WATER TREATMENT
HANDBOOK

Degrémont

WATER TREATMENT HANDBOOK

FIFTH EDITION

A HALSTED PRESS BOOK

JOHN WILEY & SONS

New York - Chichester - Brisbane - Toronto

1979

First French edition 1950
Second French edition. 1952
Third French edition 1954
First English edition 1955
Fourth French edition 1958
First Spanish edition 1959
Second English edition 1960
Fifth French edition 1963
Third English edition 1965
Second Spanish edition 1965
Sixth French edition 1966
Seventh French edition 1972
Fourth English edition 1973
Third Spanish edition 1973
First German edition 1974
First Jugoslav edition 1976
Eighth French edition 1978
Fifth English edition 1979
Fourth Spanish edition 1979

Published by Halsted Press,
a Division of John Wiley & Sons, Inc.,
605 Third Avenue, New York, N.Y. 10016

International Standard Book No. : 0470-26749-6

Library of Congress Catalog Card No. : 79.87503

Translated from French into English by
Language Consultants (France) Ltd.
68 Newington Causeway, London SEI 6 DQ

Printed in France by
FIRMIN-DIDOT S.A.
56, rue Jacob, 75006 Paris

PRESENTATION

This edition, the 18th since the first publication of the Water Treatment Handbook in 1950 and the 5th in the English language, is more exhaustive than the 1973 edition and contains much additional material. Most of the text of the last edition has been retained; the overall arrangement is unchanged.

Part I is a general survey of water and its action on the materials with which it comes into contact; it also outlines the theoretical principles of the separation and correction processes used in water treatment.

Part II describes the processes and the treatment plant, beginning with the separation processes (Chapters 4 to 12), followed by the gas-liquid exchange systems which, thanks to their dual character, effect the transition to the chemical treatment processes (Chapter 14).

Oxidation (Chapter 15) is also an ambivalent process since it has a chemical rôle (i.e. to modify the properties of the water), as well as a part to play in the destruction of microorganisms (disinfection).

Most processes used in water treatment generate liquid sludge which, according to circumstances, must be thickened, stabilized, dewatered or dried. Chapters 16 and 17 deal with this subject.

Chapters 18 and 19 (reagents, measurements and controls) develop the corresponding Chapters in the 1972 Handbook.

Part III covers treatment of the various kinds of water according to type and ultimate use. The sections on developments in municipal waste water and industrial effluent treatment have been considerably enlarged.

Part IV deals with the chemistry of water and the reagents used in water treatment, methods of analysis (with additional material on waste water and sludge) and the biology of water.

Formulae, which constitute Part V, include unchanged the concepts of hydraulics and electricity described in the previous edition and the mathematical notes from the 1965 edition. In the Chapter dealing with units of measurement, emphasis is laid on recent additions to the Système International (SI).

A new Chapter on heat is included.

The work concludes with a review, illustrated by comparative tables of the international standards, of the main legislation and regulations governing water.

The loose insert in the centre of the Handbook includes a condensed List of Contents and instructions on "how to use" the Handbook, based on reference to the general alphabetical index.

Note : all references to gallons throughout this Handbook relate to U.S. gallons.

TABLE OF CONTENTS

When read in conjunction with the 1978 French version of this Handbook, entitled « Mémento Technique de l'Eau » or with the 1979 Spanish version, entitled « Manual Tecnico del Agua », this English version can be used as an English-French or English-Spanish glossary of water treatment terms. Except for Chapter 34 dealing with legislation, the pagination is identical in all three language versions.

Part Three : Treatment Methods According to the Nature and the Final Use of the Water 597

Part One

General Aspects
of Water
and Water Treatment

WATER : PHYSICS CHEMISTRY BIOLOGY

Water is a naturally occurring substance; because, except in rare cases, it is not sufficiently pure, it cannot be used for human consumption or in industry without some form of treatment. While circulating in the ground, on the surface of the earth or even in the air, water becomes polluted and laden with solids in suspension or in solution: clay particles, vegetable waste, living organisms (plankton, bacteria, viruses), various salts, (chlorides, sulphates, sodium or calcium carbonates, iron, manganese etc.), organic matter (humic and fulvic acids, manufacturing residues), and gases.

The following table shows two typical analyses (see definitions in chapter 27) of samples of surface water and ground water of the European type:

		River water	Ground water
Temperature	(°C)	14	9.5
Turbidity	(NTU)	18	0.2
Colour	(mg/l Pt-Co)	30	10
Suspended solids	(mg/l)	25	0.35
pH		8	6.7
TAC	(French degrees)	20	35
	(milliequivalent/l)	4	7
TH	(French degrees)	22	80
	(milliequivalent/l)	4.4	16
Ca	(French degrees)	17	71
Mg	(French degrees)	5	9
Chlorides	(mg/l Cl^-)	25	70
Sulphates	(mg/l SO_4^{2-})	18	330
Iron	(mg/l Fe)	1.4	3.5
Manganese	(mg/l Mn)	trace	1.2
Ammonia	(mg/l NH_4^+)	0.7	1.5
Nitrites	(mg/l NO_2^-)	0.2	trace
Nitrates	(mg/l NO_3^-)	3	1
Free CO_2	(mg/l)	4	135
Dissolved oxygen	(mg/l)	9.5	none
Oxidizability in permanganate, heated and in acid medium	(mg/l O_2)	7.5	1.5

Because of the large number of these different kinds of impurities, water must be treated before use in order to make it suitable for its intended applications or after use in order to avoid harming the environment.

The results required, from clear, limpid water for human consumption, to ultra-pure water needed by modern electronics, are achieved by applying singly or in combination certain processes the basic principles of which will be described in chapter 3.

The treatment always produces residues (or sludge) which for the most part cannot be discharged into nature as they are, but must be treated to reduce their volume, to dewater or disinfect them etc. Water treatment is generally followed by sludge treatment.

1. WATER AND ITS PHYSICAL CHARACTERISTICS

1.1. The water molecule

Water, the result of the combination of natural hydrogen and oxygen, consists of the reciprocal combinations of two sets of isotopes:

^{1}H Hydrogen H ⎫ ^{16}O ⎫
^{2}H Deuterium D ⎬ and ^{17}O ⎬ 1
^{3}H Tritium   T ⎭ ^{18}O ⎭

It can thus theoretically contain 18 types of molecules; but in view of the extremely small proportions of ^{3}H, ^{17}O, ^{18}O, the number of molecules is reduced to H_2O, D_2O, and HDO. The mole fraction of deuterium in natural hydrogen is almost constantly equal to 1/6 000, while that of tritium is only 4×10^{-15}. **Heavy water,** the physical properties of which differ slightly from those of light water, has the chemical formula $D_2{}^{16}O$.

As a first approximation the formula for water can be written as H_2O. The water molecule has an electric moment which is reflected by its physical and electrical properties. Water is the typical liquid of the **polar** type (sometimes called a *dipole*).

To represent it a triangular diagram is used, as shown by the figure page 5. If the O-H links were exclusively covalent, the HOH angle should be 90º. Actually, owing to the electronegativity of oxygen the angle of the O-H links is about 105º.

The distance from the oxygen atom to the axis of the H$^+$ atoms is 62 pm.

---

1. For the chemical notations, see page 870.

1.2. The three states

The structure of water depends on its physical state.

The gaseous state (vapour) corresponds exactly to the formula H_2O and especially to the triangular diagram opposite.

The condensed states (water and ice) are, however, more complicated, and this accounts for their abnormal properties.

In the **solid state** the elementary arrangement consists of a central water molecule and four peripheral molecules forming the shape of a tetrahedron.

The study of the crystallographic variations, especially with the aid of the Raman spectrum, enables us to understand the transition to the liquid state from the open crystalline structure of ice.

In water in the **liquid state,** several molecules are associated by special links called **hydrogen links,** each hydrogen atom of a water molecule being linked to the oxygen atom of the neighbouring molecule. The structure is tetrahedric in the space.

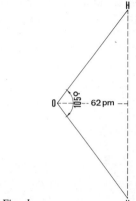

Fig. 1. —

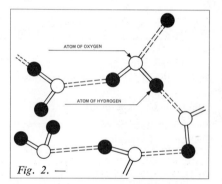

ATOM OF OXYGEN

ATOM OF HYDROGEN

Fig. 2. —

1.3. Physical properties

The following are the most important physical properties with regard to water treatment:

a) Specific mass.

Through the compacting of the molecular structure, the specific mass varies with temperature and pressure. A maximum is reached at 4.08 °C at a pressure of one bar, at 3.8 °C at 4 bars and at 3.4 °C at 10 bars.

The specific mass of pure water at 15 °C and at atmospheric pressure is 0.9990 kg/dm³. That of natural waters varies with their content of dissolved substances. Sea water with a salinity of 35 g/dm³ has an average specific mass of 1.0281 kg/dm³ at 0 °C. A variation in salinity of 1 g/l causes the specific mass to change by 0.0008 kg/dm³.

b) Thermal properties:

● **mass heat:** 4 180 J/(kg.ºC) at 0 ºC.   This varies with temperature and reaches a minimum at + 35 ºC.

● The **latent heat** of transformation is for fusion: 330 kJ/kg or 79 kcal/kg; for vaporization: 2 250 kJ/kg or 539 kcal/kg at normal pressure and at 100 ºC.

Owing to the substantial amount of mass heat and latent heat of vaporization, the large expanses of water on the earth's surface constitute veritable heat stores. This is also the reason for the use of water in industry as a heat-conveying fluid.

c) Viscosity :

This is the ability of a liquid to resist various movements, both internal and overall, such as flow.   It is the basic cause of head loss and therefore plays an important part in water treatment.   Viscosity diminishes when temperature increases.

Variations in dynamic viscosity depending on temperature :

ºC	0	5	10	15	20	25	30	35
10^{-3} Pa.s	1.797	1.523	1.301	1.138	1.007	0.895	0.800	0.723

On the other hand, it increases with a higher content of dissolved salts: sea water is therefore much more viscous than river water:

Salinity in Cl⁻ ions in g/l	0	4	8	12	16	20
Coefficient of viscosity at 20ºC 10^{-3} Pa.s	1.007	1.021	1.035	1.052	1.068	1.085

Pressure has a very special effect on the absolute viscosity of water.   Contrary to what happens with other liquids, a moderate pressure makes water less viscous at low temperatures; it somehow crushes its molecular structure.   When the pressure continues to increase, the water resumes the structure of a liquid free from any internal stresses, and again complies with the general rule that viscosity increases with pressure.

d) Surface tension.

This is a property peculiar to interfaces (boundary surfaces of two phases). It is defined as a tensile force which is exerted at the surface of the liquid and constantly tends to reduce the area of this surface to the greatest possible extent.

The surface tension γ of water is 73×10^{-3} newtons per metre (73 dyn/cm) at 18 ºC and 52.5×10^{-3} newtons per metre (52.5 dyn/cm) at 100 ºC.   It is such as to cause a capillary rise of 15 cm at 18 ºC in a tube with a diameter of 0.1 mm.

The   addition   of   dissolved   salts   generally   increases   surface   tension

($\gamma = 74.6 \times 10^{-3}$ newtons per metre or 74.6 dyn/cm for an aqueous solution of NaCl at 1 mol/l at 18 °C).

Other substances reduce surface tension: these are said to be **tensio-active.**

e) Electrical properties of water:

● **dielectric constant:** the dielectric constant of water ε, of the order of 80, is one of the highest known; this is why the ionizing power of water is so great.

● **electrical conductivity of water:** water is slightly conducting and ist behaviour is governed by Ohm's law.

The conductivity of the purest water ever obtained is:

$K = 4.2 \times 10^{-6}$ siemens per metre at 20 °C (this corresponds to a resistivity of 23.8 megohm centimetre).

The presence of dissolved salts in water increases its conductivity (see fig. 876), which varies according to the temperature.

f) Optical properties.

The transparency of water depends on the wavelength of the light passing through it. While ultraviolet light passes through it well, infrared rays, so useful from the physical and biological viewpoints, hardly penetrate it. Water absorbs a large proportion of the orange and red components of visible light; this explains the blue colour of light which has passed through a thick section of water.

This transparency is often used to measure certain forms of pollution and, consequently, the efficiency of purification treatments

2. WATER AND CHEMISTRY

The energy of formation of the water molecule, 242 kJ/mol (58 kcal/mol), is high. Water is therefore extremely stable, particularly in nature. This stability, linked with its characteristic electrical properties and molecular composition, makes water particularly suitable for dissolving many substances. Most mineral substances are in fact soluble in water, as are also a large number of gases and organic substances.

2.1. Water as a solvent

To dissolve a substance is to destroy its cohesion; its cohesion is due to electrostatic or coulombian forces which may be:

● **interatomic**
strong chemical links; covalency links (between atoms), electrovalency or ionic links (atom-electrons);

● **intermolecular**

links of cohesion between molecules (hydrogen links);

● **weak forces of attraction** (London, Van der Waals), which hold the whole substance together.

This diversity of links explains the infinite variety of states of matter.

The hydrating attraction of water (bipolar molecule) has the effect of completely or partially destroying (beginning with the weakest) the various electrostatic links between the atoms and molecules of the substance to be dissolved, which are replaced by new links with its own molecules, and creating new structures; a genuine chemical reaction **(solvation)** takes place. Complete solvation is **solution.**

2.1.1. SOLUBILITY OF THE VARIOUS PHASES

a) Gases.

The solubility of gases obeys Henry's laws, that is, the quantity of gas dissolved is proportional to the solubility coefficients α of each gas and to the concentration C of the gas in question in the gaseous phase at total pressure P in contact with the water.

The volume of gas dissolved is: $V = \alpha\, CP$.

Gas	N_2	O_2	CO_2	H_2S
Value of coefficient of solubility α at 10 º C	0.018	0.038	0.194	3.39
Value of the solubility of the gases in mg/l at 10 ºC, with water in contact with the pure gas at a pressure of 1 bar	23.2	54.3	2 318	5 112

Anhydrides (CO_2, SO_2) and various gaseous acids (HCl) dissolve and then combine. Their coefficient of solubility is much higher than that of other gases.

Oxygen is more soluble than nitrogen; the dissolved gases extracted from a water will be richer in oxygen than the initial atmosphere from which they came.

b) Liquids.

As the water molecule is polar, the solubility of a liquid in water depends on polarity of the molecules of the liquid in question. For instance, molecules containing the groups OH^- (e.g. alcohol, sugars), SH^- and NH_2^-, being very polar, are very soluble in water, whereas other liquids (hydrocarbons, carbon tetrachloride, oils and fats, etc.), which are non-polar, are very sparingly soluble.

There may be partial miscibility; for instance, two substances are miscible only above a critical temperature (a temperature above 63.5 °C for a mixture of water and phenol) or below a certain minimum (trimethylamine is soluble in all proportions only below 18.5 °C) or between two critical temperatures, one upper and one lower (the water-nicotine system).

c) Solids.

Solubility, which is determined by the maximum mass of solute that can exist in a given mass of solvent has an exact value only in the case of crystallized substances. With macromolecules there is not the precise equilibrium that exists between a crystal and the corresponding saturated solution; there is frequently no break of continuity when passing gradually from the solid substance state to the solution state. Moreover, a solute of macromolecules often includes different molecular sizes.

Water purification must take into account the size and the electrical charge of the dissolved particles. A distinction is therefore drawn between the different types of solutions and suspensions (see table fig. 3).

● *True or molecular solutions:* these are homogeneous (single phase) systems.

— crystalloid solutions: the dissolved particles are small molecules (less than a nanometre) both ionized (acids, bases, salts) and non-ionized (sugars, etc.).

— macromolecular solutions: formed from particles that are much bigger than a nanometre; may include ionized groups.

● *Colloidal suspensions:* also called micellar or pseudo-solutions, these are two-phase, distinctly heterogeneous systems, in which the dispersed particles are masses of atoms (metals) or molecules of any size.

● *Suspensions :* when the particles are visible under an optical microscope they constitute suspensions (solids) or emulsions (liquids).

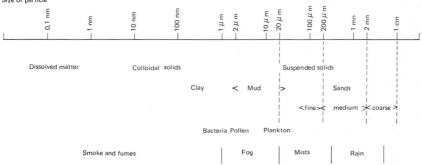

Fig. 3. — *Dimensions of various particles.*

2.1.2. HYDROPHILIZATION

The solubility of a substance may differ in various solvents: for example, sodium chloride is much more soluble in water than in alcohol, whereas paraffin is soluble in benzene but not in water.

Solubility in water depends on the nature of the substance or yet of certain of its constituent groups; the characteristic groups are therefore classified as hydrophilic (OH-CO-NH_2, etc.) or hydrophobic (CH_3-CH_2-C_6H_5).

In some cases, solvation or simple wetting take place with the aid of a third constituent called a **solubilizer** for true solutions, a **peptizer** for colloidal solutions, an **emulsifier** for emulsions, a **stabilizer** for colloidal suspensions and a **wetting agent** for surface effects.

These intermediary agents create genuine links between the solvent and substance to be dissolved, to be held in dispersion or to be wetted (they lower the surface tension).

The link on the solvent side is due to a hydrophilic group, while the link on the side of the substance to be transformed can be a chemical link (action of bases and strong acids) or a cohesion link. The latter are formed from dissymmetrical (semi-hydrophilic) molecules. One of the ends is similar to water and hydrophilic, while the other tends to associate (the action of detergents, trisodium phosphate, wetting agents) with the molecules of the substance to be stabilized or to be adsorbed on its surface. More hydrophilic aggregation or adsorption complexes are then formed.

Loss of hydrophilic properties.

The intermediary agent can break the link between the solvent and the substance which is dissolved, dispersed or wetted. Depending on the case in question, this agent will be called a **precipitant, coagulant, flocculant, thickener** or wetting **depressant**. This break can be the result of chemical action, for instance the loss of OH^- ions or of ionized groups. The intermediary agent may destroy the semi-hydrophilic cohesion link by neutralizing the hydrophilic part or by attracting the hydrophobic part on the surface either of air bubbles (flotation) or of a more or less hydrophilic insoluble adsorbant.

The break may be the result of neutralization of the electrostatic forces (by the action of polyvalent cations and ionic polyelectrolytes).

2.1.3. TRUE SOLUTIONS

The relationship between solute and solvent may be expressed in several ways:

— **concentration by weight:** number of mass units of solute dissolved in a mass unit of solvent;

— **mole fraction:** ratio of number of moles of solute to total number of moles (solvent + solute);

— **molality:** number of moles dissolved in 1 000 grammes of solvent;

— **molarity:** number of moles dissolved in 1 litre of solution.

● **Activity and concentration.**

The molecules of a solute behave in a solvent in the same way as a gas would in another gas, and indeed it has been observed that when the solute is greatly diluted, the law of perfect gases holds good. When the concentration becomes sizable, the dissolved molecules are less active than the same number of molecules of perfect gas. Concentration (c) is replaced by activity [a]:

$$[a] = (c) \, f$$

f is called the activity coefficient and tends towards 1 when the solution is very dilute.

For the reaction a A + b B $\rightleftharpoons$ c C + d D, if, in the law of mass action, we replace activities by concentrations, we obtain:

$$\frac{(A)^a \, (B)^b}{(C)^c \, (D)^d} \times \frac{f_A^a \, f_B^b}{f_C^c \, f_D^d} = K$$

$$\frac{(A)^a \, (B)^b}{(C)^c \, (D)^d} = K \quad \frac{f_C^c \, f_D^d}{f_A^a \, f_B^b} = K'$$

The thermodynamic dissociation constant K can be expressed in the form of a product $K = K' \dfrac{f_A^a \, f_B^b}{f_C^c \, f_D^d}$ where K' is the apparent dissociation constant at a given concentration, and f is the activity coefficient of the substances involved in the equilibrium.

Debye's formulae enable the activity coefficient of an ion to be calculated in terms of the ionic strength. In water with less than 1.5 g/l salinity, the following simplified expression is used for each ion:

$$- \log f_x = \frac{A \, Z^2 \, \sqrt{\mu}}{1 + 1.4 \, \sqrt{\mu}}$$

Z being the charge (valency) of ion x with molar concentration C and μ the total ionic strength $\left(\dfrac{1}{2} \, C_1 \, Z_1^2 \, 2 + \dfrac{1}{2} \, C_2 \, Z_2^2 + ... \right)$.

The value of the constant A varies with the temperature :

t °C	15	25	35	100
A	0.503	0.511	0.520	0.61

The values of K' calculated with the aid of the simplified Debye formula are valid only if the ionic strength is less than 0.1. More complicated formulae (not reproduced here) must be used for higher ionic strengths.

For the sake of convenience the expression $pK = - \log K$ is used. Tables give the values of pK for various applications of the law of mass action. The dissociation constant for a given ionic strength is calculated as follows :

$$pK' = pK + - \log \frac{f_C \, f_D}{f_A \, f_B}$$

2.1.4. IONIZATION

When certain substances, e.g. mineral salts, are dissolved, special electrical properties may be observed such as: conductivity of the liquid, possibility of electrolysis, etc.

These arise from the at least partial dissociation of the molecules into simple, electrically-charged constituents: **cation** and **anion**.

The various laws governing chemical equilibria, especially the law of mass action, must take into account the non-dissociated molecules and the various ions.

Certain acids and bases, even in relatively concentrated solution, are entirely dissociated. They are called **strong** electrolytes. For example, an ordinary solution of sodium chloride does not contain NaCl molecules, but Cl^- and Na^+ ions.

Other substances, such as acetic acid $CH_3 COOH$, are only partially dissociated in solution. These are **weak** electrolytes. In this case, we must distinguish between total acidity which comprises all possible H^+ ions, and free acidity which comprises the H^+ ions actually present.

Water itself is partially dissociated into ions according to the reversible reaction:

$$H_2O \rightleftharpoons H^+ + OH^-$$

which means that in water there are both H_2O molecules and OH^- ions (hydroxide ion) and H^+ ions (also in the hydrated form H^+, H_2O or H_3O called oxonium ion).

● **Application of the law of mass action to water: pH concept.**

Assuming that the coefficients of ionic activity are equal to 1, we obtain:

$$\frac{(H^+) \quad (OH^-)}{(H_2O)} = K$$

Since the dissociation is always weak, the concentration of the water molecules is always constant and we may write: $(H^+) (OH^-) = K_e$.

The value of the dissociation (or ionization) constant of water is of the order of 10^{-14} $(mol/l)^2$ at 23 °C. This value varies with the temperature:

Temperature °C	0	18	25	50	100
Ionization constant K_e 10^{14}	0.12	0.59	1.04	5.66	58.5
pK_e	14.93	14.23	13.98	13.25	12.24

In pure water we have: $(H^+) = (OH^-) = 10^{-7}$ mol/litre

The term 'acid medium' means a solution in which (H^+) is greater than 10^{-7} mol/l, and 'basic medium' a solution in which (H^+) is less than 10^{-7}, mol/l.

By convention the exponent of the concentration of H^+ ions or pH (hydrogen potential) is used to designate the acidity or basicity of a solution:

$$pH = - \log (H^+)$$

The pH is measured by means of coloured indicators or, preferably, by an electrometric method (glass-electrode pH meter).

● **Strength of acids and bases in aqueous solution.**

An **acid** is a substance capable of losing protons, that is, H^+ ions. A **base** is a substance capable of accepting these protons. There is thus, in an aqueous solution, an acid-base couple defined by the following equilibrium:

$$\text{Acid} + H_2O \rightleftharpoons \text{Base} + H^+$$

Applying the law of mass action and regarding the concentration of H_2O molecules as a constant, we obtain :

$$\frac{[\text{Base}]\,[H^+]}{[\text{Acid}]} = K_A \text{ and } pK_A = -\log K_A$$

K_A thus defined is called the affinity constant of the acid-base couple.

The strength of an acid is determined the byextent to which it gives off H^+ ions, that is, it is stronger the larger the value of K_A or the smaller the value of pK_A.

A base is stronger, the smaller the value of K_A.

Thus, the ammonium ion NH_4^+ is a weak acid with $pK_A = 9.2$. The corresponding base NH_4OH is a fairly strong base.

In the table on page 877 acids are classified in order of decreasing strength and bases in order of increasing strength. An acid can react on any base which is below it in table. The reaction is not complete unless the difference in pK is 5.4 units.

If, in the law of mass action, one uses concentrations (which are given by analysis) instead of activities, the apparent dissociation constants K_A' or pK_A' must be calculated taking into account the ionic strength.

The concept of pK_A makes it possible to calculate the pH of mixtures of corresponding solutions of acids, bases and salts:

— the pH of a solution of an acid with a total concentration c is:

$$pH = \frac{1}{2} pK_A - \frac{1}{2} \log c$$

— the pH of a solution of a base is:

$$pH = 7 + \frac{1}{2} pK_A + \frac{1}{2} \log c$$

— the pH of a solution of a salt is:

$$pH = \frac{1}{2} pK_1 + \frac{1}{2} pK_2.$$

K_1 and K_2 being the affinity constants of the corresponding acid and base.

● **Buffer solutions.**

In the case of a mixture of an acid of concentration (A) and the corresponding base, of concentration (B), if (A) = (B), this solution is called a buffer solution. Example: acetic acid—acetate.

A buffer solution is a solution the pH of which varies little with the addition or removal of H^+ ions. These solutions are useful when it is desired that a reaction should take place whith a constant pH.

Acetates, acid phthalates and monopotassium phosphates serve as the basis for the preparation of a whole range of buffer solutions.

● **Solubility of sparingly-soluble compounds. Solubility product.**

The ionic equilibrium state of a sparingly soluble or insoluble substance is:

$$AC \rightleftharpoons A^- + C^+$$
$$[A^-]\,[C^+] = k_s$$

The magnitude k_s or **solubility product** is constant for a given temperature and ionic strength of the solution. The value of k_s is smaller the less soluble the substance. For calcium carbonate, the solubility of which is 12 mg/l, the solubility product k_s is $10^{-8.32}$ $(mol/l)^2$. By analogy with the pH, we write:

$$pk_s = -\log 10^{-8.32} = 8.32.$$

2.2. Oxidation-reduction (Redox reactions)

Water can take part, depending on the experimental conditions, in redox reactions according to the following possible reactions:

$$2\,H_2O - 4e^- \rightleftharpoons 4\,H^+ + O_2 \nearrow$$
$$2\,H_2O + 2e^- \rightleftharpoons 2\,OH^- + H_2 \nearrow$$

In the former case, the water is a **donor** of electrons; it is a **reducing agent**: the acceptor of electrons is an oxidant. In the presence of water an oxidant releases oxygen. In the second case water is an **acceptor** of electrons; it is an **oxidant**: the donor of electrons is a reducing agent. In the presence of water a reducing agent releases hydrogen.

But reactions are very slow without catalysts and in general the action of water with respect to redox reactions may be ignored.

However, very strong oxidants and reducing agents react remarkably quickly on water: for example, chlorine easily changes into the Cl^- anion state according to the following reaction:

$$Cl_2 + 2\,e^- \longrightarrow 2\,Cl^-$$

Thus with water : $2\,Cl_2 + 2\,H_2O \longrightarrow 4\,H^+ + 4\,Cl^- + O_2$ oxygen is released and the medium becomes acid.

Water can be broken down into oxygen and hydrogen according to the following reaction:

$$2H_2O \rightleftharpoons 2\,H_2 + O_2$$

with the redox neutrality corresponding to equal pressures of oxygen and hydrogen, and a pressure of $PH_2 = 10^{-27}$ bars.

The concept of redox potential enables us to classify the various oxidants and reducing agents in relation to hydrogen and thus to each other (see chapter 3, page 95).

3. WATER AND BIOLOGY

The calorific properties of water are one of the essential factors in the thermal stability if the biosphere.

The physical, chemical and meteorological properties of water, together with those of oxygen and carbon dioxide gas, contribute to the creation of conditions favourable to the development of living beings.

Water undergoes a biological cycle in the course or which a series of exchanges takes place. For water is a vehicle which does not occur in the pure state, but contains mineral and organic substances which are sometimes useful and nutritive (both outside and inside living beings) and often harmful and polluted when this vehicle has become the receptacle for, in particular, the waste matter created by human activities.

Water plays an important part in the constitution of living beings.

Within an organism the various organs do not have the same water content. The low percentage of 22 to 34 % for bones and adipose tissue rises to 70-80 % in the various internal organs, while the highest percentage (82-94 %) is reached in nerve tissue.

Water in protoplasm becomes a colloidal solution of protein molecules made possible by the phenomenon of solvation of the side chains of these molecules.

Water constitutes the greater part of our foods:

Green vegetables and fruits............................... 78 to 97 %
Fish .... about 80 %
Meat... about 72 %

3.1. Water and cellular metabolism

It is in the external environment in which it lives that the cell seeks to find the essential substances, also called essential metabolites, which are necessary to maintain the rhythm of its activities.

Some cells are capable of synthesizing these from mineral components; they transform water, carbon dioxide gas and mineral salts into their own substance, the molecular structure of which is extremely complex. These are the **autotrophic** cells which obtain the necessary energy from the external environment (light energy or chemical energy produced by the transformation of certain mineral radicals), and synthezize reserves which are usable at any time, thus constituting potential energy.

Heterotrophic cells, on the other hand, are incapable of synthesizing all their growth factors and use nutritive substances which they split up and 'oxidize' exothermically into simpler substances: the energy released (kinetic energy) during these chemical reactions will supply the cell's needs (growth, locomotion, reproduction).

The term **metabolism** is used to cover all the energy reactions, the term **anabolism** being employed when there is a gain of potential energy (endothermic reaction).. and therefore synthesis of living matter, and **catabolism** when there is a gain in kinetic energy (exothermic reaction) and therefore degradation of nutritive substances.

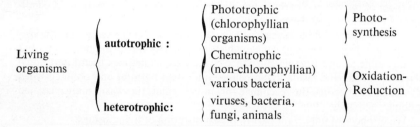

The study of energy sources therefore consists essentially of the analysis of photosynthesis and of redox phenomena in fermentation processes.

● *Anabolism:*

a) Photosynthesis: the typical case of the production of glucids (carbohydrates) by plants from the CO_2 of the atmosphere and the water in the raw sap with the aid of solar energy and in the presence of chlorophyll can be summarized by the overall equation:

$$6\ CO_2 + 6\ H_2O \longrightarrow C_6H_{12}O_6 + 6\ O_2 -2.72\ kJ/mol\ (-650\ cal/mol)$$

This biological process is quantitatively the most important in the world.

b) Mineral redox processes: other autotrophic organisms are incapable of utilizing solar energy because they have no fixing pigments; they obtain the energy necessary for their existence from the redox phenomena of mineral substances. Worthy of mention here are the Nitrosomonas which oxidize ammonia into nitrites, the Nitrobacters which transform nitrites into nitrates, ferruginous and manganiferous bacteria which oxidize ferrous and manganous bicarbonates into ferric and manganic hydroxides, sulphurous bacteria which transform hydrogen sulphide into colloidal sulphur, sulphate-reducing bacteria which reduce sulphates into hydrogen sulphide, and thiobacilli which oxidize the latter into sulphuric acid.

● *Catabolism. Fermentation process:*

Catabolism, or the process of decomposition of cellular alimentary substances, involves the formation of water or the participation of water molecules in organic oxidation and reduction reactions making use of the chemical energy contained in all nutritive substances.

The terms aerobic and anaerobic are used to characterize the type of decomposition which is in fact dehydrogenation taking place within the heterotrophic cell. If hydrogen combines with the molecular oxygen, the process is called **aerobic.** If on the other hand the process involves the transfer of hydrogen from the dehydrogenated compound to a hydrogen acceptor other than molecular oxygen, it is called **anaerobic.** Hence the idea of wholly-aerobic bacteria, wholly-anaerobic bacteria and facultative bacteria.

The motive force of decomposition (aerobic or anaerobic) of organic substances is provided by **enzymes** secreted by the organisms.

These are complex proteins (with an exact spatial arrangement) and to which the organic molecules or **substrate** can become fixed.

A distinction is made between exocellular enzymes, which break down molecular structures which are too complex to be introduced directly into the cell, and endocellular enzymes which bring about assimilation and thus form the basis of the vital phenomena which ensure the proliferation of cells. These enzymes are biological catalysts which are transformed and regenerated during the process in question.

The essential property of enzymes is their specificity linked to their structure.

The process of biological breakdown leads, depending on whether it takes place in aerobic or anaerobic conditions, to different final decomposition products. For instance, the use of glucose produces:
— in aerobic conditions:

$$C_6H_{12}O_6 + 6 O_2 \longrightarrow 6 CO_2 + 6 H_2O + 2.72 \text{ kJ/mol} (+ 650 \text{ cal/mol})$$

— in anaerobic conditions:

$$C_6H_{12}O_6 \longrightarrow 3 CO_2 + 3 CH_4 + 144 \text{ J/mol} (+ 34.4 \text{ cal/mol})$$

The heat released in anaerobiosis is only 5.3 % of the energy released in aerobiosis. As it is probable that the order of magnitude of the amount of energy necessary in order to produce new cells is the same in both cases, one deduces that it is much more economical for bacterial life to obtain its vital energy in aerobic processes than in anaerobic processes.

In other words, cellular multiplication will be much more abundant in the former case than in the latter and, as a corollary to this, the process of breakdown to the final state will be much faster, all other things being equal.

3.2. Water: the medium for microbic life

In the living world there is a group of participants, invisible but ever-present, which it is difficult to define as being animal or vegetable because they precede these two kingdoms. They play a part which is essential to the continuance of biological activities: these are the **microbes,** generally **unicellular,** the constant action of which plays a part in the recirculation of elements and in the analytical and synthetic processes without which life would stop.

Man's need to use water in ever-increasing quantities has caused him to contribute more and more to its pollution. As medical and biological science have recognized the role played by water in the appearance and transmission of certain diseases, man has had to identify and combat a number of organisms which can be contained in water. These have been given the name of **pathogens** (see page 973).

Others, however, have been found to be less harmful and man has been able to "domesticate" them and take advantage of their activities or syntheses (the useful microbes).

The pathogenic organisms are much less numerous than the useful microbes.

Among the most important microbes in water are **bacteria,** which will be described below:

Like all living cells, the bacterial cell contains a nucleus mainly composed of chromosomes massed together in the chromatin and consisting of desoxyribosenucleic acid (DNA). The nucleus controls reproduction, preserves cell lineage in a genetic code and conducts by means of messenger RNA (ribosenucleic acid) the synthesis of proteins and especially enzymes in the cytoplasm, a jelly containing RNA particles, ribosomes, as well as various organelles—mitochondria, lysosomes, etc. which play a very definite role. The bacterial cell is surrounded by a rigid membrane giving the bacterium its shape.

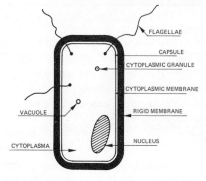

Fig. 4. —

The mobile types of microorganisms have filaments or flagella (fig. 4).

The ratio of surface area to volume is higher than in the case of other organisms. Now the metabolic rate rises with this ratio; a bacterium is more active than the more advanced organisms.

The rate of reproduction depends on the concentration of nutritive substance in the medium. In the most favourable cases a cellular division has been observed to take place in 15 to 30 minutes; sometimes it takes several days.

Bacteria only live in a medium possessing certain characteristics with regard to water content, pH, salinity, redox potential and temperature. The favourable redox potential varies fairly considerably depending on whether the bacteria are operating in aerobic or anaerobic conditions.

These conditions are closely linked with the composition of the enzyme system secreted by the bacteria. Major variations in the characteristics of the medium may result in a selection of species. Modifications of the enzyme system may occur through mutations affecting the genes of the chromosomes.

Bacteria may be classified, according to the optimum temperature for their enzymes, as **thermophilic** (temperature over 40 °C), **mesophilic** (temperature around 30 °C), **psychrophilic** (0 to 15 °C) and **cryophilic** (— 5 to 0 °C).

Some bacterial species may have a special shape owing to sporulation; the spores which they produce are cells with suspended animation possessing a structure which makes them much more resistant, for instance, to heat and dryness. When conditions return to normal the spores germinate and re-create active bacteria.

A complex bacterial culture may therefore adapt itself through selection and mutation to slow changes in the composition of the substrate on which it feeds.

In the case of heterotrophic organisms, the main nutritive substrates are protids, glucids and lipids.

3.3. Nutritive substances

Protids, the most important components of living matter, form the basis of protoplasmic and cytoplasmic matter.

Protids consist of an assembly of simple substances, the **amino-acids.** An amino-acid is a substance whose molecule contains one or more acid groups COOH and one or more amino groups NH_2 linked with the same carbon atom:

$$R-\overset{\displaystyle H}{\underset{\displaystyle COOH}{C}}-NH_2$$

Protids behave like acids or bases depending on the pH of the medium in which they are located.

The acid and amino functions can become fixed to each other and can form long-chain macromolecules the molecular weight of which can be very high (50 000 and above).

A distinction is drawn between peptids, simple proteins and compound proteins.

Glucids used to be called sugars, owing to the flavour of the simplest among them, or carbohydrates, because they correspond to the general formula:

$$C_m (H_2O)_n$$

Owing to their abundance in vegetable tissue, these are the usual foods of heterotrophic organisms.

They exist in a non-hydrolizable form (the Oses, such as glucose) or a hydrolizable form (Osides, such as starch, cellulose and glycogen).

Simple or complexe **lipids** are esters of more or less complex fatty acids and alcohols. They can emulsify in water, in which they are generally insoluble. They constitute, both in plants and in animals, an important reserve material for meeting their energy requirements.

In certain conditions, heterotrophic organisms may adapt themselves so that they can feed on other organic substrates such as phenols, aldehydes, hydrocarbons, etc.

Autrotrophic organisms may synthesise their own substance from a substrate containing mineral salts, nitrogen and phosphorus; the source of carbon may be carbon dioxide, methane, etc.

ACTION OF WATER ON MATERIALS

During the distribution or use of water, the various materials with which it is in contact can be damaged in various ways, the most common case being the corrosion of metals and of steel in particular. Other types of damage include the dissolving of calcareous materials and the degradation of concrete.

The basic processes which cause damage, and the factors common to all of them, must be known and understood so that suitable counter-measures can be adopted—such as modification of the chemical properties of the water, application of a protective coating, electrical protection or the use of a more inert material.

1. ELECTROCHEMICAL MECHANISM OF THE CORROSION OF IRON

Pure water always attacks iron because there is no area of thermodynamic stability between the two; but the kinetics of the process, which is always electrochemical, differ according to whether or not oxygen is present. In the latter case, corrosion is caused by hydrogen.

1.1. Electrochemical processes

When any metal is plunged into an electrolyte solution its ions—carrying a positive charge—tend to dissolve while the metal itself retains a negative charge. An electrode is formed with a potential which can be expressed in absolute terms by Nernst's equation:

$$E = \frac{RT}{nF} \ln \frac{P}{p} = \frac{RT}{nF} \ln \frac{C}{c}$$

where:
n : is the valency of the metal ions in question
T : is absolute temperature
R : is the molar constant of perfect gases : 8.31 J/(mol.K)
P : is the pressure at which the metal dissolves
p : is the osmotic pressure of the solution

c : is the activity of the metal ions in the solution
C : is a constant defining the concentration of ions in the metal
F : is the Faraday number.

This potential is **standard** for a standard solution of the ion in question and is then expressed as E_0, giving the general equation defining the electrode potential of a metal at 25 °C:

$$E = E_0 + \frac{0.058}{n} \log c$$

By analogy, for a hydrogen phase, the potential of the corresponding standard gaseous electrode can be expressed as:

$$E_H = 0.058 \log H^+ = -0.058 \, pH$$

In practice, the standard potentials can only be measured against a reference gaseous electrode known as a standard hydrogen electrode, whose value E_{H0} is assumed to be 0 for reference purposes.

NERNST SCALE OF STANDARD EQUILIBRIUM POTENTIALS RELATED TO THE STANDARD HYDROGEN ELECTRODE AT 25° C
(Metal immersed in a standard solution of one of its salts).

Metal	Electrode reactions	Equilibrium potential (volts)
Magnesium	$Mg = Mg^{++} + 2e^-$	$- 2.34$
Beryllium	$Be = Be^{++} + 2e^-$	$- 1.70$
Aluminium	$Al = Al^{+++} + 3e^-$	$- 1.67$
Manganese	$Mn - Mn^{++} + 2e^-$	$- 1.05$
Zinc	$Zn = Zn^{++} + 2e^-$	$- 0.762$
Chrome	$Cr = Cr^{+++} + 3e^-$	$- 0.71$
Iron	$Fe = Fe^{++} + 2e^-$	$- 0.440$
Nickel	$Ni = Ni^{++} + 2e^-$	$- 0.250$
Lead	$Pb = Pb^{++} + 2e^-$	$- 0.126$
Hydrogen	$H_2 = 2H^+ + 2e^-$	$- 0.000$ by convention
Copper	$Cu = Cu^{++} + 2e^-$	$+ 0.345$
Copper	$Cu = Cu^+ + e^-$	$+ 0.522$
Silver	$Ag = Ag^+ + e^-$	$+ 0.800$
Platinum	$Pt = Pt^{++} + 2e^-$	$+ 1.2$ approx.
Gold	$Au = Au^{+++} + 3e^-$	$+ 1.42$
Gold	$Au = Au^+ + e^-$	$+ 1.68$

For iron, therefore: $E = -0.44 + 0.029 \log Fe^{2+}$

This table classifies metals theoretically according to their potential, which is positive for the noble metals and negative for the base metals. The potentials listed are theoretically determined and the measured real potentials can be substantially different according to the quality of the metal and in particular the electrode processes. For example, in addition to dissolution of the metal as already described there is an opposing process affecting higher grade regions of the same metal immersed in the same electrolyte.

This better-grade electrode sets up a potential difference in the medium and generates an electric current which continues the dissolution of the metal electrode, or corrosion.

The process takes a very different form according to whether or not oxygen is present.

1.2. Corrosion in a deaerated medium (corrosion by hydrogen)

Electrochemical mechanism

The opposing process is a gaseous hydrogen electrode process in which hydrogen gas is formed from the H^+ ions present in the water:

$$H_2O \rightleftharpoons OH^- + H^+$$

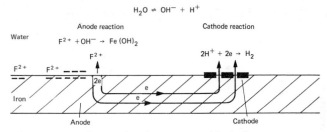

Fig. 5. — Corrosion in a deaerated medium or caused by hydrogen

This electron **capturing** process is capable of maintaining the previous electron **releasing** process.

This corrosion by hydrogen, wrongly described as chemical corrosion, is therefore basically electrochemical. It creates a positive and a negative electric pole, arbitrarily called the cathode and the anode, with an electric current circulating between them.

Metal is dissolved or oxidation takes place at the anode while the cathode is protected against corrosion. This dissolution can only affect metals whose practical potential in the electrolyte is weaker than the potential of the opposing hydrogen cathode which can form:

— on a metal which is of higher grade than the metal of the anode,
— on a foreign impurity (oxide, fouling),
— on an irregularity in the crystalline texture of the metal, in which case cold-rolled or cold-drawn zones become anodes in relation to the metal itself.

Corrosion continues to spread if it is not limited by the absence of high pH H^+ ions or by saturation of the medium with Fe^{2+} ions, which causes a protective deposit to form by precipitation of ferrous hydroxide and normally stops corrosion.

The precipitate is carried away by movement of the electrolyte; this means that in practice corrosion by hydrogen can only be stopped in stagnant water.

E_{Fe} rises because iron enters the solution, E_H drops because H_2 is given off. The reaction should stop when:

$$- 0.44 + 0.029 \log (Fe^{2+}) = - 0.058 \text{ pH}$$

i.e.:

$$\log (Fe^{2+}) = 15.1 - 2pH$$

There is no practical range of stability common to iron and water below a pH of 10.5, at which the solubility of iron is 10^{-6} mol/kg and corrosion is negligible; "acid" corrosion increases as pH rises and as the number of Fe^{2+} ions in the water falls.

In practice, this corrosion which is very heavy at acid pH values, becomes less dangerous at OpH when the concentration of H^+ is no longer sufficient to maintain the cathodic reaction and a protective film can start to form.

The Nernst graph which follows can be used to calculate the equilibrium potential of each of these two electrodes at different pH values.

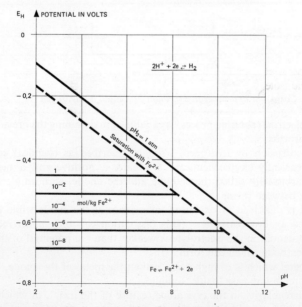

Fig. 6. — Iron-water equilibrium potential in the absence of oxygen.

The horizontal lines represent the potential of the iron electrode, independent of pH, at different concentrations of Fe^{2+} ions in the water. This concentration is limited by the solubility of the ion which tends to precipitate in the form of ferrous hydroxide Fe $(OH)_2$.

Metal is lost during this process in a deaerated medium.

Morphologically, therefore, corrosion by hydrogen appears as fairly uniform corrosion of the metal, because an infinite number of cathodes and anodes exist at the same time and are even liable to change their polarity.

1.3. Corrosion by oxygen

1.3.1. ELECTROCHEMICAL MECHANISM

In the presence of aerated water, the oxygen in the water provides the complementary electrode process in this case.

$$O_2 + 2 H_2O + 4 e^- \rightarrow 4 OH^-$$

This electrode is capable of absorbing electrons and therefore of acting as a cathode.

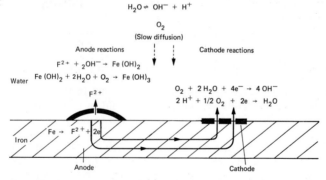

Fig. 7. — Corrosion in an aerated medium.

In this case, the equilibrium potential is governed by:
— the concentration of OH^- ions and therefore pH value,
— partial oxygen pressure.

$$E' = E_0 - \frac{RT}{F} \ln (OH^-) + \frac{RT}{4F} \ln pO_2$$

When this potential is higher than that of the metal electrode, it keeps up corrosion.

As this potential is itself more than 1 volt higher than that of the hydrogen electrode, the oxygen cannot be said to "depolarize" the hydrogen electrode by reaction with H_2 but it may be deduced from this that corrosion by oxygen is more frequent and more extensive.

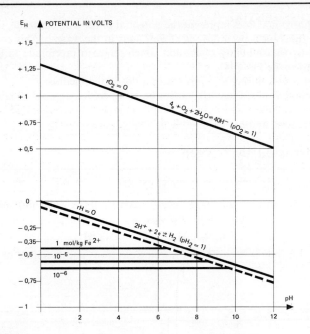

Fig. 8. — Equilibrium potentials of hydrogen and oxygen electrodes.

In this corrosion process, oxygen must be present in the dissolved state in the water; at high temperatures, the process would in fact be different.

The electrode potential is directly proportional to the quantity of dissolved oxygen.

This leads to the apparent paradox that if part of the metal is not bathed in oxygen it becomes anodic and therefore liable to corrosion in relation to the rest of the surface which is protected by the oxygen. This explains the damage caused by deposits of all kinds which prevent oxygen from reaching the underlying surfaces and set up an anodic zone.

At the oxygen cathode, the release of OH^- ions raises the pH of the water at least in the immediate neighbourhood of the metal surface. In addition, in the presence of oxygen, the Fe^{2+} ions oxidise to form Fe^{3+}. However, ferric oxide, $Fe(OH)_3$, is only very slightly soluble. Thus, instead of being washed away by the water and leaving a clean surface, as in the case of corrosion in a deaerated medium, the corrosion products collect round the anode in the case of corrosion in an aerated medium.

They form the well-known "pustules" which constitute an additional barrier to diffusion of the oxygen and strengthen the anodic character of the surface covered.

This explains why corrosion by oxygen perforates the metal.

1.3.2. DIFFERENTIAL OXYGEN CONCENTRATION

It is clear from the description of the previous mechanism, that zones in the metal containing no dissolved oxygen, such as screw threads, cavities and cracks form anodic regions while zones with dissolved oxygen form cathodic regions.

A potential difference can also appear between zones where concentration gradients of dissolved oxygen exist in the liquid film. Hence the concept, postulated by Evans, of the presence of a large number of elementary microcells formed by **differential oxygen concentration** and this concept can be extended to all irregularities in metal parts resulting from their nature, construction or degree of fouling, or just from temperature differences.

The spread of corrosion by differential oxygen concentration in the interfacial layer is governed principally by the **solubility** of oxygen and, in particular, by the **rate at which it is diffused** because slower rates promote corrosion.

The solubility of oxygen is a known function of temperature and partial pressure; it is very slight at high temperatures but never drops to nil.

The rate at which oxygen is diffused varies with the temperature, the rate at which the water circulates and the state of the surface. It is of great importance in the self-protection process which is discussed later.

2. ACTION OF CARBON DIOXIDE

2.1. CaO/CO_2 equilibrium; aggressive action on $CaCO_3$

Water in the natural state is not in fact pure and contains various dissolved chemical elements, the most common of which is calcium bicarbonate (hydrogenocarbonate).

In practice, equilibrium between this salt and carbon dioxide is governed by somewhat complex laws and any shift in equilibrium can trigger off chemical reactions such as dissolution of calcium carbonate (**aggressiveness**) or the **formation of scale** added to the simple electrochemical **corrosion** reactions specific to the metals concerned.

Calcium bicarbonate, unknown in the solid state, exists in an unstable state in aqueous solution; it tends to lose carbonic acid and precipitate $CaCO_3$:

$$Ca\,(HCO_3)_2 \rightleftharpoons CaCO_3 + H_2O + CO_2$$

$$\underbrace{}_{\substack{\text{semi-combined}\\ CO_2}} \qquad \underbrace{}_{\substack{\text{combined}\\ CO_2}} \qquad \underbrace{}_{\substack{\text{free}\\ CO_2}}$$

To maintain the calcium bicarbonate in solution, a quantity of free CO_2 known as the equilibrium CO_2 is required and this reverses the precipitation reaction.

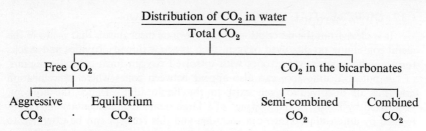

Natural water may contain a quantity of free CO_2 greater than that necessary to keep the calcium bicarbonate in solution. This excess is aggressive to limestone and is determined by the marble test.

If the free CO_2 is less than the equilibrium CO_2, the water is scaleforming. These simple definitions have become established by custom.

2.2. General study of equilibrium

All studies of the aggressive action of CO_2 are based on the following equations:
— equal electric charges

$$H^+ + 2\,Ca^{2+} + \underbrace{2\,Mg^{2+} + Na^+ + K^+}_{P^+}, \text{ etc..}$$

$$= OH^- + HCO_3^- + CO_3^{2-} + \underbrace{Cl^- + 2\,SO_4^{2-}}_{N^-} + \cdots \tag{1}$$

where P^+ and N^- are the sum of the cations or anions not involved in the CaO/CO_2 equilibrium
— dissociation of the water

$$(H^+)\,(OH^-) = K_0' \text{ with } pK_0' = pK_0 - \varepsilon \tag{2}$$

— water/H_2CO_3 equilibrium:

$$(H^+)\,(HCO_3^-) = K_1'\,(H_2CO_3) \text{ with } pK_1' = pK_1 - \varepsilon \tag{3}$$
$$(H^+)\,(CO_3^{2-}) = K_2'\,(HCO_3^-) \text{ with } pK_2' = pK_2 - 2\,\varepsilon \tag{4}$$

— solubility product of the $CaCO_3$ (see page 14) :

$$(Ca^{2+})\,(CO_3^{2-}) = K_s' \text{ with } pK_s' = pK_s - 4\,\varepsilon \tag{5}$$

ε depends on the ionic strength μ of the solution according to the equation

$$\varepsilon = \frac{\sqrt{\mu}}{1 + 1.4\sqrt{\mu}}$$

The ionic strength: $10^3\,\mu = \sum \frac{1}{2}\,C_n\,Z_n^2.$

C_n is the concentration of the ion concerned (in mmol/ kg) of valency $Z_n.e$

2.3. Graphic representations

In practice it is interesting to study the development of the six constituents H^+, OH^-, CO_3^-, HCO_3^{2-}, Ca^{2+} and H_2CO_3, and to show this by means of graphs.

Any two basic constituents are taken as co-ordinates, and the graph enabling the curves representing various calcium-carbonic constituents of water to be simply constructed is selected.

The curves (pH, HCO_3^-, CO_3^{2-} and H_2CO_3) are derived from the basic equations. The diagram obtained is valid for a family of waters with constant N and P and variable calcium-carbonic ion concentration.

A large number of diagrams are drawn using different variables. A few are summarized below.

2.3.1. TILLMANS' METHOD (1930)

This is based on the study of a pure solution of calcium bicarbonate in the presence of CO_2 and measures the equilibrium CO_2 in terms of semi-combined CO_2 in the form of bicarbonate.

This method was taken up and given general application by Guigues whose diagram uses the same system of coordinates to represent the family of curves TCa = constant and TCa — TAC = constant.

This diagram can be used in practice to determine the equilibrium of natural water, the measures required to correct its aggressive action or the progress of its scale-forming character.

2.3.2. LANGELIER METHOD AND HOOVER DIAGRAM (1936) (see page 878)

Langelier calculated the equilibrium pH or the saturation pH_s for water by reference to four parameters:
— (Ca^{2+}) calcium content;
— (HCO_3^-) total alkalinity;
— total salinity in the form of the effect of ion strength on the apparent coefficients of dissociation K_2' and K_S',
— the temperature affecting pK_S' and pK_2' (see page 14).

Substituting the value of CO_3^{2-} obtained from equation (5) in equation (4) for a water at equilibrium, we have:

$$(H^+) = \frac{K_2'}{K_S'} (HCO_3^-) (Ca^{2+})$$

and: $$pH_S = pK_2' - pK_S' - \log (Ca^{2+}) - \log (HCO_3^-)$$

Langelier also devised a **saturation index** I_s, equal to the difference between the measured pH for any given water and its calculated pH_s : $I_s = pH - pH_s$.
If $pH < pH_S$, I_S is negative and the water is corrosive.
If $pH > pH_S$, I_S is positive and the water is scale-forming.

In Langelier's graph, (HCO_3^-) is replaced by the alkalinity as $CaCO_3$ and the ionic strength by the concentration of dissolved salts (dry weight).

The concentrations are expressed in mg/l.

It should be noted that this pH_s is in fact a theoretical pH, differing from that obtained at equilibrium in practice, because the other parameters will also have changed.

In order to calculate the real equilibrium pH after treatment, changes in the values for the resultant water must first be determined.

Hoover diagram.

As Langelier's graph is fairly difficult to interpret, Hoover converted it into a parallel-axis graph which is easy to read.

This graphic resolution takes account of salinity and temperature but is qualitative only and cannot be used to work out quantities of neutralising reagents

2.3.3. HALLOPEAU METHOD

Hallopeau devised a graphic method of determining the aggressive action of water on $CaCO_3$ and of calculating quantities of neutralising reagents, by expressing saturation pH in terms of the logarithms of alkalinity and total $[Ca^{2+}]$ hardness:

$$pH_S = \log K'_S - \log K'_2 - \log 2\,[HCO_3^-] - \log [Ca^{2+}] + \log p$$

where: $p = 1 + \dfrac{2\,K'_2}{[H^+]}.$

The Hallopeau and Dubin graph (see page 56) plots the logarithm of alkalinity horizontally and the pH of the water vertically.

Free CO_2 and saturation pH are therefore represented by two sets of parallel straight lines.

Free CO_2 can be determined when the pH and alkalinity of a water are known. The graph contains two curves systems representing physical dissolution of the CO_2 and neutralisation by lime and $CaCO_3$.

Although it introduces the concepts of lime hardness and total hardness, this method, which takes no account of either total salinity or alkaline water, applies mainly to water with an average mineral content.

2.3.4. FRANQUIN AND MARÉCAUX DIAGRAM

This method is used for carbonate water with a low mineral content: the variable is the total alkalinity (TAC) of the water and the function is the sum of free and semi-combined CO_2.

The curves for constant pH and constant CO_2 are straight lines. The equilibrium curve is valid for pH values up to 10.

2.3.5. LEGRAND AND POIRIER METHOD

These authors considered the system of coordinates for (Ca^{2+}) and total CO_2 in which the equilibrium curve is plotted for a given degree of salinity, in terms of elements outside the CO_2 equilibrium.

This method can be used to define the cases of equilibrium for a given water, the conditions for reagent correction and the values obtained as a result.

3. FORMATION OF PROTECTIVE FILMS AND PASSIVATION

Corrosion can be limited by the formation of natural or deliberate protective films. These are described as natural when they are produced by the action of the constituents of the water and of temperature.

On the other hand, under certain conditions external agents such as inhibitors are used or an electric voltage is applied to produce an artificial protective film by means of controlled inhibition or controlled passivation according to the particular case.

3.1. Natural formation of protective films

This is a general inhibiting process linked with the presence of OH^- ions; it therefore takes place at neutral or alkaline pH.

Near to the boundary layer, solubility of the Fe^{2+} ions and pH value can be totally different from the values in the water itself; the difference is especially marked when the water is stagnant or very slow-moving.

The effect is that, in the cathodic regions, the H^+ ions disappear and alkalinity increases; the excess OH^- ions combine with the Fe^{2+} ions released to form ferrous hydroxide $Fe(OH)_2$ which precipitates as a fairly uniform film on the cathodic or anodic regions.

This hydroxide is unstable; according to the temperature and chemical composition of the medium it is subject to major transformations which determine the protective nature of the film formed.

3.1.1. INFLUENCE OF THE CALCIC ALKALINITY OF THE WATER (TILLMANS' MIXED FILM) IN THE CASE OF OXYGENATED WATER

Calcium bicarbonate water which contains enough oxygen and is in carbonic equilibrium causes a natural protective film to form on the cathodic regions when cold; this layer, known as the Tillmans' film, is based on a mixed precipitate of $CaCO_3$ and iron oxides.

The process takes place over a pH range of about 7 to 8.5.

If massive quantities of oxygen are diffused on to the cathodic regions, there is direct and rapid oxidation of the non-protective ferrous hydroxide into ferric hydroxide. If sufficient calcium bicarbonate is present there is simultaneous local precipitation of $CaCO_3$ which syncrystallises with the various iron oxides.

It is this naturally-formed film which protects many systems distributing preaerated well water or river water.

In practice, a protective film is formed if the three following conditions are fulfilled:

— presence of not less than 4 to 5 mg/l of dissolved oxygen;
— free CO_2 content required for CaO/CO_2 equilibrium;
— calcium bicarbonate alkalinity of the raw water sufficient to exceed the solubility product of the calcium carbonate at the cathodes (around 7 to 11 Fr°).

If there is no oxygen at all, acid corrosion develops according to H^+ ion, CO_2 and mineral content.

In practice, the probability of such protection being established can be estimated by calculating the **Ryznar stability index.**

This index, which is widely used in studies for the conditioning of industrial cooling waters, was arrived at by **experiments** on water containing calcium bicarbonate. It can be used to determine empirically the tendency of an aerated water to cause corrosion or scale.

It is expressed as follows:

$$I_R = 2\,pH_S - pH$$

where pH_S = theoretical saturation pH, calculated from the Langelier and Hoover diagram; pH = pH measured at the relevant temperature.

The table below, for temperatures from 0 to 60 °C, correlates the various positive values of the index to the tendency of the water:

Index	Tendency
4-5.................,	Heavy scale formation
5-6.................	Slight scale formation
6-7.................	Equilibrium or
7-7.5	Slight corrosive action
7.5-8.5	Heavy corrosive action

3.1.2. MAGNETITE FILM

At temperatures over 100 °C, whether or not oxygen is present, the ferrous hydroxide tends to be converted into the intermediate oxidation product, magnetite (Fe_3O_4).

$$3\,Fe\,(OH)_2 \longrightarrow Fe_3O_4 + H_2 + 2\,H_2O$$

This reduction starts at 100 °C and is complete at 200 °C.

The magnetite forms a highly-resistant film which does not redissolve at neutral or even slightly acid pH.

In the case of very pure, oxygen free-water:
— below 200 — 250 °C, pH must be raised in inverse proportion to temperature;
— over 200 — 250 °C, continuation of corrosion no longer depends on pH, above the neutral value at least.

3.2. Inhibition and passivation

Except in the two major cases just considered, steel is active in practice in all kinds of water and therefore corrodes at varying rates. This corrosion can be checked or halted by an inhibition process whereby the polarity of the electrodes is strengthened by the addition, or presence in the water, of certain quantities of soluble compounds which can have the following action:
— precipitation of insoluble products on the metal;
— buffer effect, moderating the acidity produced on the anodic regions (case already described of calcium bicarbonate waters);
— formation of an isolating mono-molecular film by adsorption of organic inhibitors.

Several of these mechanisms can work in synergy.

Inhibition relates to the solution in which the metal is immersed while « passivation » describes a condition in which the metal is covered by a protective, uncorroded film, so long as this film is not destroyed by a secondary reaction.

Oxygen can assist passivation both by the formation of oxide and by reduction of the activity of the anions which catalyse corrosion.

The degree of inhibition varies with the temperature and resistivity of the water. It is more difficult hot than cold and decreases with the resistivity of the liquid. Chloride-containing and brackish water will therefore be difficult to inhibit.

Inhibitors which act physically or by adsorption are used to combat acid corrosion while chemical inhibitors are used in cases of oxygen corrosion (see Chapter 14).

3.3. Principles of protection against corrosion

The Pourbaix diagram shows the different zones of corrosion, passivation and immunity for steel (Figure 9).

Corrosion can therefore be prevented in several ways:
— by passivating the surfaces by maintaining a sufficiently oxidizing and alkaline medium. This is the field of natural passivation;
— by adding inhibitors to the water, which slows down the corrosion process by « polarizing » the electrodes;
— by raising pH, that is by neutralization and alkalinization beyond a pH of 10.5. This is normal practice in the case of water for low-pressure boilers and certain other applications;
— by lowering electrode potential to below — 0.58 volt. This is the method of **cathodic protection** which is employed to protect buried pipes or equipment used in sea water, by immunizing the steel;
— by chemical (see page 397) or physical insulation (coatings).

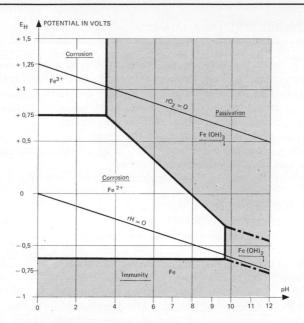

Fig. 9. — Pourbaix diagram.

3.4. Cathodic protection

By this method, any metal structure is rendered wholly cathodic and will not then corrode provided a sufficient negative electric potential is maintained so that the surface to be protected is completely polarized and remains within the passivity zone.

It can be used:

— when the electrolyte is not uniform (concentration cells in water containing various amounts of salt and in soils with varying amounts of moisture);

— when several metals are in contact with the same electrolyte.

It is generally accepted that cathodic protection is effective for applied potential values between—0.85 and—1 volt.

Current density varies with the quality of the coating on the metal structure to be protected.   The approximate figures are:

— 60 mA/m² for a bare mechanical screen immersed in sea water;

— 5 mA/m² for a mechanical screen with epoxy coating;

— 0.1 mA/m² for a pipe coated with epikote pitch, laid in wet ground which is only slightly conductive.

For protection in sea water, however, a design figure of 80 mA/m² is used to allow for long-term deterioration of coatings.

There are two methods of applying cathodic protection:

— **by using reactive, consumable anodes** made of magnesium or aluminium. An electric cell is formed by substituting a metal which carries a greater negative electric charge than steel so that an internal galvanic current is generated.

Magnesium, which has an electrolytic potential difference of 1.8 volts, is used in water with fairly high resistivity and at current strengths of a few ampères, with a consumption rate around 10 kg of Mg per ampère per year.

Aluminium, which has an electrolytic potential difference of 1.1 volts, is used in highly conductive waters, such as sea water, with a strong current and a consumption rate of about 4 kg per ampère per year.

— **by applying an external e.m.f** and setting up an external electric current. The earth which forms the anode or outlet is made of incorrodible material such as ferro-silicon or platinized titanium. The usual current densities for such anodes are 0.1 A/dm^2 and 10 A/dm^2 of electrode surface respectively.

The potential differences used are 5 to 10 volts in fresh water and 1.5 to 2 volts in sea water.

Earthing and current drainage points must be very carefully distributed.

Consumable anodes can only be used on simply-shaped structures with a small surface area or when a current outlet cannot be used for mechanical reasons (e.g. certain scraper-type settling tanks).

4. SECONDARY FACTORS IN CORROSION

The two main factors determining the conditions of corrosion are CaO/CO_2 equilibrium and oxygen content. Other factors play a part in the process of corrosion and have a major influence on its form and rate. They include the presence of dissolved salts (mainly chlorides), temperature variations, the physical cleanness of the water and the presence of microorganisms.

4.1. Influence of mineral content (sulphates, etc.)

In very pure deionised water (resistivity of 5 $M\Omega.cm$) and in the presence of oxygen, the corrosion rate of iron is negligible. Dissolution of very small quantities of salts in the water sets up corrosion in two ways:

● The total **mineral content** of water increases its conductivity and therefore lowers its resistance to the corrosion current. Corrosion can be started in this way by traces of chlorides and sulphates less than 1 mg/l.

Corrosion is stimulated by the Cl^-, Br^-, SO_4^{2-} and NO_3^- ions.

The old EVANS theory of peptization of the passivating films by chlorides is giving way to a more general hypothesis according to which overvoltages are weaker and dissolution of the steel requires less energy in the presence of certain anions. The theory is that adsorption of these anions onto the anodic regions assists migration of the iron towards the interface; it is argued that migration is much stronger with the halides than with anions of heavier atomic mass or with hydrated anions.

● **Chlorides** have a very marked influence: during the initial stage of corrosion, the concentration in the anodic region of Cl^- ions carried by the current increases, and at the same time H^+ ions appear as a result of remote precipitation of OH^- ions in the form of ferrous hydroxide. This produces a heavy local concentration of H^+ and Cl^- ions which prevents local precipitation of the hydroxides.

An increase in the chloride ion content of water increases statistically the probability that an infinite number of micro-anodes will be formed and leads to the spread of general corrosion and pitting.

This is one explanation for the appearance of pitting along the line of fluid flow inside pipes, due to the shifting of local concentrations of HCl.

One of the major factors in corrosion by increasing concentrations of chlorides is the relationship between the solubility of oxygen and the NaCl content of the water.

The oxygen content of salt or brackish water is substantially constant up to 5 g/l NaCl, at about 8 mg/l. From this point onwards, up to saturation at 310 g/l, the solubility of oxygen decreases:
— 5 mg/l at 100 g/l NaCl;
— 1 mg/l at 310 g/l NaCl.

Hache showed that corrosiveness increases up to a concentration of 10 g/l NaCl and then declines, closely in line with the decreasing solubility of oxygen. This means that brackish water is relatively more corrosive than brine. Because of their pH and the calcic TAC, natural brackish water may be only slightly corrosive but purified brackish water remains corrosive until excess alkalinity takes pH above 10.

● **Sulphates** influence corrosion in three ways:
— directly, by increasing mineral content and reducing electrical resistivity;
— indirectly, by involvement in the cycle of sulphate-reducing bacteria and in the spread of biological corrosion;
— through a specific process which degrades concrete (see page 49).

● **OH^- ions, chromates and silicates** tend, on the other hand, to form a protective film and to reduce corrosion.

● **Influence of Cu^{2+} ions:** copper in solution at the rate of less than 1 mg/l can considerably accelerate the rate of corrosion by electrical deposit of copper on the cathodes.

4.2. Influence of temperature on aerated water

● *Below 60 °C:*

In a closed vessel, heating renders the water unstable and causes scale formation followed by $CaCO_3/CO_2$ stabilization but at the same time increases corrosive action on steel. Calcification can however provide some protection.

Conversely, when water has been heated and balanced by the deposit of $CaCO_3$, it becomes aggressive and unbalanced when it returns to its original temperature.

It is also more corrosive but this time the absence of $CaCO_3$ allows corrosion to spread.

In an open industrial system two processes will take place:

a) Chemical effect:

The processes are the same as in a closed vessel but equilibrium can be shifted by CO_2 escaping, thus increasing scale-forming tendencies.

b) Electrochemical effect:

The coexistence of hot zones (where pH is lower) and cold areas (where pH is higher) produces local concentration of H^+ varying in inverse proportion and creates a heterogeneous surface with hot anodic regions and cold cathodic regions.

A temperature difference of 20 °C can result in a potential difference of 55 mV.

At high points, the formation of pockets of CO_2 gas and films of less saline condensation water result in the creation of anodic regions.

● *Above 60 °C:*

Escaping of oxygen predominates and results in the formation of differential oxygen concentration cells which are added to the previous cells to form a much higher potential. Corrosion is then very greatly increased.

Because of this, hot water systems in buildings are much more difficult to protect than industrial systems where the temperature gradients are generally 4 to 6 times less and average temperatures are below 45 °C (except at hot spots).

4.3. Influence of microorganisms

(see bacterial oxidation-reduction, page 106)

Biological corrosion is usually a secondary form of corrosion. It often shows itself by the formation of concretions in the shape of **nodules** lying very close together; these are likely to cause considerable obstructions and even underlying pitting which can develop into perforations.

These concretions consist of accumulations of fibrous ferro-bacteria, encased in partially dehydrated ferric oxide and often contain calcium carbonate.

The laminar nodular mass is often hollow. At the centre there is a liquid mass which is often very black and oxidizes rapidly in the atmosphere.

The presence of iron sulphide (release of H_2S) can often be detected by a quick quantitative analysis with a strong acid.

● *Active bacterial media:*

a) Iron bacteria:

These bacteria of the types Leptothrix, Crenothrix and Gallionella, live in an aerobic medium and synthesize their energy by consuming ferrous iron and converting it into ferric iron (see Chapter 28, page 980).

They speed up this slow oxidation by secreting enzymes, deprive the medium of oxygen and depolarize the anodes.   In this way they add further to the factors favouring corrosion.

b) Bacteria which oxidize nascent hydrogen:

These are of the autotrophic Hydrogenomonas type and synthesize their energy by oxidizing the cathodic hydrogen.

$$4 \text{ Fe} \longrightarrow Fe^{2+} + 8e^-$$

As their action depolarizes the cathode, they help to develop corrosion.

c) Sulphate-reducing bacteria:

These are of the type Desulfovibrio desulfuricans.   They are facultative anaerobic bacteria which can temporarily withstand saturation of the water with oxygen; this explains why they are found in cooling systems.   They are autotrophic and by the reduction of sulphates promote the formation of hydrogen sulphide from hydrogen from the cathodic regions.

$$SO_4^{2-} + 4 H_2 \longrightarrow S^{2-} + 4 H_2O$$

The sulphides released by reduction of the sulphates precipitate the ferrous ions in the form of black iron sulphide.

In practice, several reactions take place:

— depolarization of the cathode by the reducing bacteria:

$$SO_4^{2-} + M^{2+} + 8 H_2 \longrightarrow S^{2-} + M^{2+} + 4 H_2O$$

where M^{2+} represents an alkaline or alkaline-earth metal balancing the sulphate ion.

— formation of various corrosion products, in particular:

$$S^{2-} + Fe^{2+} \longrightarrow FeS$$

$$3 Fe^{2+} + 6 OH^- \longrightarrow 3 Fe(OH)_2$$

to give the overall reaction:

$$4 Fe^{2+} + SO_4^{2-} + 2 H^+ + 2 H_2O \rightarrow 3 Fe(OH)_2 + FeS$$

An important point to note is that **biological corrosion almost always follows electrochemical corrosion** which it only accelerates by consuming hydrogen.

The following conditions favour biological corrosion by sulphur bacteria:
— anaerobic medium;
— pH 5.5 to 8.5;
— presence of sulphate;
— presence of inorganic (PO_4 and Fe^{2+}) and organic substances which favour the spread of bacteria;
— optimum development temperature of 30 to 40 °C.

Consumption of oxygen at the surface of the nodules reduces the amount of oxygen diffused towards the interior of each nodule, which becomes increasingly anaerobic thus favouring the growth of reducing bacteria.

The formation of sulphides leads both to the precipitation of iron sulphides and to the reduction of the ferric hydroxide:

$$2\ Fe\ (OH)_3 + 3\ H_2S \longrightarrow 2\ FeS + S + 6\ H_2O$$

The interior **volume of the nodule decreases** as the sheathed ferric hydroxide is converted into denser iron sulphide.

Biological corrosion therefore consists of a strengthening of the association formed between the families of bacteria mentioned earlier. The presence of sulphate-reducing bacteria is demonstrated more by analysis of the deposit and the presence of sulphur than by the outward appearance of the nodules.

4.4. Influence of the surface condition and cleanness of the water

The presence of deposits already existing in an old, fouled system or originating from the distribution of badly-filtered water can cause corrosion in two ways:
— non-aerated zones which can generate anodic regions are created under the deposit;
— areas are formed where various strains of bacteria develop and reducing, depassivating reactions take place.
Stainless steels are specially prone to this process.

4.5. Influence of pH

The solubility of Fe^{2+} falls substantially as pH rises and an insulating film of ferrous hydroxide (and other oxides already mentioned) forms as a result; this leads to a predominant increase in the cathodic surfaces and the reduction of the anodic regions to very small surfaces. The density of the corrosion current on these anodic regions increases as their area decreases.

The risk of perforating corrosion or pitting in the presence of oxygen increases as pH approaches 10, the value at which it has been seen that corrosion decreases; it stops at 10.5 in most natural or pure water.

This explains why many cases of local corrosion have been observed in the presence of decarbonated waters with a pH between 9 and 10, which are not sufficiently alkaline to maintain the Tillmans' film.

4.6. Damage due to flow velocity

This is an extremely important but complex factor which involves physical, mechanical and electrochemical phenomena. Three types of damage can be caused in this way:

● *by cavitation:* this results from the existence of local levels of hydrostatic pressure, above and below the vapour tension of the water, which cause vapour bubbles to be released and then destroyed by implosion at very high pressures, resulting in uneven hollowing out of the solid metal.

● *by erosion-abrasion:* caused by the kinetic energy of particles of grit and other matter in the water, which cause continuous destruction of the protective layer by regular, uniform abrasion of the solid metal. Hömig maintains that the mechanism is simultaneously mechanical and electrochemical.

● *by erosion-corrosion:* due exclusively to interference with the formation of the continuous film as oxygen is diffused at a rate depending on the water flow rate. The mechanism is exclusively electrochemical: in the **absence of oxygen,** the protective film of ferrous hydroxide, governed by saturation of the boundary layer with Fe^{2+}, cannot form at high flow rates; **in the presence of oxygen,** the phenomenon is more complex because of the rate of oxygen diffusion and the possibility that a protective film of ferric oxide can form. As rate of flow increases, a zone of acid corrosion is observed, followed by increased passivation and finally no formation of any protective film.

5. CORROSION OF STAINLESS STEEL

5.1. Definitions

Stainless steel is by convention an iron-chromium alloy containing more than 11.5 % chromium. According to its crystalline structure, stainless steel is classified in one of the four categories: martensitic, ferritic, austenitic and austeno-ferritic.

● *Martensitic steels.*

They have a quadratic crystalline structure. They are very little used in boiler-making because they are difficult to form and weld. They are generally 13 % chromium steels, containing less than 0.1 % carbon.

● *Ferritic steels.*

These steels have a centred cubic crystalline structure. High chromium grades are strongly resistant to oxidation at high temperatures, particularly in a sulphur atmosphere.

● *Austenitic steels,*

They have a centred-face cubic crystalline structure. In addition to more than 16 % chromium, they contain over 6 % nickel, which promotes austenitic crystallization. These stainless steels are the most resistant to corrosion and are therefore of particular interest to water treatment engineers. Their properties vary with their carbon, chromium and nickel content.

● *Austeno-ferritic steels.*

An increase in chromium content tends to produce a ferritic structure (alpha-forming element) while a reduction in nickel content tends to produce an austenitic structure (gamma-forming element). Consequently austenitic stainless steels contain a certain amount of ferrite.

These steels have greater mechanical strength than austenitic steels and are resistant to intergranular and stress corrosion.

The compositions of the main austenitic and austeno-ferritic stainless steels used in France are listed in the table page 42.

French standard reference	Z6CN18-09	Z2CN18-10	Z6CND17-12	Z2CND17-13	Z1CNDU 25-20	Z5CNDU 21-08
Structure	Austenitic	Austenitic	Austenitic	Austenitic	Austenitic	Austeno-ferritic
Percentage composition:						
Carbon	$\leqslant 0.08$	$\leqslant 0.03$	$\leqslant 0.08$	$\leqslant 0.03$	$\leqslant 0.02$	$\leqslant 0.06$
Chromium	18 — 19	18 — 19	17	17	19 — 22	20 — 22
Nickel	9	9-10	11 — 12	12 — 13	26 — 27	6 — 9
Molybdenum			$\geqslant 2$	$\geqslant 2 - 5$	4 — 4.8	2 — 3
Copper					1 — 2	1 — 2
American equivalent	AISI 304	AISI 304 L	AISI 316	AISI 316 L		
German equivalent	X5CrNi18-09	X2CrNi18-09	X5CrNiMo18-12	X2CrNiMo18-12		
Commercial reference						
Ugine	NS 21 AS	NS 22 S	NSM 21	NSM 22 S		
Creusot-Loire					Uranus B6	UranusB50

There are even higher grades such as Hastelloy made by Union Carbide, Incoloy made by International Nickel and Monel made by Wiggin which are used for special applications only.

5.2. Corrosion of stainless steels in a liquid medium

When suitably alloyed with chromium, iron becomes passive in many types of medium.   The extent to which the alloy is stainless then depends on the stability of its passivation which is greatest in an oxidizing medium but may disappear in a reducing medium.

Passivation is achieved by the formation of a thin film of chromium oxide on the surface of the metal.   Any accidental break in this protective film will result in corrosion except in a strongly oxidizing medium when the film will reform automatically (stable passivity).   The addition of nickel increases resistance to corrosion in a slightly-oxidizing or non-oxidizing medium, while molybdenum improves resistance in the presence of reducing acids or halogen ions (chlorides, bromides, iodides).

It should be noted, however, that stainless steel can corrode in a strongly oxidizing medium, such as nitric acid containing hexavalent chromium, as a result of the phenomenon known as **transpassivity.**

The range of passivity can be widened by adding certain elements to stainless steels.   For example, the addition of silicon can prevent transpassivity while the addition of copper improves resistance to corrosion by sulphuric acid.

5.3. Various forms of corrosion of stainless steel

5.3.1. GENERAL CORROSION

Like all metals, stainless steels can in certain cases be attacked evenly over their whole surface. If the medium is not sufficiently oxidizing, the protective film on the surface of the metal can ultimately disappear, resulting in generalized attack (unstable passivity).

Furthermore, the surface condition of the metal affects the nature of its passivity and resistance will be greatest if the surfaces are not contaminated by particles of iron or scale of various kinds.

5.3.2. INTERGRANULAR (INTERCRYSTALLINE) CORROSION

This kind of corrosion occurs mainly near welding beads on stainless steel; it can also result from hot forming of the metal or inefficient thermal treatments. This phenomenon is due to local reduction of chromium content when intergranular chromium carbide, acting as a corrodible anode in relation to the rest of the metal, forms at high temperatures (400 to 800 ºC) near welding beads. In an acid medium the cohesion of the grains is then broken down and the metal becomes brittle.

This intergranular corrosion which is peculiar to austenite steels can be avoided in two ways:
— by bringing down the carbon content of the steel to a sufficiently low level ($\leqslant$ 0.03 %) in order to restrict the formation of chromium carbide;
— by using a steel stabilized with niobium or titanium, which form stable carbides with the carbon.

5.3.3. PITTING.

As opposed to what happens with ordinary steel, dissolved oxygen generally helps to passivate stainless steel except if chlorides or bromides are present, when the phenomenon of corrosion by pitting occurs. This fairly common and very dangerous type of corrosion results in the formation of perforating pits which may be almost invisible on the surface. The likelihood of pitting corrosion of stainless steel by a chloride-containing solution increases with the amount of air in the solution. Molybdenum stainless steels with a high chromium content and a low carbon content (e.g. Z2 CND 13) are relatively resistant to this type of corrosion. It is difficult to state a general rule, however, because the occurrence of pitting depends on a number of factors such as pH, degree of aeration, temperature, salt content, suspended solids content etc. of the medium; in certain cases higher grade alloys such as Uranus may have to be used.

5.3.4. CAVITY CORROSION

There are very few metals not affected by this type of corrosion. It is therefore a general phenomenon occurring in stagnant zones where diffusion is difficult if not impossible. It is found more particularly under deposits of scale, oxides, etc., under fouling, under non-metallic, non-leakproof joints, etc.

In the case of stainless steels, the development of this type of corrosion is a complex process. It is triggered off by differential oxygen concentration, leading to the formation of a small cell in which the corrosion products are trapped. If this corrosive medium consists, for example, of aerated water which is practically neutral but contains chlorides, hydrolysis of the primary corrosion products in the cell will result in the formation of hydrochloric acid which, beyond a certain critical concentration, will start the process of cavity corrosion.

Cavity corrosion is therefore characterized by an incubation period which may last several months; but when the process starts its progress can be very rapid. In such cases, corrosion is increased by the formation of local electric cells between the passive and the active metal and these quickly destroy the passivating film.

If the corrosion are washed off by mixing the water during the incubation period, the process is halted and has to start again from zero.

Nickel and molybdenum prolong the incubation period and thus increase the chance that the process will be halted during that period. However, once the incubation period is over and the process of cavity corrosion has started, its progress will generally be just as quick in high nickel and molybdenum steel as in steel containing less of these constituents.

In order to prevent cavity corrosion to the greatest possible extent, the conditions which favour development of the phenomen giving rise to differential oxygen concentration must be eliminated. For this purpose, all variations of oxygen concentration in the medium must be eliminated. It would be foolish to suppose, however, that by keeping the medium saturated with oxygen and by stirring, a saturation concentration of oxygen can be obtained in zones which are difficult to reach.

Flow rates must be high enough (if possible over 3 metres/sec) to prevent the formation of deposits and appliances must be designed so that there are no actual or potential dead zones in operation. If this is impossible, means must be provided of draining and cleaning such dead zones at intervals.

5.3.5. STRESS CORROSION.

This type of corrosion can occur in austenitic steels subjected to mechanical stresses which are either residual from a previous treatment such as stamping or welding or are set up while in service. It can occur in a non-corrosive medium but is activated by the presence of hot solutions of alkaline or alkaline-earth chlorides.

After an incubation period of varying length, corrosion appears in the form of deep cracks which spread quickly.

In order to prevent this type of corrosion, it may be necessary in some cases to relieve the stresses by an appropriate thermal treatment.

5.3.6. SPECIAL TYPES OF CORROSION

As in the case of ordinary steel, formation of the passivating film can be prevented by high velocity flow of a corrosive fluid and localized corrosion takes place as a result. The protective film can also be destroyed mechanically by abrasion from hard particles carried along in the fluid.

Galvanic corrosion can occur in mixed joints such as a weld between stainless steel and mild steel. The mild steel is then anodic in relation to the stainless steel and is subject to corrosion. On the other hand, there is then virtually no pitting or cavity corrosion of an 18-10 Mo stainless steel.

Heavy cold-forming operations can lead to the formation of work-hardened martensite which is anodic in relation to the austenite of the rest of the structure and therefore becomes a preferential zone for corrosion. This phenomenon can be avoided by using low-carbon, high-nickel steels in which the austenites are very stable.

5.4. Use of stainless steels

The general principles for the use of stainless steels can now be summarized.

At the design stage:
— surface/volume ratio must be kept as low as possible;
— sharp-angle joints and contorted layouts should be avoided;
— dead zones should be eliminated by giving the liquid a correct flowpath—drains must be provided for all unavoidable dead zones;
— zones involving heavy forming operations, where work-hardened martensite can concentrate, should be avoided;
— overlapping joints should be welded on both faces;
— mild steel supports should be secured to a stainless steel structure by means of an intermediate stainless steel plate, in order to reduce the amount of mild steel passing into the metal of the structure at welding.

At the construction stage:
— all contact with ferrous metals must be avoided;
— the welding process used should not affect the corrosion resistance of the structure. Avoid high-yield processes;
— all parts should be degreased and all ferrous inclusions should be eliminated after they are formed or machined;
— all welds should be thoroughly descaled;
— natural passivation should be completed by contact with a nitric acid solution.

5.5. Choice of a stainless steel

In the water treatment industry, a stainless steel is usually if not always chosen for its resistance to corrosion.

In view of the following facts:

— a stainless steel can always be liable to different types of corrosion;

— any element in the medium, such as oxygen, which is in contact with the steel may stabilize one type of corrosion but may activate another;

— the actual conditions of use (fluid velocity, amount of turbulence, possibility of deposits) are definite factors in the appearance of corrosion;

— the machining of stainless steel can affect its resistance to corrosion,

— it is virtually impossible to lay down specific rules for choosing a particular stainless steel since every problem of corrosion resistance is a separate case.

In general:

— for welded structures, preference will be given to very low carbon steels in order to forestall intergranular corrosion;

— it should be remembered that an increase in molybdenum content, generally accompanied by increased nickel content, improves resistance to corrosion, particularly by chlorides, and that the addition of copper enhances resistance in a sulphur medium.

General preference should therefore be given to a steel with a very low carbon content and a high alloying element content e.g. grade Z1 CNDU 25-20, but this type of steel is so expensive that it can only be used when the risk of corrosion is particularly high.

In view of the difficulty of choosing a stainless steel, it is suggested that, rather than planning from the outset to use a particular alloy, it is better to consider carefully whether the problem might not be handled with more chance of success by using a carbon steel without special protection combined with appropriate conditioning of the medium or the use of a carefully selected and applied anti-corrosion coating.

6. DEGRADATION OF CONCRETE

Concrete is a material made from various types of aggregate and cement. It frequently contains a steel reinforcement.

In theory, this reinforcement cannot corrode until degradation of the encasing concrete takes place, because of the positive electrical charge of the iron in the concrete. The potential of the iron in the concrete, which has a pH of about 11.6, is approximately $+$ 100 mV referred to the hydrogen electrode. Action to prevent degradation must therefore be directed first to the concrete, except in the case of prestressed concrete where the reinforcement, as a result of its lightness and its heavy stresses, is more often liable to stress embrittlement and chemical corrosion by percolating water. The degradation of concrete is mechanical at first and then chemical.

6.1. The *mechanical causes* of degradation.

They are of three kinds:

— excessive permeability. In the presence of aggressive water, the minimum requirement is a very compact concrete rating 300 to 400 kg/m³.

— existence of cavities and cracks due to inefficient preparation of the concrete which can be improved by increasing its plasticity by using a water/cement ratio less than 0.45 or adding a plasticizer.

— erosion caused by water velocities of over 4 m/s through pipes or by excessive thermal gradients.

Fig. 10. — A typical corroded concrete pipe

6.2. The *chemical causes*.

They are linked with the composition of the cement and the corrosiveness of the water in contact with it.

The main components of cement are silica, lime and alumina with iron, magnesia and alkalis as secondary constituents.   It generally forms a very basic medium containing a large reserve of solubilizable salts.

When cement, and particularly Portland cement, sets, large quantities of bound lime are released in the form of $Ca(OH)_2$ and tricalcic aluminate "C_3A" is formed.

According to the composition of the cement and the chemical nature of the water, this solubilization can affect the free lime and sometimes the alumina. The quantity of the main components of various cements are summarized below.

	Portland CPA	Ciment fondu (aluminous)	Supersulphated cement CSS
SiO_2	$20 - 25\%$	$5 - 16\%$	
Al_2O_3	$2 - 8\%$	30%	
CaO	$60 - 65\%$	$35 - 40\%$	50%
SO_3	$< 4\%$	$< 2.5\%$	$> 5\%$

Chemical degradation can be caused in several ways:
— aggressivity of CO_2;
— attack by strong acids;
— action of ammonia;
— action of sulphates;
— attack by strong alkalis;
— bacterial corrosion with formation of H_2S.

6.2.1. AGGRESSIVE ACTION OF CO_2

Structures are attacked by carbon dioxide in the presence of soft water or when excess CO_2 goes above 15 mg/l.

However, there is a high residue of lime alkalinity in the pores of the concrete and this allows the precipitation of a deposit of $CaCO_3$ and other salts, which temporarily slows down the destruction of the concrete by such water.

This destruction can be tolerated if the calcium bicarbonate alkalinity of the water exceeds 5 to 6° TAC and pH is over 6.5, that is around $CaCO_3/CO_2$ equilibrium.

6.2.2. ATTACK BY STRONG ACIDS

The attack increases in severity with the solubility of the calcium salts formed.   Increasing corrosivity is found in the presence of the following acids:

— phosphoric;
— sulphuric;
— nitric;
— hydrochloric, etc.
The organic acids are equally destructive.

Where the level of acidity is moderate, resistance to corrosion is improved by lowering the water/cement ratio and/or by using aluminous cements (which are very difficult to use). This applies for pH values above 2 and subject to certain precautions.

In general, however, structures are often liable to crack and can only be protected completely by using the correct type of coating.

Furthermore, several laws on the discharge of waste water recommend that the pH of water flowing in contact with walls should be between 4.5 (or 5.5) and 9.

6.2.3. ACTION OF AMMONIA

Ammonia in waste water can contribute to the destruction of concrete in two ways:
— by the development of nitrifying and therefore acidifying reactions, which can however develop only in an aerobic medium, as in atmospheric coolers for example;
— by the release of the ammonia displaced by the lime which accelerates solubilization of the latter and causes rapid degradation of the cement.

The same process can be started by the salts of magnesium or any other base weaker than lime.

Excessive concentrations of NH_4 and Mg should be avoided therefore, particularly when sulphates are also present.

6.2.4. ACTION OF SULPHATES

This action, which affects many cases, is complex; it is based on the conversion of calcium sulphate into expanding Candlot salt, also known as ettringite:
— sulphatation of the free lime in the cement by the sulphates dissolved in the water:

$$Ca(OH)_2 + Na_2SO_4 + 2 H_2O \longrightarrow CaSO_4, 2 H_2O + NaOH$$

— conversion of the aluminates in the cement into Candlot salt which expands very strongly (factor 2 to 2.5):

$$3 CaOAl_2O_3, 12 H_2O + 3 CaSO_4, 2 H_2O + 13 H_2O \longrightarrow$$
$$3 CaO, Al_2O_3, 3 CaSO_4, 31 H_2O$$

If magnesia is present, decomposition of the alkaline silicates in the cement may be added to these two mechanisms.

The German standard DIN 4030, lays down limits for assessing the aggressiveness of salt water on standard types of concrete:

	slightly aggressive water	strongly aggressive water	very strongly aggressive water
pH	6.5 — 5.5	5.5 — 4.5	< 4.5
Aggressive CO_2 HEYER's method	15 — 30 mg/l	30 — 60 mg/l	> 60
Ammonium ions mg NH_4^+/kg	15 — 30	30 — 60	> 60
Magnesium ions mg Mg^{2+}/kg	100 — 300	300 — 1 500	> 1 500
Sulphate ions mg SO_4^{2-}/kg	200 — 600	600 — 2 500	> 2 500

If the water is very strongly aggressive, coating is recommended; in the case of fairly aggressive water, use of a slag-based cement with a high hydraulicity index is possible, if the other chemical parameters of the water play no part.

Examples:
— clinker cement (CLK) containing 80 % slag aggregate
— blast-furnace cement (CHF) containing 60 — 75 % slag.

Various cements with a low "C_3A" content (see page 48) are therefore available for sea water conditions.

6.2.5. ATTACK BY STRONG ALKALIS (NaOH, KOH, Na_2CO_3)

Water with a strong alkali content is destructive to all cements because certain alumina-based constituents are liable to be solubilized; if there is no coating, therefore, it is advisable not to allow water with a pH in excess of 12 with ordinary cements or in excess of 8.5 in the case of aluminous cements.

6.2.6. BACTERIAL CORROSION WITH FORMATION OF H_2S

This kind of corrosion occurs in town sewers. The principle of corrosion in an anaerobic medium has already been described, but whereas the process follows and extends previous chemical corrosion in cooling systems, it is generally the result of the anaerobic fermentation of deposited solids in the case of domestic sewage or strong effluent.

There are two stages:
— formation and release of H_2S;
— oxidation of H_2S in the presence of water to form H_2SO_4.

These reactions are speeded up when pH falls below 6 and as the temperature of the water rises.

In a sewer, the attack takes place above the liquid surface because of the deaeration and condensation of the water.

Virtually the only way of preventing corrosion of this type is to maintain a high flow rate, after prior settling or preaeration of the water and to avoid turbulence and deaeration in the pipes.

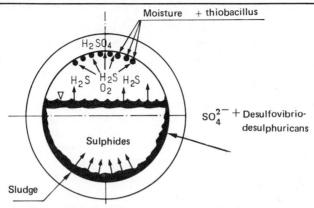

Fig. 11. — *Attack mechanism by hydrogen sulphide on a sewer pipe.*

7. CORROSION OF NON-FERROUS METALS

The principal behaviour characteristics of a number of metals, both pure and in alloys, are summarized below. Three general factors should be taken into account in all cases:

— different materials should not be used together too often as this is a source of rapid corrosion by macrocells;

— unduly high temperatures of metal element in contact with the water can cause heavy corrosion by deaeration and differential oxygen concentration; the same applies to variations of temperature;

— any stagnation causes oxygen depletion in the water. This can result in destruction of the passivating film and the release of toxic metallic ions in solution (Pb, Cu).

7.1. Aluminium

With a potential of —1.34 volts, this highly electronegative metal is very liable to corrosion in theory but in fact is passivated by oxidation in damp air and by the formation of a protective film of alumina. This passivation can also be acquired artificially by anodization.

Aluminium is sometimes used in contact with sea water, provided, first, that no impurities are deposited to form a cathodic region and secondly, that a very pure industrial grade is used.

Slightly acid water is acceptable but contact with alkaline water should be strictly avoided.

7.2. Copper

In aerated water, copper is passivated naturally by a film of cupric oxide CuO which is only very slightly soluble but can in practice be solubilized as a complex, by ammonia or cyanides.

Nevertheless, this film, unlike that formed by aluminium, is often porous and precautions must be taken during the manufacture and use of pipes.

Perforating corrosion or pitting sometimes occurs in the presence of cold, mineralized waters (type I pitting) and in annealed pipes. The same applies to hot water, with a low mineral content (type II pitting).

7.3. Lead

As lead oxide PbO and hydroxide Pb $(OH)_2$ are relatively soluble, all waters containing oxygen are highly corrosive and dangerous, unless high bicarbonate alkalinity (TAC over 12°) and a low concentration of free CO_2 (pH slightly alkaline) are present at the same time.

In that case, lead hydroxycarbonate may be formed: this is a good deal less soluble than the hydroxide and therefore protective if calcium carbonate is precipitated at the same time.

The use of lead piping for drinking water is, however, less and less recommended now and should be completely avoided for soft water.

7.4. Galvanized steels

This section deals only with coatings produced either by the electrodepositing of zinc or by hot dip galvanizing (by immersion in a bath of zinc at 450 °C or by Sendzimir's continuous process).

In these cases, a so-called "eta" (η) zinc layer, of variable thickness, is bonded strongly on the steel and its external surface is oxidized, with formation of zinc hydroxide, oxide or hydroxycarbonate according to the temperature and alkalinity of the water.

This film is of a very complex nature and slows down corrosion of the zinc by checking the diffusion of oxygen; the rate of corrosion which is high in soft water becomes very slow in water containing calcium bicarbonate.

Chlorides in sufficient concentration and particularly in an alkaline medium may precipitate insoluble oxychlorides but this simultaneously increases the risk of pitting.

The film differs from that of other metals in being a poor conductor. It does not change the potential of the zinc metal and does not raise its grade. It therefore acts more as a coat of paint or a soluble anode than as a passivating film.

When the water is corrosive and the film is destroyed, the strongly electro-negative (— 0.776 volt) zinc will itself suffer accelerated corrosion which will end by attacking the iron so that rust forms.

● *Effect of temperature:* Corrosion rate rises swiftly with temperature, reaches a peak at 60 °C and then falls back to the initial rate at 100 °C. At 60 °C, all the zinc hydroxide appears to be converted into more porous oxide which does not adhere as strongly, to form a film with a higher potential than that of the zinc. The polarity of the protective film/iron cell is reversed as a result. This causes accelerated or perforating corrosion of the bare iron surface.

● *Action of copper:* The arrival of Cu^{2+} ions in solution (exogenous copper from the system upstream) renders the zinc oxide film conductive and the iron may then corrode quickly.

The galvanizing of steel does not therefore slow down corrosion permanently except in the case of non-corrosive cold water. The temporary nature of the protection which it affords should therefore be borne in mind when the water is corrosive, particularly in the case of sea water.

7.5. Brasses

These are copper-zinc alloys with the following standard compositions:

	1	2	3
Copper	70 %	60 %	76 %
Zinc	29 %	40 %	22 %
Aluminium	—		2 %
Tin	1 %		

The third composition is recommended for use in salt water.

Dezincing or dissolution of the zinc with release of the residual Cu sometimes occurs in the first type of brass and less frequently in the others. The metal becomes porous and brittle.

Dezincing, which is increased by the presence of salts, causes a cell to be formed, with the brass as the anode, and this quickly leads to destruction of the alloy but this can be blocked by the presence of arsenic or antimony.

Grades of this alloy which eliminate this problem are now available.

8. ACTION OF SEA WATER

8.1. Corrosiveness and formation of scale

The high degree of corrosiveness of sea water is governed by its oxygen content and its temperature. Cold water pumped from great depths in the North Sea is for example much less corrosive than hot water from the surface in the Tropics. This type of corrosiveness cannot be classified by Ryznar's experimental index.

When corrosion takes place, a process of passivation is sometimes observed which slows down the actual corrosion rate more than in the case of fresh water.

At temperatures over 60 °C, sea water is also scale-forming against limestone, or non-metallic materials such as earth rocks; on the other hand, rapid corrosion can be combined with scale formation where there is contact with ferrous or non-ferrous metallic materials. The electrochemical corrosion processes increase in strength with the difference between the metallic materials.

Mg^{2+} ions in sea water act as a moderating factor capable of halving the strength of the corrosion currents.

8.2. Protection against scale formation

Below 100 °C, formation of scale is due exclusively to calcium bicarbonates. Over 100 °C, and above a certain concentration, calcium sulphate is also precipitated.

The first problem can be resolved:
— by acidifying the sea water to bring down its TAC;
— by sequestering with polyphosphates; their action is limited to temperatures below 70 °C but at higher temperatures they can be replaced by phosphonic compounds which are stable above 130 °C (taking the maximum temperature at certain hot spots).

The second problem can, however, only be dealt with by partial elimination of the calcium.

8.3. Protection against corrosion

8.3.1. STRUCTURAL MEASURES

— Use of concrete pipes or steel pipes coated with epoxy pitch, ebonite or bituminous products;
— Heads of heat exchangers and distributors coated with plastic material (rilsan, ...);

— Tubular exchangers made of naval brass or 76.22.2 brass, which is less affected by dezincing.

The use of titanium exchangers is gaining ground rapidly (power stations).

8.3.2. CONTINUING MEASURES

a) Cathodic protection:
— By means of reactive anodes made of Mg or Al, which are sometimes used in sea water (which has low resistivity). These anodes provide only minor, limited protection.
— By applying an external e.m.f. This method can be used to protect bar screens, straining drums and long pipe lengths.

b) Chemical methods :
— Protection of brass tube nests by injecting ferrous sulphate which leads to the formation of a passivating film of ferrous hydroxide. This method is used on ships' condensers.
— By adding phosphate-zinc inhibitors to the feed water for steel tube nests. This inhibition provides sufficient protection (Chapter 14).

c) Deaeration of dissolved oxygen:
— By vacuum deaeration or gas stripping (see Chapter 13). This method is designed for water injected into oil workings.
— By chemical reduction with catalyzed sodium bisulphite (see Chapter 14).

8.4. Protection against organic fouling

Fouling is a major cause of corrosion under deposits and of loss of heat exchange capacity.

There are three possible sources:
— organic mucus of bacteria and algae;
— fixing of soft organisms such as ascidia, sea anemones;
— incrustation by colonies of hard organisms such as crustaceans, mussels, oysters and barnacles which are difficult to destroy and cause perforating corrosion under deposits through differential oxygen concentration.

It is much easier to control the spread of algae than of crustaceans.

For algae, intermittent chlorination is recommended at the rate of 3 to 8 grammes of Cl_2 per m^3 of sea water, once to four times a day; the aim is a free chlorine content of 1 g/m^3 for 10 to 15 minutes at the outlet from the system.

Chlorine distribution must be very carefully controlled at pump intakes. The regulations for storing chlorine are so strict that efforts are now being made to replace it by direct electrolysis of sea water.

For crustaceans, continuous addition of 0.5 to 1 g/m^3 of chlorine for 5 consecutive weeks in spring and autumn is recommended; continuous chlorination is recommended when the water temperature exceeds 20 °C.

THE HALLOPEAU AND DUBIN CO_2 GRAPH

TEMPERATURE

Saturation pH in $CaCO_3$: $pHs = \log K'_S - \log K'_2 + 9 \cdot 2 - \log CaO + \log \left(1 + 2\dfrac{K'_2}{H}\right)$

Free CO_2 : $\log CO_2 = 0 \cdot 2 - \log K'_1 + \log Alk - pH - \log \left(1 + 2\dfrac{K'_2}{H}\right)$

Saturation $Mg(OH)_2$: $pHsMg = 2 \cdot 3 + 0 \cdot 5 \log K_1 Mg - \log K_2 - 0 \cdot 5 MgO$

Ca or MgO p.p.m.

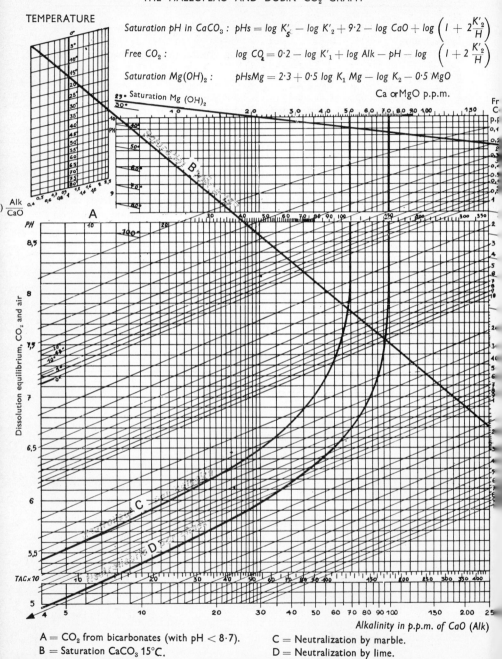

A = CO_2 from bicarbonates (with pH < 8·7).
B = Saturation $CaCO_3$ 15°C.
C = Neutralization by marble.
D = Neutralization by lime.

(1) CaO is the total calcium expressed in p.p.m. of CaO.

<table>
<tr><td>**3**</td><td># THEORY OF THE MAIN TREATMENT PROCESSES</td></tr>
</table>

In order to purify water it is generally necessary to combine a number of basic processes, which may be physical, chemical or biological in character and which have the effect of removing first of all the suspended solids, followed by the colloids and any dissolved inorganic or organic pollutants. Finally, some characteristics of the water need adjustment.

Depending on the objective in view, a variety of principles can be applied at each stage. We shall now consider the most important of these:

1. PHYSICO-CHEMICAL PROCESSES

1.1. Removal of suspended solids and colloids

1.1.1. GENERAL REMARKS

Without doubt, the largest and most evident contribution to water pollution is made by the presence of various solids.

This solid fraction needs to be removed in order to eliminate a wide range of problems, chief amongst which are: the clogging of piping and coolant systems, abrasion caused to pumps and measuring equipment and the wear of various items of equipment, which have serious implications for the running and maintenance costs of manufacturing units. Where drinking water and effluents are concerned, an additional factor is the need to comply with official regulations.

Two different principles can be applied to the separation of solid particles from water. These are:

— The direct use of gravity in the form of straightforward sedimentation, where the determining factors are the size and specific weight of the particles, or by flotation, where air bubbles systematically introduced into the suspension attach themselves to the particles. The former process can be accelerated with the aid of centrifugal force applied in cyclones and centrifuges.

— Filtration or screening.

However, the unsophisticated application of these principles comes up against the difficulty caused by the wide range of particle sizes, even for a specific type of pollution.

The table hereafter lists a number of materials and organisms with their average size and a rough indication of the time needed for these particles to settle vertically through one metre of water under the effect of gravity alone.

Particle diameter (mm)		Settling time through 1 m (roughly)
10	gravel	1 second
1	sand	10 seconds
0.1	fine sand	2 minutes
0.01	clay	2 hours
0.001	bacteria	8 days
0.0001	colloidal particle	2 years
0.00001	colloidal particle	20 years

The fact is that a litre of good quality water may contain tens of millions of particles of micron size. Despite their number, these particles will have a total weight of less than 0.1 mg.

It follows from these figures that the quality of water is a relative concept and that it is necessary, bearing in mind the requirements, to specify the acceptable level of residual pollution. In addition it is, of course, necessary to determine the methods of measurement used based on the two quality criteria normally applied, i.e. the concentration of suspended solids and the turbidity of the water, which reflects the presence of small particles which are not individually visible.

● *Colloidal suspensions and their stability in water.*

The above table underlines the very slow natural settling rate of fine, or so-called **colloidal particles,** which make up a large part of the pollution and are specific cause of turbidity. Any treatment requires as a preliminary step a change in conditions such that the particles can coalesce between themselves to form large agglomerates which are easy to remove. This coalescence does not take place naturally, since colloidal suspensions are characterized by specific forces which hold the matter in the dispersed state with a remarkable degree of stability over time.

This **stability** is due to solvation or the protective action of certain adsorbed substances as well as to the overriding effect of electrostatic forces which act to repel the particles away from each other.

In natural water colloids invariably carry a negative charge. In waste water this is often the case.

As a consequence of solvation, the colloidal particle must be regarded as being partially ionized at the surface, and with a microscope it can be seen that the application of an electrical field causes it to move.

The theoretical interpretation of these electrokinetic phenomena, and of electrophoresis in particular, has led to the hypothesis of a double ionic layer at the solid-liquid interface and to the attribution of a definite structure to this double layer (cf. fig. 12).

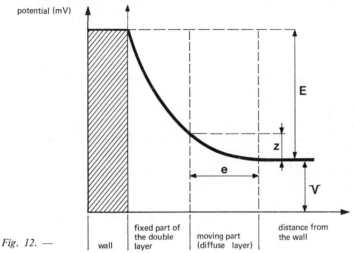

Fig. 12. —

One part of the double layer is anchored to the wall while the other is made up of a diffuse cloud of ions. V is the potential of the liquid in relation to the reference electrode.

● *Zeta potential.*

The potential difference between the surface separating the fixed and moving parts from the bulk of the liquid is traditionally called the Z (Zeta) potential.

Z is an electrokinetic potential, unlike the thermodynamic potential E calculated by Nernst's formula, which is the potential difference between the wall of the particule and the bulk of the liquid.

Z depends on both E and the thickness of the double layer. Its value determines the extent of the electrostatic forces of repulsion between the particles and hence their probability of adhesion.

The equipment used for measuring this is called the Zetameter (see page 948).

When a particle is subjected to an electrical field, it almost instantly attains a velocity such that there is equilibrium between the electrical force of attraction and the frictional force due to the viscosity of the medium. The following relationship between Z and the mobility of the particle is obtained by calculation :

$$Z = \frac{k\eta\mu}{D}$$

η = dynamic viscosity in dPa.s
D = dielectric constant of the medium
μ = mobility of the particle in μs/(V.cm)
Z = is expressed in mV.

The factor k depends on the relative values of the particle diameter and the thickness of the double layer of ions.

	k	Formula
Relatively large particle	4π	Helmholtz
Small particle, roughly spherical	6π	Huckel

The equation shows that all particles with an identical electrokinetic potential will have the same mobility whatever their radii.

The experimental curves given below show for water and sludge the percentage of particles that move in a variable potential field. The highest point of the curve gives the Zeta potential of the colloids contained in the water or sludge tested (fig. 13).

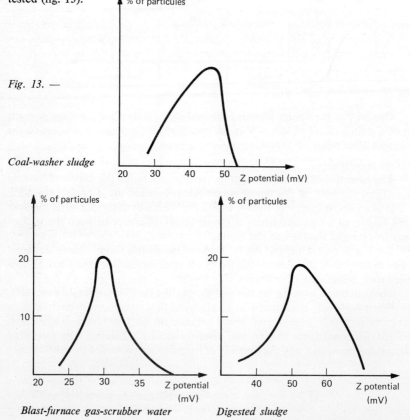

Fig. 13. —

Coal-washer sludge

Blast-furnace gas-scrubber water

Digested sludge

1.1.2. COAGULATION AND FLOCCULATION

● **Destabilization of a colloidal suspension:**

In order to bring about the separation of a colloidal suspension at a satisfactory rate, e.g. by gravity, it is necessary to agglomerate the colloidal matter to form substantially larger particles.

The suspension must therefore be converted by artificial means. Such conversion is the outcome of two separate processes:

— destabilization, usually achieved by the addition of chemical reagents which, by a bonding or adsorption mechanism, nullify the repulsive forces or act on the hydrophily of the colloidal particles;

— agglomeration of the "neutralized" colloids. This is the result of various forces of attraction acting between the particles brought into contact with each other initially by Brownian movement until they acquire a size of about 0.1 micron and then by outside mechanical agitation to bring the floc to the requisite size.

In the parlance of water treatment, it is customary to restrict the term "coagulation" to the process of destabilization and to apply the word "flocculation" to the aggregation of the neutralized colloids. The corresponding reagents are known as **coagulants** and **flocculants** respectively.

● **Coagulants and flocculants:**

a) Pattern of development:

The systematic use of polyvalent cationic inorganic salts as coagulants dates from the closing years of the last century. It was at that time that the laws governing their action were established, and it was shown that their coagulating effect depended on the valence of the ion carrying a charge opposite to that of the particles. The higher the valence, the more effective is the coagulating action (Schulze-Hardy theory).

This theory partly explains why trivalent iron and aluminium salts have been, and continue to be, widely used for all coagulation treatments applied to water.

The fact remains that, owing to hydrolysis, these coagulants have the disadvantage of modifying the physico-chemical characteristics (pH-value, conductivity) of the water so treated. When used in large doses, they cause an excessive amount of sludge, which is often troublesome. What is more, the agents do not always produce a precipitate possessing the characteristics necessary for effective separation.

These were the reasons which led to the favouring, firstly, of natural products, both inorganic (activated silica) and organic (starches, alginates), and then of synthetic products (polymers of high molar mass) known as **polyelectrolytes.**

b) Inorganic coagulants:

The most frequently used coagulants are iron and aluminium salts and especially, for economic reasons, aluminium sulphate and iron (III) chloride.

The coagulating action of these salts is the result of the hydrolysis which follows their dissolution without immediately leading to the formation of the hydroxide, e.g. $Al(OH)_3$. The intermediate aluminium compounds —hydroxy-aluminous complexes— not only provide the charges needed to neutralize the colloids but, according to some writers, are capable of polymerization, i.e. of forming bridges between the colloids and thereby initiating the flocculation process.

The amount of coagulant to be used is determined after tests on the water to be treated (see flocculation test, page 949, which can usefully be reinforced by verifying the Zeta potential of the suspension).

Depending on the water under examination, the curve showing variations in the Zeta potential as a function of the reagent dose may display widely varying characteristics.

To obtain water of good quality it is sufficient in the case of a curve approximating to curve (1) in Figure 14 to add reagent at rate A, which will make it possible to achieve a potential in the range from -3 to -5 mV. In the case of a curve of type (2), on the other hand, it is necessary to use addition rate B, which reduces the Zeta potential to zero.

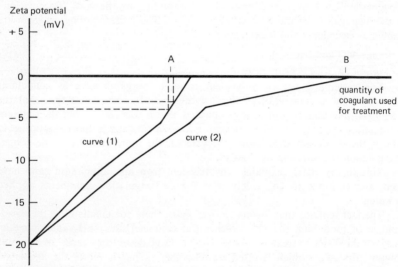

Fig. 14. — Determination of the treatment rate by the Zeta potential method.

The pH-value plays a very important part in the study of coagulation and flocculation phenomena. To illustrate: a part of the charge carried by the colloidal particles which have adsorbed OH^- ions is destroyed by an increase in the concentration of H^+ ions, and this reduces the stability of the colloidal suspension.

Similarly, it is preferable to work with a pH-value inside the range corresponding to the minimum solubility of the metallic ions of the coagulant used.

As far as the aluminium salts are concerned, the minimum concentration of Al^{3+} ions in solution is achieved with a pH-value of between 6 and 7.4. Outside this range, and depending on the mineralization of the water, a higher concentration of dissolved aluminium is liable to be found.

In some exceptional cases, the water to be treated contains a large quantity of matter rich in aluminium compounds. A simple increase in the pH-value brings about the coagulation of these compounds.

For the iron salts the pH range is much wider, and minimum solubility is reached as soon as the pH-value exceeds 5.

c) Flocculants:

— **Activated silica:** activated silica takes the form of a solution of polysilicic acid $(H_2SiO_3)_4$ obtained by the controlled polymerization of silicic acid. It lacks stability and has to be prepared at the point of use. Despite this, until the recent development of polyelectrolytes, it was for a long time the best flocculant capable of being linked to aluminium salts. It is still widely used for the processing of drinking water.

— **Organic polymers:** organic polymers are long-chain macromolecules either occurring naturally or obtained by the linking of synthetic monomers, some of which carry electrical charges or have groups which can be ionized. Natural polymers are rather inefficient, but the use of synthetic polymers has produced remarkable results. Each country has produced its own health regulations governing their use for the treatment of the water supply.

According to the ionic character of their active group, the synthetic polymers currently known as polyelectrolytes include the following:

— non-ionic polymers, almost exclusived polyacrylamides of a molar mass between 1 and 30 million.

$$\left[\begin{array}{c} CH_2 - CH \\ | \\ C = 0 \\ | \\ NH_2 \end{array}\right]_n$$

General formula of a polyacrylamide

— anionic polyelectrolytes, of a molar mass of several million, containing both groups permitting adsorption and negatively ionized groups (carboxyl or sulphuric groups) which extend the polymer. The best known is the polyacrylamide partially hydrolysed by soda, of formula:

$$\left[\begin{array}{cc} \left[\begin{array}{c} CH_2 - CH \\ | \\ C = 0 \\ | \\ NH_2 \end{array}\right]_n & \left[\begin{array}{c} CH_2 - CH \\ | \\ C = 0 \\ | \\ O - Na^+ \end{array}\right]_m \\ \text{Group} & \text{Ionized} \\ \text{permitting} & \text{group} \\ \text{adsorption} & \end{array}\right]$$

— the cationic polyelectrolytes, having in their chains a positive electrical charge due to the presence of an amine, imine or quaternary ammonium group.

$$\left[CH_2 - CH_2 - \underset{\underset{Cl^-}{\overset{H}{\mid}}}{\overset{H}{\underset{\mid}{N}}} \right]_n \left[CH_2 - \underset{\underset{Cl^-}{\overset{\mid}{NH_3^+}}}{\overset{H}{\underset{\mid}{C}}}\right]_n$$

Polyethylenimines   Polyvinylamines

● *Further possibilities for the use of polyelectrolytes:* with the production of relatively light, heavily charged cationic macromolecules, the area for the practical application of polyelectrolytes has been extended, in some favourable cases, to the coagulation process.

The positive charges of the polyelectrolyte neutralize the negative charges of the colloids and cancel out the Zeta potential. They are the better able to do this because, for a given mass of polyelectrolytes, there is a greater number of macromolecules and the chances of their encountering the colloidal particles are thereby enhanced.

Curves C_1 and C_2 below show these two types of action.

On the other hand, anionic and non-ionic polyelectrolytes do not cancel out the Zeta potential. The curves obtained with them resemble an adsorption isotherm: active radicals in the polymer chain are adsorbed on the particles, thus modifying the surface state of the particles and their Zeta potential, and causing them to agglomerate into a floc by a cross-linking mechanism.

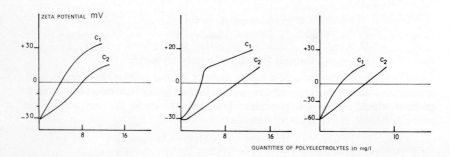

Fig. 15. — Quantities of polyelectrolytes in mg/l. Cationic.

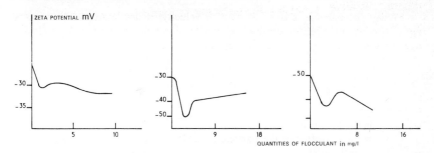

Fig. 16. — Quantities of flocculant in mg/l. Anionic.

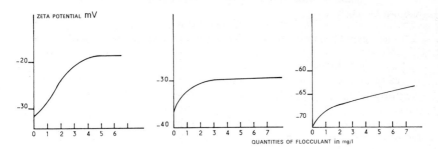

Fig. 17. — Quantities of flocculant in mg/l. Non-ionic.

1.1.3. SETTLING AND CLARIFICATION.

There are two types of substances that will settle out:
— **granular particles** that settle separately with a constant settling rate;
— **more or less flocculated particles** resulting from natural or induced agglomeration of the colloidal substance in suspension.

When the concentration of matter is low, the dispersed floc settles as if it were alone, but its settling rate increases as its dimensions increase as a result of encountering other finer particles: this is **flocculant settling**.

In the case of high concentration, the large quantity of floc causes a restrained overall sedimentation with a clearly marked interface between the sludge mass and the liquid above it: this is **zone settling**.

A. Settling of granular particles :

— **Theory.** When a granular particle is left alone in a liquid at rest, it is subjected to a motor force F_M, namely gravity, and to a resistant force F_T, the fluid drag which is the resultant of the viscous and inertial forces (fig. 18).

F_T

F_M

Fig. 18. —

When the particle is released at zero velocity, it accelerates and adopts a virtually constant velocity (terminal velocity) after a time t which is almost always negligible in comparison to the settling times in industrial operations.

This velocity is calculated from Newton's formula:

$$v^{2-n} = \frac{4\, d^{1+n} g\, (\rho_s - \rho_e)}{3\, C_{\rho_e}}$$

in which :

v   is the terminal velocity in cm/s
d   is the diameter of the particle in cm
g $= 981$ cm/s^2
ρ_s and ρ_e are the specific mass of the granular particle and the fluid
C   is the drag coefficient related to the Reynolds number Re by the formula $C = aRe^{-n}$
a and n are coefficients.

The Reynolds number $Re = \dfrac{v\rho d}{\eta}$ can be interpreted as the ratio of the inertia forces to the dynamic viscosity forces.

If the Reynolds number is small, the viscous forces are much higher than the inertial forces.   If the Reynolds number is large, the viscous forces are negligible.

The following table gives the different values of a, n and C depending on the Reynolds number.

Re	a	n	C	Formula
$10^{-4} < Re < 1$	24	1	$24/Re$	Stokes
$1 < Re < 10^3$	18.5	0.6	$18.5\, Re^{0.1}$	Allen
$10^3 < Re < 4 \times 10^5$	0.44	0	0.44	Newton

These formulae form the basis for the calculation of the movement of particles in fluids, and are used for clarification (settling of granular solids in a liquid, drops of water in air), for ascensional separation (air bubbles in water, oil drops in water), centrifuging and fluidization.

— *Correction factor—sphericity factor* Ψ.   This is the ratio between the volume of a sphere of identical total surface area and the volume of the grain.

Where Stokes' law applies, $C' = C\Psi = \dfrac{24\,\Psi}{Re}.$

Values of Ψ

Sand	2	Graphite flakes	22
Coal	2.25	Mica	170
Talc	3.25	Gypsum	4

In conclusion, it is therefore possible to calculate the sedimentation velocity of a granular particle.

In a vertical settling tank, particles with a sedimentation velocity exceeding the rising velocity of the liquid will be retained.

In a rectangular tank of depth H (fig. 19) a suspended particle on the surface when it enters the tank will settle at a constant velocity of fall V_S and reach the bottom of the tank after a period $t = \dfrac{H}{V_S}$.

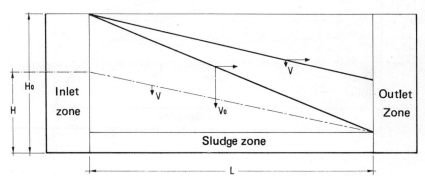

Fig. 19. —

If S_H is the horizontal area of the tank and Q the hourly flow passing through it, the capacity C of the tank will be:

$$C = HS_H = Qt$$

hence

$$V_S = \frac{H}{t} = \frac{Q}{S_H}$$

All particles with sedimentation rate exceeding Vs will be completely removed. Those with a rate Vs_1 below Vs will be removed in the proportion $\dfrac{V_{S1}}{V_S}$.

When the suspended matter contains particles of varying dimensions, the total amount removed is given by the formula :

$$(1 - C_0) + \frac{1}{V_S} \int_O^{C_0} V\, dC$$

in which C_0 is the fraction of particles having a sedimentation rate equal to or less than Vs.

Surface loading Ch is the ratio between the hourly flow and the horizontal surface area S_H.

It is a value that depends on the shape of the tank and the flow and reliably defines the sedimentation rate of the particles during a sedimentation time t_0. The upper limit of Ch in m/h consequently corresponds to the sedimentation rate V_S of these particles.

B. Flocculant settling.

Particles may join together (coalescence); they then flocculate and settle at an increasing velocity; in a tank with a horizontal hydraulic flow the settling trajectory is curved (fig. 20).

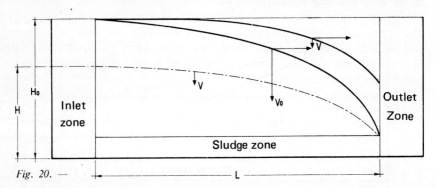

Fig. 20. —

The efficiency of flocculant settling depends not only on the surface loading but on the settling time.

There is no mathematical formula for calculation of the settling rate. On the basis of laboratory tests, percentage elimination curves p depending on depth H and time can be drawn.

The value of the velocity of fall is accepted to be of the form $V_s \simeq pt^n$.

The magnitude n is a measure of the flocculating property of the suspended solid. (For granular particles n = 0.) The curve H = f(t) on a semilogarithmic scale is a straight line from which n can be determined.

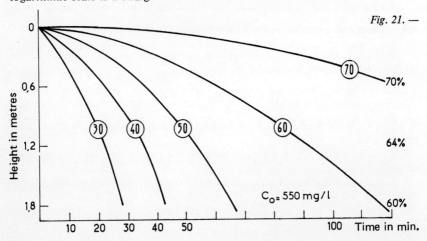

Fig. 21. —

On graph No. 21 an effective settling rate V_s can be defined, and is the ratio of the effective depth H divided by the time required to obtain a given percentage of settled particles. All particles with a settling rate not less than V_s will be removed in an ideal clarifier having a surface loading equal to V_s.

C. Zone settling of flocculated particles.

Zone settling is typical of activated sludge and flocculated chemical suspended solids when their concentration exceeds 500 mg/l. The particles adhere together and zone sedimentation occurs, thus forming an interface between the floc and the supernatant liquid.

— *Kynch's theory.*

Basic hypothesis: the velocity of fall of a particle depends solely on the local particle concentration C.

In the case of zone sedimentation in a tube of adequate height and diameter, when the depth of the top layer of sludge is measured as a function of time, the curve obtained (fig. 22) shows distinct phases:

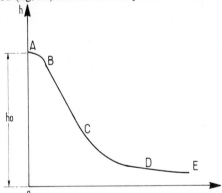

Fig. 22. —

— from A to B the interface is more or less clearly marked: this is the floccule coalescent phase. This phase does not always exist;

— from B to C, a straight section representing a constant velocity of fall V_0 (linear gradient). For a tube of given dimensions, V_0 depends on the initial concentration of solid matter and the flocculation properties of the suspension. As the initial concentration C_0 increases, the settling velocity V_0 of the mass drops. For example, for a municipal activated sludge with a concentration of suspended matter varying from 1 to 4 g/l, V_0 varies from 6 to 1.8 m/h;

— the concave section CD corresponds to a gradual reduction of the velocity of fall of the top layer of the deposit;

— from D onwards, the particles of floc come into mutual contact, and exert a compressive action on the lower layers.

Kynch's theory applies to sections BC and CD which cover the important field of settling of activated sludge in particular.

Let us take a suspended solid which has no coalescence phase when settling (fig. 23).

Calculation shows that:

— in the triangle BOC the concentration and velocity of fall are constant and are respectively equal to the initial values at B;

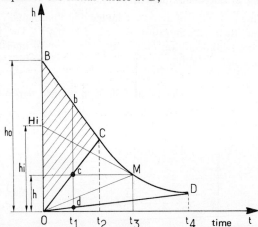

Fig. 23. —

— in the triangle COD, the equiconcentration curves are straight lines passing through the origin, which means that from the first moments of settling the layers nearest the bottom come into contact and pass through all the concentrations between the initial concentration and that corresponding to point D, where compression starts.

The sludge of depth $t_1 b$ at time t_1 therefore has three separate zones:

— a top zone bc where the concentration and velocity of fall are uniform and have retained their initial values C_0 and V_0;

— an intermediate zone cd where the concentration gradually increases from c to d and consequently the velocity of fall drops;

— a bottom zone dt_1 where the sludge floccules come into contact and are subjected to compression.

In the medium considered at time t_2, the top zone disappears and at time t_4 only the bottom zone remains.

To calculate the concentration at a point M of section CD, the tangent at M that cuts the ordinate axis at H_i can be drawn.

The depth h_i allows calculation of the concentration C_i at the interface of point M :

$$C_i = C_0 \frac{h_0}{h_i}$$

to which there corresponds a velocity of fall $V_i = \dfrac{dh}{dt}$ (gradient of MH_i).

The average sludge concentration over the full height h_u is:

$$C = C_0 \frac{h_0}{h}$$

Practical conclusion: The three sections BC, CD and DE of Kynch's curve (fig. 22) are used in the design of the following plant for zone settling:
— settling tanks and clarifiers. The section BC corresponds to installations with a vertical hydraulic flow having fixed (combined tanks, see page 216) and moving (Pulsator, see page 179) sludge blankets. The section CD corresponds to installations with a vertical or horizontal flow in which sludge thickening is desirable or essential (scraper-type settling tanks and clarifiers, see page 166);
— sludge concentrators where section DE is used (see thickeners, page 459).

— *Mohlman index.*

On Kynch's curve there is a special point used to define a sludge: it is the 30-40 minute point on the abscissa, the ordinate of which indicates the volume of settled sludge.

Determination of the mass M (in g) of suspended matter in this volume V (in cm³) defines :

— a volume index $\frac{V}{M}$ the volume (in cm³) occupied by 1 g of sludge. This is known as the *Mohlman index* I_M;

— a density index, $100 \frac{M}{V}$, the mass in g of 1 cm³ of sludge after sedimentation for 1/2 hour, or the Donaldson index I_D.

$$I_D \, I_M = 100$$

These properties cannot be related to the usual physical constants. They are easy to measure and are used to check biological sludge in waste water purification installations.

D. General design principles for settling tanks and clarifiers.

The surface area of a settling tank or clarifier is determined on the basis of two criteria:
— the surface loading in m³/ (m².d);
— the solids loading in kg of suspended solids/ (m².d).

The surface loading is related to the sedimentation rate of suspended solids. It has already been shown that this velocity can be calculated by Stokes' law for granular particles but must be measured in the case of particles that can be eliminated by flocculation.

The solids loading obviously depends on the concentration, and determines a minimum surface area to be adopted for elimination of a given quantity of such matter.

In designing a settling tank or clarifier, the larger of the areas determined on the basis of the above two criteria is adopted.

The solids loading is not generally decisive when the concentration of matter in the water to be clarified is small (e.g. free and flocculant settling) but it is important in zone settling.   In this case, the Mohlman index is a factor used to calculate the settling tank surface area from the mass loading.

— *Solids loading*: A cylindrical settling tank or clarifier of section S is fed by a flow Q_e having a solids concentration C_e.   Sludge Q_s at a concentration C_s is drawn off from the bottom.

The flow is $Q = Q_e - Q_s$.

The assessment for suspended solids is: $Q_s\, C_s = Q_e\, C_e$.

In the case of static settling tanks, Kynch's curve indicates, for any level of the sludge layer, the velocity of fall V_i and the concentration C_i. At this level the flow of solid matter is $F_i = C_i\, V_i$.

In a continuously-operating clarifier, the sedimentation velocity of solids is increased by the velocity V_f of the fluid drawn.   At the bottom of the settling tank, the flow F_s is $C_s\, V_f = \dfrac{Q_s\, C_s}{S}$.

At any level i in the sludge phase where the concentration is C_i and the individual sedimentation velocity of the particles V_i, the solid flow is equal to $F_i = C_i\,(V_i + V_f)$.

At a certain level L there is a critical concentration C_L which necessitates a maximum section S_m to prevent a bottleneck forming and to ensure that the sludge-removal conditions are met.   The flow at this level should not exceed a value F_L and the section of the clarifier should be at least $S_m = \dfrac{Q_e\, C_e}{F_L}$.

Fig. 24 shows how to obtain the critical concentration level C_L for draw-off concentration C_s by construction of Kynch's curve.

By taking the flows as equal, mathematical calculations give:

$$F = \frac{Q_e\, C_e}{S} = (V_L + V_f)\, C_L = \left(\frac{Q_e\, C_e}{S}\, \frac{1}{C_s} + V_L\right) C_L$$

where V_L is the average velocity at point L.

The solids loading is $\dfrac{Q_e\, C_e}{S} = \dfrac{V_L}{\dfrac{1}{C_L} - \dfrac{1}{C_s}}$

There are two possible alternatives:

a) L is below point B (e.g. concentrators and thickeners):

$$\frac{Q\, C_e}{S} = \frac{h_0 - h_s}{\theta_L \left(\dfrac{1}{C_L} - \dfrac{1}{C_s}\right)} = \frac{V_L}{\dfrac{1}{C_L} - \dfrac{1}{C_s}}$$

θ_L is the abscissa of point L, and V_L is the velocity of fall at point L.

The maximum loading as regards variations in the drawing off of sludge of concentration C_s is equal to that which would balance a sedimentation rate corresponding to the slope of the line $H_e L$.

b) L is above point B (e.g. clarifiers in general):

$$\frac{Q\,C_e}{S} = \frac{V_0}{\dfrac{1}{C_e} - \dfrac{1}{C_s}} \simeq \frac{C_e h_e}{\theta_e}$$

θ_e is the abscissa of point S.

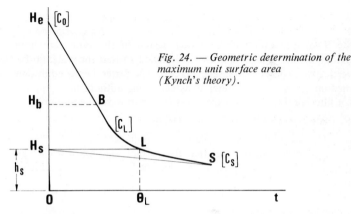

Fig. 24. — Geometric determination of the maximum unit surface area (Kynch's theory).

E. Practical design factors.

In pratice there is no ideal clarifier, for eddies can occur in the liquid, the wind may create waves on its surface, and convection currents caused by local temperature (especially those caused by the action of the sun) and density differences may affect the settling efficiency.

Every effort must be made to obtain a laminar and stable circulation with suitable ranges of value for the Reynolds and Froude numbers.

The Reynolds number calculated from the formula $\mathrm{Re} = \dfrac{vr}{v}$ must be low:

v is the water velocity
r is the hydraulic radius (sectional area divided by the wetted perimeter)
v is the kinematic viscosity of the water: 1.01×10^{-6} m^2/s at 20 °C.

The Froude number indicates the stability of a circulation process when the flow is affected primarily by gravity and inertial forces:

$$\mathrm{Fr} = \frac{v}{\sqrt{rg}}$$

v and r have the same meanings as for Re
g = 9.81 m/s^2.

The more stable the circulation, the more uniform the speed distribution over the full section of the tank, and the better the hydraulic performance. Stable circulations have high Froude numbers.

In practice, $\dfrac{H}{L}$ or $\dfrac{H}{R}$, ratios can be defined, where H is the wet depth of rectangular clarifiers of length L and circular tanks of radius R.

Taking retention time of two hours in the tank, Schmidt-Bregas gives:

$$\frac{1}{20} > \frac{H}{L} > \frac{1}{35} \text{ for rectangular tanks, and}$$

$$\frac{1}{6} > \frac{H}{r} > \frac{1}{8} \text{ for circular tanks.}$$

The operation and hydraulic design of the tanks are important factors. Sludge-blanket clarifiers must be designed so that the water to be treated passes vertically through the sludge blanket, the latter being equivalent to a porous medium constituting a filter to which some authors have applied the formulae for filtering through a porous mass (Carman Kozeni equation).

F. Plate-type settling tanks and clarifiers.

The old idea of using plate-type appliances has again become popular.

It is based on the fact that the surface loading of a clarifier with free fall does not depend on its height and that the layout in fig. 25 which was taken as a basis can be replaced by:

— the layout in fig. 26 where the flow Q can be multiplied by the number of elementary clarifiers n;

— the layout in fig. 27 where for a constant flow Q the conventional clarifier length can be divided by n.

— *Theoretical data:* let us take a system of parallel plates laid within a clarifier.

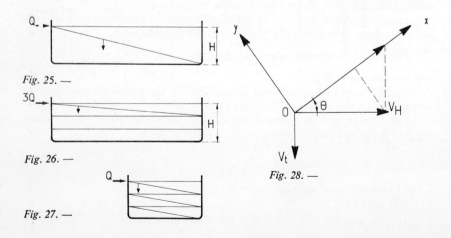

Fig. 25. —

Fig. 26. —

Fig. 27. —

Fig. 28. —

A system of co-ordinates ox, oy, oz can be defined in which :
ox is parallel to the most steeply sloping line of the plates;
oy is at right angles to the plates, and
oz is at right angles to the plane oxy (fig. 28).
θ is the angle formed by ox and the horizontal.
Let v_t be the velocity of vertical fall of a particle (Stoke's law, for example). The flow of the fluid is assumed to be laminar and uni-directional.

● 1º. The fluid moves in the direction ox positive (countercurrent settling tank) or negative (cocurrent settling tank).
The basic equations for the design of a plate-type clarifier are:

$$\frac{\mu}{v_0} = A (Y - Y^2)$$

and
$$S_c = \frac{v_t}{v_0} (\sin \theta + L \cos \theta)$$

in which
μ is the velocity of the fluid at a given point in direction x (fig. 29)
v_0 is the average velocity of the flow in the direction x
L is the ratio l/d, l being the length of the plates in the direction of the flow
d is the depth of the fluid measured at right angles to the direction of the flow
Y $= y/d$ is the ordinate of the trajectory in direction y
v_t is the settling rate of the particle having a trajectory such that it is completely
     eliminated
A is a coefficient the value of which depends on the system used
S_c is a parameter characteristic of the system

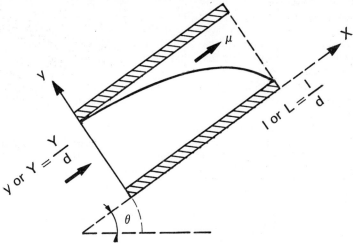

Fig. 29. — Path of a particle.

	Circular tubes	Parallel plates
A	8	6
S_c	4/3	1

The surface loading can be used as the basis for the design of a plate-type clarifier by application of the following formulae:

$$\text{Surface loading} = CK\,\frac{v_0}{L} \qquad \text{with } K = S_c\,\frac{L}{\sin\theta + L\cos\theta}$$

$C = 8.64 \times 10^{-2}$.

If v_0 in cm/s is used in these formulae, the surface loading is obtained in $m^3/(m^2.d)$.

It is possible to study the influence of the various parameters, in particular L and θ, for a sludge of given properties.

The fluid velocity may be positive (e.g. a Pulsator fitted with plates) or negative (Lamella type clarifier).

● 2⁰. The fluid moves in the direction oz.   An example is the multifloor clarifier where $\theta = 0$,

1.1.4. FLOTATION

Flotation makes use of the difference in specific mass between solids or liquid droplets and the liquid in which they are suspended.   But, in contrast to the settling process, this method of solid-liquid or liquid-liquid separation is applied only to particles whose true or apparent specific mass (the process being called "spontaneous" or "stimulated" flotation respectively) is lower than that of the liquid in which they are contained.

"Stimulated" flotation is based on the readiness with which certain solid and liquid particles link up with gas (usually air) bubbles to form "particle-gas" composites with a density less than that of the liquid in which they form the dispersed phase.

The resultant of the applied forces (gravity, buoyancy and resistance) causes the "particle-gas" composites to rise and become concentrated at the free surface of the liquid, from which they are then removed.

In order to bring about the separation by flotation of solid or liquid particles with a higher density than the liquid it is necessary that the adhesion between the particles and microbubbles should be greater than the tendency of the liquid to wet the particles.   The wetting of a solid by a liquid is determined by the magnitude of the angle of contact θ made by the surface of the solid and the gas bubble (see fig. 30).

Fig. 30. —

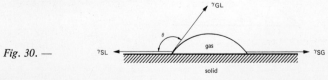

— if θ = 0, then the solid is completely wetted by the liquid, and adhesion between the solid and gas phases is impossible;

— if θ = 180°, the solid is not wetted at all by the liquid and there is optimum contact between the solid and gas phases. In point of fact, this is the extreme case which never occurs in practice as no liquid produces a θ angle of more than 110° (mercury);

— between these two values, the adhesion between the solid particles and the gas increases with the magnitude of angle θ.

This approach to the measurement of the buoyancy of particles is reasonable where their shape, be they solid or liquid (oils), is relatively simple and where their nature is well known. In the case of flocculated particles the surface phenomena are accompanied by mechanical bonds connected with the structure of the floc and, in particular, by gaseous inclusions in the floc.

Formulae for the rising velocity.

The particle-gas composite rapidly acquires a rising velocity which remains constant. This is the maximum rising speed which, as in the case of particles in the process of settling out, is calculated by applying Newton's general formula (cf. page 66) in which, under conditions of buoyancy:

d = the diameter of the particle-gas composite;

ρ_s = the effective specific mass of the particle-gas composite.

The calculations remain the same. Depending on the value of the Reynold's number, it is therefore possible to define flow systems in which the maximum rising velocity is given by the specific formulae of Stokes (laminar flow), Allen (intermediate flow conditions) or Newton (turbulent conditions).

Stoke's formula : $v = \dfrac{(\rho_e - \rho_s)\,gd^2}{18\eta}$

applied to air bubbles by themselves in water at 20 °C shows that laminar flow conditions hold good for bubbles with a diameter of less than 120 microns. Their maximum velocity is then 30 m/h. This represents an extreme case as the difference $(\rho_e - \rho_s)$ is at its maximum.

This equation reveals the influence of the various factors: the velocity v varies with d^2, with $(\rho_e - \rho_s)$ and with the temperature of the liquid, the latter varying in reciprocal ratio with the viscosity.

Another factor which needs to be taken into account is the shape or sphericity of the particle-gas composite. In the foregoing equations of Stokes and Newton the shape is taken to be spherical.

Application of this correction factor, which is easy to determine for simple geometrical shapes, leads to velocities which are lower than those which could be obtained with a sphere.

The favourable effect of the diameter, or size, of the particle-gas composite should not make us forget that, where the flotation of particles heavier than the liquid is concerned, the specific surface area, i.e. the ratio $\dfrac{\text{surface area}}{\text{volume}}$ or

$\dfrac{\text{surface area}}{\text{mass}}$, diminishes as the diameter increases. Given the same quantity of air fixed per unit of surface area, the result is a reduction of the factor $(\rho_e - \rho_s)$. The two parameters are therefore opposed to one another.

Minimum volume of gas to cause flotation.

The minimum volume of gas V_g, of specific mass ρ_g, needed to bring about the flotation of a particle of mass s and specific mass ρ_P in a liquid having a specific mass of ρ_l is given by the expression:

$$\frac{V_g}{S} = \frac{\rho_P - \rho_l}{\rho_l - \rho_g}\frac{1}{\rho_P}$$

1.1.5. FILTRATION

Filtration is a process that consists of passing a solid-liquid mixture through a porous material (filter) which retains the solids and allows the liquid (filtrate) to pass through.

If the dimensions of the suspended solids are larger than those of the pores, the solids are retained on the filter surface; this type of filtration is known as **surface filtration, cake filtration or support filtration.** When the solids are retained within the porous mass, the process is known as filtering **in volume** or **in depth.**

In both cases the flow of a liquid through a porous medium is governed by Darcy's law, which indicates that the head loss P is proportional to the filtering rate V (ratio of instantaneous flow Q to the unit of surface area), with the coefficient of proportionality k depending on the dynamic viscosity η and the resistance of the medium R.

$$V = \frac{P}{\eta R} = kP$$

A. Surface filtration.

Filtration of a sludge-laden liquid through a fabric or similar support with the formation of a filter cake of increasing thickness will now be considered (fig. 31).

According to Darcy's law, R consists of two resistances in series, the resistance R_g of the filter cake and the initial resistance R_m of the membrane.

$$R = R_g + R_m$$

with $R_g = r\,\dfrac{M}{S} = r\,\dfrac{Wv}{S}$

in which

M is the total mass of the deposited cake
W is the mass deposited per unit of volume of the filtrate
v is the volume of filtrate after a given time t
S is the filtration surface area

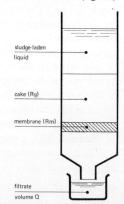

Fig. 31. —

r is the specific resistance to filtration of the cake under pressure p hence :

$$V = \frac{1}{S}\frac{dv}{dt} = \frac{P}{\eta\left(r\,\dfrac{Wv}{S} + R_m\right)}$$

integration of which produces an equation of the type $t = av^2 + bv$ where $\dfrac{t}{v} = av + b$ with $a = \dfrac{\eta\,rW}{2\,PS^2}$ and $b = \dfrac{\eta\,R_m}{PS}$.

The curve of this equation is a straight line which enables r to be defined as the slope $a = tg\theta$ (fig. 32).

$$W = \frac{W_b}{1 - \dfrac{W_b}{W_g}} \qquad \text{where} \quad W_b = \frac{\text{Weight of dry solids}}{\text{Unit weight of sludge}}$$

$$W_g = \frac{\text{Weight of dry solids}}{\text{Unit weight of cake}}$$

Note: This integration is correct only if r remains constant throughout the filtration — a condition which is completely satisfied by only incompressible sludges.

If filtration of a given volume of filter cake is continued for a fairly long period, we encounter first of all a break in the curve beyond which the dryness increases very slowly until it reaches the **limit value of dryness** of the cake.

The value of r increases with the pressure in accordance with a law given by the expression $r = r_0 + r'P^s$, in which r_0 and r' are respectively the limit specific resistance where $P = 0$ and the specific resistance where $P = 1$ bar.

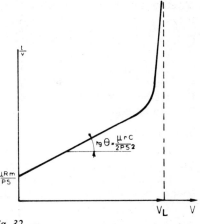

Fig. 32. —

s, known as the **compressibility factor of the sludge**, is a dimensionless number; r, the **coefficient of filterability** or specific resistance, is expressed in m/kg.

The resistance $r_{0,5}$, measured under a pressure of 49 kPa, is generally employed when comparing various sludges.

B. Filtration through a thick medium.

In the case of filtration through a porous plate, there may be two filtering phenomena:
— firstly, filtration on the surface of the plate;
— secondly, filtration inside the plate.

To determine the type of filtration (on the surface or in depth), the following dimensional criterion is considered:

$$\varepsilon = \frac{18\,P}{R_m d^2\,(s-e)}$$

in which

P is the head loss through the plate the resistance of which is R_m
d is the diameter of the particles to be retained.

Type of filtration:
— if $\varepsilon < 100$ .... membrane
— $\varepsilon > 1\,000$ .... in depth
— $100 < \varepsilon < 1\,000$ .... membrane and in depth

C. Filtration in depth.

The mathematical studies result from the following three initial equations :

— Ison's equation $\dfrac{dC}{dl} = FC$ (valid in a first approximation)   (1)

— Darcy's equation $\dfrac{dP}{dl} = kV$ (2)

in which:
C is the concentration of suspended matter in the water
V is the filtering rate calculated on the approach area at right angles to the flow
l is the distance from any section of the filter to the inlet section
k is Darcy's coefficient
F is the filtration coefficient
P is the head-loss.

— *Materials balance sheet:*

In a filter bed element of depth dl and section S, C is the inlet concentration and C + dC the outlet concentration (fig. 33);

Let q be the mass of material retained per unit volume of the bed.

The materials balance sheet is expressed by the following equation:

$$SV\,dt\,C = SV\,dt\,(C+dC) + S\,dl\,dq$$

which is reduced to:

$$-\frac{dC}{dl} = \frac{1}{V}\frac{dq}{dt} \qquad (3)$$

Equations (1) and (3) can be used to determine the clogging of the filter and the concentration of suspended matter in the filtered water. Equation (2) gives the head-loss.

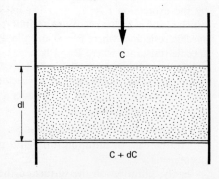

Fig. 33. —

One difficulty is that the coefficients F and k depend on the clogging and assumptions have to be made on their variations.

— *Filter clogging :*

The filtration coefficient F varies between an initial value F_0 and 0. Between these two points a linear variation $F = F_0 \left(1 - \dfrac{q}{q_1}\right)$ is assumed, in which

q is the clogging of the filtering medium (mass deposited per unit volume of bed) and

q_1 is the maximum value of q.

Integration of equations (1) and (3) gives:

$$\frac{q}{q_1} = \frac{1 - e^{-At}}{1 + e^{-At}(e^{F_0 l} - 1)} \qquad (4)$$

with $A = \dfrac{F_0 V C_0}{q_1}$

— *Quality of the filtered water :*

If L is the depth of the filter bed, then:

$$C_L = \frac{C_0}{1 + e^{-At}(e^{F_0 l} - 1)} \qquad (5)$$

If the filtered water is required to have a concentration not exceeding a fixed value C_{L_1}, t can be calculated from the above equation, the other factors being already known.

— *Loss of head*

The variation of k, Darcy's coefficient, is by assumption of the form:

$$\frac{k}{k_0} = \frac{1 + (a - 1)\dfrac{q}{q_1}}{1 - \dfrac{q}{q_1}} \qquad (6)$$

After integration, Darcy's equation becomes:

$$P = k_0 V \left[1 + \frac{a(e^{At} - 1)}{F_0}(1 - e^{-F_0 l})\right] \qquad (7)$$

which is broken down into two terms:

— one is k_0 V1 corresponding to a conventional flow in an unclogged porous medium;

— the other corresponds to the clogging of the filter bed (figs. 34 and 35).

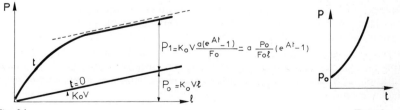

Fig. 34. —   *Fig. 35. —*

Equation (7) enables the clogging and headloss curves to be studied. It also allows calculation of the time t taken for the loss of head to reach the maximum value P_{max}.

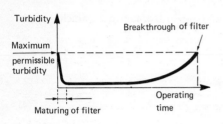

Fig. 36. —

Fig. 36 shows the quality of the filtered water during the filtration cycle. The turbidity first decreases (corresponding to the "maturing" time of the filter) then remains stable for some time before increasing again; this is the "breakthrough" point of the filter. The time t_1 during which water of the required quality is obtained and the time t_2 taken to reach the maximum design head-loss vary enormously.

The factors affecting these times t_1 and t_2 are:

1⁰ The quality and quantity of suspended matter in the water to be treated, which are defined respectively by the sludge cohesion factor K and the volume of sludge v deposited after 24 hours of sedimentation.

2⁰ The grain size of the filtering material defined by its effective size D mm.

3⁰ The filtering velocity $V = \dfrac{v}{S}$ in m/h.

4⁰ The depth L of the filter bed in m.

5⁰ The variations in head-loss expressed as m water column.

By definition, the ideal filter would have a time t_1 equal to t_2. In fact, allowance is made for a safety margin by taking t_1 greater than t_2.

1.2. Removal of dissolved matter

1.2.1. SEPARATION BY MEMBRANES

A. Semi-permeable membranes: ultrafiltration — reverse osmosis.

Techniques for separating solutes through membranes by the effect of pressure have been known for a hundred years. Notwithstanding this, we had to wait until the nineteen sixties to see the industrial application of such techniques rendered practicable by the development of synthetic membranes.

These methods, which make use of the semi-permeable properties of certain membranes (which are permeable to water and some solutes, but impermeable to others as well as to all particles), represent the development of classical filtration procedures in the direction of finer and finer separations. This, simple filtration (cf. Chapter 3, page 78) which retains particles with a diameter of more than a few microns (as in the case of in-depth filtration through sand, for example) is followed successively by:

1. microfiltration, which retains particles of over a few microns in diameter (e.g. filtration through membranes of Millipore, Sartorius and similar types);

2. ultrafiltration, which retains molecules with a molar mass of more than 10 000 to 100 000 g/mol depending on the membrane used; and

3. reverse osmosis, also known as hyperfiltration, which retains ions and molecules with a molar mass of over a few tens of grammes per mole.

There are two basic differences between filtration and microfiltration techniques on the one hand and reverse osmosis and ultrafiltration on the other:

a) Microfiltration changes none of the chemical properties of the solution whereas, with the other two procedures, separation of the dissolved species modifies the chemical potential and creates a gradient which tends to make the separated matter diffuse back in the reverse direction. To bring about a state of equilibrium, it is necessary to arrest this reverse diffusion by exerting pressure on the "filtered" liquid. In equilibrium, the pressure difference established in this way is known as the **osmotic pressure of the system.**

A simple equation relates osmotic pressure to concentration:

$$\pi = \Delta CRT$$

ΔC : the difference in concentration in $mol/m^3 = \dfrac{kg/m^3}{molar\ mass}$.

R : the molar constant of a perfect gas $= 8.314\ J/(mol.K)$.

T : the temperature in K.

π : the osmotic pressure in Pa.

Example: concentration of solute: $100\ kg/m^3$ T $= 300$ K.

for a compound having a molar mass of $0.050\ kg/mol$ $\qquad C = \dfrac{100}{0.050}.$

$$\pi = \frac{100}{0.050} \times 300 \times 8.314 = 50.10^5\ Pa,\ \text{or 50 bar.}$$

Similarly, for a compound of molar mass $0.5\ kg/mol$ $\qquad \pi = 5 \quad$ bar

for a compound of molar mass $5\ \ kg/mol$ $\qquad \pi = 0.5$ bar

for a compound of molar mass $50\ \ kg/mol$ $\qquad \pi = 0.05$ bar

Clearly, the smaller the molecule (i.e. the lower the molar mass), the greater is the osmotic pressure set up by the same difference in concentration. This explains why ultrafiltration gives rise to an osmotic back-pressure which is much lower than that experienced with reverse osmosis.

b) In the case of filtration and microfiltration, all the liquid to be treated passes through the filter plant. The suspended particles accumulate on the filter material, which, when it has been in use for a certain length of time, must either be cleaned mechanically or the "clogged" membrane must simply be replaced.

In reverse osmosis and ultrafiltration, it is no longer merely the insoluble particles which are retained by the membranes but also molecules and ions in solution. The concentration of the latter in the immediate vicinity of the membrane sets up "polarization" phenomena (cf. fig. 37) and their concentration

brings about an increase in the osmotic pressure of the solution to be treated, sometimes followed by precipitation. The measures adopted to overcome these difficulties consist in:
— passing through the membrane only a proportion of the volume to be treated, thereby setting up a continuous "reject" flow which contains the ions and molecules held back by the membrane;
— using pressures in excess of what is required theoretically, i.e. in practice:
2 to 6 bars in the case of ultrafiltration, and
20 to 80 bars for reverse osmosis.

The point must also be made that, contrary to what is the case with filtration and microfiltration, there is no theory which fully accounts for the performance of membranes used for reverse osmosis and ultrafiltration.

Be that as it may, a number of mathematical and physical models have been proposed, the most satisfactory of which are the following:
— *Ultrafiltration* involves a porous system in which the molecules dissolved in the solution to be treated are deemed to be retained when their size exceeds that of the pores through which the ultrafiltrate flows.

The concentration of the solute in this ultrafiltrate depends partly on the flow of water through the pores (which depends on the viscosity and therefore on the temperature) and partly on the concentration C_m of the liquid in contact with the membrane, which itself depends on the local concentration C_e of the solute in the liquid to be treated:

$$C_m = \Psi C_e$$

where Ψ is known as the coefficient of polarization.

This coefficient should be reduced to the minimum by scavenging as effectively as possible the surface of the membrane so as to carry off in the reject fraction the solutes and particles which tend to accumulate there (cf. fig. 37).

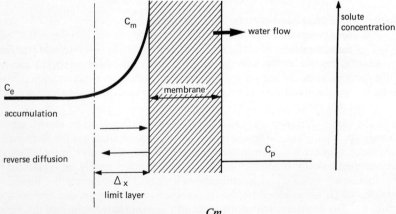

Fig. 37. — The polarization concentration $= \dfrac{Cm}{Ce}$.

— In the case of *reverse osmosis*, the porous models are no longer in any kind of agreement with the observed facts. An osmotic membrane must be thought of as a non-porous barrier to diffusion, in which transfers take place by the dissolution within the membrane of the molecular species in the solution and by the diffusion of the latter (solute and solvent) under the influence of concentration and pressure gradients which modify the chemical potential (μ) of the solution in the substance of the membrane.

The general equation for the flow J of a constituent i via a membrane of thickness e has the form:

$$J_i = - \frac{D_i C_i}{RT} \left(\frac{\partial \mu_i}{\partial C_i} \overline{\text{grad}} \, C_i + \frac{\partial \mu_i}{\partial P} \overline{\text{grad}} \, P \right)$$

where :
D_i : the diffusion coefficient of the constituent i;
C_i : the concentration of i in the membrane;
P   : the pressure.

For a water-solute system, integration of this equation, assuming that D_i is independant of C_i and that the properties of the membrane are independent of the pressure, gives for the water:

$$J_{\text{water}} = - \frac{D_e C_e V_e}{RT} \left(\frac{\Delta P - \Delta \pi}{e} \right)$$

or, for a given temperature and bearing in mind the fact that the concentrations of water are practically identical on both sides of the membrane:

$$J_{\text{water}} = A \frac{\Delta P - \Delta \pi}{e} \tag{1}$$

For the solute:

$$J_s = - \frac{D_s C_s}{RT} \frac{\Delta C_m}{e}$$

where C_m represents the concentration in the membrane.

Assuming that, under equilibrium conditions, the coefficient of partition C_p between the concentration within the membrane and the concentration at the membrane surface is constant, we have:

$$J_{\text{sol}} = B \frac{C_m - C_p}{e} \tag{2}$$

A and B are known as the permeability constants of the membrane with respect to the water and the solute.

We may express the **efficiency** R in the following terms:

$$R = 1 - \frac{C_p}{C_e} = 1 - \frac{C_p}{C_m} \times \frac{C_m}{C_e} = 1 - \Psi \frac{C_p}{C_m} = 1 - \Psi + \frac{J_e}{BC_e} \tag{3}$$

and the **salt passage** as: $SP = 100 \dfrac{C_p}{C_e} = 100 \left(\Psi - \dfrac{J_e}{BC_e} \right)$ (4)

We see again the importance of the polarization coefficient Ψ.

Equations (1), (2) and (3) or (4), which are valid whatever the type of membrane concerned, enable one to predict the direction in which the performance of an osmotic system will vary when one of the parameters affecting the system is changed. For instance, other things being equal, if the pressure rises the flow of water is increased (equation 1) while the transfer of salt (equation 2) remains unchanged (always provided that Ψ' remains constant, which generally requires that the scavenging flow be slightly increased). The quality of the treated water is thereby improved.

B. Dialysis membranes:

Unlike the semi-permeable membranes described above, the dialysis membrane is impermeable to water but lets through either all the ionized species or only those of a given sign (cations in the case of cationic, anions in that of anionic membranes) under the influence of a difference in chemical potential between the solutions impinging on its two sides.

This difference may be due to:
— a difference in concentration: **simple dialysis**
— a difference in pressure: **piezodialysis**
— a difference in electrical potential: **electrodialysis.**

Usually made from materials closely related to those used for the ion exchangers described in Chapter 10, these membranes have properties resembling those of such materials (e.g. the sensitivity of anionic membranes to dissolved organic matter).

Furthermore, where these membranes are selective in their action, we encounter once more the phenomena due to polarizing concentrations (the accumulation of the rejected ions) which we have described above and which invariably tend to counteract the desired transfer. Here, too, it is necessary to provide effective hydraulic scavenging of the membrane (cf. Chapter 12).

1.2.2. ADSORPTION

Adsorption refers to the ability of certain materials to fix at their surface organic molecules extracted from the liquid or gas phase in which they are immersed. The phenomenon is therefore one of **mass transfer** from the liquid or gas phase to the surface of the solid, for which the organic compound shows an affinity whose energy can be measured by microcalorimetric techniques.

As in all other problems relating to mass transfer, the adsorptive capacity of an adsorbent with respect to a given substance depends:
— on the developed surface area of the material; natural adsorbents (clays, zeolites etc.) have small surface areas of 50 to 200 m^2/g and their adsorptive capacity is therefore low, despite the fact that they play an important role in our natural environment. Industrial adsorbents have a surface area of at least 300 m^2/g, and that of good-quality activated carbon may attain 1 000 to 1 500 m^2/g;
— on the concentration of the organic matter in solution. Under limit conditions, an equilibrium is established between the concentration of the solution and the mass of the pollutant adsorbed per unit surface area (or per unit mass of the adsorbent). This may be expressed in the form of Freundlich's Law for equilibria:

$$\frac{x}{m} = C^{1/n} \cdot K$$

$\underline{x}$: mass of pollutant fixed per
$\underline{m}$ unit mass of the adsorbent.
C : residual concentration of
   pollutant in the liquid at
   equilibrium.
K and n : constants at a given temperature, whence the name of
   isotherm given to the corresponding curves (cf. figure
   opposite).

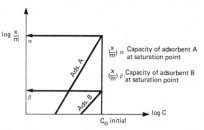

Fig. 38. —

— on the hydrodynamic characteristics of the exchange, on the relative velocity of the two phases and, in particular, on the period of contact between the solid and liquid phases;

— on the energy of the bond, i.e. on the affinity of the surface for the organic substance in question, and therefore on the nature of the substances to be absorbed. This affinity itself often depends on the pH value. In the majority of cases, an acid pH facilitates adsorption on activated carbon.

The bonding mechanisms are highly complex and have not yet been fully analysed. The forces involved are the result of purely physical phenomena (VAN DER WAALS type attraction) associated with chemical bonding mechanisms (hence the importance of the chemical groups situated on the surface of the adsorbent). This explains the absence of any satisfactory principle enabling us to predict beforehand the relative affinity of an adsorbent and an adsorbate.

1.2.3. ION EXCHANGE

Ion exchange substances are insoluble granular substances having in their molecular structure acid or basic radicals able to exchange, without any apparent modification in their physical appearance and without deterioration or solubilization, the positive or negative ions fixed on these radicals for ions of the same sign in solution in the liquid in contact with them. This process, known as ion exchange, enables the ion composition of the liquid being treated to be modified without changing the total number of ions in the liquid before the exchange.

The first ion exchange substances were natural earths; they were followed by synthetic mineral compounds (alginates, silica) and organic substances, which latter materials are today used almost exclusively under the name of resins. This term has been wrongly extended to cover any kind of exchanger.

A. Examination of the basic ion exchange reaction:

● *Use of a reversible reaction of the softening type:*

$$RNa_2 + Ca^{2+} \rightleftharpoons RCa + 2\,Na^+$$

Like any chemical equilibrium, it is governed by the law of mass action, the inverse reaction corresponding to the regeneration of the exchange substance.

If the liquid to be treated is brought into static contact with the exchange substance the reaction stops when equilibrium is reached between the liquid and the resin.

For substantially complete exchange to be achieved, it is therefore necessary to create successive equilibria stages by percolating the water through superimposed layers of exchange material. There is always a varying degree of **leakage of the ion** that it is desired to remove.

— *Laws governing a reversible ion exchange:* for each reaction involving two ions A and B, the equilibrium between the respective concentrations A and B in the liquid and in the ion exchange substance can be shown graphically (fig. 39).

Under conditions of equilibrium, and for a concentration B of X % in the solution, the exchange substance is saturated up to a concentration of Y %.

When the two ions A and B have the same affinity for the exchange substance, the equilibrium curve corresponds to a diagonal of the square. The more marked the exchange material's preference for ion B, the further the curve moves in the direction of the arrows.

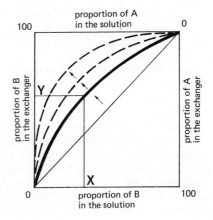

Fig. 39. —

The form of the curve for a given system of two ions depends on a large number of factors: nature and valency of the ions, concentration of ions in the liquid, type of exchange substance. In a system $Ca^{2+} \rightleftharpoons 2\,Na^+$ applied to a sulphonated polystyrene, the exchange substance always has a greater affinity for calcium than for sodium, and the more dilute the solution the more marked this will be (fig. 39).

As mentioned above, static batch treatment effected by bringing the liquid and the exchange substance into contact in a tank would reach a certain point on the curve and remain there.

If the treatment is to be continued until one ion is effectively eliminated in favour of another, the point of equilibrium must be progressively shifted by passing the liquid through a series of successive layers of the exchange substance containing fewer and fewer ions to be removed, thereby moving along the equilibrium curve almost to the zero concentration point for the unwanted ion.

If we take a layer of exchange material entirely in form A, and if a liquid containing ion B is passed through it, the successive equilibrium points between A and B give a series of isochronous concentration curves that can be represented by fig. 40 for two ions of similar affinity and by fig. 41 where the exchange material has a much higher affinity for ion B than for ion A. The "leakage point" is reached when the isochronous curve leaves the vertical right-hand axis (positive concentration of B in the last layer through which the liquid passes).

At this time, the curves for the two cases of different affinity are as follows (fig. 42).

If the area $\dfrac{B}{A + B}$ represents the fraction of the total exchanger capacity used when the leakage appears, it is clear that this fraction is much larger for B′ than for B.

This is also apparent in the treated liquid by the form of the "saturation curves" (fig. 43).

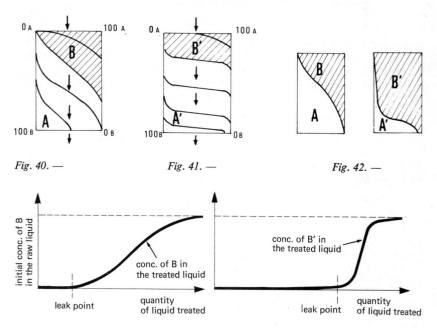

Fig. 40. —   *Fig. 41. —*   *Fig. 42. —*

Fig. 43. —

The form of the saturation curves depends not only on the static equilibrium curve mentioned above, but also on the "exchange kinetics" between the liquid and the exchange material; these kinetics involve the penetration of solutes into the exchanger, and are governed by the Donnan equilibrium laws.

These phenomena are very complex; they involve both the degree of dissociation and the ion concentration, the temperature, the nature of the exchange material/liquid interface, and the kinetics of penetration of the solid constituting this exchanger.

The total capacity of an ion exchanger, i.e. the total number of equivalents available for exchange per litre of exchange material, is only of very relative practical value; for commercial application it is the "useful capacity" defined from the isochronous graphs or saturation curves described earlier which is of importance.

A further important point in industrial applications is a direct consequence of the examination of these graphs.

Accepting the existence of an exchange material-liquid equilibrium curve, the quality of the liquid treated by an exchange layer depends on the quality of the **last layer through which the liquid passes,** whatever the qualities of the preceding layers.

If we consider a reversible reaction of the type

$$RA + B \rightleftharpoons RB + A$$

representing "treatment" from left to right and "regeneration" from right to left, it is necessary to examine the state of the exchanger at the start of a treatment cycle following a regeneration cycle.

It is clear that at the beginning of the treatment cycle the quality of the treated water, characterized by the ion leakage, will depend essentially on the degree of regeneration of the last layer of exchange material.

These factors must be borne in mind during the subsequent examination of the different qualities of exchange substances and their industrial applications.

● *Use of a non-reversible reaction:* this applies to the removal of a strong acid by a strongly-basic anion exchanger:

$$HCl + ROH \longrightarrow RCl + H_2O$$

The inverse reaction (hydrolysis) is virtually non-existent; the exchange is complete and can be obtained just as well under static or dynamic conditions. In this case ion leakage is zero, provided always that the contact time between the water and the resin is long enough.

Equilibrium reactions giving rise to an insoluble compound can be likened to this type of exchange. For example, if sea water is treated with an exchanger saturated with silver ions, the following reaction is obtained:

$$RAg + NaCl \longrightarrow RNa + AgCl$$

As AgCl is insoluble, it precipitates. Under these conditions, and according to Berthollet's law, the equilibrium shifts completely and the reaction is complete, even under static conditions.

Ion exchange is not instantaneous, and the rate of reaction depends upon the type of resin. In actual practice this type of exchange gives saturation curves similar to those in fig. 43.

The above two types of reaction can be used:

— to eliminate one or more unwanted ions from the liquid under treatment;

— to select and concentrate in the exchanger one or more ions that will later be found in the purified and concentrated state in the regeneration or washing liquid.

● *Use of a previously-fixed complex anion:* this complex ion is liable to cause secondary reactions, for example oxidation-reduction phenomena affecting the ions in the water or liquid to be treated, without itself dissolving in the liquid. Example: absorption of dissolved oxygen by oxidation of a sulphite anionic resin $R - NH_3 - HSO_3$ into a sulphate resin $R - NH_3 - HSO_4$.

Nota. — The ion exchange laws do not concern the use of ion exchangers for other purposes (catalysis, adsorption).

B. Methods of regeneration:

In the case of processes for water softening and deionization, the end of the cycle is reached when the saturation curve corresponds to that shown in figure 42 (compounds A' and B'). It can then be assumed, at least as far as the upper layers are concerned, that the ion exchanger is saturated with B' ions and is in equilibrium with the concentration of B' in the inflowing solution.

Regeneration is carried out by causing a concentrated solution of A' ions to flow through the exchanger either in the same direction as the saturation (co-current regeneration) or in the opposite direction (counter-current regeneration).

● *Co-current regeneration:* In this operation, the concentrated solution of A' ions is initially brought into contact with those layers of the ion exchanger saturated with the B' ions which are to be removed from the resin. These B' ions are then carried to those layers of the ion exchanger which are at a lower level of saturation and where the conditions are favourable to their seizure. During the first stage of regeneration it is therefore the A' ions which are eluted from the column.

It therefore appears that, in order to achieve total regeneration of the ion exchanger, it is in practice necessary to subject to a double exchange process a quantity of ions corresponding to the ration A'/B'.

Finally, if the quantity of regenerating solution is limited, the B' ions will not be completely eluted from the ion exchanger and the bottom layers will not be fully regenerated.

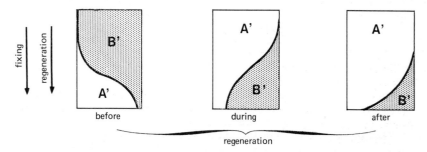

Fig. 44. — Co-current regeneration.

Consequently, during the following cycle, the B' ions will undergo automatic regeneration by the A' ions displaced from the upper layers.

● *Counter-current regeneration:* The course of events will be different when the regenerating agents are made to flow upwards from the bottom. In this case, the concentrated A' ions first of all encounter the resin layers with a low concentration of B' ions, elution of which therefore takes place in favourable conditions. What is more, the B' ions cannot be recaptured in the saturated upper layers.

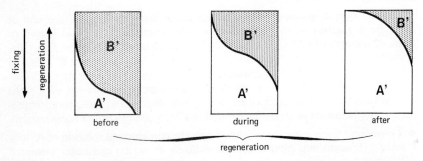

Fig. 45. — Counter-current regeneration.

Regeneration therefore takes place in far more reversible conditions than when the co-current technique is used, and from the thermodynamic point of view this means greater efficiency.

Two important advantages attaching to the principle of counter-current regeneration should be emphasized:
— the improvement in the efficiency and consequent reduction in the consumption of reagents, given equal quality; and
— the improvement in the quality of the treated water thanks to the fact that the bottom layers are regenerated with a large excess of reagent.

1.3. Liquid-liquid separation

The sedimentation theory applies to the separation of droplets of a liquid not miscible in water and considered as **free** in water. If the liquid has a specific mass below that of water (oil for example) the droplets collect on the surface.

In the case of oil-water mixtures, the Reynolds number is below 0.5 and the ascending velocity of the oil droplets is given by Stokes' law.

The American Petroleum Institute has carried out design calculations for gravity separators to remove all droplets of dimensions exceeding 0.015 cm. This can be reduced to 0.005 cm if plate-type separators are used.

Oil-water emulsions form stable suspensions that have to be broken up by suitable processes (thermal, ultrafiltration, coalescence, acidification, etc.).

1.4. Chemical processes

The addition to an aqueous solution of appropriately selected chemicals can reduce the consequences of the presence (or absence) of certain dissolved salts. This is the technique used, for example, for the total or partial removal of alkalinity, for correcting the pH value and for changing the level of oxidation of some substances in solution.

The chief possibilities are mentioned below.

1.4.1. PRECIPITATION

Chemical purification methods involve the application of Berthollet's Law and are based on the following principle:

The addition to the water of a soluble reagent which, by changing places or combining with the unwanted ions contained in the raw water, will lead to the precipitation of the product of such combination as far as the limit of its solubility.

The most widespread examples of this are:

● *The removal of carbonates using lime:* the reaction between the OH^- ions of the lime and the HCO_3 ions of the bicarbonates leads to the almost total precipitation of the calcium carbonate which is formed.

$$HCO_3^- + OH^- \longrightarrow CO_3^{2-} + H_2O$$
$$Ca^{2+} + CO_3^{2-} \longrightarrow CaCO_3 \downarrow$$

which can be expressed comprehensively by the formula:

$$Ca(HCO_3)_2 + Ca(OH)_2 \longrightarrow CaCO_3 \downarrow + 2\ H_2O$$

● *The softening of water containing calcium sulphate by adding sodium carbonate* according to the reaction:

$$CaSO_4 + Na_2CO_3 \longrightarrow Na_2SO_4 + CaCO_3 \downarrow$$

● *The removal of sulphates using barium salts:*

$$CaSO_4 + BaCl_2 \longrightarrow CaCl_2 + BaSO_4 \downarrow$$

$$Na_2SO_4 + BaCl_2 \longrightarrow 2\ NaCl + BaSO_4 \downarrow$$

● *The removal of fluorides by calcium salts:*

$$2\ F^- + Ca^{2+} \longrightarrow CaF_2 \downarrow$$

● *The removal of phosphates* as iron or aluminium phosphate, e.g.:

$$PO_4^3 + Fe^{3+} \longrightarrow FePO_4 \downarrow$$

● The *silica* content of natural water can be reduced by combining it with a complex precipitate of iron aluminate or magnesium hydroxide, although what seems to take place here is rather the adsorption of the silica than any clearly defined chemical combination.

For each of these reactions, it is vital to know:

— *The kinetics of the reaction.* The reaction often has to be performed, with a sufficient concentration and with enough time allowed, in the presence of nuclei originating from earlier reactions.

— *The limit of solubility of the product formed* at the given temperature and with the concentration of the various ions present.

— *The probable sedimentation rate of the precipitates obtained* (with or without the use of flocculants).

1.4.2. NEUTRALIZATION

This term refers to any treatment designed to adjust the pH of water either to a value approximating to neutrality or to the equilibrium pH value (see page 27 : $CaCO_3 - CO_2$ equilibrium). Initially, therefore, the water may be either acid or alkaline.

● *Acid water.*

Usually the problem is to neutralize carbonic acidity with a base or an alkaline carbonate. This is the situation encountered with water of granitic origin, in which the carbon dioxide content is excessive in relation to the level of alkaline earth bicarbonates (hydrogenocarbonates). It also applies to water obtained after clarification using a coagulant (see page 61 : coagulation and flocculation). This technique makes use of one of the following reactions:

$$2 CO_2 + Ca(OH)_2 \longrightarrow Ca(HCO_3)_2$$
$$CO_2 + NaOH \longrightarrow NaHCO_3$$
$$CO_2 + Na_2CO_3 + H_2O \longrightarrow 2 NaHCO_3$$
$$CO_2 + CaCO_3 + H_2O \longrightarrow Ca(HCO_3)_2$$

Where strong acidity is involved (e.g. adjustement of the flocculation pH, neutralization of the effluent obtained from ion exchanger regeneration or of the sludges generated by pickling processes etc.) the reactions are of the following type:

$$HCl + NaOH \longrightarrow NaCl + H_2O$$
$$2 HCl + Ca(OH)_2 \longrightarrow CaCl_2 + 2 H_2O$$

(regeneration of an ion exchanger by hydrochloric acid, hydrochloric acid pickling).

$$H_2SO_4 + 2 NaOH \longrightarrow Na_2SO_4 + 2 H_2O$$
$$H_2SO_4 + Ca(OH)_2 \longrightarrow CaSO_4 + 2 H_2O$$

(regeneration of an ion exchanger by sulphuric acid, sulphuric acid pickling etc.).

● *Alkaline water.*

This covers water which, because of its origin, is deficient in carbon dioxide. Also within this category can be placed the effluents produced by resin regeneration. Depending on the use to which the water is to be put, use is made:

— either of CO_2:

$$Ca(OH)_2 + 2\ CO_2 \longrightarrow Ca(HCO_3)_2$$

This treatment is mainly used for raising the carbonate level in the water supply.

— or of a strong acid:

$$Ca(OH)_2 + H_2SO_4 \longrightarrow CaSO_4 + 2\ H_2O$$
$$2\ NaOH + H_2SO_4 \longrightarrow Na_2SO_4 + 2\ H_2O$$

This treatment is rarely used for the water supply and is chiefly applied to industrial effluents.

1.4.3. OXIDATION AND REDUCTION

These reactions are used to modify the state of certain metals or compounds (containing nitrogen, sulphur, cyanide etc.) so as to render them either insoluble or non-toxic:

Some substances occur either in oxidized or reduced form and are converted from one to the other by acquiring electrons (reduction) or losing electrons (oxidation). A system comprising a recipient and a donor of electrons between which electrons can be transferred in such a way that one substance is oxidized while the other is reduced is known as an "oxidation-reduction" system.

$$Red \rightleftharpoons Oxid.^{n+} + ne^-$$

It should be noted here that, apart from oxygen and hydrogen which are respectively able to act only as an oxidizing and reducing agent, there are no substances which are inherently oxidizers or reducers in an absolute sense.

The possibility of such interaction is determined by electrical measurement, as described on page 578. It is defined by the concept of **oxidation-reduction** or redox potential[1], which depends on the activity of the oxidized and reduced forms according to the formula:

$$E = E_0 + \frac{K}{n} \log \frac{|\ \text{oxidized form}\ |}{|\ \text{reduced form}\ |}$$

where n is the number of electrons involved in the oxidation-reduction reaction, and E_0 the "normal" potential corresponding to the equilibrium

$$|\ \text{oxidized form}\ | = |\ \text{reduced form}\ |.$$

The various substances can be classified y comparing their E_0 potential:

A substance A having a higher normal potential than a substance B will oxidize the latter. The element B will reduce the element A.

Listed below are the normal potential E_0 values at 25 °C of a number of substances or particular relevance to water treatment. Zero potential is that of a hydrogen electrode.

---

1. The oxidation-reduction, or redox, potentia must not be confused with the "rH" (see page 876).

		E_0
Ozone	$O_3 + 2 H^+ + 2 e^- \rightleftharpoons O_2 + H_2O$	+ 2.07 V
Chlorine	$Cl_2 + 2 e^- \rightleftharpoons 2 Cl^-$	+ 1.39
Chromium VI	$Cr_2O_7^{2-} + 14 H^+ + 6 e^- \rightleftharpoons 2 Cr^{3+} + 7 H_2O$	+ 1.33
Manganese	$MnO_2 + 4 H^+ + 2 e^- \rightleftharpoons Mn_2^+ + 2 H_2O$	+ 1.23
Ammoniacal nitrogen	$NH_4^+ + 2 H_2O \rightleftharpoons NO_2^- + 8 H + 6 e^-$	+ 0.897
Nitrous nitrogen	$NO_2^- + H_2O \rightleftharpoons NO_3 + 2 H^+ + 2 e^-$	+ 0.835
Iron	$Fe^{3+} + e^- \rightleftharpoons Fe^{2+}$	+ 0.77
Sulphur	$SO_4^{2-} + 4 H^+ + 2 e^- \rightleftharpoons H_2SO_3 + H_2O$	+ 0.17

In fact, conversion of a substance from one form into the other is generally effected by means of another substance which itself occurs in both forms according to the formula:

$$aOx_1 + bRed_2 \rightleftharpoons aRed_1 + bOx_2$$

We see, therefore, that a combination of two oxidation-reduction systems is involved.

Mixing equivalent quantities of the oxidizing agent of one of the pairs and of the reducing agent of another pair (a $|\ Ox_1\ | =$ b $|\ Red_2\ |$) enables the point of equivalence to be reached. The potential E of the system is then expressed by:

$$E = \frac{b\,E_1 + a\,E_2}{a + b}$$

On a titration curve (potential against concentration of oxidizing agent) this is identified by the point of inflection.

The most widespread applications are the removal of iron and manganese from the domestic and industrial water supply and the detoxification of effluents from surface treatment plants which contain cyanides or chromates (cf. Chapter 25).

As explained below in paragraph 25, certain oxidation-reduction reactions can be brought about with the help of bacteria.

2. BIOLOGICAL PROCESSES

The relevance of biological phenomena covers an exceedingly broad field. They are introduced here with special reference to the treatment of waste water contaminated with organic matter. Treatment of this kind involves **fermentation**, i.e. chemical reactions produced by certain microorganisms. Of these, the bacteria play a vital role.

Bacteria act on the pollutants which thereby fulfil the function of a nutrient medium or substrate. The whole range of chemical reactions is catalyzed by the enzymes secreted by the bacteria, which also act as their support. The course of events can be described by following the development of the bacterial culture which exercises the purifying effect.

2.1. Growth of a bacterial culture

The growth of a bacterial culture comprises a number of phases each having a different rate (fig. 46).

1. **Latent phase** during which the microorganisms can acclimatize to the nutrient medium, by modification of the enzymatic system of the culture. This phase is specially important when the waste water has not previously been seeded with suitable microorganisms (industrial effluents).

2. A phase of growth at constant rate, generally called the **exponential phase.**

3. A phase in which the rate of growth **slows down.**

4. **A phase of zero growth.**

5. **A declining phase.**

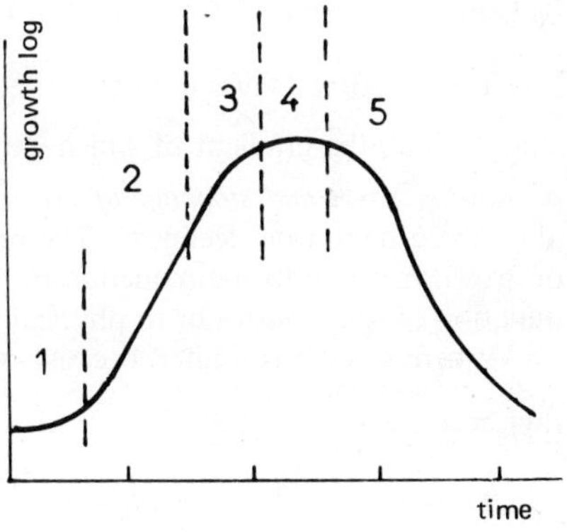

Fig. 46. — Growth curve for a bacterial culture.

Monod was one of the first to express the growth of a unicellular organism in mathematical form:

$$\frac{dS}{dt} = f(L, S)$$

in which S is the cellular mass per unit of volume and
L is the concentration of the substrate (nutrient).

● *Exponential phase:* When nutrients are present in sufficient concentration and there is nothing to inhibit growth, the culture grows at a constant rate and the above equation becomes:

$$\frac{dS}{dt} = \mu S$$

The growth rate μ is easily related to the cellular division time, t_d, that is, **the time taken** for the biomass to double:

$$\mu = \frac{1}{S}\frac{dS}{dt} = \frac{d \ln S}{dt} = \frac{0.693}{t_d}$$

$$(0.693 = \ln 2)$$

Assuming that a constant proportion of the nutrients is converted into living cells after a time Δt, the quantity of microorganisms increases by

$$\Delta S = a_m L_e$$

where a_m is expressed in mg of cells produced per mg of substrate eliminated (L_e) during the time considered.

$$\ln \frac{S}{S_0} = \ln \frac{S_0 + \Delta S}{S_0} = \ln \frac{S_0 + a m L_e}{S_0} = \ln \left(1 + \frac{a_m L_e}{S_0} \right) = \mu t$$

S_0 being the cellular concentration at time $t = 0$.

The relation between $\ln \left(1 + \dfrac{a_m L_e}{S_0} \right)$ and the time t is represented by a straight line, the gradient of which defines μ, the **exponential growth rate.**

● *Slowing down and stoppage of growth:* when circumstances become unfavourable, the growth rate declines. The reasons for the slowing down or stoppage of growth are due to the reduction in the concentration of nutrients, to an accumulation of toxic wastes or to physical modification of the environment.

Where growth is limited because of a shortage of nutrients, the basic equation becomes $\dfrac{dS}{dt} = k_1 LS$.

Taking $k_2 = \dfrac{k_1}{a_m}$, calculation gives:

$$\frac{L_f}{L_0} = \frac{1}{1 + k_2 St}$$

in which L_f is the concentration of nutrients at time t and L_0 is the initial concentration.

● *Declining phase:* During this period, the microorganisms have no more food and some of them disappear. The rate of disappearance is proportional to the concentration of microorganisms S:

$$\frac{dS}{dt} = - bS$$

b being the mortality rate.

The various phases and the equations for them are valid in aerobic or anaerobic media. Only the coefficients μ, k_2, a_m and b depend on the type of microorganisms, the type of nutrients and temperature.

2.2. Evaluation of organic pollution

In view of the diversity of the organic substances contained in effluents it would be tedious and quite pointless for the purpose of describing the biological purification process to consider the nature of each of their constituents.

The amount of organic and volatile matter contained in sludge is determined globally after incineration in a furnace at 600 °C.

For example, the chemist knows that complete oxidation of the glucose contained in sewage corresponds to the following simplified formula:

$$C_6H_{12}O_6 + 6\,O_2 = 6\,CO_2 + 6\,H_2O + 155 \text{ joules } (+\,650 \text{ calories})$$

that is to say: add $6 \times 32 = 192$ g/mol of oxygen to 180 g/mol (molar mass) of glucose, and 264 g/mol of CO_2 and 108 g/mol of water will be formed. He is therefore presented with two possible ways of following the course of this reaction:
— knowledge of the quantity of oxygen consumed;
— determination of the quantity of carbon dioxide formed.

The first method is used with:
— either a chemical oxidant which permits evaluation of the quantity consumed or;
— oxygen dissolved in water, in the presence of enzymes, catabolitic agents produced by microorganisms.

A. **Chemical oxidation:** depending on the type of oxidant and the temperature at which the reaction is effected, different values will be obtained.

Mention should be made here of the use of potassium permanganate cold (4 hours) or hot (10 minutes at boiling point) in analysing water supplies to assess their suitability as drinking water; this gives a measurement of the **liability to oxidation.**

The standard method for waste water uses potassium dichromate in a sulphuric medium; this is a powerful oxidizing agent, and gives the **chemical oxygen demand (COD).** The accuracy of measurement is of the order of 5 %. It takes two hours.

The COD covers everything likely to demand oxygen, in particular oxidizable mineral salts (sulphides, salts of metals of lower valency) and the majority of the organic compounds, whether or not liable to degradation by biological means. Mineral hydrocarbons are resistant to this powerful oxidation, and it can sometimes give misleading results.

B. **Biochemical oxydation:** the biochemical oxygen demand (BOD) is the quantity of oxygen, expressed in mg/l, consumed under test conditions (incubation at 20 °C and in darkness) during a given period to produce by biological means oxidation of the biodegradable organic matter present of the waste water.

Complete biological oxidation takes 21 to 28 days. The BOD_{final} or BOD_{21} or BOD_{28} is then obtained.

The curves showing oxygen consumption against time are of the form shown in fig. 47.

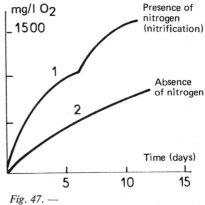

Fig. 47. —

Curve 1 was obtained with a water containing carbonaceous and nitrogenous products. It is seen that the carbonaceous products are oxidized first; this is followed by oxidation of the nitrogenous products, thus giving rise to the phenomenon of **nitrification** (see page 107).

BOD_{21}, which is too time-consuming to measure, has by convention been replaced by BOD_5, the quantity of oxygen consumed after 5 days' incubation (see page 922), and sometimes by BOD_7. BOD_5, normally represents only the biodegradable carbonaceous organic pollution.

C. **Relationship between BOD_5, BOD_{21}, and COD:** If all the organic matter in water were biodegradable, then $COD = BOD_{21}$ would apply.

For glucose, we have: $\dfrac{BOD_{21}}{BOD_5} = \dfrac{COD}{BOD_5} = 1.46$

If non-biodegradable organic matter is present, that being the case with domestic sewage (and many industrial effluents), then:

$$COD > BOD_{21}.$$

Typical examples of non-biodegradable organic substances are cellulose, coal dust, lignin, tannin, sawdust, etc.

In practice, the ratio $\dfrac{COD}{BOD_5}$ varies widely and cannot be predicted for a given water.

D. **Measurement of organic carbon:** instead of determining the oxygen consumed, the chemist can determine the carbon dioxide produced by combustion of a microspecimen. Several very expensive instruments for this purpose are available on the market. These provide a measurement of the total organic carbon (TOC) in solution. The present design of these instruments requires that the suspended solids be removed from the specimen prior to measurement. In the majority of cases, therefore, the result provides only a partial estimation of the pollution.

2.3. Aerobic treatment

When the biodegradable organic matter L_e is consumed by a mass S_a of microorganisms in an aerobic medium, the following processes occur:
— oxygen consumption by these microorganisms for their energy requirements, reproduction by cellular division (**synthesis** of living matter) and endogenous respiration (progressive auto-oxidation of their cellular mass),
— production of an excess of living and inert matter known as **excess sludge.**

It is very difficult to determine the mass of the active matter S_a experimentally; however, the mass of volatile matter S_v and total (mineral and organic) matter S_t can be measured.

— *Theoretical oxygen requirements* can be calculated from the equation:

$$O_2 \text{ consumed} = a'L_e + b'S_v.$$

Laboratory techniques (Warburg type apparatus) enable a' and b' to be determined.

— *The resulting production of the biomass* of the excess sludge is:

$$\Delta S_v = a_m L_e - bS_v$$

in which L_e is the mass of the BOD_5 eliminated in kg/day,

S_v is the mass of organic matter in kg in the fermenter (or biological reactor),

ΔS_v is expressed in kg/day.

$$\frac{\Delta S_v}{S_v} = a_m \frac{L_e}{S_v} - b$$

an equation of the type $y = ax - b$ from which a_m and b can be determined.

Note: The suspended matter (inert organic and mineral) already present in the water entering the reactor must be added to this mass of excess sludge resulting from bacterial activity on biodegradable organic matter.

A. Data relating to a biological reactor.

The **sludge loading factor** C_m is the ratio between the mass of nutrient (BOD_5 or mass of volatile matter) entering the reactor each day and the mass of sludge S_t contained in the reactor:

$$C_m = \frac{L_0}{S_t}$$

It would be more accurate to take volatile matter S_v instead of total matter S_t. This would give a sludge loading factor C_m'. In fact, if the sludge contains $\beta \%$ of volatile (or organic) matter:

$$S_v = \beta S_t$$

and
$$C_m' = \frac{L_0}{S_v} = \frac{L_0}{\beta S_t} = \frac{C_m}{\beta}$$

In conformity with normal practice, the value C_m will be adopted to define the sludge loading factor.

The **volume load or BOD loading** C_v is the mass of nutrient (BOD_5 or mass of volatile matter) per day per unit of volume entering the reactor of capacity V:

$$C_v = \frac{L_0}{V} \text{ kg/(d.m}^3)$$

With aerobiosis, if ρ is the purification efficiency in DOB_5, the oxygen consumption and production of organic sludge per kg of BOD eliminated are obtained from the equations:

$$L_e = \rho \, L_0$$

$$\frac{O_2 \text{ consumed}}{L_e} = a' + \frac{b'}{\rho \, C_m}$$

$$\frac{\Delta S_v}{L_e} = a_m - \frac{b}{\rho \, C_m}$$

Per kg of BOD eliminated, the lower the sludge loading factor the less excess sludge there will be and the higher the demand for oxygen.

The theoretical (or nominal) oxygen consumption per kg BOD, known as the **OC/load** (theoretical) therefore varies with the sludge loading factor.

Some authors take account of the **age of the sludge,** the ratio between the mass of sludge present in the reactor and the daily mass of excess sludge:

$$A = \frac{S_t}{\Delta S_t} \text{ or } \frac{S_v}{\Delta S_v}$$

The age of the sludge is inversely proportional to the sludge loading factor:

$$A = \frac{1}{a_m \, \rho \, C'_m - b}$$

Note: In anaerobiosis this ratio is called the critical solids retention time $\left(SRT = \dfrac{S_v}{\Delta S_v} \right)$ which should not be confused with the hydraulic retention time. A and SRT indicate the number of days required for the sludge to be renewed.

B. Nutrient requirements.

Like higher beings, microorganisms require balanced food. The formula:

$$C_{106}H_{180}O_{45}N_{16}P$$

has been given for the total composition of synthesized cells.

Nitrogen and phosphorus are therefore essential elements. Domestic waste water contains balanced food, but not industrial waters which often contain little N and P. In order to effect the |proper biological purification of these effluents it is therefore necessary to add to them nitrogen and phosphorus in the form of assimilable mineral salts or *nutrients.*

C. Effect of temperature.

Temperature variations affect any biological process. A temperature rise accelerates enzyme reaction rates. However, in aerobiosis, temperature has the reverse effect on the dissolution of oxygen in water. The resulting effect is defined by the equation: $\dfrac{k_2}{k'_2} = \theta^{t'-t}$

where k_2 and k'_2 are values of the constant k_2 at temperatures t and t' (see page 98) and θ varies with the biological system and the loads applied to it between 1 and 1.09.

D. *Toxicity—sepsis.*

Many substances have a toxic effect on the activity of microorganisms. Purification may be partially or totally inhibited depending on the nature or concentration of the substance concerned. **Bacteriostatic** concentration temporarily inhibits bacterial development while **bactericidal** concentration kills the bacteria. When toxic substances are introduced regularly in suitable doses, the bacterial flora becomes resistant to them. The accidental presence of heavy metals: Cu^{2+}, Cr^{6+}, Cd^{2+}, even in small quantities (0.1 mg/l), may destroy the action of the bacteria. The freshness of a waste water can be defined by its redox potential (fig. 48).

When it lies in sewers, waste water may be liable to putrid fermentation and become septic either by contamination from the overflows from septic tanks or because it remains there too long and forms deposits subject to putrefactive fermentation (see p. 976).

E. *Aeration.*

Oxygen is introduced into water by bringing it into close contact with air. At the interface of the two fluids, the monomolecular boundary layer is saturated with oxygen as soon as it is formed, and simultaneously the diffusion of gas towards deeper layers of water starts.

The diffusion rate follows **Fick's law:** $\dfrac{dm}{dt} = AK_L \dfrac{dC}{dt}$

The quantity of the substance dm that diffuses in the unit of time dt through an interface A is determined by the diffusion constant K_L and the concentration gradient $\dfrac{dC}{dt}$. If the concentrations are expressed in mg/l and the rate in mg/l.h, K_L thas the dimension h^{-1}. $K_L A$ is known as the **transfer coefficient.**

On the basis of this equation, the **oxygenation capacity** of an aeration system is defined as the quantity of oxygen expresses in g/m^3 supplied to pure water in one hour at a constant and zero oxygen content, a temperature of 10 °C and atmospheric pressure of 760 mm mercury.

The following equation gives the value of the oxygenation capacity:

$$\frac{dC}{dt} = 11.25 \frac{1}{t_1 - t_0} \log \frac{C_S - C_0}{C_S - C_t} \sqrt{\frac{k_{10}}{k_t}}$$

in which
C_s is the O_2 content at saturation in g/m^3
C_0 is the O_2 content at the beginning of the experiment, in g/m^3
C_t is the O_2 content after a time t
11.25   is saturation of O_2 at 10 °C in g/m^3
$t_1 - t_0$ is the observation period in hours
$\dfrac{dC}{dt}$ is the oxygenation capacity in $g/(m^3.h)$

k_{10} and k_t are diffusion coefficients at 10° and t °C.

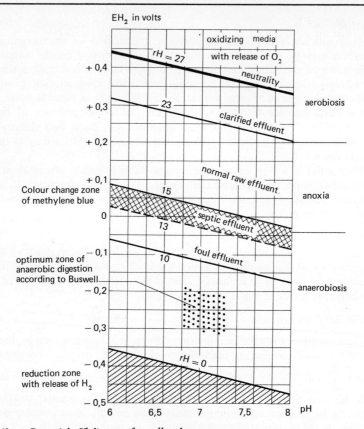

Fig. 48. — *Potential-pH diagram for polluted water.*

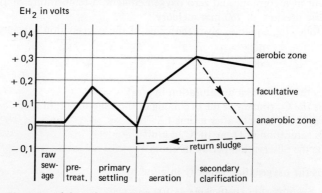

Fig. 49. — *Variation of the redox potential in a treatment plant (Hyperion).*

Taking $C_s - C_0 = D_0$ and $C_s - C_t = D_t$, O_2 deficits at the start and at time t, then the oxygenation capacity is:

$$11.25 \times 2.3 \frac{1}{t_1 - t_0} \log \frac{D_0}{D_t} \sqrt{\frac{k_{10}}{k_t}}$$

The relation between D_0 and D_t, plotted on logarithmic co-ordinates, gives a straight line as a function of time, the gradient of which tan α defines the oxygen dissolution rate :

$$\text{Oxygenation capacity} = 25.9 \tan \alpha \sqrt{\frac{k_{10}}{k_t}}$$

The quantity of oxygen introduced depends:
— on the value of the interfaces between the air and water, and their renewal,
— the oxygen gradient between the air and water,
— the time available for oxygen diffusion.

There are, however, physical and technical limits to the optimum values for these conditions.

The bubble size is important; however, it has a lower limit, as the air bubble escaping from an orifice under water has a diameter much larger than that of the pore: in practice, the bubbles formed by systems of aeration through porous substances have a diameter of about one millimetre. Finer bubbles can only be obtained by the release of air dissolved in water under pressure (a process used for flotation).

Everything else being equal, the oxygen transfer coefficient AK_L depends on the nature of the water (clean water, waste water containing suspended or dissolved matter—presence of surface-active agents), the aeration system used and the geometry of the reactor.

In general, systems of aeration are compared on the basis of their oxygenating capacity per m^3 of pure water per hour. The **specific oxygenating capacity** of a system can also be expressed in terms of the oxygen supplied per kWh.

2.4. Anaerobic treatment: digestion

Anaerobic digestion is fermentation in the absence of oxygen; the process stabilizes the organic matter by converting it as completely as possible into methane and carbon-dioxide gas. A first bacterial group consisting of **acid-producing bacteria** is responsible for converting complex organic compounds into more simple organic compounds (acetic, propionic and butyric acids) which become a source of food for a second group, the **methane-forming bacteria.**

These are the key organisms in anaerobic digestion. They are strictly anaerobic, develop slowly and are very sensitive to temperature and pH variation in their environment.

The redox potential in anaerobic fermentation is about -0.2 to -0.3 V, calculated by reference to hydrogen.

Organic matter is generally digested by mesophilic bacteria active up to 37 °C. With regard to the acid-base equilibrium, anaerobically-digested liquid can be represented as an aqueous solution of intermediate and end products.

A distinction should be made between:
— volatile acids (acetic, propionic) in equilibrium with their salts (expressed as CH_3COOH),
— carbonic acid and its acid salt: the bicarbonate (expressed as $CaCO_3$),
— ammonia.

At a given pH and a given volatile acid concentration, there is a given ratio between the acid and its salt. Likewise, for a given total CO_2 there is equilibrium between the free acid and the bicarbonate.

The various volatile acids are acids of identical strength, slightly greater than the strength of carbonic acid.

At pH 5   we have   64.3 % acetate in equilibrium with 35.7 % acetic acid
— pH 5.5   —   85   %   —   15   % —
— pH 6   —   94.74 %   —   5.26 % —
— pH 6.5   —   98.26 %   —   1.74 % —
— pH 7.0   —   99.45 %   —   0.55 % —

Total alkalinity expressed as $CaCO_3$ in mg/l or meq/l is therefore divided into bicarbonate alkalinity and acetic alkalinity.

At a pH close to 7, an index showing good digestion, all volatile acids are in a first approximation in the form of salts (partly ammoniacal).

The main criteria controlling digestion are:
— the production of gas and its composition,
— the volatile acids,
— the pH,
— total alkalinity,
— the odour of the sludge during digestion.

2.5. Bacterial oxidation-reduction

Certain oxidation-reduction reactions (see p. 95) may be brought about by the agency of bacteria:
— the oxidation of iron and manganese in water (see p. 634);
— the oxidation of sulphur compounds (see p. 978);
— the oxidation-reduction of nitrogenous compounds.

All the biological processes described so far relate to the removal of carbonated organic pollution. Effluents also contain nitrogenous compounds in the form of proteins, amino acids, urea and decomposition products as well as inorganic nitrogen, usually as ammonium salts. All these forms of nitrogenous pollution are liable, when acted upon by specific bacteria, to undergo the nitrogen transformation cycle (see p. 976).

Nitrification is the term used for the bacterial process in which the organic and ammoniacal nitrogen (total Kjeldahl nitrogen TKN) is oxidized into nitrite (by the Nitrosomonas), then into nitrate (by the Nitrobacters). Graph 50 shows the potential zone in which these changes occur.

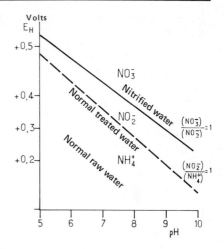

Fig. 50. — Nitrification.

During these reactions, only traces of nitrite are detected as the rate of reproduction of bacteria such as Nitrosomonas (1 day) is slower than that of the Nitrobacter under optimum environmental conditions.

In order that nitrification should occur, the age of the sludge must be such that even the bacteria which are slowest to reproduce have time to do so. If A is the age of the sludge and μ_N the growth rate of the Nitrosomonas, the following relation must be satisfied:

$$1/A < \mu_N$$

The growth rate of the Nitrosomonas is affected by temperature T according to the expression:

$$\mu_N = 0.212 \times 1.0725^{T-15}$$

It also depends on the pH and varies according to the equation:

$$\mu_N = \mu - 0.17 \, (7.2 - pH)$$

where μ is the growth rate at pH = 7.2.

Finally, the end value of the residual TKN, Ne, influences the oxidation rate K of the ammonia according to a relationship of the type:

$$K = \frac{K_m \times Ne}{K_S + Ne}$$

where :

K_m is the maximum rate of oxidation and

K_S is the ammonia concentration with $\mu = \frac{1}{2} \mu_m$.

Denitrification takes place when the dissolved oxygen concentration in the water becomes too low or drops to zero (anoxic phase): the facultative organisms use the oxygen from the nitrates, reducing them in turn to nitrites NO_2^- then to nitric oxide gas NO, nitrous oxide N_2O and nitrogen N. These changes take place at a redox potential of + 0.35 to 0.40 V at pH 7, which is the same region as nitrification.

If some external cause (anaerobic digestion) brings the redox potential to negative values (— 0.2 to — 0.3 volt), the reduction of the nitrates and nitrites gives NH_4^+.

Theoretically, denitrification is inhibited by the presence of oxygen. Actually, the bacteria participating in this process are facultative anaerobic heterotrophs which derive their energy from that liberated when electrons are transferred from the organic compounds to O_2, NO_2^- or NO_3^-.

Where the oxygen and the oxidized nitrogenous compounds are all available, the choice as final acceptor falls on the one which yields the greatest quantity of liberated energy per unit of oxidized organic matter. MacCarty[1] has published the following Table for oxygen and nitrates:

Energy source	Energy yield (kcal) by electron acceptor	
	O_2	NO_3
Domestic waste water	26.275	24.278
Methanol	27.640	26.093
Ethanol	26.267	24.720

These results show clearly that the amount of energy liberated is invariably greater with oxygen than with nitrates. Electrons are therefore transferred preferentially to oxygen. The presence of oxygen in solution inhibits denitrification.

In addition, maximum energy is liberated by the use of methanol, which explains why numerous researchers have chosen this substance as the preferred source of carbon for denitrification. On the other hand, domestic waste water provides a very similar amount of energy without cost, and is therefore very interesting as a source of carbon.

As regards the pH, it seems that the optimum value lies between 7.0 and 8.2. It follows that excessive nitrification, liable to lead to a considerable lowering of the pH in water without much of a buffer, would be prejudicial to denitrification. This would explain the difficulties encountered in the biological removal of excessive levels of TKN and $N - NH_4$.

Johnson has shown that denitrification is proportional to the quantity of substrate used. It has also been demonstrated that, whatever the carbonaceous substrate seleected, 4 to 6 mg of BOD_5 are eliminated per mg of nitric nitrogen reduced without further addition of oxygen. Denitrification therefore makes it possible to eliminate a large proportion of the carbonaceous pollution (BOD_5) by re-using the oxygen in the nitrates.

Finally, it must be borne in mind that, like any enzyme reaction, denitrification is sensitive to temperature, which at low levels (t < 10 °C) tends to reduce its velocity.

---

1. MacCarty: "Energetics of organic matter degradation in water pollution microbiology", Wiley, 1972.

Part Two

Treatment Plant and Processes

Narcisse apres auoir mesprisé plusieurs Nimphes dont il estoit aimé pour sa beauté et enrauivres Echo reuenant un iour de la Chasse s'approcha d'vne fontaine ou s'estant apperçeu il s'amouracha tellement de luy quil en secha et fut changé en vne Fleur de son nom . metam . l3.   N.langlois ex cum Pri . re

le Pautre Fin et sculp

4 PRELIMINARY TREATMENT

It is desirable for any raw water and/or sewage to be subjected to a preliminary treatment before passing to the main treatment plant. The preliminary treatment is usually of a purely physical or mechanical nature the object of which is to remove the greatest possible amount of substances whose nature or dimensions make them undesirable during later treatment processes.

Pre-treatment operations are listed below (one or several of these operations may be used, according to the size of the plant and the quality of the raw influent):
— coarse and fine screening,
— comminution,
— grit removal,
— de-silting,
— oil and grease removal,
— straining,
— treatment of grit and trash.

1. SCREENING

Bar screens are used:
— to protect the plant against the entry of large objects likely to cause blockage in different parts of the installation;
— to separate and extract more easily any bulky matter carried in the raw influent likely to interfere with subsequent operations or to create complications.

The effectiveness of coarse screening depends upon the spacing between the bars; we can distinguish between:
— fine screening, with spacings from 3 to 10 mm,
— medium screening, with spacings from 10 to 25 mm,
— coarse screening, with spacings from 50 to 100 mm.

Screening is performed either with a manually-cleaned bar screen or (necessarily, when the plant is sufficiently large) with an automatically-cleaned bar screen called a mechanical bar screen.

Mechanical bar screens are often protected by pre-screening, using a bar screen with a wider spacing (50 to 100 mm) which is customarily cleaned manually, but which may also be automatic in the case of very large installations or where the raw water carries appreciable quantities of coarse matter.

The mechanization of bar screens does not depend solely upon the size of the plant.

The need to reduce the number of manual cleaning operations may justify the choice of such apparatus even when the station is small. This mechanization becomes inevitable whenever there is a risk that a large amount of vegetable matter may suddenly accumulate on the face of the bar screen and form a tangled "mat" which can totally clog up the inlet in a matter of minutes.

When cleaning is carried out manually, the area of the screen must be designed oversize to avoid too frequent cleaning, particularly if the space between the bars is less than 20 mm.

When cleaning is carried out automatically, the equipment for recovery and removal of the screenings must be designed as a function of their weight (see paragraph 7.1, page 132). In particular the capacity of the mobile receptacle must be sufficient to take the screenings collected over a period of at least 24 hours.

When fresh sludge (from primary settlement) is treated directly (without digestion) particularly by centrifuging, after having been subjected to fine screening, this avoids the need to overdimension raw effluent screening equipment, which can be kept to an average spacing of 15 to 25 mm.

1.1. Various types of bar screens

A. MANUAL BAR SCREENS:

● These are made of straight steel bars, sometimes mounted vertically, but they are most frequently inclined at 60 to 80° to the horizontal.

● In small or medium-sized drinking-water treatment plants, the strainers on the raw water pump inlets sometimes perform the duty of bar screens. Some strainers are self-cleaning and use countercurrent flushing devices of different sorts, such as an incorporated flushing siphon.

● In small rural sewage treatment plants, and when the intake pipes are buried deep in the ground, bar screens are sometimes replaced by perforated baskets which may be lifted out for cleaning.

Fig. 51. — Curved bar-screen.

B. MECHANICAL SCREENS RAKED ON UPSTREAM SIDE:

● *Curved screens:* this type of screen is ideal for medium-sized plants and a medium pollution load.

They are especially suitable for relatively shallow installations, and an advantage of this screen is that it provides a large useful surface area. Cleaning is effected by two combs or tines, one at each end of an arm which pivots around a horizontal axis.

An extractor discharges the screenings behind the screen, either into a removable hopper, or preferably onto a conveyor belt or into the trough of an Archimedes' screw, which provides lateral removal and storage of the screenings in a receptacle of large capacity.

● *Straight screens with reciprocating cleaning:* the screen, consisting of bars of rectangular or trapezoidal cross section (reducing the risk of jamming by solid matter), is generally inclined at 80º to the horizontal and ends just above the maximum level of the liquor; it is continued by a metal or concrete apron.

The raking mechanism, with reciprocating motion, lifts the screenings. An extractor removes them into a receptacle or onto a conveyor belt situated downstream of the screen.

The raking mechanism can, for example, consist of:
— a combined rake and carriage which moves along racks. This is recommended when the height of lift and the volume of screenings are not very great;
— a rake connected to a moving carriage controlled by 2 cables. This system is used for lightly polluted influents, and in very deep installations;

Fig. 52. — Rack-and-pinion screen

Fig. 53. — Cable-type bar screen with grab.

Fig. 54. — Cable-type bar screen, with grab.
1. *Bar screen.*
2. *Grab carriage.*
3. *Tilting grab.*
4. *Grab cleaner.*
5. *Cable drum carriage.*
6. *Carriage cable.*
7. *Lateral rails.*
8. *Frame.*
9. *Carriage drive unit.*
10. *Strain limiting stop.*
11. *Slack strand, carriage.*
12. *Cable drum, grab.*
13. *Counter pulley, grab.*
14. *Grab cable.*
15. *Slack strand, grab.*
16. *Grab drive unit.*

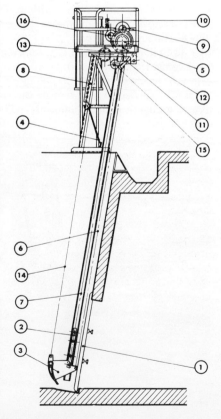

— a grab (or pivoting scoop) controlled by a separate cable and connected to a movable carriage controlled by 2 cables. This mechanism can be used for heavily polluted influents and in very deep installations.

● *Straight screens cleaned continuously:* this relatively fine screen should only be used where bulky foreign matter is infrequent (or where the water has already passed through a medium bar screen).

The straight bar screen, inclined at 80° to the horizontal, is cleaned by nylon brushes driven by an endless chain mechanism.

Screenings are removed via a trough mounted behind the bar screen.

— For very high flow rates:

When the flow rate is very high (above 30 000 m³/h (190 mgd), for example), and the water contains only small amounts of foreign matter, cleaning may be effected by a travelling mechanism which cleans only part of the bar screen at a time, and moves on laterally after each operation. A cable-operated cleaning mechanism equipped with a rake or a grab is well suited to such an arrangement.

C. MECHANICAL SCREENS WITH CLEANING MECHANISM ON THE DOWNSTREAM SIDE:

This type of screen, generally used for sewage treatment, can remove large quantities of solid matter by means of comb rakes driven by an endless chain mechanism mounted downstream of the screen.

The bar screen, which is vertical or inclined (at 60° to 80° to the horizontal), is extended to the level where removal of screenings occurs. The necessity of giving the screen sufficient rigidity limits the depth of the equipment. Depending on the type, the screenings are removed either upstream or downstream of the screen into a removable receptacle or on a conveyor.

1.2. Automatic control and protection of mechanical screens:

The generally-intermittent operation of the cleaning device may be controlled by electric timers that can be adjusted to give variable frequencies and duration of operation; or alternatively—and particularly in the case of waste water—by a differential head-loss detector; or by a combination of the two systems.

The cleaning mechanisms must be equipped with a torque-limiting device to prevent damage to the equipment in case of overloading or clogging.

Curved or straight screens cleaned with a reciprocating motion have devices which ensure that the raking mechanism stops automatically clear of the screen to avoid any risk of jamming when it restarts.

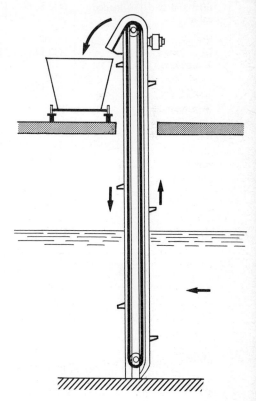

Fig. 55. — Fixed mechanical screen.

Mechanical screens.

TYPE OF SCREEN	SERVICE CONDITIONS				
	Type of influent	Range of flows m³/h (gpm)	Depth of channel m	Width of channel m	Space between bars mm
SCREENS CLEANED ON THE UPSTREAM SIDE					
Curved screens	Medium loading	10-5 000 (44-22 000)	0.40-1.70	0.30-2.00	12 to 80 special 4-10
Screens with racks	Loaded	100-10 000 (440-44 000)	1.50-5.00	0.60-2.00	12 to 80
Screens with rake and two cables	Lightly loaded	100-15 000 (440-66 000)	2.50-10.0	0.60-4.50	12 to 80
Screens with grab and 3 cables	Loaded	1 000-40 000 (4400-176 000)	2.50-10.0	1.50-5.50	12 to 100
Continuously cleaned screens with endless chain	Lightly loaded	100-15 000 (440-66 000)	1.50-8.00	0.80-3.00	12 to 25 special 4-10
SCREENS CLEANED ON THE DOWNSTREAM SIDE					
Screens with combs mounted on endless chains	Loaded	500-15 000 (2200-66 000)	1.50-4.00	0.80-4.00	10 to 60

1.3. Flow rates and headloss — Clogging

The rate of flow through the bar screen should be sufficient to carry the solids against the face of the screen without causing excessive headloss or clogging in depth between the bars. A compromise should be found to give acceptable flows between the maximum and minimum design rates.

In general, the average velocity through the bars will lie between 0.60 and 1.00 m/s (2 to 3.3 ft/sec), and may rise to 1.20 or 1.40 m/s (4 to 4.7 ft/sec) at maximum flow.

If the selected minimum flow rate is likely to permit settling out in the screen channel, stirring or other means should be provided to prevent this from happening.

Bar screens cause headlosses from 0.05 to 0.15 m (2″ to 6″ wg) for drinking water, and from 0.10 to 0.40 (4″ to 16″ wg) for sewage. (The maximum values allow for a necessary margin of safety related to possible partial blockage of the screen.)

2. COMMINUTION

This treatment applies more particularly to sewage.

It is designed to "disintegrate" the solid matter entrained by the water. Instead of being screened out of the raw effluent, these solids are shredded and pass on with the water to the following phases of treatment. The advantage of this process lies in the elimination of extra steps and problems involved in the evacuation and disposal of screenings. In practice, it has several drawbacks—in particular the need for frequent maintenance of rather delicate equipment, the risk of clogged pumps and piping through the accumulation of balls of textile or vegetable fibres bonded by grease, and the tendency to form floating scum in anaerobic digesters.

For these reasons comminution of raw sewage at the inlet to the plant is less frequently recommended when designing new plants. However, comminution is sometimes substituted for fine screening of fresh sludge (whether thickened or not) before direct treatment without digestion (heat treatment, centrifuging). In this case it is carried out by an "in line" comminutor (see below) ensuring the required fineness of comminution. Two types of comminutor are used, "stream flow" and "in line".

Both types of apparatus are especially adapted to sewage treatment and are capable of handling solid matter normally found in sewage (but only after coarse screening through bars spaced 50 to 80 mm apart, depending upon the equipment used), and of reducing the solids to particles with an average diameter of several millimetres.

2.1. Stream flow comminutors

These have the advantage of creating only a low head loss and consuming a small amount of energy.

The usual types consist of a rotating drum with a vertical axis and horizontal slots. They can be mounted either in an open channel or directly in the pipework (in the case of smaller models). The drum is made up of circular bars, and is equipped with teeth. Cutting knives are bolted to the frame.

The water flows from the outside to the inside of the drum, and the solids caught on the drum are progressively torn to shreds.

Fig. 56. — Comminutor for shredding float-   *Fig. 57. — GRIDUCTOR comminutor,*
ing debris, with lifting device.   *INFILCO DEGREMONT INC.*

2.2. In-line Comminutors

These comminutors can combine the shredding effect with a pumping effect which discharges the "disintegrated and diluted" matter. They are fixed in pipework like pumps but their pumping capacity is generally low (sometimes nil) and they may require an additional pump in series. Electric motors are considerably oversized to eliminate the risk of jamming.

These heavy-duty units usually have a screw which forces the diluted solids through a cutting screen. One (or more) adjustable blade is used to vary the grinding effect.

Type of comminutor	Flow rates m³/h (gpm)	Maximum inlet channel water depth m	Maximum loss of head m	Head at discharge m	Power of electric motor kW
Stream flow comminutor	5-8 000 (22-35200)	0.30-1.20 (1'-4')	0.10-0.35 (4"-14")		0.25-4
In-line comminutor	50-300 (220-1320)			0-2.0 (0-6'5")	7.5-20

3. GRIT REMOVAL

The object of grit removal is to eliminate from the raw water the gravel, sand and other mineral particles which would otherwise form deposits in channels and piping, to protect pumps and other equipment against abrasion, and to avoid overloading subsequent treatment phases.

Grit removal is a term used for the removal of solids with a particle size of more than 200 microns. Smaller particles are eliminated during settling and extraction operations.

The theoretical study of grit removal is related to phenomena of free-settling and involves the application of formulae developed by Stokes (laminar flow), by Newton (turbulent flow), and by Allen (transient flow). (see p. 66).

These formulae are used to calculate sedimentation rates for spherical particles. Corrections must be applied to take into account:
— the shape of the particles,
— the concentration of solids in suspension whenever they exceed approximately 0.5 %,
— the nature of the horizontal flow.

In practice, the following data (for free-falling sand particles with a density of 2.65) may be used:

d	cm	0.005	0.010	0.020	0.030	0.040	0.050	0.10	0.20	0.30	0.50	1.00
V_c	cm/s	0.2	0.7	2.3	4.0	5.6	7.2	15	27	35	47	74
$V_{c'}$	cm/s	0	0.5	1.7	3.0	4.0	5.0	11	21	26	33	
$V_{c''}$	cm/s	0	0	1.6	3.0	4.5	6.0	13	25	33	45	65
VI	cm/s	15	20	27	32	38	42	60	83	100	130	190

where:
d   diameter of the grit particle,
V_c settling rate, for fluid with negligible horizontal velocity,
$V_{c'}$ settling rate, for fluid with a horizontal velocity equal to VI,
$V_{c''}$ settling rate, for fluid with a horizontal velocity equal to 0.30 m/s.,
VI   critical horizontal fluid rate at which deposited grit particle is entrained.

3.1. Grit removal at drinking water plants

A water intake should be designed to avoid grit entrainment as much as possible. If local conditions do not permit this to be done, grit removal equipment must be provided unless the grit is to be extracted elsewhere in the course of further treatment. If the plant is to include silt extraction, it will suffice to remove particles having a diameter greater than 0.3 mm in a simple grit removal channel cleaned by hydraulic flushing.

The grit may also be removed by cycloning (hydro-cyclone on lift-pump discharge side). The hydro-cyclones give excellent grit removal in the case of grit with particle size from 100 to 500 microns, but head losses are appreciable (of several metres wg, with a minimum of 0.50 m for the more elaborate types). But there is nevertheless a risk of pump wear caused by abrasion. If the installation includes fine screening equipment (mesh openings of 1 to 2 mm, for example), sand removal must be carried out as a preliminary operation to avoid damaging the woven screens.

● *Designing the dimensions of grit removal equipment:* construction is generally of the rectangular channel type :
— the horizontal surface is calculated as a function of the settling rate V_c for the smallest particles to be removed, and of the maximum flow-rate to be handled:

$$\text{horizontal surface} = \frac{\text{maximum flow rate}}{\text{settling rate } V_c \text{ of the smallest particles to be removed}}$$

— the transversal cross-section is a function of the desired horizontal flow-rate. the flow-rate selected will be greater than the critical particle entrainment rate Vl if the design calls for hydraulic evacuation of the grit, or less than Vl if bottom scrapers are provided.

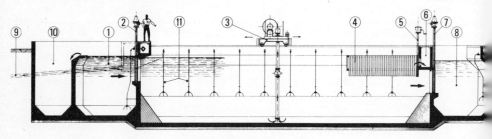

Fig. 58. — Oil and grit removal.

1. *Intake channel.*
2. *Inlet gate.*
3. *Travelling bridge for grit suction.*
4. *Stilling picket fence.*
5. *Oil removal gate.*
6. *Oil discharge channel.*
7. *Outlet gate.*
8. *Outlet channel, feeding the settling tank.*
9. *Grit discharge.*
10. *Overflow by-pass.*
11. *Submerged air blowing pipes.*

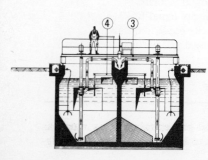

3.2. Grit removal at sewage treatment plants

Because of the many types of sewage effluents the problem here is much more difficult than in the case of drinking water plants.

It is desirable to separate the grit from other solids present in the water, particularly organic matter, so that the recovered grit will not constitute a source of trouble. Generally this is difficult. In fact, the recovered grit always contains a certain proportion of organic matter which settles out at the same time. This organic matter should be separated hydraulically using a flow-rate of about one foot per second. An improvement may be obtained by a final washing of the recovered grit in which the content of organic matter is reduced to less than 30 %.

● *Principal types of structures.*

They are listed below in the order of increasing size and efficiency:

— Simple grit-removal channels in which the horizontal velocity is proportional to the flow. These structures are used in small sewage works. The grit accumulates in a longitudinal recess in the bottom, from which it is extracted manually every 4 or 5 days.

— Constant velocity (about 0.3 m/s) channels. These units consist of 1 or preferably 2 channels equipped with a linear-equation outlet weir (depth proportional to flow). Generally provided with a manually-cleaned storage gutter, they can be designed for hydraulic transport of grit to an adjacent hopper equipped with a mechanical extraction device. Retention time about 1-2 minutes.

— Circular grit removal tanks, with tangential inlets, mechanical stirring or (better still) air agitation, in which the velocity of the cross-current at the bottom is maintained virtually constant whatever the flow. The grit is retained in a central pit, and is extracted by pump or air lift to be transferred to a gravity drying compartment. Retention time 2-3 minutes.

— Aerated rectangular grit removers, in which the aeration causes rotation of the liquor thus creating a constant-rate cross-current along the botton, at right angles to the main flow which, as it is much slower, can be varied without disadvantage. Aeration, in addition to causing the water to circulate, provides agitation which promotes the separation of organic matter that may be adhering to the grit particles.

Grit is extracted automatically :

— by a group of air lifts functioning cyclically (recovery from bottom hoppers),

— by raking (by endless chains or by scraper bridge) towards a collection pit, followed by recovery by pumping,

— direct by suction pump (or air lift) mounted on a travelling bridge. The grit is generally separated from the pumped emulsion by a mechanical method reducing the proportion of associated organic matter and transferring the grit to the storage compartment. Designed with a retention time of about 3-5 minutes, a maximum hydraulic loading of about 70 m³ water/m² of surface area/hr (28 gpm per sq ft) and a maximum length of 30 m, these structures retain about 80 % of the grit with a particle size of from 250 microns (maximum flow) to 150 microns (minimum flow).

3.3. Grit removal from industrial effluents

For effluents from the metallurgical or engineering industries, the separation of grit and other inorganic matter in the order of 100-250 microns or more can be carried out:
— either in a raked grit remover, circular in shape, with diameter of about 5-15 m (with central inlet and peripheral discharge), or square (with effluent flowing in along one side and out along the opposite side) designed with a liquor height of about 1 m, and a hydraulic loading of about 15-30 m³ effluent/m² of surface area/hr (6 to 12 gpm per sq ft). The rotating rake with central drive pushes the grit into a lateral collection pit. The mechanical extraction is carried out by Archimedes' screw or by reciprocating raked classifier,
— or by cycloning in a hydrocyclone under pressure, at reduced inlet velocity (2-3 m/s or 7 to 10 ft/sec) to limite abrasion. The equipment is about 300-700 mm (12 to 28″) in diameter. Unit flow of up to 500 m³/hr (2200 gpm). Loss of head of about 2-3 m (7 to 10 ft). Continuous extraction at a concentration of about 10 % inorganic matter. An arrangement with siphon extraction of the upper level of liquor allows the lower level to be concentrated to about 20 %, while retaining a convenient outlet pipe diameter.

Fig. 59. — Rectangular aerated grit-removal tank.

4. DESILTING

Desilting is an operation carried out before clarification of surface water containing large quantities of suspended solids. It consists of preliminary settling in which the object is to remove all the fine grit and most of the silt.

The level of concentration of suspended solids in the raw water at which desilting becomes necessary depends on the type of the principal clarifier. This level can be about 2 g/l upstream of non-scraped clarifiers, and 5-10 g/l upstream of scraped clarifiers.

Depending on the loading of suspended matter and its type, a desilter can be designed as a clarifier (this is generally the case for raw influents in which the loading does not exceed 20-30 g/l) or as a thickener when the loading of the raw influent exceeds this limit.

At low loading it is possible to use natural settling without a reagent. Then, as the loading increases, it becomes necessary to use a reagent to obtain sufficiently high clarification efficiency. As the loading rises further, the suspended solids no longer settle freely but become compacted. Generally speaking, this process is not improved by the use of a coagulant such as aluminium sulphate or ferric chloride. Only the use of specific polyelectrolytes results in adequate compaction. In fact, water treatment is becoming sludge treatment, and the dividing line between these two types of treatment varies considerably from one sludge to another.

When the desilting equipment works as a clarifier, the upward velocity varies between 2 and 6 m/h, depending on the suspended solids content required at the outlet, the type of solids and on whether or not reagents are used.

The contact time depends on the loading of suspended matter, their compaction characteristics and the final concentration required in the effluent. It generally varies between 1 and 2 hours.

When the loading is high or when the sedimentation rate is low, desilting equipment should be designed as sludge thickening equipment. The upward velocity is then lower (between 2 m/h and 0.5 m/h, sometimes even less) and the contact time lies between 2 and 5 hours.

The clarification efficiency of a desilter varies from 50 to 65 % without addition of reagent. It can reach 75 to 98 % with an adequate dose of flocculant, generally about one-third of that determined by the jar test. When the zone settling process begins, increasing doses of polyelectrolyte must be used and this very soon renders the water unsuitable for industrial purposes. In fact, only very small amounts of polyelectrolytes (about 1 mg/l) are authorized by some regulations in respect of drinking water.

Like clarifiers, desilters are circular or rectangular. The sludge is removed:
— without scraping, by gravity removal using contiguous hoppers,

— without scraping, by suction pumps mounted on a reciprocating travelling bridge moving at velocities of 1 to 3 cm/sec. The bridge can be provided with a water injection pump for washing down the sludge accumulations towards the suction pump intakes,
— with scraping, to one or more hoppers, with gravity extraction or fixed pump.

The design of desilters requires an exact knowledge of the sludge volumes produced at the most critical periods and the settlement properties of this sludge.

Fig. 60. — De-silting tanks for the preliminary settling of drinking water supplies, BANDUNG, Indonesia.

5. OIL AND GREASE REMOVAL

There are a number of oil and grease removal problems:
— removal of oil from surface waters before the main water treatment,
— grease removal from town sewage before discharge to the sewer,
— grease removal as a pre-treatment process at an urban sewage treatment plant,
— removal of oily products from waste waters at petroleum refineries and petro-chemical plants,
— oil removal from steam condensates before recycling (see condensates, page 690).

Oil removal is an operation of liquid-liquid separation, while grease removal is one of solid-liquid separation (on condition that the water temperature is sufficiently low to allow the grease to congeal).

Oils and greases are generally lighter than water, and tend to rise to the surface. Any storage tank that reduces the flow velocity and provides a quiet surface acts as an oil grease separator.

Surface recovery and removal are as far as possible to be carried out by overflowing in the case of oil (or by adherence to an endless belt) while grease is to be scraped off.

5.1. Removal of oil from surface water

The aim here is to remove any free oil that would otherwise be entrained with the water to be treated.

Natural flotation at the plant head (at the water intake or along the intake channel) is recommended. An inverted weir retains the oil, while the water flows out beneath it. The oil that accumulates on the surface is evacuated by overflowing a special fixed or adjustable weir (rotary skimmer tube).

5.2. Removal of grease from sewage before discharge to the sewer

This "at-source" pre-treatment is recommended for small plants, restaurants, small communities, etc., and is often required by by-laws. Standard grease separators (or 'grease boxes') are factory-made for flows up to 20 or 30 l/s (320 to 480 gpm). They have retention times of 3 to 5 minutes and an ascending sedimentation velocity of about 15 m/h (6 gpm per sqft).

Properly operated they can retain up to 80 % of congealed fatty matter and hold about 40 litres of light matter for each l/s of design now. Regular cleaning is essential. The water temperature should be less than 30 ºC at the separator outlet. Insofar as is possible the separators are designed to prevent bottom settling of heavy solids, but it may prove advantageous to provide ahead of the grease box an easily-cleaned settling tank with a retention time of about 1 to 3 minutes.

5.3. Grease removal in pretreatment at domestic sewage treatment plants

A primary settling tank is well adapted to the separation of grease, which rises to the surface, but is generally ill suited to the collection of large amounts of grease, liable to cause operating difficulties.

In the case of domestic sewage, grease removal is desirable (essential if there is no primary settlement) and is advantageously carried out combined with grit removal (by providing a stilling zone over part of the water surface in an aerated grit remover, grease and scum being removed by overflow or skimming). The grit-removal unit should be large enough to allow for this additional treatment.

The structure comprises an aeration zone where the air is injected into the lower level, and a stilling zone where grease rises to the surface. Any deposited sludge slides down sloping walls and returns to the aeration zone. The average retention time in the unit is about 10 to 15 minutes at average flow, with a minimum of 5 minutes. Removal of grease can be carried out either by overflow or preferably by mechanical skimming of the surface. The flow of injected air is about 0.5-2 m^3 per hour and m^3 capacity of the structure. Given the above conditions it should be possible to retain 80 % of the grease.

A separate grease remover may be recommended if the quality of the raw sewage is such that there is a large quantity of grease. In this case the stilling zone should be designed for an upward velocity of 15 to 20 m/h with a maximum of 25 m/h.

Grease removal equipment of this type is sometimes used to provide partial treatment of waste waters from certain industries such as slaughter houses and meat packing plants before discharging the water to the sewers, thus protecting the public sewer system from excessive grease deposits.

Fig. 61. — Grease removal by air injection in a slaughter-house.

In certain cases (e.g. water with strong reducing properties, or with a high ammonia concentration), it may be desirable also to use the unit for pre-aeration. It should then be designed for a retention time of at least 30 minutes with an appropriate flow of air. It should be noted that these grease removers are not intended to retain oil and hydrocarbons, for which the discharge to sewers is not permitted.

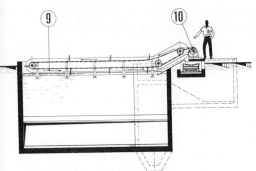

Fig. 62. — *Oil and grease removal tank.*

1. *Raw water intake.*
2. *Aeration by porous domes.*
3. *Drain-valve control pillar.*
4. *Air-lift.*
5. *Grease conveyor belt.*
6. *Sewage passing to treatment plant.*
7. *By-pass.*
8. *Drain valve.*
9. *Scraper system.*
10. *Reduction gearing and motor for the scraper drive.*

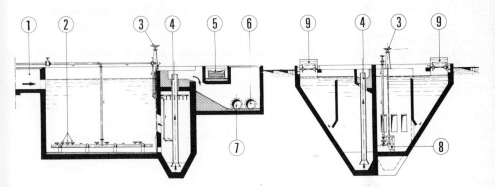

5.4. Preliminary removal of oil from effluents from oil refineries, petrochemical plants and engineering works

These effluents contain varying quantities of hydrocarbons which may be present in the free state, partially dissolved, or emulsified to a greater or lesser extent. Complete oil removal requires a two-step treatment:
— preliminary oil removal by the force of gravity without addition of reagents, reducing the hydrocarbon content to about 15-100 mg/l,
— final oil removal (dissolved air flotation, filtration, coalescence) with use of reagents (coagulation by metallic salts or by cationic polyelectrolytes), permitting complete purification.

Preliminary oil removal is achieved by natural flotation of oil globules.   It is carried out in various types of equipment.

● *Conventional longitudinal oil removal tanks:* these are rectangular tanks with longitudinal flow.   They have special inlet distribution equipment, mechanical skimmers and scrapers, and are equipped with separate outlets for treated water, surface oil and settled sludge.

The American Petroleum Institute has drawn up detailed instructions (A.P.I. standards) for designing, dimensioning and building tanks of this type on the basis of the following conditions:
— theoretical diameter of oil droplets assumed to be d $>$ 0.015 cm,
— horizontal surface calculated for an upward flow rate varying from 0.9 to 3.6 m/h (0.4 to 1.6 gpm per sq ft),
— transversal cross-section calculated for horizontal velocities from 18 to 55 m/h (1 to 3 ft/min) and equal to 15 times the upward flow rate,
— ratio of water depth to width between 0.3 and 0.5,
— width: maximum 6 m (20 ft), minimum 2 m (6 ft 8 in),
— water depth: maximum 2.50 m (8 ft 4 in), minimum 1.00 m (3 ft 4 in).

This type oil-removal equipment is best suited to heavy oil loads and wide concentration variations. In practice, it is used to reduce the hydrocarbon content to approximately 15 to 100 mg/l according to the initial percentage of emulsified products.

The use of tanks with two compartments in series results in increased oil-removal efficiency.

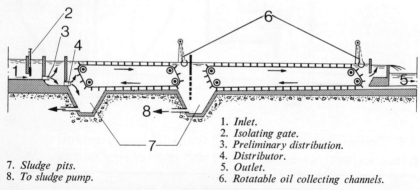

1. *Inlet.*
2. *Isolating gate.*
3. *Preliminary distribution.*
4. *Distributor.*
7. *Sludge pits.*
8. *To sludge pump.*
5. *Outlet.*
6. *Rotatable oil collecting channels.*

Fig. 63. — Longitudinal oil separator with 2 compartments working in series.

● *Scraped, circular oil removal tanks,* which are cheaper to install, simpler to operate, and equally efficient. They function at upward velocities of 5-15 m/h depending on the required degree of retention.

● *Parallel-plate oil-removal plant:* these are lamellar oil separation tanks in which oil droplets travel over a very short distance which is limited by the space between two plates (variable from 20 to 100 mm), before being trapped. They have the advantage of being very compact and of being more effective in the removal of relatively fine oil droplets.

However, they can only handle effluents carrying medium oil loads and light sludge loads, paraffin, asphalt and grease, i.e. unlikely to cause much clogging. If necessary they must be protected by the addition of a coarse preliminary treatment to avoid operating too close to the capacity limits of the equipments.

6. FINE STRAINING

Fine straining is filtration through a thin support, and is used in numerous applications in water treatment. Depending on the size of orifices in the support material, a distinction can be drawn between macrostraining and microstraining.

— **macrostraining** (through perforated steel sheet or metal wire netting, mesh size greater than 0.3 mm) is used to remove certain suspended solids, floating or semifloating matter, animal or vegetable debris, insects, twigs, algae, grass etc. sized between 0.2 and several millimetres,

— **microstraining** (through plastic or metal fabric with a mesh size less than 100 microns) is used to remove very fine suspended matter, from drinking water (plankton) or from pretreated sewage influent (discussed in Chapter 9).

Macrostraining equipment used in pretreatment can be classed as:

— stream flow equipment, with low loss of head:
• rotary macrostrainers,
• fixed scraped strainers,
— equipment requiring to be fed by pump,
• static or rotary self cleaning strainers,
• mechanical filters.

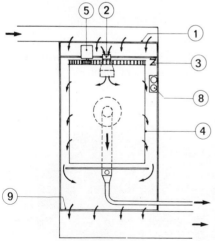

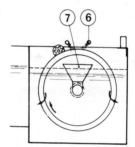

1. *Feed weir.*
2. *Raw water inlet gate.*
3. *Safety valve.*
4. *Microstrainer drum.*
5. *Drive unit.*
6. *Washing header with nozzles.*
7. *Wash-water channel and pipe system.*
8. *Head-loss detector.*
9. *Outlet weir.*

Fig. 64. — Microstraining unit.

6.1. Rotary macrostrainers:

Used for potable water, irrigation water or "thin" effluents, this equipment is available either in the form of horizontally-mounted drums when the water level is relatively constant, or rotary slat-type strainers in a continuous chain of screens when the water level varies considerably (several metres difference).

These strainers can treat flows ranging from several hundred litres per second to more than 10 m³/s (2 640 gal/sec).

Straining is performed by a number of interchangeable filter panels, consisting generally of woven metal (bronze or stainless steel wire) mounted on a rigid frame.

The mesh size varies from 0.3 to 3 mm and the wire diameter varies between 0.25 and 1 mm.

The woven macrostrainer must normally be protected by a bar screen with spacing between bars of 40 to 50 mm for mesh woven from 1 mm wire, 20 to 30 mm bar pitch for 0.5 mm wire, and 10 to 15 mm pitch for 0.3 mm wire.

The free surface area coefficient is about 50 or 60 % and the rate of filtration (ratio between the flow-rate and the free surface area of the orifices of the immersed part of the screen) is generally about 0.35 to 0.40 m/s.

The direction of filtration flow is preferably from the inside to the outside so as to facilitate screen washing and discharge of the extracted solids.

Automatic control of the wash-water jets makes it possible to limit head losses through the strainers to about 20 cm wg in normal operation. The strainers can withstand a maximum head loss of 50 cm wg.

6.2. Fixed scraped strainers:

Certain waste effluents (from slaughter houses, canning factories, etc.) must be strained before they reach the sewage treatment plant.

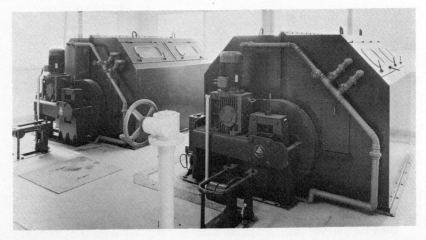

Fig. 65. — Rotating microstraining units.

This straining is effected by mechanically-cleaned screens or strainers. The suspended matter is retained on a fixed stainless-steel plate with perforations 2 to 5 mm in diameter; the entrapped solids are removed by scrapers or by brushes fixed to endless chains (for straight strainers inclined about 45° to 60° to the horizontal) or to rotating arms (for curved strainers) and discharged to a collection trough by a tipping cleaner.

If the sewage contains fatty matter which is liable to congeal, the risk of clogging of the strainers, even if the perforations are correctly chosen, may impose preliminary grease removal.

This equipment can be placed in channels 1 m - 2.5 m wide and can treat flows of the order of 100-2 000 m³/h (440 to 8 800 gpm). The loss of head must not exceed 0.5 m in normal conditions.

6.3. Static or rotary self-cleaning strainers:

Strainers with mesh size of 0.25 mm to 2 mm are also used for effluents, particularly in the food industry.

• *Static strainers* include a screen, made up of tiny horizontal bars in stainless steel, which are straight or curved and of triangular section. The effluent is distributed over the upper part of the screen whose inclination to the horizontal gradually reduces from 65° to 45° from top to bottom. In this way the effects of separation, draining and removal of the solids are obtained successively.

• *Rotary strainers* consist of a cylindrical screen with horizontal axis, with stainless steel bars of trapezoidal section, which turns slowly. The material retained on the screen is recovered by a fixed scraper and removed.

Fixed and rotary strainers can treat flows from 10 to 1 000 m³/h (44 to 4400 gpm) depending on their type and the fineness of straining required. About 2 m loss of head should be allowed for and screenings are collected at the base of the equipment. As with the other strainers (fixed scraped strainers) these strainers are liable to clogging by congealing grease.

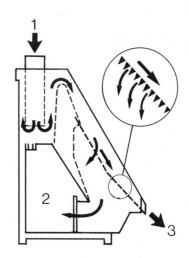

1. *Raw water inlet.*
2. *Filtrate.*
3. *Solid wastes.*

Fig. 66 — Static strainer

6.4. Mechanical filters:

When flow-rates are relatively low, suspended solids may be trapped by mechanical filters maintained under pressure by pumping.

DEGREMONT and a number of other manufacturers offer a wide range of equipment; examples are:

— **cyclone filters** with a tangential inlet at the top, a stainless steel cylindrical filter strainer with mesh orifices from several tenths of a millimetre to several millimetres, and manual or automatic washing and solids extraction using raw or filtered water. Flows vary from about 100 to 500 m³/h (0.63 to 3.15 mgd), according to the model. Head losses are up to several metres wg,

— **continuously-cleaned** rotary filters, with removable filter elements of porcelain or stainless steel (mesh size 0.1 to 1.6 mm) attached to a rotating device; cleaning is effected by passing the filter elements in front of a single orifice discharging water and the residue to the outside by counter-current flow. Filtration capacities vary according to model from 25 to 7 000 m³/h (0.16 to 44 mgd). Head losses are of several metres on a water gauge. Water consumption for the continuous washing process is 2 to 5 % of the filtered water output,

— **automatic filters** with multiple filter tubes and rotary washing arms. Flow-rates from 100 to 10 000 m³/h (0.63 to 63 mgd). Mesh size starts at 125 microns. Head losses are of the order of several metres wg. Cleaning water consumption is about 1 or 2 % of the filtered water output.

7. REMOVAL AND TREATMENT OF GRIT AND TRASH FROM WASTE WATER

7.1. Screenings

The waste recovered from the screens are extracted manually (in baskets, perforated or plain, narrow gauge trucks or dumpers), or mechanically (belt conveyor, continuous moving or reciprocating moving rake conveyor), or hydraulically (through a trough fed with pumped water).

The volume of screenings at an urban sewage treatment plant is of the order of 2 to 5 litres per head and per year for bar screens with a spacing of about 30 to 50 mm, and 5 to 10 litres per head and per year where bar spacings of about 15 to 25 mm are used.

The waste matter is often transported to a dump or buried. It can also be incinerated in a furnace provided for the purpose at the sewage plant or in a municipal garbage incineration plant. The combustion temperature must be above 800 °C to avoid emission of unpleasant smells.

To facilitate transport, the waste may be dewatered and compacted in specially-designed hydraulic presses.

7.2. Grit from grit removers

The volume of grit recovered from grit sewage during grit removal is of the order of 5 to 12 litres per head and per year.

The grit extracted manually by shovelling from small grit removal channels is generally unfit for re-use and should be buried or evacuated with the screening wastes.

In larger installations, grit recovered by pumping from the grit removal equipment may be separated from the water:

— by settling in a shallow tank; the water is then drawn off through filter floors or overflow weirs;

— by mechanical recovery (Archimedean screw for example or a reciprocating rake classifier) and stored in a fixed hopper or movable dump carts;

— by separation in a hydrocyclone and storage in a hopper equipped with an overflow weir;

— by separation in a hydro-cyclone and removal by a screw conveyor prior to storage in fixed hoppers or dump carts.

In very large installations, and before storing the grit in a hopper, it may be washed in a separate tank into which the mixture of water and grit is pumped from the grit removal equipment; the washing tank is equipped with air injection equipment to improve the washing process. Washing by addition of water can also take place in the Archimedean screw. Properly washed grit can be reused at the installation (in the drying beds).

Fig. 67. — Coarse screening with debris extraction by conveyor belt.

7.3. Grease and scum

Grease and scum recovered from the surface of grit removers, grease removers and primary settling tanks are usually not recoverable.

When these wastes are mostly organic, they may be added to the sludge undergoing anaerobic digestion (thus increasing gas production), but this has the disadvantage of promoting the formation of a floating scum layer in the digester, which must then be equipped with a particularly effective mixing system.

It is preferable to retain them in concentration tanks (with baffle outlet for the overflow with return to the inlet of the plant, to eliminate excess water), and periodically remove them by vacuum tanker trucks.

When the works include incineration of sludge or of screenings, the grease and floating scum can also be incinerated.

It is advantageous to store oily garage wastes, and oil and petrol tank flushing liquors (collected by special tankers) at the municipal sewage treatment plant. After a long period of storage, the oil that has concentrated at the surface can be extracted and burned.

1. GENERAL INFORMATION

The small size of the colloidal particles in a water, together with the fact that negative charges are spread all over their surface, means that colloidal suspensions are very stable [1].

In water treatment, **coagulation** is defined as the process by which colloidal particles are destabilized, and is achieved mainly by neutralizing their electric charge. The product used for this neutralization is called a **coagulant.**

Flocculation is the massing together of discharged particles as they are brought into contact with one another by stirring. This leads to the formation of flakes or floc, which can be settled or filtered out at a later stage of treatment. Certain products, called **flocculating agents,** may promote the formation of floc.

Separation of the floc from the water can be achieved by filtration alone, or by settling or flotation, possibly followed by filtration.

Coagulation and flocculation are frequently used in the treatment of drinking water and the preparation of process water used by industry. These techniques neutralize the colloids in the water, and adsorb them into the surface of the precipitates formed during flocculation. Certain dissolved substances can also be adsorbed into the floc (organic matter, various pollutants, etc.).

In the treatment of domestic sewage, the concentration of suspended solids is often such that flocculation can take place simply by mixing. A coagulant can also be introduced in order to help eliminate the dissolved pollution.

The composition of industrial waste waters varies widely according to the industry in question. In certain cases, flocculation of the effluent is enough because it already contains a substance capable of flocculating merely after mixing or addition of a flocculating agent. In other cases, it is necessary to use a coagulant capable of producing a precipitate which can then be flocculated.

---

1. See chapter 3, page 61, for the theory upon which the processes described here are based.

2. COAGULATION

Coagulation consists of introducing into the water a product which can:
— discharge the generally electro-negative colloids present in water;
— give rise to a precipitate.
This product is called a coagulant (see page 61).

2.1. Main coagulants

The most widely-used coagulants are based on aluminium or iron salts. In certain cases, synthetic products, such as cation polyelectrolytes, can be used.

The metal salt acts on the colloids in the water through the cation which neutralizes the negative charges before precipitation.

Cation polyelectrolyte is so-called because it carries positive charges which directly neutralize the negative colloids. Cation polyelectrolytes are generally used in combination with a metal salt, greatly reducing the salt dosage which would have been necessary. Sometimes no salts at all are necessary, and this greatly reduces the volume of sludge produced.

2.1.1. ALUMINIUM SALTS

1° Aluminium sulphate (liquid or solid form) :

$$Al_2(SO_4)_3 + 3\ Ca(HCO_3)_2 \longrightarrow 3\ CaSO_4 + 2\ Al(OH)_3 + 6\ CO_2$$

Ratio: in clarification, 10 to 150 g/m³ (commercial product) according to the quality of the raw water.
In waste water treatment, 100 to 300 g/m³ according to the type of effluent and the degree of purification required.

2° Aluminium chloride (liquid form):
This is used under exceptional circumstances only.

$$2\ AlCl_3 + 3\ Ca(HCO_3)_2 \longrightarrow 2\ Al(OH)_3 + 3\ CaCl_2 + 6\ CO_2$$

3° Aluminium sulphate + hydrated lime:

$$Al_2(SO_4)_3 + 3\ Ca(OH)_2 \longrightarrow 3\ CaSO_4 + 2\ Al(OH)_3$$

Ratio: in clarification, one part of $Ca(OH)_2$ (lime) to three parts of $Al_2(SO_4)_3$, 18 H_2O (commercial aluminium sulphate); in domestic sewage treatment, 100 to 200 g/m³ of lime to 150 to 500 g/m³ of commercial aluminium sulphate.

4° Aluminium sulphate + caustic soda :

$$Al_2(SO_4)_3 + 6\ NaOH \longrightarrow 2\ Al(OH)_3 + 3\ Na_2SO_4$$

Ratio: in clarification, the requirement of NaOH is 36 % of the dose of $Al_2(SO_4)_3$, 18 H_2O (commercial aluminium sulphate).

5⁰ Aluminium sulphate + sodium carbonate:

$$Al_2(SO_4)_3 + 3\ Na_2CO_3 + 3\ H_2O \longrightarrow 2\ Al(OH)_3 + 3\ NaSO_4 + 3\ CO_2$$
$$Al_2(SO_4)_3 + 6\ Na_2CO_3 + 6\ H_2O \longrightarrow 2\ Al(OH)_3 + 3\ Na_2SO_4 + 6\ NaHCO_3$$

Ratio: one part of anhydrous sodium carbonate to one or two parts of $Al_2(SO_4)_3$, 18 H_2O (commercial aluminium sulphate).

6⁰ Sodium aluminate alone:

$$NaAlO_2 + Ca(HCO_3)_2 + H_2O \longrightarrow Al(OH)_3 + CaCO_3 + NaHCO_3$$
$$2\ NaAlO_2 + 2\ CO_2 + 4\ H_2O \longrightarrow 2\ NaHCO_3 + Al(OH)_3$$

Ratio: in clarification, 5 to 50 g/m^3 of commercial sodium aluminate (with 50 % Al_2O_3).

7⁰ Aluminium polymers:

In some conditions, aluminium salts can be condensed in order to form polymers capable of coagulating and flocculating. **Basic aluminium polychloride (BAPC)** is formed in this way by gradually neutralizing an aluminium chloride solution with caustic soda.

Polymers can be obtained in the form $Al_6(OH)_{12}^{6+}$ to $Al_{54}(OH)_{144}^{8+}$ with exceptional coagulating and flocculating properties. BAPC needs to be prepared at the place where it is to be used (patented process). However, other products (marketed under various brand names such as WAC) can be prepared and stored for a long period because of the stability given the polymer by the addition of various products (strong acid anion such as SO_4^{2-}).

2.1.2. IRON SALTS

1⁰ Ferric chloride alone (generally liquid, sometimes crystallized):

$$2\ FeCl_3 + 3\ Ca(HCO_3)_2 \longrightarrow 3\ CaCl_2 + 2\ Fe(OH)_3 + 6\ CO_2$$

Ratio: in clarification, 5 to 150 g/m^3 $FeCl_3$, 6 H_2O (commercial ferric chloride); in domestic sewage treatment, 100 to 500 g/m^3 $FeCl_3$, 6 H_2O (commercial ferric chloride).

2⁰ Ferric chloride + hydrated lime:

$$2\ FeCl_3 + 3\ Ca(OH)_2 \longrightarrow 3\ CaCl_2 + 2\ Fe(OH)_3$$

Ratio: in domestic sewage treatment, 100 to 800 g/m^3 lime is required for 100 to 600 g/m^3 $FeCl_3$, 6 H_2O (commercial ferric chloride).

3⁰ Ferric sulphate :

$$Fe_2(SO_4)_3 + 3\ Ca(HCO_3)_2 \longrightarrow 2\ Fe(OH)_3 + 3\ CaSO_4 + 6\ CO_2$$

Ratio: in clarification, 10 to 150 g/m^3 of $Fe_2(SO_4)_3$, 9 H_2O (commercial product).

4⁰ Ferric sulphate + hydrated lime:

$$Fe_2(SO_4)_3 + 3\ Ca(OH)_2 \longrightarrow 2\ Fe(OH)_3 + 3\ CaSO_4$$

Ratio: in clarification, the requirement of hydrated lime as $Ca(OH)_2$ is 40 % of the quantity of ferric sulphate $Fe_2(SO_4)_3$, 9 H_2O.

5⁰ Ferrous sulphate alone:

$$FeSO_4 + Ca(HCO_3)_2 \longrightarrow Fe(OH)_2 + CaSO_4 + 2\ CO_2$$

Ratio: in clarification, 10 to 100 g/m³ of $FeSO_4$, 7 H_2O (commercial product); in waste water treatment, 200 to 400 g/m³ of $FeSO_4$, 7 H_2O (commercial product); in aerated waters, the ferrous hydroxide oxidizes and becomes ferric hydroxide:

$$2\ Fe(OH)_2 + \frac{1}{2}\,O_2 + H_2O \longrightarrow 2\ Fe(OH)_3$$

6⁰ Ferrous sulphate + chlorine:

$$2\ FeSO_4 + Cl_2 + 3\ Ca(HCO_3)_2 \longrightarrow 2\ Fe(OH)_3 + 2\ CaSO_4 + CaCl_2 + 6\ CO_2$$

Ratio: the requirement of chlorine is 12 % of the quantity of $FeSO_4$, 7 H_2O.
The ferrous sulphate and the chlorine can be introduced separately into the water to be treated.   A solution of ferrous sulphate can also be oxidized by the chlorine before use.   This gives a mixture of ferric sulphate and ferric chloride, widely known commercially as **ferric chlorosulphate,** by the following reaction:

$$3\ FeSO_4 + \frac{3}{2}\,Cl_2 \longrightarrow Fe_2(SO_4)_3 + FeCl_3$$

7⁰ Ferrous sulphate + hydrated lime:

$$FeSO_4 + Ca(OH)_2 \longrightarrow Fe(OH)_2 + CaSO_4$$

Ratio: in clarification, the requirement of $Ca(OH)_2$ (hydrated lime) is 26 % of the quantity of $FeSO_4$, 7 H_2O.
In waste water treatment, 100 to 150 g/m³ of lime is required for 250 to 350 g/m³ ferrous sulphate.

8⁰ Ferric chloride + sodium aluminate:

$$3\ NaAlO_2 + FeCl_3 + 6\ H_2O \longrightarrow 3\ Al(OH)_3 + Fe(OH)_3 + 3\ NaCl$$

Ratio: equal amounts of commercial sodium aluminate (with 50 % of Al_2O_3) and commercial ferric chloride $FeCl_3$, 6 H_2O.

2.1.3. OTHER COAGULANTS

1⁰ Copper sulphate:

$$CuSO_4 + Ca(HCO_3)_2 = Cu(OH)_2 + CaSO_4 + 2\ CO_2$$

Dose: 5 to 20 g/m³ (used in exceptional cases).

2⁰ Copper sulphate + hydrated lime:

$$CuSO_4 + Ca(OH)_2 = Cu(OH)_2 + CaSO_4$$

Dose: 30 grammes $Ca(OH)_2$ to 100 grammes $CuSO_4$ (copper sulphate), 5 H_2O (used in exceptional cases).

3⁰ Ozone:

Ozone is not a true coagulant as it has no action on the electric charges of the colloids in the water.

However, in specific cases when water contains complexes linking organic matter to iron or manganese, ozone can initiate a coagulation process. The complexes are destroyed by the ozone and the metal ions thus released are oxidized. With the necessary pH conditions, this may result in the formation of a small amount of a generally fragile precipitate. The density and cohesion properties of the floc so formed are inadequate to provide acceptable clarification, but may serve for filter coagulation.

2.2. Coagulation procedure

2.2.1. CHOICE OF COAGULANT

The coagulant is chosen after the water is examined in a laboratory by means of flocculation tests (see Chapter 27, page 949). When making this choice the following factors should be borne in mind:
— nature and quality of the raw water;
— variations in the quality of the raw water (daily or seasonal, especially with regard to temperature);
— quality requirements and use of the treated water;
— nature of the treatment after coagulation, (filter coagulation, settling);
— degree of purity of the reagent, particularly in the case of drinking water, where recovered substances and manufacturing by-products, etc. are ruled out.

2.2.2. INTRODUCTION OF COAGULANT

Since neutralization of the colloids is the main aim when the coagulant is introduced, it is important to diffuse the reagent as quickly as possible.

In fact, the coagulation time is extremely short (less than a second) and best results are obtained if the colloids are completely neutralized before part of the coagulant has begun to form a precipitate (for example, in the form of metal hydroxide).

Sometimes, reagents are mixed merely by the turbulence created by a weir, but it is preferable to speed up the mixing process, by using a rapid mixer or coagulator capable of achieving a velocity gradient between 100 and 1000 s⁻¹.

Rapid mixers are not essential when a sludge contact clarifier is used, but they are necessary in the case of diffused flocculation.

There are several types of mixers:

● *Static mixers:*

They are devices (screws, diaphragms or cones, etc.) installed inside pipes which cause enough turbulence to diffuse the coagulant instantly.

These mixers sometimes cause serious head loss in the pipes. They are very efficient at nominal flow, but efficiency decreases as the flow through the pipe decreases.

● *Rapid mixers:*

They are high-speed stirrers of the screw- or paddle-type (turbine) installed in a special mixing chamber (see fig. 69).

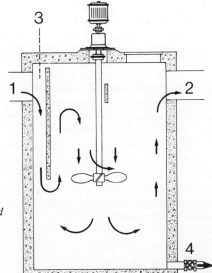

Fig. 69. — Mixing chamber with high-speed stirring device.

1. *Water inlet.*
2. *Outlet.*
3. *Reagent.*
4. *Drain.*

3. FLOCCULATION

Floc formation is initiated by the addition of the coagulant, but its volume, weight and particularly its cohesion must be increased. The floc size is improved by:
— an increase in the quantity of floc in the water; it is always an advantage to bring the water into contact with the precipitates already formed during the preceding treatment (sludge recirculation—sludge blanket) with the highest possible concentration,
— slow and even stirring to increase the chances of electrically discharged colloidal particles encountering a floc particle,
— use of products known as flocculating agents.

3.1. Flocculating agents

Flocculating agents, also known as coagulating additives, flocculation aids or filtering aids, are products which promote each of these processes. They may accelerate the reaction rate (faster flocculation) or improve the floc quality (denser, bulkier or more coherent floc).

Flocculating agents can be classified by their nature (inorganic or organic), their origin (synthetic or natural) and their electric charge (anionic, cationic, non-ionic).

3.1.1. ACTIVATED SILICA

Activated silica was the first flocculation aid used and it can still give the best results mainly when used together with aluminium sulphate.

It is generally added after the coagulant and is prepared by partially neutralizing the alkalinity of a solution of sodium silicate.

● *Preparation of activated silica:*

1. Laboratory or industrial production.

a) Baylis method:
— Take 25 ml or 35 g of sodium silicate at 41° Baumé (1.4 S.G.) (28 % of SiO_2). Dilute to 500 ml with water.
— While stirring, slowly add 170 ml water containing 2.4 ml of sulphuric acid at 66° Baumé (1.84 S.G.).
— Stir slowly for 2 hours. Dilute to 2 litres.
— The resulting solution contains 0.5 % of SiO_2.

b) Hay method:
— Take 25 ml or 35 g sodium silicate at 41° Baumé (1.4 S.G.) (28 % of SiO_2). Dilute to 400 ml with water.

— While stirring, slowly add 6.7 g of ammonium sulphate dissolved in 100 ml of water.
— Stir slowly for 2 hours.   Dilute to 1 litre.
— The resulting solution contains 1 % of SiO_2.

This solution is more stable than that obtained by the Baylis method.   The presence of ammonium salt makes it dificult to use the Hay method when there has been prior chlorination.   In both of the above methods, industrial preparation requires special precautions.

c) Sodium silicate can also be "activated" with hydrochloric acid, chlorine, carbonic acid, sodium bicarbonate, etc., in the quantity required to neutralize 90 % of the alkalinity of the sodium silicate.

2. Continuous preparation:

Industrially, however, it is preferable in large plants to regulate the feed rate of sodium silicate solution and acid solution by a continuous process. These two solutions are mixed in a tank where water can be introduced to obtain the required concentration of SiO_2; the solution is then transferred to a maturing tank where it remains for half an hour.

In order to administer 1 kg per hour of SiO_2, the following approximate porportions are necessary: 2.5 litres per hour of sodium silicate at 41° Baumé (1.4 S.G.) and 0.24 litre per hour of sulphuric acid at 66° Baumé (1.84 S.G.) or 850 g per hour of sodium bicarbonate, and water to make a total flow of 100 litres per hour.

A regular check should be made of the neutralization factor of the sodium silicate by measuring the TA and TAC of the solution of activated silica.   The TAC of the 1 % solution of SiO_2 should be about 50 to 75 French degrees.

3.1.2. OTHER INORGANIC FLOCCULATION AIDS

Other products are sometimes used for a raw water not containing sufficient suspended matter.

Among these may be mentioned:
— certain types of clay,
— whiting or precipitated calcium carbonate,
— activated carbon in powder form (when this treatment is essential),
— fine sand,
— kieselguhr (diatoms).

3.1.3. ORGANIC FLOCCULATION AIDS

With the progress made in organic chemistry, other more active flocculating agents have been developed.

1° Origin:

● Some are made from *natural products*, such as alginates (extracts of algae), starches (extracts of plant seeds), certain products derived from cellulose, certain types of gum, etc.

Alginates are used a great deal as flocculation aids with iron salts. In addition, they sometimes give good results with aluminium salts.

Alginates are obtained from alginic acid, which itself is extracted from seaweed (especially from the Laminaria genus). They are approved in all countries for use in drinking water treatment.

● Other flocculation aids are entirely *synthetic*. Polyacrylamides are in long chains and have a high molecular weight (2 to 4 $\times$ 10^6 g/mol); whereas polyamines are usually in shorter chains and have lower molecular weights (1 $\times$ 10^6 g/mol).

Polyacrylamides are used both in the clarification treatment of water and the dewatering of sludge, whereas polyamines are mainly used in clarification.

2⁰ Use:

Organic flocculation aids may be divided into two categories: solid and liquid.

There are a large number of organic flocculation aids, among which mention may be made of: Praestol, Superfloc, Magnafloc, alginates, Prosedim, Purifloc.

There is no rule for selecting the flocculation aid that will give the best results, and laboratory tests are always necessary. The time that must elapse between the addition of the coagulant and the flocculation aid will also have to be determined, as this factor is of great importance, for example, in the use of activated silica.

The manufacturer generally provides the necessary information for optimum use of his products. However, the table on page 144 gives general guidance.

Where lime is used for carbonate removal, a non-ionic or anionic flocculating agent should be selected.

In the case of clarification, a non-ionic, weakly anionic or cationic flocculating agent may be used. If the pH is close to neutral and there are large quantities of organic matter, tests should first be made with a cationic flocculating agent.

For the treatment of drinking water supplies, it is very important to refer to the legislation of the country concerned, as the use of organic flocculating agents is subject to regulations; all countries have published a list of authorized products. In France, as at 1st February 1978 no synthetic organic product has been authorized for drinking water by the Higher Council of Public Health.

TABLE SHOWING THE MAIN PROPERTIES
OF ORGANIC FLOCCULATION AIDS

Origin of the product		Starch, alginates	Solid acrylics	Liquid acrylics	Polyamines
Storage, packing		Bags	Bags or drums	Drums or containers	Drums
Handling		Weighing or dry feeders	Weighing or dry feeders	Dosing by complete drums or dosing pumps. Discharge by gravity or compressed air (containers)	
Use	Preparation of stock solution	Dispersion device useful to necessary. Concentration of stock solution 0.5-1 %	Dispersion device essential. Concentration of stock solution 0.3 to 0.5 %, exceptionally 1 %	Can be mixed with water with slow stirring device.	
	Dosing	By dosing pump	By dosing pump	Pumping as supplied or in 10 % solution (slow dosing pump because of viscosity of solution)	
	Preparation of solution for use	Dilution after dosing	Dilution after dosing	Dilution after dosing	
	Concentration on injection	1-3 $^o/_{oo}$	0.5-2 $^o/_{oo}$	approx. 1 %	
	Reaction time	30-300 seconds depending on product	20-120 seconds depending on product	20-120 seconds depending on product	

Notes:
1. To prepare the solution, a continuous dilution system can be used (e.g. hydro-ejector).
2. After injection, pumping and violent turbulence must be avoided.

3.2. Flocculation procedure

The effectiveness of flocculation is in direct proportion to the effectiveness of prior coagulation. Rapid stirring in a coagulator is followed by slow stirring in a flocculator for 5 minutes, in the case of very strong sewage, and much longer in the treatment of drinking water.

Stirring can take place in a separate flocculator or inside the clarifier proper. In the latter case either recirculating turbines or the action of the sludge blanket itself can be used (see chapter 7 on clarification).

Choosing the type of mixer and the type of flocculator is closely linked to the choice of separation process used in the next stage of the treatment. Parti-

cular care must be given to this choice when flocculation is diffusc d and is followed by static settling or flotation.

● *Flocculators:*

Flocculation takes place in tanks equipped with a stirring system which
turns relatively slowly so as not to break the flakes already formed, but fast
enough to gradually increase the floc size and to prevent deposits of sediment on
the bottom of the tank.

The capacity of the flocculation tank must allow the flocculation time, as
determined by laboratory tests, to be respected.

Stirring systems can consist of specially designed screws or a set of paddles
fixed to a vertical or horizontal rotating shaft. If possible, the flocculator should
achieve a velocity gradient between 20 and 50 s^{-1}.

It is also an advantage to have a variable speed motor drive, so that the
stirring rate can be regulated as efficiently as possible in accordance with the
quality of flocculation.

If considerable tank capacity is required for flocculation, it is preferable
to use several tanks in series. Each of these tanks is fitted with an independent
stirring system, the speed of which can be adjusted to suit the quality of flocculation.

Finally, it is important not to break the floc when it is transferred from the
flocculator to the settling zone. Depending on the degree of purification required for the treated water, the following transfer velocities must be adhered to:
— fragile metal hydroxide floc   s = 0.20 m/s
— strong metal hydroxide floc   s = 0.50 m/s
— floc from waste water   s = 1 m/s

Fig. 70. — Vertical flocculator.   *Fig. 71. — Horizontal flocculator.*

6

CHEMICAL
PRECIPITATION

1. PRINCIPLES
OF PRECIPITATION PROCESSES

Chemical precipitation means the formation of insoluble compounds of the unwanted substances contained in a water by the action of appropriate reagents; the reactions involved obey Berthollet's laws or the laws of oxidation-reduction.

The processes most commonly used in water treatment are crystalline precipitation of Ca^{2+} and Mg^{2+} ions in the first case and precipitation of metallic hydroxides in the second.

1.1. Elimination of calcium and magnesium

1.1.1. PRINCIPAL METHODS:

A. Carbonate removal by lime.

The commonest precipitation treatment is **carbonate removal by lime,** which eliminates the bicarbonate hardness (also called temporary hardness, page 896) from a water.

It will be recalled that in analyses of this type of water, hardness is expressed by the total hardness titration (in French degrees TH), the sum of the Ca^{2+} and Mg^{2+} cations, while the bicarbonate content is expressed by the complete alkalinity titration (in French degrees TAC). Non-carbonate hardness (or permanent hardness, page 896) is then expressed by:

$$TH - TAC.$$

Carbonate removal by lime therefore only partially eliminates the sum of calcium and magnesium ions because it does not affect the permanent hardness.

● *Basic reactions*: the chemical reactions for carbonate removal are as follows:

$$Ca(OH)_2 + Ca(HCO_3)_2 \longrightarrow \underline{2CaCO_3} + 2\ H_2O$$
$$\downarrow$$
$$Ca(OH)_2 + Mg(HCO_3)_2 \longrightarrow 2\ MgCO_3 + 2\ H_2O$$

Since magnesium carbonate is relatively soluble (solubility about 70 mg/1), an excess of lime will lead to the reaction:

$$Ca(OH)_2 + MgCO_3 \longrightarrow \underset{\downarrow}{CaCO_3} + \underset{\downarrow}{Mg(OH)_2}$$

If the doses of reagent are accurately measured, the alkalinity of the water is reduced to the theoretical solubility of the $CaCO_3 + Mg(OH)_2$ system, which is between 2 and 3 French degrees under normal conditions of concentration and temperature. This limit value may, however, be increased in practice by the presence of organic impurities.

If the raw water also contains sodium bicarbonate (TAC > TH), the water will retain, in addition to the amount represented by the above figure, additional alkalinity in the form of sodium carbonate or caustic soda, corresponding to the value TAC — TH.

● *Precipitation mechanism:* in order to assess the merits of the various methods of carbonate removal using lime, one needs to know the mechanism whereby calcium carbonate and magnesia are precipitated.

The reaction produced by lime in raw water is extremely slow in the absence of crystallization nuclei. In the static clarifiers formerly employed it took several days to reach chemical equilibrium. In continuous-operation clarifiers containing no sludge-contact device — these are now all but obsolete — the reaction time is still several hours.

On the other hand, if the water and lime are brought into contact with a sufficiently large volume of already-precipitated $CaCO_3$ crystals, the reaction reaches its equilibrium point in a few minutes. As precipitation takes place on the crystals, these tend to grow in volume; the sedimentation rates, governed by Stokes' law, are then increased, and the size of the equipment can be reduced.

This is true only if the surfaces of the $CaCO_3$ crystals remain sufficiently clean. Therefore, as the presence of organic colloids is liable to impede crystallization, it is common practice to add coagulant reagents such as ferric chloride, alum or aluminium polychloride to the raw waters undergoing carbonate-removal treatment in order to eliminate these colloids.

Lastly, it should be emphasized that $CaCO_3$ used alone tends to form very hard, large clusters of crystals which settle extremely rapidly, whereas magnesia used alone always appears in the form of very light flakes of $Mg(OH)_2$. If the percentage of this substance is very low, it is occluded in the calcic precipitate, but if the proportion is high, it becomes impossible to obtain bulky precipitates and the acceptable settling rate is much lower.

Assessment of carbonate-removal equipment should therefore be primarily based on its ability to produce a homogeneous mixture of raw water, reagent and $CaCO_3$ nuclei in a reaction zone of suitable size, this being followed by a settling or clarification zone the area of which depends on the conditions of formation of the precipitate and especially on the content of organic colloidal substances and magnesium salts.

It should be noted that when it is desired to obtain particularly clear decarbonated water, carbonate removal should always be followed by filtration.

B. Use of sodium carbonate.

The old processes of elimination of permanent hardness with sodium carbonate at high temperature, with barium carbonate or with tribasic sodium phosphate are now virtually obsolete and have been superseded by the more modern processes mentioned below. The cold carbonate sodium process of permanent hardness elimination, associated with lime precipitation of calcium and magnesium bicarbonates, is still sometimes used. The following reactions are involved in this elimination:

$$CaSO_4 + Na_2CO_3 \longrightarrow Na_2SO_4 + CaCO_3 \downarrow \qquad (1)$$

$$CaCl_2 + Na_2CO_3 \longrightarrow 2\ NaCl + CaCO_3 \downarrow \qquad (2)$$

This method has some disadvantages; in particular total hardness cannot at best be reduced in this way below 3 to 4 French degrees.

Preference is often given to carbonate removal by lime alone and softening on a sodium-cycle cation exchanger. In the last-mentioned method, sand filtering is essential between the clarification and softening processes.

C. Precipitation with caustic soda.

Calcium and magnesium ion elimination by precipitation with caustic soda is a variant of the combined lime and sodium carbonate treatment process described in Section *B* above.

The basic reaction is:

$$Ca(HCO_3)_2 + 2\ NaOH \longrightarrow CaCO_3 \downarrow + Na_2CO_3 + 2\ H_2O \qquad (3)$$

Precipitation of calcium carbonate is accompanied by the formation of sodium carbonate, which will react on the permanent hardness by reactions (1) and (2) above.

If caustic soda is used, therefore, the hardness of a water can be reduced by twice the amount of the reduction of alkaline earth bicarbonates. The TAC of the water can be reduced to around 3 French degrees only if there are enough calcium ions to combine with the sodium carbonate formed.

1.1.2. BASIC DATA FOR CALCULATION AND MONITORING OF PRECIPITATION

A. **Calculation of lime dose.**

Notation:

TH total hardness titration of the water,

TAC complete alkalinity titration of the raw water (measured in the presence of methyl orange),

TCa calcium hardness titration representing the total calcium salts content,

TM magnesium hardness titration representing the total magnesium salts content,

C   free carbon dioxide content of the water in French degrees, calculated as

follows: $\dfrac{\text{free } CO_2 \text{ in mg/l}}{4.4}$

TA   " simple " alkalinity titration measured in the presence of phenolphthalein, using this notation, the theoretical lime dose required for optimum calcium carbonate precipitation is:

$$CaO : 5.6 \times (TAC + C) \text{ g/m}^3$$
$$\text{or } Ca(OH)_2 : 7.4 \times (TAC + C) \text{ g/m}^3$$

To precipitate calcium carbonate and magnesia simultaneously, calculate as follows:

$$CaO = 5.6 (2\,TAC - TCa + C)$$
$$\text{or } Ca(OH)_2 = 7.4 (2\,TAC - TCa + C)$$

Measurement of the above titrations is described on page 909 and foll.

- *Monitoring of results:*

  a) Waters with positive TH — TAC.

  For precipitation of calcium carbonate only, the ideal setting is when

  $$TA = \frac{TAC}{2} \pm 0.5^o$$

corresponding to a minimum TAC of approximately 2° if the water does not contain magnesium.

If TM is greater than TH — TAC, application of this rule leads to excessive TAC values owing to the solubility of magnesium carbonate. The second method of calculation must then be used, based on $(2\,TAC - TCa + C)$, the optimum result

being obtained when: $$TA = \frac{TAC}{2} + 0.5^o \text{ to } 1^o$$

with the lowest possible value of TA.

  b) Waters in which TH — TAC is negative.

  This applies to waters containing sodium bicarbonate.   Good precipitation of calcium and magnesium is always possible by calculating the lime on the basis of TAC + TM+ C, but this gives a water containing sodium carbonate and caustic soda, and it may be desirable to use a smaller dose.

  In any case, the dose of lime must be increased (or reduced) by 5.6 grammes per cubic metre (as CaO) or 7.4 grammes per cubic metre (as $Ca(OH)_2$) per degree TA measured above or below the theoretical value.

  The values 5.6 and 7.4 apply, of course, to 100 % pure products.   In practice lime is always impure and more or less carbonated, and the actual values range from 8 to 10.

B. **Calculation of sodium carbonate dose.**

  The required dose of sodium carbonate (pure product basis) is:

  $$10.6\,(TH - TAC) \text{ g/m}^3.$$

- *Monitoring of results:* in theory, in a water containing no magnesium:

  $$TH = TAC = 2\,TA.$$

In practice, in waters part of whose permanent hardness consists of magnesium, this rule may no longer hold, and each individual case must then be considered separately.

C. Calculation of caustic soda dose

The caustic soda dose required per French degree of TAC to be precipitated (pure product basis), is: 8 (2 TAC — TCa + C) g/m³.

● *Monitoring of results:* to lower the TAC by 1 French degree, the TH must be reduced by 2 French degrees.

In practice, the dose of caustic soda introduced into the water is controlled to give minimum residual TAC.   This avoids unnecessary soda consumption leading to excessive sodium or magnesium carbonate in the water.

1.2. Silica elimination

The process used in silica elimination is more one of adsorption than of chemical precipitation proper.   However, since silica elimination is often combined with the carbonate removal reaction, it is appropriate to include it in this chapter.   It may be performed either cold or hot.

1.2.1. SILICA REMOVAL WITH SODIUM ALUMINATE.

A natural water's silica content can be substantially reduced by incorporating this element in a complex silicoaluminate of calcium and iron which forms at the temperature of the water.

For this purpose, suitable doses of ferric chloride, sodium aluminate, and lime are added to the water.

Superior silica removal is usually achieved if the water undergoes carbonate removal at the same time.   The precipitate then contains calcium carbonate as well as a complex of silica, alumina, and iron.   The higher the silica level of the water and the higher its temperature, the higher the residual silica content will be.   It is normally in the range 2 to 5 mg/l for waters containing not more than 20 mg/l silicon as silica at 20 °C.   At higher levels, 70 to 80 % reduction is achieved.

1.2.2. SILICA REMOVAL WITH MAGNESIA.

The treatment may be performed cold or hot.

● *Cold process*: the results are of the same order of magnitude as above.   Magnesia prepared in situ from magnesium oxide, which is dissolved by blowing in CO_2, is introduced into the water and then precipitated with lime.

● *Hot process :* this complements carbonate removal treatment.

If water is treated at a temperature of around 100 °C with a mixture of lime and porous anhydrous magnesia powder, the silica can be fixed by adsorption until the residual content in the treated water is as low as 1 mg/litre.

This method, combined with subsequent softening on a cation exchanger, is widely used for feed water for medium-pressure boilers.   It necessarily involves filtering of the clarified water over non-siliceous media such as marble or anthracite.

1.3. Elimination of salts from metal finishing effluents by crystalline precipitation

Various inorganic salts contained in metal finishing effluents are eliminated by precipitation by Berthollet's laws.   Chief among these are fluorides and phosphates.

In the former case, the fluorine ion treated by lime is insolubilized in the form of calcium fluoride, which then precipitates up to its solubility limit; the latter varies with the nature and concentration of the other salts present in the water, ranging from 15 to 25 mg/l expressed as F.

In the second case, phosphates are precipitated by a calcium or iron salt in the form of calcium or iron phosphates.

Precipitation of calcium sulphate in the form of gypsum ($CaSO_4.2\ H_2O$) during neutralization of a relatively concentrated sulphuric acid solution by lime is a related type of treatment.

Here again, the reactions are accelerated by the presence of pre-existing crystals and then lead to the formation of bulky precipitates.

1.4. Treatment of brines

Before brines are evaporated to produce crystallized salt or electrolysed to make chlorine, they must be purified by chemical precipitation.

In the first case, it is a matter of obtaining a salt of maximum possible purity for human consumption.   In the second case, the objective may differ somewhat depending on the type of electrolytic cell used.

If the cell is of the diaphragm type — the diaphragm separating the anode from the cathode prevents recombination of chlorine and sodium while allowing the electric current to pass — calcium and magnesium must be prevented from precipitating on the diaphragm so that it does not become clogged.

In mercury cathode cells in which the sodium is fixed in the form of an amalgam, purification must be very thorough so as to eliminate, in particular, iron and magnesium ions, as these decompose the sodium amalgam.

In both cases, wear of the graphite anodes is reduced by elimination of sulphate ions.

The main factor distinguishing a raw brine from a water is the dissolved ion concentration, which is 10 to 10,000 times greater in the former than in the latter.

The resulting incomplete ionic dissociation means that the kinetics of the precipitation reactions differs from that applicable to water treatment.

A brine passing through an electrolytic battery becomes depleted in NaCl, while it is at the same time increasingly contaminated with impurities originating from the makeup salt; these impurities must be eliminated before the brine is returned to the electrolytic battery.

● *Elimination of calcium and magnesium ions.*

Calcium and magnesium are present in the form of sulphates and chlorides.

Magnesium is precipitated by lime in the form of magnesia by the reaction:

$$Ca(OH)_2 + Mg^{2+} (SO_4 . Cl_2) \longrightarrow \underset{\downarrow}{Mg(OH)_2} + Ca^{2+} (SO_4 . Cl_2)$$

Hence the Mg^{2+} ions are replaced by Ca^{2+} ions, which remain in solution.

All the calcium is precipitated by sodium carbonate in the form of calcium carbonate. If an excess (0.5 to 1 g/l) of sodium carbonate is used, the residual solubility of the calcium will be in the range 2 to 5 mg/l.

● *Elimination of sulphate ions.*

Sulphate ions are present in brines in the form of calcium, sodium, and magnesium sulphate. As stated above, magnesium sulphate is converted into calcium sulphate in the process of magnesium ion elimination.

Fig. 72. — Brine treatment. Salt manufacturing plant, Varangeville, France.

The solubility of calcium sulphate in brine — in the region of a few grammes per litre — depends on the brine concentration; it increases with temperature up to 40 °C and then falls rapidly from 50 to 120 °C.

If Ca^{2+} ions are introduced to the brine in the form of $CaCl_2$ or $Ca(OH)_2$, sulphates are precipitated as calcium sulphate to a residual SO_4^{2-} content of 3 to 4 g/l.

Two treatment configurations are possible according to whether sulphates are or are not to be precipitated along with calcium and magnesium.

If sulphate removal is not required, lime and sodium carbonate are injected successively in two reactors through which the brine flows in series before entering a settling tank; the latter should preferably be of the sludge recirculation type with an upward flow rate of 0.75 to 2 m/h depending on the magnesium content of the brine.

Where sulphate removal is necessary, it must be carried out before elimination of Ca^{2+} ions because it involves addition of Ca^{2+} ions.

The system will then include a sludge recirculation reactor in which a high sludge content (200-300 g/l) will be maintained and where Mg^{2+} ions will be precipitated with lime and SO^{2-} ions with calcium chloride. This reactor will be followed by a reactor/sludge recirculation type settling tank into which the sodium carbonate for Ca^{2+} ion precipitation is injected, final clarification taking place in a settling tank with scraper.

1.5. Precipitation of metallic hydroxides

The main requirement here is the elimination of heavy metals: cadmium, copper, chromium, nickel, zinc, and iron which occur, in particular, in metal finishing shop effluents (page 845). Provided that they are in an ionic state and not in the form of complexes, all these metals precipitate as hydroxides or as hydrocarbonates in a pH range characteristic of each. In general, if the pH of the reaction medium is fixed between 8.5 and 9.5, the solubility of these metals remains within tolerable limits.

2. PRECIPITATION EQUIPMENT

2.1. Cold-process precipitation of calcium and magnesium

As explained in the section on the precipitation mechanism, the reaction must take place in the presence of an already formed mass of crystals if precipitation is to be fast. The equipment used for carbonate removal, whether or not combined with permanent hardness elimination, will comprise either sludge contact appliances or granular contact mass reactors.

2.1.1. SLUDGE RECIRCULATION UNITS

Combined units including a reaction zone, as described in Chapter 7 "Settling", are ideally suited to calcium and magnesium precipitation. These are the **Circulator, the Accelator, the Turbocirculator,** and **the RPS plate type clarifier.** Owing to the high sludge concentration maintained in the reaction chambers of all these types of equipment, carbonate removal is practically complete at their outlets.

In consequence, the Circulator and the Turbocirculator in particular are well suited to the precipitation of fluoride and phosphate ions. They are also used for brine treatment as reactors or reactor/clarifiers.

Particularly high sludge concentrations can be obtained in units with mechanical scrapers.

2.1.2. SLUDGE BLANKET UNITS

This equipment—especially the **Pulsator,** is highly suitable for precipitation reactions with slow kinetics—e.g., precipitation of gypsum in solutions supersaturated with calcium sulphate. This equipment allows a very close approach to the limit of the solubility product of this substance.

2.1.3. GRANULAR CONTACT UNITS

This equipment differs from the previous group mainly in the use of larger nuclei. In the sludge blanket type of units the size of the elementary crystals of calcium carbonate is 1/100 mm, but these units use a "catalyzing" mass usually consisting of grains of sand of initial size between 0.2 and 0.4 mm contained in a conical vessel. Calcium carbonate precipitates on the surface of the grains,

Fig. 73. — CIRCULATOR clarifiers.

between which the water flows upwards at high velocity, while a complete reaction and good separation of the precipitate are maintained at the same time.

These units have three advantages:

— small size;

— possibility of pressurized operation; where combined with closed filters they can be used to remove carbonates from a water without breaking the pressure to the atmosphere.

— balls 1-2 mm in diameter which dry very quickly are obtained instead of sludge.

However, they cannot be used for waters containing excessive colloids or where the magnesium content is so high that magnesia is liable to precipitate (TM > TH — TAC), as magnesia will not crystallize simultaneously with calcium carbonate.

The Gyrazur (fig. 74) consists of three cylinders of increasing diameter from bottom to top, connected by truncated cones. With this configuration the volume of the catalyzing mass can be practically twice as much as in a conventional conical unit of the same top diameter and height. The raw water is admitted to the lower cylinder at high rate through an offset pipe which imparts an upward spiral motion to it. Lime is admitted on the axis of this pipe in the form of limewater or very dilute milk of lime, so that it mixes thoroughly with the raw water.

Since the grains are continuously enlarging, it would eventually no longer be possible to keep them in fluidization; the are therefore periodically extracted and replaced by fine nuclei.

Because the Gyrazur contains a large volume of active grains, it can be operated at high rates (50 to 70 m/h).

Note that the Gyrazur can also be used for softening with caustic soda.

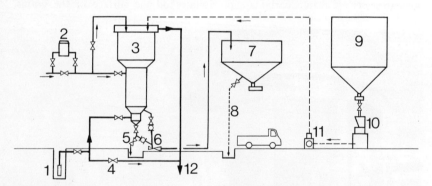

Fig. 74. Flow diagram of a GYRAZUR precipitator.

1. *Raw water lift.*
2. *Sand container.*
3. *Gyrazur.*
4. *Gyrazur by-pass.*
5. *Gyrazur drain.*
6. *Ejector.*

7. *Grain storage silo.*
8. *Grain extraction.*
9. *Lime storage silo.*
10. *Displacement feeder.*
11. *Dosing pump.*
12. *Treated water outlet.*

2.2. Equipment for hot-process carbonate and silica removal: the Thermo-Circulator

This equipment is used solely for carbonate removal with lime combined with silica removal with magnesia for medium-pressure boiler feed water. It also makes it possible to combine hot-process chemical purification with degassing.

The treatment is carried out at a low pressure corresponding to the vapour pressure for temperatures chosen between 102 and 115 °C as required.

This temperature makes possible rapid and complete reactions, which are also facilitated by sludge recirculation.

This circulation is produced by a steam emulsifier located outside the equipment, and the operation of which can easily be checked.

The raw water arrives at (1) and is dispersed by the spray nozzle (2) into the steam which occupies the upper part of the equipment. It is immediately reheated and falls into the large funnel (4) where it receives the reagents introduced via the pipe (3) and meets the sludges recirculated by the pipe (10) with the aid of the steam emulsifier (12).

This water then runs towards the lower settling zone through the central pipe (5), rises into the settling tank, where the precipitates are separated off, and then overflows to the filter unit via pipe (11).

The incondensable gases are evacuated via the stub pipe (8).

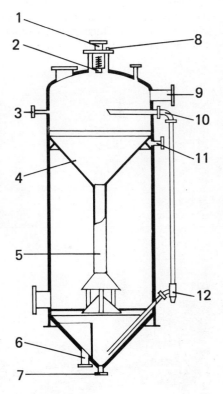

1. *Raw water inlet.*
2. *Spray nozzle.*
3. *Reagent inlet.*
4. *Funnel.*
5. *Flow pipe.*
6. *Sludge extraction pipe.*
7. *Drain.*
8. *Outlet for incondensable gases.*
9. *Heating steam inlet.*
10. *Recirculated sludge.*
11. *Condensed water outlet to filter.*
12. *Steam ejector.*

Fig. 75. — Thermo-Circulator.

Ancillary equipment, of course, includes a steam regulator and a water level regulator.

The heating steam is introduced via pipe (9).

The settled sludge mainly collects in the lower conical part, whence it is recycled, and the excess sludge is deposited in the concentrator (6), from which it is then drawn off.

This equipment may include a degassing zone fed with settled water on top. The heating steam then passes through the degassing chamber before heating the raw water.

Fig. 76. — Carbonate removal by the hot process in a thermo-circulator 7.5m (25 ft) in diameter. Output: 240 m³/h (1.5 mgd). Caltex Refinery at Kelsterbach, Federal Republic of Germany.

SETTLING
FLOTATION

The physical laws which govern solid-liquid separation processes are dealt with in Chapter 3 (pages 65 to 78). This Chapter describes the equipment (settling tanks and flotation units) in which these processes are put into use.

1. PRINCIPAL TYPES OF SETTLING

As we have seen, settling and clarifying processes are intended to enable particles suspended in the water to sink to the bottom, whether such particles already exist in the raw water, or are produced by the action of a chemical reagent added artificially (coagulation, iron removal, chemical purification) or result from physical flocculation combined with a biological treatment (municipal sewage).

1.1. Static settling

Use can be made of an intermittent batch process in which the water is allowed to remain at rest in a settling tank for several hours, and is then drained off from the top down to a point just above the layer of deposited sludge. This process is suitable for temporary installations, but on an industrial scale it is always preferable to use continuous circulation so as to avoid constant manual attention. The settling tank or clarifier consists of a rectangular or circular tank. So that the sludge will settle, the ascending velocity of the water must be less than the descending velocity of the particles. This naturally depends on the density and size of the particles.

Clarification rates can be assessed by the method described under No. 704 on page 950.

Small settling tanks are provided with bottoms sloping at 45 to 60° to permit the continuous or intermittent extraction of sludge from the lowest point.

In the case of large tanks, however, this procedure would lead to prohibitive depths; the bottom slope is therefore reduced to a minimum and the sludge is removed by a scraper system which collects the sludge in a sump from which it is easily extracted.

For sludge obtained from primary settling during the treatment of municipal sewage, the slope must be steeper than for drinking water and it is more essential to use scraper-type clarifiers.

Static settling tanks should preferably be set to operate regularly and evenly. Variations of flow cause eddy currents which raise the sludge to the surface. Moreover, any temperature variation between the raw water and the water in the tank, no matter how slight, creates convection currents which have the same effect.

The raw water inlet and the clarified water outlet must also be studied very carefully indeed to prevent the formation of stray currents and to ensure that the raw water is distributed uniformly throughout the entire useful settling zone, while providing a calm zone in which the sludge may accumulate.

To facilitate a study of the currents that form in a settling tank, the water can be coloured at the inlet. A dye that does not diffuse rapidly in the water must be used, such as Rhodamine B prepared in a concentration of one per thousand. One litre of this solution is sufficient for a clarification capacity of 60 m^3 (15 800 gal).

Where coagulation is effected by the addition of chemical reagents, clarification must be preceded by a flocculator designed for "flocculant settling" in which the concentration of suspended matter will include both that in the raw water and that introduced by the reagents.

● *Application of plate settling.*

In a static settling tank the maximum upward velocity is independent of the depth of the tank. As a result existing tanks have been fitted with plate modules, each of which constitutes a settling tank of restricted depth. For a given depth this enables the loading per unit area of an existing tank to be increased (see Chapter 3, page 74).

1.2. Sludge contact clarification

Technical progress has resulted in the improvement of flocculation by increasing the concentration of the floc, or by recycling the sludge; recycling has the effect of speeding up clarification.

In the biological treatment of sewage effluents, the final tanks in which the biological floc is separated from the purified water are called secondary clarifiers. They must allow very high recycling rates so that the sludge remains in the installation as short a time as possible before returning to the aeration tanks. Technical progress in this field has seen scraper-type clarifiers give way to suction-type clarifiers and the combined installations.

In the case of drinking water or industrial process water, flocculation and clarification can be combined in one unit such as the Circulator (with sludge circulation) or the Pulsator (with a blanket of sludge containing a high concentration of suspended matter), which enable complete reactions to be obtained with dense precipitates. The upward flow rate can then be increased considerably, from 1.5 to 6 m/h (6 to 24 gpm/sqft), depending on the type of clarifier. In this way clarified water of good and constant quality can be obtained irrespective of the degree of turbidity of the raw water and the nature of the treatment.

A special chamber or concentrator ensures the thickening of the excess sludge and enables it to be extracted automatically.

When large quantities of sludge are present, the scraper-type flocculator/clarifier should be used.

Settling systems using sludge contact improve flocculation and make better use of the quantity of reagent added by raising the concentration within the body of the sludge blanket. The phenomenon of adsorption of dissolved matter on the floc is thus improved. In the case of treatment by powdered activated carbon the concentration within the sludge blanket is such that, for the same result, a considerable reduction in the rate of dosing can be achieved. The economy can be up to 40 % of the carbon used (Chapter 11).

1.3. Application of plate settling to sludge contact clarifiers

In a sludge contact clarifier of the sludge recirculation type (Circulator, Accelator) or sludge blanket type (Pulsator), the use of plate modules improves the quality of clarified water at the same upward velocity, by trapping the residual floc which escapes from the sludge blanket. Conversely, with the same quality of clarified water the flow treated by the equipment can be considerably increased.

If inclined plates are introduced into a sludge blanket maintained in suspension an accelerated settling process is observed. The sludge settles on to the lower plate and is then subjected to a downward flow towards the base of the sludge blanket. At the same time, the water released by this process gathers under the upper plate and tends to rise rapidly towards the top of the tank. This phenomenon can be observed in the laboratory, if the settling of sludge added to two jars is compared (see fig. 77). In the left-hand vertical jar, normal sludge blanket settling takes place. In the right-hand jar, inclined at 60° to the horizontal, the combined effect of the sludge blanket and tubular settlement causes a spectacular increase in the settling rate. Fig. 78 shows this improvement in the settling rate.

If the completion of this settling process as shown by these curves is studied, it can be seen that the sludge volume of 125 ml is obtained in 6 minutes in the vertical jar, and in 2 minutes 30 seconds in the inclined jar. This equipment enables the sludge blanket effect and plate settling effect to be combined. Very

Fig. 77. — Sludge settlement in a vertical test-tube and in an inclined test-tube after 5, 10 and 15 minutes.

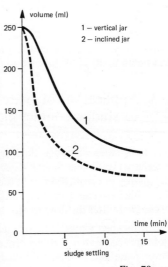

volume (ml)

1 — vertical jar
2 — inclined jar

sludge settling

time (min)

Fig. 78. —

high settling velocities can be achieved and, depending on the nature of the water and the type of suspended solids, they can reach 20 m/h (8 gpm/sq ft).

1.4. Super-accelerated clarification

The combination of the plate effect and the sludge contact effect, obtainable by adding inclined plates into the sludge blanket (Superpulsator), allows velocities twice or three times those used in normal accelerated sludge contact settling.

● *Convection currents:* any clarification process can be seriously disturbed if the specific gravity of the water does not remain constant, and this occurs if the temperature or salinity vary. A raw water temperature change of 1 ℃ per hour may be sufficient to reduce the efficiency of a clarifier. A salinity variation of 1 g/l per hour has the same effect.

2. SETTLING TANKS

When settling is carried out dynamically, that is when the liquid-solid phase separation is carried out continuously and the water in the tank is continuously moving, the design of the tank cannot be based on the upward velocity alone.

The design must take into account other parameters such as the ratio height/diameter, the Reynolds and Froude numbers which must be applied to different points in the tank to ensure that the flow is the least turbulent possible consistent with homogeneity, and the dissipation of energy where the water is distributed at the inlet to the tank. Distribution must be gradual so that no turbulence occurs to upset settling, while ensuring uniform flow into the settling zone. Particular attention must be paid to the clarified water collection system to make sure that it is uniform over the whole tank.

The hydraulic design of the system should not be limited to the liquid phase only but should include the flow, concentration and evacuation of the sludge produced whose characteristics are well-known by water treatment specialists.

Thus a settling tank should not be considered as a simple hopper, because in fact it consists of a complex piece of equipment whose efficiency is governed by its hydraulic characteristics which must be very carefully designed.

2.1. Static settling tanks

It has become the custom to use the term "static" for settling tanks which include neither sludge recirculation nor sludge blancket, although settlement in these is effected by a dynamic process.

Depending on the quantity of suspended solids, the volume of precipitates to be extracted and the slope at the bottom of the installation, the settling tank may or may not be equipped with a sludge scraping system.

2.1.1. STATIC SETTLING TANKS WITHOUT SCRAPING SYSTEMS

A. Ordinary cylindrical settling tanks with conical bottoms.

The vertical flow settling tank is used for small outputs up to about 20 m³/h (5 300 gal/h), particularly in the case of chemical purification, and also for the treatment of sewage effluents where the population does not exceed 1 000 to 2 000 people.

It is also used in larger installations, always provided that the volume of precipitates is slight and their density is high.   This settling tank may be preceded by a flocculator and even a grit removal unit if necessary.

The slope of the conical part of the installation should be between 45 and 65 ° depending on the type of water to be treated and the treatment to be used.

The average upward flow rate should be from 0.5 to 1 m/h (.2 to .4 gpm/sqft) for the clarification of drinking water, and from 1 to 2 m/h (.4 to .8 gpm/sqft) for primary settling of domestic sewage.

B. Horizontal flow static settling tanks.

In this type of settling tank, formerly used for drinking water supplies, the settling surface area in square metres must be equal to once or twice the hourly output in cubic metres of the water to be treated.

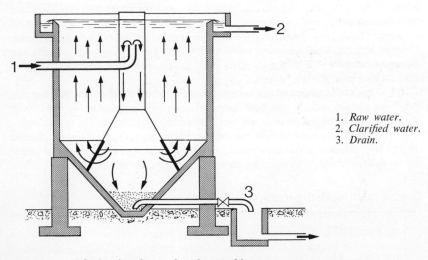

1. *Raw water.*
2. *Clarified water.*
3. *Drain.*

Fig. 79. — Cylindrical settling tank with conical bottom.

This system therefore requires large surface areas and extensive structural work.

Moreover, the tank must be drained completely from time to time to extract the deposited sludge. This method is thus only suitable for small quantities of sludge.

Static settling tanks are generally preceded by a mixing chamber for rapid diffusion of the reagents, and by a flocculator with a slow stirring device to promote flocculation (see page 71).

C. Plate-type static settling tanks.

Numerous types of plate-ytpe static settling tanks exist which are equipped with parallel plates or tubes.

The **Sedipac** shown in fig. 80 combines in one unit a mixing zone for the water and reagents added for treatment, an accelerated flocculator fitted with Superpulsator deflector plates (see page 182), and a plate settling zone. The sludge falls by gravity and is concentrated in a hopper at the bottom of the tank.

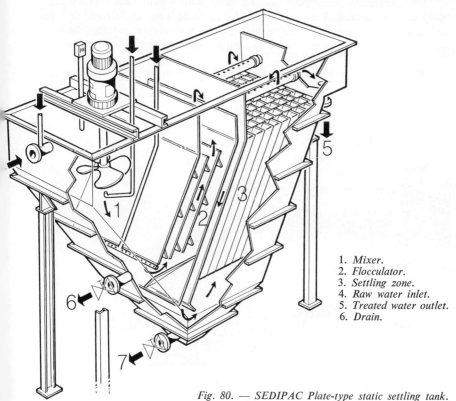

1. *Mixer.*
2. *Flocculator.*
3. *Settling zone.*
4. *Raw water inlet.*
5. *Treated water outlet.*
6. *Drain.*

Fig. 80. — SEDIPAC Plate-type static settling tank.

2.1.2. STATIC SETTLING TANKS WITH MECHANICAL SLUDGE SCRAPERS

The scraper type of settling tank is generally used for "roughing" the raw water, for primary treatment, for clarification and for chemical purification of sewage water.

It is also used for the purification of mine water, coal washing water and in general all waters containing heavy matter which will settle spontaneously.

It is important to be able to extract the sludge as and when it forms.

The use of the scraper results in a thickening of the sludge, thus reducing the volume to be extracted and keeping to a minimum the loss of water extracted with the sludge. Scraping also enables the sludge to be forced into one or more special sumps from which it is then extracted.

The travel speed of the scrapers used to collect the sludge in the concentration and extraction hopper depends on the percentage and density of the sludge which can be removed from the water by the settling process.

On average, in rectangular settling tanks with longitudinal scrapers, this speed is approximately 1 cm per second for drinking water and from 2 to 5 cm per second for sewage. In circular settling tanks with rotary scraping, the peripheral speeds of the arms are from 1 to 3 cm per second and 2 to 6 cm per second respectively.

Numerous types of settling tanks equipped with scrapers are currently available.

Fig. 81. — Primary settling tank 25 m (82 ft) in diameter with a scraper blade at the tank bottom. Output : 300 m³/h (1.9 mgd). Sewage treatment plant at Esfahan, Iran.

A. Circular settling tanks.

In circular settling tanks, the scraper is fixed to a framework rotating around the axis of the tank. It may consist of a single blade (fig. 81) or a series of scrapers in a louvre arrangement (fig. 82).

Depending on the treatment used and the quality of the water treated, it may be advisable to provide a surface scum removal system. Such systems are widely used in sewage treatment.

The scraper system may be radial or diametral. In the latter case, the bottom scrapers are doubled. They are suspended from a rotating frame driven from the periphery or from the centre.

A frequent construction is a radial bridge with peripheral drive. A reduction gear mounted on the bridge rotates a driving wheel running on the tank wall. The surface scraper is rigidly fixed to the rotating bridge and the bottom scrapers, which are generally attached by hinges, are drawn by the same bridge.

In the case of systems equipped with a central drive, the frame consists of two radial arms suspended from a central toothed crown wheel driven by a fixed reduction gear. The bottom and surface scrapers are integral with the rotating framework. The central drive can be supported either by a fixed bridge or by a central concrete column bearing on the tank bottom. The slope of the floor on which the sludge is scraped varies between 4 to 10 %. The sludge is scraped into a central hopper and is extracted by an automatic system.

Fig. 82. — Scraper type settling tank 24.5 m (80 ft) in diameter mounted in louvre formation. Waste water treatment plant at Rheims, France.

Circular scraper-type tanks can be equipped with a flocculator having a slow stirring system situated in the tank centre.

The flocculated water then flows through very wide openings without a weir (to avoid excessive turbulence) into the peripheral settling zone where the flocculated particles and suspended matter settle.

Circular scraper-type clarifiers generally have side water depths between 2 and 3.50 m (6′ 6″ to 11′ 6″).

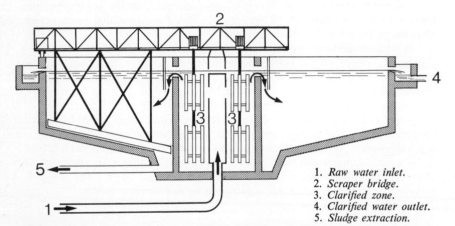

1. *Raw water inlet.*
2. *Scraper bridge.*
3. *Clarified zone.*
4. *Clarified water outlet.*
5. *Sludge extraction.*

Fig. 83. — Flocculator clarifier with scraper bridge (peripheral drive). No sludge recirculation.

Fig. 84. — Flocculator clarifier 40 m (130 ft) in diameter with scraper bridge (peripheral drive).

B. Rectangular settling tanks.

Rectangular settling tanks have the advantage of a more compact layout of the various treatment units, but on the other hand are generally more costly. A $\frac{\text{length}}{\text{width}}$ ratio between 3 and 6 is generally used. Tank depth is usually between 2.5 (8' 6'') and 4 m (13' 1''). The slope of the bottom is about 1 %.

The scraper system may be driven either by a bridge spanning the tank and moving from one end to the other or by endless chains under the water.

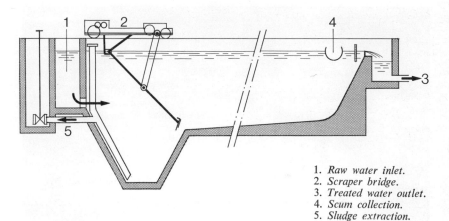

1. *Raw water inlet.*
2. *Scraper bridge.*
3. *Treated water outlet.*
4. *Scum collection.*
5. *Sludge extraction.*

Fig. 85. — Scraper-type rectangular settling tank.

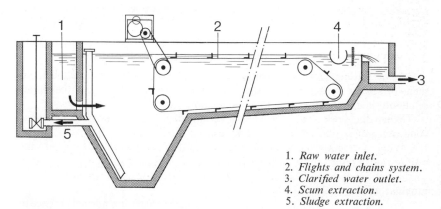

1. *Raw water inlet.*
2. *Flights and chains system.*
3. *Clarified water outlet.*
4. *Scum extraction.*
5. *Sludge extraction.*

Fig. 86. — Rectangular settling tank with flight scraper mechanism.

The sludge hoppers are immediately below the raw inlet and the scum is collected immediately upstream of the clarified water outlet. The bottom scraper works against the direction of flow in the tank, and the surface scraper in the opposite direction. The movements of the scrapers and the reversing of the direction of movement of the bridge are entirely automated.

Scraper bridges spanning and serving several tanks simultaneously are possible. The bridges can have spans exceeding 25 m (80 ft).

Where the raw water carries little suspended matter, a single bridge can be used for several tanks, the bridge being transferred periodically from one tank to the next.

Fig. 87. — Longitudinal clarifier with scraper bridge sweeping both the tank bottom and the water surface. Sewage treatment plant of SAINT-QUENTIN-EN-YVELINES, France.

2.1.3. STATIC SETTLING TANKS WITH SLUDGE SUCTION

Static settling tanks with sludge suction are used principally in the treatment of sewage by activated sludge where it is necessary to reduce the sludge retention time in the final tank to avoid degradation.

When the settling tank reaches a certain diameter, the number of sludge collection points must be increased by using a suction device.

Without creating unacceptable turbulence it is not possible to increase the rate of sludge return to the centre of the tank by increasing the speed of rotation of the bottom scraper. Suction devices can be fitted to both circular and rectangular settling tanks.

For circular tanks with a diameter less than 40 m (130 ft), it is normal practice to use a radial bridge with peripheral drive to which a horizontal channel is rigidly fixed. A number of pipes descending almost to the floor of the tank discharge into this channel.

When the diameter of the tank exceeds 40 m (130 ft), sludge suction is carried out over the whole diameter and not just the radius. The suction equipment is often driven from the centre.

In both cases, the suction effect can be obtained either by hydrostatic pressure, with the pipes discharging at a level below that of the surface level of the water in the tank, or by using an air-lift action. By adjusting the air flow it is possible separately to control the sludge flow in each suction pipe.

The removal of sludge from the movable channel is carried out via the centre by means of a siphon which is itself movable in the case of a bridge with a peripheral drive.

Fig. 88. — Clarifier with central-drive scraper bridge and sludge suction pipes, 41 metres (137 ft) in diameter. Sewage treatment plant at METZ, France.

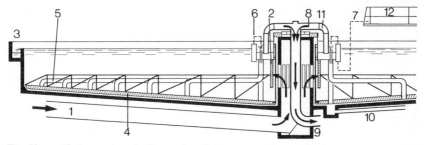

Fig. 89. — Sludge-suction clarifier with radial scraper and central drive.

1. *Intake.*
2. *Perforated baffle.*
3. *Weir.*
4. *Scraper.*
5. *Suction pipe.*
6. *Pipe-flushing air-lift.*

7. *Blown-air inlet.*
8. *Sludge extraction siphons.*
9. *Sludge outlet pipe.*
10. *Drain pipe.*
11. *Drive head*
12. *Access walkway.*

Fig. 90. — Longitudinal clarifier 21 × 29.5 metres (70 × 98 ft) equipped with a transverse bridge and sludge suction.

In a rectangular tank, the sludge sucked up is discharged into a channel which moves backwards and forwards across the tank in a direction perpendicular to the flow of water. The sludge is removed from the channel by pump or by siphon. Particular attention should be given to the design of the tank, especially to the raw water inlet, as the distribution of water must be uniform and without turbulence likely to upset settling, and to the ends of the tank to avoid "dead" zones where sludge can accumulate.

2.2. Sludge contact settling tanks

2.2.1. GENERAL REMARKS

A. Factors favouring flocculation:

To enable the particles of metal hydroxides and colloidal matter to coalesce, it is essential that they come into contact one with the other. Flocculation is considerably facilitated by slowly stirring the liquid. In addition, the chances of contact between the particles increase with their concentration in the water, and this factor led to the increasing of the concentration by retaining in the liquid a high proportion of the sludge formed by the previous treatment.

The stirring operations designed to bring into contact the water to be treated, the reagent and the sludge, must be sufficiently slow not to destroy the floc or to reconstitute a colloidal suspension.

It is, moreover, essential that the old particles brought into contact with the liquid during treatment should be in the same physical state as those particles formed by added reagents. The reintroduced sludge must not therefore have been subjected to any consolidation following a prolonged settling process which could bring about excessive dehydration.

Two processes can be used to ensure contact with the sludge :

● *Sludge circulation clarifiers:* the sludge is separated from the clarified water in a settling zone. It is then recirculated to a mixing zone provided with a mechanical **(Accelator, Turbocirculator)** or hydraulic **(Circulator)** stirring system. The raw water to which reagents have been added also enters this mixing zone.

● *Sludge blanket clarifiers* **(Pulsator** type): no attempt is made to circulate the sludge. It is only necessary to keep it in the form of an expanding mass that can be traversed by the water from the bottom upwards in as regular and uniform a manner as possible. The very mild stirring motion is obtained where the water to be treated enters the plant.

B. Sludge separation:

Within the sludge blanket or the recirculation zone, the sludge is in a state of suspension; it occupies an apparent volume that varies according to its specific gravity and the upward flow rate of the water. No appreciable sludge compacting can develop.

True separation takes place in the calm zones provided in the clarifier, though they only constitute a small part of it. The sludge concentrates in these sludge hoppers (also known as concentrators) from where it is automatically extracted by valves or siphons controlled by a programmer.

C. Application of plate settling:

The principle of plate settling is not restricted to static settling tanks but is applied also to sludge contact settling tanks in which it increases the rate of treatment.

In sludge recirculation tanks, the plate modules, which are usually in the form of tubes whose dimensions are carefully chosen to obtain the required performance without risking blockage, are placed in the settling zone where the effective surface area is thus artificially increased **(RPS plate settling tank).**

In sludge blanket settling tanks the modules are placed either in the settling zone **(Pulsator with plates or tubes)** or in the sludge blanket itself **(Superpulsator).**

In the first case, the modules play the same role as in static plate-type tanks or those with sludge recirculation.

In the second case, the modules, of elaborate design, are formed of plates and deflectors. These play a dual role: they increase the efficiency of the sludge blanket by creating slow eddying motions within it, and so noticeably reducing the time required for flocculation. In addition, they are designed to produce stationary eddies, with the maintenance between the plates of a sludge blanket subjected to high upward velocities.

D. Applications:

Sludge contact settling tanks can be used in all purification processes in which chemical aregents are added:

— coagulation of colloidal matter (clarification),

— colour and odour removal,
— precipitation of alkaline-earth salts (carbonate removal, softening),
— iron and manganese removal,
— chemical treatment of sewage.

2.2.2. SETTLING TANKS WITH SLUDGE RECIRCULATION

The basic features of these clarifiers are a reaction zone and a clarification zone. The sludge is drawn from the bottom of the clarification zone and then enters the reaction zone.

Simple in principle, these clarifiers must be very carefully designed in both their general shape and their finer details, as it is important to prevent sludge deposits and to ensure the proper circulation of the sludge without too much turbulence and to ensure satisfactory sludge contact without excessive stirring.

A. The CIRCULATOR clarifier:

This type of clarifier incorporates a particularly well-designed hydraulic system for accelerating reactions by the methodical circulation of precipitates formed with the reagents and the water to be treated.

It is of very simple construction and can be conveniently used in medium-size installations with small diameter tanks.

It is also frequently used to obtain accelerated flocculation and settling under pressure. In many small rural installations for the treatment of drinking water, where double pumping causes complications, the Circulator clarifier is of great advantage.

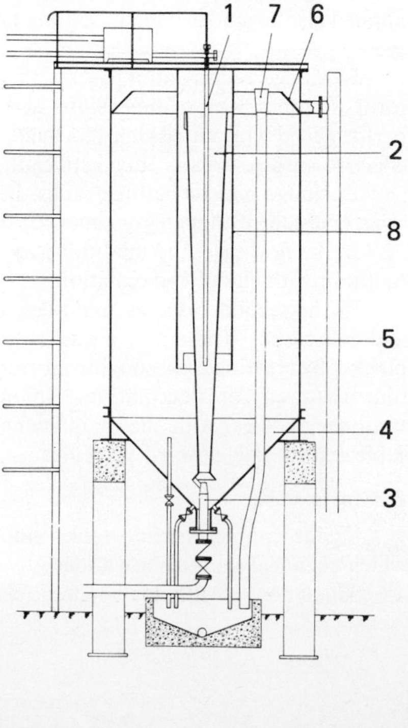

1. *Reagents.*
2. *Reaction zone.*
3. *Hydro-ejector.*
4. *Sludge concentration.*
5. *Clarification zone.*
6. *Clarified water collection channel.*
7. *Overflow.*
8. *Deflection skirt.*

Fig. 91. — The CIRCULATOR clarifier, here used for carbonate removal.

Generally, the clarifiers have a conical bottom to help the sludge to slide down to the circulating ejector. The plant has no mechanical parts (fig. 91, page 174).

For clarifiers of very large diameter that cannot have a sufficiently sloping bottom, rotating scrapers identical to those in the settling tanks shown in figs. 81 and 82 are used for continuous scraping of the sludge towards the centre.

The Circulator can normally clarify or soften water with a retention period ranging from 45 minutes to 2 hours, according to circumstances; the upward flow rate of the water must not exceed 2 m/h (.8 gpm/sqft) when used for water clarification and 5-7 m/h (2 to 2.8 gpm/sqft) when used for water softening.

B. The TURBOCIRCULATOR clarifier:

In this equipment the precipitate is circulated by a specially designed helix. This helix does not cause deterioration of the fragile precipitate of metallic hydroxide, which could not withstand recovery by the hydraulic ejector used in the conventional CIRCULATOR. It enables the equipment to be used both for clarification and softening.

The reaction zone in the centre allows complete coagulation, flocculation, softening and even oxidation reactions.

A rake system, operating continuously, rakes the sludge towards the centre where it is recovered by the recirculation system or collected in a hopper for concentration and intermittent removal.

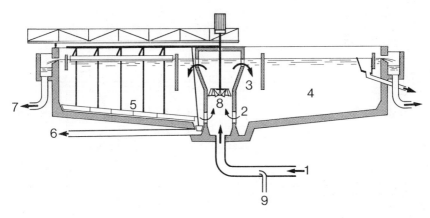

1. *Raw water inlet.*
2. *Sludge recycling.*
3. *Floc maturing.*
4. *Clarification zone.*
5. *Scrapers.*
6. *Excess sludge.*
7. *Treated water outlet.*
8. *Mixing turbine for raw water and recycled sludge.*
9. *Reagent feed.*

Fig. 92. — The TURBOCIRCULATOR clarifier.

Fig. 93. — TURBOCIRCULATOR clarifier, 53 m (176 ft) in diameter; output: 4 000 m³/h (25.3 mgd).

C. The ACCELATOR clarifier:

The **Accelator NS** clarifier has a central reaction zone surrounded by a settling zone. These two zones communicate with each other at the top and bottom.

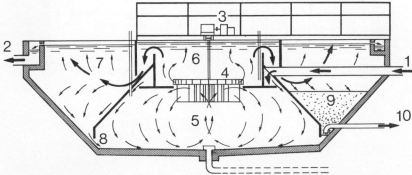

1. *Raw water inlet.*	6. *Secondary mixing and reaction zone.*
2. *Clarified water outlet.*	7. *Clarified water.*
3. *Impeller drive.*	8. *Sludge return.*
4. *Rotor impeller.*	9. *Sludge concentrator.*
5. *Primary mixing and reaction zone.*	10. *Excess sludge discharge.*

Fig. 94. — The ACCELATOR NS clarifier.

A turbine at the top of the reaction zone circulates the water towards the settling zone.

The sludge settling in this zone returns by circulation to the central zone. The resulting sludge concentration facilitates rapid flocculation and the formation of a dense precipitate.

Where necessary, bottom stirring rapidly mixes the raw water with the sludge and the reagents. It also prevents the accumulation of heavy deposits liable to clog the plant.

There are one or more sludge hoppers for extraction of excess sludge in as concentrated a form as possible.

The Accelator IS is a variation of the Accelator incorporating a rake in the lower part of the tank. This rake enables the sludge to be thickened and pushed into a sludge sump, situated at floor level, whence it is extracted.

D. The RPS plate settling tank:

The RPS settling tank is really a raked Turbocirculator, with a plate settling zone, within a rectangular structure.

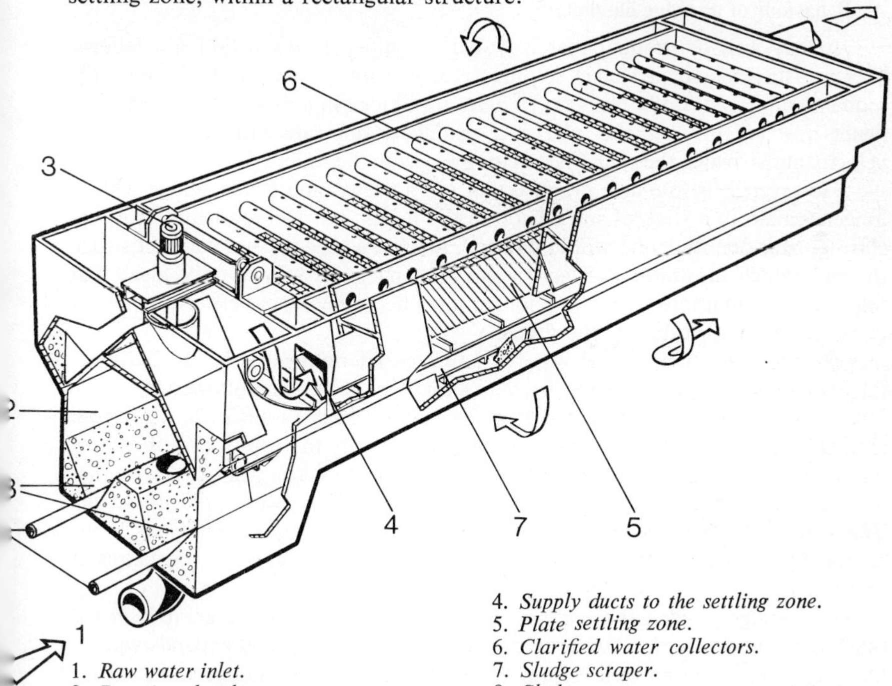

4. Supply ducts to the settling zone.
5. Plate settling zone.
6. Clarified water collectors.

1. Raw water inlet.
2. Reaction chamber.
3. Recycling impeller.

7. Sludge scraper.
8. Sludge sump.
9. Sludge extraction pipework.

Fig. 95. — The RPS plate-type settling tank.

The reaction chamber is equipped with an axial flow impeller to ensure proper circulation of the sludge, and it communicates laterally from its lower part with the plate settling zone. The floor of the reaction chamber incorporates a sludge sump. In the settling zone, a scraper continuously operates to scrape sludge from the floor into the reaction chamber where it is recirculated or concentrated before removal.

The water-sludge mixture leaving the upper part of the reaction chamber is uniformly distributed by two lateral ducts running the length of the settling zone. The uniformity of collection of the clarified water is ensured by a series of evenly-spaced perforated pipes discharging into channels situated above the distribution ducts.

When a large quantity of water is to be treated, the tank can consist of two settling zones situated on either side of the reaction chamber.

Small units can be provided without scrapers.

2.2.3. SLUDGE BLANKET CLARIFIERS

A. Behaviour of a sludge blanket:

It may well be thought that it would be sufficient, when taking a laboratory measurement of a sludge settling velocity with a view to ascertaining the requisite dimensions of a clarifier, to place the sludge in a glass tube, to circulate water upwards through the tube **at a continuous flow rate** and to measure the velocity above which the sludge is carried away with the water.

However, it is found in practice that after a few minutes the sludge will no longer remain in a state of suspension in the liquid, but will gradually heap up, starting near one wall, and will in due course form **a solid mass of compact sludge** through which the water will have made a channel. It will be quite clear that under those conditions there is no longer any effective contact between the water traversing the glass tube and the sludge.

On the other hand, if the water is admitted **intermittently**, that is, by introducing a large quantity of water in a very short period of time, followed by a prolonged period of rest, it is found that **the mass of sludge remains in a constant state of suspension.** All the sludge is carried towards the top during the short water admission period, but it then—during the succeeding rest period—settles in a regular manner as it would in a test tube of sludgy water held immobile. This is the basic principle for the measurement of the cohesion coefficient of the sludge (No. 705, page 951). This gives a sludge blanket that is homogeneous at all points.

If a sludge blanket is traversed by a vertical current of water, it is found that the volume occupied by the sludge varies with the flow and expands with it, but this is true only up to a certain limit beyond which the sludge expansion is such that the particles of which it consists are separated from each other to such an extent that the force of gravity is insufficient to maintain their cohesion. The sludge is then carried away with the water and the effect of the sludge blanket is destroyed. The limit speed so ascertained is not the maximum speed at which the clarifier can be operated. This maximum speed depends on many factors (temperature, type of water, etc.). It will be determined by test No. 705 (page 951) which is used to measure the coefficient of cohesion K of the sludge.

A sludge blanket may be likened to a coil spring which becomes compressed by the force of gravity, but which can be stretched to a greater or lesser degree by the force exerted by the water against the sludge particles constituting the spring; this counter-force naturally increases with an increase in the velocity of the water. The spring breaks if it is overstressed and this must be prevented by adopting a suitable velocity.

The resistance of the spring can be increased, and therefore the maximum possible velocity, by improving the cohesion of the sludge blanket using flocculation aids such as activated silica or polyelectrolytes. This velocity can be more than doubled if the water and sludge contact and the sludge concentration can be increased mechanically, while at the same time the amount of turbulence caused by the water passing through the sludge blanket can be reduced. This is achieved in the Superpulsator by making use of plate settling (see pages 74 and 182).

If a container without a lid is placed within a sludge mass maintained in an expanded state by an upward current of water it will be found that the sludge collects and consolidates inside the container. This is quite easily explained since the water does not circulate there, and there are no forces at work to keep the sludge in a state of expansion. The container thus constitutes a low-pressure zone (where the spring is in its naturally compressed position), surrounded by a high-pressure zone (where the spring is stretched by the friction of the water).

It is therefore quite natural that the sludge should flow rapidly away from the high-pressure to the low-pressure zone. The container in question will act as the sludge concentrator from which the excess sludge can be removed.

B. The PULSATOR clarifer (Patented in France No. 1 115 038 and principal other countries).

The Pulsator clarifier represents the industrial application of the laboratory observations and theoretical considerations expounded elsewhere. It permits the use of high upward flow-rates which, depending on the type of suspended matter, can be up to 8 m/h (3.2 gpm/sqft). See fig. 96, page 180.

● *Design and operation:* The clarifier consists of a flat-bottomed tank with a series of perforated pipes (9) at its base to distribute the raw water uniformly over the entire bottom of the clarifier. A further set of perforated pipes or channels (2) is provided at the top of the clarifier to collect the clarified water evenly and to prevent any velocity irregularity in the various parts of the tank.

Various methods are available for admitting water via the bottom pipe system at intervals, but they all require the storage of a certain volume of raw water for a certain period of time, this water then being introduced into the tank as quickly as possible.

The most economic method of effecting this operation is to introduce the raw water into a vacuum chamber (6) from which air is exhausted by a vacuum pump (7) removing an air flow more or less equal to half the maximum water flow to be treated. This vacuum chamber communicates with the bottom pipe system of the clarifier.

Under these conditions the level of raw water gradually rises in the vacuum chamber. When it reaches a value between 0.60 m (2 ft) and 1.00 m (3 ft) above the level of the water in the tank, a contact operates an electric relay for quick opening of an air inlet valve (8). Atmospheric pressure is therefore immediately applied to the water stored in the chamber and the water rushes into the tank at high speed.

These appliances are generally set so that the water in the vacuum chamber enters the settling tank in 5 to 10 seconds while the filling time for this chamber is 30 to 40 seconds.

Air is drawn from the vacuum chamber by a fan or an electric blower acting as a vacuum pump. The valve connecting the vacuum chamber to the atmosphere is opened or closed when the water reaches the upper or the lower level.

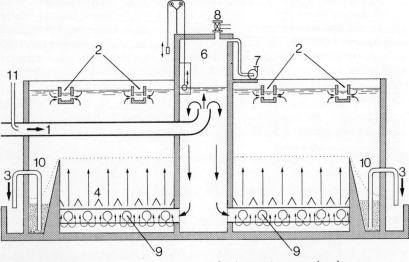

1. *Raw water inlet.*
2. *Clarified water outlet.*
3. *Sludge discharge.*
4. *Stilling plates.*
6. *Vacuum chamber.*
7. *Vacuum pump.*

8. *Automatic vacuum-breaker.*
9. *Raw water perforated distribution piping.*
10. *Sludge concentrators.*
11. *Reagent inlet.*

Fig. 96. — The PULSATOR clarifier.

The header in the bottom of the settling tank has a large cross-section to reduce head loss. The orifices in all the branch pipes are placed so that a uniform sludge blanket forms in the lower half of the settling tank. This blanket is subject to alternating vertical movement, and tends to increase in volume because of impurities brought by the raw water and the flocculation reagents added. Its level rises gradually. A particular zone of the tank is used to form hoppers (10) with inclined floors into which excess sludge discharges and is concentrated. Sludge drawoff is carried out intermittently by the discharge pipes (3)

The equipment has no mechanical sludge stirring system likely to break up the floc already formed. Due to the high concentration in the sludge blanket and its buffer action, faulty adjustment of the treatment dosing rate, or a variation in the pH of the raw water, have no immediate ill effect. A slow variation in the turbidity of the clarified water is observed, but without any massive loss of the sludge in the tank.

It is very easy to install a Pulsator clarifier in an existing tank, old filter or reservoir. In this way old installations can be modernized and their output doubled or trebled.

Such modernization schemes have been carried out at Buenos Aires (300 000 m³/d) (80 mgd), at Durban and at other plants.

C. Plate type PULSATOR clarifier:

By fitting plate modules in a Pulsator above the sludge blanket the quality of the clarified water can be improved compared with the results in a conventional tank at the same upward velocity, or the upward velocity can be increased. The plates or tubes are usually plastic and are inclined about 60° to the horizontal. Fig. 97 shows the arrangement of the modules.

Fig. 97. — Plate type PULSATOR clarifier.

Plate system.

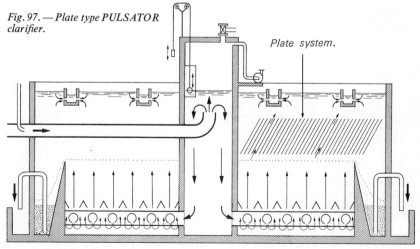

The floc particles which escape from the sludge blanket settle on the lower surfaces of the modules and accumulate there. A thin film develops until cohesion causes the film to slide back into the blanket.

D. The Superpulsator:

The Superpulsator combines the advantages of sludge contact settling, pulsation of the sludge blanket and plate settling. It has many features in common with the Pulsator, from which it has been developed, and the performances of which it improves. The principle of the raw water feed and distribution at the base of the equipment has been retained.

The mixture "coagulated water—flocculated sludge" rises vertically in parallel streams crossing the deep zone situated between the bottom distribution pipes and the inclined plates which are therefore uniformly fed with water. The stilling baffles used in the Pulsator can be omitted in the Superpulsator.

The flocculated water, distributed evenly by the distribution system, then penetrates into the system of parallel plates inclined at 60° to the horizontal and perpendicular to the concentrator. The underside of each plate is fitted with deflectors which act as supports and create slow eddying movements.

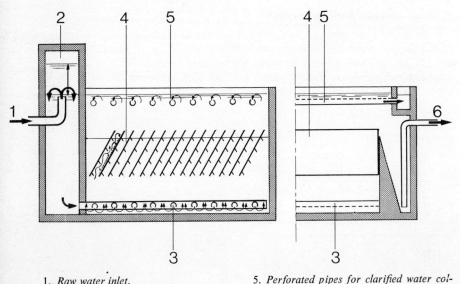

1. *Raw water inlet.*
2. *Vacuum chamber.*
3. *Perforated pipes for water distribution.*
4. *Plate system.*
5. *Perforated pipes for clarified water collection.*
6. *Sludge discharge.*

Fig. 98. — The SUPERPULSATOR clarifier.

The deflector plates in the sludge blanket enable a high sludge concentration to be maintained, twice that in a Pulsator operating at the same rate.

This high concentration in the sludge blanket, which can reach 50 % by volume, enables the Superpulsator to play the role a of filter of impurities, which is a major advantage of clarifiers with deep concentrated sludge blankets.

As in the Pulsator, the top surface of the sludge blanket is limited by decanting into the concentrator zone where no force is exerted by the upward movement of water and the recovery of clarified water is achieved by a system of collectors. The flexibility of operation of the Superpulsator permits very quick starting.

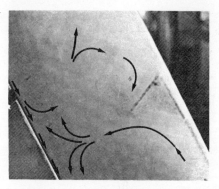

Fig. 99. —

*Fig. 100. — In the foreground, SUPERPULSATOR clarifiers. Surface area: 416 m²
(4480 sqft). Output: 3 100 m³/h (20 mgd). Morsang Water Works, Southern Paris
Region Water Supply.*

2.3. Ancillary equipment for settling tanks

2.3.1. SLUDGE REMOVAL ARRANGEMENTS

In the case of water subjected to flocculation or chemical precipitation treatment, with the exception of horizontal clarifiers without scraping systems (which have to be completely drained for cleaning), the sludge obtained during treatment is concentrated in sludge pits (or concentrators). This sludge can be extracted by a continuous system but it is preferable to draw off the deposited sludge intermittently. The frequency and duration of these sludge removal operations can be adjusted by two interlinked time switches or, better still, by an automatic instrument such as a Chronocontact 401 E of the kind shown on fig. 101.

The momentary flow is then higher and there must be a sufficient head to obtain high velocities in the pipe systems and avoid any risk of blockages or clogging.

The extraction equipment can be automatic valves, siphons or pumps.

Automatic valves are generally of the membrane or sleeve type in which closure is obtained by the application of air or water pressure around the outside of the sleeve or membrane. The opening of this valve is controlled by a three-way solenoid valve situated in the motive fluid circuit and connected to a timer.

Fig. 101. — Chronocontact 401 E.

Siphons are used in particular for the extraction of sludge from Pulsator and Superpulsator concentrators. The siphon is primed by connecting its high point to the vacuum fan or blower. It is unprimed by venting to the atmosphere.

Highly concentrated sludge is extracted by **pumps**. This is the case with scraped clarifiers with central drive treating a highly polluted water. The pumps can then with advantage be sited in the central column of the tank as shown in fig. 104, and this enables the sludge to be removed overhead.

When the flow through the clarifier is variable, and minimum loss of water and maximum concentration of the extracted sludge are required, the frequency of sludge extraction should follow the variations in flow. This can be achieved automatically by control with Chronocontact (timer) 401 E working in conjunc-

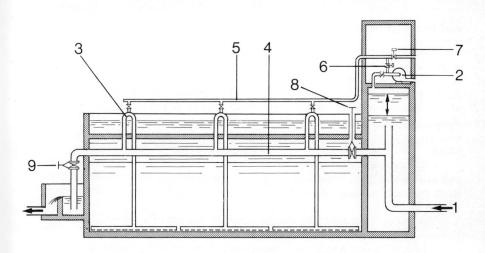

1. *Raw water inlet.*
2. *Vacuum pump.*
3. *Sludge discharge by siphon.*
4. *Main sludge discharge pipe.*
5. *Siphon negative pressure piping.*
6. *Solenoid valve.*
7. *Vent pipe.*
8-9. *Valves for automatic floor cleaning.*

Fig. 102. — Sludge extraction system using siphons for the PULSATOR and SUPER-PULSATOR clarifiers.

tion with Chronocontact 402 E (see page 565). Chronocontact 401 E causes opening of the extraction valves for a constant period but at intervals proportional to the flow through the clarifier.

In some cases, it is required to extract sludge from a clarifier only when it has reached a certain concentration. This concentration is determined by measuring the adsorption of **a gamma ray** or **an ultrasonic wave** by the sludge. A **turbidity** meter such as the DEGRÉMONT instrument described on page 574 can also be used in some cases.

The measurement of the torque developed by the drive head of the scraper mechanism also gives a good idea of the sludge concentration.

When the sludge extraction is controlled by equipment measuring the mass concentration, it is prudent to monitor the level of the sludge blanket in the clarifier using a photo-optical device. This avoids accidental overflow of the sludge when, for example, the flow increases.

In the case of manual extraction of sewage sludge, telescopic valves giving visual control are used. It will also be recalled that sludge can be extracted from static circular or rectangular tanks by suction (see page 171).

2.3.2. SCRAPERS FOR SLUDGE AND SURFACE SCUM

A. Circular clarifiers:

When the force necessary to convey the scraped sludge towards the centre of the tank is not too great, a radial or diametral rotating bridge driven at its periphery is used. The suspending rods of the scraper blade are hinged to the bridge and this avoids subjecting the latter to torsion. It rests at the centre of the tank on a roller-bearing pivot which allows deflection in the vertical plane. At the periphery it rests on a drive unit including a drive wheel coupled to a reduction gearbox and motor, and sometimes a load-bearing idle wheel.

In the case where the bridge has only one wheel, which is both load-bearing and driving and gives better adherence to the running track for a given load, its structure must be designed to resist the torsional forces induced by the motor torque.

The motor is supplied with electric power by the use of a revolving circular pickup with brushes at the centre.

With a diameter up to 20 to 25 metres and where the sludge is quite light, the floor scraper itself can consist of a single profiled blade at an angle to the radius of the tank so that the scraped sludge flows naturally towards the centre.

Above a certain tank diameter it is no longer possible to use a single blade to collect the sludge in the centre, as the angle of slip of sludge against the blade becomes insufficient. In this case a multiple blade scraper is used consisting of several louvred blades fixed on a chassis and with the angle of inclination decreasing from the periphery to the centre. This equipment is also used in small tanks when the sludge does not easily slide.

A diametral bridge is in fact two radial bridges articulated at the centre about one pivot, each with its own drive mechanism.

The use of clarifiers with peripheral drive is limited by the force on the scraper which causes them to uplift. This force increasing with the diameter of the tank, corresponds to a torque value which, related to square metre of surface area scraped, should not exceed 10 m.daN (2 400 ft pdl). When the quantity and quality of the sludge requires the use of larger torque values (40-50 m. daN/m² (900-1 100 ft pdl/sq ft) are sometimes necessary) a diametral scraper driven from the centre as shown in figure 104 should be used.

A cage on which is fixed a diametral framework supporting the scrapers surrounds the central column of the clarifier and is suspended on a drive head resting on this column. The **Centrideg** drive head, made by DEGRÉMONT and shown in fig. 103, is equipped with two reduction gearboxes coupled by a Cardan-joint transmission and driven by the same motor. Each reduction gearbox is fitted with a torque arm (patented) and this enables a high torque to be transmitted. The Centrideg heads at the top of the range are constructed for nominal torques greater than 100 000 m.daN (23,800,000 ft pdl).

Fig. 103. — CENTRIDEG drive head.

In circular tanks, the scraping of scum, when necessary, is always carried out from the centre towards the periphery. The scraper consists of a blade which is inclined to the radius in the opposite direction to that of the floor scraper. At the periphery, it ends in an articulated section which pushes the scum on to an inclined surface into a trough, the discharge level of which is situated above water level. Pipework, which must be designed to avoid any possible blockage by scum, connects the trough to the outside of the clarifier.

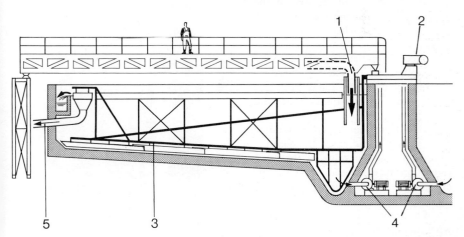

1. *Sludge inlet.*
2. *CENTRIDEG drive head.*
3. *Scraper arm.*
4. *Sludge extraction pumps.*
5. *Scum extraction.*

Fig. 104. — Thickening tank with central drive mechanism.

B. Rectangular clarifiers:

In rectangular clarifiers the sludge, which settles on the floor, is collected in a sump situated beneath the water inlet. Either a scraper driven by an endless chain (fig. 86) is used or more usually a scraper integral with a bridge spanning the tank and moving from one end to the other. In its movement upstream the bridge drags the bottom scraper, which gathers the sludge into the concentration hopper. At the end of its travel the bridge stops and the scraper is automatically lifted by a winch, either to the water surface level where during the return travel it is due to scrape scum into a channel at the downstream end, or above water level when the bridge is equipped with a separate scum scraper.

The automatic control of this type of bridge sometimes becomes quite complex when at the start of scraping, for example, it is necessary to ensure the automatic passage of the scraper under a collecting channel or, in a case of power failure, the automatic restarting must be at the point in the sequence where it stopped.

3. FLOTATION

3.1. Principal flotation systems

The definition of flotation and the theory of its kinetics are dealt with in Chapter 3, page 76.

The process is much used in the mining industry. Its object is to separate the solids on the basis of their different aptitude for being wetted by water (see page 190 on mechanical flotation).

In the field of water purification or the treatment of sludge produced from water purification, it is not usually required to "select" certain materials from the suspension and to leave other materials. If selection were required, the value of the separated products would not usually compensate for the cost of "collector" or "depressing" reagents.

Flotation can be **spontaneous** when the specific mass of the particles to be eliminated is less than that of the water.

Fig. 105. — Pilot flotation unit.

It can be **induced** by the artificial fixation of bubbles of air or gas on the particles to be removed, giving them an average specific mass less than water.

Through a similar but undesirable phenomenon, the scum within a digester can contain 20 to 40 % of dry solids even though the specific mass is only 0.8 to 0.7 kg/l.

Spontaneous thickening by flotation, as a result of fermentation, is sometimes seen in large organic sludge tanks located in desert areas.

3.1.1. SPONTANEOUS FLOTATION

Spontaneous flotation is in general use for initial oil separation from refinery process and rolling mill effluents, etc. Figure 106 shows the upward flow rates for hydrocarbon droplets of various sizes; these values are used as a basis for calculation for static oil separators.

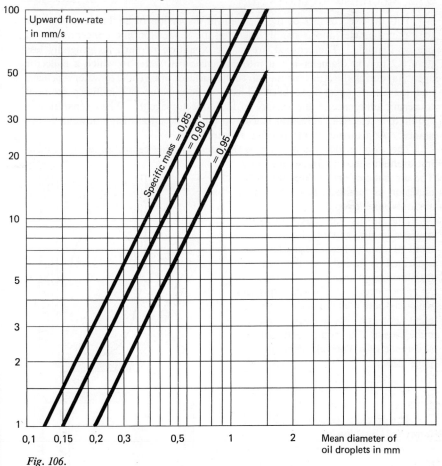

Fig. 106.

For refinery process water not containing paraffin products or other bulky waste matter, a series of parallel sloping plates a few centimetres apart can be arranged inside the oil separator (see page 128).

These plates perform two functions: (a) they improve the distribution, in a vertical plane, of the water from which oil is to be separated, and (b) they have a large contact surface area, a fact which facilitates the coalescence of the oil. The oil gathers beneath the lower surfaces of the plates and then moves upwards. The plates nevertheless often require quite complicated handling when the inevitable periodic cleaning takes place.

3.1.2. MECHANICAL FLOTATION

This process consists of the mechanical dispersion of air bubbles from 0.1 to 1 mm diameter and is mainly used for the separation and concentration of ore by foaming. The crushed ore in the condition of particles generally less than 0.2 mm in diameter is put into suspension in water to which a collector agent, favouring the attachment of air to the surfaces thus rendered water repellent, and a frothing agent have been added. A reactivator is sometimes required to permit adherence to the surface of the collector agent. When separation is required, a selective depressant is used to inhibit flotation of the unwanted substance. Adjustment of pH is often required.

The conditioned suspension is injected at the centre of a rotor turning at high speed and sucking in air. The air bubbles are sheared when the emulsion is pumped through a cage surrounding the rotor.

Due to its vigorous action, this process is not suitable for the treatment of water where the matter to be separated is usually in the form of fragile precipitate.

One mechanical flotation process which is better suited to water treatment consists of shearing the air bubbles and dispersing them with the assistance of a **Vortimix** described on page 229.

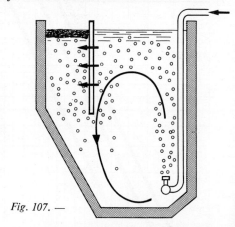

Fig. 107. —

3.1.3. FLOTATION USING BLOWN AIR

This is really spontaneous flotation improved by blowing bubbles of air (several mm in diameter) into the liquid mass.

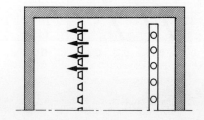

For this purpose porous devices are used to form the bubbles; if necessary, where heavily concentrated liquids are involved, medium size bubble diffusers are arranged so as to produce local turbulence designed to encourage the division of the air bubbles. The size of the bubbles must be small enough to enable them to adhere to the particles to be floated.

Two separate zones are generally provided in plants designed to remove light materials (grease, thick oil, large fibres, paper, etc); one zone is for mixing and emulsifying; the other, a calmer zone, is for flotation proper. Fig. 107 shows the operating principle.

In the emulsion zone, the suspended solids are stirred and intermixed by air. The path of the bubbles is increased by the spiral flow thus created.

In the separation and collection zone for the floating materials, the cross-flow is very slow and turbulence is reduced.

3.1.4. FLOTATION USING DISSOLVED AIR

In the water treatment field it is standard practice to reserve the term "flotation" for processes using very fine bubbles of air or "microbubbles" 40 to 70 microns in diameter, similar to those present in the "white water" running from a tap on a high-pressure main.

● *The importance of and limitations on the fineness of bubbles:* Separation by flotation of solid particles in suspension in a liquid obeys the same laws as sedimentation (see page 65), but in a " reverse" field of force. Firstly, one finds simple flotation to be governed by Stokes' law. Then again, in the case of flocculated particles or very heavy suspensions, one finds "flocculant-flotation" and flotation through a sludge blanket (see page 68). Nevertheless, it is necessary to see how this blanket is created and to what extent it can be considered as uniform.

Uniformity and continuity are linked with the diameter of the bubbles given off in the liquid mass.

Figure 108 shows the variations in the upward flow rate of the bubbles according to their diameter. 20-micron bubbles have an upward flow rate of several mm/sec, whereas bubbles several mm in diameter have velocities 10 to 30 times greater. If in a given volume an emulsion is introduced at one point with an outlet provided at the other end, the period of immersion of the air bubbles in the water and, similarly, the space filled with bubbles will be greater the smaller the upward flow rate of the bubbles and their diameter (see fig. 109).

For a flotation unit of a given cross-section, the use of bubbles several mm in diameter will result in an air flow much greater than that for microbubbles, if a satisfactory distribution of the bubbles is desired over all the cross-section.

At the same time, this increase in air flow will set up turbulent currents, disturbing satisfactory separation and setting up a kind of mechanical mixing.

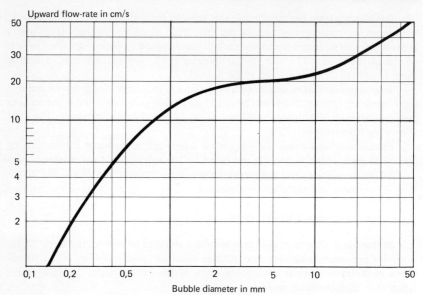

Fig. 108. —

The bubbles have a flotation effect only to the extent to which they adhere to the particles. This generally assumes that their diameters are less than the diameters of the material or of the floc in suspension.

A flotation process using means other than microbubbles can only be used in suspensions containing light and bulky materials whose surface deposits are not disturbed by whirling movements.

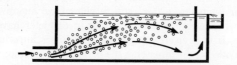

Fig. 109. —

The applications of flotation in the field of water treatment are legion:
— separation of flocculated matter in the clarification of surface water (in lieu of settling, generally for water which is cold, which a low mineral content and charged with organic matter);
— separation of flocculated or non-flocculated oil in waste water from refineries, airports and steelworks;

— separation of metallic hydroxides or pigments in the treatment of industrial waste water;

— thickening of the activated sludge (or mixed activated sludge and primary sludge) from organic waste water treatment plants.

The techniques vary according to:

— the method used for bubble formation;
— the method of feeding the flotation unit;
— the shape of the structures;
— the method of collecting the skimmings.

The most widely-used technique for producing microbubbles is **pressurization.** The bubbles are obtained by the expansion of a solution enriched with dissolved air at several bars pressure. The graph in fig. 110 shows the rate of air saturation of the water for different pressurization pressures at 20 °C. The pressurized liquid used is either raw water or re-circulated treated water. The rate of flow of the pressurized water is generally only a fraction of the rated flow for the plant; it represents 10 to 30 % of the flow to be treated, at pressures of 8 to 3 bars and on average about 60 % of excess air is allowed to dissolve in relation to the rate of saturation at atmospheric pressure. Hence, the compressed air consumption varies between 15 and 50 normal litres per m³ of water treated.

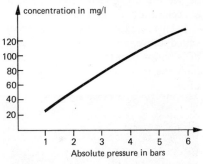

Fig. 110. —

When the quantity of material to be floated is high and thickening is particularly desired, as in the case of activated sludge, the recycled flow can reach 200 % of the nominal flow of the flotation unit. It is possible, with the use of polyelectrolytes, to obtain sludge concentration of 3-6 % for specific loadings of 5 to 13 kg of dry solids per m² per hour (1 to 3 lbs per sq ft per hour) at a downward velocity of 2 m/h (6.6 ft/hr). **Electroflotation** is another technique, in which the object is to produce bubbles of hydrogen and oxygen by electrolysis of the water by means of appropriate electrodes. The anodes are highly sensitive to corrosion and the cathodes to scaling by carbonate removal. When the protection of the anodes requires the use of protected titanium, it is not possible periodically to reverse the electrodes with the object of self-cleaning. A preliminary chemical treatment of the water or periodic descaling of the cathodes must then be allowed for.

In practice, the current densities used are of the order of 80-90 ampere hours per m² of flotation unit surface area (7.4 to 8.3 per sq ft). The production of gas is about 50-60 litres per hour per m² of area (0.16 to 0.20 cu ft/hr/sq ft). The flow rate used is about 4 m³/(h.m²) or 1.6 gpm/sq ft.

Flotation is often combined with preliminary flocculation; by incorporating a flocculation aid (see page 141) the floc can be enlarged, and the particle surface area increased. This gives improved adhesion of the bubbles and an increase in the upward flow rate of the floc.

The separation or downward flow rate of the water used in the flotation units varies according to the nature of the suspensions to be treated and also according to the method of generation and distribution of the microbubbles.

For a given flotation unit the downward velocity and the concentration of the sludge floated are strongly influenced by the value of the ratio:

$$\frac{\text{quantity of air dissolved}}{\text{quantity of material to be floated}}.$$

The greater this ratio the greater the upward force given to the particles, the higher the downward velocity, the lower the sludge density and the greater its concentration in dry solids.

This rate cannot exceed the upward flow rate of the bubbles. Although, as stated above, an extremely fine size of bubbles is particularly conducive to their distribution over the entire surface and to separation efficiency, this can sometimes limit the traverse speed and therefore limit the output of water treated by the plant.

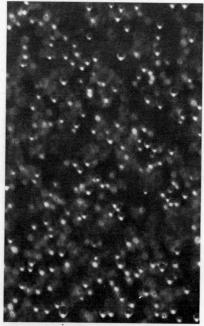

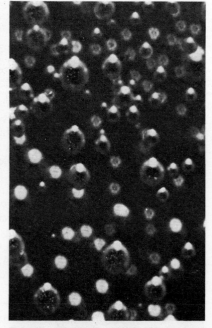

Fig. 111.— Bubbles magnified 30 times; close dispersion. Pressure: 5 bars.

Fig. 112.— Bubbles magnified 30 times; wide dispersion. Pressure: 3 bars.

In cold, low-mineral content surface water where the naturally light floc can easily be separated by flotation, it is possible to achieve upward flow rates of 5 to 8 m/h (16 to 26 ft/h). In oil separation flotation units, these rates will vary from 3 to 7 m/h (10 to 23 ft/h) according to the nature of the oil and the degree of purification required.

In the case of waste water, it is not usually possible to float all the suspended solids. Inevitably, a fairly large and very heavy part of the materials will finally accumulate on the floor of the unit. The flotation units must therefore always be fitted with a system for the elimination of bottom sludge (a steeply-conical bottom or floor scrapers).

The principal applications of the different flotation processes in water treatment are summarized in the following table.

Flotation process	Size of bubbles, microns	Energy consumption, watts/(m³.h) treated	W per gpm	Retention time, min	Principal applications
Blown air	100 to 500	20 to 30	4.5 to 6.8	2 to 5	grease
Mechanical	100 to 1 000	100 to 200	22.5 to 45	2 to 16	"roughing" of polymer and latex or elastomer suspensions
Dissolved air with 20 % recirculation	40 to 70	45 to 60	10.2 to 13.6	20 to 30	hydrocarbons solvents fibres fine suspensions flocculated particles
Electrical	50 to 70	150 to 300	34 to 68		same applications as these of dissolved air flotation in the case of hot saline water

4. FLOTATION UNITS

4.1. General technology

Flotation units can be circular or rectangular. Rectangular units are usually employed for the treatment of potable water as they can be constructed as a monobloc together with the flocculator and filters so that the land requirement is minimal.

With a flotation problem to solve, preliminary tests in the laboratory are recommended (Flotatest, see fig. 113) or tests on a semi-industrial scale (see fig. 105, page 188). It is then possible to pass on to the design calculations stage which is governed by two main parameters—the downward velocity and the quantity of solids to be floated per unit area and time.

A low downward velocity, while reducing the particle flow to the bottom of the tank and theoretically leading to a higher separation efficiency, also leads to an increase in the retention time of the floating sludge. This may cause deaeration and splitting-up, to the detriment of the sludge concentration It falls to the specialist to find the best compromise between sludge thickening and separation efficiency.

The internal layout also has a considerable influence on the performance of the flotation unit.

Fig. 113. — Flotatest.

For distribution of air bubbles, circular shaped flotation units are superior to rectangular units. The distance between the inlet column and the outlet funnel is shorter for the same capacity, and an almost uniform distribution of the bubbles can be maintained over the entire horizontal section of the unit.

4.1.1. WATER FEED

The feed system always incorporates a column or a chamber with a dual function:

— it brings into mutual contact the water to be treated (which may or may

not be flocculated) and the pressurized water. The water must, if possible, be de-pressurized immediately upstream of the unit;
— it disperses the kinetic energy of the raw water/pressurized water mixture and reduces the speed before introduction into the actual flotation zone.

The chamber also enables any large bubbles that may have formed upstream to be removed immediately.

The points and respective levels of introduction of raw and pressurized waters are of major importance in the case of prior flocculation.

Precautions to be taken in calming the mixture when it is introduced into the flotation area vary according to the size, stability and density of the flocculated or non-flocculated substances to be floated. In the case of the flotation of drinking water or slightly viscous oils, this point must be watched particularly carefully.

The emulsified water is generally introduced in the top half of the unit. Sludge is collected at the open surface and the clear liquor outlet is fitted in the lower third of the unit.

The deeper the unit and the greater the quantity of sludge that can be deposited, the farther away will be the clear liquor pick-up from the bottom. Collection is usually carried out in a peripheral chamber bounded by a siphoniform partition; in certain cases immersed headers and laterals can be used.

The more uniform the distribution of water and the microbubbles, the shallower the flotation unit can be (generally between 2 and 4 m, 6.6 and 13.2 ft.)

4.1.2. FORMATION OF BUBBLES

Figure 114 shows the principal devices used for pressure rise and air saturation of the pressurized water (which is usually re-circulated).

Several devices require the introduction of air upstream of the pressurization pump. Systems incorporating an air-cushion saturation tank are more costly, but their operation is very stable. Pressurized air dispersal devices in the pump discharge pipe have also been proposed.

For electroflotation, a bank of electrodes covers the surface area of the tank, and the water feed *must* be positioned above this grille.

Fig. 114. —

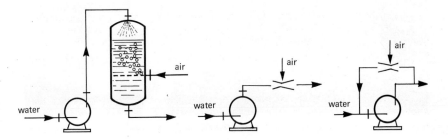

4.1.3. COLLECTION AND REMOVAL OF SLUDGE

The surface sludge layer can in certain cases attain a thickness of many inches and be extremely stable (thickening of activated sludge). In other cases, it is thinner and more fragile (flotation of floc of metallic hydroxides or of oils). When the sludge removal is not total, the layer thickens with time and acquires a degree of cohesion that facilitates the adhesion of floating particles.

The gradual and regular removal of sludge is an important point. The scraper mechanism must be especially strong as it is fitted in a flotation unit designed for sludge thickening.

On circular flotation units one or more scrapers push the sludge into a radial collecting channel with a length equal to half the radius. The access ramp must be constructed in such a way that contact with the scraper blade is always ensured.

The number of scrapers is governed by the quantity of sludge to be removed, by the rapidity with which this removal must be carried out to avoid the risk of deaeration and by the distance the sludge can be pushed without breaking up.

In rectangular flotation units the sludge is pushed by a series of scrapers driven by endless chains to a removal channel situated at one end.

DEGRÉMONT have standardized 4 types of flotation units.

Fig. 115. — Longitudinal flotators.   Output: 1 800 m³/h (11.6 mgd).

4.2. The Flotazur B

This Flotazur shown in fig. 116 is constructed in steel plate up to 6 m diameter. It has one or two surface scraper blades with central drive. The truncated cone shape of the floor makes raking unnecessary. The main flow enters vertically in the lower part of a central mixing chamber while the flow of pressurized water is introduced tangentially immediately above the main inlet.

The expansion of the pressurized water is achieved in two stages immediately adjacent to the chamber by means of two special valves.

The arrangement adopted permits both intimate contact of the material to be floated with the microbubbles without risking rupture of the floc and even distribution of flow over the whole surface of the tank.

A peripheral scum baffle causes reversal of the direction of flow without excessive acceleration likely to pull floc towards the peripheral weir.

Fig. 116. — FLOTAZUR B, 6 m (20 ft) in diameter. Output: 120 m³/h (0.8 mgd).

4.3. The Sediflotazur

The Sediflotazur is a circular flotation tank with diameter up to 20 m. The hydraulic arrangements are based on those of the Flotazur. It differs in that two sludge scrapers are provided, one for the water surface and one for the bottom. It is therefore well suited to treating water producing both sludge which can be floated and heavy sludge which can only be settled.

The floor scraper is pulled by a radial bridge driven peripherally, identical to that equipping scraped static settling tanks (see page 166). The radial surface scraper arms, the number of which varies with the amount of sludge to be removed and the diameter of the unit, are rotated by the rotation of the bridge to whose pivot the articulated arms are connected. The other end of the arms is supported on rollers which run on a peripheral track. The scraper blades, running parallel to the water surface, articulate about the arms and have a roller at each end which comes into contact with a cam. This ensures their slow gradual immersion in the water after passing above the sludge reception channel. The formation of eddies liable to disturb the floating sludge blanket is thus avoided. The sludge is raised to the level of the discharge into the channel by sliding over a ramp whose shape ensures complete contact with the scraper blade.

Tanks of large diameter, above 15 metres, are usually provided with two sludge collecting channels diametrically opposed.

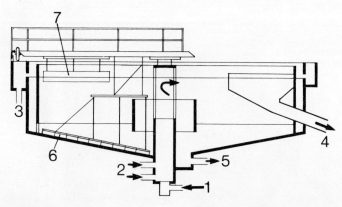

1. *Raw water inlet.*
2. *Pressurized water inlet.*
3. *Treated water outlet.*
4. *Scum discharge.*

5. *Sludge extraction.*
6. *Bottom scraper.*
7. *Surface scraper.*

Fig. 117 A. — The SEDIFLOTAZUR.

4.4. The Sediflotor

Like the Sediflotazur, the Sediflotor is equipped with a surface scraper but it is driven centrally. In some cases it is also fitted with a floor scraper. It comprises a fixed access bridge to the centre where a Centrideg or similar head (see page 187) drives a cage to which the bottom and surface scrapers are connected.

This equipment is used when the water to be treated is loaded with heavy settleable solids which need considerable torque for scraping. It is also used when severe climatic conditions make it difficult to use the peripherally-driven equipment of the Sediflotazur.

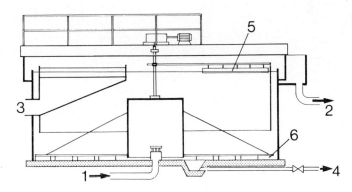

1. *Pressurized raw water inlet.*
2. *Treated water outlet.*
3. *Scum discharge.*

4. *Sludge extraction.*
5. *Surface scraper.*
6. *Bottom scraper.*

Fig. 117 B. — The SEDIFLOTOR.

4.5. The Flotazur R

The Flotazur R is in fact a combination of a flocculator and rectangular flotation unit. It is particularly suited to the treatment of potable water with a low mineral and a high organic solids content, often cold, and yielding a light, fragile floc for which the downward velocity used is 5-8 m/hr (2 to 3.2 gpm/sq ft).

After a retention time of 15 to 30 minutes in the flocculation compartment equipped with one or more slow agitators, the water passes under a baffle and enters the mixing chamber of the flotation unit where it is mixed with the pressurized water representing 6 to 15 % of the design flow. Appropriate equipment ensures the expansion of the pressurized water and its uniform mixing with the raw water over the whole width of the tank. A series of surface scrapers driven by chains periodically push the sludge towards a collection hopper situated at the opposite end to the water inlet to the unit.

In this application the production of sludge is low. Surface scraping is sometimes carried out intermittently to remove the sludge in as high a concentration as possible. The sludge can also be removed by simple overflowing. The absence of any settleable sludge makes floor scraping unnecessary.

8 | # AEROBIC BIOLOGICAL PROCESSES

The first stage in the aerobic biological purification of waste water is to stimulate the growth of bacteria which collect in films or floccules and which, by physical and physico-chemical action, capture and feed on organic waste. The second stage is usually separation by sedimentation of the sludge (called "humus" in biological filters) so produced.

1. BIOLOGICAL FILTERS

In a biological filter (sometimes referred to as a trickling filter or a percolating filter) the effluent to be treated is trickled, after primary settling through a mass of material of large specific area supporting the purifying microorganism which form a felt-like layer or more or less thick film on it. There are two types of biological filters, distinguished according to type of media:
— conventional-media biological filters, in which the media used are pozzolana, blast furnace coke, or crushed siliceous rocks;
— plastic-filled biological filters.

Whatever the material, all trickling type biological filters operate on the same principles.

The filter is aerated, in most cases by natural draught, but occasionally by forced ventilation. The purpose of this aeration is to permeate the whole of the filter with the oxygen required to maintain the microflora in a state of aerobiosis.

The pollution in the influent and the oxygen in the air spread through the biological film to the assimilating microorganisms, while conversely the by-products and the carbon dioxide are eliminated in the liquid and gaseous fluids (fig. 118).

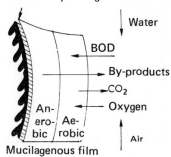

Surface of packing material

Fig. 118. —

According to the thickness of the film, an anaerobic layer may develop below the aerobic layer.

This biological film, or mucilage, consists of heterotrophic bacteria, which are normally near the surface, and of autotrophic (nitrifying) bacteria near the bottom. Fungi (Fusarium, Oospora, Geotricium) are often found in the upper layers and green algae on the surface.

There is generally an abundance of predatory fauna, both protozoa and more developed species such as worms, insect larvae, spiders, snails and slugs.

1.1. Theoretical considerations

The BOD eliminated in a biological filter depends on the nature of the influent to be treated, the hydraulic loading, temperature, and the type of filter media.

The mathematical formulation is based on the assumption that the micro-organisms in the biological filter are in the declining growth phase:

$$\frac{L_f}{L_o} = e^{-k_1 t} \tag{1}$$

in which

L_f : BOD of clarified water
L_o : BOD of the influent supplied to the filter
t : average retention time of water in the filter
k_1 : a constant depending on:
— water temperature;
— the relevant filter media;
— the nature of the relevant influent.

Research — in particular, using radioactive tracers — shows that the retention time t of the water in a biological filter follows a normal law and that its median value can be written in the form:

$$t = k_2 \times \frac{H}{Q^n} \tag{2}$$

in which

H : filter depth in m
Q : hydraulic loading in $m^3/(m^2.d)$
k_2 and n : constants.
Hence:

$$\frac{L_f}{L_o} \sim = e^{-k_1 k_2 H Q^{-n}} \tag{3}$$

According to Eckenfelder and Barnhart

$$k_1 k_2 = k_T S_s^m \tag{4}$$

in which k_T is a coefficient depending on temperature and on the nature of the media, S_s is the specific surface area of this media (m^2/m^3), and m is a positive exponent less than unity determined by experience.

Hence the final relation is written:

$$\frac{L_f}{L_o} = e^{-k_T S_m^s HQ^{-n}}$$

1.2. Conventional-media biological filters

With conventional filter media and a layer depth of 2 m, purification efficiency is relatively low (66 %) where the BOD loading (expressed in kg of BOD per m^3 of medium per day) is high. In this case, efficiency can be increased by recirculating the filter effluent back to the filter, thus diluting the feed water.

Empirical equations habe been worked out for domestic sewage effluents. For example, Rankin gives the following equations, based on a maximum hydraulic loading of 1.13 $m^3/(m^2.h)$ including recirculation:

— single-stage biological filter : $L_f = \dfrac{L_o}{2\,r + 3}$

— two-stage biological filter : $L_{f1} = 0.5\,L_o$

$$L_{f2} = \frac{L_{f1}}{2\,r + 2}$$

r = recirculation rate.

According to the applied BOD loading, a distinction is made between low-rate and high-rate filters, which have the following performance ratings with municipal waste water:

	Low	High
kg $BOD_5/(m^3.d)$ (municipal waste water)	0.08 to 0.15	0.7 to 0.8
Hydraulic loading $m^3/(m^2.h)$	< 0.4	> 0.7

In high-rate filters which normally require recirculation, the hydraulic loading is sufficient to homogenize the bacterial flora at the various levels.

Self-cleaning of the material, which then retains only a thin active film, encourages rapid exchanges and relieves the biological filter of the task of breaking down the cellular material which is formed. This process of mineralization (stabilization) is handled by other stages of the plant, such as the anaerobic digester, which necessarily involves the use of a final clarifier at the filter outlet, to collect the settled matter for transfer to the sludge treatment facilities.

In a high-rate filter, the activity of predatory agents is limited.

On the other hand, in a low-rate filter there is no continuous washing of the humus, which tends to build up in the percolating mass. The action of the predatory agents is essential, and this, together with endogenous respiration of the bacteria limits excessive proliferation of the film. In a low-rate filter, the humus is strongly mineralized and can be discharged into the outlet without final clarification if periodical discharges of humus into the final effluent are permissible. Because of the frequent risk of clogging, low-rate filters, which are uneconomic despite their high efficiency (95 %), are being used less and less and are being replaced by high-rate filters with recirculation.

Recirculation has several advantages:
— it washes the biological filter automatically;
— it seeds the effluent from the primary settling tank;
— it dilutes the high BOD sewage.

There are several possible methods of doing this (fig. 119).

Method No. 1 is the most frequently used recirculation system. The humus is continuously recycled; since the recirculation flow is taken from the bottom of the final clarifier, the surface area of the latter can be designed exclusively for an upward flow rate corresponding to the flow Q to be treated. On the other hand, the primary settling tank must be designed to take $Q (1 + R)$.

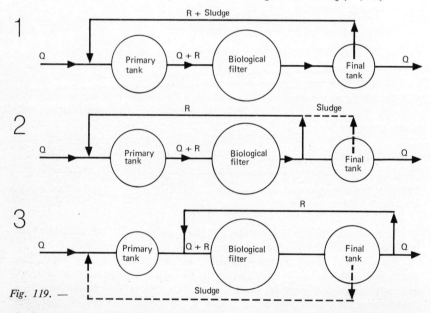

Fig. 119. —

In conventional-media biological filters, the material is pozzolana, blast-furnace coke or simply crushed siliceous rocks. It must be clean and non-friable. Grain size must be regular and between 40 and 80 mm.

Rock aggregates have a void ratio of 0.5, which means that, allowing for the biological film (0.35 m³/m³ of aggregate), the void left free for aeration is limited to 0.15. This sets a limit on the expansion of the biological mass to take high loadings.

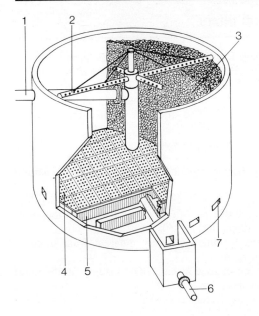

Whatever conventional material is used for the media, it is essential for the filter to be preceded by a primary settling tank to obviate clogging by coarse suspended solids in the raw influent.

1. *Influent.*
2. *Sprinkler.*
3. *Filter material.*
4. *Perforated plate.*
5. *Plate support.*
6. *Treated effluent.*
7. *Air supply.*

Fig. 120. — Sectional view of a biological filter.

Fig. 121. — Conventional-media biological filter.

1.3. Plastic-filled biological filters

Activated sludge processes have long been used for the biological treatment of high-BOD influents such as those of agroindustrial and food manufacturing plants. Conventional-media biological filters were used little, if at all, owing to the risk of clogging by excessive proliferations of filamentous biological films. However, biological filters have a number of advantages over medium- and high-rate activated sludge treatment, e.g.:
— they are less sensitive to abrupt changes in loading;
— they are easier to operate;
— they require substantially less energy, as the air is generally supplied by natural draught through the filter.

In view of these advantages, efforts have been directed towards the development of new filter materials to reduce the risk of clogging. This led to the introduction of plastic-filled biological filters in the 1960s.

The applications of plastic-media filters are very different from those of conventional types, for two main reasons:
— being less liable to clog, they can be used at high BOD loadings in the range 1-5 kg BOD_5 per m³ of media per day;
— since the material is more expensive than conventional media, the aim will be to optimize the ratio:

$$\frac{BOD_5 \text{ eliminated daily}}{\text{filter volume}}$$

For these two reasons, plastic-media biological filters are generally used at high rate. Under these conditions the BOD_5 elimination efficiency is too low to yield an effluent conforming to the specifications usually stipulated, ranging as it does between 30 and 70 % depending on the type of influent treated and the BOD loading chosen. If necessary, the plastic-media biological filter can be followed by a stage of conventional treatment — generally, activated sludge.

Equation (3) (p. 204) shows that the efficiency of a biological filter is an increasing function of the depth of the filter media. This depth rarely exceeds 7 m in practice, because of strength limitations of the materials normally used and civil engineering costs.

The "geometrical" type of plastic materials used is self-supporting, so that the filter wall is not required to withstand high mechanical stresses.

Plastic-media filters are normally used for treating concentrated effluents, so that part of the flow must always be recycled back to the filter inlet in order to maintain a minimum hydraulic loading, below which the required self-cleaning action might not be obtained. Depending on the type of media used, this hydraulic loading is in the range 1.5 to 3 m³/(m².h). Unlike the case of conventional-media filters, recycling does not improve the BOD_5 elimination efficiency here.

Effluent can be recycled either direct from the bottom of the biological filter or from the outlet of the subsequent secondary settling tank. The first approach is better because the final clarifier does not then have to be sized to handle the recycle flow in addition to the nominal flow. The recycling of a part of its excess sludge production to the filter does not entail any loss of efficiency or particular clogging problem.

Facilities must be provided for continuous sprinkling of the biological filter, if necessary using the recycled liquid from the final clarifier if the treatment flow becomes insufficient or ceases. This is particularly important, for example, at weekends or at other times when the supply of effluent declines sharply, when the relevant industrial plants are closed.

Fig. 122. — Plastic-filled biological filters.

● *Intermediate settling tank preceding activated sludge treatment.*

Plastic-media biological filters are often used without primary settling facilities. Since some flocculation of soluble matter contained in the raw waste water occurs in these filters, a considerable amount of sludge may be deposited in the intermediate settling tank. The excess sludge produced cyclically by the biological filter — which is very liable to fermentation and is not mineralized — is extracted from this tank so as not to increase the oxygen demand in the subsequent (activated sludge) treatment phase.

Since the excess sludge from biological filters settles extremely readily, the tank is sized for upward flow rates of 1.5-2.5 m/h.

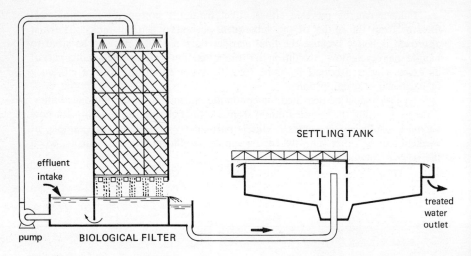

Fig. 123. — Biological filter (with recycling) and its associated clarifier.

● *Materials for plastic media.*

These materials feature high specific area (generally in the range 90-300 m²/m³) and a high voids index (about 95 %). There are two main types: "bulk" (or "random") media and "geometrical" media.

"Bulk" media seem to be more liable to clogging than "geometrical" materials and are generally used only for low pollution influents containing no suspended solids.

1.4. Specific problems of biological filters

● *Protection against cold*

Plastic-media biological filters behave like cooling towers, in which the temperature falls substantially, especially in winter. Again, the activity of the micro-organisms in a biological filter seems to be more sensitive to cold than that of the biomass of an activated sludge. This is probably due to the high rate of operation of biological filters. In cold climates, heat losses in biological filters should preferably be limited by the use of a double wall and a cover, ventilation control, etc.

● *Environmental protection*

In the treatment of certain effluents (breweries, distilleries, etc.), biological filters sometimes give rise to relatively intense smells. It may be necessary to cover the filters over and to deodorize the draught air.

Fig. 124. — Plastic-media biological filter and clarifier. Kronenbourg brewery, Obernai, France.

Flies (Psychoda) may also constitute a serious problem, massing on the filters — irrespective of media — in summer. To limit this infestation, which remains confined to the filter surface, it is important to maintain even distribution of the feed water over the entire surface.

2. ACTIVATED SLUDGE PROCESS

In this process, the growth of a bacterial culture in the form of a floc (activated sludge) is stimulated in an agitated, aerated tank (aeration tank) fed with the influent to be purified.

Agitation prevents the formation of deposits in the tank and homogenizes the mixture of bacterial floc and influent (mixed liquor); aeration means the dissolving of oxygen — either atmospheric, or from an oxygen-enriched gas, or even pure oxygen — in the mixed liquor to meet the requirements of the aerobic purifying bacteria.

After a sufficient contact time, the mixed liquor is fed to a final clarifier, or secondary settling tank, in which the clarified effluent is separated from the sludge. The sludge is recycled to the aeration tank to ensure that the latter contains a sufficient concentration of purifying bacteria. The excess (excess activated sludge) is extracted from the system and fed to the sludge treatment facilities.

2.1. Plant loading

The various types of activated sludge treatment can the classified according to their **sludge loading** C_m, which gives approximate indication of the ratio of the daily pollution mass to be eliminated to the mass of purifying bacteria employed (see page 101 and fig. 126). The following systems exist :
— high sludge loading   : $C_m > 0.5$ kg BOD_5 per day per kg sludge;
— medium sludge loading : $0.2 < C_m < 0.5$;
— low sludge loading   : $0.07 < C_m < 0.2$;
— very low sludge loading or **extended aeration:** $C_m < 0.07$.

The term "extended aeration" is preferred to "total oxidation" as the latter suggests complete conversion of all organic matter into gaseous or soluble inorganic compounds and hence a total absence of excess sludge, which is never the case.

Nitrification may occur below a sludge loading of 0.1-0.4, the exact figure depending on the water temperature and pH (see page 107).

The concept of **BOD loading** C_V (page 101), although less characteristic of the process, is frequently used. The following figures are commonly accepted, depending on the degree of prior purification and the sludge concentration of the mixed liquor:
— high BOD loading: $C_V > 1.5$ kg BOD_5 per day per m^3 aeration tank volume;
— medium BOD loading: $0.6 < C_V < 1.5$;
— low BOD loading: $0.35 < C_V < 0.6$;
— extended aeration: $C_V < 0.35$.

Aerated lagooning is an extensive process in which the BOD loading is less than 0.1 and the sludge concentration is low and variable, there being no recycling of secondary sludge. For this reason, in spite of the long retention times, the sludge loading can be high.

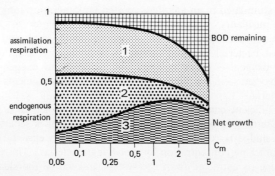

Fig. 125. — Purification parameter evolution as a function of the sludge loading C_m.

In all activated sludge treatment systems, sludge loading is the parameter which determines BOD_5 elimination efficiency (curve 1, fig. 125), the gross weight of cells synthesized from the biodegraded pollution (curve 2), and the net weight of these cells allowing for the internal catabolic reactions of endogenous respiration (curve 3).

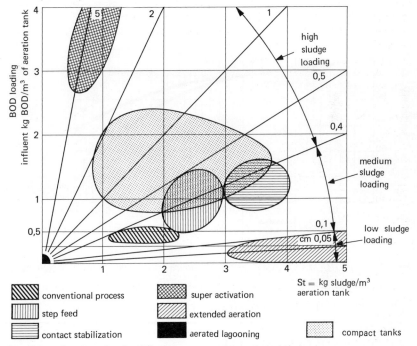

Fig. 126. — Loadings for the different systems using activated sludge.

With domestic sewage of moderate pollution level ($150 < BOD < 350$ mg/l) containing no industrial effluents, the following approximate BOD_5 purification efficiencies can be attained after perfect clarification:
— extended aeration and low sludge loading $\rho \simeq 95\%$
— medium loading $\rho \simeq 90\%$
— high loading $\rho \simeq 85\%$

Where the sewage is outside the above pollution limits or includes industrial effluents, the variation of efficiency against sludge loading may be very different. In such a case it is best to plot the individual experimental curve for the effluent under consideration, to enable the treatment facilities to be scaled accurately in accordance with the desired purification efficiency.

It should never be overlooked that the efficiency of any purification process depends not only on the biological action in the aeration tank but also on the quality of the secondary settling process which separates the biological floc from the interstitial water.

The permitted BOD loading for an activated sludge plant depends on the nature and layout of the equipment which distributes the effluent to be treated, the required air supply and the circulated sludge.

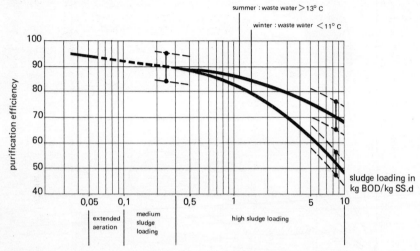

Fig. 127. — WUHRMANN experiments with Zurich sewage (90 < BOD < 150 mg/l).

2.2. The main activated sludge treatment systems

2.2.1. SEPARATE-TANK SYSTEMS.

In these systems, aeration and secondary settling take place in two different tanks, so that the recycled sludge has to be pumped from the final clarifier to the aeration tank (fig .128, no. 1).

In the conventional system, known as the **plug flow** type, the effluent to be treated Q and the activated sludge R are admitted simultaneously to the upstream end of elongated aeration tanks.

The advantages of this system are that it gives excellent quality water and encourages nitrification. However, it results in increased oxygen consumption at the entry end of the tank because the entire pollution is introduced at a single point. This disadvantage can be mitigated by **stepped feed**, improperly called step aeration, in which the effluent to be treated is distributed along the length of the aeration channel, all the recycled sludge being returned to the head of the tank (fig. 128 no. 2). The total suspended solids mass in this case is greater than in the plug flow system, for the same outlet concentration. Concentrations decrease from the inlet to the outlet sections.   The advantage of this process is that the recirculated sludge is reaerated before coming into contact with the effluent to be treated.

The system has been used by DEGRÉMONT in a number of large plants (Geneva, Metz, Tours, Montpellier, Limoges, and others) in which each aeration tank has several channels laid out side by side but operating in series.   (Sewage distribution in a stepped-feed aeration tank is illustrated on page 241).

These principles are pursued still further in the so-called contact stabilization process and its derivatives, in which the settled sewage is introduced only after substantial reactivation of the recycled sludge.   During the relatively short contact time of the activated sludge with the incoming flow, organic matter is eliminated by absorption and adsorption on the biological floc (the INFILCO-DEGRÉMONT Biosorption process).

Stepped feed and Biosorption are still medium sludge loading processes.

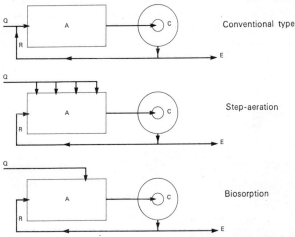

A. *Aeration tank.*
C. *Final tank.*
E. *Excess sludge.*

Q. *Raw sewage or sewage after primary settling.*
R. *Return sludge flow.*

Fig. 128. — Installation diagrams with activated sludge.

Complete mixing is a third possibility, in which the same proportion of effluent to be treated, activated sludge and oxygen — i.e., the same sludge loading — can be obtained throughout the aeration tank.   This system has superior shock resistance to those described above, but is less favourable to nitrification.   It is difficult to apply in large separate aeration tanks, especially if elongated, but comes into its own in compact units (Minibloc, see page 741, MA compact unit, page 740, and the Oxyrapid and Aero-Accelator described below).

2.2.2. THE OXYRAPID.

Oxyrapid units are compact structures enclosing both aeration and clarification compartments, where recycling rates can be very high. They can operate without any mobile underwater mechanical equipment. The length of the sludge return piping is kept to a minimum.

They constitute an excellent means of obtaining complete mixing on an economical basis in large plants, allowing high-efficiency purification at high sludge loading; they are at the same time so compact that considerable space is saved compared with the requirements of separate tanks. The Oxyrapid is suitable for populations in the range 50,000 to 250,000.

In addition to the biological advantage of complete mixing, this unit features the hydraulic advantage of sludge blanket type vertical flow settling tanks; it is also possible to vary the recirculation flow manually or automatically up to 300 %.

The Oxyrapid may be built as a circular or rectangular unit; in the latter configuration, which is most suitable for large plants, the central aeration zone is flanked on each side (or on one side only in some cases) by secondary settling zones. The construction lends itself readily to the use of prefabricated components (gantries and internal partitions). This high-performance unit can be fed with unsettled sewage if it has undergone fine bar screening.

Fig. 129. — The four OXYRAPID tanks of the FLIMS (Switzerland) sewage treatment plant.

The sludge settled in the settling zone is returned to the aeration zone through rising recirculation ducts, which house the diffusers that provide the necessary air pressure for driving the recycling process. The recirculation air is distributed by a circuit separate from the main system, so that the amount of sludge recirculated can be very easily varied, and is not affected by the air flow injected into the aeration compartment proper.

Depending on operating conditions, the continuous or intermittent recycle flow can be automatically pulsed by an adjustable programmer.

Owing to the great attention devoted to the distribution and slow introduction of the mixed liquor into the settling zones, high upward flow rates can be achieved and the maximum advantage derived from the filtering effect of the sludge blanket.

The angle of the sloping parts of the unit is normally 50-55°.

Existing Oxyrapid units are between 10 and 120 m long.

● *Construction:*

Oxyrapid types S 400 and S 450:

This type of unit, whose principle is described above, has been standardized in two depths: 4 m and 4.50 m.

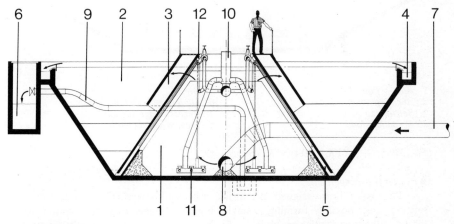

1. *Aeration zone.*
2. *Settling zone.*
3. *Feed duct.*
4. *Treated water outlet.*
5. *Sludge recycling duct.*
6. *Excess sludge extraction pit.*

7. *Raw sewage inlet.*
8. *Perforated pipe for raw sewage distribution.*
9. *Excess sludge outlet.*
10. *Blown air siphon.*
11. *Air diffuser.*
12. *Air for sludge recycling.*

Fig. 130 a. — OXYRAPID.

Oxyrapid type L (Oxyflash):

The settling compartments of this unit are fitted with submerged inclined plates giving a lamellar settling effect (see page 74), thus permitting a 50-70 % increase in average settling rate. Where the capacity of an existing type S unit is exceeded by the incoming flow, these plates can substantially improve its performance.

Oxyrapid type R:

This unit is designed for the treatment of low strength sewage and high peak storm flows. It allows a considerable increase in settling area compared with an Oxyrapid type S of the same length and with the same aeration capacity, and uses mechanical scrapers to help the sludge slide down the moderately inclined walls of the settling compartments.

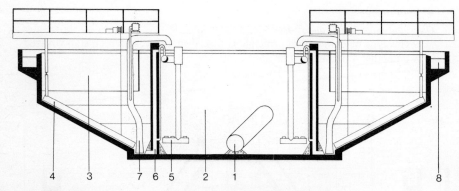

1. *Raw sewage inlet.*
2. *Aeration zone.*
3. *Settling zone.*
4. *Scraping blade.*

5. *Air.*
6. *Sludge recirculation.*
7. *Excess sludge suction.*
8. *Treated sewage outlet.*

Fig. 130 b. — OXYRAPID type R, scraper type.

● *General arrangements:*

The Oxyrapid is designed so that it can be readily combined with a primary settling tank, possibly forming a single, highly compact treatment unit, which may be either rectangular or circular. Such a complex — e.g., within the precincts of a town — can easily be totally roofed over.

Oxyrapid units are equipped with diffusers of the " medium bubble" (Vibrair) type or of the "fine bubble" (DP 230 porous disc) type.

The effluent to be treated is distributed down the centre of the plant over its whole length, either along an open channel equipped with weirs or through submerged pipes with orifices at intervals. Long units have a number of sewage inlet branches over the whole length of the plant; in some cases these are controlled by distributing siphons.

The excess activated sludge is extracted either by hydrostatic ejectors or by air lifts; extraction is always automatic (diaphragm valves and solenoid valves). The sludge extraction points are usually located at the bottom of the settling zones but may be in the aeration zone itself, if so required.

Experience confirms that, even with very widely-spaced extraction points, the concentration of activated sludge in the aeration zone varies within $\pm$ 10 % over the whole length of the unit.

2.2.3. THE AERO-ACCELATOR

The INFILCO-DEGRÉMONT Aero-Accelator is a combined tank, usually of circular shape, with a central aeration zone and an annular outer settling zone; recycling of sludge (the lowest point of the unit is situated in the aeration zone) is accelerated by the adjustable recirculation flow. The aeration zone is oxygenated and agitated by a Vortimix aerator (see page 229).

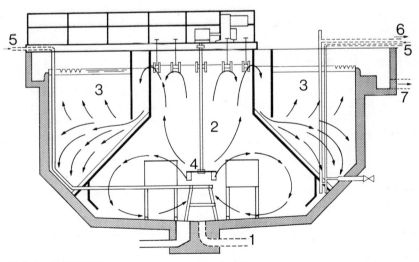

1. *Raw sewage inlet.*
2. *Aeration zone.*
3. *Settling zone.*
4. *VORTIMIX aerator.*

5. *Air*
6. *Excess sludge outlet*
7. *Treated sewage outlet*

Fig. 130 c. — AERO-ACCELATOR.

Fig. 131. — Sewage treatment unit with contiguous tanks and rectangular primary settling tank, scraper type.   HOCHDORF treatment plant, Switzerland.

2.2.4. FUNCTION OF THE SECONDARY SETTLING TANK

In separate tank systems, the treated water is separated from the sludge in **secondary settling** tanks, sometimes also called final clarifiers.   A frequent aim is to design tanks in which the sludge remains for as short a time as possible to avoid anoxybiosis, or even anaerobiosis of the sludge.

The sludge retention time depends firstly on the settling rate of the suspended particles and secondly on the method of collection of the settled sludge to be recycled to the aeration tank.

Activated sludge is flocculent and its specific gravity is very close to that of water.   Its **settleability** (see Kynch curve, page 69), which is checked on the spot by the Mohlman index (see page 71), depends on a number of factors, and is affected particularly strongly by the presence of industrial effluents in the sewage to be treated.   Settleability may also be affected during treatment by variations of the medium : variations in temperature, loading, dissolved oxygen content, accidental shutdowns of the aeration system, etc.   Sludge settleability has an important effect on the superficial hydraulic loading admissible to the tank.   In medium-sized municipal treatment plants and in correctly scaled conventional secondary settling tanks, this loading may reach a maximum peak value of 2.5 $m^3/(m^2.h)$ with activated sludge at medium loading, or 0.9 $m^3/(m^2.h)$ with extended aeration.

These maximum values are sometimes considerably reduced by poor settleability, in some cases only reaching 0.4 $m^3/(m^2.h)$ or even less.   Secondary settling capacity is determined not only by these peak values but also by daily hydraulic loadings.   Typical values are in the region of 24 $m^3/(m^2.d)$ for activated sludge at medium loading and 12 $m^3/(m^2.d)$ for extended aeration, considering the differences in sludge concentration and quality.

The sizing of a secondary settling tank —i.e., its area and volume— depends on the following parameters:

— mean and peak flows and suspended solids concentration of the activated sludge;
— mass of suspended solids introduced;
— sludge settleability;
— permissible suspended solids content of the settled sewage.

Other parameters may also be taken into account —in particular, the method of collection of the treated sewage, and the flow per unit length of the overflow weir (which should normally not exceed 15 $m^3/(h.m)$).

The **recirculation rate,** on which the suspended solids concentration of the recycled sludge depends, determines the volume and retention time of the sludge in the secondary settling tank. If it is insufficient, the volume of sludge stored (often the limiting factor for secondary settling) is excessive; the sludge blanket approaches the discharge weirs and the water quality suffers accordingly. There is a consequent risk of anaerobiosis and in some cases of denitrification of the activated sludge, causing it to rise to the surface. If the recirculation rate is excessive, settling may be impaired by the excess hydraulic energy introduced. For town sewage activated sludge of correct settleability, a recycle rate variable between 50 and 100 % of the average flow is satisfactory. For sludge which is not readily settleable, it may be as much as 200 %.

The overall sizing of the plant embraces the aeration tank, the secondary settling tank, and the recirculation system, all of which are interdepedent.

The performance of a secondary settling tank also depends on its geometry. The best results are obtained by vertical-flow settling in deep tanks with sloping floors (the angle being at least 50º), but such tanks of limited diameter are expensive to build. Because of the need for large areas, secondary settling tanks with only slightly inclined floors must therefore be used. In this case a blade or chain scraper system is required to concentrate the sludge into pits for recycling. To speed up sludge return, the number of settled sludge collection points had to be increased, leading to the design of the so-called **suction-type secondary settling tanks** (see page 170).

The sludge suction equipment is mounted on a moving bridge or scraper arms so that there are no "dead areas" which might impair the quality of the sludge. This arrangement is recommended for large-diameter tanks and for difficult conditions (hot climates, sludge of poor settleability, etc.). The suction tubes, each of which corresponds to a short scraper, continuously collect sludge from the whole of the flat floor of the tank. The suction effect may be produced by:
— hydrostatic pressure: the sludge is fed into a metal channel out of which it is pumped or siphoned;
— an airlift, allowing independent control of the flow of each suction tube;
— a combination of the two systems (in which the airlift is used partly for periodic scouring to clean the tubes).

Suction type scraper bridges standardized by DEGRÉMONT are described in Chapter 7, page 171.

2.3. Aeration systems

2.3.1. CRITERIA AND COMPARISON OF AERATION SYSTEMS

Aeration systems have two functions:
— to supply the microorganisms in the activated sludge with the oxygen they require;
— to agitate the liquor and homogenize it sufficiently to bring about intimate contact between the live medium, the pollutants, and the oxygen introduced.

These systems consist of a device or group of devices situated in a tank of set volume and shape, for dissolving a certain weight of oxygen —generally atmospheric— in the water.

A. Aeration: parameters and standard conditions.

The oxygenating ability of an aeration system can be characterized by the following parameters:
— hourly oxygen input: in kg of dissolved oxygen per hour;
— specific oxygen input: in kg of dissolved oxygen per kWh energy consumption;
— oxygenation capacity: in kg of dissolved oxygen per hour per m³ of tank;
— oxygenation efficiency: percentage of the mass of oxygen introduced which is actually dissolved with a compressed air system. The advantage of this parameter is that it eliminates the factor of the efficiency of the air blower used from the comparison, as this factor is independent of the actual air diffusers.

The aerator/tank combination is to be considered as a whole, and any operating result of the aeration system must be accompanied by a complete definition of this combination. For example, excellent oxygenation performance can be obtained under exceptional conditions: high power per m³ tank volume with surface aerators, or low air flow per diffuser in fine bubble systems.

Systems are generally compared on the basis of "standard" or nominal conditions, viz.:
— in pure water;
— at a temperature of 10 °C (20 °C in some English-speaking countries);
— at standard atmospheric pressure of 760 mm Hg;
— at a constant dissolved oxygen content of 0 mg/l.

B. Correction to convert from standard to actual conditions.

To convert from standard to actual conditions of use, a correction factor T is applied to the parameters mentioned above:

actual conditions = standard conditions × T

this factor T is itself the product of three secondary coefficients, T_p, T_d and T_t.

● T_p (often called α in the U.K. and U.S.) is a pure water/liquor exchange coefficient depending on the nature of the effluent and in particular on its content of surfactants, greases, suspended solids, etc., the aeration system itself, and the tank geometry.

● T_d: oxygen deficit coefficient.
The oxygen input is proportional to the oxygen deficit $C_s - C_x$.
C_s : oxygen saturation under actual conditions: salinity, atmospheric pressure. temperature etc.
C_x : oxygen content of the liquor.
Under standard conditions (at 10 °C), C_s is constant and equal to 11.25 mg/l, and C_x is zero. T_d is therefore equal to $\dfrac{C_s - C_x}{11.25}$.

C_s is affected by:
— the salinity of the water: the correction factor to be applied is

$$\frac{475 - 2.65\,S}{475}$$

in which S is the salinity in $1/1\,000$;
— temperature (see page 888);
— atmospheric pressure (allowing usually only for altitude).
C_x is usually assumed to be 1-2 mg/l.

● T_t : transfer velocity coefficient. As the temperature rises, the gas-liquid transfer velocity increases. The correction is $T_t = 1.024^{t-10}$, t being expressed in °C.

Note that while the coefficients T_d and T_t may be independent of the aeration system, this is by no means true of the exchange coefficient T_p. For this reason, the overall oxygenating ability of different types of aeration systems does not vary in the same way when the standard conditions are corrected to take account of the actual conditions; any objective comparison must be based on the latter. Unfortunately, this calls for a knowledge of T_p, which can only be determined accurately by means of precise measurements taken in a pilot biological purification plant fed with the particular effluent under consideration.

Note in particular that the coefficient T_p may be substantially lower with "fine bubble" air diffusion than with a "coarse bubble" system or surface aeration, largely owing to the effect of surface-active detergents.

C. Other criteria of comparison.

The comparison extends to other ancillary features which are difficult to quantify and can only be assessed qualitatively, as for example:
— **mixing**, which must allow a high enough sweep velocity to prevent deposits and ensure an even consistency of the mixed liquor;
— **possibility of adapting** to different requirements while maintaining a good specific oxygen input;
— **reliability** of all components, such as reduction gear, blower, diffuser, porous element, pipes, etc.

For example, there is no point in having an aerator with excellent oxygenation ability if this is achieved at the cost of insufficient hydraulic agitation or liability to clogging, the result of which would be a fall in oxygenation capacity and the occurrence of anaerobic deposits in the tank.

2.3.2. SURFACE AERATION

A. Different types of aerator.

Surface aerators are divided into three main groups:

— *Slow, vertical-shaft* turbine aerators of peripheral speed generally in the range 4-6 m/s. This type takes in water through the base either with or without a central draft tube, and throws it radially.

— *High-speed vertical-shaft* aerators running at 750-1 800 rev/min, driven directly by the motor without intermediate reduction gearing; in consequence the impeller, usually housed in a tube, is of small diameter. This type of aerator is more suitable for lagooning than for activated sludge tanks.

— *Horizontal-shaft* aerators, mounted either lengthwise or crosswise. Their performance is better at low specific powers.

The following pages deal with the operating conditions for the first type, to which the most commonly used types of surface aerators belong.

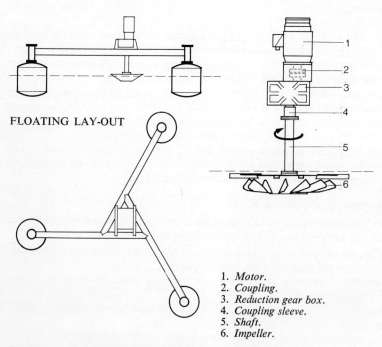

FLOATING LAY-OUT

1. *Motor.*
2. *Coupling.*
3. *Reduction gear box.*
4. *Coupling sleeve.*
5. *Shaft.*
6. *Impeller.*

Fig. 132. — ACTIROTOR surface aerator.

B. Power.

The power absorbed on the rotor shaft of a vertical-shaft surface aerator is given by the equation:

$$p = KN_p D^5 N^n$$

in which D is the rotor diameter, N is the rotational speed in rev/min, N_p is a coefficient depending on the rotor geometry, the respective dimensions of the rotor and the tank, and on immersion depth, and n is an exponent between 2 and 3. It should also be noted that for a given immersion depth, power normally varies with speed raised to the power 2.4 to 2.7. Again, after the initial starting period, movement of the water, and in particular its rotation, causes as it were a relative speed slip between the turbine and the liquid mass, thereby reducing power; this fall is especially marked in circular tanks and when specific power is high.

C. Specific oxygen input.

In any given system, specific oxygen input is influenced by a number of factors:

— **specific power,** i.e., power consumed per m³ of tank volume; up to an upper limit of 70-80 W/m³, an increase in specific power increases specific oxygen feed;

— **speed;**

— **immersion depth:** the effect of immersion depth on efficiency at a given speed varies from aerator to aerator;

— **tank shape:** a circular tank makes it easier for the mass of liquid to rotate and thus slightly reduces the specific oxygen feed (unless radial baffles are installed);

— **area/depth ratio** of the tank: optimum performance is obtained when the side of the square or the diameter is twice the depth.

The specific oxygen inputs of most slow-speed aerators are between 1.5 and 2.5 kg of oxygen per net kWh (mechanical energy measured on the aerator shaft).

Fig. 133. — Actirotor R 8020 (75 kW) mounted on a reinforced-concrete walkway.

Fig. 134. — Surface aerator testing by INFILCO-DEGREMONT Inc. at Tucson, Arizona.

D. Mixing.

The mixing effect in a tank, which is characterized by the speed at which the bottom is swept, is influenced by two main parameters:
— **specific power:** in a given tank, velocity increases with power;
— **gyration radius:** by analogy with flow through a pipe, the gyration radius is defined as the ratio of volume to wetted surface area. Mixing improves as this radius increases.

E. ACTIROTORS (fig. 132).

DEGRÉMONT has standardized a range of surface aerators rated at 3 to 100 kW to meet the different requirements mentioned. These are slow, vertical turbine aerators with a hollow hub carrying thin-profile blades; the assembly forms a completely open wheel, eliminating all risk of clogging (matting).

Actirotor units can be mounted on a fixed support (bridge or platform) or on a floating rig (the three floats are positioned far enough apart not to impede the turbine flow).

Owing to their high pumping capacity, Actirotor units are normally installed without a draft tube, but can be so equipped to cope with very dense sludge (aerobic stabilization).

● **Oxygenation:** the standard specific oxygen input ranges between 1.8 and 2.3 kg of oxygen per net kWh depending on installation conditions, the specific power being 40 W/m³ tank volume.

Fig. 135. — Two aeration tanks (capacity: 3 750 m³ or 990,000 gal each) equipped with 12 37 kW Actirotor, R 6016 type.

2.3.3. COMPRESSED-AIR AERATION

With this process, air is injected under pressure into the mass of liquid by appropriate means classified under three main headings according to the size of the bubbles produced:
— **coarse bubbles** (dia > 6 mm) : the air is injected either directly through vertical pipes or through wide-mouth diffusers;
— **medium bubbles** (dia 4 to 6 mm) : bubble size is reduced by various means, such as spargers, small orifices, etc;
— **fine bubbles** (dia < 3 mm) : produced by diffusing air through porous media (50 micron pores).

A. Oxygenation efficiency.

The standard oxygenation efficiency of a system is determined by several factors:
— **type of diffusers** and their layout (isolated, in line, on a floor);
— **depth of air blowing.** Within the limits of 2.5 to 8 m, it is accepted that efficiency is proportional to depth of immersion;
— **air flow.** With coarse and medium-bubble systems, efficiency is improved by increasing flow. In fine-bubble systems, with an increase in flow the bubbles coalesce more and efficiency generally tends to fall;
— **tank cross section.** Oxygenation efficiency may fall if the area/depth ratio of the tank is excessive.

B. Mixing.

In general, mixing in the aeration tanks causes less problems when they are equipped with compressed-air facilities than with surface aerators, provided the air is diffused near to the bottom.

Fig. 136. — Porous diffusers in an aeration tank.

Fig. 137. — Porous diffuser at work.

C. Air diffusers.

● *DP 230 porous discs.*

The porous disc consists of grains of artificial corundum (alumina α) of specified particle size bonded by a ceramic adhesive vitrified at high temperature. The material has good resistance to most corrosive chemicals, except for fluorine and its derivatives. The discs are secured to PVC or stainless steel bases mounted on submerged longitudinal feeders grouped or distributed throughout the bottom of the tank. The discs give high oxygenation efficiency under standard conditions (see fig. 138) in the region of 20-25 % at an immersion depth of 4.00 m. The air blown in must first be thoroughly filtered (dust content $<$ 15 mg per 1 000 m³).

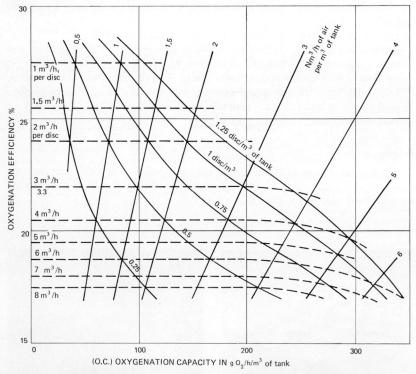

Fig. 138. — Efficiency curves. Type DP 230 porous diffuser disc.

● *VIBRAIR diffusers.*

Vibrair diffusers distribute medium-sized air bubbles; they consist of a moulded polythene body carrying a vibrating valve. Like the disc diffusers they are mounted on submerged feeders (see fig. 140).

The oxygenation efficiency of a VIBRAIR diffuser (8 to 12 % at 4 m) is lower than that of the disc type, but VIBRAIR diffusers are rugged enough to allow cheaper air conditioning, production and transportation; they are also especially suitable for small and medium-size plants. They can work with a discontinuous air supply.

Fig. 139. — Medium bubble diffusers, moulded polythene body, ribbed vibrating cover in aluminium bronze.

Fig. 140. — Aeration tank equipped with Vibrair diffusers.

● *Other diffusers:*

Tubular diffusers made of porous ceramics (fine bubbles), by winding plastic thread on a cylindrical support (medium bubbles), or from plastics (rigid or flexible polythene and polyurethane).

Negative-pressure systems, in which air is fed at low pressure into the neck of an ejector through which the contents of the tank are recirculated.

2.3.4. MIXED AERATION: *the Vortimix aerator*

Some equipment combines the effects of mechanical agitation and air blowing. An example is the INFILCO-DEGRÉMONT *Vortimix* aerator which comprises:
— a mechanical agitator with submerged inclined-blade impeller setting up a downward circulation of liquid;
— a compressed air distribution device underneath the impeller.

The medium air bubbles emitted by this device are atomized by the rotor blades and intimately mixed with the pumped water. The water-air mixture is directed diagonally towards the tank bottom.

This system gives continuous effective mixing as the agitator has good hydraulic efficiency, and permits separate control of oxygenation by variation of the introduced air flow. The intense agitation produced by the impeller in the air introduction zone is favourable to oxygenation efficiency, which may reach 30 %.

Overall energy consumption is about the same as that of a slow surface aerator. It may be less where mixing problems are preponderant. This type of equipment is particularly suitable for sewage treatment in regions with cold winters and where aerosol formation is to be avoided.

Fig. 141. — Aeration tanks equipped with 16 50 HP Vortimix aerators at Palo Alto, California, U.S.A. (Courtesy of INFILCO-DEGREMONT Inc.)

2.3.5. USE OF PURE OXYGEN

The use of pure oxygen is a recent biological purification technique for waste waters.

A. Advantages.

The two main advantages of pure oxygen are as follows:

— owing to the very high partial pressure of oxygen in the diffused gas, many times more oxygen can be introduced to the activated sludge liquor for a given energy cost than from atmospheric air alone; conversely, the energy cost is lower for equal oxygenation capacity. Hence far more microorganisms may be present per unit volume of the reactor;

— for the same reason, the dissolved oxygen content of the activated sludge liquor or water may easily reach or even exceed 6 to 8 mg/l, which would entail a prohibitive energy cost with conventional aeration. This very high dissolved oxygen content favours oxygenation within the biological floc, even in the most concentrated suspensions.

These advantages allow the size of biological reactors to be reduced, because the concentration of the activated sludge is increased and its activity is somewhat enhanced. At equal sludge loading, the excess sludge formed has lower putrescibility, and there is often less of it than with traditional aeration facilities.

B. Applications.

Pure oxygen can be used for various purposes:
— continuous-duty oxygen type activated sludge plants: treatment plants for concentrated industrial effluents whose pollutants are mainly biodegradable, and large-scale municipal waste water treatment works;
— variable-loading activated sludge plants: purification plants with highly variable pollution levels, mainly in tourist areas, with much higher oxygen requirements in the high season than the average demand for the year:

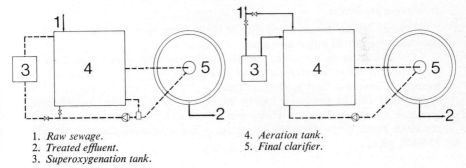

1. *Raw sewage.*   4. *Aeration tank.*
2. *Treated effluent.*   5. *Final clarifier.*
3. *Superoxygenation tank.*

Fig. 142. — Flow-sheets for additional oxygen input systems using a separate cell.

— preoxygenation of waste water or liquors with oxygen contents of 10 mg/l or substantially higher:
— for deodorization;
— for oxygen enrichment prior to enclosed-chamber biological filtration (DEGRÉMONT patent), whereby biological filtration unit sizes can be appreciably reduced;
— for doping in conventional plants, by increasing the dissolved oxygen content of an aeration tank still operated with atmospheric air.

C. Practical application.

The commonest system using oxygen-enriched gas comprises an air-tight covered tank in which the "gas blanket" is maintained at a high partial pressure of oxygen. This arrangement is also environmentally favourable. In exceptional cases where the discharged gases may have to be deodorized, the gas scrubbing equipment can be much simpler with pure oxygen than with atmospheric air systems.

Pure oxygen purification of highly polluted effluents in enclosed tanks has the disadvantage that it hinders elimination of the free CO_2 formed by bacterial respiration. To limit CO_2 production the alkalinity of the liquor can be increased or atmospheric stripping can be carried out. For this reason two stages of treatment —oxygen followed by air— are often advantageous. If pure oxygen is used in the initial stages of a reactor, both energy costs and reactor volumes for elimination of most of the pollution can be limited. If air is then used in the later stages, free CO_2 can be eliminated and degassing and foam formation in the final clarifier can be reduced. The air helps to eliminate residual pollution and can effect nitrification.

To minimize losses, the multistage biological reactor normally has a number of compartments, through which the gaseous mixture passes in succession. With municipal waste water, the oxygen utilization rate in the reactor is about 90 %.

D. Oxygen production.

In small plants, oxygen is supplied in liquid form, stored in efficiently insulated containers fitted with evaporators.

For larger plants, economic considerations generally dictate local oxygen production, except in favourable cases where a nearby oxygen pipeline gives access to supplies from a central production unit.

Oxygen can be produced locally by two technologies: molecular sieves (pressure swing adsorbers or PSAs) for capacities up to 30 tonnes/day, and cryogenic plants above 10 or 20 tonnes/day. The oxygen produced is usually 95-99 % pure.

3. BIOLOGICAL TREATABILITY OF WASTE WATER

Whenever statistics regarding suitability for biological treatment are inadequate, laboratory or semi-industrial pilot studies must be carried out. This type of study is particularly advisable in the case of industrial effluents or municipal waste water containing industrial waste.

The composition of the waste can, of course, affect the possibility of biological treatment and the nature and metabolism of the bacterial strains which can adapt to the waste.

For example, the temperature of the waste must be known, because a biological treatment can be inhibited by both high and low temperatures. The ability of purifying bacteria to exist and spread is influenced by the pH value and the presence of assimilable nitrogen and phosphorus. Lastly, it is essential to know whether or not the waste water contains toxic matter. Some matter is absolutely **toxic;** other types are only relatively so, and some strains can be adapted to their presence.

In order to determine parameters for the essential biokinetic relationships which are needed to design the plant, such specific strains can be adapted in continuous-running pilot laboratory plant.

3.1. Pilot laboratory plant

Figure 143 shows one type of plant developed in the laboratory for this type of study; the aeration and clarification zones are separate. After aeration, the activated sludge is delivered to the lower part of the clarification cylinder and is recirculated by an air lift after being separated from the treated water. This unit works according to the complete mixing principle, thus simplifying the mathematical relationships which can be established from the observed operating parameters.

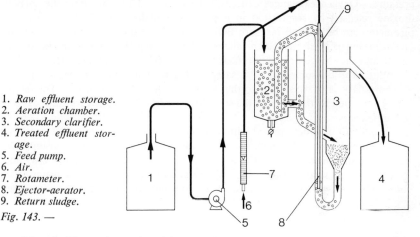

1. *Raw effluent storage.*
2. *Aeration chamber.*
3. *Secondary clarifier.*
4. *Treated effluent storage.*
5. *Feed pump.*
6. *Air.*
7. *Rotameter.*
8. *Ejector-aerator.*
9. *Return sludge.*

Fig. 143. —

The air lift can be replaced by a peristaltic pump giving precise control of the recycle flow and making the aeration system independent of the return sludge pumping.

The normal method of acclimatizing an activated sludge to an industrial waste effluent is to start with a sludge from a town treatment plant and then to feed it, in the pilot plant, with a mixture of domestic waste and the effluent to be tested. Over about ten days, the proportion of industrial waste is gradually raised in steps every one or two days. If 100 % industrial waste is reached and the sludge remains active, the effluent may be regarded as treatable; if not, it is possible to determine by how much it must be diluted with town sewage for it to remain assimilable by the micro-organisms. This test must, of course, be carried out on a medium which is naturally or artificially balanced as regards assimilable carbon, nitrogen and phosphorus.

Once the sludge has been acclimatized, variable quantities of sewage can be fed in to determine a number of parameters, such as growth rate, from which the purification efficiency to be expected for this type of water can be calculated by reference to the sludge loading applied to the sludge in question.

Oxygen requirements and sludge production can then be determined very accurately by manometric measurements.

3.2. Manometric methods—Respirometers

The Warburg respirometer is the standard instrument for studying the respiratory activity of activated sludge.

Its principle is based on the fact that, at constant volume and temperature, any variation in the quantity of a gas can be measured by the variation of its pressure. Oxygen exchanges can easily be measured by absorbing the carbon dioxide released by respiration in a potassium hydroxide solution. The respirometer can be used to measure BOD_5, and to test the toxic effects of certain products, but it is of particular value for determining the respiratory factors a' and b' and the sludge production factors a and b (Chapter 3, page 102).

When the consumption of a sample of activated sludge is measured in a Warburg flask, a curve similar to that shown in fig. 144 is obtained. The factor b' is represented by the slope of the straight line representing this constant oxygen consumption.

The factor a' is determined with adapted sludge taken from a plant working with a known load. A comparison is made between the oxygen consumption curves obtained from two sets of flasks, one filled with activated sludge only and the other with a mixture of the same sludge and of a sample of waste water to be compared. After a time, the slope of the consumption curve for the second set

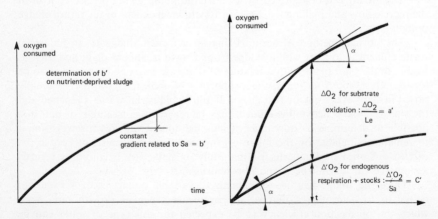

Fig. 144. — Warburg method of determining respiratory factors.

moves to a value exactly the same as that obtained for the first set; it may then be assumed that all the BOD brought in by the waste water tested has been purified and, at that moment, the difference in oxygen consumption between the two sets of flasks, referred to the quantity of BOD introduced, gives the value factor a'.

When the ratio $\dfrac{\text{final BOD}}{\text{BOD}_5}$ is known, a and b can be calculated and give a very accurate measurement of oxygen requirements and sludge production; the figures are very much more accurate than those obtainable from a small pilot plant for which calculations of air flow and weight of sludge are liable to substantial error.

All the data obtained from experiments on pilot plants and with the Warburg respirometer can be built into complex **mathematical models** which can be handled by a computer. These models make it possible to calculate and design the component elements of a biological purification plant.

This method is currently being employed systematically. The results show that plants calculated on this basis do indeed achieve the performance expected.

However, Warburg respirometers have certain disadvantages: their oxygenation capacity cannot easily be checked, and the operations involved necessitate shutdowns which may mask certain phenomena when the pollution is very rapidly assimilable.

The present tendency is to replace these instruments by larger reactors of the industrial fermenter type, in which oxygen consumption is measured continuously by probes.

Fig. 145. — Warburg respirometer.

4. CONTROL OF BIOLOGICAL PROCESSES

In theory, biological processes can be controlled automatically by means of mathematical models in which data obtained on a pilot plant operated at a steady rate (growth rate, respiration coefficients, etc.) are applied on a transient basis. It is, however, insufficient to control the oxidation stage alone. Operation of the final clarifier, which plays just as essential a role, must also be controlled.

4.1. Purpose of control

Before defining any kind of control system, the aims and limits of the action which can be taken must be clearly stated.

From the functioning standpoint only, the theoretical aim should be to hold the sludge loading at its optimum value. It would then be possible:

— to maintain high purification efficiency as pollution increased;

— to reduce energy consumption when pollution decreased, by not retaining in the system more activated sludge than the amount strictly necessary for the required level of purification efficiency;

— lastly and most important, to ensure regular, even growth of the biological flora, thus guaranteeing its vital balance and therefore a constant quality of purification.

From the operation standpoint, any automating device must measure and adjust some or all of the parameters needed to operate the purification plant:

— amount of dissolved oxygen in the mixed liquor;

— excess sludge flow;

— return sludge flow;

— in some systems, distribution of the effluent to be treated along the tank sides;

— degree of pollution of entering influent;

— quality of purified effluent.

In most sewage purification plants, the sludge loading cannot be regulated to a strictly constant value, owing to the following biological and hydraulic factors: because of variations of flow and the degree of pollution, the hourly **BOD** loading can vary by a factor as great as five or even ten in a single day.

To maintain a constant ratio between the cellular biomass and the pollution load removed, it would be necessary to have :

$$\frac{dS_t}{S_t} = \frac{dL_e}{L_e}$$

Since the second term can reach very high values, ranging from 20 to 50 % per hour, a comparable growth rate can only be attained with a very high loading (exponential phase), i.e. with a very low purification efficiency. A minimum bacterial mass must therefore be maintained at all times, with an inevitable drop in the sludge loading at off-peak periods.

From the hydraulic standpoint, heavy extraction of excess sludge dilutes the activated sludge until it is not cohesive enough and becomes difficult to settle. Variations in recirculation flow have only a limited effect on the weight of sludge available. Experience shows that the proportion of sludge transferred from the final clarifier to the aeration tank by varying the rate of recirculation cannot easily be raised above one third of the amount of sludge in the latter tank.

Efforts must therefore be concentrated on controlling oxygen input and reducing variations in the sludge loading by spreading. This can only be done by relating oxygen requirements to the amount of sludge.

The measurement of oxygen requirements (or their variations) alone would be too inaccurate and would involve a major risk of control loop error.

4.2. Principe of control

The mass of activated sludge can be assessed with varying degrees of accuracy, and the method best suited to the particular problem and to the size of the plant should be selected in each case.

The percentage volume does not represent weight exactly because the Mohlman index of the sludge varies with the sludge loading. However, this index varies so slowly that percentage volume can be used, as a first approximation, as a basis for regulation.

If more precise control is required, the measurement of the percentage of sludge can be combined with the measurement of the oxygen requirement.

Indeed, any increase in the ratio $\dfrac{\text{requirement } O_2/m^3}{\text{percentage of sludge}}$ is immediately reflected in an increase of the sludge loading; to keep this within acceptable limits, the volume of sludge must be increased as quickly as possible and, while this considerably reduces daily variations in the sludge loading, the length of time for which the mass of microorganisms is 'represented' by the volume of sludge will be substantially increased.

It will be appreciated that if measurement of the percentage of sludge is replaced by measurement of its weight, the checks linking the volume and mass characteristics of the sludge will be eliminated. This is a first improvement.

A further step towards greater accuracy is to estimate the intrinsic respiratory activity of the sludge by measuring it with a continuously-operating **industrial respirometer** instead of estimating the oxygen requirement of the tank (which includes the oxygen required to oxidize the substrate).

This is so because, as long as the loading limits within which all pollution is sure to be 'sorbed' are not exceeded, the amount of waste matter L_e extracted from the purified water is the sum of the immediately-assimilable waste matter L_i and of stored waste matter L_s.

$$L_e = L_i + L_s.$$

Therefore: $O_2 = a' (L_i + L_s) + b'S_a$ (see chapter 2, page 101).

It is possible to measure $\dfrac{O_2}{S_a}$ = respiratory activity in relation to the mass of sludge

$$\frac{O_2}{S_a} = a' \frac{(L_i + L_s)}{S_a} + b' = \frac{a'L_i}{S_a} + c' \text{ with } c' = \frac{L_s}{S_a} + b'$$

But the fraction of stored waste matter itself depends on the sludge loading:

$$\frac{L_s}{S_a} = kC_m$$

giving $c' = a'kC_m + b'$ which can be written

$$c' = K C_m + b'$$

c' is then defined as the specific respiratory activity of the sludge in the absence of any fresh substrate input.

The Wuhrmann curves (fig. 146) show that with a sludge loading of less than 1, the value of K is substantially constant, so that the curve expressing the relationship between sludge loading and specific respiratory activity is almost a straight line. In practice, this is the zone used because, beyond it, respiratory activity increases more slowly than the loading, thus tending to show that the stock accumulation potential is reaching a limit and that 'sorption' is no longer complete; purification efficiency therefore drops.

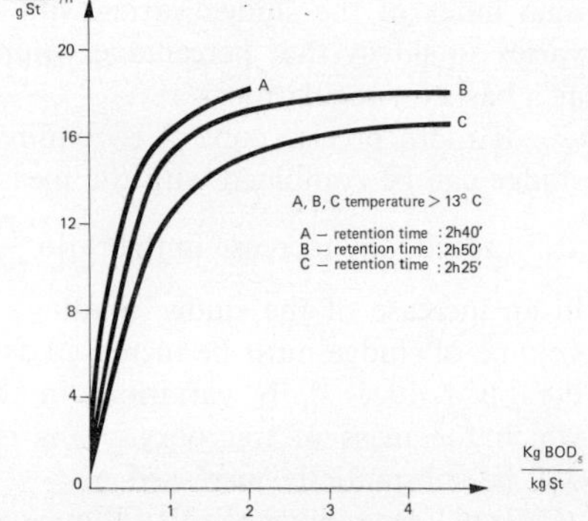

Thus, the specific respiratory factor c' of the sludge also represents the sludge loading, and thus the level of efficiency.

Fig. 146. —

A, B, C temperature > 13° C
A — retention time : 2h40'
B — retention time : 2h50'
C — retention time : 2h25'

The factor c' can be measured on the sludge taken from the aeration tank at a point where elimination of the substrate is known to be complete.

Lastly, when a device to control biological purification is installed at a plant, the ultimate aim is, of course, to be able to link the amount of sludge directly to the amount of waste matter entering the plant, this without intermediate measurement.

Measurement of the oxygen demand of the reactor cannot on its own represent the degree of pollution because of variations in the growth rate and respiratory activity of the bacterial flora in the tanks. Reactor oxygen demand can do no more than indicate a tendency. This is not enough for controlling the operation of a large plant. A more accurate measurement of the pollution load must be obtained with a COD-meter, or better still, with an industrial respirometer receiving a mixture of the effluent to be treated and activated sludge, which is held in a separate reactor in a very slow stage of development with a low sludge loading factor.

An industrial respirometer of this kind can be designed for separate measurement of rapidly assimilable pollution and pollution which is slowly degradable after extracellular or intracellular storage (suspended solids, colloids, large molecules). The data thus collected can be fed to a computer which can control the process of biological purification using a mathematical formulation of the Lefort-Jacquart type.

Such a direct measurement of the pollution load is of unquestionable value as a direct check on waste matter entering the plant and also a means of direct regulation in the two following cases:
— in a plant with a primary settling tank, the control of biological purification can then be anticipated more clearly;
— in a plant which is overloaded or is designed to treat only a fraction of the maximum loading, it is possible to assess the optimum amount of pollution which can be admitted while remaining compatible with the required degree of purification.

4.3. Practical applications

4.3.1. CONTROL OF OXYGEN INPUT

Signals from the **oxygen analyser** actuate the air blowers or the surface aerators. There are several alternatives, depending on the degree of accuracy required:
— the oxygen content of the tanks is held within a given minimum/maximum range;
— oxygen production is only varied if the dissolved oxygen content remains above or below the set limits for a specified minimum time;
— the quantity of oxygen introduced is increased or decreased in very small stages with a minimum possible interval between pulses. This eliminates the risk of 'hunting' and makes use of the regulating effect of the biological reactor.

In the case or blowers, the air flow can be varied in different ways:
— startup or shutdown of a standby feed unit;
— startup or shutdown of various single-or multiple-speed units in cascade;
— introduction or elimination of successive head losses on the delivery side of centrifugal blowers;
— continuous speed variation of blowers;
— continuous variation of blower feed guide vanes.

Fig. 148. — Gallery housing centrifugal blowers.   On the left is the control console.

With surface aerators, the following alternatives are possible:
— syncopated startup or shutdown with time delay relays;
— operation at different speeds;
— variation of blade submersion depth;
— continuous speed variation.

With compressed air oxygenation at great depth (over 3 m) it is normally possible to maintain high energy efficiency over a wide range of air flows. This is not the case with shallow depth oxygenation. The problems arising here are just as difficult with surface aerators, whose specific inputs vary widely according to the immersion depth and speeds used.

Furthermore, while the dissolved-oxygen content can be very closely regulated in complete mixing tanks because the oxygen is evenly spread throughout the liquid mass, the same is not true in the case of aeration tanks designed for stepped aeration or stepped feed.

It then becomes necessary to place a number of detectors in the successive compartments of the tank and to bring about changes in oxygen production on the basis of all the data received from these detectors. The energy gain obtained by regulation is therefore uncertain if the dissolved oxygen content varies too widely between compartments. The position can be improved by automatic equalization of the oxygenation rates, by progressive remote control of the inlet distributors for the water to be treated over the whole length of the tanks.

In the example illustrated in figure 149, which depicts a stepped-feed activated sludge tank, the flow of primary settled effluent is distributed by the three distributors A_1, A_2 and A_3, the liquor discharges over weir B, and the secondary return sludge is recycled to the beginning of channel 1. A number of different combinations for distribution of the flow of sewage can be programmed. If the influent loading varies in such a way that the deviation in the dissolved oxygen contents measured by probes S_1 and S_2 exceeds a fixed set value, the system switches automatically to the flow distribution combination which will give a deviation lower

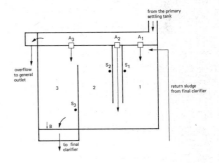

Fig. 149. —

than the set value. The overall air flow introduced is controlled automatically by the dissolved oxygen probe S_3 which actuates the air blowers.

4.3.2. CONTROL OF THE SLUDGE MASS

This is achieved by varying the extraction of excess sludge and the sludge return flow. The former is the more important because of the range of regulation which it allows.

A. Extraction of excess sludge

One of the following systems can be adopted, according to the nature of the effluent, the size of the plant, the required purification rate and the daily variation of pollution:

a) constant-flow extraction;

b) programmed variable-flow extraction.

These very simple methods can be combined as follows with measurement of the volume or weight of sludge:

c) constant-flow extraction, with a lower limit on the amount of sludge;

d) programmed and timed variable-flow extraction with upper and lower limits on the quantity of sludge.

Extraction can be linked directly to the amount of sludge present:

e) extraction of a constant amount of sludge at large city plants. This arrangement has the disadvantage that sludge is also extracted when pollution is at its peak;

f) programmed extraction of a variable quantity of sludge. The amount of sludge in the system is automatically varied according to the time of day.

None of the above methods provides an effective and immediate check on the sludge loading. This can only be achieved by incorporating the oxygen requirements of the bacterial flora. The following solutions are possible:

g) extraction with the quantity of sludge governed by the supply of oxygen. The amount of sludge in the tank (by percentage or by weight) is regulated to as constant a set ratio as possible with the oxygen requirement represented by the output of the oxygen generators (see fig. 150);

h) extraction with constant specific respiratory activity. In this case also, a minimum quantity of sludge is always retained in the system;

i) extraction governed by the pollution load entering the system. The mass of sludge is regulated by the degree of pollution.

B. Recirculation rate

By varying the recirculation flow it should be possible:

— to reduce the volume of sludge in the final tanks at peak times;

— to increase the volume outside peak hours.

But:

— a minimum retention time in the final tanks must be observed;

— recirculation flow must not be excessive at peak flow times because, for hydraulic reasons, the volume of sludge in process of settling can increase and the quality of the purified water will deteriorate.

The various alternatives are as follows:

— recirculation flow programmed over a period,

— constant or slightly-variable rate of recirculation, with limit on maximum recirculation flow;

— variable recirculation flow governed by intake flow.

4.3.3. CONTROL OF THE POLLUTION LOAD ADMITTED

Pumps or flow dividers controlling intake flow can be regulated by reference to the amount of waste matter entering the system. It may be necessary to treat:

— either a constant pollution load in the plant, by increasing or reducing feed flow according to the concentration of the raw sewage;

— or a pollution load varying with the self-purifying capacity of the receiving waters (expressed by their instantaneous flow). The pollution load to be eliminated is then automatically calculated and displayed on the 'pollution regulator';

— or a pollution load varying with the time of day.

Regulation of the pollution load entering the system can be combined with regulation of the amount leaving the system; or at least, automatic monitoring of the quality of purification is feasible.

This check can be used, for example, to govern recirculation of the purified water to the intake of the plant, or regulation of the flow entering with an upstream storage tank, or even a modification of the sludge loading.

Sludge concentration measuring units (Sedimometers).

Fig. 150.

Fig. 151.

4.3.4. CONTROL OF A PURE OXYGEN TREATMENT SYSTEM

The problem is complicated here by the need to maintain a minimum partial pressure of oxygen in the atmosphere of the reactor if the latter is closed, while at the same time allowing some leakage so as to prevent gradual enrichment of this atmosphere with carbon dioxide and, above all, with nitrogen.

The most commonly used systems control oxygen introduction on the basis of measurement of the total pressure of the gases in the atmosphere of the first reactor chamber, the leakage being controlled by measurement of the partial pressure of the oxygen in the final chamber.

The above systems can be supplemented, as in conventional air units, by having the aerators' operation controlled by the dissolved oxygen content. However the attention is drawn to the fact that, especially when pure oxygen only is used, mixing may be a more important consideration than oxygenation.

For safety purposes a hydrocarbon detector completes the reactor monitoring equipment.

WATER FILTRATION

1. GENERAL

Filters are designed to retain the particles contained in a liquid, either on the surface of the filter media or "in depth". Surface filtration requires a supporting media, whereas in-depth filtration is performed through a filter bed.

Surface filtration:
— through mesh media: by screening, straining, micromesh straining, filter presses or vacuum filters, etc.;
— through a thick porous supporting media: filters of sintered materials or cartridge-type filters;
— through a pre-coated supporting media: candle filters, frame filters, tray filters or drum filters, etc.

"In-depth" filtration through a single or multi-layer filter bed:
The theoretical aspect of these two main methods of filtration has been covered in Chapter 3, page 78.

In order to proceed from the theoretical aspect to the practical use of the most suitable filter for the operation concerned, it is well to be familiar with the various filtering mechanisms.

A. Filtering mechanisms.

Depending on the nature of the particles to be retained and the filtering material to be used, one or more of the following three mechanisms may be involved: deposit mechanism, fixation mechanism, detachment mechanism.

1. — Deposit mechanisms.

There are two main types:
— *Mechanical straining:* This retains all particles larger than the mesh size of the filter or the void between the particles already deposited, which themselves form a filter material.

The finer the mesh of the filtering material, the more marked this latter phenomenon will be: it is of little significance in a filter bed composed of relatively coarse material, but is of paramount importance in filtration through a fine-mesh medium: strainer, filter sleeve, etc.

— *Deposit on the filter material:* the suspended particle follows a current in the liquid; depending on its size in relation to the pores, it may be able to pass through the filter material without being retained. However, various phenomena cause its path to change and bring it into contact with the material.

The following phenomena may be identified:
— direct interception;
— diffusion by Brownian motion;
— attraction by van der Waals forces;
— the inertia of the particle;
— settling; particles may settle on the filter material by gravity, whatever the filtering direction;
— gyrational movement under the action of hydrodynamic forces;
— coagulation due to the action of enzymes (e.g. biological membrane in slow filters).

These deposit mechanisms occur mainly during the process of filtering in depth.

2. — Fixation mechanisms.

The fixation of particles on the surface of the filter material is promoted by a slow rate of flow, and is due to physical forces (wedging, cohesion) and to adsorption forces, mainly van der Waals forces.

3. — Detachment mechanisms.

As a result of the mechanisms referred to above, a reduction occurs in the space between the walls of the material covered with particles that have already settled. Consequently the velocity increases and the flow can change from laminar to turbulent. The retained deposits may become partially detached and be driven deeper into the filter material or even carried off in the filtrate.

The solid particles in a liquid and the colloidal particles that are flocculated to a greater or lesser degree do not have the same properties and do not react to the same extent to the above mechanisms. Direct filtration of a liquid in which the suspended matter retains its original state and electrical charge will therefore be very different from filtration of a coagulated liquid.

B. Clogging and washing of the filter material.

Clogging is the gradual blocking of the interstices of the filter material.

If a constant inlet pressure is maintained, the flow of filtrate will decline. To keep out-flow constant, the initial pressure must be increased as the filter becomes clogged.

The clogging rate depends on:
— the matter to be retained: the more suspended matter there is in the liquid, the greater the cohesion of this matter, and the more liable it is to proliferate, then the greater will be the clogging rate;
— the rate of filtration;
— the characteristics of the filter material: size of pores, uniform grain size, roughness, shape of the material.

The effect of these factors is studied in Chapter 3.

When the filter becomes clogged, it must be restored to its original condition by efficient and economic washing; the method used depends on the type of filter and the matter it retains.

C. Choice of filtering method.

Several different criteria govern the choice between the different types of surface filtration and filtration through a filter bed:
— properties of the liquid to be filtered, its impurities and the way they evolve with time;
— the quality of the filtrate to be obtained and the permissible tolerances;
— the quality of the mass of retained material when the object is to recover it;
— installation conditions;
— facilities available for washing.

The various methods involve different capital and running costs, which in turn depend on the prior conditioning of the liquid to be filtered, the washing method, the degree of automation and control, etc.

In selecting a filter, the possibility of easy, efficient and economic washing is as important as obtaining the best filtrate quality, since this quality will only be maintained if the washing process allows the filter material to remain intact.

2. SURFACE FILTRATION

Depending on the required conditions, this type of filtering can be carried out through a thin medium, a thick porous supporting medium or a precoated supporting medium.

2.1. Filtration through thin media

There are many types of filters using thin supporting media. They differ in their method of operation, and can be either open filters operating at atmospheric pressure or pressure filters.

The only filters that will be examined here are those providing fine filtration, as the other types are considered elsewhere in this handbook in connection with coarse screening of water or other liquids. This fine filtration is generally called "microfiltration" as opposed to "macrofiltration", a process in which particles larger than 150 micrometres (μm) are retained and "ultrafiltration", where the cut is between 0.4 and 0.004 micrometres (see Chapter 12, p. 361).

2.1.1. MICROSTRAINING AT ATMOSPHERIC PRESSURE

The primary objective in microstraining is to remove the plankton contained in surface waters. The process will, of course, also remove suspended matter of large size and plant or animal debris in the water.

Various control systems are used to regulate the speed of rotation of the drum, and one or more banks of washing jets are used, depending on the variations in the flow-rate and the clogging capacity of the water.

Optimum efficiency is obtained by maintaining a more or less constant head-loss resulting from partial clogging by the particles to be retained.

However, the plant efficiency will always be limited by several factors:
— the washed filter fabric does not carry an effective deposit at the start of the filtering cycle, and filtering is then limited to the size of the mesh alone.
— plankton elimination is never complete. The plankton can grow again, particularly when the temperature rises.
— certain very small eggs can easily pass through the filter fabric and hatch in the downstream tanks, where crustaceans visible to the naked eye may develop.
— because of the risk of corrosion of the microstrainer fabric or its supports, it cannot be used for continuous treatment of heavily prechlorinated water.
— microstrainers have to be fairly large to cope with peaks of plankton growth occuring several times a year. If too small, the output of the station could be significantly reduced during these peaks and during alluvial high-water periods.

Fig. 152. — Microstrainer filtering elements shown out of water.

Fig. 153. — Microstrainer filtering elements. Cleaning the filter fabrics.

The metal or plastic filter fabrics generally have mesh sizes of 20 to 40 micrometres, and exceptionally 10 micrometres. The smaller the mesh, the greater must be the straining area. Thus, with a mesh size of 35 micrometres, the filtration rate should, at the most be 35 m/h (14.3 gpm per sq ft) calculated over the total area of the strainer (50 m/h (20.5 gpm per sq ft) on the real submerged area) for a maximum clogging capacity of 10 (see analysis 712, page 756).

The reduction in the clogging capacity of the water by microstraining varies from 50 to 80 %, with a mean of about 65 %. For comparison purposes, a good settling tank gives a reduction of 80 to 90 % without prechlorination and 95 to 99 % with prechlorination.

The microstrainer must be considered only as a device of limited effectiveness, applicable to water containing few suspended solids. It has no effect on colour and on dissolved organic matter, and only removes the coarsest proportion of the suspended particles.

For the really effective disposal of plankton, clarification preceded by prechlorination is essential.

2.1.2. MICROSTRAINING UNDER PRESSURE

This is normally carried out with one of the following three types of filter:

● *Filters with stacked discs* in which filtering is performed by the clearances left between the discs. These filters are particularly sensitive to algae and fibrous matter which may cause matting and irreversible clogging.

For particle sizes down to 10 μm, filters are available with capacities of 10 to 100 m³/h (44 to 440 gpm). 250 m³/h (1 100 gpm) per filter can be attained for 150 micrometres.

● *Filters using trays, candles, basket or other filter elements covered with:*

— metal gauze, either plain or folded to increase the filter area. This type is particularly vulnerable to matting by fibrous elements and to the wedging of hard particles such as grains of fine grit. For mesh sizes of 2 to 40 micrometres filters are available with ontputs of 100 l/h to 150 m³/h (0.44 gpm to 660 gpm);

— metal or plastic wire wound on a frame, also affected by the wedging of grains of hard materials. A fineness of 3 micrometres and outputs of from 10 to 1 000 m³/h (44 to 4 400 gpm) with mesh sizes of 5 to 125 micrometres can be obtained with such filters;

— a specially shaped section wound on uprights. The disadvantages of the two previous types are reduced with this method. For mesh sizes of 80 to 125 micrometres, outputs ranging from a few m³/h to more than 5 000 m³/h (2 200 gpm) per filter can be obtained.

● *Filters using cartridges or thin filtering plates:* These cartridges or plates can be made of paper (folded or plain), made of cellulose or synthetic fibres, and can be discarded when they become clogged. This type of filter is often used as a finishing safety filter when the water must be permanently free from any matter that might escape the preceding treatments, e.g. for making beer, carbonated drinks, etc. Extremely fine particles of only a few micrometres, and even certain bacteria, can be removed in this way. On the other hand, a filter of this kind must not in normal operation receive water likely to have a relatively high content of suspended solids, as it would become clogged immediately.

2.2. Filtration through a thick porous medium

In this type of microfiltration, the medium provides not only surface retention, but also filtration in depth, though this depth is never very great.

Depending on circumstances, porosity is obtained by:
— yarn made from cotton, glass, polypropylene, etc., specially wound on a rigid core, with the porosity decreasing towards the filtrate outlet. The output ranges from a few gallons per hour to 500 m³/h(2 200 gpm) per filter for pore sizes from 0.3 to 75 micrometres;
— porous sintered products, metal, sand, porcelain, plastic, etc. The Chamberland filters make use of porcelain candles of very fine porosity giving a high headloss (several bars) for a small flow of water. When clogged, the candles can be removed and, in certain cases, steam-cleaned, but cleaning is never complete and irreversible clogging occurs quite frequently.

These sintered products can have pore sizes down to 5 micrometres and filters are available with outputs ranging from a few gallons per hour to 400 m³/h (1 800 gpm) for pore sizes of 40 micrometres.

2.3. Filtration through precoated supporting media

When large flows have to be treated by microfiltration without irreversible blocking of the filter element, precoat filters are used. Filtering is no longer carried out through fixed units, but through a filter material introduced into the appliance at the beginning of each operating cycle to form a microporous filtering layer on a fixed medium. When the filter becomes clogged, this precoat layer is removed by the washing operation. The texture of the precoat consists of a large number of very small-diameter channels, generally giving a filtrate of low turbidity.

2.3.1. CANDLE FILTERS

A. Description—operation.

Fig. 154. — Supporting plate equipped with candles.

DEGREMONT precoat filters consist of a cylindrical vessel with dished ends, fitted internally with a number of candles fixed to a support plate and arranged vertically. These candles are perforated hollow cylinders of stainless steel onto which a thin layer of synthetic fibre is wound in the form of a sleeve, to the outside of which the precoat layer is applied in the initial form of a dilute suspension.

The finest particle sizes for the precoat material are in fact smaller than the sleeve mesh sizes (this is necessary to prevent subsequent clogging) but these particles are nevertheless retained by the sleeve, since they become deposited on the arches formed between the fibres by the largest particles of the precoat layer. Care must be taken in applying this precoat. Before starting filtration, it is advisable to apply the precoat in closed-circuit operation in order to ensure that the fines which initially pass through the sleeve are subsequently deposited on the precoat as it forms.

When the filtrate runs clear, filtering can be started at a rate varying from 1 to 15 m/h (0.4 to 6 gpm per sq ft) depending on the type of treatment, on the nature of substances to be retained, on their concentration, on their behaviour when compacted and on the duration of the filtration cycle. The filtrate obtained is generally of high quality, although there are exceptions when the liquid to be filtered contains extremely fine colloidal particles. To remove these it is necessary to choose either a very fine precoat material at the risk of rapid clogging of the filter and short cycles, or a very adsorbent one.

Fig. 155. — Pre-coat filter.

The cycles can be lengthened or the filtration rate increased by starting with a smaller precoat and by injecting extra precoat material throughout the filtering cycle; the suspended solids deposited on the precoat do not then form a "cocoon" that is difficult to permeate but a porous layer, since they then mix with the added precoat material. This new layer then takes part in the filtering operation without a rapid increase in the loss of head.

Washing is carried out when the appropriate maximum degree of clogging is reached.

B. Washing.

If washing is to be effective, the suspended solids retained on the precoat material must be completely detached from all the candles; this can be done by ordinary air-water washing or the "Cannon" washing process that gives the DEGREMONT filter its name.

● **Ordinary washing.** After the filter is stopped and opened to atmosphere at the top, partial drainage allows an air cushion to form. The air inlet valve is then closed and this air is compressed at the pressure of the feed•pump. The valve at the base of the filter is opened sharply: the sudden decompression of the air drives out the water which passes through the candles from the inside and detaches the deposits. The filter is then drained and the candles rinsed.

● **"Cannon washing"** (DEGREMONT patent). As before, an air cushion is formed at the top of the filter; then, with the filter isolated, compressed air is

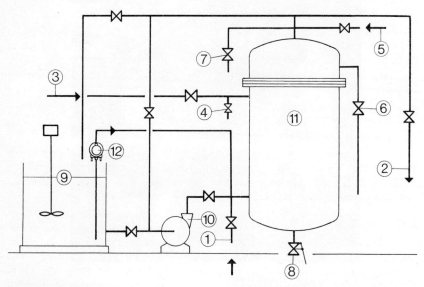

1. *Water to be filtered.*	8. *Filter drain valve.*
2. *Filtered water.*	9. *Preparation of the pre-coat media*
3. *Compressed air.*	10. *Pre-coat pump.*
4. *Decompression valve.*	11. *Filter.*
5. *Wash-water.*	12. *Dosing pump for continuous pre-coat*
6. *Bell drain valve.*	*maintenance.*
7. *Vent.*	

Fig. 156. — Diagram of a pre-coat filtration system with "Cannon" washing.

forced under the plate supporting the candles. This air drives a certain amount of water through the candles towards the top of the filter, while compressing the top air cushion. The air under the plate is then suddenly decompressed by opening the vent to the atmosphere: the water travels from the inside to the outside of the candles at very high speed. This "Cannon" effect instantaneously detaches all the deposits, which fall to the bottom of the filter. The operation is completed by drainage.

This type of washing allows very long candles to be used, rules out any risk of irreversible clogging and ensures minimal water loss.

C. Precoat materials.

Depending on the application, the main precoat materials used are diatomaceous earth, cellulose, powdered activated carbon and powdered ion-exchange resins.

● **Cellulose,** used in the form of high-purity fibres, has a filtering capacity comparable to that of a slow filter paper, but has only a very low adsorbent capacity. It is insoluble in cold or warm water, but starts to hydrolyse at 85 °C.

● **Diatoms** are fossilized siliceous shells of marine origin; they are very fine (5 to 100 micrometres) and have some adsorbent capacity. In the presence of water laden with colloids they provide better clarification than cellulose.

They are also able to adsorb emulsified impurities such as oils or hydrocarbons. The silica of the diatoms is slightly soluble in demineralized water.

● Because of its very high adsorbent capacity, **activated carbon** can be used on a supporting bed of cellulose or diatoms for colour removal and for almost complete elimination of organic matter of vegetable origin.

● **Powdered ion-exchange cation and anion resins,** mixed in varying proportions, provide filtration with thorough elimination of colloidal iron or complete demineralization of condensates in thermal and nuclear power stations. In this case, special arrangements must be made for coating after washing.

2.3.2. DISCS OR TRAY FILTERS

Filtration through a layer of precoat material can also be effected in disc or tray-type filters (fixed or rotating) arranged horizontally or vertically. The trays or discs are covered with a fabric supporting the precoat material. There are many different models, with plain or complex trays designed to obtain an even distribution of the deposits. Back-wash devices, using water alone, even if combined with movement of the filtering media supports, are not as efficient as washing with air and water.

3. FILTRATION THROUGH A FILTER-BED

3.1. General

Filtration through a filter bed is used when the quantity of suspended solids is large but their particle size is relatively small.

If this type of filtration is to be effective, the suspended solids must be able to penetrate the bed to a sufficient depth and not clog its surface. In addition, the grain size for the material or materials constituting the bed, and also the depth of the layer, must be carefully selected to ensure that the filtrate meets the quality requirements.

This filtration process can be used without prior treatment of the water, and is suitable for slow or rapid filtration of water when it is desired merely to reduce the concentration of suspended solids without acting upon the colour or organic matter content.

If optimum clarification of the water by rapid filtration is required, prior treatment by the addition of reagents, possibly combined with a settling process, will be necessary.

Any filter clogs up as its bed becomes laden with retained matter. When clogging becomes excessive or the filtrate quality begins to suffer, the filter bed must be washed. This washing operation must restore the original properties of the bed, or otherwise the filter would become less and less effective and the filter medium would have to be removed for complete cleaning or replacement.

A filter will only operate effectively if there is a perfect distribution through the filtering mass of the water to be filtered, of the wash-water and of the air (if used) utilized for washing. The method of collecting the filtered water and of distributing the wash-water (related to the type of support for the filter bed) is a factor of crucial importance for any filter.

Depending on the properties of the particles to be retained, it may be best to use one layer of varying depth of a uniformly sized material, two or more layers of different grain size (each level being uniform), or one or more layers of materials of completely heterogeneous and graduated grain size.

The effectiveness of a filter is also greatly affected by the method of controlling its throughput; where control facilites exist for individual filters, they must be secure against hunting phenomena, while the general control system for the station must not cause sudden surges in the flow to each filter at times when the general flow-rate is being changed or when washing operations are being carried out; otherwise the matter retained by the bed will pass rapidly through the filter, thus causing premature "break-through".

It will therefore be obvious that many conditions must be met by a filter if good filtration is to be achieved. There is no "universal" filter, but suitable types are available to deal with each of the problems to be overcome.

3.2. Slow filtration

The object of slow filtration is to purify surface waters without prior coagulation or sedimentation. The colloidal matter is coagulated by the enzymes secreted by algae and by microorganisms which are retained on the sand (biological membrane).

For satisfactory results three stages of filtration are generally necessary:
— coarse filters working at a rate of 20 to 30 m^3/m^2 per 24 hours (0.34 to 0.51 gpm/sq ft);
— prefilters working at a rate of 10 to 20 m^3/m^2 per 24 hours (0.17 to 0.34 gpm/sq ft);
— filters working at a rate of 3 to 7 m^3/m^2 per 24 hours (0.03 to 0.12 gpm/sq ft).

The slow filtration rate ensures a fairly low head-loss at each stage, and the filters are washed on an average once a month. Coarse filters and prefilters are washed more often according to the turbidity of the raw water.

Immediately after washing the quality of the filtrate is not satisfactory and the filtrate must be allowed to discharge to waste until such time as the biological membrane forms; this takes several days.

Slow filtration gives good clarification results provided that the water does not contain large quantities of suspended solids, and that a low final filtration rate is maintained. However, when the suspended solids in the water increase, coarse filters and prefilters are not sufficiently effective, and the turbidity of the treated water is likely to rise well above the values permitted by the appropriate standards unless the filtration rate is further reduced.

These filters are also particularly sensitive to a high plankton growth which may clog their surface.

Moreover, if slow filters are used for surface water with a high content of organic matter and chemical pollutants, the filtered water may still retain an unpleasant taste.

High hopes were once placed in slow filters as final or "polishing" purification plant. However, their biological action is not effective in eliminating all micropollutants (phenols, detergents, pesticides); they can, for example, only remove about 50 % of organochlorinated pesticides. Slow filtration is therefore a limited polishing process.

3.3. Rapid filtration

During rapid filtration, the water passes through the filtering layer at rates of from 4 to 50 m/h (1.6 to 20 gpm/sq ft). There is practically no biological action; at the most there is some nitrification in certain cases when the velocity is limited, when the oxygen content is adequate and when the nitrifying bacteria find favourable nutritive conditions in the water.

3.3.1. METHODS OF RAPID FILTRATION

The principal methods are:
— direct filtration, in which case no reagents are added to the water to be filtered;
— filtration with coagulation on the filter of water not previously clarified;
— filtration of coagulated and clarified water.

A. Direct filtration.

With surface filtration it is customary to define the quality of the filtrate by the percentage removal of particles according to their sizing. This criterion is not applicable to direct filtration, as it does not make allowance for the different filtration mechanisms which vary throughout the year depending on temperature, the size and type of particles, their concentration, their clogging capacity, their colloidal state, their content of microorganisms, etc.

However, when the quality of the water to be filtered is well known throughout the year, it is possible to predict a maximum content of suspended matter in the filtrate.

When the trend in the percentage by weight of the retained matter is observed, it is found that it is not constant when the quality of the liquid to be filtered varies; for example, it may range during the year from 50 to 95 %.

The length of the filtration cycle varies even more according to the abovementioned factors; it also depends on the possible accidental presence of matter liable to clog the filter in its topmost layers. In some cases clogging time variations of from 1 to 10 or even more may be noted, depending on the type of filter.

The widest range of filtration rates depending on the application is found with direct filtration. They commonly range from 4 to 25 m/h (1.6 to 10.2 gpm/sq ft), with peaks exceeding 50 m/h (20 gpm/sq ft) in some cases.

That is why a decision on direct filtration and its characteristics cannot be taken lightly on the basis of a single analysis and a single test on the liquid to be filtered. It is essential to know how it may vary throughout the year.

Only experience or a detailed study of the case can enable a correct choice to be made from the wide range of possibilities.

B. Filtration with coagulation on the filter.

Granular filtering materials do not retain colloidal matter; to obtain a perfectly clear filtered water, coagulation must be effected prior to filtration.

In most cases there is no question of a coagulant dose sufficient to ensure the complete neutralization of the electronegative charge of the particles in the water, since the resultant quantity of sludge would be too large and would rapidly clog the filters.

When the colour and suspended and organic matter contents are low, it is sufficient to add a small dose of coagulant (generally 2 to 10 g/m³ with a maximum of 15 g/m³ of aluminium sulphate for example), and if necessary a neutralizing agent to correct the pH and a flocculation aid. The coagulant dose determines the quality of the filtrate; the flocculant, when necessary, improves it slightly and its main purpose is to extend the length of the filtering cycle by increasing sludge cohesion and facilitating fixation of the sludge on the filter material grains. The filter mass, within which there is a laminar flow, acts as an excellent flocculator for the microscopic floc formed in this way. Too high a dose of the flocculation aid must be avoided, as it might cause surface clogging and consequently rapid blocking of the filter.

Cationic polyelectrolytes act both as coagulants and as flocculants, because of their positive charge and molecular structure. They may be regarded as supplanting both the coagulant (aluminium sulphate, iron(III) chloride) and the flocculant (activated silica or polymer, either anionic or non-ionic), and they have the advantage of generating a smaller volume of sludge.

Essentially, the filtration rate is linked to the pollution load of the water to be filtered and to the desired result. Where the aim is to retain reasonably long filtration cycles and guard against the eventuality of a temporary falling off in the quality of the water to be filtered, the rate is normally between 4 and 10 m/h (1.6 to 4 gpm/sq ft). This rate may, however, be very greatly increased when the water being treated is only slightly polluted or the standard of final turbidity permits. In treating the water for swimming pools, for instance, it is possible to work at rates of 40 to 60 m/h (16 to 24 gpm/sq ft) using very small doses of coagulant.

This process cannot, however, be applied universally, and must be restricted to water whose content is known to remain low throughout the year, and only slightly coloured (if colour removal is required) as otherwise it might become impossible to operate the plant when turbidity reaches a peak, which is precisely the case where effective purification is most necessary; the availability of a clarifier guards against such dangers.

When river water with a variable content has to be treated to supply certain industrial systems, it is sometimes possible to combine coarse filtration with coagulation on the filter.

C. Filtration of a coagulated and clarified water.

The floc resulting from total coagulation of the water is largely removed at the clarification stage; the water to be filtered contains only traces of floc, and floc cohesion depends on the reagents used. With good settling, the filters are ideally placed to receive water of almost constant quality and with a low content of suspended solids. Filtration then becomes the polishing safety treatment necessary when the water is intended for public consumption, sophisticated industrial treatment or the manufacture of high-quality industrial products.

The filtering rates depend on the required filtrate quality; they may range from 5 to 20 m/h (2 to 8 gpm/sq ft) depending on the quality of the clarified water and the type of filters used.

D. Wastewater filtration.

Two filtration techniques are applied to the tertiary treatment of wastewater:
— traditional filtration, in which physico-chemical methods only are employed to eliminate the pollution, and;

— biological filtration, in which physico-chemical purification is backed up by biological treatment aimed at reducing the soluble BOD.

The two types of filtration can be applied following biological or physico-chemical treatment.

● *Traditional filtration:* by means of this type of treatment, which brings about a limited measure of improvement in the effluent discharged from a biological purification plant, it is possible to achieve standards of 20 mg BOD_5 and 20 mg of suspended solids, or less, per litre of discharged water.

Traditional filtration is also the method applied to allow the recycling of treated effluent through some cooling circuits.

Where the aim is to bring about a marked reduction in the concentration of suspended solids contained in a good-quality, purified effluent, it is generally less costly to add a filtration stage than to scale up the biological purification (and especially the clarification) plant.

Filtration is carried out through a bed of homogeneous sand with an effective grain size of 1 to 2 mm at a rate of 10 m/h (4 gpm/sq ft) or more. Washing is by air injection and back-wash.

Given good-quality effluent from a biological purification plant, it is possible to expect a reduction of about 70 % of the suspended solids and about 40 % of the BOD_5

● *Biological filtration:* in this case, use is made of a filler material, such as Biolite, which has an open porous structure suitable for the anchorage and development of a bacterial film.

Aeration is carried out prior to filtration in order to promote the bacterial activity. As with the sand filter bed, filtering rates are in the 8 to 12 m/h range (3.2 to 4.8 gpm/sq ft), giving removal rates of 80 % for the suspended solids and better than 60 % for the BOD.

Fig. 157. — Tertiary filtration on Biolite. Sewage treatment plant at La Tremblade (France).

A further advantage of this method is that, for a given head-loss, it enables a greater amount of suspended solids to be retained per unit volume of filter material than can be achieved with sand having the equivalent effective grain size.

3.3.8. THE POROUS MEDIA

A. Physical properties.

A filtering material is generally defined by the following factors; precise definitions and methods of measurement will be found in Chapter 27, page 939:
— **grain size,** defined by a curve representing the percentage by weight of the grains passing through the meshes of a series of standard sieves;
— **effective size:** corresponding to the percentage of 10 in the above mentioned curve and—in combination with the following two factors—enabling the filtrate quality to be predicted to a large extent;
— **coefficient of uniformity:** this being the ratio of sizes corresponding to the percentages 60 and 10 in the above-mentioned curve;
— **grain shape:** this being either rough (crushed material) or smooth (river and sea sand).

To obtain a water quality similar to that obtained with a given round grain size, a smaller effective size must be used when working with coarse grains.

Given an equal size, the head-loss increase is less with coarse grains than with round grains for, contrary to what might be expected, coarse grains bed down less easily than round grains, and leave larger spaces for the water to pass through;
— **friability:** the friability test allows suitable filter materials to be selected without the risk that the washing operations will produce fines. Its importance varies according to the mode of operation of the filter.

The friable material is generally unacceptable, especially when the filtration direction is downwards, and when the filter is washed with water only, as the fines formed clog the filter surface;
— **loss by acid attack:** obviously a high loss caused by acid attack cannot be tolerated when the water is likely to contain corrosive carbon dioxide gas;
— **the density of the grains** making up the filtering media;
— **their apparent bulk density in air and water** which provides a measure of the volumes occupied in air and water by a given mass of the material.

There are other properties specific to adsorbent materials such as activated carbon; they will be examined when the use of these materials is discussed.

B. Nature of the porous media.

Quartz sand was the first material used for filtration, and it is still the basic material in many existing filters.

Anthracite or marble can be used instead when any trace of silica must be avoided in industrial processes or when they are easier to obtain.

For some methods of treatment such as final purification, tertiary treatment of effluents etc., it is expedient to use materials with a large specific surface area, e.g. expanded schists, Biolite, pozzolana or other similar material.

In multi-layer filters, sand may be combined with anthracite, garnet, schists of varying porosity, etc. provided that these materials have low friability and suffer little loss by acid attack.

Finally, filtering may be effected through physically-strong activated granulated carbon in the following cases:

— to replace sand after clarification treatment to eliminate the residual floc and to combat pollution by adsorption at the same time;

— in the second filtering stage for "polishing" treatment only or for dechlorination.

Numerous experiments have shown that under the same operating conditions all filtering materials that are not porous in themselves and do not react chemically with the water to be filtered or the matter dissolved in the water behave in the same way if they are of the same effective size and the same shape; the filtering output is the same and the quality of the filtered water is identical.

However, if an adsorbent material such as activated carbon is used, certain dissolved substances are additionally adsorbed, and the quality of the filtered water is no longer the same from the chemical aspect.

C. Choice of grain size for a single filter layer.

This choice must be made while also considering the depth of the layer. Assuming suitable depths, the fields of application of the different effective sizes can be broadly outlined as follows, for a coefficient of uniformity generally comprised between 1.2 and 1.6 or even 1.8:

● *Effective sizes from 0.3 to 0.5 mm:* very rapid filtration under pressure up to 25 m/h (10 gpm/sq ft) and even 50 m/h (20 gpm/sq ft) for swimming-pool water. Direct filtration of water with a low impurity content. Filtration in mobile installations, with coagulation on the filter, of raw water having a turbidity of less than 100 mg/l of silica. The head-loss may attain several bars. Clogging, generally fairly rapid, naturally depends on the filtration rate and the depth of the bed. Cleaning must be effected by water alone at a sufficient rate to cause expansion of the filter bed.

● *From 0.6 to 0.8 mm:* filtration wihtout prior clarification, with or without coagulation on the filter, provided the water is not very dirty (turbidity less than 50 mg/l of silica).

Filtration of clarified water at a limited rate (7 m/h or 2.8 gpm/sq ft)) in open filters and at higher rates in closed filters where the head-loss can be higher; it is not generally possible to exceed 0.6 bar with a clogged filter without spoiling the quality of the filtered water. This size of filtering material can be used in a heterogeneous layer resting on supporting layers and washed by water only; it can also be used in a uniform layer on a false floor with wash-water and air scour.

● *From 0.9 to 1.35 mm:* this standard grain size in Continental Europe is used in a homogeneous layer to filter clarified water or water with a low turbidity with coagulation on the filter; maximum head loss is 0.3 bar. It is also used for direct filtration of water not containing large quantities of suspended solids and intended for industrial use. This grain size is ideally suited for false floor filters washed with water and air, and allows filtration rates of up to 15 and 20 m/h (6 to 8 gpm/sq ft) depending on the required filtrate quality.

● *From 1.35 to 2.5 mm:* for the removal of coarse contaminants from industrial effluents or for tertiary effluent treatment. Used as a supporting layer for 0.4 to 0.8 mm material.

● *From 3 to 25 mm:* used almostly exclusively for supporting layers.

D. Depth of a single filtering layer.

With a filter media of a given grain size, if the depth of the filtering layer is gradually increased, it is found, after the filter "matures", that the turbidity of the filtrate decreases until at a certain depth it reaches a stable value that shows no further improvement even if the depth is increased. This depth defines the minimum layer to be used with a clean filter to obtain the best filtrate according to the grain size of the material used. It gives a minimum head-loss for the filtration rate considered.

To leave an acceptable clogging margin, the depth of the layer must be increased; the time during which a clear filtrate is maintained (time t_1, see fig. 158) is proportional to the depth of the layer.

In order to benefit from the clogging of almost the full depth of the layer, there must be a head-loss such that the time t_2 taken to reach it is little shorter than the time t_1. This head-loss value represents the maximum limit beyond which the filtering mass would rapidly "break through". The finer the sand, the higher is this maximum value.

As a general rule it may therefore be said that the finer the sand the shallower need be the depth of the filtering bed, but also the higher the mean and maximum loss of head. These considerations are easily interpreted by means of standard clogging curves. Let us take the example of a filter which by design has a maximum head-loss in the sand of $P_2 = 1.50$ m (5 ft) on a column of water.

The top curve in fig. 158 shows the variation of the head-loss with time; the maximum head-loss P_2 is attained after a time t_2.

The bottom curve shows the variation in turbidity with time; if the maximum tolerated turbidity is "e", it is reached after a time $t_1 < t_2$. This shows that the filter is badly designed and that is bed is of insufficient depth to provide the maximum head-loss; the depth should be increased until t_1 is greater than t_2.

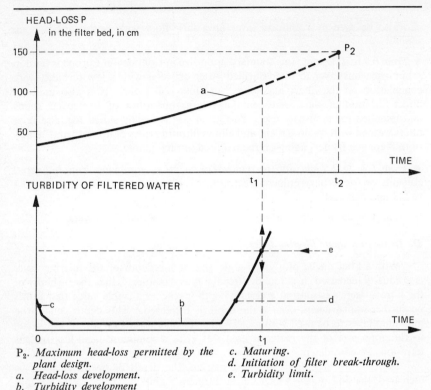

P$_2$. Maximum head-loss permitted by the   c. Maturing.
   plant design.   d. Initiation of filter break-through.
a. Head-loss development.   e. Turbidity limit.
b. Turbidity development

Fig. 158.

The graphs in fig. 159 represent an open filter, with a sand depth BD and a water depth AB. The ordinate of the right-hand graph represents the levels of the pressure take-offs A, B, C, D measured from the floor D of the filter, and the abscissa show the pressures represented in depth of water on the same scale as the ordinate. Thus at point B of the filter, at the top of the filter bed, the pressure is always equal to the water depth AB, plotted at B'b. At point C of the filter bed, the pressure takes the value AC, plotted at C'c$_0$, when the filter is shut down. Likewise the static pressure at floor level is equal to AD plotted at D'd$_0$. All the points representing the static pressure at different levels of the filter are on the 45° straight line A'd$_0$.

With the filter in operation, the head-loss in clean sand is, according to Darcy's law, proportional to the depth of the sand and to the flow-rate, taken as constant for this purpose. The pressure at point C of the filter becomes equal to C'c$_1$, with the value c$_0$c$_1$ representing the loss of head of the sand between

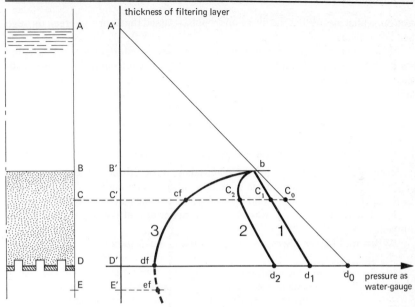

1. *Curve representing pressures in clean sand.*

2. *Curve representing pressures during clogging.*

3. *Curve representing pressures after filter break-through.*

Fig. 159. — *Downward filtration (distribution of pressures).*

levels B and C; likewise, at floor level, the pressure at D becomes equal to $D'd_1$, the head-loss in clean sand being d_0d_1.

The line bc_1d_1 is a straight line since c_0c_1 and d_0d_1 are proportional to the depth of sand (Darcy's law).

When the sand is completely matured, the plotting of the pressures $C'c_2$ and $D'd_2$ at the various levels of the sand gives the curve bc_2d_2 representing the pressures in the filter; it has a curving section and a straight section parallel to the straight line bd_1 representing the head-loss with a clean filter. Point C_5, showing the start of the linear head-loss, indicates the level C reached by the impurities in the sand; below C, as the head-loss is linear, the sand is clean. The point c_2 defines the depth BC of the **"filtration front"** at the time concerned.

The minimum sand depth and minimum head-loss anticipated before clogging are therefore BC and c_0c_2 respectively.

The shift of point c_2 during clogging represents the advance of the filtration front. In fig. 159, where the filter no longer gives clear water once the maximum head-loss P_2 is reached, the curve representing the pressures at different points in the filter is given by $bc_fd_fe_f$; it reaches the floor without having a straight section, which means that the filtration front has passed the floor and break through has taken place.

If a filter with a greater depth of sand had been used, the curve representing the pressure at the different points of the filter for the maximum available head-loss would have become linear at point e_f: this immediately gives the minimum depth DE of sand that should have been added to make $t_1 = t_2$.

Finally, experience shows that the values of t_1 corresponding to different depths of a specific sand are reasonably proportional to the corresponding thicknesses.

E. Grain size and depth of material in multi-layer filters.

The effective grain sizes of the materials making up the two or three layers of a multi-layer filter must be in a certain ratio to one another, and this ratio itself depends on the nature and size range of the particles to be retained, the respective densities of the grains of filtering material and the washing technique applied.

When multi-layer filters started to be used, much importance was attached to effecting a clear-cut separation between the individual layers. However, a certain degree of interpenetration of the layers is not harmful, although it is important that the coefficient of uniformity of each layer should be as small as possible (1.5 maximum) in order to prevent very fine grains being concentrated at the surface by the hydraulic grading effect of the washing operation. Such grading would result, for each layer, in the same troublesome clogging of the surface as is observed in single-layer filters washed either with water alone or first with air then with water.

In the case of two-layer filters, the effective size of the sand used for the bottom layer is normally between 0.4 and 0.8 mm, while that of the anthracite or pumice varies between 0.8 and 2.5 mm.

The depth of the layer depends on the previous treatment applied to the water to be filtered, the filtration rate and on the nature and quantity of the particles of floc to be retained.

As a first approximation, we may say that, other things being equal, the overall depth of the two layers corresponds to about 70 % of the equivalent depth of a single layer made up of homogeneous material which remains so after washing.

If full advantage is to be taken of a two-layer filter, it should be provided with 1/3 sand and 2/3 anthracite, or some other material lighter than sand.

Where filters comprising more than two layers are concerned, everything depends on the arrangement of the three, or even four or five layers. Clearly, the greater the number of layers, the more critical does the choice of media become and greater is the importance attaching to the method of washing.

3.3.3. THE USE OF POROUS MEDIA

A. Downward filtration through a single heterogeneous layer.

For the treatment of drinking water, use is occasionally made of sand filters with an effective grain size of about 0.55 mm and washed by a reverse current of water at a velocity of 35 to 40 m/h (14 to 16 gpm/sq ft) to bring about the necessary expansion of the layer. After washing, the sand which was homogeneous throughout when first charged into the filter is found to have undergone hydraulic grading, with 0.3 mm fines at the surface and 0.9 mm grains at the bottom of the layer.

The filter material is thereby rendered heterogeneous and unconducive to the use of the full bed depth. What happens is that the retained impurities are arrested within the first few centimetres where they set up very large local head-losses which are likely to shorten the filtration cycle and cause degassing of the water by lowering the pressure to below atmospheric level.

Figure 160 shows graphically the pressure pattern inside a 0.60 m deep filter bed surmounted by 1.50 m of water and with a 2 m head-loss in the sand due to clogging.

The shaded portion denotes sand which is under vacuum at the end of the cycle, only the depth BC of the layer becoming clogged.

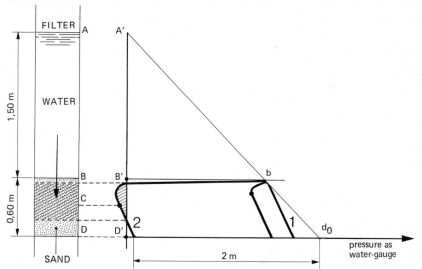

1. *Curve representing clean sand pressures.*   2. *Curve representing sand depth clogged*
BC. *Depth of clogged sand.*   *under a 2 m (7 ft) WG head-loss.*

Fig. 160. — Downward filtration through a single heterogeneous layer (distribution of pressures). The cross-hatched portion of sand is under vacuum.

Filters with homogeneous layers of larger grain sizes and multi-layer filters were developed to overcome these disadvantages.

B. *Downward filtration through a uniform layer.*

Uniform-layer filters are those in which the effective grain size of the filtering material is uniform throughout the whole depth of the bed, both initially and after washing. Filters of this type are washed simultaneously with air and water, and are then rinsed without expansion of the filter medium.

In the first washing stage, the backwash is combined with the air scour and there is no expansion of the filter bed; on the contrary, there is often a certain degree of sand compacting when the backwash flow is small; the air ensures complete agitation of the sand which, after the air scour, is completely uniform, as it was originally. During the second rinsing phase (intended to remove from the filter the dirt already extracted from the sand and collected in the surface water), there is virtually no expansion; this is desirable in order to prevent hydraulic grading of the sand that has already been uniformly blended in the preceding stage.

During filtration, therefore, the impurities penetrate deep into the sand nstead of clogging the surface as is the case with filters with a heterogeneous or non-uniform bed. In addition, the use of a coarser sand reduces the risk of the formation of a vacuum.

The graph in fig. 161 shows the pressure curve for an open Aquazur filter, type V (see page 291) operating at a filtration rate of 15 m/h (6 gpm per sq ft) with 1.50 m (5 ft) depth of homogeneous sand of an effective size of about 0.95 mm, and with 1.20 m (4 ft) of water above the sand; the head-loss reserved solely for the clogging of the sand is 2 m (6.56 ft).

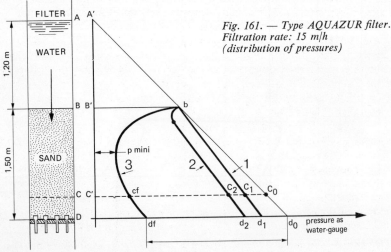

Fig. 161. — Type AQUAZUR filter.
Filtration rate: 15 m/h
(distribution of pressures)

1. *Curve representing pressures in clean sand.*
2. *Curve representing pressures during clogging.*
3. *Curve representing pressures with clogging under a 2 m (7 ft) WG head-loss.*

The curve bc_fd_f represents the state of the pressures in the filter bed with a clogged filter; at no point does the pressure drop below atmospheric pressure. Point C represents the final advance of the filtration front, which leaves some 30 cm (1 ft) of clean sand, a necessary safety margin.

C. Filtration through a multi-layer filter bed.

This filtration can be performed in either direction, downwards or upwards, the purpose always being to avoid the shortcomings inherent in filters with a heterogeneous filter layer, i.e. the likelihood of surface clogging and restricted filtering rates.

● *Downward filtration:* to increase the filtering rate and prolong the operating time of these filters, part of the fine sand is replaced by a lighter material with an effective grain size greater than that of the sand beneath. This lighter material is normally anthracite, but it may also be a porous schist, a volcanic or plastic medium or similar.

The effective grain size of the anthracite is 2 to 3 times greater than that of the sand.

The idea, of considerable antiquity having been patented in 1880 by Smith, Cuchet and Monfort, is that the impurities should penetrate throughout the whole body of the coarser upper layer, leaving the bottom layer to perform a finishing and safety function. For some kinds of raw water, each of the layers can also act to retain suspended solids of a size corresponding to that of the filter media employed.

The grain size of each of the layers is chosen in such a way that they undergo similar expansion with the same flow of wash-water, thus enabling them to be regraded prior to the recommencement of filtration.

The uniformity coefficient of the various media (cf. page 259), and especially that of the material constituting the upper layer, must be as low as possible (not more than 1.5) to prevent the surface of each layer from becoming clogged with impurities.

The velocity of the backwash should increase in proportion to the grain size and the temperature of the water. It must be possible to expand each of the layers by at least 10 to 15 %.

In some instances provision must be made for adapting this velocity to the temperature of the water so as to maintain permanently an adequate degree of expansion without the risk of filter material being lost to the drain.

The complete equipment of a filter of this kind includes a device for monitoring the water level prior to washing with air alone as well as apparatus for monitoring and controlling the flow of wash-water. The edges of the channels used to recover the wash-water can be shaped so as to minimize the losses of light media.

Filters comprising 3 and 4 layers also exist and these improve the in-depth penetration of the impurities, although this imposes a variety of conditions affecting grain sizes and the washing technique applied.

The adoption of this type of multi-layer filter has in fact enabled filtration rates and cycle durations to be increased when compared to the filters with a heterogeneous layer of fine sand. However, it has not eliminated the washing disadvantages inherent in these filters.

Again, owing to the loss of material to the drain during washing, it is necessary annually to replace 5 to 7 % of the medium making up the top filter layer.

Compared with the filter with a single uniform layer of sand, the advantages of the multi-layer equipment begin to dwindle. The fact of the matter is that a filter with a single uniform layer can operate at the same filtering rates, with the same cycle duration and with the same final head-loss, and this by using a uniform and slightly finer grade of sand than the anthracite of the double-layer equipment combined with a greater depth to allow the retention in depth of the same quantity of impurities per m² of filter area. Where washing is concerned, the advantage is on the side of the single-layer sand filter, washed with water and air, in which the existence of mud-balls is unknown and where the loss of sand is very slight.

On the other hand, the multi-layer filter is particularly attractive for the retention of very fine precipitates in the form of floc of low cohesive strength, such as is encountered in a number of processes for the removal of iron and manganese.

● *Upward filtration:* in this system, filter beds of grain size decreasing from the bottom to the top are used, once again with the purpose of enabling the impurities to penetrate deep into the bed so as to make maximum use of the filter mass and increase the length of the filtration cycles.

The first problem that arises with this type of filter, which is generally washed by water and air, is the need to ensure uniform distribution of the flow to be filtered; this is done by passing the water through orifices or nozzles not liable to become clogged by the impurities and the plankton growth in the raw water.

In addition, the sand is subjected to a lifting action by the water passing through it upwards from the bottom; this action increases with the loss of head, and creates local areas of expansion of the fine sand in the top part of the filter bed, which breaks through for a few minutes before resuming its filtering role To overcome this disadvantage, efforts have been made to stabilize the fine sand by embedding at the top a horizontal grid consisting of flat bars set on edge.

This grid system does not completely eliminate the undesirable sudden expansion that occurs mainly whenever the filtration rate is accelerated as a result of rapid or large increases in the flow to be filtered. To reduce these disadvantages further, greater sand depths should be used, merely to provide weight. In addition, rate of flow measurement devices and metering of the filtered volume on each filter must be provided so that the need for washing can be determined on the basis of this volume rather than the loss of head; strict operating instructions must be given to ensure that rates of flow are only varied slowly.

All these considerations rule out the use of this type of filter in practice if filtering results are to be really reliable.

● *Reverse-current filtration:* another way of stabilizing the fine sand layer is to discharge the filtered water through a submerged header pipe located near the centre of the finer top layer and to introduce above this layer the prefiltered water from the coarser bottom layer; this means that a pressure is applied above the sand equal to the pressure tending to lift it (see fig. 280).

With a non-uniform sand layer, clogging occurs chiefly at the level corresponding to the size of the impurities, a zone of very restricted depth, since below it the sand is too coarse to filter efficiently and above it the sand is too fine. Moreover, the flow filtered in the top zone is very small, since all the fines collect on the surface.

Only with a uniform sand, therefore, does a dual-flow filtration system make it possible to obtain regular penetration in depth by the matter to be retained. This sand then provides proper filtration in both directions within the homogeneous top layer, almost doubling the flow filtered by a single filter.

Finally, the washing system must be carefully designed so as to carry all the sludge concentrated in the bottom coarse-filtration layer up through the top sand layer and so as to maintain the uniformity of this layer in order to prevent the fines coming to the surface and blocking the downward filtration.

3.4. Washing

Washing is an extremely important operation, which, if inadequate, leads to the permanent clogging of some areas and to a consequent reduction in the section available for the passage of the water. The head-loss increase becomes more rapid, and filtration is locally accelerated and less effective. The filter bed can then become a focus for the proliferation of microorganisms which are harmful to the quality and taste of the water.

3.4.1. METHODS OF WASHING

The filtering material is washed by a current of water, generally flowing from the bottom upwards, the purpose of which is to dislodge the impurities and convey them to a discharge channel. Simultaneously, the filtering material needs to be agitated in the current of water. A number of methods can be employed to achieve this result.

A. Washing with water alone to expand the filter bed.

The current of water must be sufficient to expand the filtering material, i.e. to bring about an apparent increase in its volume of at least 15 %.

As the viscosity of water varies according to temperature, it is desirable that a system should be provided for measuring this and for regulating the flow of wash-water so as to keep the degree of expansion constant over time.

The expanded layer then becomes subject to convection currents; in certain zones the filtering material moves downwards and in others upwards, which means that portions of the compact layer of sludge encrusting the filtering material surface are carried deep down to form hard and bulky mud balls as a result of the whirling action of the currents.

This is partly overcome by breaking up the surface crust with powerful jets of high-pressure water ejected from fixed or rotating nozzles (surface washers). This method calls for considerable care and makes it necessary to measure expansion of the filtering material exactly. Its greatest drawback is that it results in a size grading, concentrating the filtering material fines on the surface, and is therefore unsatisfactory for downward filtration.

B. Simultaneous air and water washing without expansion.

A second method, now becoming more widespread, is to use a backwash velocity which will not cause expansion of the sand, and at the same time to disturb the sand by air scour. The sand thus remains stable, and the surface crust is completely broken up by the air; in this way, mudballs do not have a chance to form; they are in fact unknown with this type of washing process.

During air scour, the wash-water flow-rate can be varied over a wide range, but must not fall below a figure of 5 m³/h per m² (2 gpm per sq ft). The higher this flow-rate, the more rapid and effective will be the washing. The maximum figure will depend on the material and the filter parameters.

When the impurities have been removed from the filtering material and collected in the layer of water between the sand and the sludge channel, "rinsing" must take place, i.e. the layer of dirty water must be replaced by clear water.

Rinsing may be carried out by various methods after the air scour has stopped:
— continue the backwash at a constant rate of flow until the discharged water runs clear. The time this takes is inversely proportional to the rate of flow of water (which must never drop below 12 m³/h per m² (4.8 gpm per sq ft) and proportional to the depth of the layer of water above the filtering material;
— increase the rate of flow of water during rinsing to at least 15 m³/h per m² (6 gpm per sq ft);
— cleanse the surface of the filter with a horizontal current of raw or clarified water combined with the backwash;
— drain off the dirty water above the sand and cleanse the filtering material surface as above.

C. Washing with air and water in succession.

This method of washing is used when the nature of the filtering material is such that it is impossible to use air and water simultaneously without running the risk that the wash-water will carry off the filter media to the drain. This applies to filter beds composed of fine sand or low-density materials such as anthracite or activated carbon.

In the first stage of the washing operation, air is used by itself to detach the retained impurities from the filtering material. In the second stage, a back-wash of water with a sufficiently high flow-rate to bring about the expansion of the filtering material(s) enables the impurities detached during the first stage to be removed from the bed and carried away.

In the case of impurities which are heavy or particularly difficult to remove, this sequence may be repeated several times (pulsation washing).

D. Washing wastewater filters.

The washing of these filters presents special features only in the case of biological filtration through expanded materials, where the presence of organic sludge, which is more difficult to eliminate than inorganic impurities, may necessitate repeating the washing sequence several times. The flow-rates of the air scour and washwater must be adjusted to suit the filtering materials.

The washwater containing the impurities is then recycled back upstream in the plant for further treatment.

3.4.2. FREQUENCY OF WASHING

The frequency of washing depends on the nature of the water to be filtered. In practice, the loss of head is taken as a criterion, and washing is carried out as soon as it reaches a certain limit, incorrectly called maximum clogging. In fact, this loss of head depends on both clogging and flow. This method permits a check to be kept upon the clogging of the filter only if the flow is constant.

However, it is not necessary to consider the flow provided that the maximum flow is not exceeded and does not vary too much during the cycle; a maximum head-loss is determined on the basis of the required quality provided it remains within the prescribed limits.

If the rate of flow is essentially variable, the best method is to arrange washing after filtration of a certain volume of water determined on the basis of the quality obtained at the end of the cycle under running conditions.

3.4.3. CONSUMPTION OF WASH-WATER

The amount of washwater consumed depends essentially on the character and weight of the particles retained per cubic metre of filtering material. The combined use of air scour and clarified water makes it possible to reduce water consumption by some 20 to 30 % as compared with washing with water alone.

The consumption of washwater is greater:
— the deeper the layer of water above the filtering material;
— the lower the flow-rate of the backwash (water alone);
— the greater the distance separating the individual sludge discharge channels;
— the larger the quantity of sludge to be removed; and
— the greater the cohesion and density of the sludge.

Water consumption is also increased by high-pressure surface washing.

3.4.4. CHOICE OF NOZZLES FOR WASHING RAPID FILTERS

Nozzles for floor fixing are available in two types according to the washing method:

● **nozzles for washing with water alone:**

These nozzles can be distinguished by their shape, slot width and the material from which they are made.

● **nozzles for wash-water and air scour:**

— either the air is distributed by a branched perforated header under the floor, in which case D.13 nozzles (fig. 162) discharge the water-air mixture;
— or the air is distributed by an air cushion; in the latter case longstem nozzles figs. 164 to 167) specially designed for the purpose, provide perfectly uniform distribution of the air and water.

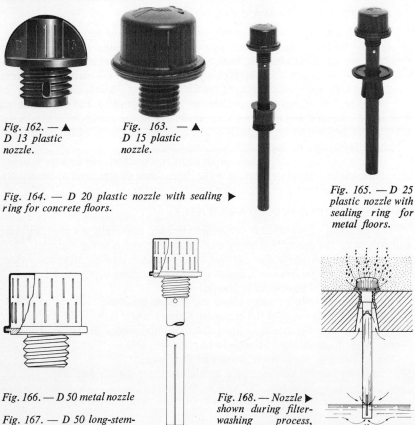

Fig. 162. — ▲ D 13 plastic nozzle.

Fig. 163. — ▲ D 15 plastic nozzle.

Fig. 164. — D 20 plastic nozzle with sealing ▶ ring for concrete floors.

Fig. 165. — D 25 plastic nozzle with sealing ring for metal floors.

Fig. 166. — D 50 metal nozzle

Fig. 167. — D 50 long-stem-med metal nozzle ▶

Fig. 168. — Nozzle ▶ shown during filter-washing process, using air and water.

Fig. 168 is a cross-section of a long-stem nozzle set in a concrete floor, shown here during a washing cycle using air and water.

This nozzle has a head with slots fine enough to prevent the ingress of the filter material, and a stem consisting of a tube with a hole at the top and a slot near the bottom.

The air forced under the floor forms a cushion which, once formed, passes through the holes and slots in the nozzles and provides a good air-water mixture evenly distributed over the entire filter area.

This particularly efficient cleaning system means a saving in water. It can be used successfully even with filter beds exceeding 2 m (6.56 ft) in depth.

3.4.5. QUANTITIES OF SUSPENDED SOLIDS WHICH CAN BE REMOVED BY FILTRATION

The suspended solids lodge between the grains of the filter material. Since sufficient space must always be left for the water to percolate, the sludge must not, on average, fill more than one quarter of the total volume of voids in the material.

Since, irrespective of grain size, one cubic metre of filtering material contains about 450 litres of voids, the volume available for the retention of particles is about 110 litres, provided that the effective grain size of the filtering medium is suited to the nature of the particles.

When the suspended solids are based on colloidal floc, their dry matter content does not exceed 10 g/l; the quantity that can be removed per m^3 of filter material is therefore no more than $110 \times 10 = 1\ 100$ g.

This figure is increased when the floc contains dense mineral matter (clays, calcium carbonate). For a sludge containing 60 g/l dry matter it can be:

$$110 \times 60 = 6\ 600 \text{ g}$$

Finally, much higher values can be obtained in the filtering of industrial impurities (rolling mill scale, barium and lead salts, etc.).

Fig. 169. — Four of a total of eight triple 3 m diameter towers used for the removal of rolling-mill scale. Total output: 3,000 m^3/h (19 mgd). The SIDELOR steelworks at ROMBAS (France).

These values indicate the maximum permissible content of suspended solids in the raw water entering a filter once its filtration rate and the length of the cycle between two washing operations have been determined.

For example, a filter with a bed one metre (3 ft 4 in) deep operating at a rate of 10 m/h (4 gpm per sq ft) and requiring washing every eight hours (80 m³ water per m³ filter bed (600 gal per cu ft) between washing operations) cannot cope with more than $\dfrac{1\,100}{80} = 13.75$ mg/l flocculated suspended solids

or $\dfrac{6\,600}{80} = 82.5$ mg/l mineral suspended solids.

For suspended solids in river water, the figure will be midway between the above two values.

4. PRESSURE FILTERS

These filters are usually made of metal.

4.1. Vertical filters washed by water alone

These filters contain filter materials the grain size and density of which must be compatible with the back-wash rates of flow that must be provided to expand them. The filter layer is supported by successive beds of materials that become progressively coarser from the top to the bottom and the filtered water is collected by a branched perforated header embedded in the coarsest layer.

These filters are generally loaded with a single layer of sand or anthracite.

1. *Filter body.*
2. *Filter material.*
3. *Collection system.*
4. *Raw water inlet.*
5. *Filtered water outlet.*
6. *Air vent.*
7. *Drain cock.*
8. *Hand-hole.*
9. *Wash-water outlet.*
10. *Rewash if required.*

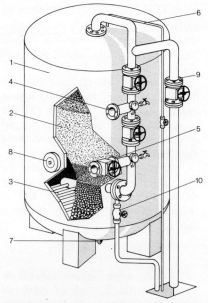

Fig. 170. — *Closed pressure filter with water backwash.*

However, multiple layers may be used: e.g. fine sand topped with coarser anthracite. Depending on the grain size of the filter bed, the filtration rate may vary from 5 to 50 m/h (2 to 20 gpm per sq ft).

For this type of filter the maximum loss of head at the end of the cycle depends primarily on the fineness of the filter layer and the filtration rate; it may vary from 0.2 to 2 bars.

The wash-water rate of flow, also dependent on the grain-size, must cause expansion of the filter bed so that its depth increases by 15-25 %.

For a sand filter bed, the rate of flow is given by the following table:

Rate of flow	25 to 35 m/h 10 to 14 gpm per sq ft	40 to 50 m/h 16 to 20 gpm per sq ft	55 to 70 m/h 22 to 28 gpm per sq ft	70 to 90 m/h 28 to 36 gpm per sq ft
Effective grain size	0.35 mm	0.55 mm	0.75 mm	0.95 mm

The wash-water flow-rate can easily be checked (an essential procedure) by fitting a flow measuring weir to the sludge pit. At the same time the change in the quality of the discharged water can be observed, and the length of the washing operation adjusted accordingly.

It will vary from 5 to 8 minutes depending on the depth of sand and the nature of the matter retained.

4.2. Vertical filters designed for washing by air and water

4.2.1. FILTERS WITH A SINGLE UNIFORM LAYER, WASHED SIMULTANEOUSLY BY AIR AND WATER

The filter bed, homogeneous throughout its depth, is placed on a perforated metal floor with rings to which metal or plastic nozzles are screwed, depending on the type and temperature of the liquid to be filtered.

1. *Filter body.*
2. *Filtering mass.*
3. *Floor with nozzles.*
4. *Feed basin.*
5. *Raw water inlet.*
6. *Filtered water outlet.*
7. *Wash-water inlet.*
8. *Wash-water outlet.*
9. *Air scour inlet.*
10. *Air blow-off.*
11. *Drain and air release.*
12. *Man-hole.*
13. *Hoisting lug.*

Fig. 171. — FV 2 B filter.

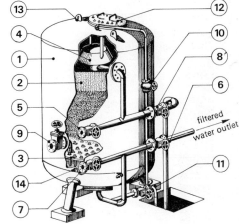

These are ordinary sand filters; they are washed by reduced flow backwash water and air scour followed by a full-flow rinse with filtered water.

The normal characteristics of this type of filter are as follows:

— Effective grain size................... 0.7 to 1.35 mm
— Air requirement .... 50 m³/h per m² of filter
                                      (2.2 cu ft per min per sq ft)
— Air rate of flow during injection ...... 5 to 7 m³/h per m² of filter
                                      (0.24 to 0.38 cu ft per min per sq ft)
— Rinse water flow .... 15 to 25 m³/h per m² of filter
                                      (6 to 10 gpm per sq ft)
— Loss of head at end of cycle .... 100 to 400 mbar

Fig. 172. — The ten filters 6 m (19.6 ft) in diameter operated by S.A. ESPERANCE-LONGDOZ, CHERTAL (Belgium). Views inside and outside the building.

The depth of the layer is adjusted to suit the rate of filtration and the quantity of solids to be retained.

Filtration rates are generally between 4 and 20 m/h (1.6 to 8 gpm per sq ft). In industrial operations, this filter can be used with bed depths of 1 to 2 m made up of sand with an effective grain size ranging from 0.65 to 2 mm. Velocities are sometimes well in excess of those applicable to clarified water: 20 to 40 m/h (8 to 16 gpm per sq ft) with a deep bed of sand with an effective size of 1.8 to 2 mm used for water contaminated with metallic oxides, and 30 to 60 m/h (12 to 24 gpm per sq ft) through a layer of fine, 0.65 mm sand for the treatment of deep sea-water.

These filters, which are suitable for use in batteries made up of large-diameter units, offer the following major advantages: simplicity of operation; complete functional reliability; and a low instantaneous rate of flow of wash-water, the consumption of which is reduced.

4.2.2. SINGLE AND DOUBLE-LAYER FILTERS WASHED SUCCESSIVELY WITH AIR AND WATER

The vertical filters described above can also be adapted to the use of a layer of light filtering material (e.g. anthracite, or granular activated carbon) or to several layers of different media.

To provide for the proper expansion of the layers, these materials necessitate a method of washing which differs from that applied to filters with a single uniform layer. Once the level of the water has been lowered to the top of the uppermost layer, the washing operation begins by blowing through with air alone, causing the impurities and fines of the bottom layer to be diffused throughout the whole depth of the filter bed. As a second stage, the filtering material must be washed with a high-velocity reverse current of water, the purpose of which is to expand the filter bed thereby expelling the impurities and regrading the filter media.

The velocity of the return flow of filtered water is chosen to suit the nature of the filtering material, its grain size, the temperature and the degree of expansion desired. Where the fine material of the double layer is made up of sand, the flow-rates of the reverse flow of water should be as shown in the table for filters with a simple backwash for the same size of sand as that constituting the bottom layer. As the rates of flow are greater than those encountered with filters with a single uniform layer, the piping, the valves and the capacity of the wash pump must be scaled up accordingly. In addition, the expansion of the filter bed means that it is necessary to raise the channel used for removing the wash water from the filter.

4.2.3. U.H.R. FILTERS

These are ultra-rapid two-layer filters, for which the various combinations of anthracite and sand grain sizes are selected to suit the kind of filtration required.

These U.H.R. filters have been specially designed to work at rates of 25 to 50 m/h (10 to 20 gpm per sq ft) and to retain very dense and granular materials or large quantities of suspended solids.

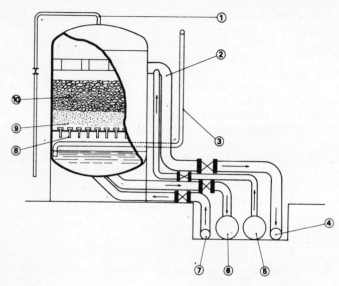

1. *Vent.*
2. *Wash-water outlet.*
3. *Air inlet pipe.*
4. *Wash-water discharge.*
5. *Raw water inlet.*

6. *Filtered water outlet.*
7. *Wash-water inlet.*
8. *Floor complete with nozzles.*
9. *Sand.*
10. *Anthracite.*

Fig. 173. — MEDIAZUR U.H.R. type filter.

With this object in mind:
— sand/anthracite combinations with the optimum grain sizes are used;
— the filters operate with filter beds of 2 m depth and more; and
— the applied technology, particularly as regards the air scour and sludge removal, is designed for the evacuation of large quantities of retained matter.

U.H.R. filters are manufactured as vertical cylindrical units, of metal or concrete construction, which may be of the open or pressurized variety. These filters are able to operate with long intervals between washes. Despite their high instantaneous wash-water consumption, they are particularly favoured for the treatment of rolling mill and oily refinery effluents. Another suitable area of application is the treatment of storm runoff water.

4.2.4. DOUBLE AND TRIPLE TOWERS

These consist of two or three filters each with a single filter layer or two filter layers superimposed one on top of the other in the same cylindrical casing. These towers behave in exactly the same way as would separate units of the same number and size as their unit components. Compared to a single filter of the same total area, they have the advantage of occupying less floor space provided sufficient

headroom is available. Because they are divided into several units, the instantaneous quantity of wash-water used is also reduced; this is a great advantage:
— when the wash-water has to be taken from the filtered water, because the wastage is small;
— and when on the other hand the flow of filtered water has to remain constant, and the rate of filtration on the other units in service has to be stepped up; the necessary increase in speed then is not so great.

Because it is necessary to wash the different units in turn, fully-automatic filter control is advisable; and the valves can remain small.

Fig. 174. — *Superimposed filters 2.80 m (9 ft 32 in) in diameter for process water. Total output: 135 m³/h (0.85 mgd).*

4.2.5. D2F TYPE REVERSE-CURRENT (DUAL FLOW) FILTERS

This type of filter makes it possible to double the filtration rate per m² of horizontal section by the expedient of using a filter bed which is traversed both in the downward and upward directions, the filtrate being drawn off in the middle. The filter (fig. 175) comprises chiefly:
— the raw water intake. This is divided and made to follow two paths, one (1) leading to the dished top of the vessel, and the other (2) to the dished bottom section;
— the floor fitted with special long-stemmed nozzles and provided with wide flow apertures. These are embedded in a layer of supporting gravel;
— a deep layer of filtering sand of uniform grain size throughout its depth;

— the branched header (3) for drawing off the filtered water. This is embedded in the middle of the filter bed;
— the bowl for collecting the wash water, which then passes through a pipe to the drain.

In fact, this filter amounts to no more than two similar filters mounted one on top of the other which operate with opposite directions of flow but at the same filtration rate and with the same head-loss. These two currents hold the sand in its initial position, so there is no danger of sand flurries during operation. This mode of operation is rendered possible only by the homogeneity of the filtering bed throughout its depth. If the fine sand were concentrated in the upper part, this would lead to very rapid clogging of the surface and almost the entire flow would be upwards, thus nullifying the initial advantage of the process and bringing about the rapid breakthrough of the lower filter element.

While the washing operation must be effective, it must not cause expansion of the sand bed. This is achieved by a simultaneous air scour and backwash, the velocities of which are chosen to suit the grain size of the sand and the nature of the retained particles.

In particularly difficult cases, where the retained matter is at once heavy, sticky and abundant, pulsation washing (DEGRÉMONT Patent) can be applied to this type of filter. This ensures that the filter layers are kept in perfect condition without the risk of impurities accumulating at any point during the filtration cycle. This washing operation uses standard equipment, to which is added a butterfly valve which opens the compressed air circuit to the atmosphere. This valve can be operated automatically by a timing device.
— air flow: 60 to 80 m³/(h.m²) (3.2 to 4.3 cu ft per min per sq ft).
— backwash flow during blowing: 10 to 15 m³/(h.m²) (4 to 6 gpm per sq ft).
— rinse water flow-rate: 20 to 30 m³/(h.m²) (8 to 12 gpm per sq ft) in the case of filters retaining relatively light contaminants, and 30 to 50 m³/(h.m²) (12 to 20 gpm per sq ft) for filters used to separate heavy precipitates.

1. *Raw water inlet.*
2. *Upper water inlet.*
3. *Lower water inlet.*
4. *Filtered water outlet.*
5. *Wash water inlet.*
6. *Wash water outlet.*
7. *Air vent.*
8. *Blown air inlet.*
9. *Blown air by-pass.*
10. *Wash-water by-pass.*
11. *Rinsing of filtered water header.*
12. *To drain.*

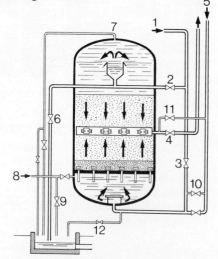

Fig 175. — D2F type enclosed filter for *reverse-current operation.*

This D2F filter can be equipped with all the accessories necessary for semi-automatic or fully-automatic operation.

This type of filter is to be recommended particularly for direct filtration of river water or relatively turbid water, and in all cases where long cycles between washing operations are essential.

Fig. 176. — D2F filters 2.50 m (8 ft 3 in) indiameter, with a single header pipe system (semiautomatic control) for factory process water. Output: 300 m³/h (1.9 mgd).

Fig. 177. — Three D2F process-water filters 2.80 m (9 ft 3 in) in diameter with a two-header pipe system each. Output: 300 m³/h (1.9 mgd).

4.3. Horizontal filters designed for water backwash and air scour

The closed filter washed by water and air may also be constructed as horizontal cylindrical vessel. This ensures economy in construction when large filtering areas are required, as it is possible to increase the length of the vessel without altering its diameter. The operating principles of these filters are identical with those of vertical filters.

These filters are used for raw water with a low or medium suspended solids content, as the depth of sand layer is limited by their design. They are also suitable for filtering clarified water, and can operate either under pressure or at atmospheric pressure.

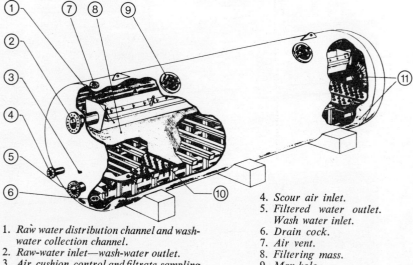

4. *Scour air inlet.*
5. *Filtered water outlet.*
   Wash water inlet.
6. *Drain cock.*
7. *Air vent.*
8. *Filtering mass.*
9. *Man-hole.*
10. *Floor complete with nozzles.*

1. *Raw water distribution channel and wash-water collection channel.*
2. *Raw-water inlet—wash-water outlet.*
3. *Air cushion control and filtrate sampling valve.*

Fig. 178. — Horizontal filter with filter floor designed for washing by back-wash water and air scour.

5. OPEN FILTERS

The majority of filter plants engaged in the supply of drinking water, as well as a good number of industrial water and effluent clarification plants with a large throughput, make use of open, generally concrete-built, filters.

Depending on circumstances, either no reagent at all is added to the raw water beforehand, or it is simply coagulated without any settling stage, or, and this is what usually happens, the water is coagulated, flocculated and clarified. The method of treatment influences the technological design of the filters and, more especially, the overall concept of the filter battery.

Leaving aside slow filters (see page 255), these open filters normally operate at filtration rates between 4 and 20 m/h (1.6 to 8 gpm per sq ft). Looked at from the technological point of view, we can classify these units basically as:

● Rapid filters of the traditional type, working at filtration rates between about 5 and 10 m/h (2 to 4 gpm per sq ft).

Within this category, DEGRÉMONT manufactures the Aquazur T and N, the Mediazur T and N, and the Mediazur G types.

● High-rate filters, operating at filtration rates of 7 to 20 m/h (2.8 to 8 gpm per sq ft). Included in this category are the:

Aquazur V, Mediazur V and Mediazur GH, as well as the reverse-current filters. hese various types are described in detail below.

5.1. Rapid filters — 5 to 10 m/h (2 to 4 gpm/sq ft)

5.1.1. AQUAZUR T AND N FILTERS WITH A SINGLE UNIFORM LAYER

The characteristic features of these filters are:
— a filter bed of uniform grain size, which stays homogeneous after washing;
— backwashing simultaneously by air (with a high flow-rate) and water (with a low rate of flow), followed by a medium-flow rinse, which does not expand the filter bed;
— a shallow (0.50 m) depth of water above the sand; and
— a reduced head loss, usually of 2 m, to avoid the possibility of excessive clogging causing serious deaeration of the water.

Depending on the nature of the water and its susceptibility to deaeration, the maximum filtration rate may be from 7 to 10 m/h (2.8 to 4 gpm per sq ft).

Types T and N differ only as regards the kind of floor used to support the filtering material.

Fig. 179. — AQUAZUR V filters: simultaneous air and water backwash.

T type filters are fitted with long-stemmed D20 nozzles (fig. 164) which are screwed to a removable floor made up of slabs of concrete (fig. 181) or of reinforced polyester (fig. 180).

Fig. 180. — *Slab of reinforced polyester, equal in length to the width of the filter cell, (V type filter).*

Fig. 181. — *T Type AQUAZUR filters in course of installation. From left to right: the supporting members, the floor made up of concrete slabs, and the floor fitted with nozzles.*

In T type filters with a small filtering area, the scour air is distributed below the floor via a feed pipe (fig. 182).

In the larger filters the air is distributed through a concrete conduit located below one of the sludge extraction channels (fig. 183).

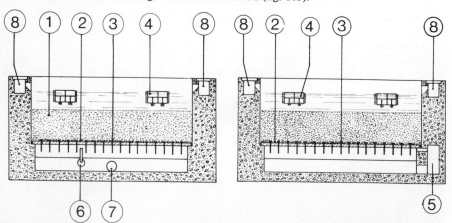

Fig. 182. — *Model T AQUAZUR filter with concrete floor and scour air feed pipe.*

Fig. 183. — *Model T AQUAZUR filter with concrete floor and air-water channel.*

1. *Sand.*
2. *Concrete floor.*
3. *Nozzles.*
4. *Water inlet clack-valves.*
5. *Air water distribution and filtered water outlet channel.*
6. *Scour air feed pipe.*
7. *Wash-water inlet and filtered water outlet*
8. *Sludge extraction channels.*

In both cases, the air is distributed uniformly over the whole surface area of the filter thanks to the creation of an air-cushion by the long-stemmed nozzles (the density of which must be at least 50 per m² of floor area) (42 per square yard).

The N type filters are fitted with small D 13 nozzles with no stem (fig. 162) which are screwed into the floor made up of fibrocement slabs at the rate of about 80 per m² (67 per square yard) of floor area. The scour air is distributed via a branched perforated feed pipe located under the floor.

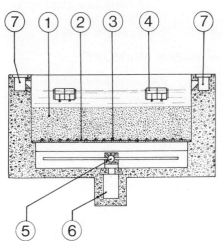

1. *Sand.*
2. *Fibrocement floor.*
3. *Nozzles.*
4. *Water inlet clack-valves.*
5. *Blown air header.*
6. *Wash-water inlet and filtered water outlet.*
7. *Sludge extraction channels.*

Fig 184. — *Type N AQUAZUR filter with fibrocement filter floor.*

These T and N filters have only three valves, one for the filtrate, one for the wash-water and one for the air.

The unfiltered water inlet is controlled by a clack-valve which closes automatically during the washing process when the water level in the filter becomes higher than the level in the inlet channel (figs. 185 and 186).

Sludge extraction is ensured by overflow into longitudinal troughs.

Fig. 185. — *Water inlet valve. Open (during filtration).*

Fig. 186. — *Water inlet valve. Closed (during the washing process).*

The water level in these filters is controlled by a siphon with "partialization" control, by a butterfly valve or by a seat valve.

Their characteristics are as follows:

— filtered water output......................... 5 to 10 m³/ (h.m²)
        (2 to 4 gpm per sq ft)
— wash water requirement during air scour........ 5 to 7 m³/ (h.m²)
        (2 to 2.8 gpm per sq ft)
— air scour requirement........................ 50 to 60 m³/ (h.m²)
        (2.7 to 3.2 cu ft/min sq ft)
— rinse water flow rate......................... 20 m³/ (h.m²)
        (8 gpm per sq ft)

DIMENSIONS OF THE AQUAZUR MODEL "T"

With scour air feed pipe			*With air-and-water channel*		
Width in m	Length in m	Surface area in m²	Width in m	Length in m	Surface area in m²
2.46	2.58	6.35	3.00	8.18	24.5
—	3.42	8.5	—	9.34	28
—	4.26	10.5	—	10.5	31.5
—	5.10	12.5	—	11.6	35
—	5.94	14.5	—	12.82	38.5
—	6.78	16.5			
—	7.62	19			
—	8.46	21	3.50	8.02	28
—	9.30	23	—	9,01	31.5
—	10.14	25	—	10.01	35
			—	11	38.5
			—	12	42
3.07	7.62	23.5	—	13	45.5
—	8.46	26	—	13.99	49
—	9.30	28.5	—	14.98	52.5
—	10.14	31			
—	10.98	33.5	4.00	11.66	46
			—	12.82	51
			—	13.98	56
			—	15.14	60.5
			—	16.30	65
			—	17.46	70

The above dimensions relate to single-cell filters. Double-cell filters comprise two cells identical to the above.

These filters may be arranged either as single filters (containing one control system for each filter unit) or as double filters (two filter units connected at the top and bottom, with a single control system).

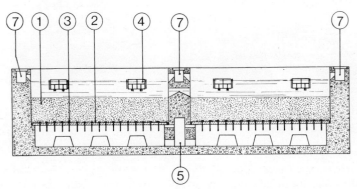

1. *Sand.*
2. *Concrete floor.*
3. *Nozzles.*
4. *Water inlet clack-valves.*

5. *Scour air wash water distribution and filtered water outlet channel.*
7. *Sludge extraction channels.*

Fig. 187. — Model T AQUAZUR dual filter with concrete floor and air water channel.

● *Washing Aquazur filters with a shallow depth of water:*

— **Manual washing:** to wash a filter equipped with a single layer of sand, close the filtrate outlet valve, start the air blower, open the scour air inlet valve and start the wash-water pump. As soon as the air is well distributed beneath the full area of the filter, with the water level below the sludge weir level, slightly open the wash-water inlet valve.

After 10 minutes, close the air inlet valve, switch off the blower, and fully open the wash-water inlet valve. When the water discharged to drain runs clean, close the wash-water inlet valve.

The filter is ready to be returned to service by opening the filtrate outlet valve.

— **Semi-automatic or automatic washing :** the washing operations can be partially or fully automated.

The instantaneous power necessary for washing (blower and pump) is about 1.5 kW per m^2 (1.2 kW per sq yd); washing takes about 15 minutes, not counting dead time. Wash-water consumption depends primarily on the quality of the water being treated, and generally varies between 1 and 2 % of the filtered volume.

Fig. 188. — Control gallery for four AQUAZUR filters each having a surface area of 30 m² (323 sq ft). At the extreme right is the wash console. Output: 1 250 m³/h (7.9 mgd). Water works of BAYONNE (France).

● *Conversion of a horizontal metal filter to an Aquazur filter:*

Horizontal metal filters can be used as open filters, with the water level during filtration just above the washing channel. They are then fitted with a control unit (siphon, butterfly valve) identical to those used for the standard Aquazur filters, and are washed with air and water in a similar fashion.

This is particularly advantageous in cases where a short building time for a water treatment plant is required; the time for the concrete works may thus be reduced.

Fig. 189. — Gallery housing horizontal pressure filters. Output: 1,800 m³/h (11.4 mgd). Water works at ROME (Italy).

5.1.2. MEDIAZUR T AND N FILTERS WITH A SINGLE OR DOUBLE LAYER AND WITH SUCCESSIVE AIR AND WATER WASHING

As a method of washing, simultaneous air scour and backwash with extraction of the impurities by an overflow system can be applied only when permitted by the density of the media employed. While it is feasible with sand, it is not practicable with activated carbon or anthracite, the low density of which would cause serious losses of material if washed by air and water.

With this in mind, the T and N type filters, which are filled with:
— either a single layer of a low-density media such as anthracite or activated carbon with a coarse grain size,
— or two layers of different materials, e.g. sand and anthracite,
have been designed to solve this problem, as regards the filter feed and the washing system.

In the Mediazur filters, the top layer of filtering material is covered only by a shallow depth of water.

Fig. 191. — Diaphragm-type obturator controlling untreated water feed.

The same washing technique is applied to these filters as is described on page 277 in reference to the same type of vertical metal pressure filter, i.e. drainage followed in succession by washing with air alone and then with water alone, accompanied by expansion.

The raw water can no longer be admitted via clack-valves as the washing sequence starts with a drainage stage. The inlet comprises diaphragm-type valves inflated by compressed air and a thick-wall weir at the point of entry of the water to prevent any erosion of the filter bed.

Operation of these filters requires careful supervision. It is necessary, in particular, to adjust and monitor the washing conditions so as to avoid excessive losses of filtering material.

5.1.3. MEDIAZUR G FILTERS WITH SMALL SIZE CHANNELS AND A SINGLE FILTER BED, WASHED SUCCESSIVELY BY AIR AND WATER

Where the filtering material is both light and of small grain size, it is necessary to collect the sludge along long overflow weirs so as to prevent any of the filter media from being carried off to the drain. This applies especially to the activated carbon with an effective grain size of about 0.55 mm which is ordinarily used for second-stage filtration and which does not stay homogeneous after washing.

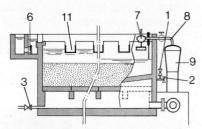

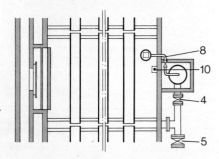

1. *Air scour valve.*
2. *Air vent valve.*
3. *Drain cock.*
4. *Filtrate valve.*
5. *Wash-water valve.*
6. *Diaphragm valve.*
7. *Partializing control box.*
8. *Connection between siphon and partializing box.*
9. *Siphon.*
10. *Level probe.*
11. *Untreated water inlet—washwater outlet.*

Fig. 192. — MEDIAZUR filter with activated carbon bed for second stage filtration.

The Mediazur G type filter is a variant of the Mediatur T and N models which has been adapted to the use of this kind of material. Like these other models, it is fitted with:
— a raw water intake via a diaphragm valve and a thick-wall weir, and;
— a washing sequence which comprises preliminary drainage followed by air scour on its own and then by washing with water alone accompanied by expansion of the filtering material.

It differs from the others in that it is fitted with several transverse sludge extraction channels.

The Mediazur G type filter can be filled with several layers of different media.

5.2. High-rate filters — 7 to 20 m/hour (2.8 to 8 gpm/sq ft)

The higher the filtering rate, the more sophisticated does the applied technology become, especially as regards:
— the choice of filtering material(s);
— the layer depth;
— the washing technique, and
— the overall hydraulic system.

5.2.1. V TYPE AQUAZUR FILTERS WITH A SINGLE LAYER WHICH REMAINS UNIFORM AFTER WASHING

A layer of homogeneous material (usually sand) in an Aquazur filter is
capable of satisfying completely the requirements imposed by high-rate filtration.

Simultaneous air scour and backwash accompanied by surface sweep,
followed by a water rinse with the same rate of flow (without expansion of the
filter bed and keeping up the surface sweep), makes it possible to maintain the
filter bed in its initial condition without erosion.

Given equal pretreatment of the raw water, the quality of the filtrate produced by a filter with a single deep layer is practically indistinguishable from that
obtained with a two-layer filter.

● *Characteristics and design features.*

The V type Aquazur filter is equipped with the same kinds of floor and
the same control mechanisms (siphon or butterfly valve) as the model T Aquazur
filters, from which it differs in respect of:
— the considerable depth of water above the filter bed, which is never less than
1 m and is usually 1.20 m;
— the deep homogeneous filter bed (sand, or Biolite in the case of effluent treatment); and
— the feed and washing systems, the latter employing surface sweep with raw water.

*Fig. 193. — MORSANG water works (France). Water supply of the Southern Paris
Region. V type AQUAZUR filters. Filtration rate: 12.3 m/h (5 gpm per sq ft). Output:
20 mgd.*

Where filtration is through sand, the main parameters are as follows:
— effective grain size of the filtering material, usually 0.95 mm to 1.35 mm (limit values 0.7 mm to 2 mm);
— filtration rate: 7 to 20 m³/h per m² (2.8 to 8 gpm per sq ft);
— flow-rate of filtrate backwash: 13 to 15 m³/h per m² (5.2. to 6 gpm/sq ft);
— flow-rate of air scour: 50 to 60 m³/h per m² (2.7 to 3.2 cu ft/min sq ft).

The standard dimensions of the cells of V type filters are given in the following Table:

V TYPE FILTERS
DIMENSIONS OF SINGLE-CELL FILTERS WITH CONCRETE FLOOR

Width m	Length m	Surface area m²	Width m	Length m	Surface area m²
3.00	8.18	24.5	4.00	15.14	60.5
—	9.34	28	—	16.30	65
—	10.5	31.5	—	17.46	70
—	11.6	35			
—	12.82	38.5			
			4.66	12	56
			—	13	60.5
3.50	8.02	28	—	13.99	65
—	9.01	31.5	—	14.96	69.5
—	10.01	36	—	15.98	74.5
—	11	38.5	—	16.97	79
—	12	42			
—	13	45.5			
—	13.99	49	5.00	13.98	70
—	14.96	52.5	—	15.14	77
			—	16.30	81.5
			—	17.46	87
4.00	11.6	46	—	18.62	93
—	12.82	51	—	19.78	99
—	13.98	56	—	20.94	105

Two-cell filters comprise two cells identical to the above.

Filters with a floor made up of reinforced polyester slabs have the same dimensions as those with a concrete floor but are limited to widths of 3, 3.5, 4 and 5 metres.

The filter feed can be designed in different ways according to the type of control system required, i.e. reacting to the flow-rate or water level.

The diagram (fig. 194-195), shows a two-cell V type filter with an upstream control system. Passing from the common raw water channel (1), the water to be filtered is equally distributed between the different filters by a combination of the diaphragm (2) and the weir (3). The water enters the filter through the two side apertures (4) and passes through the filter bed (5), the depth of which depends on the filtering rate. The filtered water is then collected in the channel (6), governed by the control mechanism (7), and is then discharged into the common filtrate channel (8). Item (9) indicates the head-loss in the filter and may also transmit an appropriate signal.

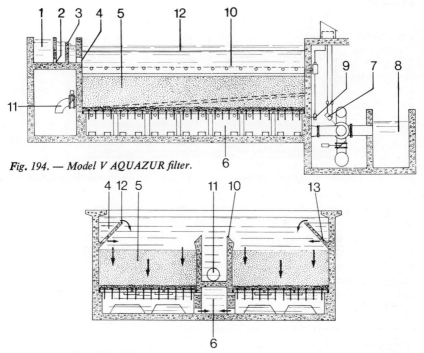

Fig. 194. — Model V AQUAZUR filter.

Fig. 195. — Model V AQUAZUR filter shown during the filtration phase.

When the maximum degree of clogging compatible with good filtrate quality is reached, washing is carried out (fig. 196).

As a first stage, the level of the water is lowered to the upper edges of the washing channel (10). This lowering of the water level is usually effected by shutting down the filtering operation and opening the wash water discharge valve (11). It can also be done by continuing to filter while cutting off the filter feed with the aid of diaphragm shut-off valves (fig. 191) built into the raw water intake system.

The second washing stage consists of combining the conventional washing in an Aquazur filter with surface sweeping using the water to be filtered. The latter continues to flow in through the orifices (4), runs into the V-shaped channels and enters the top of the filter through the orifices (13). At the same time air scour is effected plus a backwash of filtered water of 13 to 15 m³/h per m² (5.2. to 6 gpm/sq ft) made possible by the special form of the wash-water weirs. This stage lasts 4 to 5 minutes.

The final washing stage consists or rinsing at the same backwash flow of filtered water as before, while maintaining the surface sweeping. It lasts 3 to 4 minutes.

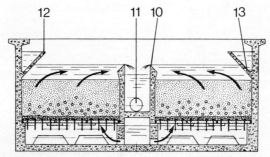

Fig. 196. — Model V AQUAZUR filter; air-water scouring phase with surface sweeping.

Taking account of the valve operating times, washing takes 10 to 12 minutes and is followed by refilling of the filter to its normal working level.

The washing of a battery of V model filters therefore requires no more than a blower and a wash-water pump operating at the same rate of flow during both the blowing and rinsing stages. In view of the fact that washing is performed without expansion of the filter bed, there is no need to adjust the rate of flow to suit periodical changes in the water temperature.

● *Advantages inherent in this type of filter.*

Filters of this kind combine all the features necessary to high-quality filtration and efficient washing:

— throughout the whole washing operation, the filter continues to be fed, wholly or in part, by the raw water needed for surface sweeping. During this period, the other filters in the same battery undergo little or no violent increase in flow-rate or filtration rate;

— these filters are specially suitable for high filtering rates, for which sand beds 1 m to 2 m in depth can be used;

— throughout the entire filtration cycle, a positive pressure is maintained over the whole depth of the sand;

— the washing technique without expansion avoids any hydraulic grading of the filter bed (cf. page 266);

— the washing operation calls for only a small filtrate backwash flow, and this cuts down both the power consumption and the amount of equipment required;

— the water backwash, which continues throughout the air scour, is backed up by surface sweeping. The wastage of water is identical to that of the T and N model Aquazur filters;
— filtration is restarted by raising the water level, which brings about a gradual startup after washing irrespective of the control system used. Where required, this gradual startup can be extended over a period of 15 minutes;
— finally, the method of washing by means of a pump with a constant rate of delivery means that it is unnecessary to install a raised water tank with all the complications which result from such an arrangement.

Fig. 197. — Model V AQUAZUR filters with a central control gallery. ORLY drinking-water plant for the city of PARIS.

5.2.2. SINGLE AND MULTI-LAYER MEDIAZUR V TYPE FILTERS WASHED SUCCESSIVELY WITH AIR AND WATER

As has already been noted in relation to Mediazur T and N model filters, V type filters can be adapted to the use of a filter bed comprising:
— either a single, deep layer of light material such as anthracite or coarse-grained activated carbon;
— or two layers of different materials, such as anthracite and sand.

The configuration of the Aquazur V model filters working with a considerable depth of water has been retained, as has also the type of floor and the control system. Only the systems of filter feed and washing have been modified, as it is clearly impossible to wash light materials by simultaneous air scour and back-

wash without inviting the risk of loss of a large quantity of the filter media.
The washing sequence comprises three stages:
— drainage of the raw water until its level is just above the top of the filter bed.
This lowering of the level can be achieved either by filtration or by discharge to
the drains,
— air scour alone to detach the impurities from the grains of the filtering mate-
rial(s). During this stage, there is no surface sweep with raw water;
— high-velocity washing with water alone, the purpose of which is to expand
the filter bed, to remove the impurities which have been diffused throughout its
depth by the air scour and, in the case of a two-layer filter bed, to effect the more
or less complete hydraulic regrading of the two different media.
During this final stage, the raw water is once more admitted via the side
channels to sweep the surface and thereby accelerate the removal of the impurities.
This calls for:
— one or several gate valves or diaphragm-type shut-off valves (see fig. 191)
so as to cut off completely the raw water feed during the drainage and air scour
stages and to isolate it partially also during the final stage of washing with water
alone, thereby setting up a low-velocity surface sweeping action;
— an electrode located above the top of the filter bed to arrest drainage prior to
washing;
— an air scour flow-rate of 50 to 60 m³/h per m² (2.7 to 3.2 cu ft/min sq ft) via
nozzles at a density of numbering more than 50 per m² of filter surface area
(42 per sq yd);
— a high backwash flow-rate adjusted to the material(s) making up the filter
bed and designed to keep the degree of expansion of the filter bed constant during
the washing period.
It follows that, given the same filtration rate and the same filtration cycle
time, a filter of this kind requires considerably more accessory equipment than a
unit with a deep, homogeneous sand bed.

5.2.3. MEDIAZUR GH FILTERS WITH SMALL SIZE CHANNELS AND A SINGLE FILTER BED WASHED SUCCESSIVELY BY AIR AND WATER

The aforementioned filters, working with a considerable depth of water and
filtering material, cannot be applied without modification when the filter media
used is both light and finely divided, such as the activated carbon of low effective
grain size used for second stage filtration. To avoid being carried off down the
drain, media of this kind require a low approach velocity to the weirs during
washing.

In these circumstances, it is necessary to use filters in which the sludge is
collected by a number of transverse channels (fig. 198) which ensure a low rate
of flow per metre length of the weir lip.

These GH model filters have the same cell dimensions as the Aquazur V
type, and they can be fitted with the same floors and control systems. The
method of feed is adapted to the control system used, and the washing technique

is identical to that employed for Mediazur V model filters, i.e. it involves washing with air alone followed by water backwash accompanied by expansion of the filter bed, the sole difference being that surface sweep is no longer necessary.

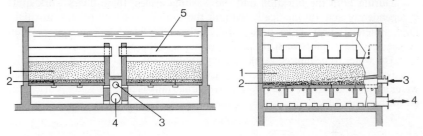

1. *Filter media.*
2. *Supporting layer.*
3. *Blown air inlet.*

4. *Wash-water inlet.   Filtered water outlet.*
5. *Raw water inlet.   Wash-water outlet.*

Fig. 198. — Model GH MEDIAZUR filter.

5.2.4. D2F TYPE REVERSE-CURRENT FILTERS

The principle of this kind of open, reverse-current filter is similar to that incorporated in the reverse-current pressure filter described on page 279. The filter may, or may not, be preceded by a flocculation/clarification stage.

A control system monitors the outflow rate of the filtrate in relation to the level of water in the filter. Prior provision is made for distributing the water equally between the filters.

Open D2F type filters give very satisfactory results and, more especially, they make it possible to operate at high filtration rates while retaining the advantage of simultaneous air/water washing.

This kind of filter can be designed as a model T unit for filtration rates up to about 12 m/h (4.8 gpm per sq ft) or as a model V filter for a greater depth of water (see fig. 199) and higher filtration rates.

1. *Upper raw water intake.*
2. *Bottom raw water intake.*
3. *Filter bed.*
4. *Header fitted with nozzles for drawing off filtered water.*
5. *Layer of gravel.*
6. *Floor fitted with long-stemmed nozzles.*
7. *Sludge extraction channel.*
8. *Filtrate outlet channel.*
9. *Air/water washing channel.*

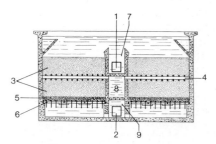

Fig. 199. — Reverse-current AQUAZUR V model filter.

5.3. Valveless, self-washing filters with automatic siphon

During both the filtration and the washing cycles, these filters work quite independently and automatically without the use of electricity or any auxiliary fluid systems.   They have been supplied by DEGRÉMONT for more than 50 years.   Many improvements have been made to them.

The water to be filtered is taken from a buffer tank.   After filtration through a bed of small grain size, the water rises from the bottom of the filter to the filtered water storage tank at the top.   When the storage tank is full, the water is discharged to the point of use.

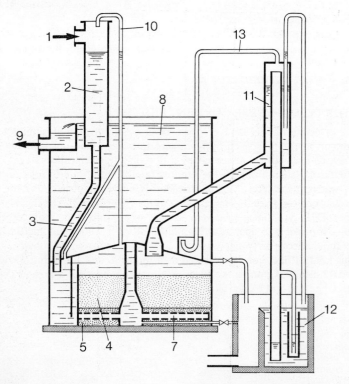

1. *Raw water inlet.*
2. *Header tank.*
3. *Raw water pipe.*
4. *Filter layer.*
5. *Supporting layer.*

Fig. 200. — Valve-less, self-washing filter.

7. *Filtrate pipe.*
8. *Filtrate storage tank.*
9. *Filtrate outlet.*
10. *Vent pipe.*
11. *Wash-water siphon.*
12. *Wash-water return tank.*
13. *Unpriming pipe.*

When the filter bed becomes clogged, the level rises in the buffer tank and in the upstream arm of the siphon (the downstream end of which is submerged). When the maximum design head loss is reached, the compressed air in the siphon is discharged suddenly: the siphon is immediately primed, and causes the water in the filtered water storage tank to flow back through the filter; the capacity of the tank is designed to ensure adequate washing of the filter.

With this type of filter, the bed can never become excessively clogged, as washing is carried out automatically at a predetermined fixed head loss; no mechanical parts or auxiliary fluid are needed.

This filter can be used, for example, when compressed air and electrical power are not available. It is suitable for waters with a low or medium suspended solids content when the water distribution system can tolerate a stoppage long enough for the storage tank to refill.

The maximum filtration rate is 10 m³/h per m² of filter area as a general rule (4 gpm/sq ft).

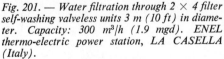

Fig. 201. — Water filtration through 2 × 4 filter self-washing valveless units 3 m (10 ft) in diameter. Capacity: 300 m³/h (1.9 mgd). ENEL thermo-electric power station, LA CASELLA (Italy).

5.4. Dry filters

These are filters that are not submerged in water; they are, in fact, filters in which the water is sprayed above the surface of the sand as in biological filters.

They are used in some countries, especially Holland, to treat clear underground water containing both manganese and ammonia.

Air circulation through the sand is accelerated by drawing the air from under the floor with an air pump or ejector.

Nitrifying bacteria grow in the sand and convert the ammoniacal nitrogen to nitric nitrogen. Manganese generally precipitates well in the form of MnO_2 on the surface of the sand, but the presence of ammonia appears necessary, probably because it encourages the growth of oxidizing bacteria.

If any difficulty is encountered in removing manganese, its precipitation is always facilitated by adding a little potassium permanganate before filtration.

Dry filters clog up gradually and must be washed when pools of water form on the surface of the sand, as these indicate that the air can no longer penetrate into the filter mass. It is then impossible for oxidation of manganese and ammonia to take place.

To obtain good results, there must be a low iron content and the water must not contain organic matter.

6. FILTER CONTROL

Three main types of filters can be distinguished according to their control system: variable-level constant-output filters, constant-flow filters (the output being matched to the total flow treated), and variable-flow filters operating without equal distribution or individual control.

6.1. Variable-level constant-output filters

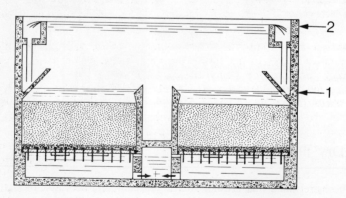

1. *Low level, clean filter.*   2. *High level, clogged filter.*

Fig. 202. — Variable-level constant-output AQUAZUR V filter.

Equal distribution of the total flow to be filtered is obtained at the inlet to the filters where the water falls from a variable height depending on the degree of clogging. When the filter is clean, the sand is just covered with water; the level of water is maintained by the level of the weir of the outlet siphon. At maximum clogging the level reaches the supply water level.

Generally, this water level is at a height of 1.50 (5 ft) to 2.00 m (6 ft 8 in) depending on the grain size of the filter medium. It is not so high (0.80 m (2 ft 8 in) to 1.00 m (3 ft 4 in) when, instead of a filter medium, the filter contains a mass of a neutralizing product whose primary purpose is not filtration. This applies to Neutralite filters, for which the variable-level type of control is commonly used.

6.2. Constant-output filters with compensation for clogging

The level of water in the filters is either stationary or subject to only slight variations. The filtered water is discharged 2-3 m lower down at a constant rate of flow equivalent to the total feed-rate divided by the number of filters.

The maintenance of this constant flow-rate irrespective of the degree of clogging of the filters is ensured by a control mechanism located at the outlet of each filter which responds either to the water level or to the flow-rate. When the filter is clean, this device creates a large supplementary head-loss which is cancelled when the filter is completely clogged, i.e. the control system compensates for the clogging of the filter bed.

6.2.1. FILTER CONTROLLERS

A. *Hydraulic controller.*

DÉGRÉMONT concentric siphon and its "partialization box" (fig. 203) are used for level control, the partialization box being the detection and control unit and the siphon the regulating unit.

● *Siphon:* the siphon consists of two concentric tubes, with the flow passing from the inside branch to the outside branch. It operates in the same way as an ordinary siphon but it is far more stable.

If air is introduced at the top, this air is driven by the water into the descending branch where the density of the water-air mixture is reduced, thus reducing the vacuum at the neck. Without partialization air, the vacuum at the neck is equal, ignoring the head loss in the downstream branch, to the height H of fall between the water level on the filter and the water level in the downstream chamber. With partialization by air, this vacuum is reduced to height h_1 equal to the product of H and the density of the water-air mixture. The difference $H - h_1 = h_2$ represents the loss of head created by introduction of air (fig. 204).

If h_1 represents the head loss with a clean filter due to the flow to be filtered passing through the filter bed, the floor and the filtered water discharge pipe to the siphon neck, h_2 represents the clogging height available for the filter bed.

With a clean filter, it is therefore only necessary to introduce a quantity of air sufficient to create this head loss h_2 and, as the filter bed gradually clogs up, to reduce the air flow until it becomes zero to bring h_1 to H.

● *Partialization box* (siphon control valve) (fig. 205): this is the control device that introduces air at the top of the siphon to control its flow. It may be defined as a flap C suspended from a spring D fixed to a point F (fig. 204).

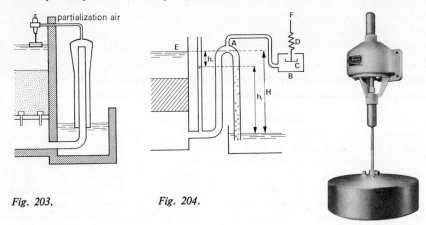

Fig. 203.   *Fig. 204.*

Fig. 205. — Partialization box.

In the first instance, with a constant flow, F is fixed. The filter gradually clogs; its flow drops, causing a reduction in the density of the water-air mixture and hence in the vacuum h_1 at the neck and in the container of the partialization box; the air inlet flow and section are then reduced by the action of the spring; the density of the water-air mixture increases, providing a height h_1 greater than that before clogging. When the filter is completely clogged, no more air enters; the filter discharges at its maximum geometric fall. If the filter is not washed at that time, its flow will start to drop.

Consequently this partialization box provides automatic head loss compensation. It also enables the filter flow to be adapted to the total flow to be filtered, provided the level of point F is connected to the level of the box float. An increase in flow corresponds to an increase in the height of point F and a reduction in the quantity of air entering the siphon. The head loss h_2 drops, causing an increase in the flow discharged by the siphon.

● *Vacuum gauge indicating the head loss.* If a vacuum gauge is placed in the neck of the siphon, it measures the vacuum h_1 representing the head loss through the filter and its pipes.

● *Priming the siphon.* When a filter is started, to avoid a sudden increase in the flow before the water on which the float of the partialization box rests reaches its normal level, it is sufficient to provide for a gradual movement of point F or an auxiliary air inlet that gradually shuts down.

With upstream control of an AQUAZUR filter with a shallow depth of water, this auxiliary air inlet is controlled by a clack-valve fitted to the box.

With downstream control, the float on the partialization box gradually reaches its equilibrium level by hydraulic drainage.

B. Electronic controller.

To function properly, compressed-air controllers need a perfectly pure air supply, and this is very often not available in water treatment plants. To complement their range of controllers, DEGRÉMONT have therefore developed an electronic unit which acts hydraulically or pneumatically on the mechanism being controlled and is specially suited to overcoming problems associated with the control of their filters. The unit works on the following principle (see fig. 206):

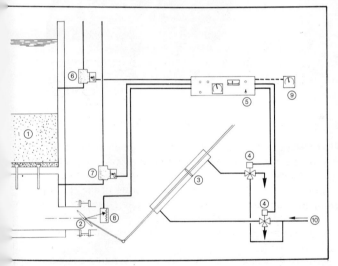

1. *Filter bed.*
2. *Filtered water outlet valve.*
3. *Valve control jack.*
4. *Solenoid valves controlling supply of transmission fluid to jack.*
5. *DEGREMONT electronic control unit.*
6. *Pressure sensor for measuring the level in the filter.*
7. *Underfloor pressure sensor.*
8. *Potentiometer coupled to valve axis.*
9. *Clogging indicator (extra).*
10. *Supply of transmission fluid for jack.*

Fig. 206. — Diagram illustrating the electronic filter control mechanism.

A pressure sensor (6) equipped with strain gauges emits an electric signal proportional to its depth of immersion. This signal is compared to a reference datum representing the level to be maintained constant. After determination of the direction of the imbalance, any deviation between the signal and the datum which exceeds a threshold value intrinsic to the system triggers the electronic control unit (5) which opens one of the two solenoid valves (4) governing the supply of transmission fluid to the jack (3) controlling the butterfly valve (2) for the filtered water, which is thereby fractionally opened or closed until a state of equilibrium is re-established.

A potentiometer (8) coupled to the axis of the butterfly valve sets up an adjustable reaction in the control loop, and over a period this gradually cancels itself out, thereby restoring without hunting the water to the desired level.

This entirely transistorized equipment incorporates a number of ancillary devices which make it possible to adjust the control band, the level of reaction, the amplification factor and possibly also the rate of opening after washing so as to effect a gradual startup over the desired period.

Water or compressed air which has undergone simple filtration can be used optionally as the transmission fluid for the jack controlling the butterfly valve.

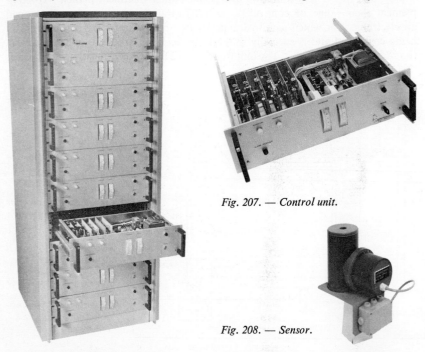

Fig. 207. — Control unit.

Fig. 208. — Sensor.

Fig. 209. — Cabinet housing the entire range of DEGREMONT electronic filter control units.

6.2.2. CONTROLLING A BATTERY OF FILTERS

Two types of control are generally used: control with flow measurement and constant-level control.

A. Control with flow measurement.

Each filter is equipped with a controller on its filtered water outlet, designed to discharge a constant and identical flow for all filters. The flow of filtered water is measured by a pressure differential system (venturi, nozzle, etc.) which sends a signal to the controller; the controller compares the signal with the set flow control point.   According to the difference, the controller closes or opens the component governing the discharge (butterfly valve, diaphragm valve, siphon) until the measurement and the set point coincide.

This control method is used on batteries of both pressure filters and gravity filters.

In the latter case (described in more detail below), nothing maintains the water level on the filters.   Consequently another controller must be added to adjust this level in accordance with the general control system of the plant.

● *With overall upstream control*, a general instrument detects the intake flow and maintains the water level on the filters with a variation of 10 cm (4 in) to 30 cm (12 in) by adjusting the individual set flow of the filters.   If the intake increases, the level upstream of the filters rises and the general detector increases the set flow of the filters until the general upstream level is stabilized, indicating that the flow of filtered water is equal to the flow entering the plant.   Fig. 210 represents a control system of this kind with pneumatic controllers; it would be the same with electric or hydraulic controllers.

The flow to be treated is therefore set before the water enters the plant and the overall control system discharges an equal flow of filtered water, distributing it evenly among the filters.   The flow to be treated can be fixed by programme or on the basis of the level in the filtered water tank.

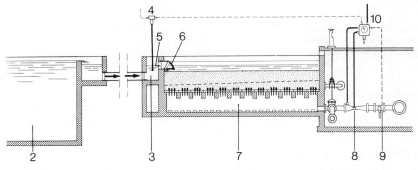

2. *Clarifier.*
3. *Water inlet channel.*
4. *Pneumatic transmitter.*
5. *Water inlet orifice.*

6. *Automatic water inlet valve.*
7. *Filter.*
8. *Venturi.*
9. *Control valve.*
10. *Flow control.*

Fig. 210. — Upstream control with flow measurement.   The flow to be treated is pre-set, and the control system must draw off this pre-set quantity of water and distribute it equally over all the filters currently in operation.

● *With overall downstream control*, a general instrument detects the level in the filtered water storage tank and adjust the individual set flow of each filter accordingly.   Another general controller on the channel supplying the filters detects the level in it and maintains it with a variation of 10 cm (4 in) to 30 cm (12 in) by acting on the instrument controlling the flow entering the station, so as to supply the filters with a flow equal to that which they have been set to discharge. Fig. 211 shows this type of cascade regulation with downstream control.

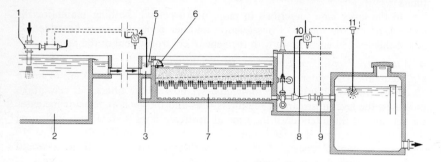

1. *Pneumatically-controlled raw water in-*   6. *Automatic intake valve.*
   take non-return valve.   7. *Filter.*
2. *Clarifier.*   8. *Venturi.*
3. *Clarified water channel supplying filters.* 9. *Control valve.*
4. *Pneumatic transmission or control unit.*   10. *Filter flow control.*
5. *Intake aperture for water entering filter.* 11. *Pneumatic device for signalling level in*
                                                     filtrate tank.

Fig. 211. — Downstream control with flow measurement. The flow of water to be treated must be equal to the flow of discharged water, the latter being variable. This flow must be equally distributed among all the filters currently in operation.

B. Control with maintenance of a constant level.

To obtain a constant flow from each filter, a constant level can be used. In this case, first of all the total flow is equally distributed among the filters, whose outlet systems are controlled by the constant level (up- or downstream) taken as a reference point.

● *Upstream control* (see fig. 212): the flow entering the station is first distributed equally at the inlet to each filter, which therefore receives a flow equal to the incoming flow divided by the number of filters.

Each filter has a control element detecting the upstream level which it maintains constant by acting on the control regulating the output.

As the upstream level is kept constant, the output is equal to the intake and clogging is compensated until it reaches a maximum value related to the available head.

— *Equal flow distribution* (fig. 212): the partialization box (7) shown in the diagram of the T model Aquazur filter (cf. page 284) forms part of the equipment of each filter and maintains a level which is both constant and identical in all

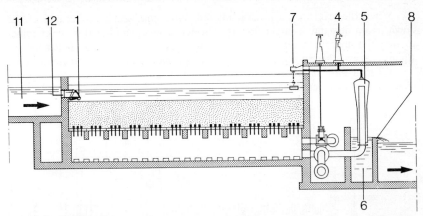

1. *Water inlet valve.*
4. *Head-loss indicator.*
5. *Concentric siphon.*
6. *Filtered water discharge chamber.*

7. *Siphon control valve (partialization box).*
8. *Filtered water outlet weir.*
11. *Clarified water channel.*
12. *Calibrated orifice.*

Fig. 212. — Upstream control ensuring equal distribution to AQUAZUR filters (with maintenance of a constant level).

the units. The filters are fed from a combined header in which, due to the low design velocity, the level of the water is virtually horizontal. The head-loss set up by the calibrated apertures (1) is therefore the same for all the filters, which thus receive the same raw water flow.

— *Control and compensation of clogging head loss :* the partialization box (7) acts on the outlet device (siphon 5) to deliver a flow equal to the intake. If the flow of water arriving at the filter increases, the level tends to rise and the siphon control valve increases the flow delivered by the siphon and vice versa.

Likewise, when a filter becomes clogged, its upstream level tends to rise; the partialization box acts as above. Consequently the head loss is compensated.

When a filter is stopped, the total intake is automatically distributed over the filters remaining in service (except in filters with a surface-sweeping system when the water continues to feed the filter being washed) and there is no risk of the filter bed becoming dry, as if the water level drops the partialization box opens fully and breaks the siphon.

● *Downstream control:* figure 213 shows how this type of control is used; the partialization boxes (7) fitted to each filter are designed to maintain a constant level, identical for all filters, in the filtered water discharge chambers. The level in the general filtered water channel is horizontal and the calibrated orifices (9), all discharging under the same head, deliver an identical flow from each chamber, equal to the total flow divided by the number of filters. Consequently the downstream demand is equally distributed.

When the level in the general filtered water channel drops because of the downstream demand, a larger flow is delivered from each filtered water chamber so that the level in it tends to drop. The partialization boxes (7) act on the siphons (5) and thereby increase the flow until it matches the downstream demand.

When a filter is stopped, the total flow demanded is automatically spread over all the filters remaining in service.

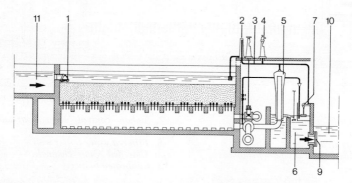

1. *Raw water intake non-return valve.*
2. *Loop to avoid sand emersion.*
3. *Slow start-up pipe.*
4. *Clogging indicator.*
5. *Concentric siphon.*

6. *Filtrate return chamber.*
7. *Partialization box.*
9. *Calibrated aperture.*
10. *Filtrate channel.*
11. *Clarified water channel.*

Fig. 213. — *Downstream control of AQUAZUR filters ensuring equal distribution (with maintenance of a constant level).*

C. Comparison of the various methods of control.

This comparison must be made in the context of optimum filtration. It is universally recognized that the quality of the effluent will be better if the flow is kept stable or if its variations are slow at each change in the running conditions of the plant.

The best control system, therefore, is one with controllers that are simple to maintain and adjust and that operate without hunting and controls that refer to the largest possible areas of water so that the set point variations are slow.

From this point of view, upstream control referred to the total area of the filters is certainly the type giving smoothest operation. In the same way, control by partialization-type siphons or by electro-hydraulic controller is preferable to that by pneumatic controller, which is known to be sensitive to the least degree of clogging, causing hunting.

6.3. Filters without equal distribution or individual control (declining-rate filters)

There are in existence plants with filters that do not operate on either the variable-level or the individual-control system.

This applies to some filter batteries operating under pressure. Batteries of gravity filters can also operate with variable flow.

In this system, all the filters receive their raw water from the same pipe or channel, without any fall since there is no need to worry about equal distribution; this is an advantage, as it avoids breaking down flocculated particles; such breakdown sometimes makes filtration more difficult when the particles are fragile.

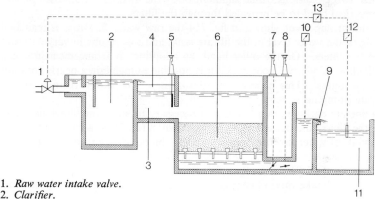

1. *Raw water intake valve.*
2. *Clarifier.*
3. *Filter feed channel.*
4. *Overflow weir.*
5. *Intake yalve.*
6. *Declining-rate filter.*
7. *Filtrate outlet valve.*
8. *Valve for controlling auxiliary head-loss.*
9. *Individual filter outlet weir.*
10. *Individual flow-meter.*
11. *Joint filtrate tank.*
12. *Measurement of level in tank.*
13. *Raw water flow controller.*

Fig. 214. — *Diagram of a general downstream control system for filters with declining flow-rate.*

The filtered water flows into individual chambers, where the level of the weir (9) ensures that the filter bed remains covered during periods of filter shutdown or reduced-flow operation. Each filter outlet is equipped with a filtrate valve (7), open or closed, backed up by another valve (8) which sets up the auxiliary head-loss. The raw water feed rate (1) is regulated according to the water level in the filtrate tank (11) via a water level detector (12) and a general control unit (12).

In an arrangement of this kind, the auxiliary head-loss "p" set up by the valve (8) is adjusted in such a way that, with the maximum flow Q being treated by the plant:

— the individual flow-rate of the filters varies according to the degree of clogging by $\pm$ m% of the mean rate of flow, which is equal to $\dfrac{Q}{N}$, where N is the number of filters in service. So, the flow through a filter which has just been washed is $\left(1 + \dfrac{m}{100}\right)\dfrac{Q}{N}$, while that through a filter prior to washing $\left(1 - \dfrac{m}{100}\right)\dfrac{Q}{N}$.

The value of "m" is currently put at between 20 and 40 %, depending on the predetermined mean velocity.

— The head-loss due to the clogging of the filter prior to washing should, when restored to the value corresponding to the mean rate of filtration, acquire a normal value as determined by the nature of the water and the type of treatment carried out, i.e. from 1.75 to 2 metres (5'9" to 6'7").

These two conditions define simultaneously the auxiliary head-loss "p" and the geometrical head to be provided through the filters.

In the arrangement shown in fig. 214 the raw water flow-rate is regulated according to the water level in the filtrate tank, and this results in varying water levels in the filters. On the other hand, by adopting an arrangement in which the intake flow is otherwise controlled it is possible to maintain a nearly constant level in the filters. This necessitates adjustment of the downstream level and the filtrate pumping system.

To operate declining-rate filters, it is necessary to know the individual flow-rates of the filters, and in the arrangement shown in fig. 214 this can be measured by the depth of water flowing over the weir at the outlet from the filtrate return chamber.

This method of filter control implies:

— a raw water intake valve of considerable cross section to avoid any appreciable head-loss;

— a substantial depth of water above the filter bed;

— a greater filter depth, and consequently a larger structure, than is required for a filter operating at the same mean velocity;

— a geometrical head which is less than that of a filter battery operating at a filtration rate equivalent to the average velocity in a battery made up of declining rate filters, for the same increase in head-loss;

— a less good filtrate quality at the start of the cycle owing to the accelerated filtration rate at the beginning;

— generally speaking, a superior filtrate quality at the end of the cycle thanks to the deceleration of the filtration rate, which limits the risk of breakthrough;

— the lengthy isolation of filters for washing. It is necessary first of all to drain away by filtration the considerable depth of water above the filter bed. This is followed by washing and gradual restarting of the filter. This sequence of operations may take more than an hour per filter and usually requires that two filters be shut down simultaneously, one being drained while the other is being washed, which increases by one the requisite number of filter units as compared with the traditional system of control;

— relatively simple operation, provided that the overall flow-rate and the quality of the raw water remain constant;
— much more painstaking supervision, on the other hand, in the following circumstances:
● where the overall raw water flow-rate varies, any such variation calls for an appropriate adjustment of the auxiliary head-loss set up by the valve (8) so as to provide an adequate incremental head-loss, failing which the volume of water filtered per m² would be very appreciably reduced;
● where rapid changes occur in the quality of the raw water, in which situation the level in the clarified water channel rises quickly as the filters cannot be washed with sufficient speed. This entails the risk of serious loss of water to the overflow system which must be provided upstream of the filters.

7. MONITORING AND AUTOMATION OF FILTERS

7.1. Monitoring equipment

Depending on the type of filter and the control system used, the following functions can be monitored:
— the level of clogging of the filter bed by means of a "head-loss gauge" which may measure either the positive or negative pressure (in the case of control by siphon). Where the user wishes to locate all the visual displays and possibly a head-loss recording device in a central control room, this equipment must be able to relay its signal remotely. Where the washing cycle is triggered by the clogging of the filter bed reaching a certain level, provision must be made for a suitable adjustable setting mechanism on the head-loss gauge;
— the amount by which valves are open or closed; contact switches on the filter valves supply this information;
— the flow of filtered water delivered by each filter; this is useful for filters equipped with flow controllers, but not for those operating on the level-control system with equal distribution in advance;
— the flows of scouring air or wash-water; this is not always necessary. If a volumetric blower is used, the air flow will be correct. As for measurement of the wash-water flow, it is only necessary in filters where this flow has to be controlled to give a precise degree of expansion according to the grain size of the filter material: this applies to filters washed with water alone or filters washed with air and water when the two fluids are used separately. On the other hand, in Aquazur filters with a sand bed where the main washing is done by simultaneous backwash and air scour, it is not necessary to measure the rinsing water flow as it does not have to expand the sand. A correct pump capacity is the only require-

ment. However, it is still useful to measure the wash-water flow in order to determine the total quantity of water used in this operation;
— the turbidity of the filtered water. It is sometimes considered sufficient to measure the turbidity at the general filtered water outlet.

This measurement makes it possible to correct the treatment or general washing instructions when necessary, in accordance with the variations in the properties of the raw water.

It would be ideal to check the turbidity at the outlet of each filter.

7.2 Automation

The operation of a filter plant can be made easier by motorizing the valves incorporated in the equipment and linking them to a system of remote push-button control. This is what is known as an **assisted control system** and does not constitute an automated system in the true sense as the make-up of the various washing sequences, the order in which they are performed and their duration remain the responsibility of the operator.

The three types of automation described in Chapter 19 (page 592) are applicable to the control of the washing operations carried out in a filtration plant and cover:

● *Remote manual control of washing sequences:* in this case, the filter control unit, which may be in the form of a **cyclomatic** (a system of push-buttons connected to relays each one of which relates to a given washing sequence, the order of which is governed by interlocking electrical controls), is usually mounted on a console placed near the filter.

● *Automatic sequential control with manual startup:* in this instance, it is possible either to provide a separate console housing the control equipment for each filter (as above) or to bring together all the control systems for the various filters in a centralized control room.

● *Automatic sequential control with automatic startup:* where this system is used, the automatic washing process is generally triggered by the head-loss reaching a maximum level, although other parameters can also be applied such as the volume of filtered water, the turbidity of the filtrate or the duration of the filtration cycle, in which case the automated control system is governed by one or more of these parameters.

As certain items of washing equipment, e.g. the pump and air-blower, are common to a group of filters, simultaneous washing of several filters belonging to the group is impossible. Therefore, while one filter is being washed, any requirements for washing originating from the units in operation are stored in the memory and then fed back into the automated control system either in the order in which they were committed to memory or according to the numbering of the filters.

10 | ION EXCHANGE

1. PROPERTIES OF AN ION EXCHANGE MATERIAL

An ion exchange substance for industrial use must meet the following specifications:

— **Its chemical structure, generally macromolecular,** must be such that its molecule contains one or more acid or basic radicals.

To clarify the exchange phenomena, the presence of these radicals enables a cation exchanger to be assimilated to an acid of form H-R and an anion exchanger to a base of form R-OH. The strength of the acid or base depends on the nature of the molecular nucleus and the radicals attached to it, HCO_2, HSO_3, NH_3OH, etc.

The exchange material is monofunctional if only one variety of radicals is present, e.g. HCO_2, or HSO_3, and polyfunctional if the molecule contains simultaneously radicals of different natures and consequently different ionic strengths, for example:

— **The product must be insoluble under normal conditions of use.**

In practice, all the exchange materials in current use meet this requirement, and their true solubility at ambient temperatures, disregarding the initial period, is not detectable by the usual methods of analysis under normal conditions of flow and temperature. This is no longer true of certain exchange materials once a certain temperature is reached.

— **The product must be in the form of granules of maximum homogeneity and of dimensions such that their loss of head in percolation remains acceptable.**

— **The changes in state of the exchange substance must not cause any deterioration in its physical structure.**

The exchange material may be required to fix ions or ionized complexes of very varied dimensions and weights.

In some cases this causes an appreciable swelling or contraction (up to 100 % for some carboxylic resins (HCO_2R) between the H and NH_4 phases). This swelling and contraction must obviously not cause the grains to burst. The design of the apparatus must, in the most difficult cases, allow for this expansion without causing excessive stresses in the bed.

It should never be forgotten that there are certain all-too-often disregarded limitations on the use of ion exchangers:

— Ion exchangers can function only in the presence of a liquid phase of limited concentration and cannot solve every problem.

— Ion exchangers are made to fix ions and not to filter substances in suspension, colloids or oily emulsions. The latter substances can only shorten the life of the exchange products. The complex problem of soluble organic substances must be reserved for a more detailed study.

— The presence of large quantities of dissolved gases in the water can cause serious disturbances in the activity of the exchangers.

— The energetic oxidizers Cl_2 and O_3 affect some resins.

— Generally speaking, great caution should be used in the practical application on an industrial scale of laboratory results, and when reading the documentation produced by ion exchanger manufacturers.

The rules for the design and use of the appliances are just as important as a knowledge of the theoretical performances of the exchange substances themselves.

Ion exchange substances used for the applications described below take the form of granules 0.3-1.2 mm in size. Powdered resins of between 5 and 30 microns, known as "microresins", are available for certain special uses (treatment of condensates, waters from nuclear circuits).

2. ION EXCHANGE VOCABULARY

● **Exchange capacity of an ion exchange material.**

This is the weight of ions that can be fixed per unit of volume (or exceptionally per unit of mass) of the exchange material concerned. The capacity is expressed in gramme-equivalents per litre of compressed resin or in degrees per unit of volume; the value of the degrees in terms of the gramme-equivalent varies from one country to another (French degree, German degree, etc.).

A distinction is drawn between:

— *Total capacity*, which is the maximum weight of ions that can be exchanged and which is the property of a given resin.

— *Usable capacity*, which is the usable fraction of the above, depending on the hydraulic and chemical conditions in each individual application.

● **Bed volume.** Ratio of the volume of liquid to be treated per hour to the volume of resin.

● **Ion flow.** Bed volume multiplied by salinity of water (number of milliequivalents treated per litre of resin per hour).

● **Regeneration rate.** Weight of reagent used to regenerate one unit of volume of the ion exchange material.

● **Regeneration efficiency.** This is expressed by the relation:

$$\frac{\text{Gramme-equivalents of regenerating reagent used}}{\substack{\text{Gramme-equivalents of reagent corresponding} \\ \text{stoichiometrically to the ions eluted}}}$$

In this case, the term efficiency is incorrectly used, as in fact it expresses the reverse of the conventional meaning of efficiency.

● **Leakage.** This is the ratio (expressed as %) of the concentration of the ion to be fixed in the liquid, after and before treatment.

● **Attrition.** Mechanical wear of the granules during their use.

3. MAIN TYPES OF ION EXCHANGERS

There are two main groups of ion exchangers:

— *cation exchangers*, in which the molecule contains acid, sulphonic or carboxylic radicals of the HSO_3 or HCO_2 type able to fix mineral or organic cations and to exchange them either with each other or with the hydrogen ion H^+;

— *anion exchangers*, containing basic radicals, for instance tertiary amine or quaternary ammonium functions, able to fix mineral or organic anions and exchange them either with each other or with the hydroxyl ion OH^-.

3.1. Cation exchangers

● *Mineral cation and sulphonated carbon exchangers:* these products are of little more than historical interest.

● *Synthetic cation exchangers:* these products can be divided into two groups: **strongly-acid exchangers,** with HSO_3 sulphonic radicals and acidities close to that of sulphuric acid, and **weakly-acid exchangers** with HCO_2 carboxylic radicals and similar in strength to organic acids such as formic or acetic acid.

● *Sulphonated polystyrenes:* the great majority of cation exchangers in current use consist of these products.

They are all obtained in the same way:

— copolymerization of styrene and divinylbenzene in emulsion form to obtain perfect spheres on solidification;

— sulphonation of the beads thus obtained.

The products obtained by this process are virtually monofunctional. Their physical and chemical properties vary depending on the percentage of divinylbenzene to styrene, known as the degree of crosslinkage, which generally varies from 6 to 16 %.

A large number of firms supply products of this kind under various trade names.

The first list below contains a number of products in common use for fixed beds with a moderate percolation rate and for the treatment of waters of average properties.

These products are not suitable for treatment (continuous or batch processes) at high rates or with frequent cycles, or for the treatment of oxidizing water.

There are on the market specially-designed resins, generally **with a high degree of cross-linkage** and frequently of the "macroporous" type. There are many of these and the best known are those in the third column hereunder.

Supplier	Name of product	
	SULPHONATED POLYSTYRENES IN CURRENT USE	SULPHONATED POLYSTYRENES WITH A HIGH DEGREE OF CROSS-LINKING
Bayer (Germany)	Lewatit S 100	Lewatit SP 120
Dia-Prosim (France) and Diamond Shamrock (U.S.A.)	Duolite C 20	Duolite C 26, C or CI
Dow Chemical (U.S.A.)	Dowex HCR — S and W	Dowex MSC — 1
Imacti (Holland)	Imac C 12	Imac C 16 P
Montedison (Italy)	Kastel C 300	Kastel C 300 AGR — P
Resindion (Italy)	Relite CF	Relite CFS
Rohm & Haas (U.S.A.)	Amberlite IR 120	Amberlite IR 200
Zerolit (G.B.)	Z 225	Z 625

● *Carboxylic exchangers:* these products, of the general formula $HCO_2 - R$, are of the weakly acid type. In water treatment, they can free carbonic acid by fixing Ca, Mg, Na, etc. cations, corresponding to the bicarbonates, but cannot exchange cations in equilibrium with sulphate, chloride or nitrate anions.

Here again a wide variety of resins is available, and some are listed below:

Supplier	Name of product
Bayer	Lewatit CNP 80
Dia-Prosim	Duolite CC 3
Dow Chemical	Dowex CCR 2
Montedison	Kastel C 100
Resindion	Relite CC
Rohm & Haas	Amberlite IRC 50
	IRC 84
Zerolit	Z 236

3.2. Anion exchangers

Anion exchangers can be divided into two main groups:
— weakly- or moderately-basic anion exchangers;
— strongly-basic anion exchangers.

The two types can be distinguished in practive as follows:
— the weakly-basic types do not fix very weak acids such as carbonic acid or silica, but the strongly-basic types fix them completely;
— the strongly-basic types alone are able to release the bases from their salts by the following standard reaction:

$$R\ OH + NaCl \rightleftharpoons R\ Cl + NaOH$$

— the weakly-basic types are more or less sensitive to hydrolysis, in the form of the displacement by pure water of the anions previously fixed to the resin:

$$R\ Cl + H_2O \rightleftharpoons R\ OH + HCl$$

whereas the strongly-basic types are more or less insensitive to this phenomenon.

● *Weakly- or moderately-basic anion exchangers:* all these products consist of a mixture of primary, secondary, tertiary and sometimes quaternary amines. The nucleus of the molecule is very varied in nature and may be aliphatic, aromatic or heterocyclic.

There follows a non-exhaustive list of resins of this type.
Some of these resins are macroporous in structure.

Supplier	Name of product
Bayer	Lewatit MP 62, MP 64
Dia-Prosim	Duolite A 30 B, A 368, A 369
Dow Chemical	Dowex WGR
Imacti	Imac A 20 S
Resindion	Relite 4 MS
Rohm & Haas	IRA 93, IRA 94
Zerolit	H (ip)

● *Strongly-basic anion exchangers:* the existence of quaternary ammoniums in the molecule is typical of these products.

All the strongly-basic resins used for demineralization purposes belong to two main groups commonly known as type I and type II.

The former consists of simply quaternary ammonium radicals, the latter of alkylated quaternary ammonium radicals.

Each type has its own field of application, depending on the nature of the water to be treated and the conditions applying to the regeneration cycle.
The two types differ in the following respects:
— in type I, the basicity is strong and the capacity low; the regeneration efficiency is poor;
— in type II the basicity is weaker and the capacity higher; the regeneration efficiency is also better.

The following list of resins used for ordinary applications is not exhaustive:

Supplier	Type I	Type II
Bayer	Lewatit M 500 M 504	Lewatit M 600
Dia-Prosim	Duolite A 101 D	Duolite A 102 D
Dow Chemical	Dowex SBR	Dowex SAR
Montedison	A 500	A 300
Resindion	Relite 3 A	Relite 2 A
Rohm & Haas	Amberlite IRA 400 IRA 402	Amberlite IRA 410
Zerolit	FF (ip)	N (ip)

A wide range of **high porosity resins** is now available. Called by the manufacturers macroporous or macroreticular, isoporous, homoporous, they are difficult to define in terms of method of manufacture. Experience has proved these types of structure to have a twofold advantage:
— Better treatment results and better resistance to fouling in the presence of

waters containing organic colloids. The percentage of organic matter fixed is higher than with conventional resins, but the percentage of elution on regeneration is also higher.
— Better mechanical resistance both to physical stresses (pressure variations) and to chemical stresses (changes in state of ionic saturation).
These products include:

Suppiler	Type I	Type II
Bayer	Lewatit MP 500	Lewatit MP 600
Dia-Prosim	Duolite A 161	Duolite A 162
Dow Chemical	Dowex MSA 1	—
Imacti	Imac S 5 40	Imac S 5 42
Montedison	Kastel A 500 P	Kastel A 300 P
Resindion	Relite 3 AS	Relite 2 AS
Rohm & Haas	Amberlite IRA 900	Amberlite IRA 910
Zerolit	MP F	MP N

4. USE OF ION EXCHANGERS

All the techniques described in this chapter relate to ion exchange processes, and should not be used unless the raw water has been subjected to a form of pretreatment suited to its type, which must include removal of suspended solids, organic matter, residual chlorine, chloramines, etc. The pretreatment varies with the type of ion exchanger used.

4.1. Fixed-bed with downward flow regeneration

The operating cycle of an exchanger is limited by the exchange capacity of the layer related to an exchangeable mass of ions and consequently to a certain volume of water treated between two regeneration operations.
The cycle has four phases:
— Fixation: the volume of water defined above passes through the layer from top to bottom.
— Back-flow expansion of the resin layer.
— Regeneration: the suitably diluted regenerant is passed through the bed from top to bottom.
— Rinsing: removal of the regenerant by washing with water from top to bottom.

● *General characteristics of an ion exchange unit:* whatever the type of exchange, whether for softening, carbonate removal or deionization, each appliance normally consists of a vertical closed cylindrical container holding the resin. The latter can be placed in direct contact with the device collecting the treated liquid. The device may consist either of nozzles evenly distributed over a tray or of a system of perforated tubes of a suitable number and size. The resins may also be supported by a layer of inert granular materials (silex, anthracite or plastic beads), the layer being drained by the draw-off system (fig. 215).

Sufficient free space is left above the resin bed to allow it to expand normally (between 30 and 100 % of the compressed volume depending on the type of resin) during back-flow expansion.

Both the water to be treated and the regenerant are admitted at the top of the container by a distribution system of varying complexity.

The appliance has an external set of valves and pipes for the various operations of fixing, expansion, regeneration and rinsing. The valves may be manually or automatically controlled, or can even be replaced by a central multiport valve.

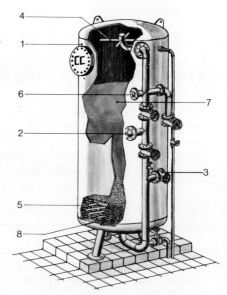

1. *Exchanger body.*
2. *Raw water inlet.*
3. *Treated water outlet.*
4. *Raw water distributor.*
5. *Treated water collector.*
6. *Regenerant discharge.*
7. *Resin.*
8. *Support layer.*

Fig. 215. — Softener.

The three main uses of ion exchangers in water treatment are:
— softening;
— carbonate removal;
— total deionization.

N.B. : to simplify matters, the reactions in the following description are taken as being complete. In practice a slight ion leakage always occurs.

4.1.1. SOFTENING

A cation exchanger regenerated with a sodium chloride solution is used for this purpose.

All the salts in the treated water are transformed into sodium salts.

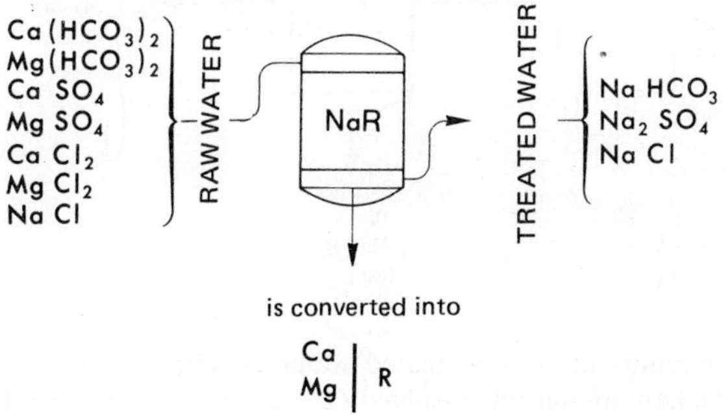

Ca(HCO$_3$)$_2$
Mg(HCO$_3$)$_2$
Ca SO$_4$
Mg SO$_4$
Ca Cl$_2$
Mg Cl$_2$
Na Cl

NaR

Na HCO$_3$
Na$_2$ SO$_4$
Na Cl

RAW WATER   TREATED WATER

is converted into

Ca
Mg   R

Fig. 216. —

The hardness of the treated water is virtually nil. Its pH and alkalinity values remain unchanged.

Softening can be done after preliminary purification with lime; this purification process removes the bicarbonates and reduces the TAC to a value generally between 2 and 4 French degrees.

In this case the water obtained is both free from carbonates and softened.

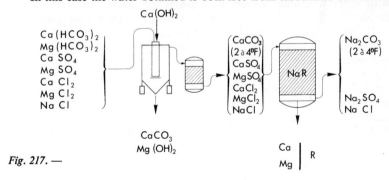

Ca(HCO$_3$)$_2$
Mg(HCO$_3$)$_2$
Ca SO$_4$
Mg SO$_4$
Ca Cl$_2$
Mg Cl$_2$
Na Cl

Ca(OH)$_2$

CaCO$_3$
(2 à 4°F)
Ca SO$_4$
MgSO$_4$
CaCl$_2$
MgCl$_2$
NaCl

NaR

Na$_2$CO$_3$
(2 à 4°F)

Na$_2$SO$_4$
Na Cl

CaCO$_3$
Mg(OH)$_2$

Ca
Mg   R

Fig. 217. —

4.1.2. CARBONATE REMOVAL BY ION EXCHANGE RESINS

The process uses a carboxylic resin which is in the HR form, having been previously regenerated by an acid. It has the property of fixing metallic cations and releasing the corresponding anions in the form of free acid, until the pH of the treated water reaches a level of between 4 and 5, at which point all the carbonic acid from the bicarbonates is released. The cations associated with the anions of strong acids (chlorides, nitrates, sulphates) are not fixed by the resin.

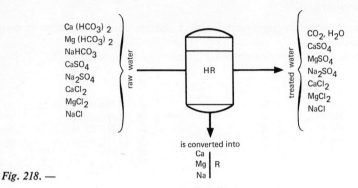

Fig. 218. —

In these circumstances, the treated water contains all the original salts of strong acids and an amount of dissolved CO_2 which is proportional to the bicarbonates in the raw water. The alkalinity of this water may be nil, and its hardness equal to the TH-TAC value of the raw water; the hardness value may therefore fall to zero if the TH is equal to or less than the TAC, since the resin exchanges alkaline-earth ions in preference to alkaline ions.

In the opposite case a zero hardness can be obtained by combining in the same vessel a layer of carboxylic and a layer of sulphonic resin, regenerated in turn with a strong acid and a solution of sodium chloride.

With water containing sodium bicarbonate the fixation efficiency of carboxylic resins is poor, and the H/Na method is sometimes used instead: a sulphonic resin in H form is placed in parallel with another in Na form, and while the former fixes all the cations and releases the corresponding acids, the latter produces softened water.

A mixture of decationized and softened water in suitable proportions provides treated water of the same composition as the first method. It has, however, the disadvantage of requiring the acid water to be kept strictly proportional to the bicarbonated water, as otherwise an acid, hence a corrosive, mixture is obtained.

With these systems it is generally advisable to remove the dissolved CO_2 produced by the ion exchange process.

4.1.3. TOTAL DEIONIZATION

In the simplest system, the water flows successively through a cation exchanger regenerated with an acid and an anion exchanger regenerated with caustic soda.

After passing through the cation exchanger, all the cations are retained by the resin, and there is no longer anything in the water but the acids of the salts it contained initially.

The anion exchanger intended to retain these acids may comprise:
— either a weakly-basic resin which retains the strong anions but not the weak ones such as CO_2 or silica (fig. 219).

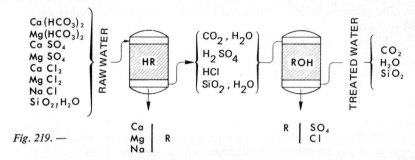

Fig. 219. —

— or a strongly basic resin which simultaneously retains both the strong and weak anions, including carbon dioxide and silica.

It is, however, advisable in this instance to ensure that the CO_2 is physically removed, in order to reduce the consumption of the reagent used to regenerate the anion exchanger (at least where the CO_2 content represents an appreciable proportion of the salinity).

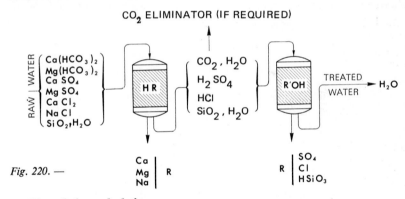

Fig. 220. —

A. Plant design calculations.

The necessary data for these calculations are as follows:

— TAC concentration of the raw water in French degrees;

— SAF concentration of the raw water in degrees ($SO_4 + Cl + NO_3$);

— silica content as $TSiO_2$ ($1^o = 12$ mg/l SiO_2);

— TCO_2: content of carbon dioxide in the water after passing through the cation exchanger and where appropriate after elimination of carbon dioxide;

— volume V of water to be supplied between regeneration processes, in m^3, including service water if appropriate;

— hourly output Q in m^3;

— exchange capacity C of the resins expressed in degrees/litre per litre of consolidated resins. (The degrees may be replaced by milli-equivalents, where 1 milli-equivalent = 5 French degrees.)

The anion exchanger is calculated first: the volume to be used is given by the formulae:

$$V_a = \frac{V \times SAF}{C}$$

for a weakly-basic exchanger, and

$$V_a = \frac{V \times (SAF + TCO_2 + TSiO_2)}{C}$$

for a strongly-basic exchanger.

Then the cation exchanger is calculated, allowing for the additional water αV_a necessary to rinse the anion exchanger, where α may vary from 5 to 20 depending on the type of resin.

$$V_c = \frac{(V + \alpha V_a)(SAF + TAC)}{C}$$

The volumes calculated must then be compared with the hourly output to be treated. There are upper limits to the flow rate or to the bed volume.

If V_c or V_a are too low, they should be adjusted, possibly by increasing the cycle volume V.

B. Deionization flow diagrams.

For total deionization, there are a number of possible alternatives arising from the fact that an absolutely complete exchange cannot be obtained with the simplified layouts described above and which require varying amounts of excess reagent.

Various combinations of exchangers have therefore been studied as a means of either increasing the purity of the demineralized water or reducing consumption of regenerating reagents.

In the diagrams which follow:
— Cf = weakly-acid cation,
— Cf = strongly-acid cation,
— Af = moderately or weakly-basic anion,
— Af = strongly-basic anion,
— CO_2 = physical removal of carbon dioxide,
— Lm = mixed bed.

For each combination, the general ranges of use and performance obtained are given below.

● **Systems with a single run over a cation exchanger:**

a) *Strongly acid cation exchanger + weakly basic anion exchanger:* this system consists of an apparatus containing strongly acid cation exchangers regenerated with a strong acid, operating in series with an apparatus containing weakly (or moderately) basic anion exchangers regenerated with caustic soda or ammonia. The water produced is either used without further treatment, if the carbonic acid content is not harmful, or passed through a CO_2 eliminator located before or after the anion exchanger.

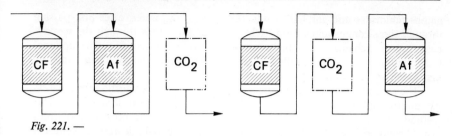

Fig. 221. —

The treated water now contains only a few mg/l of ionized salts, but all the silica and, if degassed, a few mg/l of carbon dioxide. According to the regeneration rate selected for the cation exchanger, the conductivity may vary from 2 to 20 microsiemens per cm. The pH is in the range 6 to 6.5 if the carbon dioxide is properly removed.

b) Plant comprising a strongly-acid cation exchanger and a strongly-basic anion exchanger: all ions, including silica, are removed from water passed in turn over a cation exchanger and a strongly-basic anion exchanger.

In most cases it is advisable to reduce the flow of ions passed to the anion exchanger by installing between the anion and cation exchangers a carbon dioxide eliminator intended to reduce the CO_2 content to a few mg per litre.

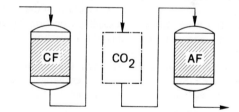

Fig. 222. —

The quality of the demineralized water depends mainly on the rate at which the cation exchanger is regenerated. Ion leakage takes the form of a trace of caustic soda (or lime if the raw water contains no sodium), corresponding to ion leakage from the cation exchanger. Reduction of the silica content itself depends on the sodium content remaining in the demineralized water.

In practice, the water obtained by this method generally has a conductivity of 3 to 20 microsiemens per cm, a silica content of 0.05 to 0.5 mg/l and a pH between 7 and 9.

c) Plant comprising a strongly-acid cation exchanger, a weakly-basic anion exchanger and a strongly-basic anion exchanger: this combination is a variant of the previous one and provides exactly the same quality of water while offering economic advantages when the water to be treated contains a high proportion of strong anions (chlorides and sulphates).   In this system, the water, after first passing through the strongly-acid cation exchanger, passes in turn through the weakly-basic anion exchanger and the strongly-basic anion exchanger.   The optional CO_2 eliminator may be installed either between the cation exchanger and the first anion exchanger or between the two anion exchangers.

The anion exchangers are regenerated in series, with the caustic soda solution passing through the strongly-basic and weakly-basic resins in that order.   This method requires much less caustic soda than the previous one, because the surplus soda remaining after normal regeneration of the strongly-basic resin is sufficient to regenerate the weakly-basic resin completely.

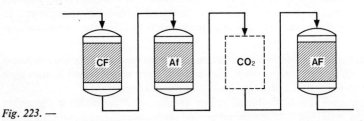

Fig. 223. —

d) *Plant comprising a weakly-acid cation exchanger and a strongly-acid cation exchanger in series:* this combination is a new variant of the previous ones.   It is advantageous when the water contains a high proportion of bicarbonates.   In this system, the water passes in turn through a carboxylic exchanger and a sulphonic exchanger.

Regeneration is effected in series through the sulphonic and carboxylic exchangers in that order.   Since the carboxylic exchanger is regenerated almost stoichiometrically from the surplus free acid remaining after sulphonic regeneration, the total regeneration rate is considerably reduced.

Fig. 224 shows one of the possible combinations for this system.

Fig. 224. —

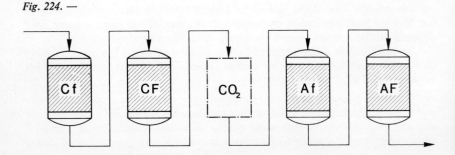

● **Systems with two runs over cation exchangers:**

As already noted, the quality of the water leaving the four combinations described above is determined by ion leakage from the initial cation exchanger. As a result of this leakage, which varies with the composition of the raw water and the rate of regeneration, the quality of the demineralized water obtained is not sufficient for certain uses such as feeding very high-pressure boilers and various chemical, nuclear or electronic applications.

The demineralized water, therefore, has to be further treated in a finishing plant. The ion leakage normally consists exclusively of free bases or alkaline bicarbonates, according to whether the primary system terminates with a strongly or weakly-basic resin. Consequently, if the water from the primary system is passed over a second cation exchanger, it emerges with a very high proportion of its cations removed and, therefore, in a very pure state. A number of combinations based on this pattern are used in practice :

1° — primary cation exchanger:
— CO_2 eliminator if fitted;
— weakly or moderately basic anion exchanger;
— secondary cation exchanger;
— strongly-basic anion exchanger;

In this case, the water leaving the second cation exchanger still contains silica and carbon dioxide which are retained in the strongly-basic anion exchanger. The water acquires a very low conductivity, and the silica content is between 20 and 100 micrograms per litre.

2° — primary cation exchanger;
— deaerator, if fitted;
— strongly-basic anion exchanger;
— secondary cation exchanger;
— second strongly-basic anion exchanger.

With this arrangement, the water treated in the second cation exchanger contains only traces of caustic soda, the second anion exchanger does very little work and the water is of excellent quality (conductivity 0.05 to 1 microsiemens per cm, silica content between 5 and 20 micrograms per litre).

In a useful variant of the above arrangement, the primary strongly-basic anion exchanger is replaced by a combination of a weakly-basic and a strongly-basic exchanger in series.

The resultant arrangement (CF—Af—AF—CF—AF) is preferred for feeding very high-pressure boilers and is extremely reliable. Caustic soda consumption is very low because the two anionic resins are regenerated in series. Finally, experience has shown that this combination gives the best results and the best resin performance when the raw water contains organic matter.

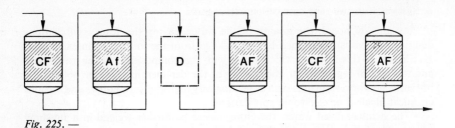

Fig. 225. —

Another variant of this arrangement (Cf-CF-Af-AF-CF-AF) uses less regenerating acid than the previous combination when the raw water has a high content of bicarbonates.

Water from which the cations have been removed and with a slightly acid pH is perfectly suitable for certain applications in the chemical industry. In such cases, the following can be used:
— primary cation exchanger;
— anion exchanger with elimination of carbon dioxide;
— secondary cation exchanger.

The water so obtained has a very low dry extract, contains almost no Na and Ca cations, has a pH between 6 and 7 and a conductivity exceeding 1 microsiemens per cm.

4.2. Back-flow regeneration; use of stratified beds

It was long since realized that it was illogical to pass the regenerant through the resin bed in the same direction as the liquid to be purified. The regenerant cannot be systematically used up if it encounters layers of ion exchanger in a state of decreasing saturation (see general aspects, Chapter 3 page 92).

However, various difficulties relating to the design and technology of ion exchange treatment hindered the introduction of back-flow regeneration.

If the regenerant is to pass through in an upward direction it must be introduced at the bottom of the column and a system installed to collect it at the top of the bed of ion exchange resins. Technological improvements have made it possible to control the expansion of the resin bed and to ensure that the regenerant is distributed as efficiently as possible. The various processes may be divided into three categories:

A. Bed retention by water injection.

The regenerant solution is injected at the bottom of the bed of ion exchange resins, while water is at the same time introduced at the top of the column. The fluids are removed from discharge points located in the upper part of the resin bed.

Evaluations of the regeneration cycle have shown that because of the instability of the hydraulic system it is difficult to make optimum use of the regenerant with this method. Results have nevertheless demonstrated an appreciable improvement in the quality of the treated water, in comparison with downward flow regeneration, particularly when a pulsed system of regeneration is used and the rate of upward flow restricted to between 2 and 2.5 $m^3/(m^2.h)$.

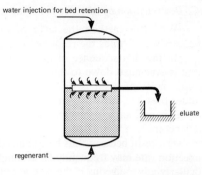

Fig. 226. —

B. Bed retention by mechanical means.

Various mechanical methods of bed retention are used, such as inflation of a rubber or plastic membrane during regeneration, or the use of an inert material to fill the empty space above the resin bed, etc.

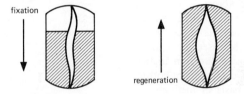

Fig. 227. —

In the "floating bed" process the flow during the exchange cycle is from bottom to top and during the regeneration cycle from top to bottom.

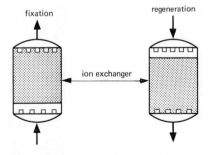

Fig. 228. —

These processes need suitable additional equipment to remove the resin fines and suspended solids brought in by the raw water and the reagents.

C. Bed retention by air injection or air exhaustion.

With this method the layer of ion exchange resin is partially drained, a process which holds the resin bed in place during regeneration.

In the first sequence of the regeneration cycle the top of the ion exchange unit is evacuated down to the level of the eluate collector, which ensures that the upper layer of resin is drained. This process involves either injection of air under pressure, or exhaustion of air by means of an external device. The flow of air is maintained throughout the cycle of injection and displacement of the regenerant (see figure 230).

The resin bed can thus be stabilized and its expansion prevented; the reagent injection rate may frequently be as much as 10 m³/(m². h), a high rate being particularly advantageous if the regenerant is sulphuric acid.

● **Use of stratified beds.**

In certain cases it is possible to combine in a single plant strong and weak resins of the same polarity, provided they are sufficiently different in density. In these circumstances the resins are graded by back-flow decompaction, so that during the fixation cycle the liquid requiring treatment passes in turn through the weak resin and then the strong resin, a design which is in line with the various systems described above.

If the system is to operate with maximum efficiency the regenerant must, of course, be passed through the resin in the opposite direction to that of the exchange cycle, i.e. from bottom to top. On the basis of methods developed to ensure efficient upward regeneration, stratified bed systems of the basic design shown below can be used efficiently:

Fig. 230. — Back-flow regeneration. Bed retention by air exhaustion.

Fig. 229. —

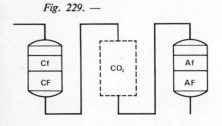

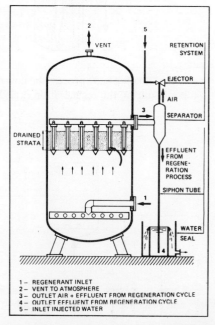

1 – REGENERANT INLET
2 – VENT TO ATMOSPHERE
3 – OUTLET AIR + EFFLUENT FROM REGENERATION CYCLE
4 – OUTLET EFFLUENT FROM REGENERATION CYCLE
5 – INLET INJECTED WATER

The difference in density between carboxylic and sulphonic resins means that they separate less efficiently than moderately and strongly basic anions, and cationic stratified beds are therefore rarely used.

● **Performance of resins regenerated by back-flow methods.**

The quality of the treated water varies with the regeneration rate selected for the ion exchange resins. Water of average salinity and silica content has a conductivity in the range of 0.5 to 5 microsiemens per cm after primary treatment, and the silica level is generally below 50 μg/l. Such primary treatment systems can therefore be used to provide feed water for boilers of medium pressure without any need to include a polishing process using secondary cation/anion resins.

If water of still greater purity is required, systems containing several cation exchange stages can be used (see fig. 231).

The mixed-bed technique is very frequently used in polishing units.

SYNOPSIS OF STANDARD COMBINATIONS OF DEIONIZATION UNITS

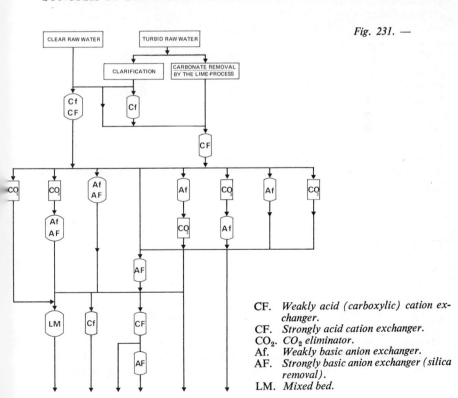

Fig. 231. —

CF. *Weakly acid (carboxylic) cation exchanger.*
CF. *Strongly acid cation exchanger.*
CO_2. *CO_2 eliminator.*
Af. *Weakly basic anion exchanger.*
AF. *Strongly basic anion exchanger (silica removal).*
LM. *Mixed bed.*

4.3. Mixed-bed installation

This differs essentially from the separate bed system in that both the anion and cation resins are used in a single vessel. The two resins are intimately mixed by agitation with compressed air. The resin beads are thus arranged side by side, and the whole bed behaves like an infinite number of anion and cation exchangers in series.

To carry out regeneration, the two resins are separated by decompacting them with a stream of water. As the anion resin is the lighter, it rises to the top, while the heavier cation resin falls to the bottom.

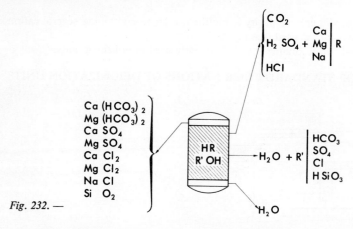

Fig. 232. —

When the resins have been separated, each of them is regenerated separately the anion portion with caustic soda and the cation part with a strong acid. Any excess of regenerant is removed by washing each bed separately.

After partial emptying of the vessel, the two resins are remixed with compressed air. Washing is completed, the vessel is then ready for a fresh cycle.

The advantages of mixed-bed deionization as compared with the separate-bed system are as follows:
— the water obtained is of very high purity and its quality remains constant throughout the cycle;
— the pH is almost neutral;
— rinsing water consumption is very low.

The disadvantages of mixed beds are a lower exchange capacity and a more complicated operating procedure because of the need for separation and remixing to be carried out absolutely correctly.

Mixed-bed exchangers are most frequently used either for finishing, following a primary deionization system, or when the raw water contains very few ions, as in the case of condensed water or nuclear tank water deionized in a closed circuit.

In the latter cases a complex system of ion exchangers can be replaced by a single mixed bed.

Special layouts have also been used as follows:
— cation exchanger—CO_2 eliminator—mixed bed,;
— softener—mixed bed;
— cation exchanger—weakly-basic anion exchanger—mixed bed (advantageous for water with a high content of strong anions).

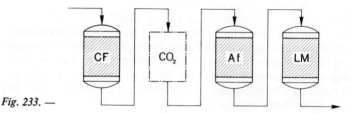

Fig. 233. —

When the water has a high silica content or if silica removal must be as complete as possible, one of the following layouts may be used:

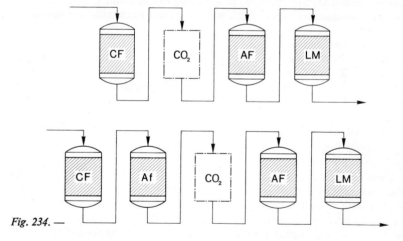

Fig. 234. —

4.4. Moving-bed ion exchange processes

All the processes described above use fixed resin layers operating in batch cycles and contained in vertical vessels.

In such apparatus, each unit has a fixing, expansion, regeneration and washing cycle, following which the ion exchanger is again in its original state, ready to start a new cycle.

This system has several disadvantages:
— use of resin masses generally calculated not on the basis of the hourly flow but on the required autonomy between two regenerations, so that where salinities are high vast exchanger volumes have to be immobilized;
— interruption of the output of purified water while regeneration is in progress; this means that either the processing plant has to be duplicated or extensive storage facilities for treated water have to be provided;
— complexity for the regeneration operations;
— high consumption of bed-expansion and rinsing water.

Moreover, since it is essential to stop the cycle as soon as ion leakage appears at the bottom of the resin layer, which is well before all the resin is saturated with ions to be fixed, the fixing and regeneration performance of the resin is well below the theoretically possible limit. Ion exchange experts have therefore long been considering the possibility of substituting a continuous back-flow process for the conventional method.

The difficulties to be overcome included:
— regular and controlled circulation of the resin;
— separation of the exhausted resin from the treated liquids;
— correct distribution of the fluids in a moving bed of resin;
— use of a circulation method which would no exert mechanical stresses on the resin beads;
— regulating and control devices.

All these difficulties have been overcome with the development of the DEGREMONT ECI Continuous Ion Exchange process, the current version of which incorporates various improvements and simplifications to the systems described in previous editions. The ion exchange unit comprises (see figure 235):
— a fixing column, equipped with a washing/fines removal hopper;
— a resin/raw liquid separation column (optional);
— a regeneration column.

These columns are generally of the compacted-bed type. However the fixing or the regeneration column may be of the fluidized-bed type, e.g. for the treatment of liquids with a high SS content. The resin circulates semi-continuously at programmed intervals, moving from the bottom of the fixing column to the regeneration column and then to the washing hopper before being reinjected into the fixing column. Because all the liquids circulate in the opposite direction to the resins, the various exchanges (fixing, regeneration, washing) show an optimum efficiency, smaller quantities of resin are required, and regeneration performance is greatly improved. The flows of untreated liquid, regenerant, dilution water and washwater are set in advance, and the resin circulation is predetermined by adjusting the emptying interval of the metering hopper on top of the regeneration column. If the composition of the untreated liquid varies, the regenerant injection rate and resin circulation rate are adjusted.

The above description applies to a single ion exchanger used for:

— softening (cation resin regenerated with sodium chloride);

— cation removal (cation resin regenerated by an acid).

If demineralization of the conventional type is required, two identical systems can be combined, each comprising three columns arranged in series, one containing a cation resin regenerated with acid and the other an anion resin regenerated with caustic soda or ammonia.

The quality of the water obtained is limited by the ion leakage from the cation exchanger, which is dependent on the regeneration rate and the salinity of the fluid requiring treatment. If very high-grade water is required (conductivity less than 1 microsiemens per cm, silica down to less than 50 µg/l) an advantageous new method is to use a continuous cation-anion mixed bed, the layout and essential cycles of which are described below.

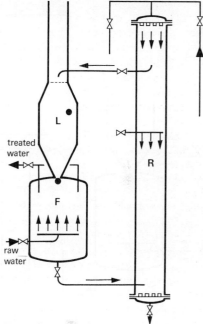

F = *fixing column.*
R = *regeneration column.*
L = *washing/fines removal hopper.*

Fig. 235. —

The installation comprises (fig. 236):

— a mixed-bed fixing column;

— a resin separation column;

— two regeneration columns, one for the cation and the other for the anion exchanger.

In a mixed bed it is not necessary to continue washing until the last traces of salts are completely eliminated. Use may be made of feed hoppers of a smaller size than the washing/fines removal hoppers.

The functions of the main column are fixation and discharge.

The separation column is supplied with mixed resins carried hydraulically from the bottom of the fixing column, and it provides for a clear separation of the cation and anion resins, the former being removed from the bottom and the latter at a suitable level higher up the column.

A level detector actuates an automatic valve which starts or stops the resin

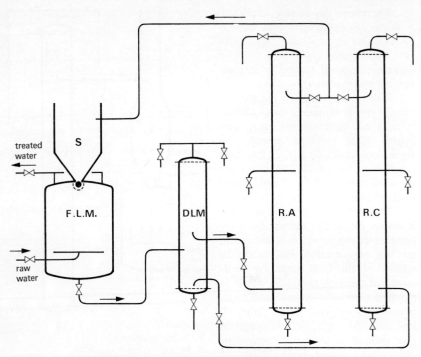

Fig. 236. — *Mixed bed system of continuous deionization.*

F.L.M. = *Fixing column.*   R.A. = *Anion regeneration column.*
D.L.M. = *Separation column.*   R.C. = *Cation regeneration column.*
                                   S.   = *Resin storage.*

supply according to the rate of transfer to the regeneration columns, which operate as described earlier.

The regenerated and washed resins are transferred hydraulically to a mixing hopper which feeds the fixing column.

The resin flow rate is determined very simply by multiplying the volume of each metering compartment by the number of times it is emptied per hour.

This type of plant is used for water which does not contain too high a percentage of bicarbonates, for the design of the mixed bed makes physical elimination of the CO_2 from the bicarbonates impossible. The anion exchanger therefore has to fix all this carbon dioxide, a fact which involves high resin costs and unnecessary reagent consumption.

The most economic layout for treating water containing bicarbonates in order to obtain a high degree of purity consists of continuous carbonate removal over carboxylic resin on the pattern of the single exchanger described earlier, followed by physical elimination of carbon dioxide and a continuous mixed bed as just described.

Regeneration of the sulphonic regenerating column followed by regeneration of the carboxylic column gives a saving in acid consumption; the system has an intermediate tank and allows secondary dilution if the raw water is rich in calcium, and if one wishes to avoid too high a calcium salt concentration in the carboxylic regeneration effluent.

Finally, with this process acid consumption can be reduced by the combined regeneration described above, and soda consumption can be reduced by combining physical removal of the carbon dioxide from the bicarbonates with systematic back-flow regeneration.

● *Applications of the continuous process in the chemical industry:* for treatment of a valuable liquor that must be recovered without being diluted (sugar liquor, glucoses, solutions containing uranium, chemical solutions to be purified), the process must not involve any appreciable dilution. Batch processes always involve a high degree of dilution, as the change from the fixing to the regeneration phase and vice versa necessitates having zones in which the concentration is increased and reduced, and relatively extensive washing. With the continuous process, dilution can be frequently limited to negligible values provided the liquid carrying the resins is recovered in a special column. This column is designed so that the resin is separated from the concentrated liquid in a zone where the very slight dilution only introduces a very small volume of additional water into the treatment circuit.

● *Special advantages of the DEGREMONT-ECI (Continuous Ion Exchange) process:* the types of disadvantage that the continuous processes are designed to overcome have already been mentioned. Compared with other processes, which have mostly remained in the pilot stage, the one described above has the following specific advantages:

1⁰ the resins and fluids are always cycled in the opposite direction to each other, and in the most logical direction, since the regenerant solutions are always injected into fully compacted resins;

2⁰ the resin is transported hydraulically, and no mechanical stresses are therefore imparted;

3⁰ the fixing, regenerating and washing columns are calculated independently of each other and may be suited to any special cases of flow, salinity and regeneration performance;

4⁰ the distribution and compacting devices have been carefully designed so as to reduce the H.T.U. (height of transfer unit) to a minimum and to come as close as possible to the efficiency ratings derived from the equilibrium curves.

A further advantage of all the continuous processes is that the acid and alkaline regeneration effluents are discharged at low constant flow-rates and are therefore much easier to neutralize than in the case with batch processes, nor do they require the installation of large neutralization tanks.

5. PERFORMANCE CHECKS AND ADDITIONAL TREATMENTS

5.1. Checking the performance of a deionization plant

The checks to be made on the performance of a deionization plant essentially include the following measurements:
— conductivity (or its inverse, resistivity);
— silica concentration;
— hardness where necessary;
— sodium concentration;
— pH.
These measurements can be made in the laboratory:
— silica (method No. 204, page 906);
— hardness (methods Nos. 302B, and 303B, page 910);
— sodium—with flame spectrophotometer;
— pH (electrometric methods).

The maximum reliability can only be obtained by continuous automatic checks—see Chapter 19: conductivity, page 575; silica and hardness, page 581; pH, page 577.

For the correct interpretation of the conductivity measurement and the consequent deduction of the ion leakage value, it should be borne in mind that normally, in a properly-designed installation, the deionized water contains only traces of caustic soda if the system ends with a strongly-basic exchange, and traces of sodium bicarbonate or sodium carbonate if it ends with a weakly-basic exchanger.

The following graphs can be used to show the equivalence between measured conductivity and ion leakage in mg/l.

5.2. Removal of organic matter by adsorbent resins

In all the foregoing discussions it has been assumed that the raw water contains mineral substances only.

In reality, all natural waters contain varying quantities of organic substances which react differently with resins according to their nature. Some pass through the resin bed unaffected, others are reversibly fixed and are eliminated on regeneration, while yet others are irreversibly fixed and tend to contaminate the resins.

In practice, the last disadvantage applies particularly to strongly-basic anion exchangers; cation exchangers are virtually unaffected while weakly-basic anion exchangers fix such substances in a more or less reversible manner.

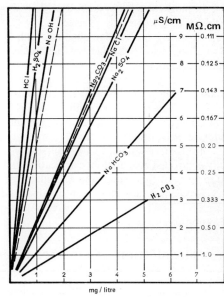

Fig. 237. — Specific conductivity of certain electrolytes art 20 °C.

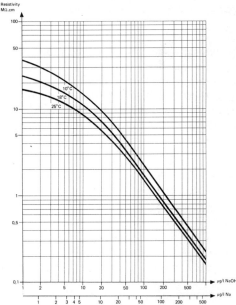

Fig. 238. — Resistivity of water as a function of soda content.

In most cases the systems described above give satisfactory industrial results if the resins are correctly selected.

However, the results obtained with conventional combinations have not always been wholly satisfactory, particularly in the case of raw waters containing humic compounds.

For such raw waters, highly porous and adsorbent anion exchangers can be used either for finishing at the end of a standard system or at the head of the system to protect the demineralizing resins proper.   In the latter case, the organic substances are fixed directly by the resin and eliminated by treatment with sodium chloride or caustic soda.

For surface waters, it is of course advisable to precede deionization by carefully controlled coagulation, clarification and filtration, which will often eliminate 50 to 90 % of the soluble and colloidal organic substances.

Excellent results have been obtained with humic acids by passing the water over finely powdered activated carbon used on a pre-coat filter after coagulation/filtration treatment.

5.3. Disinfection of ion exchangers

Operating difficulties sometimes occur due to the presence of micro-organisms. These difficulties are of two types:

— clogging of the bed invaded by bacterial colonies (especially on carboxylic resins);

— internal contamination of the resin pores (especially on anion exchangers).

The remedies, which must not be used without consulting an expert, are of two types:

— preventative, by prior continuous or batch chlorination of the raw water;

— curative, by disinfection of the resin bed with formol, or with a solution of a product with a quaternary ammonium base.

ADSORPTION AND ADSORBENTS

The term adsorption was defined in Chapter 3 (page 86) and an indication was given as to the main factors influencing the transfer of organic matter from a solution into the surface of an adsorbent.

This chapter describes the main adsorbents, of which activated carbon is the most widely used, and gives an account of the use and main applications of activated carbon in granular and powder form.

1. MAIN ADSORBENTS

1.1. Activated carbon

It was mentioned in Chapter 3 that specific surface area is one of the factors determining the quality of an adsorbent. Inexpensive techniques, such as chemical or thermal activation, can be used to make the specific surface area of various carbon substrates very large (700 to 1 500 m²/g). This is why activated carbon is by far the most widely used adsorbent in the water treatment industry.

The common types of activated carbon are prepared from anthracite, bituminous or soft coal, petroleum coke, peat, wood and coconut.

Experience shows that activated carbon has a broad spectrum of adsorptive activity, as most organic molecules are retained on its surface; hardest to retain are the shorter molecules (especially those containing less than three atoms of carbon: ordinary alcohols, primary organic acids), and the least polar. However, heavier molecules, aromatic compounds, substituted hydrocarbons, etc. are firmly retained.

It is obvious that the molecules mentioned above as being hard to retain are conversely the easiest to break down biologically; this means that adsorption treatment is complementary to biological treatment.

Activated carbon is therefore used in the following processes:
— the final polishing treatment of very pure industrial process water and drinking water, particularly when produced from surface water, in which case the activated carbon will retain the dissolved organic compounds not broken down by natural, biological means (self-purification of waterways): micropollutants, substances determining the taste of the water; it will also adsorb traces of certain heavy metals (see Chapter 20, page 650);
— the "tertiary" treatment of municipal and industrial waste water. The carbon retains dissolved organic compounds which have resisted upstream biological treatment, thereby eliminating a large part of the residual COD;
— industrial waste water treatment, when the effluent cannot be decomposed biologically or when it contains organic toxic elements ruling out the use of biological techniques. The use of activated carbon often allows the selective retention of toxic elements and the resultant liquid can then be decomposed by normal biological means.

Another widely exploited property of activated carbon is its **catalytic action,** particularly on the oxidation of water by free chlorine:

$$Cl_2 + H_2O \longrightarrow 2HCl + \frac{1}{2}O_2 \nearrow$$

This is the method used for the **dechlorination** of water subjected to excess chlorination treatment. This dechlorination action is characterized by the **half-dechlorination depth,** that is, the depth of the filter bed which, for a given velocity, causes the reduction by one-half of the amount of chlorine in the water measurement (cf. Chapter 27, page 947). The pH has considerable influence on this depth. According to the temperature, the free chlorine content and the tolerance allowed on the residual chlorine, loads are used of between 5 and 15 volumes of water per volume of activated carbon per hour.

The same type of catalytic action is used to break down chloramines into nitrogen and hydrochloric acid. However, the kinetics are slower than in the case of free chlorine (the depth of half-dechlorination is very much greater); therefore the load must be greatly reduced in volume if comparable results are to be obtained.

The dechlorination capacity of a carbon is affected by any factor that might interfere with the contact between the carbon and the water to be treated, such as deposits of calcium carbonate, surface saturation through adsorption of various pollutants, etc.

1.2. Other adsorbents

Apart from a few natural adsorbents already mentioned in Chapter 3, in recent years attempts have been made to develop new adsorbents:

— inorganic adsorbents: alumina and other activated inorganic oxides; they can have a very large specific surface area, but only a few substances are compatible with them, making them very specific adsorbents;

— organic adsorbents: macromolecular resins with specific surface areas of between 300 and 500 m²/g; their adsorptive capacity is poor compared with activated carbon; however, these resins have better adsorptive kinetics (range of use between 5 and 10 bv/hr) and are often easier to regenerate (low binding energy).

They are recommended in a few cases such as the protection of ion exchange systems in metal finishing processes where non-ionic detergents are present, or the retention of phenol compounds for recovery.

Here should also be mentioned the "scavengers", which are highly-porous anionic resins, as described in Chapter 10. However, these resins have a smaller specific surface area and their action on polar substances (such as humic acids, anionic detergents) is partly due to their ionic charge, which distinguishes them from other adsorbents.

2. USE OF ACTIVATED CARBON

Since activated carbon is by far the most important, we shall limit our study to this one adsorbent. It is available in two forms: powdered carbon and granular carbon.

2.1. Powdered carbon

Powdered carbon takes the form of grains between 10 and 50 μm and its use is generally combined with clarifying treatment. It is added continuously to the water together with the flocculating reagents. The carbon enters the floc and is then extracted from the water with it. It is sometimes recommended to carry out this extraction by direct filtration, but it is better to use a recirculation clarifier (Turbocirculator) or better still, a sludge blanket clarifier (Pulsator). These clarifiers considerably increase the time during which the water and carbon are in contact, thereby making it easier to attain equilibrium. Thus, by using a Pulsator instead of a static settling tank, a saving of 15 to 40 % of carbon can be achieved, while still obtaining the same result.

Powdered carbon can also be used as a precoat for a candle filter for the final polishing of very pure industrial process water (treatment of condensates, rinsing water in the electronics industry).

2.1.1. ADVANTAGES

— Activated carbon in powder form is 2 to 3 times less expensive than granular carbon;
— extra quantities of powder may be used to handle pollution peaks;
— investment costs are low when the treatment only involves a single stage of flocculation-settling (a single feeder for activated carbon is all that is necessary);
— adsorption is rapid, since the surface area of the powder is directly accessible;
— the activated carbon promotes sedimentation by making the floc heavier.

2.1.2. DISADVANTAGES

— The activated carbon cannot be regenerated when mixed with hydroxide sludge and must then be regarded as expendable. When carbon is used on its own (without inorganic coagulants), modern techniques of regeneration in fluidized beds can be used; but even these lead to very heavy losses;
— it is difficult to eliminate the final traces of impurities without adding an excessive amount of activated carbon.

Therefore powdered carbon is mostly used when intermittent or small quantities are required (smaller than 25 to 50 g/m³ depending on the case).

Fig. 239. — Unit for the dosing and dilution of activated carbon. Orly water works. Paris water supply.

2.2. Granular carbon

2.2.1. CHARACTERISTICS

The table below shows examples of a few physical characteristics of several types of granular carbon usable in water treatment. Considerable variations can be observed and it is important to bear these in mind when selecting the carbon and the equipment necessary for its use (grain size, density, friability).

Raw material	A Peat	B Peat	C Coal	D Coal	E Wood	F Petroleum coke
Grain size:						
• Effective size mm	0.95	0.66	0.45	0.6	1.7	0.4
• Uniformity coefficient	1.5	1.4	2.2	1.6	1.4	1.4
Form	crushed	extruded	crushed	crushed	crushed	pellets
Friability[1] :						
• 750 strokes	31	22	20	11	2.2	25
• 1 500 strokes	54	38	30	23	6.7	45
Bulk density (compacted)	0.29	0.42	0.67	0.45	0.69	0.58
Specific surface area (m²/g)	650/750	1 100/ 1 200	800/900	1 050/ 1 200	1 000/ 1 100	1 100/ 1 300
Ash content %	4 to 6	5.5	8	8	12	0.2

1. Cf. Analysis n°. 502, page 939.

2.2.2. CARBON BED TECHNOLOGY

Granular carbon is used as a filter bed through which the raw water passes, leaving behind its impurities which are extracted methodically. The water, as it progressively loses its pollutants, encounters zones of activated carbon which are less and less saturated and therefore more and more active.

Functions of a carbon bed.

A compact bed has four functions:

● *Filtration:* this must often be reduced to a minimum in order to avoid clogging of the bed, which is unavoidable without efficient cleaning systems to break up the layers completely after each cycle. In addition, the carbon tends to extract adsorbable products from the floc with which it is in contact, causing premature saturation. This is why it is often advisable to use sand filtration as a preliminary step.

● *Biological media :* the surface of carbon offers ideal conditions for bacteriae growth. This phenomenon can assist purification, but can also be very dangerous if not properly controlled (anaerobic fermentation giving off odour, clogging of bed, etc.).

● *Catalytic action* (see Chapter 11, page 342).

● *Adsorption:* this remains the principal role of the carbon. Good water-carbon contact must be maintained and whenever the capacity of the carbon (see paragraph 2.2.3. below) is important for the economics of the process, a countercurrent system should be set up. There are two possible arrangements: moving beds and fixed beds.

— *The moving bed* makes use of the countercurrent principle for the flow of water and activated carbon.

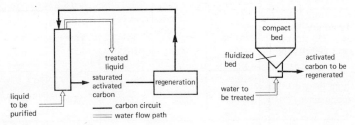

Fig. 240. —

The base of the bed can be a fluidized bed which, inter alia, facilitates the extraction of the carbon (cf. figure 240).

— *Fixed beds.*

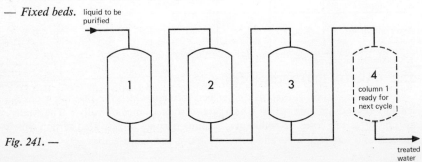

Fig. 241. —

Generally speaking, a series of 2 or 3 columns is used so that the first column can be thoroughly washed without losing the efficiency of the methodical extraction system and, in the same way, only the most saturated carbon (first column) is regenerated. Thus, a countercurrent system can be organized by changing around the order of the columns — 1, 2, 3 → 2, 3, 1 (after regeneration), etc.

The technology used here is similar to that used in sand filters. Special attention should be paid to washing systems, due to the risk of carbon loss through flotation.

2.2.3. ADSORPTIVE CAPACITY OF CARBON

Whether treatment using activated carbon is economical or not largely depends on **the adsorptive capacity of the carbon,** expressed in grams of retained COD per kilogram of activated carbon, which characterizes the "consumption of carbon" in obtaining a given result. For a given system, containing polluted water and carbon, this capacity depends on:

● *the depth of the bed:* the deeper a bed, the easier it deals with extended adsorptive fronts without excessive leakage (a similar principle to that of ion exchange described in Chapter 3 on page 88) while still ensuring thorough saturation of the upper layer;

● *the exchange rate:* experience shows that 3 volumes of water per volume of carbon per hour can seldom be exceeded when treating high levels of pollution. In the treatment of drinking water, in which the content of adsorbable products is very low, any decision as to the economic optimum must take account of the high investment costs, with the result that higher rates are used (5 to 10 bv/h) and incomplete saturation of the carbon must be accepted.

The theory gives only an indication of the trend of the laws of adsorption. It still remains essential to call upon the experience of the expert and to carry out dynamic tests on columns of sufficient size so that results can be extrapolated.

2.2.4. REGENERATION

Activated carbon (like artificial adsorbents) is an expensive product. In most cases the cost of replacing the saturated carbon would be prohibitive. It must therefore be regenerated, and three methods have been developed for this purpose:

● *Steam regeneration:* this method is restricted to regenerating carbon which has only retained a few very volatile products. However, steam treatment can be useful in unclogging the surface of the grains and disinfecting the carbon.

● *Thermal regeneration:* by means of pyrolysis, burning off adsorbed organic substances. In order to avoid igniting the carbon, it is heated to about 800 °C in a controlled atmosphere. This is the most widely used method and regenerates the carbon very well, but has two disadvantages:

— it requires considerable investment in either a multiple-hearth furnace, a

fluidized bed incinerator or a rotary kiln. The furnace must have monitoring devices for atmosphere and temperature, a dewatering system at the inlet and a carbon quenching system at the outlet;
— it causes high carbon loss (7 to 10 % per regeneration), so that after 10 to 14 regenerations, the starting volume of carbon, will, on average, have been entirely replaced.

● *Chemical regeneration:* DEGREMONT have developed a process based on the action of a solvent used at a temperature of approximately 100 °C and with a high pH. The advantage of this process is that, for the same capital outlay, only minimum carbon loss occurs (about 1 % of the quantity treated).

However, the use of chemical reagents for regeneration (alkaline reagent and solvent) leads to the formation of eluates from which the solvent must be separated by distillation. The pollutants are then destroyed by incineration unless they can be recovered.

This process is less widely used than thermal regeneration.

● *Biological regeneration:* this method of regeneration has not yet been applied on an industrial scale.

2.3. Combined use of powdered and granular carbon

When surface water with variable levels of pollution is being treated, it may be advantageous to use both powdered carbon in the clarification process to tackle pollution peaks and granular carbon to adsorb normal pollution.

2.4. Main applications

As was emphasized above (p. 341), activated carbon should be considered every time dissolved organic pollutants need to be eliminated. Examples are:
— detergents;
— synthetic soluble dyes;
— chlorinated solvents;
— phenols and hydroxyl derivatives;
— aromatic derivatives, substituted or unsubstituted, especially chlorine or nitrate derivatives;
— tastes and smells.

SEPARATION BY MEMBRANES

The present chapter covers the three membrane-separation techniques which have proved to be of practical use in the treatment of water: reverse osmosis, ultrafiltration, electrodialysis.

The physical phenomena involved and the most useful properties of the semi-permeable and dialysis membranes which constitute the "core" of the treatment plant are described in Chapter 3, page 82.

The operating conditions and the main applications of these techniques are described below. It should be noted that they have only recently appeared on the market and that, like all new arrivals, they are still capable of extensive adaptation which will open up numerous new areas of application and reduce their cost.

1. REVERSE OSMOSIS

1.1. Osmosis and reverse osmosis

Reverse osmosis derives from the fact that direct or natural osmosis is a reversible process.

In fig. 241, compartment A contains an aqueous solution of inorganic salts, whereas compartment B contains pure water. When direct, or natural, osmosis occurs, pure water passes from compartment B into compartment A. The level in compartment A rises until the pressure set up by the liquid column counterbalances the flow of pure water. A state of osmotic equilibrium is thus achieved and, as has been explained in Chapter 3, page 83, the magnitude of this hydrostatic pressure is termed the osmotic pressure of solution A.

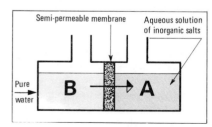

Fig. 242. — Direct osmosis.

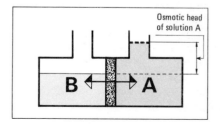

Fig. 243. — Osmotic equilibrium.

If we now apply to the saline solution a hydrostatic pressure in excess of the osmotic pressure (see fig. 244), it is found that pure water flows in the direction opposite to that described above and that the salts are retained by the membrane. This phenomenon has become known as reverse osmosis.

Reverse osmosis makes use of semi-permeable membranes which allow water to pass through while retaining 90-99 % of all the inorganic substances in solution, 95-99 % of the organic constituents, and 100 % of the most finely divided colloidal matter (bacteria, viruses, colloidal silica etc...).

N.B. The efficiency of membranes in the removal of salts is stated to vary from 90 to 99 %. In other words, their **salt passage (S.P.)** lies in the 10 % to 1 % range.

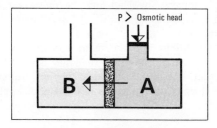

Fig. 244. — Reverse osmosis.

1. *Raw water.*
2. *Purified water (permeate).*
3. *Concentrate (waste water).*
4. *High-pressure pump.*
5. *Reverse osmosis module.*
6. *Semi-permeable membrane.*
7. *Discharge valve.*

Fig. 245. — Simplified reverse osmosis flow diagram.

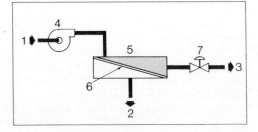

Membranes are assembled in units called **modules,** and the simplest practical arrangement is to connect together in series a high-pressure pump and a module. The pure water passes through the membrane under the action of the pressure, while the concentrated reject is extracted continuously via a flow control valve (fig. 245). **The ratio between the flow-rate of the purified water and that of the raw water feed is called the conversion factor (Y) and is expressed as a percentage.**

Lastly, as has already been pointed out in Chapter 3, page 84, the rate of flow through a reverse osmosis membrane is directly proportional to the effective pressure (i.e. to the difference between the applied and osmotic pressures). It must be remembered here that the concentration in a single module ranges from the inlet concentration, C_e, to the maximum concentration $C_r \times \Psi$ (where C_r is the concentration of the reject and Ψ is the polarization coefficient). To obtain a satisfactory flow of purified water it is therefore advisable to apply a pressure well in excess of the osmotic pressure of the reject (cf. Table 1).

In practice, the pressures used vary between 25 and 80 bar (at least where large plants are concerned).

TABLE 1

THE OSMOTIC PRESSURE OF VARIOUS SALTS IN SOLUTION

Salt	Concentration mg/l	Osmotic pressure bar
NaCl	35 000	27.86
NaCl	1 000	0.79
Na_2SO_4	1 000	0.42
$MgSO_4$	1 000	0.25
$CaCl_2$	1 000	0.58
$NaHCO_3$	1 000	0.89
$MgCl_2$	1 000	0.67

1.2. Reverse osmosis membranes

When membranes are prepared in the laboratory, they are usually flat and scant attention is paid to the choice of materials. On the other hand, when it is applied in an industrial context, the economic viability of the process becomes the overriding objective. This is determined mainly by the configuration of the membrane and by the stability in time of the material chosen.

There are today two types of membrane on the market:
— cellulose acetate (a mixture of mono-, di- and triacetate), and
— aromatic polyamide membranes.

Cellulose acetate membranes give a high flow-rate per unit surface area and are used in the form of tubes, as spirally-wound flat sheets and, more recently, as hollow fibres.

Polyamide membranes, on the other hand, have a lower specific rate of flow. They are manufactured in the form of hollow fibres so as to achieve the maximum surface area per unit volume. This is about 15 times that of the spirally-wound membranes (cf. figs. 246 and 247).

It is important to note that polyamide membranes possess outstanding resistance to the action of chemical and biological agents and that they therefore have a longer life than acetate membranes, which are inevitably subject to hydrolysis (although this hydrolysis can be cut down to a minimum by adhering to strict control of the operating pH and temperature).

The differing characteristics of the two types of membrane made respectively of cellulose diacetate and polyamide are listed in Table 2.

TABLE 2
SPIRALLY-WOUND ACETATE
AND HOLLOW FIBRE POLYAMIDE MEMBRANES

Membranes	Applied to the treatment of water with a salinity of < 15 g/l	
1. Material	B-9 aromatic polyamide	cellulose acetate
2. Configuration	hollow fibres	spirally-wound hollow fibres
3. Physical data: — Normal working pressure — Maximum back-pressure of treated water — Maximum operating temperature — Maximum storage temperature	28 bar 3.5 bar 35 °C 40 °C	30 — 42 bar 30 °C 30 °C
4. Chemical characteristics: — pH acceptable — Hydrolysis — Bacterial attack — Free chlorine: maximum acceptable continuous dose — Other oxidizing agents	4 — 11 unaffected unaffected pH ≤ 8 : 0.1 mg/l pH > 8 : 0.25 mg/l highly resistant	4.5 — 6.5 highly sensitive highly sensitive } 0.5 — 1 mg/l moderately resistant
5. Operating life	3 — 5 years	2 — 3 years
6. Salt passage (NaCl)	5 — 10 %	5 — 10 %

Also available are membranes intended specifically for the treatment of sea-water. Based on the same polymers, these have a closer texture which enable them to desalinate in a single stage solutions, like sea-water, which contain some tens of grammes of salt per litre. Since 1975, the application of B10 hollow fibre polyamide membranes has formed the basis of several sea-water desalination plants, notably for shipboard use.

1.2.1. SPIRALLY-WOUND MODULES (cf. Fig. 246)

In this arrangement, pairs of membranes are wound round a central tube which collects the permeate. The solution to be demineralized flows parallel to the central tube through the gaps made by spacers (usually of plastic mesh)

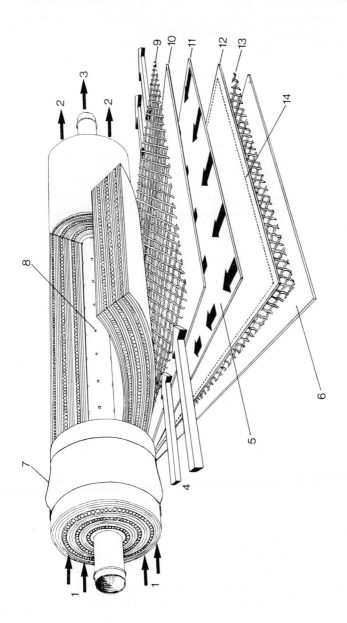

1. *Raw water.*
2. *Reject.*
3. *Permeate outlet.*
4. *Direction of flow of raw water.*
5. *Direction of flow of permeate.*

6. *Protective coating.*
7. *Seal between module and casing.*
8. *Perforated tube for collecting permeate.*
9. *Spacer.*
11. *Membrane.*

11. *Permeate collector.*
12. *Membrane.*
13. *Spacer.*
14. *Line of seam connecting the two membranes.*

Fig. 246. — Spirally-wound module.

between the two active faces of the membranes. The permeate is collected in a porous material, through which it flows to the central tube.

Using flat membranes, this technique makes it possible to construct modules which are markedly more compact than those which first appeared on the market (i.e. flat membranes on top of porous plates or supported by tubes).

1.2.2. HOLLOW FIBRE MODULES, E.G. THE PERMASEP B-9 * MODULE (cf. Fig. 247)

A hollow fibre can be likened to a thick-walled, porous cylinder, the strength of which depends on the ratio of the outside and inside diameters. Provided that this ratio remains constant as the two diameters are decreased, the mechanical strength of the cylinder is maintained despite the reduction of the wall-thickness (which increases the flow-rate of the water passing through it). This fact makes it possible to achieve a membrane with the maximum surface area per unit volume which is at the same time capable of withstanding high pressures without mechanical support.

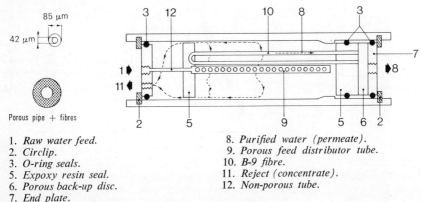

1. *Raw water feed.*
2. *Circlip.*
3. *O-ring seals.*
5. *Expoxy resin seal.*
6. *Porous back-up disc.*
7. *End plate.*

8. *Purified water (permeate).*
9. *Porous feed distributor tube.*
10. *B-9 fibre.*
11. *Reject (concentrate).*
12. *Non-porous tube.*

Fig. 247. — *Basic diagram of B-9 module.*

Hollow B-9 fibre has an outside diameter of 85 μm with an inside diameter of 42μm. Of the 21 μm wall-thickness it is only the outside layer (about 0.1 μm thick) of more compact texture which provides the selective action. A membrane of this type is said to be "asymetrical". The fibres are able to withstand working pressures of 28 bar without damage. They are assembled into modules in the way shown diagrammatically in Fig. 247. Several hundred thousand fibres are arranged in a U-shaped configuration inside a fibreglass pressure vessel. Raw water under pressure is distributed radially inside the module by a porous or perforated axial header which runs the whole length of the module.

---

* Registered trademark of Du Pont de Nemours.

Under the effect of the pressure outside the fibres, the pure water passes through the fibre wall and into the central duct, by which it is carried out through the epoxy resin sealing plate in which the free ends are supported. It is then collected in a porous disc and extracted from the module.

The concentrated reject is collected on the outside of the fibre bundle before being extracted through an orifice located at the same end as the raw water inlet. This type of module is made in a number of sizes with varying output capacities.

TABLE 3

CHARACTERISTICS OF THE PERMASEP B-9 0840 MODULE

Length (m)	1.2
Diameter (cm)	25
Output capacity * (m³/d)	53
Salt passage * (%)	< 10
Maximum working pressure (bar)	28
Permissible back-pressure (bar)	3.5
pH range	$4 - 11$
Maximum operating temperature (°C)	35

* Output capacity and salt passage are measured at 25 °C with a 1 500 mg/l sodium chloride solution at 28 bar with a conversion factor of 75 %.

By way of example, the chief characteristic data relating to the most widely used module are given in Table 3. Noteworthy is the high output capacity (53 m³/d or 9.7 gpm) of the module in relation to its modest volume of 60 dm³ (2.12 cu ft).

1.3. Reverse osmosis plants

Irrespective of the nature of any pre- or post-treatment which may be applied, a reverse osmosis plant comprises a number of elementary modules juxtaposed in a particular geometrical arrangement. This fact underlies the ease with which units can be expanded and explains how it has been possible, in the space of a few years, to build up from pilot plants with an output capacity of just a few cubic metres per hour to industrial installations with capabilities of the order of thousands of cubic metres per hour, such as the MANFUHA and SALBUKH plants at RIYADH (Saudi Arabia) with output capacities of 38,500 and 46,000 m³/d respectively (10 and 12 mgd).

In the simplest plant layouts the modules are **arranged in parallel.** All the modules then operate under identical pressure conditions and with the same conversion factor. This is the system adopted for most low-capacity plants (see fig. 248). A cartridge-type filter protects both the high-pressure pump and the membranes against the penetration of suspended solids. Two pressure-gauges located upstream and downstream of the modules provide the means of monitoring continuously the head-loss within the system. Two flowmeters, measuring respectively the treated and the reject water, indicate the conversion ratio, which is controlled by the two regulating valves shown in fig. 248.

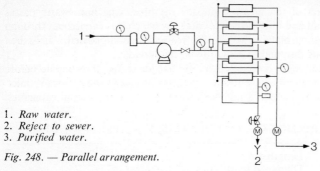

1. *Raw water.*
2. *Reject to sewer.*
3. *Purified water.*

Fig. 248. — Parallel arrangement.

Often other plant layouts are chosen. For example, a series arrangement (see fig. 249) may be used to increase the conversion ratio. The reject solution from the first stage provides the feed for the second stage. An intermediate pump is not necessary as the pressure available at the outlet from the first stage differs very little (2 — 3 bar head-loss) from the inlet pressure of the second stage. A system of this kind, usually referred to as the "**reject staging**", is easily capable of achieving conversion ratios of 70 to 90 % (in a two or three-stage plant) without any increase in the polarization coefficient.

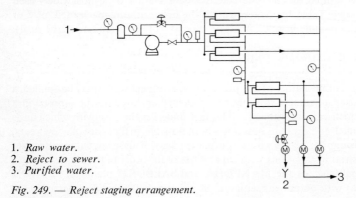

1. *Raw water.*
2. *Reject to sewer.*
3. *Purified water.*

Fig. 249. — Reject staging arrangement.

For other applications, e.g. for the production of very high-quality water, a two-stage treatment of the purified water ("**product staging**") may be used. The output from the first stage is then pumped to the second stage, where the purification process is repeated. The low-concentration reject from the second stage is recycled upstream of the plant to effect some dilution of the feed water.

To satisfy the requirements most usually encountered, DEGRÉMONT have developed their standard **OSMOPAC** range of monobloc reverse osmosis units, which includes equipment comprising from 3 to 30 modules (**OP 3** to **OP 30**), (see fig. 250).

Fig. 250. — Monobloc OP 30 osmosis unit.

● *Plant operating conditions.*

Fundamentally, osmosis is a process of concentration. It is therefore necessary to avoid the precipitation of any of the solutes contained in the reject, failing which the system will be exposed to the danger of rapid scale formation. Now the concentration of the reject is directly linked to the chosen conversion factor (Y), so that, if we ignore the salts which pass through the membrane, the concentration of the reject C_r is given by the expression:

$$C_r = \frac{100\,C_e}{100 - Y}$$ where C_e is the concentration of the feed water

Hence, if $Y = 75\,\%$, $C_r = 4\,C_e$.

If the aim is to achieve conversion factors in excess of 20 to 30 %, some pretreatment of the raw water is generally necessary. The optimum conversion ratio can only be determined in the light of a complete analysis of the raw water (the higher the value of Y, the more intensive is the pretreatment likely to be, but the less energy will be consumed in producing a cubic metre of treated water).

Such pretreatment is designed to eliminate or lock up either the anion or the cation of the compound which is likely to be precipitated. With this object in view, the solution is subjected to one of a number of treatments which include: acid vaccination, carbonate removal, softening on ion-exchange resin, silicon removal, iron removal etc.

All the suspended particles rejected by the reverse osmosis membrane can accumulate on its surface if the conversion ratio becomes excessive (resulting in inadequate sweeping of the membrane). The optimum conversion factor can be determined by applying a test to measure the fouling index (see analysis 712 E, p. 956). Where the water contains an excessive quantity of colloidal particles, clarification and filtration treatment is required to prevent rapid fouling of the membrane.

It is necessary, in any event, to provide for periodical cleaning of the membranes to remove all traces of deposited matter. It is here that the advantages of a chemically stable membrane become particularly apparent, as it is then possible to employ a wide range of cleaning solutions suited to the various impurities contained in the water.

1.4. Applications of reverse osmosis

Reverse osmosis is appropriate to all cases where the object is to demineralize the water or to bring about a concentration of ions or organic molecules. This having been said, it remains true that at the present time two highly important applications account for the lion's share of the market.

1.4.1. THE SUPPLY OF DRINKING AND PROCESS WATER to communities whose natural resources are limited to brackish water.

In this situation, demineralization is the principal objective. Tables 5 and 6 contain, by way of example, data relating to two plants handling water with levels of salinity of about 1,500 and 6,000 mg/l. Clearly, the degree to which the various ions are removed is determined not only by their valency but also by the chosen conversion ratio, which varies widely in the two examples cited.

1.4.2. THE PRODUCTION OF VERY HIGH-QUALITY WATER, e.g. water for boilers and the ultrapure water needed by the electronic, pharmaceutical and nuclear industries as well as by laboratories and hospitals.

Here, use is made of the ability of the membranes to arrest not merely the dissolved salts but also the organic molecules with a molar mass in excess of about 70 g/mol and, more especially, the most finely particulate matter such as viruses.

Applied to the treatment of surface water, reverse osmosis frequently removes very many pollutants which have eluded the self-purifying action of rivers. Very high-quality water can be produced in this way if the sequence of treatment is completed by mixed-bed ion exchange to eliminate the final traces of salts which have passed through the membrane.

Tables N⁰. 4 provide an example of the performance achieved against micro-pollutants contaminating surface waters.

TABLES 4

THE REMOVAL OF ORGANIC MATTER (O.M.)

Removing the oxygen consuming capacity (of water taken from the mouth of the Loire river).

$$P = 28 \text{ bar}$$

Test No.	Y (%) (salt passage)	O.M mg/l $KMnO_4$	
		raw water	treated water
1	75	10.1	0.6
2	75	12.3	0.5
3	75	10.4	0
4	75	10.4	0
5	75	12.0	0
6	75	51.5	3.8
7	50	51.5	0.3

Removal of micropollutants (water taken from the Seine river above Paris)
P = 28 bar.

Product	Concentration in raw water µg/l	Salt passage, Y = 75 %	Salt passage, Y = 50 %
HCB — PM 285	9.8	2	1
Lindane PM 291	10	2.4	2.1
Parathion PM 252	2	0.4	0.3

TABLE 5

Operating conditions:

P = 28 bar
Y = 90 % (3-stage reject staging)

Pretreatment: carbonate removal, filtration, acidification to pH 5.5.

	Raw water	Water after carbonate removal	Water after reverse osmosis
Ca^{2+} mg/l $CaCO_3$	425	40	4.5
Mg^{2+} —	200	200	22.5
Na^+ —	478	763	127
HCO_3^- —	160	0	7.5
CO_3^{2-} —	0	40	0
SO_4^{2-} —	521	521	49.5
Cl^- —	422	442	97
T.S. * mg/l	1 470	1 270	200

* Total salinity.

TABLE 6

Operating conditions:

— P = 28 bar.
— Y = 50 % (single stage osmosis).

— Polyphosphate: 5 mg/l.
— Acidification H_2SO_4, pH = 5.

	Water after pretreatment mg/l	Water after reverse osmosis mg/l
Ca^{2+}	750	10
Mg^{2+}	850	14
Na^+	970	48
HCO_3^-	85	50
SO_4^{2-}	600	14
Cl^-	3 280	160
PO_4^{3-}	4.6	—
Total salinity	6 540	296
pH	6.2	6.2
Conductivity µS/cm	12 000	550

1.4.3. SEA-WATER DESALINATION

As has been pointed out above, some membranes are capable of removing over 98.5 % of monovalent ions. Using these, drinking water is obtained in a single desalination stage, and this enables reverse osmosis to compete with distillation techniques. A further point is that the plants for applying reverse osmosis are far simpler and easier to operate.

Electricity consumption is about 8 kWh per cubic metre. If suitable turbines were used to recover the potential energy in the reject (70 % of the water pumped is available at a pressure of over 50 bar) it would be possible to reduce the power consumption to below 5 kWh per cubic metre of desalinated water. Such a performance is greatly superior to that of even the most sophisticated distillation units.

Table 7 below shows the quality of water obtained with a reverse osmosis demineralization plant equipped with B-10 modules.

TABLE 7
PERFORMANCE OF A B-10 MODULE IN
TREATING SEA-WATER

Operating conditions :

P = 56 bar   Conversion factor Y = 25 %   Acidification H_2SO_4

	Feed water mg/l	Purified water mg/l	Salt passage %
Ca^{2+}	329	1.5	0.5
Mg^{2+}	1 031	3	0.3
Na^+	9 419	77	0.8
K^+	355	3	0.8
Sr^{2+}	13	0.2*	—
So_4^{2-}	2 200	1*	0.05
PO_4^{3-}	9	0	—
HCO_3^-	68	26	38
Cl^-	15 825	120	38
F^-	0	0	—
NO_3^-	3.4	0	—
SiO_2	0	0	—
Total salinity	32 680	185	0.6
Conductivity μS/cm	37 250	390	1
pH	6.3	6.3	—

* Value on borderline of instrument sensitivity.

1.4.4. TREATMENT OF INDUSTRIAL EFFLUENT.

Reverse osmosis warrants consideration when economic advantage can be derived from the concentration of certain industrial residues. For example :

— the recovery of metals from the effluents generated by electroplating plants (engaged in nickel, copper, brass and cadmium plating);

— the recovery of ammonium nitrate from the condensates produced in plants, manufacturing nitrogenous fertilizers.

It is, of course, generally impracticable to contemplate the recovery processing of the whole of the effluent discharged by an entire works. In order to prevent the discharge of liquid mixtures adversely affecting the recovery of desirable products, reverse osmosis must be applied at the point of discharge from workshops which generate effluent rich in recoverable ions.

1.4.5. RECLAMATION OF HIGH-QUALITY WATER FROM THE OUTPUT OF TRADITIONAL PURIFICATION SYSTEMS

By means of reverse osmosis it is possible to remove the final traces of salts and dissolved organic matter which persist in water which has undergone conventional treatment (physico-chemical, biological or adsorptive.) This is exemplified by a case in the United States where it has proved possible, by applying reverse osmosis to sewage which has undergone biological treatment followed by advanced physico-chemical processing, to replenish water resources used for human consumption. In an application of this kind, the removal rates may exceed: 92 % of the residual COD, 95 % of the phosphates, and 90 % of the nitrates, accompanied by thorough disinfection of the water. It is important to ensure that the water is subjected first of all to highly reliable and intensive physical pretreatment.

2. ULTRAFILTRATION

2.1. Principle

As has been explained in Chapter 3, ultrafiltration membranes are able to separate colloids and macromolecules whose molar mass exceeds the "cutting limit" of the membrane. As the osmotic pressure to be overcome is negligible low pressures of 2 — 5 bar are used.

2.2. Ultrafilters

The character of the membranes used for ultrafiltration varies very widely. As very many synthetic polymers and copolymers possess satisfactory properties it is impossible to provide an exhaustive list. The user should therefore contact the specialists to find out the characteristics of a given membrane, e.g. its cutting limit, chemical resistance, thermal stability and so on.

The sweeping of ultrafiltration membranes calls for even greater care than do those used for reverse osmosis, because:

— the specific flow-rate is higher (in the case of pure water subjected to a pressure

of 3 bar, it exceeds 2000 1/d per m² — 50 gal per day per sq ft of membrane surface area);
— compared with ions, the macromolecules and colloids deposited on the surface show very little tendency to diffuse back into the liquid, and it can only be
relied on sweeping of the membrane wall, which means that very low conversion
factors are used: 0.5 to 5 % in the case of all heavily contaminated liquids.   This
calls for recycling systems such as those shown in the following diagrams, and
implies a high level of energy consumption: 4 to 12 kWh/m³ (15 to 45 wh per gal)
of ultrafiltrate.

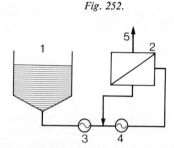

Fig. 251.   *Fig. 252.*

1. *Pressurized feed (e.g. water + salt).*
2. *Regulating valve.*
3. *Ultrafiltrate (water + salt).*
4. *Macromolecule (protein).*   1. *Storage tank.*
5. *Macromolecule or colloid.*   2. *Ultrafilter.*
6. *Salt.*   3. *Feed pump (4 bar).*
7. *Water.*   4. *Circulating pump (1.5 bar).*
8. *Concentrate (e.g. proteins).*   5. *Ultrafiltrate.*

Depending on the manufacturer, membranes may be of the following types:
— flat with filter-press type support;
— flat, spirally-wound;
— tubular inside a supporting cylinder;
— tubular, in the form of hollow fibres whithout a supporting structure; the
fibre diameters are much greater than those used for reverse osmosis (0,5 to 1.5 mm
O.D.) so as to allow the raw fluid to flow *inside* the fibres.
    The advantages and shortcomings of these different variants depends on the
fluid to be treated and call for a variety of pretreatment.   Least prone to clogging
are, of course, the tubular membranes, followed by the flat design and the spirally-
wound and hollow-fibre membranes.   The last-named require thorough pretreatment (filtration through pores in the 10 — 20 μm range) to avoid premature
clogging of the membranes.

2.3. Applications

Current applications of ultrafiltration include the following:
— clarification of wines and fruit juices (in which the target products pass through
the membrane);

— protein/salt separation (blood plasma);
— protein/salt and sugar separation from milk and serums (to obtain liquid cheeses);
— enzyme recovery;
— concentration of macromolecular suspensions (polyvinyl alcohols etc.).

Other applications are concerned with the treatment of concentrated industrial effluents, where the concentrate is sometimes recycled. Examples are:
— treatment of the rinse-water originating from electrophoretic painting booths (recycling of the pigments and paint resins);
— treatment of spent soluble oils for the purpose of:

• destruction: by ultrafiltration the waste emulsion containing between 2 and 5 % oil can be concentrated up to about 40 % to 50 % and then burnt off with possible recovery of the heat energy;

• recycling: this applies to the treatment of so-called "synthetic" oils, where it is possible to remove from the concentrated ultrafiltrate the contaminants which have been introduced into the oil during use (metal dust, abrasive pastes, free oils and grease) and to recycle a clean machining fluid (cf. Chapter 23, page 709).

3. ELECTRODIALYSIS

Principle.

If a liquid rich in ions is subjected to an electrical field by means of two electrodes with a continuous (d.c.) potential difference applied between them, the cations will be attracted to the negative electrode (cathode) and the anions will be attracted to the positive electrode (anode).

If nothing impedes their movement, they will each become discharged on the opposite-sign electrodes, and electrolysis thus takes place.

On the other hand, if a specially-arranged series of selective dialysis membranes (cf. Chapter 3 p. 86) is placed between the electrodes:
— some being negative and permeable only to the cations,
— and the others being positive, and permeable only to the anions, and arranged alternately as shown in diagram (fig. 253), the migration of ions as described above is restricted, as the anions cannot pass through the negative membranes and the cations cannot pass through the positive membranes.

In the case, therefore, of the cell of 2 × 3 membranes in diagram 1, where compartments 1, 2, 3, 4 and 5 are fed by a flow consisting of a sodium chloride solution, the ions in compartments 1, 3 and 5 pass into compartments 2 and 4 under the influence of the electrical field by the electrodes.

It is easy to see that in this way the water in compartments 1, 3 and 5 becomes weak in salt, and thereby becomes demineralized, while the water in compartments 2 and 4 becomes concentrated by the quantity of salt given up within the other compartments.

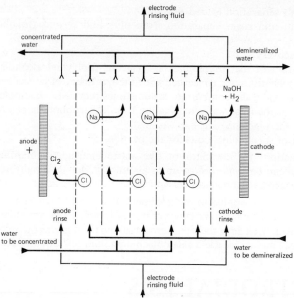

Fig. 253. —

The introduction into the system of each coulomb will therefore result in one valent weight in grammes of anion and cation leaving each of the demineralizing compartments (1, 3 and 5). This valent weight in grammes will be added to the others already present in the concentration compartments (2 and 4).

As the potential difference is proportional to the number of cells, the power consumption per kilogramme of salt removed is more or less constant (i.e. 0.6 to 0.8 kWh/kg - 0.27 to 0.36 kWh/lb of salt removed).

It is therefore possible to demineralize water by this process. At the same time, the non-ionized molecules (in particular, organic compounds) and the colloids remain behind in the treated water.

The main **limitations of this method** are due to:

● *The impossibility of obtaining fully demineralized water;* as the corresponding compartments would have an excessive electrical resistance leading to ohmic losses (generally speaking, it is unreasonable to try to reduce the salinity of the treated water below 300 mg/l); and

● *The cost of the treated water which increases rapidly with the salinity of the feed:*
— on the one hand, as we have already seen, the power consumed is proportional to the quantity of salts removed;
— on the other hand, if we wish to avoid a fall-off in selectivity and a reverse diffusion of ions caused by an excessive chemical gradient between the two faces of the membrane, the efficiency of the desalination must be restricted. Depending

on the internal hydraulic conditions in the electrodialysis units, the optimum level of salt removal which can be achieved is between 40 and 66 % per processing stage (i.e. the salt passage is between 60 % and 34 %).

It is for these reasons that most units are built up of several stages arranged in series, whereby the sought-after reduction in salinity is achieved by successive steps (see fig. 254 showing a "2-stage" installation).

It is necessary to pretreat the raw water before it is passed into the electrodia, lysis units :

— turbidity must be removed, and

— the iron content must be reduced to a fraction of a milligramme per litre.

● *All the salts which are liable to be precipitated* in the concentration compartments *must be eliminated* beforehand. Account must be taken of the phenomenon of polarization (see Chapter 3, page 84) which, in the case of electrodialysis, tends not only to cause excessive concentration of the ions present in the water to be treated but also to change the pH value (due to the local over-concentration of OH^- or H^+ ions), which may reinforce the tendency of some compounds to precipitate.

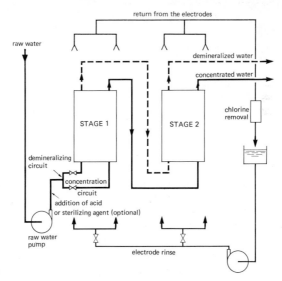

Fig. 254. —

The main **area of application for electrodialysis** is the production of drinking water from brackish water with a low salt content (0.8 to 2 g/l) and here it remains competitive with reverse osmosis. It also offers advantages for the desalination of colloidal and organic solutions (e.g. for demineralizing whey). In these circumstances, the use of reverse osmosis would involve the attendant concentration of all the species present and would produce demineralized water, whereas electrodialysis removes only the ionized species.

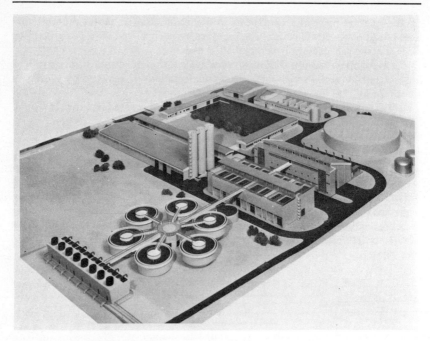

Fig. 255. —Water supply of the City of Riyadh, Saudi Arabia. Salbukh Plant. Output : 60 000 m³/d (16 mgd).

<table>
<tr><td>

13

</td><td>

GAS-LIQUID
EXCHANGES

</td></tr>
</table>

Water treatment in its three aspects — physical, chemical, and biological — makes widespread use of gas-liquid exchange phenomena.

The aim is often to dissolve a gas such as air, oxygen, ozone, or chlorine in the water, for example, in biological aeration, iron removal, disinfection, and other treatments.

In other cases the opposite problem is involved — that of eliminating dissolved gases: carbon dioxide contained in a water undergoing deionization, oxygen in a boiler feedwater, etc.

The phenomenon discussed in this chapter thus has these two opposing facets. However, the aeration systems used mainly in the treatment of waste waters will be mentioned only for the sake of completeness, as they are described in detail in Chapter 8, which deals with aerobic biological processes.

The liquid-gas system follows the laws of mass transfer from one phase to the other until an equilibrium state is ultimately reached.

Absorption is the operation by which optimum conditions are created for dissolving in the liquid phase the soluble gases contained in the gas phase. **Desorption** is the name given to the opposite operation, whereby volatile gases dissolved in the liquid phare transferred to the gas phase.

When conditions favour the occurrence of these two phenomena, they take place simultaneously. For example, aeration of a water rich in CO_2 and poor on O_2 gives a water depleted in CO_2 and enriched in O_2.

1. THEORY OF GAS-LIQUID EXCHANGE

The theory of absorption and desorption phenomena is relatively complex, and an exhaustive discussion would exceed the scope of this volume. The fundamentals only will be presented here.

Under equilibrium conditions, these phenomena are governed by **Henry's law,** which states that at a given temperature the concentration of a gas dissolved in a liquid phase is proportional to the pressure of this gas in the contiguous gas phase. Consequently, in order to bring about a gas-liquid exchange, the partial pressure of the gas in the gas phase must be either increased or reduced, depending on the required direction of the exchange.

The partial pressure of this gas, where the latter is not alone, follows **Dalton's law** which, together with the **law of ideal gases,** indicates the relative proportions of the molar fractions of the different gases at their partial pressures, whose sum constitutes the total pressure of the gas phase.

Hence, for a mixture of gases occupying a volume V at temperature T under pressure P and consisting of m_1, m_2, ... m_n grammes of gases of respective molar masses M_1, M_2, ... M_n exerting partial pressures P_1, p_2 ..... p_n:

$$P = p_1 + p_2 + ... + p_n$$

$$\frac{p_1 \, M_1}{m_1} = \frac{p_2 \, M_2}{m_2} = ... = \frac{p_n \, M_n}{m_n}$$

In dynamic processes, according to **Whitman's and Lewis's** theory, mass transfer takes place between the two phases (liquid and gas) through two films on either side of the interface, in accordance with the following diagrams.

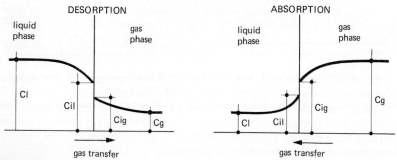

Fig. 256. —

Cl and Cg are the gas concentrations in the liquid and gas phases respectively; Cil and Cig are the same concentrations at the interface. Only Cl and Cg are accessible to measurement.

The following relation applies under steady-state conditions (N being the flow of gas transferred and assuming no accumulation at the interface):

$$N = k_1 \, (Cl - Cil) = k_g \, (Cig - Cg).$$

In the case of desorption, the concentration at the interface Cil must be kept below the required residual concentration Cl_r.

The difference between the concentrations Cl and Cil can be likened to a potential difference under which the gas transfer takes place; $\frac{1}{k}$ can in this case be regarded as a transfer resistance. The same argument can be used in the case of desorption. Under dynamic conditions, therefore, the following two parameters can be defined: *number of transfer units NTU*, and *equivalent height of transfer unit HTU*.

While the concept of number of transfer units is general, the HTU concept applies only to continuous contact systems in which the two phases circulate in countercurrent.

The number of transfer units as it were expresses the difficulty of obtaining a given exchange. It is given by the expressions (same notation as before):

$$\text{for the liquid phase} \quad NTU_l = \log \frac{Cl - Cil}{Cl_r - Cil}$$

$$\text{for the gas phase} \quad NTU_g = \log \frac{Cg - Cig}{Cg_r - Cig}$$

The equivalent height of a transfer unit is connected with the transfer coefficient, and therefore characterizes the transfer resistance. It depends on the physicochemical nature of the relevant phases, the hydrodynamic characteristics of the system (turbulence), the type of packing of the exchange column, and other factors.

The HTUs are given for different gases dissolved in water by experimental curves applicable to different types of contact masses constituting the packing.

The height of contact mass to be used to solve a given problem is associated with the NTU and the HTU by the following relation applied to the liquid phase:

$$H = HTU \times NTU$$

It may be concluded from the foregoing that to bring about gas **depletion or enrichment** of the liquid phase, it is essential to optimize the conditions necessary to facilitate the transfer of matter from the liquid to the gas phase or vice versa.

The more severe the conditions laid down, the greater the care that must be devoted to the design and construction of the equipment.

2. GAS-LIQUID EXCHANGE EQUIPMENT

The types of industrial equipment for performing gas-liquid exchanges in water treatment can be classified as follows:

— static aerators and mixers;
— mechanical aerators and mixers;
— pressurized aerators;
— entrainment (or stripping) type deaerators, also known as atmospheric deaerators;
— thermal deaerators;
— vacuum deaerators;
— combinations of the above processes.

2.1. Static aerators and mixers

As their name implies, aerators use atmospheric air as the only gas phase. They can eliminate from the water all gases other than those contained in the air, as well as enriching the water in oxygen. The following processes are used:
— spraying of water in air;
— spraying or trickling of water in an open structure with natural or forced air draught;
— ditto with the addition of trickling through a layer of suitable materials;
— air distribution in the water by means of coarse, medium, or fine bubbles.

2.1.1. SPRAYING

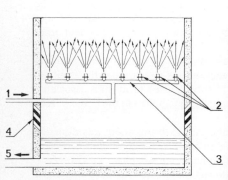

1. *Water to be aerated.*
2. *Spray nozzles.*
3. *Header pipe and laterals.*
4. *Ventilation louvres.*
5. *Aerated water.*

Fig. 257. — Spray tower.   *Fig. 258. — Spray nozzles.*

The water is sprayed through nozzles mounted on one or more manifolds, thus increasing the area of the water-air interface, thus equally facilitating elimination of excess gases and oxygen enrichment. The water may be sprayed at low pressure through a large number of nozzles, or at 0.8 bar, in which case the number of nozzles may be smaller.

This treatment is appropriate for up to 70 % CO_2 elimination or the elimination of smells due to other substances contained in the water. It is effective for oxygen enrichment in iron removal and manganese removal or to increase the dissolved oxygen content of waters containing little or no dissolved oxygen.

2.1.2. TRICKLING OF WATER IN AIR

There are two possibilities, depending on the degree of aeration required:
— plate-type aerator with or without spraying and with natural air draught (fig. 259 a);
— the same aerator with a blower to improve efficiency by setting up a more or less powerful forced draught (see fig. 259 b).

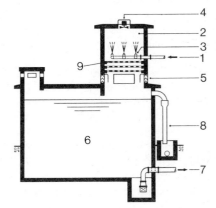

Fig. 259. —
Aerator designed for both spray and trickle.

a) Natural aeration.

1. *Inlet for water to be treated.*
2. *Spray chamber.*
3. *Spray nozzles.*
4. *Fan.*
5. *Aeration louvres.*
6. *Treated water tank.*
7. *To treated water pumps.*
8. *Overflow.*
9. *Trickle plates*

b) Forced aeration.

Usually, the aerators are of the " weir" or the "plate column" type, of varying degrees of complexity.   The water falls from one stage to another in thin sheets.

The number of cascades required and the height of fall can easily be determined by a simple test carried out "in situ": methods Nos 701 and 711 (see page 955).

The efficiency of this type of aerator is moderate and is dependent on its method of manufacture and the characteristics of the water to be aerated.

2.1.3. TRICKLING OF WATER THROUGH A CONTACT MASS — BIOLOGICAL FILTER

This aeration system is mentioned here for the sake of completeness; it is examined in detail in Chapter 8 (page 203).

2.1.4. DEEP TANK BUBBLE AERATION

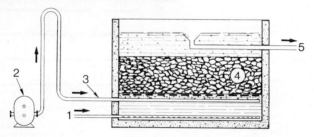

1. *Water to be aerated.*
2. *Air blower.*
3. *Injection of blown air.*
4. *Contact mass.*
5. *Collection of aerated water.*

Fig. 260. — Bubble aerator in deep water.

The water to be treated and the blown air are introduced in cocurrent at the bottom of the tank. The aerated water is drawn off at surface level. The treatment is improved by the presence of a contact mass which considerably increases the areas of liquid and gas in contact (fig. 260). This type of aeration is more particularly reserved for waters with a low suspended solids content.

Aeration rates of 65 to 75 % can be achieved, provided that a high rate of air flow is used (50 to 100 $m^3/(h.m^2)$; 2.8 to 5.6 cu ft per min per sq ft) and water flow-rates from 10 to 30 $m^3/(h.m^2)$ (4 to 12 gpm per sq ft).

This type of aerator is recommended for the removal of corrosive carbon dioxide coupled with oxidation of the ferrous iron in bore-hole water. The aerator is also suitable for nitrification treatment or for deaeration of supersaturated waters.

The bubble method can also be used whenever it is required to bring a gas into contact with water : e.g. carbon dioxide for recarbonation of previously lime-treated water, ozone for oxidation and disinfection, etc.

2.1.5. BUBBLE AERATION IN A SHALLOW DEPTH OF WATER

Water can also be aerated by introducing air bubbles under a **shallow depth** of water. Figure 261 shows an aerator based on this principle. It is equipped with a horizontal floor in finely-perforated stainless-steel sheet, over which the water to be treated flows at a maximum depth of 25 to 30 cm. The water is highly emulsified by a strong current of air blown under the floor by a single fan. Contrary to what has for a long time been assumed, this arrangement can also remove part of the equilibrium CO_2 in addition to the corrosive CO_2. The required precautions must therefore be taken to prevent the aerated water from becoming scale-forming.

The depth of water can be controlled in relation to the rate of flow, and the air flow can be adjusted to conform to the accepted residual carbon dioxide content.

Rates of aeration of 85 to 100 % can be attained with water flowrates of 20 to 40 m^3/h per m^2 (8 to 16 gpm per sq ft) of floor area, and air flowrates 30 to 60 times the water flow-rate.

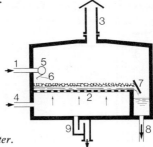

1. *Water inlet.*
2. *Perforated steel-sheet.*
3. *CO_2 outlet.*
4. *Air inlet.*
5. *Distributing pipe.*
6. *Deflecting baffle.*
7. *Foam shield.*
8. *Treated water outlet.*
9. *Siphon.*

Fig. 261. — Bubble aeration in shallow water.

These types of aerators should be chosen where a raw or filtered water is to be brought to its carbonic equilibrium without increasing its mineralization.

2.1.6. GAS DIFFUSION (ADMISSION OF BLOWN GAS: AIR, O_2, O_3)

In this treatment, a pressurized gas is dispersed in a liquid mass which may or may not contain suspended solids.

The equipment used is often named in accordance with the size of the gas bubbles produecd, viz.:

— *coarse-bubble* equipment: gas is injected either direct through vertical pipes or through large-orifice diffusers;

— *medium-bubble* equipment: Vibrair units;

— *fine-bubble* equipment: porous diffusers.

The first two types of equipment are used almost exclusively in waste water aeration. The third, in the form of domes, plates, tubes, or discs, is used in the same field and also for the injection of other gases such as ozone, CO_2, etc. (see page 227).

A detailed description of the various types of gas diffusers is given in Chapter 8, on page 228 foll. For ozone diffusion, see Chapter 15, page 420.

2.2. Mechanical aerators and mixers

This equipment group includes:

— surface aerators;

— deep tank mixers with blown air injection.

Like blown gas diffusers, these facilities are used mainly in waste water treatment; they are described in Chapter 8, on page 224 foll.

2.3. Pressurized aerators

These aerators are often used for iron removal from deep underground waters. They are comparable with those described in section 2.4.   The closed oxygenation tower includes a bed of volcanic lava supported on a floor.   The water, at pump pressure, feeds a mixer which receives blown air.   The water-air mixture is fed to the bottom of the layer of packing material.   Any excess air is discharged to the atmosphere through an air-valve.   The water is collected at the top of the column (figs. 262).

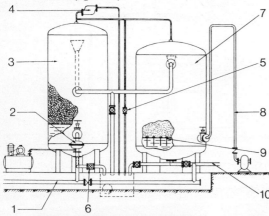

1. *Raw water inlet.*
2. *Offset baffle type mixer.*
3. *Oxidation tower.*
4. *Air vent.*
5. *Air discharge valve.*
6. *Drain valve.*
7. *Sand filter.*
8. *Scour air.*
9. *Floor with nozzles.*
10. *Iron-free water outlet.*

Fig. 262. — *Iron removal plant.*

Fig. 263. — *Oxidation column and filter for iron-removal from drinking water; output : 50 m³/h (220 gpm).*

Fig. 262. — Iron-removal unit. Output: 150 m³/h (660 gpm).

2.4. Gas entrainment (stripping) type deaerators

In this type of equipment, gases or volatile substances dissolved in the liquid phase are entrained (stripped) by passing another gas through the liquid phase in countercurrent; this other gas (the entraining gas) may be steam. The content of the gas to be eliminated in the entraining gas must be virtually nil. At a given temperature, the concentration of a gas in the liquid phase is proportional to its partial pressure in the gas phase — i.e., to its concentration in the latter. As long as any disequilibrium persists, gas will be transferred from one phase to the other.

In this type of treatment, therefore, the liquid becomes depleted in gases not contained in the entraining gas and enriched in gases which the latter does contain. The final gas concentrations in the liquid phase will therefore be proportional to the partial pressures of the same gases in the gas phase.

A unit of this kind must have the following characteristics:

— spraying or very good distribution of the water throughout the surface area of the column;

— large contact area between the liquid and gas phases;

— high-purity entraining gas, which is scrubbed by the water being treated and which must also contain virtually zero level of the gas whose concentration in the water is to be reduced;

— perfect distribution of entraining gas at the bottom of the column;

— water and entraining gas flows carefully calculated to give the desired residual dissolved gas concentration.

The name applied to this type of equipment varies according to its function; hence the terms CO_2 eliminator, H_2S eliminator, etc.

2.4.1. CO_2 ELIMINATOR

In this equipment, the water is either sprayed or finely dispersed and uniformly distributed above a packing layer normally consisting of Raschig or similar rings, volcanic lava, coke, etc. A powerful draught from a fan is blown on the underside of a perforated plate supporting the packing material. The water and air circulate in countercurrent. After degassing, the water is collected in a tank positioned underneath the contact column.

As stated in the section "Theory of Gas-Liquid Exchange ", the residual CO_2 concentration in the liquid phase (i.e., in the water) depends on the temperature and transit rate of the water, the type of packing and its volume, and the air flow.

The orders of magnitude for **DEGREMONT** equipment are as follows:
— water flow: 30 to 50 $m^3/(m^2.h)$: 12 to 20 gpm per sq ft;
— air flow: about 50 times the water flow;
— packing height : 1.5 to 2.5 m (5 to 8 ft).

A residual concentration very close to the equilibrium concentration can be obtained with a correctly calculated CO_2 eliminator.

This equipment can also be used under very similar conditions for cold H_2S elimination by air stripping.

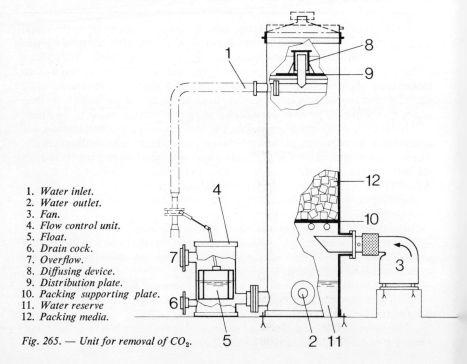

1. *Water inlet.*
2. *Water outlet.*
3. *Fan.*
4. *Flow control unit.*
5. *Float.*
6. *Drain cock.*
7. *Overflow.*
8. *Diffusing device.*
9. *Distribution plate.*
10. *Packing supporting plate.*
11. *Water reserve*
12. *Packing media.*

Fig. 265. — Unit for removal of CO_2.

Fig. 266. — CO_2 removal unit. SEAT factory, Barcelona, Spain.

2.4.2. STEAM STRIPPING (ammonia elimination)

The solubility of gases declines with increasing temperature, so that they can be more easily eliminated the higher the temperature to which the liquid phase is heated. The solubility and stability of ammonia in water are so great that this gas can only be eliminated by steam stripping. The entraining gas phase, enriched with noncondensable substances, including ammonia, is eliminated continuously. It then undergoes other treatment for the reclamation or destruction of ammonia compounds.

Whenever hazardous gases arise in the treatment process, special precautions must be taken to avoid such risks as corrosion, explosion, respiratory troubles, etc.

2.5. Thermal deaerators

This type of deaerator, which constitues a specific application of stripping, is used mainly to eliminate dissolved oxygen and carbon dioxide from steam-generator feed water in order to protect the boiler, the upstream heaters and all the water-steam circuits from corrosion.

Two fundamental laws governing the solubility of gases form the basis of the physical deaeration of water. The first, Henry's law, states that the concentration of gas by weight dissolved in a liquid at a given temperature is proportional to its partial pressure in the atmosphere above the liquid (cf. page 367). The second law states that the solubility of a gas in water is a non-explicit decreasing function of temperature.

According to the first law, therefore, water can theoretically be deaerated by maintaining the appropriate pressure and saturated steam temperature in the vessel containing the water; the dissolved gases will then automatically join the vapour phase. If there is a continuous flow of water through the vessel, the gases must be released into the atmosphere as they pass into the vapour phase; this takes place naturally if the pressure within the vessel is higher than atmospheric, but an ejector or a vacuum pump must be used if the vessel is at less than atmospheric pressure.

Whether vacuum or pressure deaeration is adopted, the appliance used must fulfil the following three conditions:
— maximum water-steam interface;
— water temperature must be held very close to saturated steam temperature at deaeration pressure;
— the partial pressure, in the vessel, of the gas to be eliminated must be lower than that corresponding to the final required content, in accordance with Henry's law.

2.5.1. DEAERATING TANK

The DEGRÉMONT deaerating tank is designed to work at pressures of 300 millibars (5 psi) and over. Starting from cold water it delivers water containing less than 7 µg/l of oxygen, at the saturated steam temperature corresponding to the working pressure.

It consists (fig. 267) of a horizontal, three-compartment tank:
— a heating compartment (1) surmounted by an atomizer chamber;
— a boiling compartment (2) with a steam distributing pipe system.
This compartment communicates with the first one at two points: at the bottom, through a set of calibrated apertures in the water (3) and at the top, between off-set baffles (4) in the steam phase;
— a deaerated water outlet compartment (10) separated from the previous compartment by a partition and communicating with it at the bottom.

The water atomized in the first compartment is heated as it condenses the steam escaping from the boiling compartment through the off-set baffles (4). The mixture of untreated water and condensed steam enters the boiling compartment through the apertures (3) and comes into close contact with the steam. In compartment (1), the water separates from the steam, and a natural circulation of water between this compartment and the boiling compartment is set up by the difference in density between the water and the water/steam mixture. Before leaving the tank, the water is thus brought into contact with the live steam several times which completes the deaeration process.

The gases entrained by the steam are evacuated through the annular space (11) round the atomizer chamber where the reduction in temperature gives rise to partial condensation of the steam. The gases, mixed with a volume of steam equal to their own volume, are then discharged outside the unit through a special pipe (6) fitted with a diaphragm designed so that the working pressure is maintained within the unit.

The water flow to the deaerator is controlled by a valve (7) according to the level in the tank and thus according to consumption. Constant pressure is maintained in the deaeration circuit by a valve (8) which more or less throttles the steam as it enters. This deaerator is also fitted with the following compulsory safety devices:

— a vacuum breaking valve (9), which prevents evacuation of the deaerator due to sudden condensation of steam;

— an excess pressure protection device consisting either of a hydraulic siphon which also acts as an overflow in units with working pressure not exceeding 0.3 bar, or of one or more safety valves for higher pressures. In the latter case, the unit includes an overflow pipe closed off by a valve whose opening is controlled by the level controller.

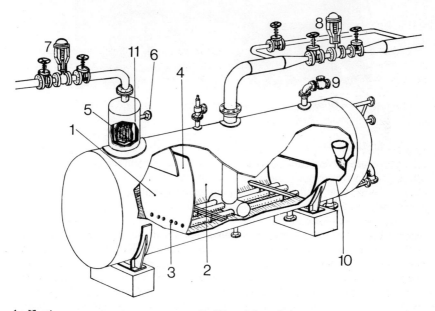

1. *Heating compartment.*
2. *Boiling compartment.*
3. *Calibrated orifices.*
4. *Offset baffles.*
5. *Atomizer chamber.*
6. *Outlet of gases mixed with steam.*
7. *Water inlet valve.*
8. *Valve maintaining a constant pressure in the unit.*
9. *Vacuum breaking valve.*
10. *Deaerated water outlet compartment.*
11. *Annular space round the atomizer chamber.*

Fig. 267. — Deaerating tank.

Fig. 268. — Thermal deaerator for boiler-feed water.   Output: 10 m³/h (44 gpm).

2.5.2. DOME-TYPE   DEAERATOR   MOUNTED   ON   TANK

This model is very suitable for service involving frequent starting and stopping, and its high performance is virtually independent of variations in the throughput. A dome, which can easily be mounted on an existing tank, combines the heating and bubbling functions of the types already described.

It operates as follows (fig. 269): the water, which is dispersed by the atomizers (1) at the top of the dome, comes into contact with part of the steam from the boiling compartment (2) and is heated as a result.   It collects on the tray (3) and runs down to the bottom of the boiling compartment through the pipe (4) which is concentric with the steam distribution pipe (5) ending in a terminal diffuser (6). The water in close contact with the steam overflows from the boiling compartment and runs through pipe (7) to the deaerated water storage tank.   The same pressure is maintained in the dome and the tank by means of an equalization tube (8).

The boiling compartment has a density flow recirculating device which considerably enhances deaeration efficiency.

As on the previous models, the gas/steam mixture is discharged through a pipe at the top of the atomizer chamber in the coldest zone.

Fig. 269. — A dome-type deaerator.

1. *Atomizer chamber.*
2. *Boiling compartment.*
3. *Heated water collection.*
4. *To the bottom of the boiling compartment.*
5. *Steam inlet pipe.*
6. *Diffuser.*
7. *Deaerated water outlet.*
8. *Equalization tube.*

2.5.3. VERTICAL TANK DEAERATOR

This equipment can be used advantageously for deaerating water flows not exceeding 25 to 30 m³/h (110 to 132 gpm) at a pressure of 0.3 bar. It can be shut down frequently and for extended periods (e.g., at weekends).

Its design (fig. 270) is based on the principles used in the dome type deaerator.

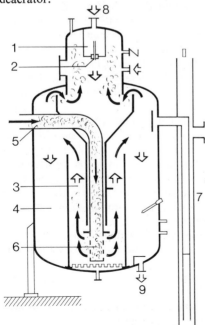

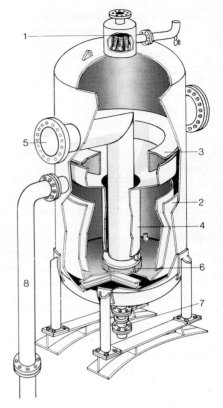

1. *Atomiser chamber.*
2. *Atomizer.*
3. *Boiling compartment*
4. *Storage compartment*
5. *Steam inlet pipe.*
6. *Bubbling compartment.*
7. *Hydraulic siphon.*
8. *Water to be deaerated.*
10. *Deaerated water.*

Fig. 270. — Vertical tank deaerator.

2.6. Vacuum deaerator

Apart from the thermal deaerator, the vacuum deaerator is the only type of equipment whereby gases dissolved in the water can be eliminated without enrichment with other gases, because both units tend to create a gas phase composed either partially or entirely of steam.

There is no alternative to the use of the vacuum deaerator where the requirement is as stated and where treatment must be carried out at less than 100 °C.

The unit operates at a total gas phase pressure equal to the steam pressure plus the partial pressures of the gases, all at the temperature of the water undergoing treatment.

The **DEGREMONT** vacuum deaerator consists of a hermetically sealed vertical tank, inside which are one, two or three floors each carrying a packing of Raschig rings or similar material.   Each of the resulting chambers constitutes a deaeration stage.   Assuming one and the same flow rate, the number of stages to be used depends on the quantity of gas to be eliminated and the final residual gas content; each stage represents a fixed number of transfer units for the relevant gas.   In the three-stage deaerator, the water is sprayed on to the surface of the first-stage packing (upper stage), passes through the packing, and is collected by a device which distributes it to the second-stage packing layer.   This process is repeated for the third stage.   Each distributor is fitted with a siphon device to allow the use of decreasing pressures from the upper to the lower stages.   The water then falls into a storage tank kept at the same vacuum level as the last stage, from which it is pumped to the point of use.

The vacuum is set up by one of the following types of facilities:

— vacuum pumps (one per stage).   The pumps operate at the same flow rate and are therefore interchangeable; the pressure of each pump corresponds to that of the stage to which it is assigned;

— ejectors (one per stage).   In this case, a single pump feeds the ejectors, and the latter are calculated to furnish the vacuum required by the stage to which they are assigned;

— barometric condenser type steam ejector, usually of two-stage construction.

The entire equipment must be of very high quality construction and must be perfectly airtight as any penetration of air would impair the treatment.

These deaerators are fitted with control and monitoring devices suitable for the particular conditions of vacuum operation.

The vacuum deaerator is sometimes equipped with facilities distributing steam evenly below the floor supporting the mass of packing material; the higher temperature thus obtained makes it possible to work at an absolute pressure higher than the otherwise very low, and difficult to maintain, bottom stage pressure.

Ambient temperature vacuum deaerators are used particularly in treatment systems for waters (fresh or sea water) to be injected into oil wells.

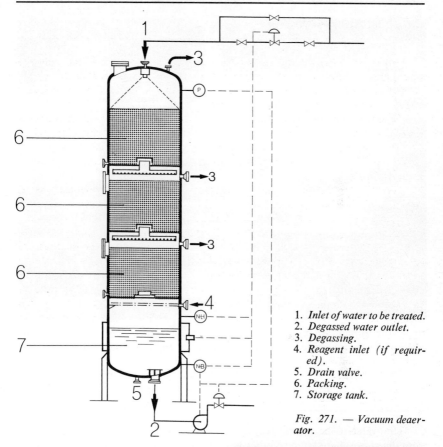

1. Inlet of water to be treated.
2. Degassed water outlet.
3. Degassing.
4. Reagent inlet (if required).
5. Drain valve.
6. Packing.
7. Storage tank.

Fig. 271. — Vacuum deaerator.

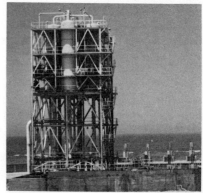

Fig. 272. — 3 vacuum deaerators. Output: 700 m³/h (3 100 gpm) each.

GULF OF SUEZ PETROLEUM Co., Ars Shuk Meir (Egypt).

Fig. 273. — Oil rig at DUBAI (Persian Gulf) equipped with three vacuum deaerators each with an output of 750 m³/h (3 300 gpm). (Courtesy of Infilco Degremont Inc.)

2.7. Combined deaeration

The water to be treated may sometimes contain not only dissolved gases from the atmosphere but also other gases; at the same time, treatment may have to be carried out at ambient temperature. This is the case, for example, with sea water acidification to obviate deposits of scale from bicarbonates. The carbon dioxide thus formed may reach levels of as much as 150-200 mg/l, thus considerably increasing the volumes of gas to be extracted and consequently also capital and operating costs. It is then advantageous to eliminate gases in two stages, viz.:

— elimination of gases (CO_2) by means of an atmospheric deaerator — e.g., a CO_2 eliminator;

— elimination of remaining dissolved gases in a vacuum degasser down to the desired residual concentrations.

14

THE CHEMICAL CORRECTION AND CONDITIONING OF WATER

The general purpose of the types of treatment examined in previous chapters is to **separate** undesirable substances from water. This chapter however, will deal with all those agents which can be **added** to water to adapt its quality to its purpose by means of chemical correction and conditioning. While these terms are synonymous the latter is usually applied to industrial applications.

The object of the addition processes to be considered may be either:
— to confer new properties on the water, making it suitable for the use for which it is intended. This category covers the addition of salts to excessively soft water (see remineralization, page 390), oxygen introduced by aeration (see chapter 13), fluorides (see chapter 20), scale and corrosion inhibitors, and biocides;
— or by adding suitable agents to eliminate the harmful effects of a substance already present in the water. This category covers both **pH correction reactions,** a number of which may be the secondary consequence of another type of treatment (though such phenomena lie outside the scope of the present Chapter as they are incidental only), and the **chemical reduction of oxygen.**

1. APPLICATIONS

1.1. Corrosion prevention

A. *Drinking water and miscellaneous industrial systems.*

The corrosion mechanisms described in Chapter 2 (see page 21 et seq.) are such that, even if a water satisfies the conditions of calcium carbonate equilibrium in not attacking or forming scale, it may nevertheless have a corrosive effect on unlined cast iron or steel pipes. There are various ways of protecting metal against corrosion and these include in particular insulating surface layers formed in one of two ways:
● either naturally by deposition of Tillmans **protective layer** (see page 31) comprising a mixture of calcium carbonate, iron hydroxides and iron carbonates. In order that this protective layer may form, it is necessary firstly that the pH should be compatible with a condition of carbonate equilibrium in the water at its service

temperature; secondly that the TAC and calcium TH should be sufficiently high (at least 7 French degrees, or 70 mg/l of $CaCO_3$, for waters with low concentrations of sulphates and chlorides although these minima should be increased in the presence of strong acid salts, especially chlorides) and thirdly that the level of dissolved oxygen should be at least 4 to 5 mg/l.   Many waters do not satisfy all these conditions and depending on circumstances, some or all of the following corrective treatments are then applied:

— neutralization of carbonic acidity until the equilibrium pH value is reached;
— remineralization of water containing no calcium bicarbonate (soft water from granite regions; desalinated water);
— increasing the level of dissolved oxygen (see Chapter 13).

● or by **film-forming** treatment (see page 396: corrosion inhibition). Where the aforementioned conditions for the formation of a protective carbonate layer are not met, such treatment deposits on the pipe an isolating layer made up basically of phosphates, silicates, chromates, zinc salts or organic substances.   These compounds can be applied equally to open or closed circuits or to mains systems.

B. *Boiler and heating-circuit water.*

This major field for the application of conditioning techniques is described in Chapter 22.

1.2. Scale prevention

The process described above, leading to the formation of a protective carbonate layer, is self-limiting: it stops when there is no more bare iron.   However, when the pH of the water is higher than the equilibrium value (due to a deficiency of "balancing" CO_2) continued deposition of "scale "occurs (calcium carbonate), and this may proceed until the pipe is completely blocked.   The preventive techniques falling within the scope of this chapter (i.e. other than softening, carbonate removal and deionization) mostly involve the controlled addition of acid, called "vaccination", (in the case of drinking or industrial circuit water), or the addition of CO_2 (restoration of the carbonate level of softened water with excess lime), or chemical conditioning (this applies particularly to industrial systems, see paragraph 2.3 of the present Chapter).

Acidification of drinking water to overcome its scale-forming tendency is rarely practised, though this technique is often applied to industrial systems (cooling circuits) and is also used as pretreatment before desalination.

The formation of scale may be caused by salts other than calcium carbonate, e.g. by calcium sulphate in supersaturated solution (sulphuric regeneration effluents in ion exchange; cooling circuits with high levels of concentration; equipment for distilling sea water; ore leaching liquors in hydrometallurgy). Irrespective of the precipitation techniques using seeding, chemical conditioning may also produce a satisfactory solution in these circumstances.

1.3. Prevention of organic growths and fouling

Certain conditioning agents are introduced into industrial water systems to prevent the growth of miscellaneous microorganisms (bacteria, moulds, algae, etc.) and to combat fouling (see page 701).

1.4. Neutralization of various effluents

Before discharge into the natural environment it is important to adjust the pH value of the effluent to conform to the regulations in force relating to:
— acid and alkaline industrial effluents;
— acid effluents generated by mine drainage.

Certain industrial effluents also require neutralization (correction of excessively acidic or alkaline pH) prior to a further treatment stage (biological or physico-chemical).

2. MAIN PROCESSES

2.1. pH correction (neutralization)

Firstly, it should be remembered that it is sometimes possible to correct the pH value by physical techniques involving gas/liquid exchange (this is especially true of the removal of carbon dioxide; the techniques are described in Chapter 13). Here we shall considerer only those instances where a chemical reaction is effected in the water to be treated.

2.1.1. ADDITION OF BASIC REAGENTS

In the case of water intented for human consumption, the reagents used are **caustic soda, lime** and **sodium carbonate.** These materials are also used in the treatment of industrial water, for which other reagents are also employed for specific applications:
● **lithia** in the nuclear industry;
● **neutralizing amines** (ammonia, cyclohexylamine, ethanolamine, morpholine, etc.) in boiler-feed water: when the steam condenses these agents combine with the dissolved carbon dioxide to form an amine bicarbonate. The coefficients of distribution of the CO_2 between the steam and water phases are such that the doses can be much smaller than the stoichiometric quantities calculated on the basis of the CO_2, actually released in the boiler. At low and medium pressures, the dose is of the order of 1 g per gramme of CO_2 released. At high pressure, after thermal degassing, the doses are about 1 g/m^3 of water.

● **Calcium carbonate,** in powder form, for neutralizing industrial effluents.

A further point is that, in the case of water for industrial purposes, the re-agents do not have to be as pure as those used for drinking water treatment, where the quality of the materials must be carefully checked; specifically, when caustic soda is used, it must be free from mercury.

In the presence of carbonic acid, these reagents produce bicarbonates. With the strong acids contained in certain industrial effluents, neutral salts are obtained.

Because of its low cost, lime is the most widely used reagent in conventional applications.

The reagents may be added to the water :

● in powder form (lime, sodium carbonate) by dry feeders (see page 555);

● in solution or suspension, by dosing pumps or gravity feeders (see page 551); when lime is used for the final adjustment of the pH of a clear water before it is delivered, it is advisable to use a lime saturator, since it retains the unburned impurities and supplies a transparent lime water. Milk of lime on the other hand always gives water a degree of cloudiness which varies with the purity of the commercial product and the dose of lime required.

The effectiveness of the treatment depends on how well the neutralizing reagent is mixed with the water to be treated: it is therefore important to obtain an even mixture in reaction vessels equipped with stirrers or, when necessary, in Turbactors which provide the best mixing conditions.

2.1.2. FILTRATION THROUGH ALKALINE-EARTH PRODUCTS

This type of treatment, which uses materials with a base of calcium carbonate mixed, where appropriate, with magnesium carbonate or magnesia, is most often applied to the neutralization of corrosive carbon dioxide which forms bicarbonates during the filtration process.

In the past it was common practice to use marble for this purpose, but because of its slow rate of reaction and the impossibility of obtaining perfect equilibrium, other products known under their commercial names of Neutralite, Magno, Akdolit, etc. are now preferred. The reaction kinetics of these agents give complete efficiency with a relatively small contact mass.

These granular reagents are mostly used in pressurized closed filters. Sometimes the open type of filter is used, when sufficient head is available.

In principle, filtering is possible in both directions. In the downward direction, the neutralizing medium also acts as a mechanical filter, becoming compacted in the process with a tendency to clog which varies with the quality of the water being treated. Backflow washing is therefore necessary and the standard filter is designed to permit expansion of the filter media using only water at a high flow-rate delivered by a backwash pump or from a water tower. If the aim is to reduce the quantity of water consumed by each expansion operation, water and air can be used together for the backwash.

Where the water is percolated upwards, expansion and washing are theoretically not necessary, but whithout them this type of treatment should be restricted to clear water, free from suspended solids.

Some of the products on the market are calcined during manufacture and therefore contain a high proportion of alkaline-earth oxides. When they are first put into service they give the water a high degree of alkalinity, which gradually diminishes in the course of time.

Neutralite, with no free alkali, does not have these disadvantages and is stable indefinitely. It is therefore widely used and will be given particular attention here. Neutralite is available in various grain sizes. It consists of calcium and magnesium carbonates and its special structure ensures a rapid and uniform solubility which is always in proportion to the amount of CO_2 to be neutralized.

1 m³ (35 ft³) of Neutralite is sufficient to treat 3 to 10 m³ (800 to 2 600 gal) of water per hour, depending on the concentration of CO_2.

Neutralite is used in the form of a filtering layer 0.8 to 1.5 m (2.6 to 5 ft) thick. The head loss caused by clean Neutralite varies, according to filtration rate and the depth of the bed, from 0.2 to 0.5 m (0.65 to 1.6 ft) However, allowance must be made for clogging and a head-loss of 1 metre through the filters must be allowed for.

Fig. 274. — Neutralization by means of Neutralite. The chlorinator is shown on the right. Throughput: 25 m³/h (110 gpm).

Consumption varies according to circumstances from 1.6 to 2.2 g/gramme of neutralized carbonic acid and partial reloading of the filter from time to time is sufficient to compensate for the quantity which disappears by dissolving in the water.

CONSUMPTION OF NEUTRALIZING REAGENTS

Reagent	Consumption of pure product per g of corrosive CO_2	Increase of hardness per g of corrosive Co_2 in French degrees	Reaction
Lime as $Ca(OH)_2$	0.84	0.11	$2\,CO_2 + Ca(OH)_2$ $= Ca(HCO_3)_2$
Sodium hydroxide as NaOH	0.91	0	$CO_2 + NaOH = NaHCO_3$
Sodium carbonate as Na_2CO_3	2.4	0	$CO_2 + Na_2CO_3 + H_2O$ $= 2\,NaHCO_3$
Marble as $CaCO_3$	2.3	0.22	$Co_2 + CaCO_3 + H_2O$ $= Ca(HCO_3)_2$
Magnesium oxide as MgO	0.45	0.11	$2\,CO_2 + MgO + H_2O$ $= Mg(HCO_3)_2$
Neutralite	1.6-2.2	0.12-0.22	$3\,CO_2 + CaMgOCO_3 + 2\,H_2O$ $= Ca(HCO_3)_2 + Mg(HCO_3)_2$

2.1.3. ACIDIFICATION

The main applications of this technique are as follows: correction of scale forming water, vaccination of industrial systems, treatment prior to desalination, neutralization of alkaline effluents and pH adjustement after softening with lime.

When CO_2 is used, the plant comprises: storage tanks or cylinders, a gas flowmeter and a dissolving tower. Otherwise, sulphuric acid and sometimes hydrochloric acid are used and these are added by measuring pumps.

2.1.4. RECIPROCAL NEUTRALIZATION

In some special cases, a chemical reagent can be dispensed with, by using the interaction of two or more waters of opposite characteristics:
— corrosive waters and scale-forming waters (however, additional reagent often has to be added to achieve the exact calcium carbonate equilibrium conditions);
— acid and alkaline effluents.

We can also include in this category those cases where acid and alkaline waters are passed alternately through carboxyl resins.

2.2. Remineralization

This treatment is applied when it is necessary to give an excessively soft water a certain amount of calcium bicarbonate, i.e. a sufficient TAC and calcium

TH to bring about the deposition of a protective carbonate layer on unlined cast iron or steel pipe.

There are several methods for obtaining the recommended TAC and TH values:

A. *Carbon dioxide and lime or calcium carbonate.*

In most cases, carbon dioxide and lime are used. (The lime is prepared in a saturator when a clear water is being remineralized.)

8.8. g of CO_2 plus 5.6 g of CaO per m^3 of water should be added per degree TAC. The carbon dioxide can be taken from engine exhaust or a flue and washed in a trickle water column before being introduced by blowers into the water to be treated. It is also possible to use submerged burners to burn a liquid or gaseous hydrocarbon contained in the liquid itself.

The quantities of fuel required to generate 1 kg of CO_2 are:

Coke .... 350 g
Fuel oil .... 450 g

The use of commercial liquid CO_2 ensures a higher-purity product.

Lime injection can be replaced by filtration through Neutralite and carbon dioxide consumption is thereby appreciably reduced (to about 5 g/m^3 per degree of remineralization).

B. *Sodium bicarbonate and calcium salt.*

HCO_3^- bicarbonate ions (as sodium bicarbonate) and Ca^{2+} calcium ions (generally as calcium chloride, though sometimes as calcium sulphate) are introduced into the water simultaneously.

The equilibrium pH must allow for the new ionic strength of the water.

To obtain an increase of 1 French degree in 1 m^3 of water, 16.8 g of sodium bicarbonate must be used with either

— 11.1 g of calcium chloride (as $CaCl_2$)

— or 13.6 g calcium sulphate (as $CaSO_4$).

These techniques are used with small and medium-size plants.

C. *Sodium bicarbonate and lime.*

Under method B above, it may also be necessary to add lime to bring the pH to the equilibrium value in line with the new mineralization of the water.

Sometimes, the TH of the water is high enough and it is only the alkalinity that needs to be increased. If so, sodium bicarbonate is added to the water to raise the TAC, and lime to adjust the pH.

D. *Sodium carbonate and carbon dioxide.*

Depending on the availability of local reagents, the TAC can also be increased by a combination of $Na_2CO_3 + CO_2$. Here, the flow of CO_2 may be controlled by the pH required and there is no need to add another reagent if the TH is high enough (if not, a calcium salt is added as in B above).

E. Sulphuric acid and calcium carbonate.

The water and crushed chalk are introduced into a settling tank or contact tank and sulphuric acid is added in a quantity equivalent to the quantity of bicarbonate to be added to the water.   The water may also filtered through a layer of marble or limestone after addition of the same quantity of sulphuric acid.

To raise 1 m³ of water by one French degree, 9.8 g of H_2SO_4 and a minimum of 20 g of $CaCO_3$ are required.   It is essential to end with the addition of lime, since marble alone is not sufficient to bring the pH to the equilibrium value.

F. Neutralite and/or Mineralite.

In some special cases and where only small flows are involved (sea-water distilled on board ship; offshore drilling platforms; small reverse osmosis units, etc.), the treated water is remineralized by percolating it in closed filters through Neutralite, or Mineralite (consisting mostly of $CaSO_4$ anhydrite) or through both at the same time.

2.3. Precipitation inhibition

As a result of changes in the medium (pH, temperature, concentration, etc.), a number of substances dissolved in the water are likely to become partly insoluble and form hard and adherent scale.

The best known example of this is the precipitation of calcium carbonate, the mechanism of which is explained in Chapter 2, page 31.

Until recently, the precipitation of calcium sulphate and magnesium compounds was referred to only in relation to boilers and sea-water distillation plant, but it is now found in cooling circuits operating with high pH values and with high levels of concentration.

Apart from water treatment, these precipitation phenomena are encountered in hydrometallurgy, sugar refining, etc.

To prevent such precipitation, a number of processes have been proposed, for example:

— softening, page 321;
— carbonate removal on resin, page 321;
— carbonate removal with lime, page 147;
— vaccination, by dissolving carbonated salts through a strong acid, page 386.

Certain chemical compounds are capable of a double effect on metallic ions, particularly calcium and magnesium:

1. — *They can form very soluble complexes* which possess all the characteristics of sodium salts, so that water treated in this way has the attributes of a softened water.

This process consumes about 50 g of polyphosphate per French degree of calcium, or 1 000 g per m^3 of water at 20° TH.

2. — When used in very much lower concentrations, *they inhibit crystal nucleation* and thereby maintain the solution in a state of supersaturation: this is called the "threshold effect".

For economic reasons, these compounds are used according to the latter method.

The most widely used of these substances are those referred to by the general name of **polyphosphates.**

The phosphoric ánion has in fact the property of producing linear or cyclic condensation products.

Polyphosphates proper have a linear structure and the following general formula:

$$M_4P_2O_7, \ n(MPO_3)$$

The first members of the series of sodium salts are pyrophosphate $Na_4P_2O_7$, followed by tripolyphosphate, $Na_4P_2O_7$ $NaPO_3$ or $Na_5P_3O_{10}$.

The **polymetaphosphates** have a cyclic structure and correspond to the general formula $(NaPO_3)_n$.

The best-known members are trimetaphosphate $(NaPO_3)_3$ and hexametaphosphate $(NaPO_3)_6$.

The commercial products sold under these names are often mixtures, the prefix of which represents the average degree of condensation.

The inhibiting action is completely effective on calcium carbonate, and a little less so on magnesia and calcium sulphate.

Usually about 2 g/m^3 of polyphosphate is used to stabilize a water with a TH and TAC of the order of 20°. This dose increase with the TH, the TAC, the turbidity and the temperature of use. The rules of the Comité Supérieur d'Hygiène de France for drinking water lay down a maximum of 5 g/m^3 of polyphosphates, expressed in terms of P_2O_5.

Polyphosphates are available on the market in two main forms:

— Crystalline polyphosphates, which are readily soluble and are introduced into the water in the form of a solution by means of a dosing pump. They are suitable for very accurate dosing.

— "Vitreous" polyphosphates, which dissolve very slowly and are put into solution by means of simple percolation. This device is more suitable for small flows, but it is difficult to achieve constant dosing with a variable flow.

Polyphosphate have the disadvantage that they decompose progressively by hydrolysis, forming PO_4^{3-} orthophosphate ions. The rate of hydrolysis increases with the temperature and acidity of the medium, although one cannot say there is any precise threshold beyond which polyphosphates are destroyed. Their effectiveness becomes dubious beyond 60 °C.

Therefore, at high temperatures, not only do polyphosphates become ineffective against scale, but the PO_4^{3-} ions that are formed precipitate in the form of tricalcium phosphate, which is very sparingly soluble and this in turn leads to the formation of calcium phosphocarbonate.

For higher temperatures and stricter stabilization requirements, use is made of elaborate organic substances, whose behaviour is completely identical with that of polyphosphates.

The action of EDTA (ethylene diamine tetraacetic acid) and its sodium salts is not due to a threshold effect; they have been used particularly in boiler water containing traces of hardness :

$$COOH - CH_2 \diagdown \qquad\qquad CH_2 - COOH$$
$$N - CH_2 - CH_2 - N$$
$$COOH - CH_2 \diagup \qquad\qquad CH_2 - COOH$$

EDTA is usually sold as a 50 % (approx.) solution.

In cooling circuits, organic derivatives of phosphorus—and in particular **phosphonates**—are widely used today.

There are two main types of phosphonates:

— AMP—tri (phosphonic methylene) -amino acid

$$N \equiv \left\{ CH_2 - \overset{\displaystyle OH}{\underset{\displaystyle O}{\overset{|}{\underset{\parallel}{P}}}} - OH \right\} 3$$

— HEDP—diphosphonic hydroxyethylidene acid

$$CH_3 - \overset{\displaystyle OH}{\underset{}{\overset{|}{C}}} = \left\{ \overset{\displaystyle OH}{\underset{\displaystyle O}{\overset{|}{\underset{\parallel}{P}}}} - OH \right\} 2$$

These compounds are generally marketed in the form of concentrated solutions. They are stable well beyond 100 °C and are sensitive to the presence of free chlorine. Although non-toxic, their use in drinking water is subject to approval by the competent authorities.

Their dosages vary both with the characteristics and conditions in which the water is used, and also from one commercial product to another. On average, they are of the order of 1 g/m^3 as P_2O_5, or about 10 g/m^3 of commercial product.

If, despite the presence of phosphate, slight precipitation of calcium carbonate does occur, the precipitate so formed is generally flocculent and will not form scale.

N.B. — The use of polyphosphates, EDTA or organic phosphorus compounds, sometimes advocated in order to stabilize ferruginous waters, is a very controversial question; iron removal is the only safe treatment.

These compounds hamper the formation of the crystalline structure, acting both at the molecular level and on the crystalline nucleus.

In pipe systems and cooling and boiler circuits the formation of scale and cumulative deposition can also be prevented, by using substances called *dispersants* which hold or put solid particles in suspension in the water. Among these the tannins—or preferably sodium tannate (see page 396)are used in low- and medium-pressure boilers; dosage: 2 g/m³ per degree of calcium.

Lignosulphonates and other vegetable extracts were widely used to protect cooling circuits; but now synthetic substances such as polyacrylates and poly-acrylamides partially hydrolysed from sulphonate derivatives, etc. are preferred.

The doses and conditions of use vary appreciably. It is advisable to consider in particular the cost-effectiveness achieved.

To facilitate even distribution, all these substances are generally introduced into the water in the form of fairly dilute solutions of a few percent strength. A further point which is particularly relevant when dealing with open circuits, is that the reagent dose should be related to the flow of water. A dosing unit must therefore comprise at least a dissolving tank, a stirrer and a feeder.

2.4. Reduction of oxygen

Three main substances can be used:
— sodium sulphite, with or without catalyst;
— hydrazine;
— sodium tannate.

For each of these the reaction rate increases with temperature and pH.

A. Sodium sulphite. This acts according to the reaction:

$$2\ Na_2SO_3 + O_2 \longrightarrow 2\ Na_2SO_4$$

that is, the theoretical consumption is 7.88 g of anhydrous sodium sulphite per gramme of dissolved oxygen.

From 80 °C upwards the reaction rate of natural sulphite is generally sufficient, but at lower temperatures it is necessary to use sodium sulphite in conjunction with a metal salt catalyst. The catalysed sulphite must be protected from becoming damp when stored in crystallized form and from contact with the air when in the form of a solution.

For each gramme of oxygen reduced, about 9 grammes of sodium sulphite are produced; thus the total salinity of a water saturated with air when cold is increased by about 100 mg/l by reduction of the oxygen with sulphite. This increase is often prohibitive for medium-pressure and high-pressure boilers, and it not infrequently necessitates an increase in the blow-down processes in low-pressure boilers.

B. Hydrazine. This reduces oxygen according to the reaction:

$$N_2H_4 + O_2 \longrightarrow 2\ H_2O + N_2$$

or, theoretically, 1 g of N_2H_4 per gramme of oxygen. Hydrazine is generally sold commercially in the form of hydrazine hydrate ($N_2H_4 \cdot H_2O$) containing 24 % of $N_2H_4 \cdot H_2O$ or 15 % of N_2H_4.

The reaction is very slow when the water is cold. At 80 °C the necessary dose is still several times the theoretical quantity. Catalytic methods for this reaction have been proposed.

Hydrazine and its decomposition products are all volatile, and thus this method does not add any fixed salinity to the boiler water. At temperatures above 270 °C, however, allowance must be made for the decomposition products, especially ammonia and nitrogen.

C. Sodium tannate. This is extracted by a specific process.

In a boiler, 2 g of sodium tannate completely reduce 1 g of oxygen. The reaction rates are similar to those of hydrazine; the decomposition products are nearly all volatile. The use of tannate is not, however, recommended above 35 bars (510 psi).

Sodium tannate promotes the holding in suspension of all solid particles; it also forms a protective film over steel surfaces.

After thermal degassing it is usually recommended that an excess of 0.1 to 0.3 g/m³ of N_2H_4 or 20 to 30 g/m³ of sodium sulphite should be kept in the water of low-pressure and medium-pressure boilers. In relation to the feed water, this corresponds to rates of consumption of the order of 0.1 g/m³ of N_2H_4 or 1 g/m³ of Na_2SO_3.

2.5. Corrosion inhibition

We shall examine under this heading the main chemical products which can be added to water in order to eliminate its corrosive effect on metals. These are also called passivating agents.

Unlike the methods described in detail at the beginning of this chapter, corrosion inhibitors do not act on the chemical constituents of water but generally form over the metals a thin protective film which, by eliminating the metal/water contact, prevents the corrosion which this contact would produce.

A very large number of products are used for this purpose.

2.5.1. SIMPLE INHIBITORS

A. Physical inhibitors:

These are compounds which have a strong affinity for the solid surfaces on which they are adsorbed.

Carried in the water in suspension or very dilute solution, they cover the walls with a film similar to a coat of paint.

Best known of these are the **film-forming amines:** fatty amines containing 4 to 18 carbon atoms, in which one end of the molecule is hydrophilic and the other hydrophobic. Running parallel to each other and perpendicular to the metal walls, the molecules form a continuous, waterproof film which prevents contact between the water and the metal and thus eliminates all causes of corrosion.

The dosage adopted is between 2 and 20 g/m^3, but the method of application and requisite analytical checks require great care.

For the protection of condensate return systems, the use of these amines is particularly advisable when a large amount of CO_2 is released, i.e. when the consumption of neutralizing amine would also the substantial.

When, as in most cases, the structure of the system impedes the formation of a truly continuous film, this method must be avoided, since all the corrosion currents will concentrate on the unprotected surfaces, thus producing cavitation and eventually even perforations.

B. Chemical inhibitors:

The electro-chemical mechanism of corrosion is described in Chapter 2. Chemical inhibitors make use of the electrical potential difference (cell effect) forming in the metal in order to produce the protective layer and immediately to inhibit that "cell" effect. The protection is applied selectively where actually needed, thus affording better protection.

Certain inhibitors act on the corrosion anodes (chromates), others on the cathodes (zinc salts) and others on the anodes and/or cathodes, depending on how they are used (phosphates). Their action consists of changing either the potential or the polarization curves of the electrodes.

— **Polyphosphates:** These substances are described on page 393; owing to their increasing instability with rising temperature, their use is restricted to circuits in which the maximum temperatures reached by the walls are moderate.

They are generally used in aerated waters with a pH of the order of 6.5 in doses of 5 to 50 g/m^3; these doses increase with the hardness of the water. Their action against corrosion is dubious.

— **Silicates** have the advantage, where drinking waters are concerned, of protecting ordinary steel against corrosion by soft and hot water when used in doses compatible with legislation (in France the maximum dose is 10 g/m^3 SiO_2).

They are available in either liquid or vitreous form.

— **Chromates** are anode inhibitors with very high thermal stability and which, however, cannot be used for drinking water.

Dosage: Several grammes per litre when used alone; in this case their use is restricted to completely closed circuits where only very small quantities of topping-up water are required.

— **Nitrites** give good protection to steel, but copper alloys and aluminium are less well protected.

The dosage is again of the order of one gramme per litre of water.

2.5.2. COMPOUND INHIBITORS

The action of corrosion inhibitors often needs to be stabilized, reinforced or supplemented by the use of other chemical products of the type of buffer salts, catalysts or dispersants. Corrosion inhibitors are in fact usually presented commercially in this compound form.

Furthermore, the combined use of two inhibitors gives protection against corrosion which is much greater than the sum of their individual effects: synergy takes place. This makes it possible, among other things, greatly to reduce the doses required.

This type of inhibitor is particularly advantageous for conditioning water in cooling circuits involving air-cooling in which the consumption of water, and therefore of inhibitor, is heavy.

For "closed" circuits where the amount of water added is very small, it is often an advantage to use a large dose of inhibitor, as this simplifies checking.

A. Inhibitors for closed circuits:

— Inhibitors with a base of buffered chromates provide a remarkable degree of protection for all the usual metals up to temperatures above 150 °C. Typical is the RD8 complex which is added at a rate of 2 to 5 g/l, increasing according to the chloride content of the water.

Use demineralized water, softened water or, at worst, water of low hardness. Chromates are incompatible with the anti-freezes in current use.

— Inhibitors with a base of buffered nitrites provide good protection for steel and cast iron. They usually contain substances which protect copper and aluminium. They are compatible with anti-freezes. The RD11 complex is typical of this class;

Doses required: 2 to 20 g/l in demineralized or softened water, or, at the worst, water of low hardness.

B. Inhibitors for semi-closed circuits:

Without its being possible to give an exhaustive list, mention may be made of some types of these inhibitors on the basis of their main constituents; the additives may vary from one commercial product to another.

— **Chromate-phosphates.**
Appearance: yellow powder.
Dose required: 30 to 80 g/m³ of the commercial product or 15 to 40 g/m³ expressed in terms of CrO_4.
pH of the water of the order of 6.5.
These give good protection against corrosion but combine the disadvantages, described below, of chromates and phosphates.

— **Zinc phosphates** (type Z 106) are also very effective, especially with soft waters containing chlorides.
Appearance: white powder.
Dose required: 12 to 60 g/m³.
pH of the water: 6.5 to 7.
Advantages: The blow-down does not generally create any poisoning problems.

Disadvantages:

● Like all polyphosphates, they hydrolyse faster the higher the temperature.

● They introduce into the circuits assimilable phosphorus which encourages the development of algae and bacteria.

— **Zinc chromates** provide excellent protection for steel and all the usual metals.

● Appearance: yellow powder.

● Dose required: 25 to 40 g/m³ or 6 to 10 g/m³ of CrO_4.

● pH of the water: 6.4 to 6.8.

Advantages: they inhibit the development of algae and bacteria. Low cost of use.

Disadvantages: direct discharge of the deconcentration blow-down is not possible.

— **Organozinc compounds**—by this term we mean various organic compounds, but especially those which include phosphorus in their molecule.

● Nature: liquid, acid.

● Dose required: 30 to 100 g/m³.

● pH of the water: 6.5 to 8.5.

Advantages: good high-temperature stability. No blow-down toxicity problems.

Disadvantages: slightly less effective against corrosion and a little more costly to use.

C. Inhibitors for distribution systems:

Even for the protection of industrial water distribution systems, the list of inhibitors is greatly reduced by the following two requirements:

— they must contain no toxic substances at all;

— they must be effective in doses small enough not to render their use prohibitively costly.

For water intended for human consumption, whether hot or cold, inhibitors are necessarily subject to approval.

At present, only certain substances of the zinc phosphate type can be used, in doses of the order of 10 to 20 g/m³.

2.6. Use of miscellaneous agents

A. Biocides:

Under this heading are grouped all those substances which act against the growth of algae, fungi, bacteria, or even shell-fish and insects, etc.

In practice, their use is restricted to industrial uses, particularly cooling systems.

Leaving aside the most toxic substances, especially heavy metal compounds, mention may be made of the following biocides:

— chlorine and its inorganic derivatives (see page 401);

— quaternary ammonium compounds;

— organic sulphur derivatives.

Depending on the nature of the product and the application in question, the dose used varies widely, from 1 to 100 g/m³.

As regards biocides, the method of use is at least as important as the choice of product. Of foremost importance is the need to avoid the phenomenon of tolerance, that is, the ability of certain members of a species to survive in the presence of a biocide and gradually to continue development by creating resistant strains.

In order to prevent this phenomenon, it is advisable not to use biocides continuously, but rather to administer so-called "shock" or "impact" treatments, varying their frequency, as appropriate, from once a day to once every three months. Equally, the same biocide should not be used all the time.

Very serious consideration should be given to the effect of biocides on the environment. The majority of hydraulic installations are not totally closed systems and involve discharging effluent into sewers or rivers.

Biocides are generally stable and methods of destroying them are not widespread. It is therefore often advisable to employ systematic treatment based on chlorine, alternating or supplementing it with biocides.

Lastly, it should be noted that all methods of prevention are much more effective and less costly than remedial treatment.

B. Cleaning agents.

Reference is made here to the numerous materials and composite media that are used in particular to maintain or re-establish the cleanliness of cooling circuits equipped with an atmospheric cooler.

The **biocides** described above often have a cleansing effect, particularly quaternary ammonium compounds because of their detergent properties.

Dispersants have the property of holding in suspension in water particles that would otherwise tend to become deposited in the circuits.

Of these, without giving an exhaustive list, the following may be mentioned:

— **tannates** and **lignosulphonates** whose efficiency is well known,

— **organic polymers**—polyacrylates and polyacrylamides—of average molecular weight and used to an increasing extent.

But here again, operating conditions are at least as important as the agents themselves: firstly, it is better to keep a circuit clean from the moment it is put into operation than to have to clean it later. Secondly, even though dispersants hold particles in suspension, there comes a time sooner or later when suspended solids have to be removed from the circuit. Hence the advantage of combining the action of dispersants with by-pass filtration.

As for the doses required, although these are about 5 to 10 g/m^3 for continuous treatments, they may reach figures of 200-300, or even 500 g/m^3 in cleaning operations. Actually, wherever possible, it is better in these circumstances to relate the level of the dispersant medium consumption to the weight of the solids to be put back into suspension, or at least, to the surface area of metal to be cleaned.

OXIDATION
DISINFECTION

In order to achieve objectives such as the removal of certain undesirable inorganic substances in solution (e.g. iron or manganese compounds), the elimination of foreign tastes and smell and the destruction of pathogenic microorganisms (disinfection), water treatment makes use of chemical (oxidation) and physical (ultra-violet) techniques.

In the present chapter, oxidizing agents are described from the point of view of their application (chlorine and its derivatives, ozone and bromine) and production (chlorine dioxide, ozone). Where and how these oxidizing agents and disinfectants are applied during the various treatment sequences is dealt with in Chapter 20, which considers in detail the treatment of water of varying provenance.

1. OXIDATION AND DISINFECTION BY CHLORINE

1.1. The action of chlorine

Chlorine is the most widely used reagent for disinfecting water. It possesses very considerable residual oxidizing capacity and is therefore useful for the destruction of organic matter. Its lethal effect on bacteria is due to the destruction of the enzymes essential to the survival of pathogens.

Chlorine dissolved in water reacts with its solvent according to the reaction:

$$Cl_2 + H_2O \rightleftharpoons HClO + HCl$$

which is accompanied by the secondary reaction:

$$HClO \rightleftharpoons ClO^- + H^+$$

The direction of these equilibrium reactions depends on the pH value of the medium. If the pH is below 2, all the chlorine occurs in molecular form. At pH 5, the molecular chlorine has entirely disappeared and recurs as hypochlorous acid (HClO). At pH 10, the chlorine is combined in the form of hypochlorite ions (ClO$^-$).

If the pH value lies between 5 and 10, which is usually the case with water subjected to chlorination, we encounter a mixture of hypochlorous acid and hypochlorite ions the relative proportions of which vary according to the pH value, as shown by the following curve:

As its bactericidal effect is more marked when it is in the form of HClO, chlorine is more efficient in an acid than an alkaline medium. Its action increases with the time of contact between the water and the reagent. A short contact time can be compensated by using a larger dose of reagent.

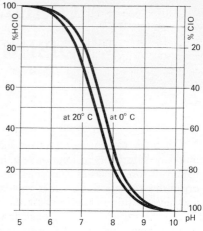

Fig. 275.

1.2. The use of chlorine

In the field of water treatment, chlorine and its derivatives are chiefly used for prechlorination, disinfecting drinking water, treatment against algae and mollusca in cooling circuit water and for the final treatment of municipal sewage.

1.2.1. PRECHLORINATION (BREAK POINT)

If increasing doses of chlorine are added to the water and the quantity of residual chlorine is measured at the end of one hour, it is often found that instead of increasing regularly, this residual quantity reaches a maximum and then decreases, reaches a minimum and then increases regularly in accordance with the curve in the Figure 276.

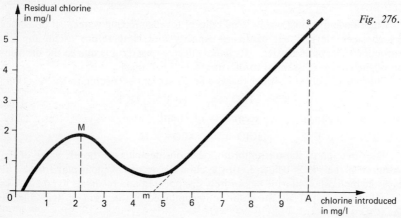

Fig. 276.

This is due to the fact that, in the initial stage, the chlorine combines with the organic matter in the water and with the ammonia, either free or in the form of amines, to produce chlorinated compounds (chloramines) which, in the second stage, are destroyed by the increased concentration of chlorine. The critical or "break" point is reached when the amount of chlorine added corresponds to the minimum "m" on the curve, at which point the water contains only free chlorine and traces of chloramines.

As addition compounds containing chlorine are the primary reason for water tasting unpleasant, the break-point corresponds to that amount of chlorine which combines the weakest taste with the maximum discolouring effect.

It should be noted, however, that there are some kinds of water which do not have a break-point.

● *Monitoring the prechlorination process:* the process is monitored by detecting the presence of free chlorine in the water. The measurement must not be affected by the presence of combined chlorine in the form of chloramines. In the laboratory, amperometric techniques are applied in conditions which preclude the chloramines from affecting the measurement. Industrially, the measurement can be performed automatically with the aid of an automatic photocolorimeter of the king described on page 581 and by using syringaldazine as the specific reagent for free chlorine.

A rapid manual check can be carried out with the help of a comparator with tinted screens.

1.2.2. THE DISINFECTION OF DRINKING WATER

Water intended for human consumption must be free from putrid bacteria and pathogens. This result is achieved by disinfecting the water by the injection of chlorine water, either in the treated water tank or, if the plant is pressurized, in the rising main to the storage reservoir.

In both cases, means must be provided for the efficient mixing together of the chlorine solution and the water being treated. A contact time of at least 30 minutes is required, at the end of which the residual chlorine concentration must still be between 0.1 and 0.2 mg/litre.

As a rough approximation, the amount of chlorine to be added is that which will ensure that traces of free chlorine remain after 30 minutes of contact time. The measurement is carried out by pouring increasing doses of chlorine into a series of 1 litre flasks containing the water under examination. After a contact time of 30 minutes at the same temperature the quantity of residual free chlorine is measured.

● *Disinfection control:* disinfection can only be completely checked by bacteriological tests. However, frequent measurements of residual chlorine should be made, and it is desirable to use a method of analysis which establishes the level of both the free and the combined chlorine (see Method of Analysis on page 954).

Where the water to be disinfected has not undergone prior break-point treatment, it is essential to use one of the above two methods of measurement so as to determine the free chlorine separately from the combined chlorine present in the chloramines.

The excess chlorine can be measured automatically by galvanometric or photocolorimetric techniques (see page 581).

1.2.3. CONTROLLING THE AMOUNT OF CHLORINE GAS |USED AND STORING LIQUID CHLORINE

Conventional chlorine feeding equipment (see page 560) is based on the normal principles applied in the distribution of gases.

The equipment contains an expansion device, a flow regulator, a flow meter and a unit for dissolving chlorine,

For safety reasons, virtually all this equipment works under a partial vacuum. The pressure in the chamber located immediately behind the inlet valve which reduces the pressure of the compressed gas is slightly lower than atmospheric pressure.

There is always a danger of leaks, even with equipment working under vacuum, because chlorine is always introduced under pressure. To prevent serious corrosion of neighbouring apparatus (electric motors and switchgear), storage and distribution equipment should always be housed in separate premises and installed in accordance with regulations (p. 1160).

For recommended installation procedures see Chapter 18,.

Physiologically, chlorine is a powerful respiratory irritant, but gives warning of its presence in concentrations well below the danger level. In an enclosed area, low concentrations sometimes cause nausea and coughing.

Special safety measures must be taken to deal with emergencies. A suitable gas mask, with a special filtering cartridge, must be kept easily accessible and within reach at a point outside the area likely to be affected in the event of an accident.

As regards the **toxicity** of chlorine, the concentrations producing various physiological reactions in human beings are generally accepted to be as follows:

	ppm
Permissible harmless concentration in air breathed during 8 hours' work.	1
Perceptible odour	3.5
Throat irritation from	15
Coughing induced from	30
Maximum for short period of exposure	40
Dangerous, even for short period of exposure	40 to 60
Quickly fatal	1 000

2. OXIDATION AND DISINFECTION BY CHLORINE DERIVATIVES

2.1. Chloramines

These are very stable antiseptics that act more slowly than chlorine but remain active for longer in water.

They are generally prepared from chlorine and ammonia (one-quarter to one-half as much ammonia as chlorine) or ammoniacal salts.

These chlorine compounds are not very widely used at the present time.

2.2. Chlorine dioxide

Chlorine dioxide is a yellow gas with an acrid odour. In a concentration of more than 10 % by volume in air it is explosive, but it is quite harmless in solution in water. It is a highly effective oxidizing agent with powerful deodorizing and bleaching properties. Its action on pathogenic substances is at least equal to that of chlorine. It should be used in preference to chlorine whenever the water to be treated contains traces of phenols which are able to combine with the chlorine giving the water the unpleasant taste of chlorophenol. It quickly oxidizes iron salts, which it converts into insoluble ferric hydroxide. Similarly, when used in excess of a concentration which varies with the pH of the water, it precipitates manganese salts in the form of manganese dioxide. Used at the pre-chlorination stage, it therefore enables these metals to be eliminated.

Chlorine dioxide is invariably produced on the spot by reacting chlorine or hydrochloric acid in solution with sodium chlorite.

In the solid state, sodium chlorite must be handled with care and must be kept away from damp, heat and combustible materials.

● *Preparation from hydrochloric acid:* the reaction producing chlorine dioxide is complex but may be broadly summed up as:

$$5 \ NaClO_2 + 4 \ HCl \longrightarrow 4 \ ClO_2 + NaCl + 2 \ H_2O$$

10 g of pure sodium chlorite require 3.2 g of hydrochloric acid to produce 6 g of chlorine dioxide.

The reaction takes place slowly and reaches completion only in the presence of excess acid. In practice, the proportion of hydrochloric acid used is 3 to 4 times the stoichiometric amount.

The equipment for producing the chlorine dioxide and metering the reagents can be constructed as shown in the diagram below (fig. 277).

The hydrochloric acid and sodium chlorite solutions are prepared in separate tanks with a concentration of, say, 8.5 % and 7 % respectively. The two solutions are extracted by a dual dosing pump which delivers them into a mixing vessel, the capacity of which is equivalent to a contact time of some twenty minutes when the pump is operating at full capacity. Given an identical rate of flow through each head of the pump, this results in a 20 g/l chlorine dioxide solution at the outlet from the mixing vessel.

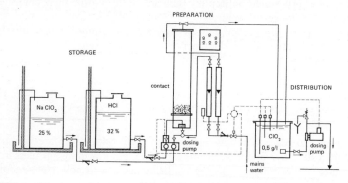

Fig. 277. — *Diagram of equipment for producing chlorine dioxide from hydrochloric acid.*

1. *Chlorine cylinder.*
2. *Chlorinator.*
3. *Mains water.*
4. *Ejector.*
5. *Solenoid valve.*

6. *Chlorite preparation tank.*
7. *Water for dilution.*
8. *Dosing pump.*
9. *Reaction chamber.*
10. *Constant-flow distribution.*
11. *Solution storage tank.*
12. *Level sensor.*
13. *Electric control cabinet for chlorinator and dosing pump (8).*
14. *Solution delivery pump.*
15. *Variable-flow distribution.*
16. *Syncopated or variable-speed control for pump.*

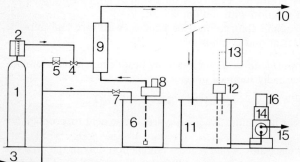

Fig. 278. — *Diagram of equipment for producing chlorine dioxide from chlorine.*

● *Preparation from chlorine:* in this case, the reaction can be written:

$$2\ NaClO_2 + Cl_2 \longrightarrow 2\ ClO_2 + 2\ NaCl$$

in which 10 g of pure sodium chlorite react with 3.9 g of chlorine gas to form 7.45 g of chlorine dioxide (fig. 278).

A solution of chlorine water with a pH of less than 2.5 is prepared and mixed with the required volume of 10 % sodium chlorite solution. To prevent sodium chlorite not converted to chlorine dioxide from being injected into the water to be treated, the mixture is in fact never made in stoichiometric proportions and a fairly large excess of chlorine is always used. At the cost of an additional complication, this excess can be reduced by lowering the pH value of the medium in which the reaction takes place to below 2 by the addition of sulphuric acid.

2.3. Sodium hypochlorite

Sodium hypochlorite solutions, commonly known as "Javelle water" or bleach, are characterized by their active chlorine content, which is measured in **chlorometric degrees** (Gay-Lussac). This is the quantity of free chlorine in litres which, in normal conditions (0 °C and 760 mm mercury column), has the same oxidizing power as 1 kg of product.

1 chlorometric degree = 3.17 g Cl_2 per kg

The sodium hypochlorite solution is added to the water in its concentrated commercial form or, if this method results in excessively small volumes of reagent, in a dilute state.

The sodium alkalinity of the sodium hypochlorite precipitates the hardness of the dilution water in preparation tanks; it can result in the furring of pipes and feeding equipment. This can be remedied either by preparing the solution 24 hours in advance in order to allow the precipitates sufficient time to settle, or by placing in the tank on the occasion of each filling approximately 50 g of hexametaphosphate of sodium to 100 litres of water. With very hard waters, a preliminary softening process may be necessary.

Measuring the addition rate of sodium hypochlorite solution can be easily performed by means of the simplest feeder is the DRC model (described on page 551) which is a simplified instrument allowing the surface collection of the disinfecting fluid which discharges regularly and evenly either at a constant rate of flow or in proportion to the flow to be treated if the latter is variable.

When the disinfecting fluid has to be fed into a pipe under pressure, at constant or variable flow, proportional feed pumps (described on page 548) are generally used, as they allow injection under pressure.

● *Disinfection control.*

As in the case of treatment with chlorine gas, the process is monitored (apart from bacteriological tests) by verifying the presence in the treated water of a residual quantity of reagent.

Where chlorine dioxide prepared from chlorine is used, the residual Cl_2 is determined (Method of analysis, page 708).

3. ELECTROCHLORINATION

3.1. Principle

This involves manufacturing sodium hypochlorite on the spot by the electrolysis of a sodium chloride solution. The application of this method is particularly attractive when the salt solution is available in the form of sea-water. Electrochlorination makes it possible to eliminate the constraints imposed by considerations of safety when large amounts of liquid chlorine have to be stored. There is no need to provide large reagent storage tanks since the manufacture of sodium hypochlorite proceeds continuously to meet the demand as it occurs. Where the starting solution is an artificially prepared sodium chloride solution, the problems of handling and transport are greatly simplified by the fact that the reagent is harmless.

The ionic dissociation of sodium chloride subjected to electrolysis gives rise to the following reactions:

at the cathode $2 Na^+ + 2e^- \longrightarrow 2 Na$

$2 Na + 2 H_2O \longrightarrow NaOH + H_2$

at the anode $2 Cl^- - 2e^- \longrightarrow Cl_2$

$Cl_2 + H_2O \longrightarrow HCl + HClO$

In the tank, the caustic soda thus formed reacts with the hydrochloric and hypochlorous acid to produce respectively sodium chloride and sodium hypochlorite, so that the whole reaction may be written:

$$2 NaCl + 3 H_2O \longrightarrow NaClO + NaCl + 2 H_2O + H_2 \nearrow$$

In order to obtain the equivalent of 1 gramme of chlorine it is necessary to produce 2.1 g of sodium hypochlorite with the consumption of 3.3 g of sodium chloride and the liberation of 0.056 g of hydrogen.

3.2. Electrolyzer design

An electrolyzer generally comprises a plastic tank which is immune to attack by sodium hypochlorite and is provided with an inlet and an outlet for the saline solution. Special provision is made at the outlet for the removal of the hydrogen liberated.

The electrodes are mounted in series inside the tank. Those at the end are coupled respectively to the negative (cathode) and positive (anode) poles of the current generator. The electrodes are usually made of titanium, and the anode as well as the anodic surfaces of the intermediate bipolar electrodes are plated with platinum to protect them against corrosion.

3.3. Operating conditions

Electrolyzers are arranged in series or parallel depending on the characteristics of the electrolyte and on the desired concentration of the sodium hypochlorite solution. Recirculation of the saline solution is possible in both cases. Using a solution of the optimum concentration, i.e. about 30 g of NaCl per litre, it is possible in a single pass through the electrolyzer to obtain an equivalent chlorine concentration of about 250 mg/l.

By using electrolyzers arranged in series and recycling the solution, this concentration can be raised to about 3 g/l where the starting solution is sea-water and to 6-8 g/l in the case of an artificial sodium chloride solution.

Where the purpose is to disinfect drinking water, a high concentration is advisable so as to reduce as far as possible the quantity of sodium chloride added to the water.

Electricity consumption is about 4 to 4.5 kWh per kg of chlorine equivalent produced.

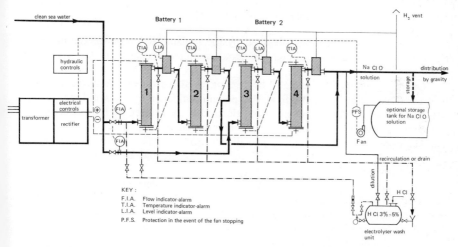

Fig. 279. — Diagram of a plant for the production of sodium hypochlorite from ordinary sea-water.

4. OXIDATION AND DISINFECTION BY OZONE

4.1. Characteristics and production of ozone

4.1.1. PHYSICAL CHARACTERISTICS

Ozone is an allotropic variety of oxygen with the formula O_3. It is a gas which is blue in colour and possesses the following main physical characteristics:

— molar mass: 48 g/mol;

— density in relation to air: 1.657;

— specific mass at O °C and a pressure of 760 mm mercury column: 2.143 ′kg/m³;

— heat of formation of a mol given constant volume: 143 kJ (34.2 kcal).

4.1.2. PRINCIPLE OF OZONE PRODUCTION

Ozone is an unstable gas obtained by the ionizing action on oxygen of an electric field derived from a high potential; a violet discharge is the visible result.

Ozone is obtained industrially by passing a current of air, or oxygen, between two electrodes subjected to an AC potential difference. To avoid the formation of an arc, one or sometimes both electrodes are covered with a dielectric of uniform thickness forming an equipotential surface.

The potential difference applied to the electrodes obviously depends on the type and thickness of the dielectric and the width of the ionization gap; in practice it is between 10,000 and 20,000 V.

For a given potential difference, ozone production depends mainly on the geometry of the components of the ozonizer, the dielectric properties of the insulator, the current frequency, the dryness of the air, the pressure and the desired ozone concentration in the air or oxygen. It also depends on the temperature of the cooling water in the ozonizer.

The usual concentration of ozone in ozonized air averages from 10 to 20 g/m³.

For this concentration, and assuming that intensive drying is carried out (dew point between—40 and—60 °C), the productive capacity of present-day ozonizers varies, depending on type, from 50 to 100 g per hour per square metre of dielectric surface area with a current frequency of 50 Hz. In certain circumstances, higher figures can be achieved by raising the frequency.

Depending on its size, the total energy consumption of the complete unit varies from 20 to 30 Wh per gramme of ozone produced. The ozonizer itself consumes 14 to 18 Wh per gramme of ozone. Owing to losses in the electrical discharge, a very large proportion of this energy is converted into heat which gives rise to a considerable temperature increase. As the ozone yield is reduced when the gas temperature rises, it is necessary to provide a cooling system which generally takes the form of a cooling-water circuit.

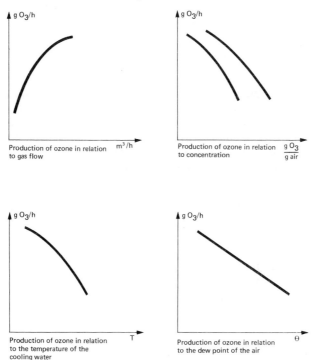

Fig. 280.

4.1.3. TYPES OF OZONIZERS

There are at present two types of apparatus available for ozone production: plate-type ozonizers and tube-type ozonizers.

A. Plate-type ozonizers.

These consist of flat dielectrics and metal electrodes. The whole unit is generally housed in a sealed chamber and incorporates a water cooling system.

B. Tube-type ozonizers.

Tube ozonizers consist of two concentric electrodes and a dielectric tube.

Tube ozonizers differ in the horizontal or vertical positioning of the electrodes and in the location of the dielectric with respect to the cooling water.

● *Two-stage low-pressure drying:* in large installations, this is the preferred method of drying.

In this method, the air passes initially through a cooler (1st stage) where a large part of the water vapour which it contains is condensed out.

The air leaves the cooler at a temperature between 2 and 5 °C and in a condition of near-saturation. It then flows through a drier (2nd stage) which, as in the previous case, contains an adsorbent material such as activated alumina. This drier, which also comprises two drying units, works at low pressure. The adsorbent medium is here regenerated by the passage of hot air.

C. DEGRÉMONT ozonizers.

DEGRÉMONT ozonizers are available in a variety of standard sizes and are of two types: single and double. In the single ozonizer, each metal tube in the bundle contains one dielectric tube. In the double ozonizer, two dielectrics are accommodated, back to back, in the same metal tube.

These items of equipment cover a range of production capacities varying from 0.25 kg/h to 10 kg/h. Within this range, the monobloc units produce up to 2.5 kg/h (OZONAZUR A and MB).

The monobloc units comprise in the same cabinet (OZONAZUR A) or mounted on a single chassis (OZONAZUR MB): the ozone producing cell, the air conditioning equipment, the low-voltage electrics needed for the power supply and for the automatic and safety appliances and the high-voltage transformer. Since the "heatless" method of drying is used, the air has to be supplied at a pressure of between 5 and 10 bar.

For laboratory requirements, the OZOLAB T_1 and OZOLAB T_2 units provide a source of ozone which can be freely adjusted between 0-8 g/h and 0 — 16 g/h respectively (when fed with air) or between 0-15 and 0-30 g/h (with an oxygen feed).

Fig. 282. — OZOLAB laboratory ozonizer type T2S.

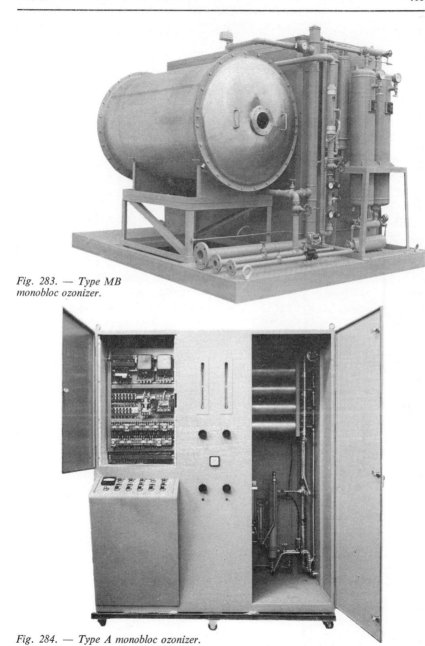

Fig. 283. — Type MB monobloc ozonizer.

Fig. 284. — Type A monobloc ozonizer.

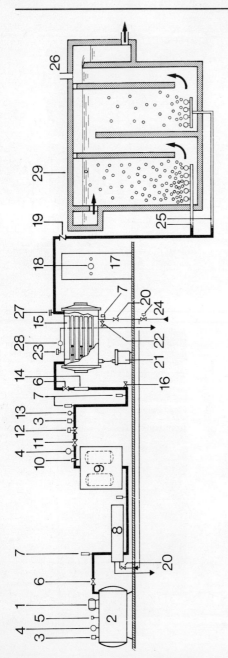

1. *Compressor unit.*
2. *Air tank.*
3. *Calibrated valve.*
4. *Pressure gauge.*
5. *Pressurestat.*
6. *Air regulating valve.*
7. *Thermometer.*
8. *Cooler/exchanger.*
9. *Automatic drier.*
10. *Expansion valve.*

11. *Stop valve.*
12. *Solenoid valve.*
13. *Hygrometer.*
14. *Air flow meter.*
15. *Ozonizer.*
16. *Dry air sampler.*
17. *Electrical cabinet.*
18. *Voltage switch.*
19. *Check valve.*
20. *Cooling water control valve.*

21. *Transformer.*
22. *Drain valve.*
23. *Thermometer/thermostat.*
24. *Automatic valve.*
25. *Diffusion tube.*
26. *Vent.*
27. *Ozonized air sampler.*
28. *Flow monitor.*
29. *Contact towers.*

Fig. 285. — Disinfection with ozone. High-pressure operation. Single-stage drying.

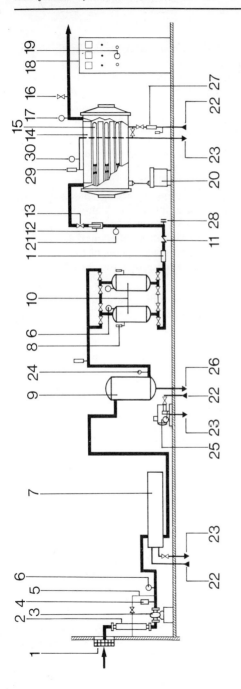

1. Air filter.
2. Silencer.
3. Blower.
4. Calibrated discharge valve.
5. Air discharge valve.
6. Pressure gauge.
7. Exchanger/cooler.
8. Thermometer.
9. Refrigerating set.
10. Driers using adsorbent medium.
11. Check valve.
12. Air flow meter with electrical contact.
13. Regulating valve.
14. Ozonizer.
15. Ozone generating tubes.
16. Ozonized air sampling point.
17. Pressurestat.
18. Electric cabinet.
19. Voltage switch.
20. Step-up transformer.
21. Hygrometer.
22. Cooling water inlet.
23. Cooling water outlet.
24. Thermostat.
25. Coolant compressor.
26. Condensation water outlet.
27. Cooling water flow meter.
28. Dry air sampling point.
29. Thermometer/thermostat.
30. Flow monitor.

Fig. 286. — *Disinfection with ozone. Low-pressure operation. Two-stage drying.*

D. *Electrical supply to ozonizers.*

There are two possibilities for supplying electricity to ozonizers depending on the quantity of ozone to be produced:
— where the quantities are small, e.g. less than 10 kg of ozoner per hour per unit, the supply frequency to the ozonizers is that of the mains, i.e. 50 or 60 Hz;
— where the quantities are larger, individual study of each particular case may show that it is more economical to supply the ozonizers at medium frequency using a static frequency generator with an output in the lower audio range.

● *Supply at industrial frequency:* here, the ozone output is controlled by varying the voltage applied to the ozonizer. This can be done either in steps, using a switch to connect the fixed mains voltage to taps on the transformer corresponding to several transformer ratios or continuously, using a varying autotransformer to supply a voltage capable of continuous variation to the high-voltage transformer.

As the power factor of an ozonizer is low, with a lead of about 0.4, an inductive compensator can be fitted so as to bring the power factor of the plant to a value close to unity, thereby reducing the installed power (expressed in terms of kVA) demanded by the equipment.

● *Medium frequency supply:* medium frequency ozonizers are supplied through rotating machines or static frequency generators called "inverters".

The rotating machines concerned are motor-driven alternators.

Inverters are thyristor-type static frequency generators which first rectify the frequency of the three-phase mains supply and then set up a medium frequency controlled by thyristors connected in a bridge configuration.

Producing ozone with the aid of these static frequency generators offers a number of advantages, including in particular:
— an increase in the specific capacity of the apparatus, i.e. operating with a frequency in the 300 to 600 Hz range makes it possible, with the same equipment, to achieve capacity and output figures equal to more than twice those obtained at industrial frequency;
— at nominal cost, a power factor close to unity;
— a reduction of the operating voltage;
— flexibility in service, the capacity —and therefore the output level— being varied by regulation of the thyristors.

Fig. 287. — Internal view of an ozonizer.

Fig. 288. — Front view of three Type GL ozonizers installed in the city of NANTES (France). Total output capacity: 30 kg of ozone per hour (66 lbs per hr).

4.2. Application of ozone

4.2.1. PRINCIPLE

The purpose to be achieved by the ozone treatment needs to be clearly defined.

Where **drinking water** is being treated, the primary objective may be to improve the organoleptic qualities of the water (colour, flavour threshold). In this case, where the water is produced from raw water with a low pollution load, use can be made of a single contact apparatus to bring about an ozonation reaction appropriate to the circumstances. The contact time ranges from 4 to 6 minutes.

Where dependable virucidal action is required, it is generally considered necessary to maintain for 4 minutes a residual ozone level of 0.4 g/m³. It is essential in these circumstances to effect contact in two compartments.

The function of the second compartment is to bring about the virucidal action proper. At the inlet into this compartment, the residual ozone level must be at least 0.4 g/m³ and sufficient ozone must be introduced at the bottom of the tower to maintain this

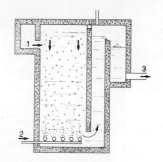

1. *Water inlet.* 2. *Ozonized air inlet.* 3. *Ozonized water outlet.*

Fig. 289. — Contact tower (single contact chamber).

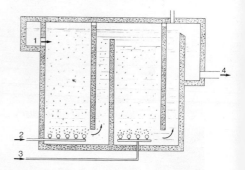

1. *Water inlet.* 2. 3. *Ozonized air inlet.* 4. *Ozonized water outlet.*

Fig. 290. — Two compartment contact tower.

level for 4 minutes. The amount of ozone to introduce in this compartment remains more or less constant and equivalent to between 0.4 and 0.6 g of ozone per cubic metre.

The first compartment is intended to satisfy the chemical ozone demand of the water. The period of contact and the amount supplied in this compartment are strictly dependent on this demand. Where the demand is low (limited to reducing colour or taste), treatment may be based on a usage rate of 0.4 to 1 g of ozone per m³ and a contact time of 4 to 6 minutes. Where the water is liable to contain micropollutants which have to be oxidized, it may be thought advisable to increase considerably both the amount supplied and the period of contact. Depending on the quantity of micropollutants to be eliminated, the usage rate may be increased to 5 g/m³ or even higher. The duration of contact may vary from 4 to 12 minutes according to the oxidation kinetics of the contaminants to be oxidized. In each case, every effort should be made to determine the potential level of chemical pollution. The size of this demand can be assessed in the laboratory.

In some circumstances it is only the oxidizing action of the ozone which is called for (in order to remove iron or manganese). Use is then made either of a single-chamber contact unit or of a system for recycling the residual ozonized air (cf. fig. 291).

For the treatment of **industrial effluents,** several contact units may be arranged in series, and here a counter-flow system can be adopted if required so as to make the maximum use of the residual ozone.

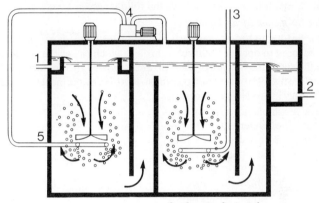

1. *Water inlet.*
2. *Ozonized water outlet.*
3. *Ozonized air inlet.*
4. *Recovered ozonized air compressor.*
5. *Recycled ozonized air intake.*

Fig. 291. — Contact towers in series with recycling of residual ozonized air.

For the treatment of **waste water (industrial or domestic),** where the water to be treated contains flocculated matter and is required to be disinfected, it is wise to employ an impeller type contact system. The point here is that the bubbles produced by porous diffusers cannot set up a sufficient turbulence to break up the agglomerated matter and thus enable the ozone to bring about the complete oxidation of the bacteria and viruses. Ozone usage rates may go up to 10-20 g/m^3. Wherever the effluent is heavily polluted, a preliminary laboratory test is essential in every case.

4.2.2. WATER/OZONE CONTACT

Careful attention must be given to the way in which the ozonized air and water are brought into contact, as it is the choice of the gas/liquid interface factor and the concentration of ozone in the gas phase which determine the efficiency of the operation by which the ozone is put into solution. Considered only from the point of view of the solubility of ozone in water, it is expedient to maximize the ozone concentration in the ozonized air which is injected. But when, given a constant rate of ozone usage, the concentration is increased, this results in a rise in the power consumption of the ozonizer and in a considerable decrease in the area of the water/bubbles interface due to the reduction in volume of the latter. It should also be borne in mind that the efficiency with which the ozone is passed into solution increases with the pressure at which the gas is injected and, in particular, with the increasing depth of the contact chamber. With tanks 7 to 8 m deep, it is easy to achieve 95 % efficiency in dissolving the ozone.

Of the various methods of bringing the ozone into contact with the water to be treated, mention may be made of the following:

A. Contact by means of an injector.

● Where a head equal to at least two metres water gauge is available, this pressure can be used to operate an injector through which passes the whole flow to be treated and which draws in ozonized air and supplies the contact column from the bottom (see fig. 292). It is not possible with this arrangement to achieve a high evel of efficiency in dissolving the ozone.

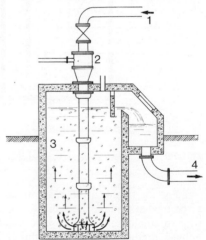

● When the available head is less than two metres, the water flow to be treated must be divided into two unequal parts. The smaller part is pumped to increase its pressure and to operate the ozonized air injector. The remainder of the flow is introduced by gravity at the bottom of the contact column. This method is not very efficient because of the poor blending of the ozone concentration in that portion of the water which has not been passed through the injector.

B. Contact by porous diffusers.

Porous diffusers at the base of a contact column ensure that the ozonized air is divided up into very small bubbles. The inlet for the water to be disinfected is located at the top (see fig. 289). This counter-flow process ensures the intimate contact of the two fluids. Contact columns with several compartments can be produced, with the

1. *Intake of water to be disinfected.*
2. *Ozonized air mixing injector.*
3. *Contact tower.*
4. *Disinfection water outlet.*

Fig. 292. — *Operating principle with ozonized air injector.*

counter-flow system also preferably being used for partial injections of ozonized air. The ozonized air is introduced into the water via porous tubes or discs.

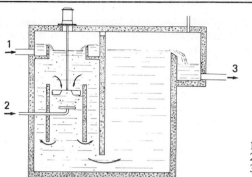

1. *Water inlet.*
2. *Ozonized air inlet.*
3. *Ozonized water outlet.*

Fig. 293. — Contact tower equipped with impeller.

C. Contact by special impeller.

The water to be disinfected is introduced into the suction area of an impeller which delivers it downwards to meet a flow of ozonized air injected beneath the impeller. A very fine emulsion (ozonized air—water) diffuses throughout the upstream part of the contact column and is picked up again by the impeller, which recycles several times the flow of water to be treated (see fig. 293).

A special profile impeller shears the bubbles of ozonized air and ensures that the gas mixture is effectively distributed throughout the body of liquid. The system described on page 229 (Vortimix) can be used to achieve this result.

4.2.3. DESTRUCTION OF RESIDUAL OZONE

After is has been in contact with the water to be treated, the air released by the vent of the contact tower still contains a certain amount of residual ozone. Depending on the method by which the ozone is brought into contact with the water and on the residual ozone level which is to be maintained in the treated water, this loss of ozone can vary from 1 to 15 % of the total quantity of ozone produced.

It is important that air too rich in ozone should not be released outside. Occasionally atmospheric dilution is considered sufficient, but the usual method involves the systematic destruction of the excess ozone (by thermal, chemical or catalytic means). It is also possible to recycle the residual ozone to the inlet into the contact system, although this technique does entail a sizeable consumption of energy (to repressurize the air containing the residual ozone or to drive a recycling impeller) for a modest return—particularly in the case of drinking water. The fact is that the air contains only a low proportion of residual ozone (0.2 to 1 g/m^3) and that, at this concentration, the efficiency with which the ozone can be dissolved is very poor (50 %). A further point is that this recycling does not solve the problem of disposing of the final ozone residue, the concentration of which in the air should not exceed 2 mg per m^3.

4.2.4. PRECAUTIONS WHEN USING OZONE

Consideration must be given to the distance at which the closest users of ozone-treated water are located. Ozone is unstable in solution in water, but when the concentration in the water at the outlet from the contact tower is 0.4 g/m^3, traces can still be detected after more than half an hour. It is for this reason that, when the contact time in the tank for treated water is short, corrosion is sometimes found to occur in the installations of those consumers situated close to the treatment plant. It is then advisable to neutralize the excess ozone contained in the water to be fed into the mains.

It can also happen that the water remains for some time in the tank for treated water and that the users are located at some distance from the treatment plant. In these circumstances, the residual ozone contained in the mains water is zero. The absence of residual disinfectant may lead to the development of plankton or bacteria in the supply system.

To avoid the proliferation of these microorganisms, the method generally adopted is to inject, after ozone treatment, a very small proportion of disinfectan with a long-term effect. Chlorine or chlorine dioxide can be used for this purpose without running the risk of giving the water any extraneous flavour, as the ozone has already oxidized the organic matter which might give rise to this.

The effect of ozone may vary depending on the temperature. It is sometimes found that, at temperatures below $5 \, ^\circ\text{C}$, the power of ozone to act on those constituents which lend a flavour to the water is reduced.

This is the reason why, when dealing with water of variable temperature, it is often advisable to combine the action of ozone with that of activated carbon.

4.2.5. METHODS OF OZONE DETERMINATION

Determining the ozone present in water

● *Colorimetric method:* see method of analysis 214 B on page 908.

● *Volumetric method:*

Ozone liberates iodine from a solution of potassium iodide, and the iodine is then titrated using a solution either of potassium arsenite or of sodium thiosulphate.

To 1 litre of the water to be analyzed add 10 cc of 10 % potassium iodide solution plus a pinch or 4 cc of starch paste. Homogenize the mixture by stirring. With a burette, add drops of N/35.5 sodium arsenite or sodium thiosulphate solution until decolorization of the titrate occurs. If n is the number of cc of titrant used, then the ozone concentration is equal to: $n \times 0.676 \text{ mg/l}$.

Note: Determination of the liberated iodine by means of a reducing solution indicates the total oxidizing agents present in the water. Should the specimen of water also contain chlorine in solution, the determination is carried out in two stages, the first embracing both oxidizing agents and the second following destruction of the ozone by glycocoll. The difference between the two readings is the ozone concentration in the water.

Determining the ozone present in air

● *Spectrophotometric method:* this involves the use of apparatus specially designed for the purpose, which measures the absorption of ultraviolet radiation by ozone. The apparatus is capable of continuous operation and gives the concentration of ozone in the air or oxygen supplied to it.

● *Chemical method:* in this method, a volume of gas which is measured by a counter at the bubbler outlet, is passed through a potassium iodide solution, and this is followed by titration in an acid medium of the iodine liberated by a sodium thiosulphate solution.

If V is the volume in standard litres of the dry gas which has passed through the counter, T is the normality of the sodium thiosulphate solution and x is the volume of this reagent introduced to decolorize the iodine solution, then the concentration of ozone in the gas in mg/l is given by the expression:

$$C = 24 \, T \frac{x}{V}$$

5. OTHER OXIDATION AND DISINFECTION PROCESSES

5.1. Use of potassium permanganate

This relatively expensive reagent (it costs 2.5 to 3 times as much as chlorine) is today used especially for the pretreatment of water to remove the manganese in solution. It acts more efficiently than chlorine on iron and manganese, and its effect is independent of the concentration of these two metals in the water.

Because of its poor performance as a disinfectant, potassium permanganate is not used in this role in water treatment plants. Despite this and the difficulty of putting it into solution, it is however sometimes used at a concentration of 30 g/m³ and with a contact time of at least 24 hours to disinfect tanks and water mains before these are placed, or brought back, into service.

5.2. Use of bromine

Bromine possesses antiseptic and algicidal properties which suggest its use for disinfecting swimming pool water. The minimum residual concentration to be maintained is 0.4 g/m³. At this dosage, irrespective of pH, it does not cause the water to smell or irritate the eyes. However, it has been found that, with high and rapidly fluctuating levels of pool usage it is necessary to raise the level of residual bromine to as much as 2 g/m³ to guarantee thoroughly dependable disinfection.

5.3. Use of chlorine/bromine mixtures

Some writers consider that, for the treatment of swimming pool water, the use of a mixture of 95 % chlorine and 5 % bromine facilitates the elimination of coliform bacteria. According to this hypothesis, the bromine combines with the organic matter introduced by the swimmers while the chlorine is entirely devoted to the destruction of the coliforms.

The technique has also been suggested for the disinfection of effluents.

5.4. Disinfection by ultra-violet radiation

Ultra-violet radiation is emitted by very low pressure mercury vapour lamps, the power rating of which may be as much as 200 watts and whose average life is between 2 000 and 4 000 hours. The wavelengths lie in the 200-300 nm range (2 000-3 000 Angström units) with a maximum microbicidal action around 250 nm,

The water to be treated must be made to flow close to the lamp and in a layer which should be as shallow as possible as ultra-violet rays are quickly absorbed by the water, which should be perfectly clear.

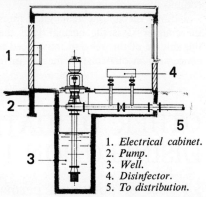

Fig. 294. — Disinfection by ultra-violet rays.

1. *Electrical cabinet.*
2. *Pump.*
3. *Well.*
4. *Disinfector.*
5. *To distribution.*

Given a depth of water of 15 to 20 cm, a 36 watt lamp is capable of disinfecting 3 m³/h (13 gpm). In actual practice, it is calculated that, to carry out disinfection with sufficient speed, an energy consumption of 40 Wh/m³ is required.

The water to be treated is usually pressurized. It is made to flow through a pipe, at the centre of which is located a quartz tube accommodating the radiation source. In this way, a thin layer of water is exposed to the germicidal action of the rays.

5.5. Disinfection by silver

This electrolytic (oligodynamic) disinfecting process, which has not yet been satisfactorily explained, requires that the silver should be present in the ionic state. The application of the process is still the subject of argument, but it is occasionally used for treating swimming pool water.

5.6. Disinfection by ionizing radiation

The development of this process is currently receiving a certain amount of attention in connection with the disinfection of sewage.

As the source of ionizing radiation, use is generally made of cobalt 60, which has a specific activity of 25 Ci/g. It is used in the form of sticks 4 cm in diameter and of varying lengths according to the activity. The water to be treated is made to flow through sheaths which encase the radiation sources and which are sealed to prevent any radioactive pollution.

According to published data, radiation intensities should be as follows:
— for disinfection purposes: 450 kilorad at 10^5 Ci/(m³.h);
— for total sterilization: 4.5. megarad at 10^6 Ci/(m³.h).

The average efficiency of the plant is said to be between 70 and 90 %.

16 NATURE, STABILIZATION THICKENING AND CONDITIONING OF SLUDGE

1. ORIGIN AND NATURE OF THE DIFFERENT TYPES OF SLUDGE

The solid pollutants, and the converted substances they produce, removed from the liquid phase during treatment of any kind of water finally collect as suspensions which vary in concentration and are known as "sludge ".

All types of sludge form an extremely liquid waste product of little or no value. Some sludge is chemically inert, but those types originating from biological treatment processes are fermentable and often have an offensive smell.

All sludge requires some form of treatment, whether stabilization, thickening or dewatering possibly followed by drying and incineration, or a combination of one or more of these processes, before being discharged into the natural environment.

Chapters 16 and 17 will review the main treatment processes to which water treatment sludge from all sources may be submitted, including surface water, domestic sewage and industrial waste water.

1.1. Classification

It is essential to classify a sludge in order to select the treatment method applicable to it, and to forecast the performance of the equipment to be used. The following table endeavours to classify the various types of sludge according to their origin and also according to their hydrophilic colloidal matter content as this plays a major role in the behaviour of the sludge during dewatering.

CLASSIFICATION OF SLUDGE

Principal characteristic of the sludge	Origin — Industry	Water treatment Sludge pretreatment	Constituents of sludge
hydrophilic organic	1. Domestic waste water treatment plants (D.W.W.)	Primary settling (raw primary) Primary settling + Anaerobic digestion (digested primary) Primary settling + Biological treatment (raw primary + act. secondary) Primary settling + Biological treatment + Anaerobic digestion (primary + act.; Anaerobic dig.) Extended aeration + Aerobic stabilization Physico-chemical (Flocculation-Settling)	Predominantly volatile solids: VS/DS: 30 to 90 % — Protein matter, often very fermentable — Vegetable or animal wastes — Animal and sometimes mineral oils and fats — Hydrophilic hydroxides (Al, Fe) in physicochemical treatment — Hydrocarbons (petro-chemistry)
	2. W.W. from agriculture and foodstuffs industry ● breweries ● abattoirs ● potato processing ● dairies ● canneries ● stock rearing (piggery manure)	2. Settling Biological including extended aeration, aerobic stabilization or anaerobic digestion	
	3. W.W. from textile industry, organic chemical industry (including petrochemistry)	3. Physico-chemical (Floc.-Settling) Biological	
	4. Any polishing biological treatment	4. Biological	
hydrophilic oily	1. W.W. from refineries 2. W.W. from engineering works (soluble oils)	— Oil removal — Flocculation-Settling /Flotation	— Mineral oils and greases — Hydrocarbons — Hydroxides (Al, Fe)
	3. W.W. from cold rolling, metallurgy	— Biological (refineries)	— (possibly) biological V.S.
hydrophobic oily	1. W.W. from steel rolling mills	Settling	— Dense and readily settleable DS (Scale-Fe oxides) — Considerable mineral oil and grease

Principal characteristic of the sludge	Origin — Industry	Water treatment Sludge pretreatment	Constituents of sludge
hydrophilic inorganic	1. Drinking water and industrial make-up water (river water or groundwater) • clarification • partial carbonate removal • deionization eluates 2. W.W. from metal finishing treatment • pickling • anodizing • galvanizing • painting 3. W.W. from inorganic chemical industry 4. W.W. from colouring agents; dyeworks 5. W.W. from tanneries 6. Total final treatment of W.W. for recirculation	1. Physico-chemical (Floc.-Settling); Neutralization (eluates) 2. Neutralization + Flocculation-Settling Decontamination (cyanides, Cr $^{6+}$) + Flocculation-Settling 3. *id.* 1 4. *id.* 1 + possible biological 5. *id.* 4 6. *id.* 1 + filtration	Predominantly hydrophilic metallic hydroxides (Fe, Al, Cr...) + V.S. ($<$ 30 % of DS) + $CaCO_3$ (carbonate removal) or $CaSO_4$ 2 H_2O (neutralization H_2SO_4) 4. Mineral + organic 5. + animal fats and organic matter
hydrophobic inorganic	1. Industrial make-up water—carbonate removal (river water or groundwater) 2. Iron and steel industry—Steelworks—Foundries—Gas scrubbing 3. Coal washing 4. Incineration of refuse Flue gas scrubbing	Neutralization — Flocculation Settling	Dense inorganic solids Low content of hydrophilic hydroxides (Fe, Al, Mg $<$ 5 % of DS) Low V.S. content ($<$ 5 % of DS)
fibrous	1. W.W. from paper mills 2. W.W. from paper pulp 3. W.W. from board mills	— Settling/Flotation (fibre recovery) — Flocculation — Settling — (Possibly) biological	Cellulose fibres + possibly sawdust and shavings + hydrophilic hydroxides (in varying amounts) + possible biological V.S.

VS = Volatile solids   W.W. = Waste Water
DS = Dry solids   D.W.W. = Domestic Waste Water

The composition of a sludge depends both on the nature of the initial pollution of the water and the treatment processes to which that water has been submitted, whether physical, physico-chemical or biological.

● *Hydrophilic organic sludge* : this is one of the largest categories. The difficulties encountered in dewatering this sludge are due to the presence of a large proportion of hydrophilic colloids. All types of sludge resulting from the biological treatment of waste water and whose volatile solids content may be as much as 90 % of the total dry solids content (waste water from the food industry, organic chemical industry, for example) are included in this category.

Hydroxides of a hydrophilic nature such as iron or aluminium hydroxides, derived from inorganic flocculants used during water treatment, may be present.

These organic types of sludge require conditioning before mechanical dewatering.

● *Hydrophilic inorganic sludge* : this sludge contains metal hydroxides formed during the physico-chemical treatment processes as a result of the precipitation of metallic ions present in the raw water (Al, Fe, Zn, Cr...) or due to the use of inorganic flocculants (ferrous or ferric salts, aluminium salts).

● *Oily sludge* : this is characterized by the presence in the effluents of small quantities of mineral (or animal) oils or fats. These oils are in emulsion or adsorbed onto the hydrophilic or hydrophobic sludge particles. A proportion of biological sludge may also be present, in cases of final activated sludge treatment (e.g. treatment of refinery effluents).

● *Hydrophobic inorganic sludge* : this sludge is characterized by a preponderant amount of particulate matter with little or no bound water (sand, silt, slag, rolling-mill scale, crystallized salts, etc.).

The dewatering, which is easy in principle, of this type of sludge may be hindered by the presence of hydrophilic inorganic matter from the flocculating agents used in water treatment.

● *Fibrous sludge* : this sludge is generally easy to dewater, except when the intensive recovery of fibres makes it hydrophilic, because of the presence of hydroxides or biological sludge or both.

1.2. Factors characterizing the nature of a sludge

A. Dry solids content (DS) : this is generally expressed in grammes per litre or as a percentage by weight and is determined by drying at 105 °C to a constant weight. In the case of liquid sludge, it generally approximates to the suspended solids content, determined by filtration or centrifugation.

B. Volatile solids content (VS) : this is expressed as a percentage by weight of the dry solids content. It is determined by gasification in an oven at 550-600 °C.

In the case of hydrophilic organic sludge, in particular, it is often near the organic matter content and is characteristic of the nitrogenous matter content.

C. Weight of contents (especially in the case of organic sludge) :

— C and H to assess the degree of stabilisation or to deduce the net calorific value;

— N and P to evaluate the agricultural value of the sludge;

— other contents (e.g. heavy metals). In the case of inorganic sludge, the Fe, Mg, Al, Cr, calcium salts (carbonates and sulphates), and silica contents are often useful.

D. Composition of the interstitial water:

— dissolved substances;

— TAC, TA;

— COD, BOD_5, pH, etc.

1.3. Factors characterizing the structure of the sludge

A. Apparent viscosity in relation to the rheological behaviour:

Sludge suspensions are not Newtonian liquids: the value found for viscosity is quite relative and depends on the shearing stress applied.

In the case of some sludges, and with certain precautions, a viscosity known as the Bingham viscosity may be deduced for a characteristic stress T_B (see fig. 295).

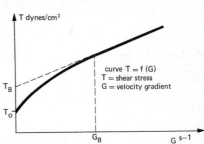

Fig. 295. — Determination of the BINGHAM viscosity.

Viscosity may be considered as a measurement of the intensity of the inter-particulate forces. It also permits evaluation of the thixotropic character of a sludge (the capacity of a sludge to form into a gel when motionless and to return to the fluid state when only lightly stirred). This property is very useful for assessing the possibility of collecting, transporting and pump-ing sludge.

B. Particle size distribution.

C. Nature of the water contained in the sludge:

This water is the sum of:

— free water which can be fairly easily eliminated,

— bound water comprising: colloidal hydration water, capillary water, cellular and chemically bound water.

The release of bound water requires considerable energy; for instance, cellular water, in particular, is only separable by heat treatment (thermal conditioning, drying or incineration).

The proportion of free water and bound water is therefore decisive in the suitability of a sludge for dewatering. An approximate value can be obtained by **thermogravimetry**, i.e. by drawing the weight loss curve of water at constant temperature in a thickened sludge sample in specified handling conditions (fig. 296). The point where the thermogram bends may be determined by drawing the curve $V = f(S)$ where V is the drying rate and S the dryness of the sample (fig. 297). A dryness S_L is read for each sludge, corresponding to the first critical point: S_L is considered to be the dryness of the sludge after loss of the free water: for the sake of practical interpretation, the free water is defined, in thermogravimetry, as the quantity of water capable of elimination at constant drying rate.

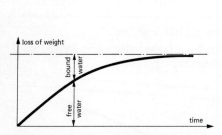

Fig. 296. — Dewatering curve for sludge dried by evaporation at constant temperature.

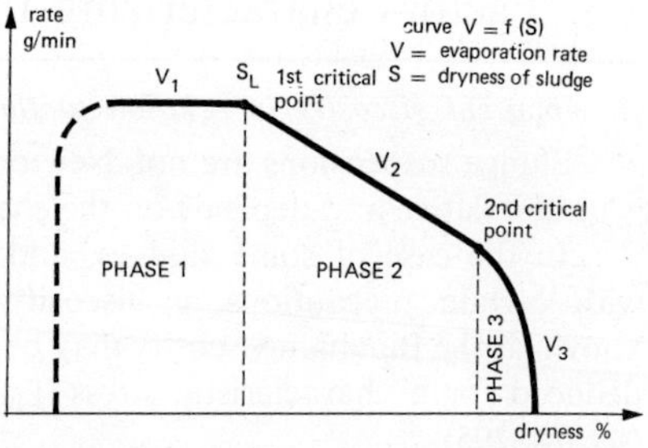

Fig. 297. — Thermogravimetry of sludge.

With the thermogravimetric method, it is possible to estimate the hydrophilic tendency of a sludge, and also to:

— follow the evolution of the proportion of bound water as a function of the various conditioning processes.;

— evaluate fairly accurately in the laboratory the performances of the dewatering units;

— determine, by comparative studies, a maximum dryness for each type of treatment;

— calculate, if required, the liaison energy of the various types of water with the sludgy matter.

The interpretation of hydroxide sludge thermograms is more complex because of the presence of rather a large proportion of chemically bound water.

1.4. Factors characterizing the behaviour of a sludge during dewatering

These factors arise from the dewatering techniques to be used and are dealt with in detail later in the chapter:
— ability to undergo thickening (see page 460);
— numerical characterization of filtrability; specific resistance to filtration (see page 481);
— numerical characterization of the compressibility of a sludge under the effect of an increase in pressure (see page 483);
— determination of a maximum dryness under a given pressure;
— centrifugability, characterized by the separation rate of the dry solids under the effect of a high acceleration field.

2. DISPOSAL OF TREATED SLUDGE BY-PRODUCTS

Sludge is often difficult to dispose of economically, and its removal is almost always a heavy item in operating costs. From the economic standpoint, the real aim is to limit the cost of sludge treatment (labour, reagents, electric power of heat, transport, disposal etc.). On the other hand, the protection of industrial health and the environment calls for methods which will cause the least nuisance, sometimes in spite of higher costs.

Sludge and its by-products are disposed of as follows:

A. Soil improvement

Sludge from the treatment of municipal sewage and certain kinds of industrial effluents can be used for this purpose. The table in page 434 gives the principal agronomic properties of such sludge according to the type of treatment. Its value lies more in the humic matter which it provides and in the improvement of the water-retention properties of the soil, than in its nutrient matter content. It is particularly suitable for the cultivation of flowers, lawns and trees; the addition of stabilized or digested sludge to the soil can help to promote the growth of the autotrophic microbial flora by acting directly on the mineral nutrition of the plants. It can be spread:
— in liquid form (after preliminary thickening);
— in a form suitable for shovelling (dewatered to give a dry matter content

of less than 10 % in the case of colloidal biological sludge, and more than 50 % in the case of very dense inorganic sludge);
— in powder form (partial drying to between 65 and 90 % dryness).

AGRONOMIC PROPERTIES OF DOMESTIC SEWAGE SLUDGE (PRIMARY + ACTIVATED SLUDGE) AS A PERCENTAGE OF DRY MATTER

	N	P_2O_5	K_2O	Organic matter
Raw sludge	3-5/4-5	2/3	0-5/1	60/80
Digested sludge	2/2-5	1/2	0-2/0-5	40/65

The hazards due to bacteria should not, however, be overestimated as microbial action in the soil is important. The restrictions of use arise from the risk of the release of odours from insufficiently stabilized sludge or from too high a concentration of heavy metals (Zn, Cd, Cu, Pb, Ni, Cr_6, Hg, ...). The optimum annual quantities of sludge to be spread per hectare can be determined in each case by a suitable agronomic survey. The products added for conditioning the sludge before dewatering (inorganic or organic flocculating agents) should then be taken into account.

The addition of waste water treatment sludge to domestic refuse enriches the product in active and humic matter; the C/N ratio of the mixture, about 25 to 30, is very suitable for thermophilic aerobic fermentation. Composting facilitates subsequent assimilation by the soil and considerably reduces the content of pathogenic germs. The sludge, which is usually stabilized first by biological treatment, must be as dry as possible in order to be incorporated with the refuse, the water content of the mixture being about 50 %. Preliminary dewatering is therefore usually necessary.

Direct composting of the sludge can also be envisaged, but after addition of a material containing carbon (sawdust, straw, etc...), and preliminary dewatering of the sludge. In the case of sludge with a final high moisture content, the cost of the added material becomes prohibitive, because of the quantity required.

B. *Recovery of products*

Only some of the constituents of the sludge are recoverable. They include in particular:
— fibres in the case of the paper and timber industries;
— coagulants in the cas of the treatment of river water for use as drinking or industrial process water. (When this treatment makes use of aluminium sulphate, a substantial part of the alumina can be recovered by subsequent acidification of

the sludge; after recarbonation, possible chemical conditioning and dewatering, the final sludge can be disposed of in a " shovelable " form);
— lime or calcium carbonate in the case of a massive lime treatment (paper-mill liquors, waste water, decarbonated water). The simultaneous treatment of carbonate-removal sludge and predominantly organic sludge can be recommended when mineral reagents are used; carbonate-removal sludge can also be used for the neutralization of acid effluents;
— zinc, copper, chrome in the case of the purification of water used in the treatment of metal surfaces and in various industrial processes.

But even when certain constituents of the sludge have been recovered, a residue of unusable substances always remains.

C. *Recovery of energy*

The use of sludge to generate energy is generally an intermediate stage only. Sludge is very rarely used as a fuel, and this application is confined to previously-dried sludge obtained from the sedimentation of water containing a very large proportion of coal dust (e.g. Emschergenossenschaft in the Ruhr (Germany), territory) and to oily or greasy suspensions recovered by flotation.

Energy will of course only be recovered from sludge with a high organic-matter content, and will only be a by-product of sludge treatment, not the main purpose.

Energy is recovered in two main forms:
— production of methane gas by fermentation (anaerobic digestion); this gas is used for heating or to generate electricity or for the thermal conditioning of the sludge itself.
— use of the calorific value of the dry matter in incinerators. The energy generated is used essentially to predry the sludge. When the sludge is autocombustible, it is possible to recover the thermal energy, possibly transformed into electrical energy, in the combustion gases.

It will be noted that, in addition to the partial or total reduction of the organic matter in the sludge, the recovery of energy eliminates pathogenic germs and destroys the fermenting power of the product.

D. *Sale of by-products*

The ash could, for example, be used for making road surfaces, soil stabilizers or cements, but manufacturers have not yet moved in this direction, and prefer to use better-quality materials which are easier to handle.

Very fibrous sludge which is unlikely to ferment can, after dewatering, be used to stabilize landfill which is to be grassed over.

E. *Disposal on land*

Water or sewage treatment sludge is most often disposed of in this way. The amount of residue varies with the sludge treatment process, but a substantial residue remains even after incineration.

The sludge can be simply discharged into a sludge lagoon, which takes months or years to drain and evaporate, or it can be used to fill in excavations or depressions with dry sludge which can be compacted much more rapidly.

An approach which is sometimes considered is to incorporate products like silicates, cements, etc. with the liquid sludge before discharge. After a rest period, the mixture solidifies. This method of treatment — apparently suitable for toxic sludge — has the disadvantage of permanently condemning a large area of ground at the site of discharge. Furthermore, the risk of leaching by runoff water cannot be altogether discounted.

F. Disposal at sea

This quick method of disposal may take the form of discharging the sludge from barges or lighters at intervals or may involve the use of a long and very deep underwater pipe.

Disposal at sea involves and long detailed prior investigation of surface and depth currents, as well as very thorough bacteriological, biological and fish ecology studies. The destruction of pathogenic germs and the breakdown of organic matter are slow in sea water. The germs survive for several days.

All floating matter must be removed from sludge disposed of at sea; the sludge must also be physically, chemically and biologically stable. Preliminary mixing with sea-water assists immersion. In this way the city of Los Angeles disposes of a large part of its sewage sludge in the Pacific Ocean after prior digestion; the sludge pipe-line is laid at a depth of 70 m (230 ft) over more than 10 km (6 miles) of sea bed.

This method has also been adopted in the Mediterranean, where red sludge containing a large amount of sodium ferro-silico-aluminates from the production of alumina is disposed of into deep water.

G. Reinjection into the ground

With this technique, sludge is injected in the liquid state into porous sub-soil pockets separated by continuous strata of clay. A thorough geological survey is essential. The bore-hole must be sealed completely to avoid contaminating the water-bearing strata through which it passes.

The depth of injection varies from 100 to 4 000 m (330 to 13 200 ft) and the pressure used can reach 70 bars (1 000 psi).

This technique must be strickly limited to sludge which is very difficult to treat. Its use does not always seem to have been associated with the necessary safeguards.

3. OBJECTIVES OF SLUDGE TREATMENT AND THE PROCESSES USED

The best method of treating the sludge will vary according to how much waste land is available, the extent to which the sludge is fermentable or can be dried and the economic factors involved (cost of land, labour, power, reagents, amortization of investment, health requirements, etc.); but the final objectives will be the same in all cases:

● **Reduction of volume.** This can be achieved either by straightforward thickening (when the dryness rating of the product can reach 10 % or even 20 % in very exceptional cases although not dry enough to be removed by shovel) or by dewatering by natural drainage, by mechanical means or by heating or, finally, by incineration.

● **Reduction of fermenting power** (stabilization) which can be achieved by:
— anaerobic digestion;
— aerobic stabilization;
— chemical stabilization;
— pasteurization;
— heating;
— incineration as a final stage.

The table on page 438 shows the most usual sludge treatment process chains.

4. SLUDGE STABILIZATION

4.1. Anaerobic digestion

Methane fermentation (see page 105) is one of the most powerful means of destroying cells known to biology, and this process can be used to eliminate a large quantity of organic matter.

Anaerobic digestion takes place in two stages—liquefaction followed by gasification.

During liquefaction mainly volatile acids are produced.

During gasification, the strictly anaerobic methane bacteria produce methane gas from the volatile acids or from the alcohols formed during the first stage. These slowly-developing microorganisms are sensitive to variations in pH (maximum activity between 6.8 and 7.2).

When large quantities of volatile acids are produced, the pH value decreases, and this checks the biological digestion process. A high bicarbonate alkalinity can have a beneficial buffer effect.

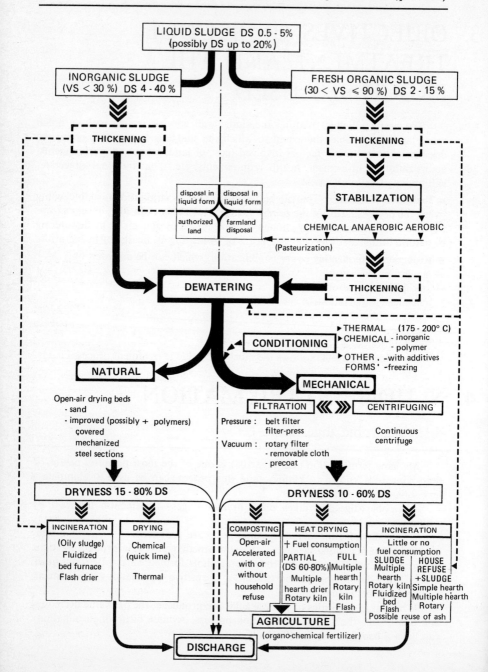

4.1.1. GAS PRODUCTION — TEMPERATURE AND RETENTION TIME

The gas generated is essentially methane (CH_4) and carbon dioxide (CO_2) in the following proportions by volume:

$$CH_4: \quad 65 \text{ to } 70 \ \%$$
$$CO_2: \quad 25 \text{ to } 30 \ \%$$

Small quantities of other substances may be present: oxygen O_2 (0 to 0.3 %), carbon monoxide CO (2 to 4 %), nitrogen N_2 (1 %), hydrocarbons (0 to 1.5 %), hydrogen sulphide H_2S.

The quality of digestion is best expressed by the quantity of gas produced: his depends basically on two factors:

— temperature;
— retention time.

Fig. 300 shows the maximum quantity of gas obtained from the digestion of 1 kg of organic matter at various temperatures. Plants generally aim at reducing the volatile matter content by digestion by 45 to 50 %.

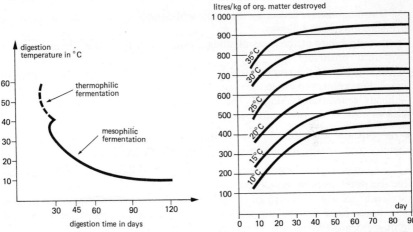

Fig. 299. — *Time needed for the total digestion of municipal sewage sludge at different temperatures.*

Fig. 300. — *Variation of gas production as a function of time and temperature.*

Since 1 kg of **destroyed** organic matter in domestic sewage gives some 900 to 1 000 l of gas (1 lb gives 14 to 16 ft³), well-balanced digestion generally produces about 400 to 500 l (6.5 to 8 ft³/lb) of gas per kg of organic matter **entering the plant.**

Temperature is a decisive factor in efficient digestion, and affects the speed at which the process starts, the stability of fermentation and the output of gas. Sludge heating units are essential on all but very small plants (serving a population of a few thousand) in which the sludge is retained in unheated digesters for a very long time.

Fig. 299 shows how the time required for complete digestion varies with the temperature in the two cases of mesophilic and thermophilic fermentation (see page 19); only mesophilic fermentation, below 35 °C, with greater stability of operation, is discussed below.

Retention time depends on the rate at which the microorganisms reproduce, and minimum theoretical digestion time can be put at 3 to 4 days. Below this figure, the loss of microorganisms due to the quantity of sludge evacuated from the digester is higher than the growth of new bacteria due to the transformation of the matter fed in.

The net calorific value of the gas produced by digestion is between 22 600 and 25 100 kJ/Nm^3 (607 to 675 BTU/ft^3).

This figure varies with the proportion of methane; the proportion is often related to the rate of digestion (net calorific value is higher when digestion is slower).

4.1.2. PARAMETERS AFFECTING THE EFFICIENCY OF ANAEROBIC DIGESTION

The main factors favouring the stable and complete development of methanic fermentation are as follows:

— **the temperature** of the mass of sludge, which must be sufficiently high and as constant as possible;

— **the capacity of the reactor,** which must be designed to allow the matter to remain long enough to attain the required degree of decomposition;

— **highly concentrated** sludge. The sludge fed into the digester should be as thick as possible for two main reasons:

● the capacity (and therefore the price) of the digester can be reduced for a retention time corresponding to a given level of efficiency,

● the concentration of methane-producing bacteria is increased; the biochemical reaction develops more rapidly and is much easier to start. For industrial operation, a concentration of 15 g/l (15 000 ppm) of organic matter may be regarded as a lower limit for the digestion of sewage sludge. This is an important point, especially for starting up a digester;

— **vigorous stirring.** This increases the chances of bringing the micro-organisms into contact with the matter to be broken down and gives the sludge an even consistency. The difference observable between the digestion performance of a small laboratory apparatus and that of an industrial plant is due to the much more vigorous stirring which is easy to achieve in a small tank. Powerful and effective stirring eliminates differences in temperature and in organic matter concentration (raw sludge and sludge in process of digestion) in different parts of the digester;

— **regularity of feed.** Both the input of raw sludge and the extraction of digested sludge must be very regular in order to keep the ratio of organic matter to micro-organisms as constant as possible, and thus avoid any sudden variation in the regular development of the microorganisms.

The main factors which inhibit or interfere with digestion are as follows:
— the presence of certain toxic substances:
- heavy cations (copper, nickel, zinc);
- excess NH_4^+ ions;
- sulphides;
- certain organic compounds (e.g. cyanides, phenols, phthalates);
- too high a concentration of detergents;
— sudden variation of pH beyond the optimum range (6.8 to 7.2), caused by the introduction of alkaline or acid solutions;
— any sudden variation of temperature or load.

Fig. 301. — Overall view of sludge digesters at the sewage treatment plant of TOURS, France.

2 primary digesters, capacity 2 000 m³ (70 600 cu ft).
2 secondary digesters, capacity 1 000 m³ (35 300 cu ft).

4.1.3. RESULTS OF TREATMENT—ADVANTAGES OF DIGESTION

Fresh domestic sludge is grey or yellowish in colour; it contains faecal matter, paper, vegetable waste, etc. It has a foul smell. After complete digestion, the sludge is black (iron sulphide) and smells of tar. The original constituents are virtually indistinguishable (except for human and animal hair and some seeds). Most of the pathogenic germs have been destroyed, because alkaline digestion has a very powerful bactericidal effect (but the destruction of certain types of virus and of Koch's bacillus is disputed).

Sludge drains on drying beds in the open air without difficulty and without any harmful effects.

Taking:

— m_1 as the percentage of mineral matter in the raw sludge,
— m_2 as the percentage of mineral matter in the digested sludge,
the reduction of volatile matters obtained by digestion and expressed as x, is given by the following formula:

$$x = 1 - \frac{m_1 (100 - m_2)}{m_2 (100 - m_1)}$$

In the case of domestic sewage, a 45 to 50 % reduction in volatile matter content means substantially the straightforward elimination of one third of the dry matter and a 30 % reduction in the volume of sludge, allowing for thickening of the raw and the digested sludge respectively. The sludge dewatering plant can be very much smaller.

Anaerobic digestion has the advantage of providing energy in the form of a valuable product: methane gas. More gas is produced than is needed for the thermal energy requirements of the digestion plant itself (i.e. for heating and electric motors). Methane can easily be stored in bell-type gasholders or in pressurized spheres.

In addition, a digestion unit provides a safety margin advantageous to the operation of any treatment plant. With the buffer capacity provided by the digester or digesters, the downstream sludge treatment units (dewatering units or incinerators) can be stopped for a while for repair or overhaul.

It is also possible to store digested sludge or dry it in the open air, without odours.

4.1.4. ONE- OR TWO-STAGE DIGESTION— DESIGN PRINCIPLES

The digestion plant often consists of two stages: the digestion stage; the thickening stage.

If both stages take place in the same tank, the sludge cannot be stirred very vigorously, because a still zone is required for thickening.

If the two stages take place in two different tanks, the second tank is intended to thicken the sludge and to complete the digestion process. However, digested sludge is often difficult to thicken mainly because of the presence of large quantities of occluded gas. Lengthy retention periods are therefore necessary.

If thickening of the digested sludge is not required, high-rate digestion may be carried out with a secondary tank of small capacity (with a retention time of a few days) or even without a secondary tank. On the other hand, preliminary thickening of the fresh sludge is valuable as it allows a longer retention time for the same tank capacity.

High-rate digestion is not the result of using two stages but of designing the

primary digester to ensure rapid and complete digestion. The features necessary for this purpose are as follows:
— powerful stirring to give an even consistency and prevent the formation of a scum layer;
— reliable heating of the sludge to about 35 °C under all running conditions (make-up fuel required on starting);
— regular feed of raw sludge;
— means of controlling the pH of the sludge;
— adequate thickening of the raw sludge;
— multiple sampling points;
— several extraction points.

Domestic sewage sludge can then be digested with loadings of up to 5 kg of volatile matter per day per cubic metre (0.31 lb/ft³ d) of digester capacity (at a pilot plant).

The table below gives the size limits for medium-rate (single stage only) and high-rate (1- or 2-stage) digesters. In the latter case, it is assumed that the primary digesters are equipped with the best stirring and heating systems currently available and are fed with thick sludge.

	Primary digester			Secondary digester
	Load of volatile matter		Retention time in days	Retention time in days
	kg/m³ d	lb/ft³ d		
Digestion with medium loading (heated to 25 °C)	0.8	0.05	37	—
Digestion with medium loading (heated to 35 °C)	1.2	0.07	25	—
Digestion with heavy loading (heated to 35 °C)	3-4	0.18-0.25	10	2-4

The factors determining the size of **unheated** digesters vary very widely. They depend to a large extent on climatic and operating conditions. Even in the temperate regions of Western Europe, the process of digestion is practically halted for 3 to 6 months of the year. Digestion capacity must allow for a minimum retention of 90 days. It can be reduced somewhat in the case of an Imhoff tank in which contact between the sludge and the raw effluent always keeps the temperature of the sludge above the minimum of 11 to 12 °C, below which methane fermentation is halted.

For a quick calculation of domestic sewage plant dimensions, the following figures for volume of sludge and weight of dry matter per head of population can be used; the figures relate to effluents from separate-sewage systems in which the suspended solids are two thirds organics; the values indicated between brackets are those otbained after intermediate thickening of the fresh sludge.

	Dry matter		Dry	Quantity of sludge	
	g/hd d	lb/hd d	matter %	litre/hd d	gal/hd d
Primary settling units					
Fresh sludge	54	0.12	5 (10)	1.08 (0.54)	0.29 (0.15)
Digested sludge	34	0.075	10	0.34	0.09
Complete plant					
a) settling + biological filter:					
fresh sludge	74	0.16	5(7)	1.48 (1.05)	0.40 (0.27)
digested sludge	48	0.105	8	0.80-0.60	0.21-0.158
b) settling + activated sludge:					
fresh sludge	85	0.185	4.5 (6)	1.87 (1.41)	0.49 (0.37)
digested sludge	55	0.12	4-6	1.38-0.92	0.36-0.24

4.1.5. DESIGN OF THE DIGESTERS

The technique of the IMHOFF or two-storied tank, combining primary settling in the upper chamber and unheated digestion in the lower chamber, is virtually no longer used. It was mainly used in small plants. The processes of extended aeration and aerobic stabilization, although more expensive in energy consumption, are less sensitive to temperature change and require less capital expenditure, and are now preferred for small communities.

A. Medium-rate digestion

Medium-rate digestion takes place in a single digester, which should preferably be heated.

The digester must be equipped with powerful means for breaking up the scum, and this system must act effectively whatever the consistency and depth of the layer.

The scum-breaking system must be free of any accumulation of fibre or paper. An advantageous approach is to use rotary sprinklers supplied by external sludge recirculation pumps (fig. 304).

In medium-rate digestion, where the mass of sludge cannot be stirred very vigorously, the fresh sludge is fed into the upper part of the digester.

A normally-operated medium-rate digester comprises three main zones:
— floating scum;
— an intermediate layer (water band), where the concentration of dry solids is lowest;
— a bottom layer where the sludge already digested or in process of digestion gradually thickens.

The digester must have an overflow system which will remove only the least concentrated sludge. The overflow is usually returned to the head of the plant.

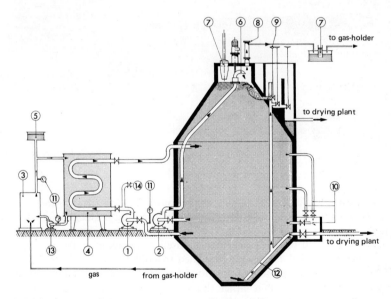

1. *Sludge-heating circulation pump.*
2. *Sludge scum-breaking circulation pump.*
3. *Hot-water boiler.*
4. *Heat exchanger.*
5. *Expansion tank.*
6. *Scum-breaker.*
7. *Vacuum-breaking and pressure-limiting device.*

8. *Gas outlet.*
9. *Scum extraction.*
10. *Sampling pipes.*
11. *Thermometer.*
12. *Extraction of digested sludge.*
13. *Hot-water circulation pump.*
14. *Fresh sludge inlet.*

Fig. 302. — Single medium-rate digester.

The system may consist either of a telescopic valve or of a series of outlets at different levels; the digester overflow can thus be set at the optimum level.

The digester floor must slope as steeply as possible because stirring is not sufficient to displace the accumulations of sludge which settle when the slope is only slight.

Fig. 303. — Medium rate digester and its associated gasometer. Sewage treatment plant at SENS (France).

The digested sludge is drawn off at the lowest point of the digester.

The gas produced by digestion is normally used in the sludge-heating boiler. A gasholder is by-pass mounted on the gas line to the user circuits.

The gas stored in this holder can be used to maintain a steady supply.

The gasholder bell is normally designed to hold 5 to 10 l (0.18 to 0.36 cu ft) per head of population, amounting to 20 to 40 % of the daily output. If the gas is not recovered it is either discharged direct into the atmosphere or, preferably, it is run to a waste gas burner.

The digester is fitted with vacuum-breaking and pressure-limiting devices in the form of water seals or mechanical valves.

B. High-rate digestion

High-rate digestion is usually in two stages. The second-stage digesters can be open or covered. Open digesters must be of small diameter (maximum 8 m (27 ft)) because they have no scum brakers, and because it must be possible to remove the floating scum by hand.

If the secondary digester is covered, it must, like the primary digester, have a gas recovery system, if the surplus gas which continues to be released at the end of the methane fermentation process is not to be lost. In that case, the scum is broken up by the same means as in the primary digesters.

C. Mixing systems in the digesters

The sludge in the digesters may be mixed by recirculating large quantities of sludge from the bottom to the upper part of the digester by means of external pumps, until the entire volume of sludge is turned over. Because of the inevitable limitations imposed by the pump delivery obtainable, this technique requires the construction of tanks with steeply-sloping floors in order to avoid large dead sludge zones and to facilitate methodical mixing of the contents of the digesters. Very deep tanks are expensive and it is for this reason that this system has, in many cases, been superseded by the **gas mixing** system (fig. 306).

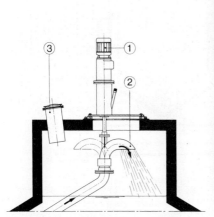

Fig. 304. — *Hydraulic scum-breaker.*
1. *Motor and reduction gear.*
2. *Rotary sprinkler.* 3. *Sight-glass.*

Fig. 305. — *Boiler-house, hydraulic-mixing digesters and gasometer. SAINT-BRIEUC (France) sewage treatment plant.*

With this method, the sludge is mixed by feeding gas under pressure into the digester.

Various layouts and points for introducing the gas have been suggested; one very efficient method is to concentrate the gas flow at the centre and base of the digester. This concentration of gas sets up a powerful stirring "swirl" from the centre to the periphery when the $\dfrac{\text{diameter}}{\text{height}}$ ratio is appropriate.

The bottom of the digester is swept by a sludge-flow in the opposite direction to the surface flow.

The injected gas taken from the gasholder or the dome of the digester is forced through diffusers or vertical pipes at the rate of about 0.8 m³/ (h. m²) (2.6 ft³/ft² h).

Because of its powerful recirculating effect, the gas mixing method allows the construction of digesters with a gradually sloping floor.

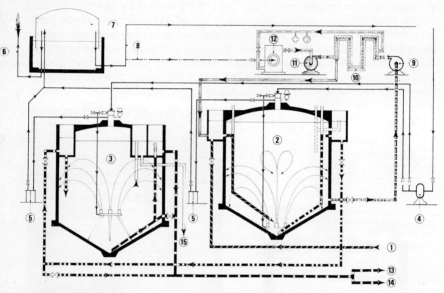

1. *Fresh-sludge inlet.*
2. *Primary digestion tank.*
3. *Secondary digestion tank.*
4. *Gas compressor.*
5. *Water traps.*
6. *Waste gas burner.*
7. *Gasometer.*
8. *To boiler-room and blowers.*

9. *Sludge circulation pump.*
10. *Heat exchanger.*
11. *Hot-water pump.*
12. *Boiler.*
13. *To final sludge treatment.*
14. *To lagoons.*
15. *Overflow.*

Fig. 306. — Diagram of two-stage digestion process with gas mixing equipment.

The cover often takes the form of a fixed roof. Floating covers can also be used, but they are fairly costly to install and maintain.

The sludge is heated in the same way as in sludge-recirculation digesters.

D. Heating of digesters

The sludge is sometimes heated by the direct injection of steam. This method is not used much because of the necessity of guarding against the risk of overheating the sludge near the injection points. The use of steam boilers,

Fig. 307. — Internal view of a digester 33 m (108 ft) in diameter. Achères IV sewage treatment plant, Paris, France.

which are dear, requires special precautions. Hot-water sludge heaters are most frequently employed. It is possible to use coil exchangers in which the sludge recirculates slowly or tubular exchangers (fig. 308) with fast circulation of both the sludge (1 to 1.5 m/s; 3 to 5 ft/sec) and the hot water (1.5 to 25 m/s, 4.9 to 82 ft/s) giving exchange factors of up to capable of 4500 kJ/(^{o}C m^2 h) or 450 BTU/ (^{o}C sq ft h).

In addition to heating the raw sludge, the heating system also has to make up external losses of heat by radiation and convection.

The heat transfer coefficients depend on the materials used and on the thermal conditions of the site on which the structure foundations are laid.

Heat losses can be very considerably increased if the base of the digester is below the underground water level.

For temperate countries, external heat losses can be estimated, as a first approximation, at 2,100 to 2,500 kJ/(m^3. d) (57 to 68 BTU/ft^3 d) for capacities below 1 000 m^3 (35 000 ft^3), and at 1 250 kJ/(m^3 d) (44 BTU/ft^3 d) for capacities over 3 000 m^3 (100 000 ft^3).

In many cases, the thermal insulation of the digester walls includes, outside the actual concrete wall, a rendered hollow-brick or hollowblock wall, about 20 cm (8 in) from the main wall.

Increasing use is now being made of mineral wool and expanded materials, foam, etc., for thermal insulation.

The sludge to be heated is collected at the foot of the vertical wall and, after passing through the heat exchanger, is pumped into the digester at a point at least 10 m (35 ft) from the collecting point.

Before it enters the digester fresh sludge should, for preference, be mixed with the recirculated sludge undergoing digestion.

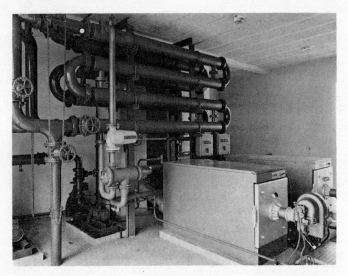

Fig. 308. — Heating system for digestion sludge: boilers, pumps and heat exchangers.

The temperature of the hot water used in the high-rate tubular exchangers can reach 90 °C. In low-rate exchangers it must be limited to 60 °C to avoid heating the sludge and clogging the exchangers.

Boilers fed with the gas produced by digestion must also have fuel oil or natural gas burners so that the sludge can be heated for plant start-up, or if the digestion gas supply fails.

Devices to prevent flame blow-back are often fitted on the boilers and at various gas user points.

As the amount of gas produced by digestion greatly exceeds the in-plant heating requirements, the surplus can be used either to feed gas engines or to supply a unit for sludge drying, incinerating or heat conditioning.

4.1.6. STARTING AND OPERATING A DIGESTION UNIT

The difficulties sometimes experienced in operating a digestion unit are very often due to negligence or mistakes when starting up. A careful and methodical starting procedure is the key to regular operation and satisfactory treatment

A digestion unit is not simply a tank to be filled with sludge but a reactor for biochemical reactions, the balance of which must be carefully monitored and which can easily be guided by regular observations.

The optimum conditions for the correct starting of digestion are as follows:

— heating of the digester to 35 °C;

— maintenance of the sludge pH between 6.8 and 7.2 (in a high-rate unit is often useful to inject about 8 kg (18 lb) of lime per day per 1 000 inhabitants, until the pH stabilizes at 6.8);

— feed to be kept as regular as possible, and not exceeding about 30 % of rated load in a high-rate unit;

— stirring of the sludge in the digester when the concentration in the digester reaches a minimum of 25 g of dry matter per litre (0.21 lb/gal).

If these simple instructions are followed, the process of digestion can be started in 2 or 3 weeks in a high-rate unit.

One difficulty is sometimes encountered at start-up. Heavy foaming of the sludge, with formation of a gas rich in CO_2, occurs when the process of digestion begins. This foaming appears fairly suddenly and the whole sludge mass starts to ferment vigorously. It is further accelerated by the presence of protein matter or certain surfactants which have not been completely broken down. This problem can often be overcome by adhering strictly to the optimum temperature and monitoring the feed.

Once the process of digestion has begun, the digester load can be **steadily** and quickly increased.

The main ways of monitoring a digestion unit are as follows:

— measuring the amount of gas produced;

— measuring the pH;

— measuring the volatile acid content. This figure should normally be kept below 500 mg/l (it can go up to 1 200 mg/l in the case of highly-organic sludges treated in high-rate units, and a closer watch is then necessary);

— measuring alkalinity (bicarbonate and acetic—see page 106). This measurement is only significant when taken together with the figure for volatile acids. The degree of alkalinity must always be much higher than the equivalent volatile acid content, to provide an adequate buffer and ensure stable digestion. The volatile acid/TAC ratio (expressed in milli-equivalents) must be under 0.5, with an optimum value around 0.2;

— analysis of the gas. Continuous measurement of CO_2 and CH_4 content is the quickest method of revealing any anomaly in the process of digestion. In particular, it reveals the presence of toxic or inhibiting substances in the reactor. This type of check can be used with advantage at very large plants, because the operating staff are warned immediately.

The essential data needed for compiling the operating results for a digestion unit are: daily quantity of dry solids introduced into the digester (or, if absolutely necessary, the daily flow of sludge entering the digester), the daily volume of gas produced and the reduction of volatile solids expressed as a percentage.

4.2. Aerobic stabilization of sludge

This process is also often referred to as "aerobic digestion". With this method, extended aeration of the sludge is used to stimulate (in the case of primary sludge) or continue (activated sludge) the growth of aerobic microorganisms beyond the period of cell synthesis to reach the stage of auto-oxidation. This is the mechanism of endogenous respiration, which is represented at a first stage by the elimination of cellular matter.

$$C_5H_7O_2N + 5 O_2 \longrightarrow 5 CO_2 + 2 H_2O + NH_3$$

The NH_3 then oxidizes biologically to become NO_3^-. Figure 309 shows how the extended oxidation of a mass of activated sludge leads to a final reduction in the weight of "sludge".

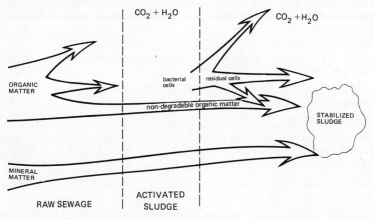

Fig. 309. — Simplified diagram of aerobic stabilization.

4.2.1. ANAEROBIC DIGESTION AND AEROBIC STABILIZATION

As a result of the greater amount of thermal energy released by oxidation, the aerobic bacteria work faster than the methane bacteria. However, the fraction of the cells which cannot be broken down biologically is greater in the case of aerobic stabilization.

The advantages of the two processes are compared in the table which follows:

	Aerobic stabilization	Anaerobic digestion
Breakdown products	CO_2, H_2O, NO_3^-	NH_4, H_2O_4, CH, CO_4
Energy available for the bacteria (for 1 mole of dextrose)	650 cal	35 cal
Rate of breakdown	+	—
Final reduction of volatile matter	—	+
BOD_5 of supernatant	50-500 mg/l	500-3 000 mg/l
BOD_5 per g of volatile matter	—	+
VS/DS ratio	+	—
Odour	=	=
Filtrability	—	+

Like all biological processes, aerobic stabilization is strongly influenced by temperature, and a change from 20 °C to 10 °C can extend stabilization time by 50 % for an equal percentage reduction in volatile matter. But this loss of performance caused by a fall in temperature does not have the same serious disadvantages as those encountered in the same conditions with unheated anaerobic digestion, i.e. a putrid smell, grey sludge which drains badly and an acid pH.

The rate of reduction of volatile matter obtained by stabilization is, in normal weather conditions, considerably lower than that obtained with heated anaerobic digestion. The elimination of pathogenic germs is also less effective (about 85 %) and the destruction of worms' eggs more doubtful.

However, because sludge stabilization is a sturdy process which is easy to operate and can withstand variations of load, of pH value and even, to some extent, of temperature, it is used on an increasing scale at medium-sized and rural treatment plants.

Aerobic stabilization is a more flexible process than anaerobic digestion in which the methane bacteria are affected by the ecological conditions (presence of Cr^{6+} and heavy cations); it is thus well suited for the treatment of industrial effluents.

It requires higher energy costs than anaerobic digestion, but the latter process costs more to install.

Fig. 310. — Aerobic stabilization tanks. ÉLANCOURT, France.

4.2.2. CRITERIA FOR ASSESSING GOOD STABILIZATION

Correctly-stabilized sludge has no smell and is easy to drain. On leaving the stabilization tank, it appears as a brown liquor of even consistency, in which the coarse organic matter from the raw water has been dispersed and reduced in volume.

It is, however, preferable not to store this type or sludge in liquid form for too long a period.

This qualitative definition of stabilized sludge, which is easily understood by anyone who has experience of operating an aerobic digestion unit, has been considered insufficient by a number of specialists. Criteria based directly on analysis have been proposed:

— Kehr defines the degree of stabilization in terms of the maximum rate of sludge respiration. This limit, corresponding to the endogenous respiration rate, ranges from 0.10 to 0.15 kg O_2/kg of organic matter per day.

— In France, the following test is often applied: "after 120 h of continuous aeration, unfed stabilized sludge must have a dissolved-oxygen content of 2 mg/l in the liquor and must not have lost more than 10 % of suspended solids by weight".

— The organic-matter reduction rate is a debatable basis of assessment, because aerobic stabilization is very often applied to activated sludge which has not been mixed with primary sludge and comes from direct aerobic treatment of the raw water. According to the age of this activated sludge, the organic matter may have been substantially reduced in the aeration tank before stabilization begins.

Fig. 311 shows how a mixture of primary sludge and activated sludge originally containing 65 % of volatile matter evolves during stabilization. This reduction of volatile matter S can generally be expressed in the form:

$$S = a \log t + b$$

Stabilization affects only the active organic matter which, in excess activated sludge, will account for at most 40 to 45 % of the total sludge mass, whereas total organic matter accounts for 75 %. Only the biodegradable matter will be reduced by aerobic action. Since this active biodegradable matter in any case leaves a residue of 25 % of its weight, it will be seen that perfect stabilization of such sludge will mean a reduction in volatile matter of about 0.75 (40 to 45 %) = 30 to 35 %. The corresponding "mineralization" of the sludge will therefore raise the mineral matter content from 25 % to barely 35 %.

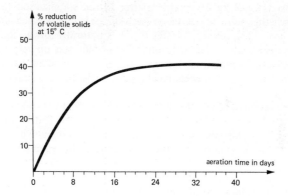

Fig. 311.

4.2.3. DESIGNING A SLUDGE-STABILIZING UNIT

The real retention time (or age) of the sludge is the basic factor in determining size.

The percentage reduction of volatile matter increases with retention time, but the ratio of volatile matter destroyed to volatile matter in the stabilizing unit generally falls.

A stabilization time of 12 days is most often adopted in temperate climates for activated sludge obtained by direct aerobic treatment and containing about 70 % volatile matter.

Under the same conditions, this time has to be raised to 15 days for fresh sludge (primary and activated sludge) containing 75 % volatile matter. The stabilization period is closely linked with the sludge loading in the aerator.

It should be noted that temperature has a considerable effect and that, with the same sludge loading and retention time, a 40 % reduction in volatile matter at 20 °C drops to 30 % at 12 °C.

However, retention time may not be the only factor limiting the rate of stabilization.

The sludge loading expressed as dry volatile matter fed in, expressed in kg/d m³ or in lb/ft³ d of stabilization tank is also important. It would appear that for good stabilization a loading of 2 kg/m³ d (125 lb/1 000 ft³ d) must not be exceeded with activated sludge or of 3 to 4 kg/m³ d (187 to 250 lb/1 000 ft³ d) with a mixture of fresh sludge. When the sludge is too thick, it becomes difficult to ensure perfect homogeneization of the liquor and rapid diffusion of the oxygen within the biological floc.

The installed aeration power depends not only on oxygen demand but also on the need for stirring and turbulence.

The demand for oxygen **actually supplied** to the bacterial flora is about 0.1 kg O_2 per day per kg of volatile matter in the stabilization tank, varying according to whether activated sludge or fresh sludge is to be treated and according to retention time. The oxygen transfer factor is naturally low as compared with the figure for clean water. An aeration system which sets up very heavy turbulence seems to be best for aerating fairly viscous sludge where mechanical dispersal plays an important role. In mechanical aeration, the power required for oxygenation only is about 20 W/m³ of tank approximately (570 W/1 000 ft³).

The power required to stir the mixture and prevent the formation of deposits can easily reach 25 to 30 W/m³ of tank per hour (710 to 850 W/1 000 ft³) and, for this reason, is usually the only figure taken into account, when aerators are designed. It is for this reason that the introduction of oxygen into stabilization tanks is usually syncopated.

When compressed air is used for aeration, an air flow of 5 to 6 m³/h . m² (28 to 32 cu ft per min per 100 sq ft) of tank is often used.

4.2.4. AEROBIC STABILIZATION PROCEDURE

To make the best use of the available tank capacity, sludge concentration must always be kept as high as possible (to increase stabilization time).

It is generally easier to thicken the sludge before stabilization than after. The optimum concentration (20 to 25 g/l; 0,20 to 0,25 lb/gal) in the stabilization tank is rather difficult to attain with sludge from the direct biological treatment of raw sewage.

On the other hand, this concentration can easily be achieved with primary or fresh sludge. However, it is advisable not to exceed this figure as the resulting increased liquor viscosity would overload the motors which drive the surface aerators or air compressors. Stabilized sludge is always very slow to thicken.

It is always best to stop the aeration system in the stabilization tank before feeding in **fresh** sludge, in order to free a clear supernatant liquid. Similarly, the concentration of the sludge drawn off can be improved by stopping aeration before extracting.

The aeration equipment used is either surface aerators (ACTIROTOR) or diffusers delivering large or medium-sized bubbles (VIBRAIR G.M.). While surface aerators are better for diffusing the oxygen and breaking down bulky products, they can, on the other hand, destroy the biological floc mechanically to a certain extent in the long run, with the result that thickening and subsequent conditioning of the sludge then becomes more difficult.

The supply of fresh sludge into the tank should also be as regular as possible and the aim should be at least one feed daily under normal running conditions.

A large amount of foaming is often observed during the first few weeks or even months of operation. The thick layer of scum may reach a depth of 20 to 30 cm (8 to 12 in) and is linked with nitrification-denitrification phenomena and with inadequate degradation of proteins or surfactants. After this period has elapsed, foaming usually decreases and no longer becomes troublesome.

Aerobic stabilizers must be so shaped that the floor is completely swept by the movement of the liquor (suitable rotation radius—see page 226).

The recirculating pumping capacity of mechanical aerators must be high.

In the case of small or medium-sized plant (for a population of less than 20 000), the stabilizing tank is usually of the complete mixing type. This process has, however, the disadvantage of delivering to the drying beds sludge containing a certain, if small, proportion of matter which has only been in the stabilization tank for a short time.

For larger plants, it is better to have two or more stabilization tanks working in series, with the last of the series receiving no fresh sludge.

4.3. Chemical stabilization of sludge

The stabilization of sludge by (anaerobic or aerobic) biological means requires large-capacity tanks. When reducing the investment costs is a prior aim, the fermentable power of the sludge can be considerably decreased simply by adding chemical reagents.

The introduction of reagents does not alter the quantity of biodegradable organic matter but has an essentially bactericidal action. Lime, because of its low cost and its alkalinity, is the reagent most often used, but its effect differs according to whether it is used on liquid sludge or on dewatered sludge.

In the case of **liquid sludge,** the addition of lime raises the pH and temporarily halts evil-smelling acid fermentation. Lime is of common use in sludge thickeners, with raw sludge in particular. It also often improves the initial filtrability of the sludge. The amount of lime used for raw sludge is about 10 % of the amount of dry solids; it will vary according to the septicity of the raw waste water, the treatment process, temperature and thickening time. However, even with lime treatment, liquid sludge may not be stored for too long because of the dilution effect and the renewed fermentation in the course of time of the sludge particles which escaped alkalinization.

In the case of **dewatered sludge,** lime stabilization has a much more lasting effect, and the lower the water content the more stable the sludge, as acid fermentation can then only develop with difficulty. However, it is less easy to mix the lime and sludge and very powerful mixing equipment is required. The use of quicklime allows advantage to be taken of the exothermic effect of the hydration reaction to increase the dryness of the sludge. Storage in the fresh air further reduces the number of pathogenic germs, especially if antiseptics have been used during dewatering.

By raising the pH of the sludge above 11, the coliform content is reduced to about 10^3 germs per gramme of dry solids (against 10^9 in the case of sludge which has not received lime treatment) and Salmonellae practically disappear. The amount of lime to be used may exceed 30 % of the initial dry solids content of the sludge. The effect of alkalinization alone on the mycobacteria is uncertain but it is effective vis-à-vis the enteroviruses.

The bactericidal action of flocculants such as iron salts used without lime appears to be negligible and polyelectrolytes sometimes have a negative effect.

In the case of particularly putrid liquid sludge, powerful oxidizing agents such as chlorine or hydrogen peroxide may be necessary.

4.4. Pasteurization

Pasteurization is not, strictly speaking, a stabilization process. Its sole purpose is to render the sludge aseptic so that there is no risk in using it in liquid form in agriculture, even on fodder or vegetable crops.

Under this process, the sludge is held at a temperature of 70 °C for about twenty minutes. The treatment is applied to the raw liquid sludge, whether thickened or not; it also appears to destroy viruses.

A pasteurization unit (fig. 312) is usually run intermittently and may be fully automated. Heat recovery is hardly possible because of the low temperatures used. It would be of little economic interest. Consequently, additional heat is required if there is no digester.

Pasteurization can be applied to fresh, stabilized or digested sludge. In the

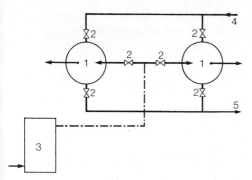

1. *Reactor.*
2. *Automatic valves.*
3. *Steam boiler.*
4. *Sludge to be pasteurized.*
5. *Pasteurized sludge.*

Fig. 312. — The operating principle of an alternating pasteurization plant.

case of fresh sludge, there is still a fairly considerable risk of evil-smelling fermentation after spreading.

Pasteurization has very little effect on the colloidal structure of the sludge because the matter is only heated to a moderate temperature.

4.5. Irradiation

Gamma-ray irradiation is also essentially a decontamination process. Results vary to a considerable extent according to the dose of radiation used. Irradiation at about 800 krad seems to give results comparable with liming; at a level of 1 200 krad, the faecal flora would appear to be reduced by 3 to 5 log 10.

5. SLUDGE THICKENING

This is the first and simplest stage in reducing the volume of sludge without high energy consumption.

Thickening has several advantages:

— at equal capacity, it increases retention time in the aerobic and anaerobic digesters. Digestion overflows may even be cut out completely by thickening the fresh sludge;

— more reliable performance. In particular, at plants treating municipal sewage, the sludge from the primary settling tanks can be pumped in very liquid form and even continuously. Any risk of clogging in the extraction pipes is virtually eliminated. The thick sludge extraction problems are then concentrated in a

single unit, **the thickener,** which can be fitted with all the necessary preventive and monitoring devices, such as displacement-type extraction pumps, flushing and clearing devices, additional screening and degritting units, concentration analysers;
— reduction in volume of the conditioning reactors (particularly in the case of thermal conditioning);
— improved performance of the dewatering units, such as drying beds, vacuum filters, belt filters, centrifuges and particularly filter-presses;
— establishment, in the case of thickening by settling of a buffer capacity between the water treatment and sludge treatment systems.

In no case will the increase in the DS content achieved by thickening produce a sludge which can be shovelled.

The disadvantage of thickening is that, apart from the additional cost, which can frequently be offset elsewhere, it can leave organic sludge with a bad smell or an unpleasant appearance.   The smell can be dealt with fairly easily by prior lime addition, which is in fact often necessary for the direct dewatering of fresh sludge or during the start-up of anaerobic digesters.

Another method of smell abatement is to cover the tanks, set up a negative pressure in the enclosed chamber formed and treat the polluted air aspirated.

Two techniques are used in gravity thickening:
— settling;
— flotation.

The second method ensures a higher specific production than the first, but requires higher energy consumption.

5.1. Thickening by settling

The sludge-laden suspension is fed into a tank (thickener), where it remains for a long time so that the sludge can become compacted and then be extracted from the bottom, while the interstitial liquid is drawn off at the top.

The sedimentation curve for the sludge is a good guide to the dimensioning of the thickening tank.   This curve is plotted in the laboratory using a vessel of adequate size, fitted with a very slow stirring device if necessary.   This curve represents the variation against time of the separation depth of the mass of sludge in the test tube.

Kynch's theory (page 69) can be used to calculate the thickeners and, in particular, to determine the surface corresponding to the critical concentration.

It is possible to determine the maximum flow F_L corresponding to a selected extraction concentration C_u. To do this, it is merely necessary to draw the tangent to the curve $F_s = f(C_i)$ from the abscissa C_u; the slope u defines the rate of extraction flow and the intersection with the axis of the ordinates determines the maximum flow F_L (fig. 313).

It is more difficult to determine the depth of the thickening tank, which must be the sum of the compression height required to reach the desired concentration plus a certain amount of "freeboard". The latter corresponds to a clarification zone where a flocculant settling process takes place. The upward flow-rate of the water in this zone should always be less than the settling rate of the individual particles in the sludge suspension.

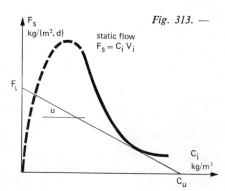

Fig. 313. —

The amount of "freeboard" should also provide an operating safety margin if extraction conditions vary and should ensure satisfactory hydraulic distribution of the liquor entering the tank. Freeboard depth is generally between 1 and 2 m (3.2 to 6.5 ft) and may be more when the sludge is highly fermentable.

Thickening tanks should always be built deep enough to allow the sludge to become compacted under the weight of the solid mass alone. Taking the storage capacity into account, 3.50 m (12 ft) is a minimum for domestic sewage sludge.

In the case of well-digested sewage sludge, or hydroxide sludge, tanks may be deeper in order to obtain an efficient compression effect.

Factors favourable to thickening include:

— the presence of dense, coarse-grained particles which compress the underlying layers better than a hydrophilic biological floc;

— the influence of thermal conditioning and chemical flocculation on the structure of the sludgy matter;

— prevention of the fermentation of organic sludge as the gases produced hinder consolidation. Lime treatment, maintaining the sludge at an alkaline pH value, is the most usual method.

When thickening non-stabilized organic sludge, the sludgy material must not be kept in the tank longer than about a day.

When precise experimental data on sedimentation curves are not available the following specific loading can be used, as a first approximation, to calculate the surface area of thickening tanks:

Type of sludge	Specific loading		Possible concentration of S.S.
	kg S.S./m² d	lb/ft² d	g/litre
Fresh primary sludge	80-120	16-24	100
	(according to V.S.S. content)		
Primary sludge + fresh activated sludge	50-70	10-14	50-70
	(according to V.S.S. content)		
Activated sludge only	25-30	5-6	50-70
	(according to V.S.S. content)		
Sludge from carbonate removal process	400	80	150-250
Sludge from the flocculation of drinking water using metallic hydroxides	15-25	3-5	30-40

The overflow rate is not an essential criterion for calculating the size of a thickener. If the tank is deep enough, this rate can reach 1 m/h (3.2 ft/h) with a mixture of primary and activated sludge.

Thickeners may be either without scrapers or mechanized.

5.1.1. NON-SCRAPED THICKENERS

These are generally simple, cylindrical tanks tapering at the bottom and designed in the manner of static settling tanks. The slope of the floor must be adequate (50 to 70º to the horizontal). This type of thickener, which is rarely buried, can be up to about 5 m (16 ft) in diameter.

5.1.2. MECHANIZED THICKENERS (fig. 315)

The round tank, with a floor slope generally between 10 and 20º to the horizontal, is equipped with a dual-purpose rotating mechanism:

Fig. 314. — Thickeners equipped with central drive bridge, scrapers and vertical fences, 14 m (46 ft) in diameter. RHEIMS, France, sewage treatment plant.

— it transfers the settled sludge to the central sludge-collecting hopper by means of scrapers moving immediately above the floor;
— it helps to free the interstitial water and gases held in the sludge by using a vertical "picket fence" mounted on the rotating mechanism.

The whole mechanical system usually has a strongly-constructed centre drive cwith a double radial arm. High torque values may be needed in the case of large-diameter tanks handling heavy sludge (for example, with sludge from washing beet, the drive torque for thickeners 40 m (130 ft) in diameter can exceed 100 000 m daN (23.8 × 10⁶ ft pdl) (see page 187).

The scraping system should consist of a series of "louvre-type" scrapers to ensure that the sludge does not clog during transfer to the centre, and to reduce the physical links between particles, by " ploughing" the mass.

In certain cases (lack of space, risk of smells), thickening may be carried out in the raw sewage settling tank; the settling tank/thickener then contains a large thickening zone at its centre, equipped with many radial arms. Tanks of this type are very deep, varying from 6 to 9 m (20 to 29.5 ft) at the centre.

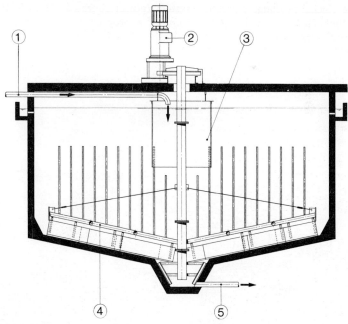

1. *Supply pipe.*
2. *Drive unit, with motor and reduction gear.*
3. *Centre well.*
4. *Scraper arm.*
5. *To extraction pump.*

Fig. 315. — Scraper-type thickener. Centre drive.

5.1.3. OPERATION OF THICKENERS

The sludge for thickening can be fed in under the floor, or from the top if there is serious risk that the pipes may become clogged.

The thickened sludge is usually extracted through a pipe which is also below the floor. In some difficult cases, the sludge can be removed from the top. Fig. 104 is a cross-section of a thickener in which both the sludge for thickening and the thickened sludge are carried through overhead pipes. This is a very reliable system and is recommended for heavy types of sludge (containing clay or sand).

The extraction valves or pumps must be located as near as possible to the centre sludge hopper. Displacement pumps (reciprocating, eccentric rotor or diaphragm types) can handle more concentrated sludges than simple centrifugal pumps.

An access tunnel under the floor is often very useful.

Thickeners with an unscraped sludge hopper occupying only one sector, and thickeners with sludge extraction from a number of peripheral hoppers are not recommended.

The thickened-sludge extraction system (power-operated valves or pumps an be controlled automatically by means of a programmer or a gamma-ray oncentration analyser.

Fig. 316. — PARIS-ACHÈRES sewage treatment plant. One of the eight scraper-type circular sludge thickeners: diameter 60 m (196 ft). Total capacity: 104,000 m³ (3,700,000 cu ft).

5.2. Thickening with elutriation

Elutriation is the process of washing the sludge with clean water to improve the physical and chemical properties of the sludge-laden suspension:
— elimination of fine and colloidal matter;
— reduction of alkalinity (anaerobic digested sewage sludge).

The elimination of fine particles accelerates thickening of the sludge and increases the output of mechanical dewatering systems (vacuum filters, filter-presses, centrifuges).

The lowering of alkalinity should reduce consumption of the reagents, such as ferric chloride and lime, used to condition the sludge, particularly when a vacuum or pressure filter is used.

In the case of highly fermentable sludge, elutriation with well-aerated, purified water also reduces the risk of septic fermentation.

It is sometimes necessary to use soft water for the reduction of the salt content in the sludge, in order to avoid fusion of the ash produced in calcination, e.g. oil refinery sludge containing large concentrations of salts (10 to 20 g/l of NaCl.).

Elutriation tanks are similar to thickeners, with the difference that a large volume of water is added to the sludge at the tank intake.

The concentration of sludge obtained by elutriation is similar to that obtained by thickening. The overflow from the elutriator generally contains a fair amount of fine and colloidal matter which, after recirculation to the head of the plant, normally has to be readsorbed by flocculated sludge (biological or chemical).

5.3. Thickening by flotation

This method, which was examined in Chapter 7, has two advantages:
— smaller surface area and volume of the thickening units,
— colloidal sludge (such as activated sludge) is here more highly concentrated than with the static thickening method. This is very advantageous for the entire sludge treatment system.

Against this, running costs are higher, and large quantities of thickened sludge cannot be stored in the tank.

Until now, flotation has been mainly used to thicken activated sludge and recover cellulose fibres, but it can also be applied to hydrophilic inorganic sludge (e.g. of the aluminium hydroxide type) formed by floc of low particulate density.

The most widely-used process in sludge treatment consists of expansion of the fluid which has previously been in contact with compressed air at pressures of 3 to 6 bar (bubble size from 15 to 100 microns). Pressurization may be of two types:

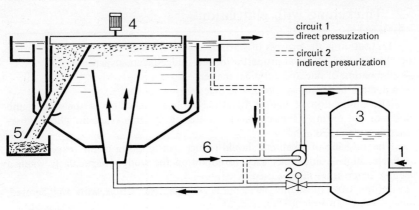

circuit 1
= direct pressuzization

=== circuit 2
indirect pressurization

1. *Compressed air.*
2. *Expansion valve.*
3. *Saturation vessel.*

4. *Flotation unit.*
5. *Thickened sludge.*
6. *Sludge to be thickened.*

Fig. 317. — Principle of sludge thickening by flotation.

— direct: total or partial pressurization of the sludge itself;
— indirect: pressurization of water (often clarified water), followed by its injection, immediately after expansion, into the sludge suspension.

The main parameters for flotation may be summarized as follows:
— concentration of the sludge, temperature, surface condition of the particles, size of the floc (modifiable by the addition of a polyelectrolyte, Mohlman index...);
— rate of dissolution of air, presence of surfactants (which have a favourable effect), expansion without desorption, recirculation rate in cases of indirect pressurization;
— surface loading and, above all, sludge loading.

Fig. 318. — Pilot flotation unit

It is preferable to thicken mixed, fresh or digested sewage sludge by settling. On the other hand, flotation is very suitable for the flocculated structure of activated sludge.

Hydrophilic sludge with light floc, even when treated in large settling tanks (loading: 15-20 kg DS/(m².d.), does not become highly concentrated (thickened sludge: 15-25 g/l).

The following values have been obtained as a result of experience acquired with thickening by flotation of excess activated sewage sludge:
— the concentration of the sludge fed in to the flotation unit should be less than 6-8 g/l DS;
— previous flocculation with a polyelectrolyte considerably increases the capacity of the flotation unit and the concentration of floated sludge;
— the size of flotation thickeners is generally calculated on the following basis:
• specific loading: 5 to 13 kg DS/(m².h) (1 to 2.7 lb dry solids/sq.ft.h);
• surface loading: less than 5 m³/(h.m²) (2 gpm/sq.ft);
— concentration of the floated sludge:
• without a polyelectrolyte: 3 to 4.5 % DS;
• with a polyelectrolyte: 3.5 to 6 % DS according to dosage and to loading;
— separation efficiency: generally higher than 90 % (> 95 % with polymers);
— in cases of indirect pressurization, the recirculation rates are generally high (50 to 150 %, sometimes more).

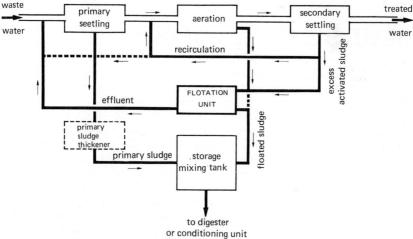

Fig. 319. — Thickening by flotation of activated municipal sludge.

The sludge storage capacity is sufficient for several hours, after which time the air bubbles disappear and the sludge regains its normal specific weight; during the first few hours, however, drainage improves with time, and this is reflected in the markedly higher upward concentration of the floated layer.

This beneficial effect is used to advantage in post-flotation units, which are specially-designed tanks for the storage of floated sludge.

Fig. 320. — Sediflotor, 7.5 m (25 ft) in diameter. DALLAS Airport, Fort Worth, U.S.A.

6. SLUDGE CONDITIONING

Even after thickening, the sludge still contains a very high proportion of water: 97 to 98 % if the dry solids are highly organic and, in particular, proteinic; 70 to 80 % if the dry solids are mineral, heavy and granular.

To obtain sludge which can be " trucked" or " shovelled", additional dewatering is necessary. Conditioning is generally essential to prepare the sludge for the various equipment. Difficulties are often encountered in the processes used for reduction of the volume of liquid sludge; these difficulties are closely linked with the hydrophilic (or sometimes hydrophobic) colloidal mass, of which sludge suspensions are mainly formed.

In practice, additional **flocculating** energy will be needed to modify the internal cohesive forces of the sludge, to "break down" colloidal stability and to artificially increase the size of the particles. In the case of hydrophilic sludge, part of the bound water must be freed and transformed into free water (see page 431).

Of the various conditioning processes, thermal treatment is by far the most effective **in reducing particle hydrophily**; chemical flocculation, which makes use of mineral electrolytes, and above all, lime, also reduces the bound water, but to a lesser extent.

Conditioning, which is intended to render the sludge drainable, filtrable or centrifugable, uses physical (in particular, thermal) or chemical processes or a combination of the two. Adequate sludge conditioning is the basis of efficient operation of the dewatering plant.

6.1. Chemical conditioning

By application of coagulation and flocculation phenomena (see Chapter 5), chemical conditioning leads to the agglomeration of the particles in the form of a three-dimensional system or "floc".

Inorganic or organic flocculating agents are used. Each of these types has its own special properties, especially insofar as concerns reducing particle hydrophily, mainly observed in the presence of lime, and the size of the floc formed (voluminous with polyelectrolytes).

Inorganic reagents seem to be mores uitable for drying by vacuum or pressure filtration and organic reagents for centrifuge drying or pressure belt filtration.

6.1.1. INORGANIC REAGENTS

Polyvalent cationic mineral electrolytes and lime give a relatively fine and stable floc.

The most effective and cheapest metal salts are:
— ferric chloride, $FeCl_3$ at a concentration of about 600 g/l;
— ferric sulphate, $Fe_2(SO_4)_3$, 9 H_2O;
— ferric chlorosulphate, $FeSO_4Cl$;
— ferrous sulphate, $FeSO_4$, 7 H_2O;
and, to a lesser degree, the various aluminium salts.

Ampholyte electrolytes (which behave like acids vis-à-vis bases and vice-versa) have a dual action:
— a coagulant action: their charge is generally opposite to that of the sludge particles;
— a flocculating action: forming complex hydrated hydroxides, such as $[Fe(H_2O)_6, (OH)_3]_n$ which play the part of a mineral polymer.

The most effective electrolytes have proved to be iron salts (mainly the trivalent salts). The introduction of lime (in the form of 5-10 % milk of lime) after the introduction of the electrolyte solution, is always advantageous because of:
— the alkalinization obtained (pH $>$ 9, which corresponds to a correct flocculation pH);
— precipitation of a number of organic calcium salts;
— decrease in the rate of bound water (reduction of specific resistance and compressibility, with production of a drier and more consistent cake);
— the introduction of a dense inorganic loading, causing dilution of the colloidal medium.

In the case of hydrophilic hydroxide sludge, lime alone is generally sufficient to improve the filtrability of the sludge to an acceptable level; but is sometimes needed in large quantities: about 50 % of the weight of DS, expressed as CaO.

In the case of hydrophilic organic sludge, lime is often used in combination with iron salt. The rates are generally between 3 and 12 % $FeCl_3$ and between 6 and 30 % CaO in relation to the dry solids in the sludge.

Ferric chloride is one of the most widely-used reagents.

The amount of ferric chloride and lime required increases with the volatile matter content of the sludge.

Genter has proposed graphs (fig. 321) from which the quantity of $FeCl_3$ required can be calculated on the basis of the alkalinity and volatile matter content of the sludge (particularly anaerobic digested sludge) in the case of vacuum filtration. It is nevertheless advisable, in each separate case, to carry out some simple laboratory tests to determine the amount of reagent needed by measuring specific resistance (see page 481).

The amount of lime expressed as CaO is generally between 150 and 250 % of the dose of $FeCl_3$ in the case of non-elutriated sludge.

The Sub-Chapter on Dewatering (pp. 493, 500, 506, 517) gives figures for the conditioning rates used with various types of sludge. These rates should be reduced when the sludge contains large quantities of fibres or mineral matter and increased when it contains protein organic matter.

A high proportion of the added reagents remains in the dewatered sludge in solid form, as a result of the precipitation of metal hydroxide ($Fe(OH)_3$) and calcium carbonate ($CaCO_3$). At the same time, 20 to 40 % of the mass of reagent goes into solution in the form of chlorides, in the liquid phase (filtrate).

According to local availability and in order to economize on operating costs, flocculating agents other than $FeCl_3$ may have to be used, particularly:

— ferrous sulphate: doses of Fe^{2+} are often double those of Fe^{3+} and efficiency is limited with very organic and very hydrophilic sludge. The advantage of this reagent lies in its low cost, due to the fact that it is a by-product of steel pickling and of the titanium industry. To increase its efficiency, ferrous sulphate is sometimes oxidated (usually with chlorine gas) to form ferric sulphate;

— aluminium salts: these are much less effective than ferric salts and frequently clog the filter cloths; they give good results (sometimes in combination with polyelectrolytes) in filter-presses with cycles of 12 to 24 hours.

● *Use of mineral reagents:* although certain grades of sludge can be flocculated in 1 or 2 minutes with very well-designed mixing equipment and under given flow conditions, it is generally preferable to use flocculation tanks in which the sludge can remain for 5 to 10 minutes, so that the floc can develop sufficiently.

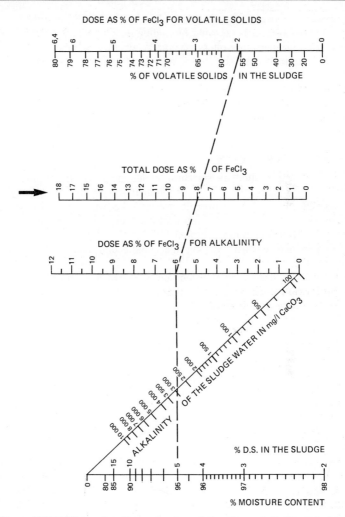

Fig. 321. — GENTER graph. Determination of the dose of FeCl₃ calculated as a percentage of the weight of dry matter.

These tanks generally have two compartments, one for the metallic salt, the other for the lime (see the diagram for organic sludge, page 509).

It is advisable to allow a period of 15 to 30 minutes for the sludge to mature. This period is, in fact, usually allowed for, as the conditioned sludge is retained in the mixing tanks and in the trough of the vacuum filter or in the buffer tank preceding the filter press.

The reagents must be mixed completely with the sludge, and their diffusion throughout the mass can be assisted by adding diluting water. Complete automation of conditioning units offers very great advantages, reagent dosing being automatically controlled by the sludge flow and possibly by the sludge concentration (gamma ray densimeters and ultrasounds).

To avoid destruction of the floc formed, transfer of flocculated sludge should be limited to short distances by gravity, or, if absolutely essential, by slow displacement pumps (piston-membrane, eccentric rotor), or by intermittent pneumatic pressure.

6.1.2. POLYELECTROLYTES

Natural products such as starches, gums, polysaccharides and alginates have been used for many years for conditioning inorganic industrial sludge. Very much more effective synthetic organic products known as polyelectrolytes (see Chapter 5), are used in smaller quantities than inorganic reagents, but they cost more and have to be much more diluted.

These products consist of long macromolecular chains. They give the following results:
— extremely well-differentiated flocculation by the formation of bridges between particles as a result of the long ramified chains, followed by a coagulating action in the case of cationic polymers;
— a very considerable reduction in the specific resistance of the sludge;
— little or no influence on particulate hydrophily: the compressibility of the sludge remains stable (sometimes increasing).

Many polyelectrolytes are currently available on the world market, including, among the best-known, Prosedim, Sedipur, Praestol, Magnafloc, Drewfloc, Superfloc, etc. The final choice of a product is often made only after the plant has begun operations. Cationic polymers are mainly effective in the conditioning of sludge with a high content of colloidal organic matter while anionic polymers are widely used with suspensions of a preponderantly mineral nature. A combination of anionic and cationic polymers is sometimes necessary when flocculating a mixture of several different types of sludge (e.g. biological sludge + hydroxide sludge). Sludges with a high protein content (over 80 % volatile matter) often require strong cation polymers of high molecular weight. The tables of performances on pages 500 and 517 show the most usual conditioning rates: 0.3 to 2 kg anhydrous polymer per tonne of DS with mineral sludge, and 2 to 7 kg in the case of organic sludge.

● *Use of polyelectrolytes:* polyelectrolytes are supplied in powder or very viscous syrup form. They have to be greatly diluted before being injected. The satisfactory concentration of the initial solution is generally between 0.3 and 1 % with an optimum of 0.5 % for the majority of products.

The preparation of relatively concentrated and viscous initial solutions avoids the storage of large quantities of dilute solutions and the use of high feed-rate

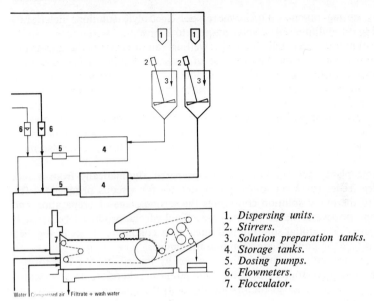

1. *Dispersing units.*
2. *Stirrers.*
3. *Solution preparation tanks.*
4. *Storage tanks.*
5. *Dosing pumps.*
6. *Flowmeters.*
7. *Flocculator.*

Fig. 322. — *PressDeg filtration unit with polyelectrolyte preparation.*

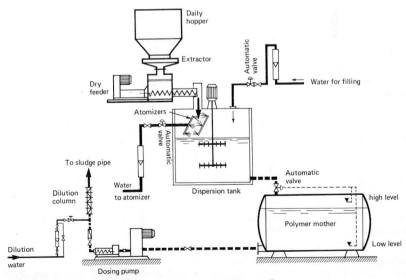

Fig. 323. — *Automatic dosing of polymer with "Jet spray".*

and costly dosing pumps. Furthermore, polyelectrolyte solutions deteriorate very quickly if too dilute (0.1 % solutions in 1 to 2 days).

The solution is prepared, firstly by wetting (or dispersing) the anhydrous product (this is not necessary in the case of a viscous liquid where the content is often limited to 15 or 20 % of active product). This dispersion, which often has to be done by hand, should be carried out with care to avoid the formation of agglomerates. It is effected with specially adapted ejectors or, in the case of large flows, by "jet spray" dosage of the powder. When injecting the polyelectrolyte in liquid form, the liquid is, for preference, pumped from the tank of the initial solution, by means of piston-type or eccentric-rotor-type displacement pumps.

In large plants, the solution may be prepared automatically from a silo, using a dry feeder (or a pumping system suitable for viscous products). The maturing of the initial solution constitutes the second stage of preparation and requires one or several hours (in the case of anhydrous products) before it is ready for use. The initial solution is diluted on a continuous basis in the feed pipe, just before the point of injection into the sludge.

With polyelectrolytes, flocculation is generally immediate; the floc formed is bulky but somewhat fragile. The injection points are therefore usually located almost immediately ahead of the dewatering unit, the mixing being violent but of short duration.

6.2. Thermal conditioning

The bond between the water and the colloidal matter can also be broken by thermal methods:
— by raising the temperature;
— by lowering the temperature.

6.2.1. HEATING THE SLUDGE

The idea of heating sludge dates from the start of this century; the actual technical process was first introduced by the British engineer Porteous in 1935. Many improvements have since been made to the equipment used, and other thermal conditioning processes have been developed.

The physical structure of the sludge is irreversibly transformed by heating it to a sufficiently high temperature (particularly if it contains a high proportion of organic and colloidal matter). Heating takes place, at a temperature varying between 160 and 210 °C, and held for 30 to 60 minutes according to the type of sludge to be treated and the subsequent method of drying.

During heating, the colloidal gels are destroyed. Two processes take place simultaneously:
— certain types of suspended solids are solubilized (for example, starch is hydrolysed to form sugars);

— matter in solution is precipitated (e.g. glucose).

The fats remain relatively stable. Cellulose is only slightly decomposed. The table page 476 gives variations in the composition of one type of sludge (municipal sewage) heated at 180 °C for widely-differing times and shows how transformation of the main components is affected by heating.

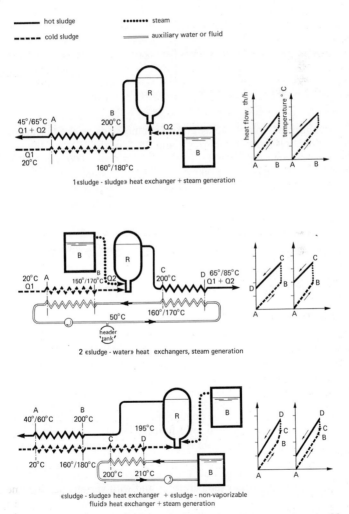

Fig. 324. — Operating diagrams for a sludge heat-conditioning unit: possible circuits.

The proportion of volatile acids in the interstitial water of the heated sludge (filtrate) is between 15 and 35 % of the volatile matter; the figure is higher for fresh sludge than for digested sludge. At the same time, the proportion of amino-acids varies between 60 and 65 % and is higher in digested sludge.

Heating temperature and time	% of volatile matter			
	lipids	glucides	protides	other
180 °C — 10 minutes	51.2	44	3.35	1.45
180 °C — 75 minutes	47.4	47.6	3.7	1.30

The quantity of volatile acids and of amino-acids in the filtrate rises with the temperature. Heating solubilizes about 25 % of the COD of the raw sludge and produces filtrates with a BOD_5 of around 2 000 to 3 000 mg/l with digested municipal sewage sludge, and 5 000 mg/l with fresh sludge of the same origin. The amount of nitrogen present in the liquid phase is relatively high (0.5 to 1 g/l as NH_4^+); phosphorus, on the other hand, remains in the sludge. This is also usually the case with heavy metals.

The main advantages of thermal conditioning are:
— it is universally applicable to all predominantly organic sludges;
— it is easy to control;
— it disinfects the sludge;
— the sludge produced can be quickly thickened in a static thickener and is easy to dewater;
— no chemical reagents are necessary.

With municipal sewage sludge heated at 180 to 200 °C, concentrations of 150 g/l and over can be achieved simply by sedimentation. The structure of the sludge is so much improved that filtration without the use of reagents is possible. In addition, the cake produced will be drier than that obtained by chemical conditioning.

The filtrability of the sludge is improved by a higher temperature or increased heating time, with temperature the dominant factor (substantially the same results are obtained if the sludge is heated for 1 hour at 180 °C, for 30 minutes at 190 °C and for 15 minutes at 200 °C). Short heating times improve the quality of the filtrate.

The curves in fig. 325 show how the specific resistance (page 481) of a particular sludge evolves at various temperatures and with various heating times.

The gain in filtrability obtained by heating varies with the organic matter content of the sludge. The after-heating filtrability of fresh municipal sludge is generally better than that of digested sludge heated in the same way (but the volume of sludge to be treated is generally greater, by about 30 %). The association of an anaerobic digestion process with thermal conditioning is, nevertheless, a very advantageous combination, as the excess gas produced by the digesters supplies most of the heat required by the thermal conditioning unit. Furthermore, the fluidity and very homogeneous nature of the digested sludge and its improved stabilization rate means that the heat exchangers can be technologically simpler and less expensive. Finally, the large capacity of the digesters ensures regular operation of the heating unit.

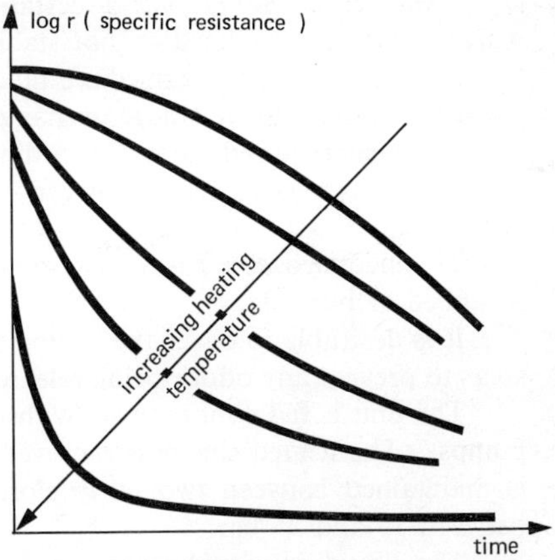

Fig. 325. — *Relative influence of heating time and temperature on the specific resistance of the sludge.*

It must be pointed out that the heat treatment of sludge has the following disadvantages:
— production of highly-polluted liquor;
— protection is required against smells (mainly due to the incondensable gases evacuated from the reactor and vapour from the heated sludge thickeners);
— the high investment costs of a unit using very advanced technology.

6.2.2. SLUDGE HEATING PROCEDURE

Figure 324 gives the principal operating diagrams for a heating unit. In all cases, every effort is made to recover as much heat as possible from the heated sludge to preheat the feed sludge, so that the amount of heat actually required is equivalent to that necessary for raising the temperature by about 50 °C. The heat is generally recovered in tubular exchangers, operating in reverse flow, although attempts have been made to preheat the feed sludge by direct contact with successive expansion stages of the treated sludge.

It is always advantageous to have a separate reactor ensuring a minimum heating time at an adequate temperature which can be controlled.

The required heat can be added either by direct injection of live steam into the reactor, or by an additional sludge/steam or sludge/superheated water exchanger. Although it is not, theoretically, the most economical in energy, the method of direct injection of steam into the reactor has the great advantage of guaranteeing the heating temperature however clogged the exchangers/sludge heaters may be and without making the latter too long.

With sludge containing a certain proportion of heterogeneous material which is poorly stabilized or not stabilized at all, the sludge should preferably be preheated with an intermediate fluid; in all other cases, e.g. digested sludge, the use of direct sludge/sludge exchangers is very suitable.

The heat required varies, according to the layouts and the condition of the exchangers as regards scale, between 3 300 and 6 300 kJ/kg DS (1 450 to 2 700 B.T.U./lb).

The incondensable gases have to be evacuated from the reactor; they can be scrubbed or burned.

It is desirable to cover the heated sludge storage tanks and/or to cool them, so as to prevent any odour being released.

The unit is fed continuously with raw sludge by high-pressure displacement pumps. The heated sludge is removed so that the level of sludge in the reactor is maintained between two fairly close points. Boiler test pressure is usually between 15 and 25 bar.

The liquid extracted from the dried heated sludge can be disposed of in the following ways; it can be:

— spread on farmland;

— returned to the head of the plant with or without a buffer storage tank whence it can be reinjected at the head of the biological treatment plant during a low-pollution period;

— given separate biological treatment: this approach, requiring tanks of limited capacity only, nevertheless takes considerable time, because of the highly-concentrated liquor.

Either activated-sludge or plastic-packing biological filters can be used. In the case of activated sludge, aeration time is about 24 hours, and the reduction in BOD_5 can reach 96 to 98 %, with some phosphorus being added, either in the form of nutrient salts or by diluting the liquor with pre-settled or raw effluent. In the latter case, larger aerators are required.

6.2.3. WET COMBUSTION

This process originally consisted of heating the sludge in the presence of air under pressures which could go as high as 200 bar. The organic matter was thoroughly oxidized and the colloidal matter physically transformed at the same time. Very much lower pressures have gradually been introduced, the air supply then having little influence. The process produces the same results as thermal conditioning.

The heating process and wet-combustion process produce desinfected sludge in which all pathogenic germs, eggs, worms, viruses, etc., have been destroyed. When disposed of to an outdoor tip this sludge does not ferment or give off a putrid smell.

6.3. Other methods of conditioning

6.3.1. FREEZING

The quantity of water associated with the dry matter included in the sludge can be reduced and the particles grouped in consequence, by total solidification of the sludge by freezing for a suitable time.   This grouping leads to the formation of thin lamellae which remain relatively stable after the ice has melted.

The filtrability of the sludge is improved and it is much easier to drain. After freezing, 20 % dryness can be achieved in 24 hours by spreading organic municipal sludge in 5 cm (2 in) layers.

The sludge in frozen for 1 to 4 hours at temperatures around — 10 to — 20 °C. Figure 326 shows one possible layout for a freezing unit.

In the case of organic sludge, the effectiveness of the process is greatly increased by prior chemical conditioning.

The freezing method has so far had few applications, and its use is mainly confined to predominantly mineral sludge from the treatment of drinking and industrial waters.

Despite the low initial cost, this technique is still somewhat expensive.   It would appear to be suitable for medium-sized plants, and is often combined with a mechanical filtration system.

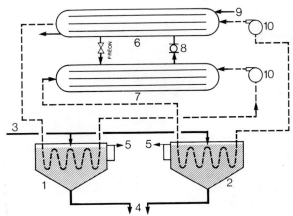

1. *Heating tank.*
2. *Freezing tank.*
3. *Inlet for sludge to be frozen.*
4. *Supernatant evacuation.*
5. *Extraction of frozen sludge to dewatering plant.*

6. *Refrigerant.*
7. *Condenser.*
8. *Freon compressor.*
9. *Glycol pump.*
10. *Circulation pumps.*

Fig. 326. — A typical operational diagram of sludge freezing plant.

Fresh sludge from domestic sewage very quickly becomes the focus of evil-smelling fermentation when it melts after freezing.

6.3.2. CONDITIONING WITH INERT ADDITIVES

The cohesion of the sludge, and hence its filtrability, is improved by adding dry matter which is generally inert. Compressibility is also decreased. This has the effect of increasing the dry-matter content of the dewatered cake and making it easier to transport, but it does not add substantially to the output of the dewatering system proper.

The additive may be ashes (returned from the incinerator), sawdust, kieselguhr, paper fibres, calcium carbonate, etc.

Ashes have some disadvantages:
— the proportion of incombustible matter in the cake is considerably increased;
— risk of solidification within the pipes carrying the sludge;
— risk of abrasion.

Fine particles are not as rule separated when fibres are added, and the degree of pollution in the interstitial liquor remains high.

6.3.3. CONDITIONING BY MEANS OF SOLVENTS

Some solvents of the aliphatic amine type have very variable solubility in water with slight variations in temperature. It is then possible more easily to separate the dry solids contained in the water by mechanical means, while recovering most of the solvent.

Fig. 326 B. — Continuous sludge pasteurization unit at ROMONT, Switzerland.

SLUDGE DEWATERING, DRYING AND INCINERATION

17

1. DEWATERING

1.1. Filtration

1.1.1. NUMERICAL CHARACTERIZATION OF FILTRABILITY

Filtration is so far the commonest method of dewatering sludge produced by water or waste water treatment. Filtration may consist solely of draining through beds of sand or it may be "mechanical", under vacuum or under pressure, which requires more complicated equipment.

Experience shows that although sludge suspensions are usually highly complex mixtures, theoretical work on the laws of membrane filtration and flow conditions in cakes yields mathematical laws which are well borne out by practical results with real sludge.

A. Specific filtration resistance of the sludge (see page 79).

The specific resistance r is defined as the resistance opposed to filtration (or passage of the filtrate) by a quantity of cake deposited on 1 m² of filter surface and containing 1 kg of dry product. The general laws covering surface filtration lead to the following main equation:

$$\frac{dv}{dt} \neq \frac{PS^2}{\eta\, rCv}$$

Integration at constant pressure of the above relation gives the simplified equation:

$$\frac{t}{v} = \frac{\eta\, Cr}{2PS^2} \times v = av$$

in which: t = time in seconds
  v = volume of filtrate obtained after time t (in m³)
  η = dynamic viscosity of filtrate (at room temperature η is approximately equal to 1.1×10^{-3} Pa.s)
  C = dry solids concentration of sludge suspension (kg/m³).
  S = filtering area (m²)
  P = pressure gradient (Pa)
  r = specific resistance under pressure P (m/kg)

Several assumptions have been made to arrive at this simplified equation; in particular, that:
— the resistance of the filtering material (R_m) is negligible in the case of sludge compared with the specific resistance r of the cake (except when clogged);
— with most sludges (except very dense suspensions), W (see page 78) can be replaced with only very slight error by C, the concentration of the sludge liquid.

A simple laboratory test of filtrability, applying a known differential pressure to a small surface, gives the "filtration curve" shown in figs. 32 and 330. This curve is a straight line of slope a. The filtrability coefficient of the sludge under pressure P can then be expressed as $r = \dfrac{2\,aPS^2}{\eta\,C}$

The specific resistance (or filtrability coefficient of the sludge at a differential pressure of 0.5 bar) is an essential magnitude which is highly representative of the sludge. The purpose of conditioning is to lower this resistance in order to accelerate filtration.

It is affected by the percentage of hydrophilic colloidal matter in the sludge; in the case of sewage, this proportion of colloidal matter is fairly close in line with the percentage of volatile matter, and fig. 327 shows relatively characteristic "zones" for fresh and digested sludge.

The specific resistance of a sludge is theoretically independent of most of the conditions of dewatering; it is characteristic of the matter itself, regardless of its concentration.

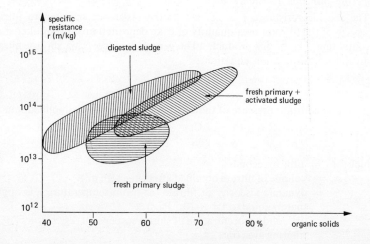

Fig. 327. — Specific resistance of municipal sewage sludge.

B. Compressibility of the sludge (see page 79).

When the differential pressure is increased, the pores of the cake are closed up, thus increasing resistance to filtration. The compressibility coefficient of the sludge is incorporated in the definition of the coefficient of filtrability as follows:

$$r = 2^s \, r_{0.5} \, P^s = r' \, P^s$$

According to whether the value of s defined graphically by the slope of the straight line $\log \dfrac{r}{r'} = s. \log P$ (see figure 328) is less than, equal to, or greater than unity, the filtration flow will increase, remain constant, or fall as P increases.

Insoluble crystalline and morphologically related substances — e.g., sludge made up of water-repellent particulate matter — are usually not readily compressible (s close to 0 or less than 0.3). Suspensions with hydrophilic particles (organic sludge, gel-like hydroxides) have compressibility coefficients exceeding 0.5, approaching or sometimes even exceeding unity (e.g., high-protein pure activated sludge).

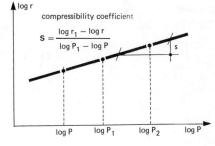

Fig. 328. — *Determination of compressibility coefficient.*

With many kinds of organic sludge, there is even a "critical pressure" beyond which the pores in the cake close up so much that drainage becomes impossible. In the case of municipal sewage sludge, for example, a filtration pressure of over 15 bars (210 psi) serves little purpose.

This is why a gradual increase in pressure has some advantage in theory because it delays this compacting of the cake.

Fig. 329 illustrates diagrammatically the structures of different types of sludge.

C. Maximum dryness.

When sludge is filtered at a high differential pressure (from 5 to 15 bars; 70 to 210 psi), the filtration curve $\dfrac{t}{V} = f(V)$ has the characteristic form shown in fig. 330.

The curve will be seen to have three sections:
— A straight line AB, over which the general law of filtration applies as in the case of a lower differential pressure (e.g. 0.5 to 1 bar; 7 to 14 psi). The slope of this straight line is as defined earlier.

— An asymptotic section CD, over which an increase in filtration time gives no further increase in the volume of filtrate. This "blockage" is the result of complex phenomena such as differential compaction of the cake, an increase in the viscosity of the filtrate, clogging of the filter medium and deformation of compacted matter by the pressure.

— An intermediate section BC, varying in length.

The dryness of the cake after an indefinite time of filtration is referred to as "maximum dryness" S_L. The maximum filtrate volume V_L is a logarithmic function of the pressure P. Carman's curve is supplemented by an asymptotic section, as follows:

$$\frac{t}{V} = av + \frac{K \times V}{(V_L - V) \, V_L}$$

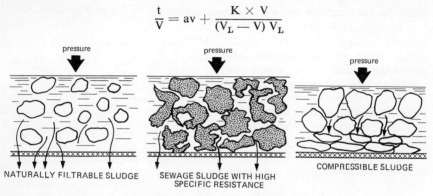

NATURALLY FILTRABLE SLUDGE   SEWAGE SLUDGE WITH HIGH   COMPRESSIBLE SLUDGE
                                      SPECIFIC RESISTANCE

Fig. 329. — Diagrams illustrating the structure of different types of sludge.

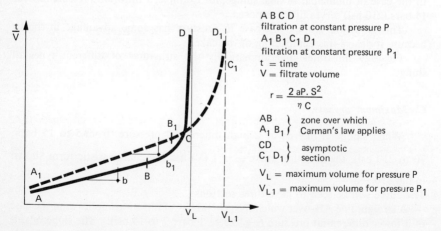

A B C D
filtration at constant pressure P
$A_1 B_1 C_1 D_1$
filtration at constant pressure P_1
t = time
V = filtrate volume

$$r = \frac{2\,aP.\,S^2}{\eta\,C}$$

AB   }   zone over which
$A_1 B_1$ } Carman's law applies

CD   }   asymptotic
$C_1 D_1$ } section

V_L = maximum volume for pressure P
V_{L1} = maximum volume for pressure P_1

Fig. 330. — Filtration at different pressures.

By plotting curves for $S_L = f(P)$, it is possible to determine, in the case of high-pressure filtration, the pressure which must be used to obtain the required degree of dryness. The selected pressure must be such that the operating point is below the point C in order to maintain acceptable operating conditions (adequate rate of filtration).

Maximum dryness is inversely proportional to the colloidal matter content of the sludge.

1.1.2. DRYING BEDS

Until a few years ago, sludge was most commonly dried on drained sand beds. This method is still employed at moderate-sized plants, despite the large amount of space and labour required, and at certain large plants under favourable climatic conditions.

For reasons of environmental hygiene, this natural dewatering is appropriate only for well stabilized, non putrescible sludge. Owing to the weather, especially in temperate countries, efficiency is usually low.

Drying beds are usually made up of a 10 cm (4 in) layer of 0.5 to 1.5 mm $\left(\dfrac{1}{32} \text{ to } \dfrac{1}{16} \text{ in}\right)$ sand, spread over a 20 cm (8 in) bed of 15 to 25 mm $\left(\dfrac{5}{8} \text{ to } 1 \text{ in}\right)$ gravel. The drains, which are laid under the gravel, are usually unjointed pipes made of cement, or of stoneware when the sludge is aggressive. There must be enough pipes, with the correct amount of slope, to drain the whole mass of sludge evenly (fig. 331). The layer of spread sludge is about 30 cm (12 in) thick. If it is too thick the top layer of sand quickly becomes clogged.

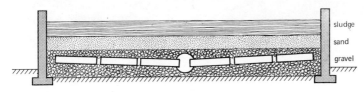

sludge

sand

gravel

Fig. 331. — Typical cross-section of a drying bed.

When the bed is fed at one point only, its width cannot easily exceed 8 m (30 ft) and its length 20 m (70 ft).

Drying beds operate on the following principle:

— first stage: dewatering by drainage or filtration at very low pressure. Free water is eliminated from the sludge suspension in this way down to a content of approximately 80 % in the case of domestic sewage sludge;

— second stage: evaporation of part of the bound water. Atmospheric drying can then produce dryness values of up to 65 % dry solids depending on retention time, weather conditions, and sludge characteristics.

The dried sludge is most often removed by hand. This can be avoided by mechanizing the beds (fig. 332) by means of motorized bridges equipped with a scraper and evacuation mechanism. With mechanization, larger drying bed areas are possible. The motorized bridges spread the liquid sludge over the whole drying area. On the other hand, the initial outlay is noticeably increased by the need to reinforce the supporting walls and to add fairly heavy equipment.

The following output figures can be adopted for the drying of sewage sludge on well-drained conventional beds (in the Mediterranean area).

Fig. 332. — 20-hectare (50-acre) drying beds with motorized bridges for sludge spreading and removal. PARIS-ACHÈRES II sewage treatment plant.

	Output in kg D.S./ m² d	lb/ft² d	Drying time
Digested sludge	0.4 - 0.6	0.8 - 0.12	1 month
Stabilized sludge	0.3 - 0.5	0.6 - 0.10	1 month

Output is greatly affected by the climate. The rate of filling can be as low as 3 to 4 times a year in certain temperate zones with an oceanic climate; filling is sometimes halted throughout the winter. Bed efficiency can be enhanced by chemical conditioning (using mainly polyelectrolytes). The drainage rate can be considerably increased by this easily applied process of flocculation with polymers.

Very cold weather may be beneficial for drying beds, while creating a form of sludge conditioning by freezing, but the solidification of sludge must affect the whole mass. This very rarely leads to an increase in efficiency, because the natural conditions do not often correspond to the successive periods of freezing and melting which are theoretically desirable.

Other processes whereby efficiency can be improved subject to certain conditions (in particular, where the initial moisture content of the sludge is less than 90 %) and which have been applied to a limited extent should be noted for the sake of completeness—for example, balanced drainage and electroosmosis. Drying on specially

Fig. 333. — Drying bed with polymer (foreground) and without polymer (middle ground). Time: 8 days.

shaped steel drainage sections (higher capital cost) is sometimes used where there are small quantities of industrial sludge (metallic hydroxides from metal surface finishing treatment, flocculated with polyelectrolytes).

1.1.3. FILTER BAGS

Filter bags of simple design impose only limited constraints and can be used in small plants to facilitate dumping or spreading on land. The polyelectrolyte-flocculated sludge is loaded into synthetic fabric suspended bags fitted with an additional central drainage column. These bags (2.5 m high and 1 m in diameter) give fast and considerable sludge concentration by simple drainage. The resulting filtrate is of good quality.

Depending on the initial solids content, 5 to 15 m³ of sludge per cycle can be placed in a bag.

The degree of thickening obtained depends on the nature of the sludge and on the draining time (between 6 and 24 h).

Sludge type	*Initial concentration %*	*Dryness after thickening %*
Fresh primary	6 — 10	18 — 23
Digested primary	6 — 8	17 — 22
Fresh mixed	3.5 — 6	13 — 17
Digested mixed	3.5 — 6	13 — 16
Stabilized aerobic	1.5 — 2.5	8 — 13
River water clarification (aluminium sulphate, little silt)	0.1 — 0.2	5 — 8
Carbonate removal from well water	3 — 6	35 — 45

This technique of high-rate thickening by drainage is also valuable for winter emergency service of small plants or to minimize liquid sludge volumes to be transported.

Fig. 334. — Thickening plant using filter bags.

1.1.4. VACUUM FILTRATION

This is the oldest continuous type of mechanical dewatering technique.

The vacuum filters most commonly used to dry water treatment sludge are of the rotary drum and open trough type. Other types of filters, used in the chemical, paper, or coal industries, may also occasionally be used — e.g., disc filters, vacuum belt filters, horizontal table filters, and paper filters.

A. Description and mode of operation.

A rotary drum filter consists basically of a revolving cylinder (or drum) partially submerged in a tank (or trough) containing the sludge to be filtered. The drum is formed by the juxtaposition of a number of compartments (or sectors) which are sealed off from each other and covered with a filtering fabric. Each compartment is connected by a pipe to an essential device known as the "distributor".

● *Filtering sequence* (fig. 335)

In zone A the compartments are linked to the vacuum source through the distributor. The liquid drawn off is discharged to the outside through a separating vessel. The matter trapped on the fabric accumulates to form a thickening cake. **Zone B** is the dewatering zone where drainage of the cake continues without addition of matter and with the compartments connected to the vacuum pump (if necessary, at a differential pressure higher than that in the preceding cells).

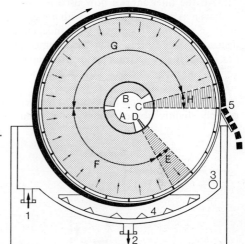

1. *Feed.*
2. *Drain.*
3. *Overflow.*
4. *Stirring fence.*
5. *Discharge by scraper blade.*
E.   *Cloth washing.*
F.   *Filtration zone.*
G.   *Dewatering zone.*
H.   *Detachment of cake.*
A, B. *Sectors under vacuum.*
C, D. *Sectors subjected to air pressure.*

Fig. 335. — The main phases of a vacuum-filtration cycle.

Zone C is the discharge zone. Methods of discharge include:
— compressed air blowing and scraper;
— parallel coiled wire or chains;
— a pressure roller which also completes dewatering by applying additional pressure;
— removable filter cloth.

In the last case, the filter fabric is completely detached from the drum (fig. 336). The change of curvature frees the cake and this is followed by scraping as an added precaution.

Sludge is filtered continuously with an industrial vacuum of 300 to 600 mm (12 to 24 in) of mercury. The thickness of the cake ranges from 5 to 20 mm ($\frac{3}{16}$ to $\frac{3}{4}$ in). Filters with removable fabric can be operated with the thinnest cakes, that is, on sludge with the highest specific resistances; this is made possible by continuous washing of the fabric, by water at a pressure of 3-4 bar, which greatly reduces the risk of clogging.

On an industrial-scale filter the cake forms in a few minutes. The drum rotates from 8 to 15 times per hour.

In the case of greasy, oily, or very thin sludges which are very liable to clog, the addition of a pre-coat layer is often essential (fig. 337). The drum then has only one compartment in communication with the vacuum pump. Before filtering starts, the filter is covered with a 50 to 80 mm (2 to 3 in) coating of some material of suitable porosity such as diatomaceous earth, wood flour, fly-ash, etc. A very rigid scraper, fitted with an adjustable micrometer advance, removes a film from the pre-coat together with the cake and keeps the filtering surface clean at all times.

All the filter fabrics now used are made of synthetic fibres. They should be woven as evenly as possible with a mesh void of up to 100 microns. A double layer of metal coil springs, with independent external washing, has been used instead of filter fabrics. This arrangement, which is ideal for preventing clogging, has the serious disadvantage of producing very thick filtrates because the mesh void is too large whereas filtrates obtained with filter fabrics are clear (less than 300 mg/l of dry solids).

These spring type filters are, however, difficult to operate with nonfibrous or only slightly fibrous sludge, when their productivity becomes unpredictable.

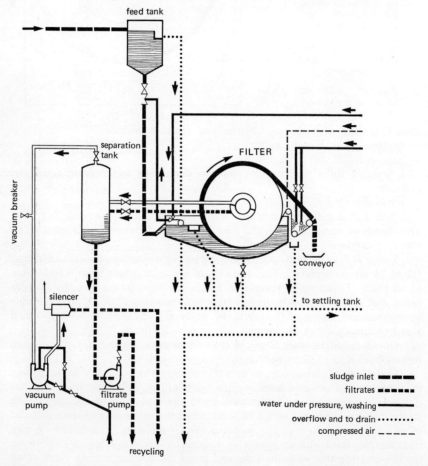

Fig. 336. — Diagrammatic description of a vacuum-type rotary filter with removable cloth.

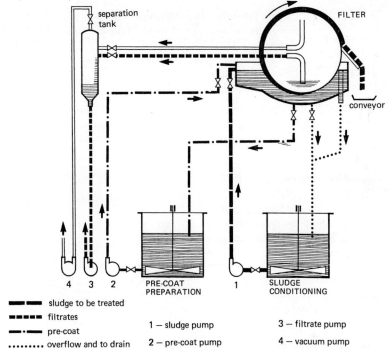

Fig. 337. — Operating diagram of a vacuum-type rotary pre-coat filter.

Legend in figure:

■■ ■■ sludge to be treated
■■■■■ filtrates
■—·— pre-coat
.... overflow and to drain

1 — sludge pump   3 — filtrate pump
2 — pre-coat pump   4 — vacuum pump

B. Output capacity.

In practice, the concept of filtration capacity is used; this is determined by laboratory experiments and expressed in kg of dry solids retained per m² and per hour.

Usable filtration capacity ranges between 10 and 30 kg/ (m².h) for example for sewage sludge, corresponding to values of $r_{0.5}$ of between 10^{11} and 5×10^{11} m/kg for common sludge concentrations of 3 to 8 % DS.

The output capacity L of a vacuum filter may be theoretically approached by integration of Carman's equation for the effective filtration time $t_f = nT$ (where T represents total drum rotation duration and n is the immersion fraction):

$$L = k \times \left[\frac{2\,P \times C \times n}{\eta \times r \times T} \right]^{\frac{1}{2}} \times \frac{1 - w_b}{1 - \dfrac{w_b}{w_g}}$$

in which k = correction factor depending in particular on the filter medium's own resistance (k is normally taken as between 0.75 and 0.85);

P = vacuum applied, usually about 49 kPa (0.5 bar);

r   = specific resistance of conditioned sludge at pressure P (which
            can easily be determined with a Büchner funnel in the labo-
            ratory);
W_b ,W_g = see definitions in page 79 (expressed as fractions);
C   = DS concentration of conditioned sludge;
n   = drum immersion fraction (0.25 to 0.40).

This relation is well borne out under practical conditions. It indicates the
value of a high sludge concentration (L = K $\sqrt{C}$) and of keeping the filter medium
clean (factor k); the latter is easy with filters having continuously-washed removable cloths.

C. Calculating the rate of chemical conditioning.

The rate of conditioning can be
roughly calculated by performing a
series of quick tests on a Büchner apparatus or a special laboratory cell, connected to a vacuum source—a water
pump (the cake must be dewatered in
2 to 4 minutes). A more accurate calculation can be made with the same
equipment by plotting filtration curves,
$\frac{t}{V} = f(V)$, and specific resistance curves

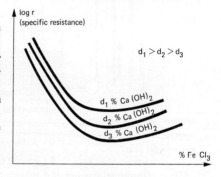

Fig. 338. — Variation of specific resistance with sludge conditioning.

(r) for various rates of conditioning.
Figure 338 gives for a particular sludge
the set of curves log r = f (FeCl$_3$) for
different doses of Ca(OH)$_2$.

● *Capillary suction filtrability test:* this is a rapid evaluation test, devised by
Gale and Baskerville, which can readily be substituted for specific resistance determination in, for example, the industrial operation of a filtration plant. The
principle of measurement is simple: a sludge sample is poured into a bottomless
cylindrical tank resting on absorbent paper, which absorbs the filtrate by capillary
suction. The suction time (CST) is the number of seconds taken by the filtrate
ring to travel 1 cm. The higher the filtrability of the sludge, the shorter this
time; for a sludge of given concentration, it can be readily correlated with the
specific resistance $r_{0.5}$ provided that the latter is not too low. The correlation
with $r_{0.5}$ is better if CST is compared with $r_{0.5} \times C$. Certain precautions
(sampling, floc shearing, etc.) must be taken when performing this test.

The test is particularly suitable for the study of chemical conditioning processes using inorganic salts, thermal conditioning of a given sludge, and monitoring filter operation. For the latter purpose, a rapid test is more useful than
a universal but laborious method of measurement.

D. Performance.

The following table summarizes a large number of operating results obtained with removable-cloth filters fed with different types of domestic sewage and industrial sludge.

PERFORMANCE OF VACUUM FILTERS

Sludge type	*Origin*	*Nature*	*Conditioning* $FeCl_3$ [1] %	*Conditioning* CaO [1] %	*Filter capacity kg DS/(m².h)*	*Cake dryness* %
Organic hydrophilic	DWW [2]	Primary	2-4	7-11	30-40	26-32
		Fresh mixed	3-6	13-19	20-30	23-27
		Digested mixed	5-7	13.5-21	20-25	24-28
		Stabilized aerobic	7-12	22-37	15-20	18-22
Organic hydrophilic	DWW [2]	Fresh mixed	Thermal conditioning at 180 °C for 40 min		25-30	33-40
		Digested mixed			15-30	25-40
	Brewery	Biological + 10 % carbonate removal sludge	7	19	20-30	20-25
Fibrous	Paper pulp	Sawdust and shavings	—	—	40-50	30
	Paper mill	Fibres 20 % + hydroxides	—	19	15-20	25
Inorganic water-repellent	Makeup water	Carbonate removal on lime, Fe < 1 % [1]	—	—	50-70	40-50
	Steelworks	Converter gas scrubbing	—	—	60-70	60
	Coal mining	Coal washing	Polymer: 0.3 kg/t DS		25-30	35
Oily	Refinery	First stage settling sludge	—	15-22 (CaO)	5-10 with precoat	35-45

1. Percentage referred to weight of dry solids
2. Domestic waste water.

Vacuum filters are not normally suitable for very hydrophilic colloidal sludge because even with intensive conditioning the filtrability of this sludge remains greater than 10^{12} m/kg.

With chemical conditioning, inorganic reagents are preferable to organic products. The following are recommended:
— for organic sludge: iron salts combined with lime;
— for inorganic sludge: mainly lime.

These inorganic reagents have a favourable effect on the compressibility, consistency, and dryness of the cake produced.

If polyelectrolytes are used, there is a risk of break-up of the cake: owing to the very substantial decline in its specific resistance — not readily controllable in practice — the cake becomes so thick that it can no longer withstand its own weight when it emerges from the sludge bath; it then falls back into the through. In this case the mechanical cohesion of the cake must be measured in the laboratory.

E. Vacuum filtration practice.

Drum-type vacuum filters may have a unit area of 50 m² (500 ft²) or more.

Vacuum filtration is a continuous process which can easily be automated. Raw sludge feed is governed by the sludge level in the filter trough or the conditioning tanks. This control can be continuous but is most frequently of the on-off

Fig. 339. — Filtration of digested sludge through rotary vacuum filters with washed removable cloths at the LILLE sewage treatment plant (France).
Unit filtration area: 40 m² (400 sq ft).
Output capacity: 3,5 to 4,5 t/h dried sludge.

type through pumps or pneumatic valves. The dosing system is also controlled by the level.

Further automation is possible on the basis of variations in the concentration of the raw sludge. The amount of conditioner is then governed by sludge flow multiplied by concentration. For this purpose an electromagnetic flowmeter and a concentration analyser (e.g., of the gamma-ray type) must be installed on the raw sludge feed pipe.

The vacuum pumps used are of the fluid ring type. If the make-up water is very hard and the plant is large, provision may be made for softening the water and removing its carbonate content.

Filter fabric can give up to 3 000 hours of service. After a time, the conditioning lime clogs the fabric by the deposit of carbonates. The fabric should be cleared from time to time; this can easily be done by filling the filter trough with a passivated hydrochloric acid solution and brushing the fabric.

The filters can be made of a wide variety of materials. They are most often of steel but, if the sludge is very corrosive or creates special problems, they may be made of stainless steel, vulcanized steel, titanium, lead, wood, resins, etc.

The power absorbed by vacuum filtration is about 1.5 kW/m^2 of drum area.

Fig. 340. — Sludge dewatering by vacuum filtration.

1.1.5. PRESSURE BELT FILTRATION

This filtration technique is particularly suitable for small or medium-sized plants, for several reasons:
— continuity of process and continuous washing of filter medium;
— ease of operation and robustness (low velocity);
— low capital, labour, energy, and maintenance costs;
— sludge obtained can be shovelled.

A. Description and operation

The development of filterbelt presses is due largely to the evolution of synthetic organic polymers. The fundamental condition for the application of filterbelt presses to sludge dewatering is the "superflocculation" given by appropriate polyelectrolytes which provide a suspension of a coarse floc in clear interstitial water. The specific resistance $r_{0.5}$ of the sludge is thus reduced to approximately 10^{10} m/kg; the flocculated sludge then dries naturally by simple drainage on a fabric with relatively large mesh size (0.2 to 0.5 mm), or in drainage drums in some filters. This natural drainage doubles or quadruples the dry solids content of the sludge in less than a minute, giving a cake of sufficient consistency to withstand progressively increasing pressure. The pressure applied is limited by the risk of sludge creep: the pressing zone is not sealed laterally, and the sludge itself provides the seal under the pressure which it can withstand at each instant. The maximum permissible pressure remains relatively low (about 1 bar), so that this process cannot yield cake dryness values as high as plate type filter presses.

The preliminary drainage phase is therefore of vital importance, as it imparts greater cohesion and better resistance to the sludge for the subsequent pressing phase. Reagents must be selected judiciously on the basis of the following criteria:
— production of a floc of maximum drainability;
— minimum consumption.

In addition to the principles applied in the three functional phases of a belt filter—adequate sludge flocculation, free drainage, and progressive pressing— the effect of a fourth principle should also be mentioned: namely, cake deformation. This deformation is more or less sudden and gives rise to shear stresses which release the interstitial water and give rise to new flow channels in the cake. It begins in the drainage zone when the cake is handled by combing rakes and the levelling roller, and continues in the pressing zone by alternate bending of the cake trapped between the two filter cloths.

Many different types of belt filters in which these principles are applied have been constructed. Some of them have a filter cloth and a watertight pressure cloth. A good example is the Flocpress (fig. 341).

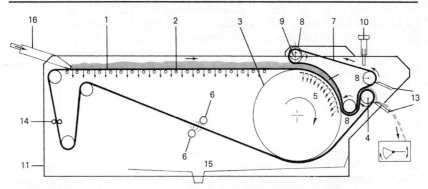

1. *Filter belt.*
2. *Support rollers.*
3. *Pressure drum.*
4. *Outlet roller.*
5. *Grooves and perforations.*
6. *Washing nozzles.*
7. *Pressure belt.*
8. *Pressure rollers.*

9. *Variable-height shaft.*
10. *Jack (pressure adjustement).*
11. *Frame.*
12. *Pressing gap (air space).*
13. *Scraper.*
14. *Belt guide.*
15. *Filtrate + wash water discharge.*
16. *Feed channel.*

Fig. 341. — Diagrammatic cross-section of the FLOCPRESS.

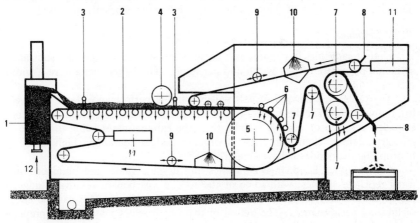

1. *Flocculator/mixer.*
2. *Drainage zone.*
3. *Combing rakes (movable).*
4. *Levelling roller (movable).*
5. *Filter drum.*
6. *Additional pressing rollers controlled by pneumatic jacks.*

7. *Idler rollers of different diameters and drive rollers.*
8. *Scrapers.*
9. *Pneumatic cloth centring device.*
10. *Enclosed washing system.*
11. *Pneumatic cloth tensioning jacks.*
12. *Polymer injection.*

Fig. 342. — Diagrammatic cross-section of the PRESSDEG.

Other filters have a vacuum chamber at the end of the free drainage zone, but this entails higher electricity consumption.

In the search for higher dryness values, a trend has emerged towards the adoption of dual filter belts associated with elaborate pressing and cake deformation systems.

Figure 343 shows one of these types of filters: the PressDeg, in which water is eliminated from both sides of the cake. The dewatering pressure is obtained mainly by the tension of the filter cloths; the pressures developed along the cylinders are inversely proportional to the radii of the rollers.

The main feature of this type of belt filter is its great simplicity of operation: because all the components involved in the different phases of operation are immediately visible, the operator can quickly correct any irregularities arising and readily adjust the unit in accordance with the varying characteristics of the sludge. Optimum efficiency for each type of sludge is obtained by simple adjustements of:

— cloth tension (using pneumatic jacks) (even tensioning is necessary);
— pressure;
— linear velocity of the two cloths (usually in the range 0.5 to 4 m/min).

Some filters have additional pressing modules and feature increased (but still limited) pressures. The corresponding gains in dryness are highly variable depending on the type of sludge, while the capital cost is heavily increased.

B. Performance.

Output capacities in pressure belt filtration are quoted in kg of dry substance extracted per metre of belt width and per hour. The largest units have a width of 3 m.

The table on page 500 reveals widely varying performance data: relatively low output rates and cake dryness values are obtained with biological sludges from extended aeration or aerobic stabilization, the extreme cases being in the food industry (dairies, canneries, etc.). Inorganic sludges containing high levels of hydrophilic hydroxides behave similarly.

On the other hand, high capacities and dryness values are obtained with water-repellent sludges; with some of these —e.g., carbonate removal sludge— the simultaneous presence of hydrophilic hydroxides (Fe, Al,Mg) has the effect of limiting performance. The action of these hydroxides may become preponderant above a certain level (25-30 %).

Filterbelt presses should not be used with oily sludge because it has the effect of "waterproofing" the filter cloths (the oil bridges over the mesh openings) and owing to the risk of creep (lateral flattening-out of cakes) on this equipment, whose sides are not sealed. Polymers are used very widely; inorganic reagents are seldom employed except with sludge containing very high levels of fibrous matter.

The dryness values obtained with filterbelt presses are similar to those given

Fig. 343. — Type 762 PRESSDEG sludge filtration unit.

Fig. 344. — Dewatered sludge leaving the filter belt press.

PERFORMANCE OF FILTERBELT PRESSES

Sludge type	*Nature and origin*	*Sludge concentration %*	*Capacity kg DS / (m.h)*	*Cake dryness value %*	*Polymer consumption kg (anhydrous) / t DS*
	Domestic waste water:				
	Primary fresh	5-10	250-400	27-35	0.9-2
	Primary digested	4-9	250-500	27-36	1-3
	Mixed fresh				
	(VS < 75 %)	3.5-8	130-300	21-28	1.5-5
	Mixed digested	3-7	120-350	20-28	2-5
Organic hydrophilic	Extended aeration or aerobic stabilization	1.5-3.5	80-150	15-25	2-5
	Physicochemical (FeCl₃ 150 mg/l, lime 200 mg/l)	4-8	200-300	20-27	2-4
	Industrialneffluents:				
	Dairies-extended aeration	2-3.5	50-90	11-16	3-5
	Carbonate removal				
	1. $\frac{Fe}{DS} \leqslant 1\%$				
Inorganic water-repellent	$\frac{Mg}{DS} \leqslant 2\%$	15-30	500-1 000	55-70	0.2-0.5
	2. $1\% < \frac{Fe}{DS} < 5\%$	10-20	300-700	45-65	0.5-1
	Refuse incineration — flue gas scrubbing	15-25	800-1 000	40-50	0.3-0.5
	Iron and steel industry — blast furnace gas scrubbing	15-25	400-700	38-55	0.5-1
	Semichemical pulp manufacture				
Fibrous	1. Fibres + sawdust	4-7	200-400	25-40	not applicable
	2. Shavings + sawdust	8-15	600-1 000	35-45	not applicable
	Paper mills — Al salt flocculation	2.5-4	100-350	22-30	1-2
	Drinking water—flocculation (Al or Fe salt)	3-6	80-150	16-23	1.5-3
Inorganic hydrophilic	Partial carbonate removal	5-8	150-200	25-33	1-2
	Organic dyestuffs plant — flocculation (iron salt)	7-10	150-250	25-30	2-3

by vacuum filters of centrifuges, but electricity consumption varies substantially between the different techniques:

Filterbelt	: 5-20 kWh/t DS;
Filter press	: 15-40 kWh/t DS;
Continuous decanter	: 30-60 kWh/t DS;
Rotary vacuum filter	: 50-150 kWh/t DS.

1.1.6. PRESSURE FILTRATION

This technique is coming to be used increasingly widely although it is an intermittent process and in spite of its high capital cost, for the following reasons:

● Increasing mechanization of equipment, so that only minimal labour and supervision is required for filter opening for cake removal or filter cloth washing (automatic in-place washing);

● the requirement of high-dryness cakes to permit:
— sludge autocombustion on incineration;
— reduced fuel consumption in the case of heat drying;
— easier disposal;
— use of sludge as a filling material;
— limitation of transport costs for dewatered sludge.

Only the filter press provides the very high effective pressures (15 bar and above) which give maximum cake dryness values (normally exceeding 30 %).

A. Description and mode of operation.

The filter presses most often used for dewatering sludge are of the plate type; these plates are easy to mechanize for cake discharge. A plate filter consists basically of a set of vertical, recessed plates pressed hard against each other by hydraulic jacks at one end of the pack (fig. 345).

Filter fabrics are applied to the two grooved surfaces of these plates. The sludge to be filtered is forced into the spaces formed between adjacent plates. The filtrate recovered through the grooves at the back of the fabric is carried away by ducts running through the plates (this method of removal is hygienic and reduces the risk of smells).

The plates also have apertures which are aligned to allow the sludge to be fed through. These apertures are at the centre or at the corners of the plates. A centre feed seems to give a

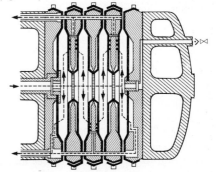

Fig. 345. — Simplified diagram of a filter press.

more even distribution of pressure over the whole plate, thus improving drainage.

The filtrate-removal ducts can similarly be joined from plate to plate or can be separate for each plate. With the second arrangement the quality of the filtrate and, thus, a tear in any filter can be checked visually.

The largest filter press units have plate side lengths of 1.80-2 metres and include up to 130 plates. Total filtration area may be as much as 800 m² (8600 sq ft).

The filter fabric is generally made of synthetic fibres. The correct choice of fabric has a great influence on the performance of a pressure filtering unit. In some cases, the filter fabric is not fitted directly to the plate but is applied on a coarser sub-fabric, to improve the distribution of pressure over the whole filtering surface, facilitate discharge of the filtrate, and ensure more efficient washing. The complete filter press is sealed by the very powerful pressure forcing the fabric-covered plates against each other. Filtering pressure can reach 25 bars, but there is rarely any point in going above 15 bars for dewatering sludge.

The space formed by the central recess between two adjacent plates determines the thickness of the cake. This will be set at between 30 and 20 mm (1 1/4 and 3/4 in) according to the specific resistance of the sludge and the time selected for the complete sequence.

For sludges which filter rapidly (e.g., carbonate removal sludge), thicknesses of 40 and even 50 mm are used in order to prolong cycle times between filter openings.

The plates may be made of different materials:
— cast iron;
— steel-reinforced rubber;
— steel lined with moulded rubber;
— glass fibre reinforced polyester, and polypropylene (lighter in weight);
— stainless steel (for highly corrosive sludge).

There are two main types of frames:
— plates suspended from an upper rail (fig. 346);
— plates supported laterally by two longitudinal guides (fig. 347).

B. Filtration cycle :

The different operating phases of a filter press are as follows:

● *Filling.* During this relatively brief phase (3 to 10 % of the complete cycle), the filtration chambers enclosed by the outer surfaces of the cloths of two contiguous plates are filled with the sludge to be filtered. Filling time depends on the flow of the feed pump.

● *Filtration.* Once the chambers are filled, continuous introduction of the suspension to be dewatered causes an increase in pressure inside the chambers due to contraction of the pores of the sludge as it thickens.

● *Filter opening.* In this operation, the filter is "opened". The plates are separated from each other and the cakes formed between pairs of plates are successively ejected.

● *Cleaning.* This operation is carried out occasionally only; it entails rinsing and brushing the cloths to restore their original drainability.

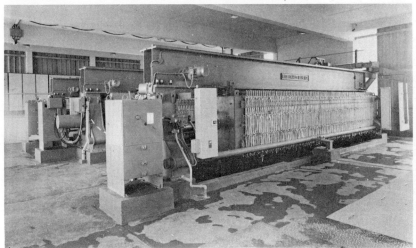

Fig. 346. — View of the battery of 3 filter-presses installed at the Water Pollution Control Works of CANNES, France.
Total filtration surface area: 660 m² (7 100 sq ft).

Fig. 347. — Partial view of the filter-press installed at the Shell-Chimie plant at Fos-sur-mer, France.

Total automation of filtration sequences involves mechanization of plate transport and reduction of the substantial downtime involved in periodic filter cloth cleaning. This time has been considerably shortened by the development of a washing method using water at very high pressure (80-100 bar) synchronized with displacement of the plates.

Fig. 348. — Filter-press. Mechanical and semi-automatic cloth washing.

Fig. 349. — Filter-press. Automatic cloth washing.

Cleaning of the filter cloths of a 500 m² filter press with 130 plates requires only 2 to 4 hours' work by a single operator; the cloths are cleaned at intervals of 20 to 50 cycles in the case of well conditioned domestic sewage sludge.

The filtration phase may be terminated by the following means, as appropriate: manually, by timer, or (most commonly with large-scale filters) by a filtrate flow sensor, usually set for and end-of-filtration flow in the region of 10 to 20 l/(m².h.).

C. Filtration capacity.

The output capacity of a filter press is generally in the region of 2 to 10 kg DS/(m².h); high values of $r_{0.5}$ (up to 8×10^{12} m/kg) may be appropriate. Pressing times range between 1 and 6 hours, often being limited to 1 to 3 hours. Pressures applied vary from 6 to 15 bar. One of the advantages of the filter press is that it tolerates higher sludge specific resistances than a vacuum filter, thus giving:

— reduced consumption of conditioning reagents (dose reduction is limited in fact by the clogging of filter cloths);

— possibility of using less efficient reagents, which are also less expensive (e.g., $FeSO .7H_2O$);

— possibility of filtering sludge without prior conditioning where $r_{0.5}$ is between 5×10^{11} and 8×10^{12} m/kg, in which range the vacuum filter cannot be used with raw sludge.

Using the parameters of specific resistance $r_{0.5}$ (i.e., at a pressure $P = 0.5$ bar) and compressibility s (the factors can be determined in the laboratory on conditioned sludge by means of a steel cell with piston), a formula to calculate the pressing time has been deduced from the laws of surface filtration:

$$t_f = k \times \frac{\eta \times r_{0.5} \times 2^s \times e^2 \times C}{P^{1-s}} \left(\frac{S_f d_g}{C} - 1\right)^2$$

in which

t_f = pressing time
η = filtrate viscosity
$r_{0.5}$ = specific resistance of conditioned sludge at $P = 0.5$ bar
s = compressibility coefficient of conditioned sludge
e = cake thickness
C = DS concentration of conditioned sludge
P = filtration pressure
S_f = percentage dryness value of cake
d_g = volume mass of final cake
k = correction factor for cloth clogging ($k = 1.2$-1.3 if the filter has automatic cloth washing facilities).

The output capacity L of the filter is given by the formula:

$$L = \frac{e \times d_g \times S_f}{2 \times t_{cy} \times 100}$$

in which t_{cy} = total cycle time = $t_f + t_d + t_r$
where t_d = opening, dropping and closing time
t_r = filling time.

t_r depends on the flow of the feed pump and is typically between 5 and 15 minutes. With a mechanized filter, t_d is 10 to 20 minutes, provided that the cakes come away cleanly.

The following important points on the operation of a filter press should be noted:
— the pressing time t_f is proportional to the square of the chamber thickness and inversely proportional to the concentration of the conditioned sludge;
— pressing time also depends on the specific resistance and compressibility coefficient of the sludge.

Nevertheless, the filter press allows greater latitude with the conditioning rate, and hence greater operational security, than the vacuum filter.

It is therefore advantageous to bring care to sludge preparation to obtain maximum thickening and to reduce specific resistance and compressibility. Most high protein organic sludges have a compressibility coefficient of between 0.8 and 1.5 even after chemical conditioning; relatively little gain on pressing time can be achieved in this case by increasing the pressure, but higher pressure will give a greater final dryness value of the cake. Curves 1, 2 and 3 in p. 508 show how the final dryness S_L and filtration time vary with the pressure, the thickness of the cake and the type of conditioning.

D. Performance.

The following table sets out the results to be expected from filter press treatment of the main types of sludge.

Sludge	Nature and origin	Conditioning FeCl₃ [1] %	CaO [1] %	Filtration capacity kg DS / (m².h)	Cake dryness %
Organic hydrophilic VS/DS 30 to 90 %	Fresh domestic sewage sludge (mixed)	3-7	11-19	2-4	40-50
	Digested domestic sewage sludge (mixed)	4-7	11-22.5	2-4	35-50
		Thermal conditioning		2.5-5	45-60
	Domestic sewage sludge, extended aeration	6-10	15-30	1.5-3	33-38
	Domestic sewage sludge, physicochemical (FeCl₃ + lime)	—	15-22.5	1.5-2.5	33-45
	Brewery, biological sludge	5-7	22.5 (with 10 % carbonate removal sludge)	2.5-3	35-38
	Amino acid synthesis biological sludge	7-12	30	1.5-2	30-35
Inorganic hydrophilic Hydroxides of Fe, Al, Cr, etc.	Partial carbonate removal, Fe/DS ≤ 10 %		7.5-11	5	50-55
	6 % ≤ Fe/DS ≤ 10 %		—	6	55-60
	Surface water clarification, aluminium salt (little silt)		19-30	1.5-3	30-40
	Aluminium pickling (HCl)		—	5-6	35-40
	Lime neutralization Aluminium anodization, neutralization		15	5	40
	with caustic soda		—	1.5	30
	Pickling of steel, lime neutralization		—	3-5	45-50
	Electroplating, galvanizing		—	2	30-35
	Chromate removal		19	2	30

1. Percentage referred to weight of dry solids.

Sludge type	Nature and origin	Conditioning		Filtration capacity kg DS / (m².h)	Cake dryness %
		FeCl₃[1] %	CaO [1] %		
Inorganic water-repellent	Carbonate removal Fe/DS ≤ 2%		—	10-20	60-70
	2% ≤ Fe/DS ≤ 5%		—	8-13	50-60
	Refuse incineration, flue gas scrubbing		—	10-15	55-60
	Gas scrubbing		—	7-15	60-70
Oily	Cutting oil or soluble oil effluents — Acid breaking + lime neutralization	11	3		50-60
	— Flotation (Al salt + lime) of hydro-carbons (30 % referred to DS)	22.5	4		50-55
Inorganic, oily	Rolling mill effluents, high levels of oils and fats		—	15-20	85

1. Percentage referred to weight of dry solids.

The table on pages 506 and 507 shows that the filter press is applicable to almost all types of sludge; the following comments may be made on the different sludges:

● *Hydrophilic organic sludge:* thermal conditioning, which is virtually confined to fresh or digested sludge in large-scale domestic sewage treatment plants, yields a sludge which can be readily dewatered by release of the water bound to the hydrophilic fraction.

Inorganic chemical conditioning is recommended to avoid the difficulties often met with in polyelectrolyte conditioning: risk of cake sticking (cakes are usually more spongy), and waste of time due to frequent cloth washing.

● *Inorganic hydrophilic sludge:* the filter press gives high dryness values and normally calls for addition of lime, which has a very favourable effect on the hydrophilic structures (especially those of metallic hydroxides).

● *Inorganic water-repellent sludge:* these high-density sludges can be readily dewatered owing to their low compressibility and their very crystalline particles.

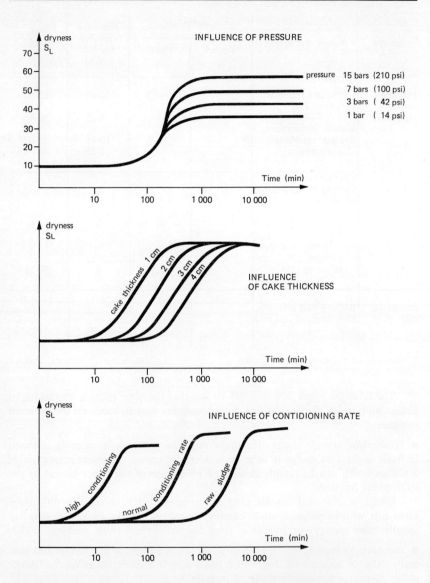

Fig. 350. — Parameters affecting filtration time in pressure filtration.

Grain size structure is a crucial factor in determining the dewatering characteristics. Again, filtration is carried out without a conditioning agent.

● *Oily sludge:* oils are present in emulsified form or are adsorbed on the particles. The filter press can be used with light oils owing to its high operating pressure. The presence of animal or plant fats can sometimes impair the satisfactory working of the filter, whose cloths must then be degreased at frequent intervals.

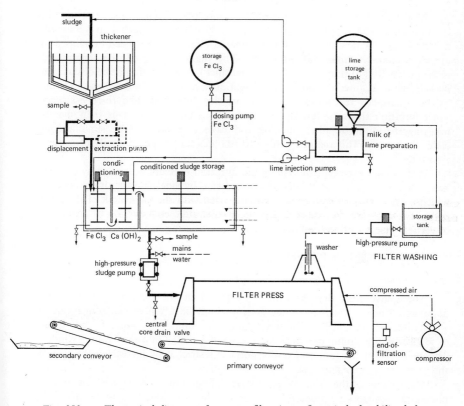

Fig. 351. — Theoretical diagram of pressure filtration. Organic hydrophilic sludge.

E. Pressure filtration practice.

Close attention must be devoted to the method of feeding filter presses. The following facilities may be used for this purpose:
— diaphragm pumps with automatic flow adjustment;
— displacement pumps incorporated in a bypass line connected to a buffer tank with air cushion (eccentric screw or piston pumps);
— a double set of tanks under air pressure with cyclic filling and emptying.

It is virtually impossible to control chemical (or thermal) conditioning by a closed loop system in accordance with the inlet flow into the filter presses, owing to the substantial variations in this flow. The sludge is conditioned in tanks preceding the filters (see fig. 351).

Cloth life is highly variable, 500 pressing cycles being a minimum.

1.1.7. AUTOMATED THIN-CAKE PRESSURE FILTRATION

Mechanized plate type filter presses require the presence of an operator to monitor cake discharge for a period of 20 to 30 minutes at each 1 to 3 hour filtration cycle. A new type of fully automated filter press has been developed in order to eliminate these labour costs. The main feature of this equipment is the production of thin cakes (3 to 10 mm thick), which could not be discharged without manual intervention in a conventional plate type filter press. Owing to the square-law relationship between pressing time and cake thickness, very short filtration times (a few minutes) can be achieved, and it is then merely necessary to minimize filter opening and cake removal time by simultaneous cake discharge. This method gives 5 to 10 times the output capacity of a plate type filter.

According to the type of system, the filtration pressure is applied by a membrane compressing several plates positioned side by side or on a single plate to which a cyclic motion is imparted, synchronized with the movement of the filter belt. The pressure may also be exerted by hydraulic compression of a number of vertical plates with one solid surface and one filtering surface.

1.1.8. CONTINUOUS PRESSES

A. Continuous screw type presses.

These give high dryness values (35-45 %) in the treatment of partially dewatered sludge (which has undergone vacuum filtration or centrifugation).

These presses consist of a screw (single or double) rotating at low speed (a few revolutions/minute) which compresses the sludge in a perforated cylinder. The clearance at the sediment outlet point is restricted by increasing the size of the screw or by a closing cone.

On account of the principle of this type of equipment (wide-mesh perforated plate and high working pressure), only coarse, long-fibre sludge can be effectively treated. One of the main applications is the dewatering of primary paper mill sludge previously thickened to 15-20 % and containing fibres, sawdust, wood shavings, bark, etc. Operational problems frequently arise with some sludges in connection with cleaning of the perforated cylinder.

B. Continuous disc type presses.

The same considerations apply to the use of these presses as to screw presses, and their applications are also the same. They consist of two large-diameter circular discs perforated on the inside and rotating very slowly. The two discs are inclined to the axis so as to form a compression "wedge" in which dewatering takes place.

1.2. Centrifugal separation

1.2.1. CENTRIFUGATION AND CENTRIFUGABILITY (fig. 352)

In a cylindrical vessel turning at an angular speed of ω (radians/sec) or N (rev/min) and containing a liquid ring of mean radius R (in metres), the centrifugal acceleration a in m/sec^2 to which the particles are subjected is given by the expression:

$$a = \omega^2 R = 0.011 N^2 R$$

The centrifugal fields obtained with machines built on an industrial scale can reach values more than 1 000 times that of the earth's gravitational field that is normally used for all settling processes. The acceleration generated by centrifugal separation is always expressed by reference to the earths' field as a multiple of g ($g = 9.81$ m/s^2).

The force exerted on a particle of unit mass is expressed by:

$$F_C = 0.011 N^2 R (d_S - d_L) \times \frac{1}{g}$$

in which d_S = specific mass of particle
d_L = specific mass of interstitial liquid

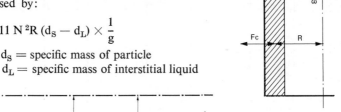

Fig. 352. —

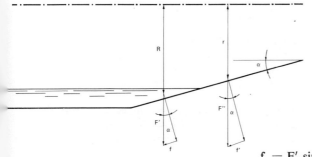

Fig. 353. — *Effect of bowl conicity and rotational speed on reflux force f'.*

$$f = F'_C \sin \alpha (d_S - d_L) = F' \sin \alpha$$
in which $F'_C = 11.2 \times 10^{-4} RN^2$
$$f' = F''_C \sin \alpha \, d_S = F'' \sin \alpha$$
in which $F''_C = 11.2 \times 10^{-4} rN^2$

The forces so applied to the fine particles of a sludge-laden suspension give centrigufal accelerations of 1 000 to 3 000 g and settling rates of about 10 m/h, which is something like 50 times faster than that of the natural thickening of the sludge. In addition, in the case of concentrated suspensions, they break the bonding forces and thus separate particles which would not be moved by ordinary static sedimentation.

When a sample of sludge containing a high proportion of colloidal matter is subjected to a centrifugal field (for example, 1 000 *g* for 1 to 2 minutes) in a laboratory centrifuge, the following appear in the test tube (fig. 354):
— a very cloudy supernatant liquid, containing 0.5 to 1 g/l DS;
— a bottom layer of sediment which can be subdivided into 2 zones :

● a concentrated lower zone of dense matter
    DS content: 30 to 35 %
    $\dfrac{VS}{DS}$ ratio: 65 %   (VS = volatile solids)

● a less concentrated upper zone of non-cohering matter
    DS content: 10 to 17 %
    $\dfrac{VS}{DS}$ ratio: 85 to 90 %

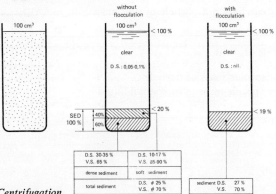

Fig. 354. — Centrifugation of a municipal sewage sludge.

When the experiment is repeated with a conditioned sludge, especially if an organic polyelectrolyte has been used, the following are formed, as appears from fig. 354:
— a clear suspernatant liquid containing very little DS;
— a completely homogeneous bottom layer, less concentrated than the denser layer but more concentrated than the mean of two layers from non-conditioned sludge.

This capacity of the sludge suspension to separate, in a laboratory centrifuge into two distinct phases—one very clear (containing approximately 0.1 g/l of DS) and the second sludge-laden and substantially homogeneous—is the "centrifugability" of the sludge, and conditioning aims at improving it. At a given acceleration (around 1 000 *g*) and a given centrifugation time (a few minutes), "centrifugability" will be inversely proportional to the DS content of the supernatant liquid. Another important parameter is cake consistency, which can be measured by penetrometry.

1.2.2. CONTINUOUS DECANTERS

The machines most commonly used for sewage sludge are continuous horizontal decanters.

Both feed and discharge of centrate and sediment are continuous in these decanters, which consist basically of a horizontal-axis cylindrico-conical bowl rotating at high rate with a helical extraction screw rotating at a slightly different rate inside it. Some centrifuges even have a dual inner screw, for progressive acceleration of the sludge.

Apart from its absolutely continuous operation, the main reasons for selection of this type of centrifuge for sludge dewatering are as follows:

● The phases are separated by centrifugal decanting, followed by screw transportation of the sediment, which also passes through an additional drying zone, out of the liquid phase. There is no risk of clogging because the liquid phase does not pass through a filter medium as in the case of centrifugal driers.

● Unlike disc type separators (see page 520), which have narrow orifices and channels inside the bowl, continuous decanters are not liable to clogging of orifices provided that the sludge particle size is appropriate.

● A homogeneous sediment and high separation efficiencies are obtained where the sludge is conditioned with polyelectrolytes.

Recent progress in centrifugal decanting has enabled this technique to be applied on a large scale to many different types of sludge, and it has also been adapted for the dewatering of hydrophilic colloidal sludges, as most commonly encountered in waste water treatment:
— better equipment geometry;
— better knowledge of operational parameters (speed, depth of liquid ring, etc.);
— very low risk of clogging;
— improved anti-abrasion lining;
— increased specificity of synthetic polymers.

A. Construction parameters.

Continuous decanters are distinguished first of all by the respective directions of travel of the sludge suspension and of the sediment: cocurrent or countercurrent (fig. 355 et 356). Central feed is generally used; tangential feed, although more favourable to progressive acceleration of the flocculated sludge, has been abandoned because of operational difficulties (frequent clogging due to heterogeneous sludge).

The countercurrent centrifuge is suitable for dense sludge, while the cocurrent type is better suited to colloidal hydrophilic sludge because of its lower internal turbulence.

1. *Feed.*
2. *Sediment outlet.*
3. *Centrate outlet.*
4. *Bowl.*
5. *Helical scraper.*
Fig. 355. — Continuous uni-directional flow centrifuge fed at the centre.

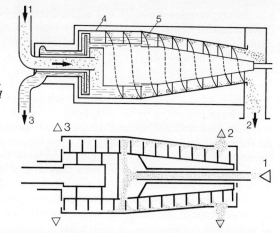

Fig. 356. — Continuous counter-flow decanter fed at the centre.

The caption to fig. 357 indicates the construction parameters important to the proper operation of a continuous decanter.

1. *Geometry of the bowl: Ratio* L/D_1
Cone angle α.
2. *Bowl velocity* Va *(N) or acceleration Fc.*
3. *Relative velocity of scraper* V_R.
4. *Scraper pitch P.*
5. *Depth of liquid ring* $0.5 (D_1 — D_3)$.
Fig. 357. — Centrifuge constructional parameters.

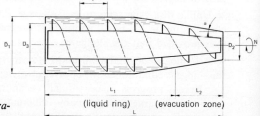

The bowl shloud be a tapered cylinder, with separation taking place basically in the cylindrical section, and the conical section carrying the sediment to the point of evacuation. At the point of transition from the cylindrical to the conical section, the sediment is subjected to back-pressure which becomes weaker as the angle of the bowl reduces. This pressure must not destroy the cohesion of the thickened sludge, otherwise extraction will be poor and a substantial amount of suspension will be reformed (see fig. 355).

This pressure may also be reduced by reducing the rotating speed of the bowl; this should not be reduced too much, however, otherwise the sediment will not be dry enough.

Now that hydrophilic colloidal sludges are more and more frequently encountered, it has been necessary to develop decanters more suitable for the extraction of sediments which are less coherent even after flocculation:

Their main features are the following:

— ration $\dfrac{L}{D_1}$ between 2.5 and 3.5;

— cone angle α: 6 to 8°. For very difficult sludges (e.g., biological sludge from extended aeration), α may be reduced to 4°;

— facilities for varying liquid ring depths, first in countercurrent machines and then in cocurrent decanters. Reduction of the emergence zone facilitates sediment extraction, although its dryness is impaired. The ability rapidly to modify the depth of the liquid ring becomes an important parameter of operational flexibility where the sludge to be treated is of varying consistency;

— in sewage sludge treatment, the helical scraper should preferably be of the single-thread type. P generally lies within the range 0.15 to 0.22 D_1;

— the centrifugal forces applied range between 500 and 2 500 g; variable-speed motors are now increasingly being used, so that the operation of the decanter can be rapidly matched to the characteristics of the sludge suspension;

— The relative velocity V_R of the scraper may vary from 3 to 30 rev/min depending on the required performance (dryness, flow). It is this parameter that determines the sediment compacting time in the machine.

Fig. 358. — 43 cm (17") diameter centrifuges. Sludge dewatering at Biarritz, France.

Fig. 359. — 63 cm (25") diameter centrifuges. Intercommunal treatment plant at Vevey-Montreux, Switzerland.

B. Operating parameters.

These parameters are:
— feed flow;
— DS content of the raw suspension;
— the nature of the interstitial liquor (density, viscosity varying with the temperature);
— the properties of the sludge and, in particular, the way it has been conditioned.

The effect of these parameters can be roughly expressed as follows:

	DS output capacity	Extraction efficiency	Dryness of sediment
Feed flow	+	—	+
DS content	+	+	+
Conditioning	+	+	— or =

The quality of the conditioning is vital. With many types of sludge and, in particular, sewage sludge, quality is the determining factor in obtaining high extraction efficiency (95 % and over).

The optimum conditioning of a sludge which is to be centrifuged should lead to the formation of a bulky, heavy floc rather than a granular floc (which is better for filtration). Polyelectrolytes are therefore the best reagents for the purpose.

An initially high DS content is a favourable factor, but is must not exceed a certain limit beyond which the sludge cannot easily be moved by the screw. This limit would appear to be around 8 to 10 % dryness of the feed sludge. In the case of sewage sludge, too high a DS content may hinder diffusion of the flocculants because it is too viscous.

As in the case of the belt filter, it is vital to find the optimum polyelectrolyte. Laboratory studies of the sludge behaviour (estimation of floc size, measurement of shear resistance, and penetrometry) are seldom sufficient and it is usually necessary to conduct various tests at industrial scale.

C. Performance.

The table below gives the cake dryness values which can be expected with different types of sludge treated in continuous decanters.

The polymer consumptions and performance data given in the table correspond to an extraction efficiency exceeding 95 %. Higher dryness values than those given in the table can easily be obtained by limiting extraction to the densest materials (see fig. 354). This procedure entails a fall in extraction efficiency and an increase in the solids content of the centrate, which complicates the recycling of the latter.

Extraction efficiency is expressed in the form:

$$P = 1 - \frac{C_F (C_S - C)}{C (C_S - C_F)}$$

in which C $=$ DS concentration of sludge introduced
    C_S $=$ DS concentration of sediment extracted
    C_F $=$ DS concentration of centrate
The capacities stated refer to decanters with 40 or 45 cm bowl diameter.

Sludge type	Nature and origin	Polyelectrolyte kg (anhydrous) / t DS	Mass flow kg DS / h	Dryness value %
Organic hydrophilic	Mixed sewage sludge, fresh, VS $\leqslant$ 75 %	2-5	300-500	19-26
	Mixed [1] sewage sludge, digested, VS $\leqslant$ 60 %	2.5-5.5	250-500	17-26
	Sewage sludge from extended aeration or aerobic stabilization	3-6	200-300	14-22
	Biological sludge - dairy - VS $>$ 80 %	3.5-6	100-200	9-16
	Potato processing—lime-treated mixed fresh sludge	5-7	300-350	20-25
Oily hydrophilic	Refinery—flotation with polymer—oils 30 % [2]	1-2	350-600	30-40
	Treatment of soluble oils—acid and physicochemical breaking	0.5-1	400-700	40-60
Fibrous	Paper mill white liquors —flocculation (iron salt + lime)	1-2	200-300	20-30
Inorganic hydrophilic	Surface water clarification (Al salt + silts)	1-2	300-400	22-27
	Organic dyestuffs industry (iron salt + lime)	2-4	300-450	19-25
	Tannery (flocculation—Al salt)	1-2	200-300	18-22
	Distillery, prior physico-chemical treatment P $>$ 90 %	—	300-400	22-26

1. **Mixed** : mixture of primary sludge and excess activated sludge
2. Percentage referred to mass of DS.

The technique of centrifugation is particularly well suited to difficult sludges, which it dries well while giving almost total separation of solids. However, the resulting dryness values are never very high. The equipment is very compact and, because processing takes place in a sealed enclosure, very hygienic. Operation does not require either direct supervision or full-time labour. The major disadvantage is the inevitable wear of the helical scraper if abrasive material is present. However, considerable progress has been achieved in protective linings for components subject to erosion (in particular, by the use of tungsten carbide or ceramic materials), and screw life may exceed 4 000 hours before the worn parts have to be relined.

Nevertheless, the continuous decanter is not recommended for grainy and abrasive oxide-based sludge (e.g., water-repellent, inorganic sludge) from iron and steel plants. In addition, it is always worth while as a precaution to remove grit in a cyclone from sludge thought liable to contain it.

The continuous decanter is well suited to the treatment of oily sludge, although it is impossible to avoid the entrainment of a certain amount of oil into the centrate.

D. *Continuous decanter practice.*
A basic layout is shown in fig. 360.

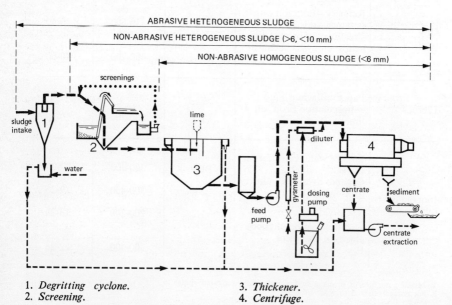

1. *Degritting cyclone.*   3. *Thickener.*
2. *Screening.*   4. *Centrifuge.*

Fig. 360. — *Diagram showing the working principle of an industrial-scale centrifuging plant.*

There are several types of machines; the bowl diameters of the commonest types may be as large as 900 mm (35"). In domestic sewage or industrial effluent treatment plants in which heterogeneous sludge is liable to arise, special precautions should be taken if only small machines are available (bowl diameters 25 to 45 cm). In particular, preliminary bar screening and grit removal must be thorough; bar screens and cyclones should precede the thickeners, where the sludge is not yet dense. Bar screening of the sludge may be avoided where the raw sewage or effluent undergoes additional very fine bar screening (8-10 mm).

Decanters are fed by variable-flow displacement pumps. Flocculation is carried out in-line, i.e., by direct introduction of the dilute polymer solution into the sludge line just upstream of the centrifuge.

The plant may be fully automated, and include the following features:
— safety devices sensitive to excessive friction between bowl and screw;
— automatic washing;
— closed loop control of reagent pumps by sludge pumps;
— closed loop control of proportioning pumps by influent sludge concentration.

The centrate is usually recycled to the plant inlet. This raises no problems if the sludge is properly conditioned, as the effluents then usually contain only 0.05 to 0.2 % suspended solids.

Centrifugation without polymer results in the recycling of centrates containing a high proportion of solid matter (1 to 3 % dry substance, very fine and very colloidal), eventually leading to clogging of the plant with fine particles, even if small seasonal clearances are undertaken. In this case the centrate must be treated before recycling. Failing this, it must be discharged.

1.2.3. ACCELERATED SLUDGE THICKENING BY CENTRIFUGATION

Some types of centrifuges can be used for sludge thickening without addition of polyelectrolytes. This type of treatment is applicable to very hydrophilic, organic or inorganic sludge whose DS content, even after long-period static thickening, reaches only 2 to 3 % and is only moderately improved by treatment in dewatering equipment.

Centrifugal thickening gives a sludge of 5-8 % DS content, resulting in a substantial reduction of the area and volume of treatment facilites (especially when compared with static thickening). However, it is important to note that this technique consumes a great deal of energy and that the buffer volume of the static thickener is eliminated. Several types of centrifugal machines are used for thickening:

● The *continuous decanter* with very small cone vertex half-angle (4°). The level of the liquid ring is adjusted until the sediment emergence zone is totally eliminated; semiliquid sludge can then be extracted from the conical part of the machine.

The disadvantage of these machines when used without polymers and where extraction efficiencies exceeding 85 % are desired is their low output capacity. In his field of application, the velocities used are very slow.

● The *disc type separator* (fig. 361) with thickened sludge discharge by evacuation through very small diameter (1-2 mm) calibrated orifices or by periodic opening of the bowl. This type of centrifugal machine can be used only with homogeneous sludge consisting solely of fine colloidal particles (drinking water treatment sludge or biological sludge from effluents containing only dissolved pollutants). Other types of sludge require very fine (less than 1 mm) double screening, which is frequently not very effective. The satisfactory operation of the equipment depends to a great extent on the proportion of coarse particles liable to clog the disc or nozzles. These machines operate at high rotational speeds.

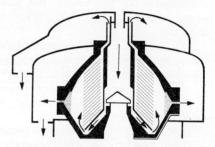

Fig. 361. — Disc type separator. Diagrammatic cross-section.

● The *basket centrifuge*, vertical or horizontal, with scraper and draw-off tube. This fully automated machine operates intermittently with alternate filling and emptying cycles. As with the continuous decanter, mass flows are very low where acceptable extraction efficiencies are to be maintained.

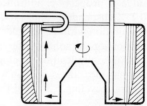

Fig. 362. — Basket centrifuge. Diagrammatic cross section.

The basket centrifuge is also used for dewatering, but in this case it is preferable to add polymers in order to obtain a worthwhile price/capacity ratio. These machines are less subject to wear than the continuous centrifuge because there is no helical scraper.

1.3. Mobile dewatering units

Filterbelt presses less than 2 m wide, continuous decanters (bowl diameter less than 52 cm), and small filter presses can be mounted on road trailers. The trailer then also accommodates the sludge feed pump, reagent station, and control console.

The advantage of this system is that it can serve a number of small treatment plants. It is essential for the latter to include sludge stabilizing facilities with sufficient storage capacity to cover the interval between visits by the mobile unit. Capital and operating cost studies show that a single mobile unit can satisfy the requirements of three or four plants serving a population equivalent of 2 000 to 10,000 within a radius of 50 km.

Fig. 363. — Mobile sludge-dewatering unit.

2. DRYING AND INCINERATION

Drying, a term generally reserved for heat drying, comprises the evaporative elimination of the interstitial water in the sludge. In the case of total drying, the final product is practically reduced to the "dry solids" level (both organic and inorganic).

Incineration not only totally eliminates the interstitial water but also involves combustion of the organic matter contained in the sludge. Incineration gives the least weight of residues of all processes: only ash remains, consisting of nothing but the inorganic content of the sludge.

Incineration obviously includes a drying phase, but since it uses the calorific value of the organic substance of the sludge, it always requires less energy than a process confined to heat drying.

For this reason heat drying is only worth considering if the end product can be reclaimed and marketed as fertilizer (waste water organic sludge) or within an industrial manufacturing process.

Heat drying and incineration are generally used only with sludge which has already been dewatered mechanically (by filtration or centrifugal separation), because it costs much less to remove the water by mechanical processes than by evaporation. In some cases, however, the water from liquid sludge can be evaporated directly (by drying or incineration). For example:
— when cheap heat (or heat otherwise lost) is available as in the case of certain organic chemical industries (oil refineries, petrochemicals, carbochemicals);
— when there is only a small volume of liquor to be treated, containing a very high proportion of pollutants, so that the elimination of any kind of treatment (biological or otherwise) of the liquid phase makes the relatively high cost of evaporation or direct combustion economically acceptable;
— in the case of mixed combustion with other waste, such as domestic refuse, acting as make-up fuel;
— when it is profitable to recover by-products by drying or incineration from liquors which cannot be concentrated by mechanical means. For example: black liquors from paper-mills and kraft pulp mills.

As thermal energy becomes more and more expensive in many countries, these processes are coming to be used less and less frequently.

Some fundamental thermal concepts are outlined in Chapter 33.

Drying may be direct, the sludge being brought into contact with the combustion gases, or indirect by means of hot air (combustion gas/air exchangers, hot water/air exchangers with boilers, etc.).

2.1. Heat balance

In the case of straightforward drying, all the heat comes from an outside source and fuel has to be consumed. In the case of incineration, the substantial amount of heat provided by combustion of the organic matter in the sludge, may, in some circumstances, be sufficient to supply all the thermal units needed to sustain combustion; the sludge is then referred to as **auto-combustible.** With sewage sludge containing 70 % volatile matter in the dry matter, the limit of auto-combustibility lies between 60 and 70 % humidity according to the type of incinerator.

The total heat of the evaporated water, expressed in kJ/kg, can quickly be calculated from the formula:

$$2\ 700 + 2.1\ (t - 100)$$
or
$$[1\ 150 + 0.9\ (t - 100)]\ BTU/lb$$

where t is the temperature in °C at which the hot gases leave the system. The heat balance is determined by estimating the following items:
● thermal units consumed:
— vaporization and superheating of the water;
— losses in the ash:

— sensible heat (combustion air of organic matter and make-up fuel);
— furnace radiation;
● thermal units provided by:
— the organic matter;
— the make-up fuel:
The flow of air must be as near as possible to the comburivorous power, but it is difficult to get below 20 to 30 % excess air.

The heat to be supplied for evaporation alone will range between 4 600 and 6 700 kJ/kg of water (1 980 and 2 870 BTU/lb) depending on the technology of the drying and incineration units (and in particular on heat recovery from the hot gases).

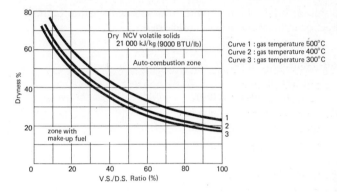

Fig. 364. — Auto-combustion curves.

Figure 364 shows the autocombustibility zone of the sludge according to its water content and $\frac{VS}{DS}$ ratio, for industrial-scale operation. The upper limit of humidity corresponds to a gas outlet temperature of 300 °C and the lower limit to a temperature of 500 °C.

The initial cost of a drying and incinerating unit is directly linked to the weight of water to be evaporated. The cost of constructing and operating the incinerating unit alone will therefore depend very largely on the extent to which the sludge has been previously dewatered.

2.2. Main components of a drying and incineration plant

The drier and the furnace are not the only components of a drying and incineration plant. The principal units are as follows:
— the sludge feeder (see page 766) :
● with or without regulating storage facilities with extractor,
● with or without crusher for breaking up cake lumps,

- with conveyor, Redler conveyor (endless chain type trough conveyor), displacement pumps, etc.;
— the drier and furnace proper, or one of these units only;
— the blower system or systems:
  - for operation at positive excess pressure or negative pressure,
  - conveying drying gas or air, combustion gas or air, or fluidization air,
  - with one or more admission points to the drying/incineration unit;
— the heat recovery unit, if necessary, for the combustion gases (comburent air heater or waste heat boiler);
— auxiliary heat source:
  - outside or inside the unit (with make-up and starting burners);
  - direct or indirect (in the case of drying);
— the drying/incineration unit control facility;
— the dust control facilities:
  - dry type (cyclone);
  - wet type (spraying, scrubber);
  - electrostatic;
— ash elimination:
  - continuous or intermittent;
  - hydraulic (by pumping suspensions of less than 200-300 g/l concentration);
  - dry, in enclosed containers;
  - wetted, in open containers.

2.3. Phases and methods of drying

When the interstitial water is removed from sludge in a drier at constant temperature, there are basically two drying stages (fig. 365):
— **A rapid drying** phase at constant speed (zone 1), during which the partial pressure of the liquid evaporating on the surface of the material is equal to the vapour pressure at the temperature concerned. The water migrates from the inside to the surface. All the capillary water is removed.
— **A slower drying** phase (zone 2), marking a change in the deep vapour pressure, due to the difference in temperature from the surface to the core of the material. The vapour formed is diffused through the top surface layer. During this phase, the extracted water has a much higher bonding energy.

With hygroscopic materials, for which water content is determined essentially by adsorptive or osmotic forces, zone 2 represents the typical drying pattern. This is frequently the case with sewage sludge (particularly if it has been pre-dewatered).

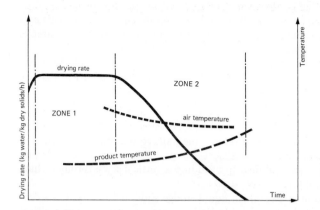

Fig. 365. —

When drying is complete, the material has the same humidity as the ambient medium in the drier. It is difficult, in practice, to reach a residual humidity of less than 8 % at the outlet of industrial driers. If the dried material is then stored in the open, even when protected from rain and snow, it is likely to take up more water according to the humidity of the atmosphere. The concept of "dry material" is always relative.

The degree of vaporization increases:

— as partial vapour pressure in the surrounding air decreases (and therefore as the temperature rises);

— as the surface area of the material increases;

— as the renewal of the contact of this surface with the air increases.

The temperature curves for the gas and the material vary differently according to whether drying is against the flow or with the flow; the same applies to the partial vapour pressures. Counter-flow drying brings the coolest gases into contact with the wettest material; its exchange principle and thermal efficiency are better than for parallel flow (with-flow) drying.

With-flow drying is not recommended when the wet material is sensitive to temperature (formation of crust, as on many sewage sludges); in the case of counter-flow drying, the dried material may not withstand the radiation from the hot gases. This is one of the reasons why, for sewage sludge, the hot gases in contact with the material at the end of a drying process must be at a temperature below 450 °C and the product at about 200 °C.

Because the available temperature gradients are higher, drying with the gases of combustion enables the size of the driers to be reduced as compared with those needed for hot air drying. This is the most widely used method for sewage treatment sludge; indirect drying is reserved for valuable products and those that are sensitive to sudden temperature variations.

Pre-drying of sludge is almost always essential in the case of incineration, if only so that the dry product can be raised to ignition temperature. This part of drying can be carried out in a unit separate from the incinerator (for example, using the gas of combustion as hot fluid) or in the same unit.

Most incineration furnaces can also be used for drying. Some furnaces, however, are designed for drying only. The following types of driers may be mentioned:

— Ball type driers, in which a current of hot air at not more than 150 °C passes upwards through a bed of balls through which the sludge is conveyed downwards. A screw conveyor returns the balls to the top of the bed.

— Flash driers, in which the influent sludge, after mixing with already dried sludge, is crushed and then sprayed into a vertical stack through which the hot (combustion) gases pass at 600-700 °C.

— Hot-plate type driers, in which liquid sludge slides along a (flat or cylindrical) surface heated on the inside by hot gases or steam. These driers require an effective scraping system to avoid fouling of the heating surface and for sludge evacuation.

This type of drier only partially reduces the water content.

In highly specific cases—concentrated solutions rather than suspensions or sludges proper—other concentration processes commonly used in industry may be applied, e.g., multiple-effect evaporation by pressure release or vacuum or by thermocompression.

Finally, **lyophilization** (freeze drying) would in principle be suitable for high-grade products—in particular, ones used in the food industry. In this process, the ice contained in a previously frozen material is sublimated direct in vacuo.

2.4. Principal types of furnaces

2.4.1. MULTIPLE-HEARTH FURNACE (fig. 366).

This has hitherto been the most commonly used type of furnace for sewage sludge. It consists basically of a set of plates (or hearths) over which the product descends in succession. It is moved from hearth to hearth by a set of rotary scrapers driven by a vertical centre shaft coupled to a drive unit outside the furnace.

These furnaces operate in the counter-current mode an therefore have high thermal efficiency. The outlet temperature of the gases is approximately 400 °C, while that of the wet sludge at the upper drying levels barely exceeds 70 °C, so that deodorisation (by afterburning using an additional burner at the gas outlet) is not normally necessary.

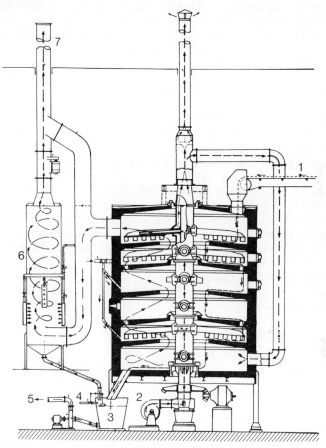

1. *Sludge input.*
2. *Cooling air.*
3. *Ash.*
4. *Added water.*

5. *To lagoon.*
6. *Cyclonic washer.*
7. *Combustion gases to atmosphere.*

Fig. 366. — Diagram of a multiple-hearth furnace.

The sludge is dried very gradually in the upper levels (to a water content of 50 to 60 % for sewage sludge) and then drops on to the combustion hearth where it is totally ignited in an oxidizing atmosphere at a temperature of 760 to 870 °C.

Combustion is completed in the lowest hearths, where a high proportion of the ash is recovered; the ash is cooled in contact with the cool combustion air and is normally tipped as dust into a water-filled extinction tank. Hence the combustion gases have a very low dust content, which can be reduced to less than 200 mg/m³ STP by simple wet scrubbing.

These furnaces, which have a very big heating and exchange surface, can be up to 7 m (23 ft) in diameter; the number of levels can vary from 4 to 12.

The fly ash content of the fumes normally requires only a small consumption of washing water.

This type of furnace can also be used as a drier, with the products taken off at the side, level with the lowest drying hearth.

A hearth furnace used as an incinerator can take fats, skimmings or bar screen debris which are then fed in at constant flow on to one of the hearths upstream of the combustion stage.

These furnaces have considerable heat inertia, with the corresponding advantages and disadvantages: heavy fuel consumption at start-up after a long stoppage, but low consumption after being stopped for only a day or two.

The sludge is fed in at the top from dosing hoppers preceded by conveyors. As for all other types of furnace the cakes from the filter presses must first be broken up, to give a steady feed into the incinerator.

Hearth furnaces may also be used:
— for regeneration of granular activated carbon (operating in a reducing atmosphere with added steam);
— for recycling calcic sludge (quicklime can be reclaimed by cooking sludge originating from massive lime treatment).

2.4.2. ROTARY KILNS

This is industry's most popular type of unit for handling the problems of separate or combined drying and incineration (mining industries, iron and steel, cement-making, etc.).

These driers are easy to operate and can be used for anything from 1 to 10 tons per hour of sludge which has previously been dewatered.

The cylindrical rotary drier, which is set at a slight angle to the horizontal, usually works against the flow (fig. 367); the combustion zone is lined with refractory material.

With a combustion temperature of 900 to 1 000 °C and 50 % excess air, the outlet temperature of gases from sewage sludge is around 300 °C.

The kiln can be used as either a drier or an incinerator.

The sludge is dried in the first half of the drier by a set of louvres and lifting angles.

Most of the ash cooled by the air of combustion is recovered at the lower end of the kiln and tipped into a skip for transport. The fly ash is recovered from the dust-extracting cyclone on the gas outlet. The gases are washed as they leave the cyclone.

The whole unit works at a negative pressure produced by the combustion-air

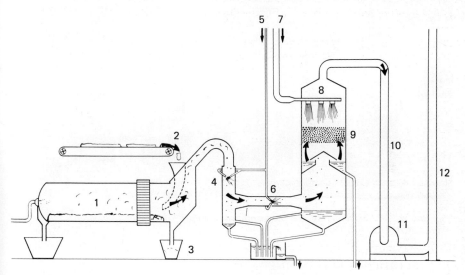

1. *Kiln.*
2. *Sludge loading.*
3. *Dust.*
4. *Saturation and wash sprinkler.*
5. *Wash water.*
6. *Venturi jet.*

7. *Cooling water.*
8. *Sprinklers.*
9. *Concentration pack.*
10. *Washed and cooled flue gases.*
11. *Fan.*
12. *Chimney.*

Fig. 367. — Principle of a rotary kiln incinerator.

fan. The drier is controlled basically by reference to the temperature of the outlet gases.

Bar screen refuse can be burnt in this type of kiln, preferably after being cut up or shredded.

With-flow operation is best for the combustion of greasy or oily sludge. In such cases, the kiln cannot be used as a drier.

Rotary kilns are simple and cheap both to install and to run as driers and incinerators.

The intrinsic thermal efficiency of these units is excellent, but external losses by radiation are high; with very intermittent operation, heat consumption increases substantially.

Linings should preferably be made of refractory steel owing to the rotation of the kiln.

Fig. 368 A. — Sludge drying and incineration in a rotary kiln. Water Pollution Control Works of BIARRITZ, France.

2.4.3. FLUIDIZED BED INCINERATORS

The fluidized bed incinerator offers the major advantage of having virtually no moving mechanical parts in contact with the hot gases. It removes all smell from the fumes but at the cost of a not insignificant loss of heat. The operation of the PYROBED furnace is illustrated in fig. 369.

It requires finer control than the hearth or kiln type of furnace because the chamber is smaller.

The fluidized bed technique, which is widely used in the chemical, mining and coal industries, has recently been adopted for the combustion of water treatment sludge. Combustion is complete, even in the presence of a small excess of air.

The sludge is fed into a bed of inert material—usually sand—held in suspension by a rising flow of air injected at the bottom through a distribution grating with a number of baffles. Bed depth at rest ranges between 0.50 and 0.80 m. Bed agitation facilitates sludge breakup.

Fluidization must be sufficiently intense to allow rapid heat dispersal, but must not cause sand entrainment. For this reason the air flow through the grating can be varied only within narrow limits.

Combustion of a high proportion of the dried fines and volatile solids is completed in the freeboard zone above the fluidized bed, where the temperature is in the region of 900 °C—i.e., 100 to 200 °C higher than the bed temperature.

The higher the moisture content of the sludge to be burnt, the more evenly the sludge flow, and any additional heating, must be distributed over the bed surface. However, the fluidized bed method requires:

— recovery of all the burnt matter in the form of fly ash. This is achieved by cyclone treatment followed by washing, or by washing alone. The cyclone, which works under pressure, must be sealed. The wet washer delivers a suspension of ash containing 2 to 4 % DS which must be lagoon treated. Bulky non-combustible matter must not be introduced into the furnace; bar screen refuse is, for example, very difficult to dispose of;

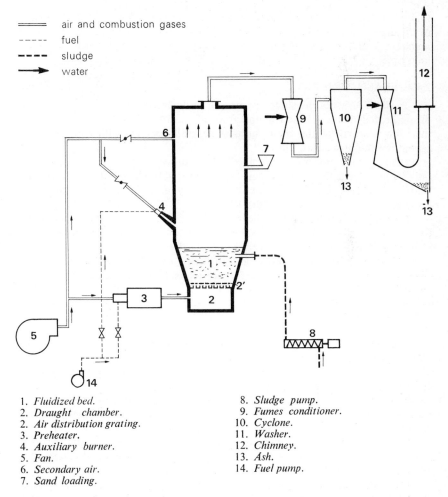

1. *Fluidized bed.*
2. *Draught chamber.*
2. *Air distribution grating.*
3. *Preheater.*
4. *Auxiliary burner.*
5. *Fan.*
6. *Secondary air.*
7. *Sand loading.*

8. *Sludge pump.*
9. *Fumes conditioner.*
10. *Cyclone.*
11. *Washer.*
12. *Chimney.*
13. *Ash.*
14. *Fuel pump.*

Fig. 368 B. — Diagram illustrating the principle of a fluidized-bed furnace without heat recovery from flue-gases.

— injection of air blown at sufficiently high pressure to offset head losses in the fluidized bed;

— recovery of the heat from the hot gases if high thermal efficiency is also required. A first-class heat exchanger is necessary; but when the outlet temperature of the hot gases is below 500 ºC, installation of the heat exchanger becomes uneconomical. The heat recovered is used to heat the furnace air feed which is brought to around 400 ºC, or in a boiler outside the system.

Fluidized bed furnaces can easily be shut down at night; they use very little fuel when restarting, partly because the heating chamber is small and partly because of the high calorific inertia of the refractory mass and of the mass of sand which maintain a high temperature over the whole unit when shut down even for a long time.

2.4.4. FLASH DRIERS

In these furnaces the sludge is sprayed in finely divided form into a current of air heated to the ignition temperature of its dry solids content. Owing to their low thermal inertia, these furnaces generally operate with high excess air quantities and at constant flow.

Being of only moderate thermal efficiency, this type of furnace is appropriate only for low sludge flows and where heat is available cheaply (e.g., oil refinery sludge). To reduce make up fuel costs, spray units have been designed to disperse sludge containing 15 % dry solids.

Direct incineration is also applicable to liquors containing highly concentrated or toxic pollutants; physicochemical or biological treatment of these liquors is sometimes expensive and may produce a considerable volume of sludge. This treatment may even prove impossible. In such a case, the solutions to be incinerated are often so saline that the furnace technology becomes very costly and exacting.

2.5. Pyrolysis

In pyrolysis, the sludge is first dried and then heated in a reducing atmosphere. The volatile solids are next distilled in the form of a highly combustible gas. The "fixed carbon", which is normally not volatile, can also be gasified, usually with the aid of make-up steam which facilitates the reaction.

The pyrolysis gases not used to maintain the temperature in the pyrolysis zone are burnt in an afterburner chamber; the waste heat recovered can be used for steam generation or energy reclamation.

Calorific efficiency can normally be increased where pyrolysis is used, although the installation is more complex. The air excess necessary in practice for combustion of the gases is less than that required for combustion of solid substances, for which as much as 100 % air excess may be required in conventional furnaces. This technique is potentially very valuable as energy costs continue to increase.

2.6. Flue gas treatment

The aim of treating the flue gases produced by a drying or incineration unit is to eliminate dust, smells, and the smoke plume.   The lower the initial dust content of the combustion gases, the more readily the **dust can be eliminated.**

Depending on the individual situation, the process involves one or more of the following phases: dry cycloning, wet scrubbing, or electrostatic dust removal. Wet scrubbing is effected by venturi scrubbers, water curtain devices, or countercurrent plate systems.

Electrostatic dust removal facilities are usually required, where the dust content of the gases discharged through the flue must be reduced to less than 100 mg/m^3 STP.

Odour elimination takes account of a subjective factor: the lower the gas temperature and the water content of the sludge at the outlet of the drying zone, the higher the risk of smells.

Driers or furnaces giving smooth and progressive drying and with properly ducted gas and sludge circuits are much less liable to cause smells than less sophisticated units such as countercurrent type rotary kilns.   In these more elaborate units, the water can be evaporated in the coldest zones, whilst volatile solids are evaporated in a zone in which they can be ignited before discharge because the gas temperature is still high enough for this purpose.

In the absence of sophisticated drying facilities, all the gases leaving the drier or furnace must be heated to at least 650 °C to eliminate odours.   In fluidized bed furnaces, this criterion is always satisfied without additional equipment. From the point of view of deodorization, pyrolysis also has evident advantages, since the gases as a whole can be raised to a high temperature in the afterburning chamber.

The reasons for smoke **plume elimination** are purely aesthetic.   It is sometimes demanded under very exacting environmental conditions, but is superfluous from the hygiene or health point of view.

A number of factors are responsible for the occurrence of a smoke plume:
— brightness of ambient air;
— air turbulence around the chimney;
— gas outlet velocity;
— absolute humidity of ambient air.
If all the gases are heated intensively before discharge, the flue gas temperature in the atmosphere will only reach its critical value (below which water droplets condense) when the gases have substantially dispersed, in which case the plume will not be visible.   However, this involves high energy consumption.

Additional preliminary cooling of the flue gases, to condense part of the water contained in them, is usually preferred; the gases are then heated again slightly so that the gases discharged have a relative humidity substantially below the satu-ration concentration at the relevant temperature.

2.7. Sludge and domestic refuse

Apart from the possibilities of composting for the combined treatment of sewage sludge and domestic refuse, sludge incineration (or drying) can usefully be associated with domestic refuse incineration.

If mechanically thickened or dewatered sludge is to be dried only, the refuse incineration furnace can directly or indirectly supply the heat necessary for the drier.   In this case the wet gases from the drier can perfectly well be fed through the furnace for complete deodorization.

If the sludge is to be incinerated, predried sludge (dried as above) can be fed into the refuse furnace, or alternatively combustion can be used.

As a rough guide it should be noted that with domestic refuse having an NCV of 7 500 to 8 400 kJ/kg (3 200 to 3 600 BTU/lb) for the crude product after removal of ferrous material, the quantity of fresh sludge arising from the corres-ponding population can theoretically be burnt without addition of fuel oil if the moisture content of the sludge is in the region of 90 % (assuming a $\frac{VS}{DS}$ content of 70 %).

Thickened sludge and domestic refuse have already been burned together directly, but adaptation of the incineration process raises problems.   If a sludge incinerator is used (for instance, a multiple hearth furnace), the refuse must first be finely crushed.   If a garbage incinerator is used, the sludge must be powdered so that it can be mixed with the refuse and the incinerator must be designed to prevent the unburnt sludge from passing too quickly into the ash-pit.   In order to burn refuse and mechanically dewatered sludge in the same unit, the matter fed into the incinerator must also be of uniform consistency.

In general, the operation of a plant designed for the simultaneous treatment of sludge and refuse will be more flexible if the refuse incineration unit is sepa-rate from the sludge drying (and/or incineration) unit.   However, as much heat transfer as can be required should be provided for.   The separated operation of the two facilities may prove useful in emergency or where they run at very different rates.

18

REAGENT STORAGE AND FEEDING

1. PRINCIPAL REAGENTS USED IN WATER TREATMENT

Many chemical reagents are used in the treatment of water. They are of two types:
- *specific:* coagulants (aluminium sulphate, ferric chloride, etc.), oxidizing and disinfecting agents (chlorine, ozone, etc.), aids (activated silica, polyelectrolytes);
- *general:* bases (soda, lime, etc.) and acids (sulphuric and hydrochloric acid).

For the purposes of this handbook, only a few data concerning the principal reagents will be given.

1.1. Specific reagents

- *Aluminium sulphate.* Used as a coagulant.

This reagent is used in either solid or liquid form. The solid forms are crushed slabs, grains and powder, with a theoretical formula of $Al_2(SO_4)_3 \, 18H_2O$. It is generally defined by its alumina content, expressed as Al_2O_3, which is approximately 17 %. The apparent density of powdered aluminium sulphate is in the region of 1,000 kg/m³.

Like the solid form, the liquid form, is defined by its alumina content, Al_2O_3; this concentration is usually between 8 and 8.5 %, or 48 to 49 % powder equivalent, or 630 to 650 g of $Al_2(SO_4)_3 \, 18 \, H_2O$ per litre of aqueous solution.

Since aluminium sulphate is the salt of a weak base (aluminium hydroxide) and a strong acid (sulphuric acid), its aqueous solutions are very acid; their pH varies between 2 and 3.8 depending on the sulphate/alumina molar ratio. This acidity has to be taken into account for the purpose of storage, preparation and distribution.
- *Basic aluminium polychloride* (B.A.P.C.) Used as a coagulant.

B.A.P.C. is a coagulant prepared by progressively neutralizing aluminium chloride with soda. During neutralization, aluminium polymers are formed homogeneously.

Aluminium chloride $AlCl_3$ has a solid or liquid form; it is generally used in its liquid form (density: 1.29 kg/dm^3; Al_2O_3 content: 11.4 %).

Since the solution is acid, plastic materials must be used for its preparation and distribution.

B.A.P.C. is prepared intermittently, cannot be kept and must be used quickly.

● *Ferric chloride* ($FeCl_3$). Used as a coagulant.

Ferric chloride has a solid and liquid form; it is mostly used in the latter orm.

The solid form has the appearance of a yellowish-brown deliquescent crystalline mass, with a theoretical formula of $FeCl_3$ 6 H_2O; it must be kept away from heat, because it melts in its water of crystallization at 34 °C.

The commercial liquid form contains about 40 % of pure $FeCl_3$. In order to avoid any confusion between the contents of the pure and commercial products, it is better to express the proportion of coagulant in Fe equivalent, i.e. 20.5 % for the solid form and about 14 % for the commercial aqueous solution. The aqueous solutions of ferric chloride are rapidly reduced to ferrous chloride, $FeCl_2$, in the presence of iron. This explains the product's strong corrosive effect on steel, and the consequent need to protect the tanks used for storage, preparation and distribution.

● *Ferrous sulphate* ($FeSO_4$ 7 H_2O). Used as a coagulant.

Ferrous sulphate takes the form of a green powder with an apparent density of about 900 kg/m^3. It is completely soluble in water and preparation of the aqueous solution presents no problems.

It contains approximately 19 % iron. The pH of a 10 % solution is about 2.8. Therefore, tanks for storage, preparation and distribution must be protected when they are made of metal. Tanks made of a plastic material are also of frequent use.

● *Ferric sulphate* $Fe_2(SO_4)_3$. Used as a coagulant.

This takes the form of a white powder, very soluble in water, with an apparent density of 1,000 kg/m^3. In aqueous solution, hydrolysis occurs and sulphuric acid forms. As in the case of ferrous sulphate, precautions must be taken against this acidity.

● *Ferric chlorosulphate* ($FeClSO_4$). Used as a coagulant.

This has the form of a dark, brownish-red concentrated solution, with a density of about 1,500 kg/m^3.

The commercial liquid form contains approximately 13 to 14 % of iron, or about 200 kg/m^3. This solution is acid (370 kg/m^3 of SO_4^{2-} and 125 kg/m^3 of Cl^-). It is therefore necessary, as with ferrous sulphate, to protect metal tanks used for storage, preparation and distribution, or to have them made of a plastic material.

● *Sodium silicate.* (Na_2SiO_3). Used as a floccuclant and conditioner.
This reagent is available in solid or liquid form, the latter being the more widely used because it is easy to prepare.
Liquid sodium silicate is defined by its mass ratio Rma of its molar ration Rmo and its density. The mass ratio is the ratio between the mass os silica and the mass of sodium oxide

$$Rma = \frac{SiO_2 \; mass}{Na_2O \; mass}$$

Alkalinity increases as the ratio falls: silicates are generally classified as neutral (Rma $<$ 3) and alkaline (Rma $\geqslant$ 3).
The molar ratio is the ratio between the number of moles of silica and of sodium oxide:

$$Rmo = 1.0323 \times Rma$$

The concentration of the solutions is determined by the percentage mass of n SiO_2Na_2O in relation to the total mass: n $SiO_2Na_2O + H_2O$.
The viscosity of a solution of sodium silicate increases with the $\frac{SiO_2}{Na_2O}$ ratio or with a drop in temperature.
The principal commercial sodium silicates used are defined by their concentration in degrees Baumé.

	35/37 °Baumé silicate	38/40 °Baumé silicate	40/42 °Baumé silicate
% silicate	24 to 26	26.5 to 28.5	27.5 to 29.5
Mass ratio	3.3 to 3.4	3.2 to 3.3	3.1 to 3.2
Density (kg/m³)	1320 to 1340	1350 to 1380	1380 to 1410
Viscosity at 20 °C (in mPa.s or cp)	40 to 70	100 to 200	180 to 300

It is very important to know the percentage of silica and the mass ratio for each type of sodium silicate in the manufacture of activated silica, because the mass of activated silica and the mass of acid, respectively, can then be established for the partial neutralization of the sodium silicate.
Sodium silicates must never be stored in galvanized steel containers, because the zinc is attacked and hydrogen is given off.

● *Sodium carbonate* (Na_2CO_3). Used to adjust the alkalinity titration.

Used as a white anhydrous powder, soluble in water, with an apparent density varying between 500 and 700 kg/m^3, according to the extent to which it is compacted. Its solubility is rather poor: about 100 g/l at 20 ºC. It is readily decomposed by most acids.

● *Sodium bicarbonate* ($NaHCO_3$). Used to adjust the complete alkalinity titration.

This is used in the form of a powdered solid, having a density of 800 to 1 200 kg/m^3, depending on the extent to which it is compacted. Its solubility is rather poor (96 g/l at 20 ºC).

● *Polyelectrolytes.* Used as flocculation aids.

These are used in solid (powder) or liquid form, and are classified anionic, cationic (very variable polarity) or neutral polyelectrolytes.

Owing to the large number of polyelectrolytes and their specific nature, flocculation tests should be carried out before any selection is made.

Powdered polyelectrolytes are suspended at concentrations of 2 to 10 g/l during a minimum contact time of 30 to 60 minutes, and fed as suspensions at 0.1 to 1 g/l. The length of time that suspensions can be kept is less than a week.

Liquid polyelectrolytes are distributed in the same concentrations, expressed as dry products.

Their solubility varies a great deal and they are highly viscous (up to 10 Pa/s (100 poises) at concentrations of 5 g/l). Apparent density varies from 300 to 600 kg/m^3.

Polyelectrolytes generally attack unprotected steel.

● *Potassium permanganate* ($KMnO_4$). Used as a disinfectant and oxidant of manganese in water.

It is used in the form of a powdered solid, having a density of 800 to 1,200 kg/m^3, depending on the extent to which it is compacted. Its solubility is very poor: 5 g/l at 20 ºC after a contact time of 15 minutes and 30 g/l at 20 ºC after an hour.

At 8 ºC, it takes 3 1/2 hours to reach the latter concentration (recommended figure for average mineralization, TAC = 20 º Fr. approx.).

It attacks ferrous metals and tanks must be made of a plastic material, or lined with ebonite if made of steel.

● *Sodium chlorite* ($NaClO_2$). Used in conjunction with chlorine or hydrochloric acid to form chlorine dioxide (disinfection).

It is used in the form of a powdered solid, containing 50 to 80 % $NaClO_2$, or in liquid form, containing about 24 to 25 % $NaClO_2$ (300 g/l). Its solubility is good: approximately 550 g/l at 20 ºC.

Both liquid and solid forms must be stored in inert materials: polyvinyl chloride, polythene, glass, stoneware, porcelain or molybdenum stainless steel. The powdered product must not come into contact with reducing materials, especially organic materials: fabric, paper, wood, etc. (risk of explosion or combustion). Non-diluted sodium chlorite must never be mixed with a concentrated acid.

● *Sodium hypochlorite* (NaClO). Used as a disinfectant.

Commercial Javel extract measures 47 to 50 ° on the Gay-Lyssac chlorometric scale, corresponding to about 150 g/l of active chlorine equivalent. It must be diluted with soft water in order to prevent precipitation of insoluble carbonates. Javel extract is an unstable product which is decomposed by heat and light; it must therefore be protected from all sources of light.

Many metals, even in trace form, act as catalysts in its decomposition. It is absolutely necessary to prevent it coming into contact with ordinary or stainless steel, copper, nickel, manganese, cobalt and their alloys.

● *Calcium hypochlorite* (Ca(OCl)$_2$). Used as a disinfectant.

This is generally used in solid (powder) form. It may contain as much as 92 to 94 % of Ca(OCl)$_2$, corresponding to about 650 to 700 g/kg of active chlorine. Its apparent density is approximately 1,000 kg/m^3. Its solubility is very poor.

● *Sodium hyposulphite* (Na$_2$S$_2$O$_3$). Used to reduce chlorine and its oxygenated derivatives.

Used in the form of an anhydrous powder, having a density of 1,000 to 1,200 kg/m^3, depending on the extent to which it is compacted. It is very soluble in water: about 700 gl/l at 20 °C.

It aqueous solutions are slightly alkaline; it is decomposed by acids, releasing SO$_2$ and depositing sulphur.

● *Sodium bisulphite* (NaHSO$_3$). Used as a reducing agent, transforming cupric salts into cuprous salts, and acting on many organic substances which it destroys and discolours.

Supplied industrially as an aqueous solution, emitting an odour of sulphur dioxide, it is defined by the percentage of sulphur dioxide which it is capable of releasing. Usually, typical solutions contain 300 g/l of SO$_2$. Its corrosive action attacks iron and many grades of steel. Therefore, lined steel or plastic materials must be used for containers employed in its preparation and distribution.

● *Caro's acid* (H$_2$SO$_5$) (permonosulphuric acid). Used in cyanide removal.

Commercial Caro's acid is a colourless liquid with a density of 1.3 kg/dm^3; its H$_2$SO$_5$ concentration is 200 g/l and it contains more than 15 % NH$_4$. It crystallizes at around —25 °C. It is stored and distributed in plastic containers.

1.2. General reagents

● *Sulphuric acid* (H_2SO_4). Used for the regeneration of cation exchangers, pH correction and the preparation of activated silica.

Commercial sulphuric acids are solutions of varying concentrations of H_2SO_4 in water. They are defined by this concentration.

The acid generally used in water treatment is the 65/66° Baumé technical acid, which has an average density of 1 830 kg/m³ and contains between 92 and 98 % H_2SO_4 by weight. Its viscosity at 20 °C is 25 mPa/s (25 centipoises).

When diluted with water, a fairly considerable amount of heat is generated, and precautions must be taken when this is done. In particular, the acid must always be added to the water, never vice versa.

Sulphuric acids diluted in water attack metals and steels. Special steels or plastic materials should therefore be used.

Above a certain degree of concentration, sulphuric acid no longer acts as a strong acid, and may be stored in tanks made of ordinary low-carbon steel, and protected against the entry of damp air.

● *Hydrochloric acid* (HCl). Used for the regeneration of cation exchangers, pH correction, and the preparation of chlorine dioxide.

The acids used in water treatment are technical acids containing between 31 and 36 % HCl by weight and with a density of 1 170 kg/m³ at 15 °C.

A solution of hydrochloric acid is a strong acid which in the hot state attacks all common metals (giving off hydrogen) and in the cold state, the majority of them. Steel tanks must therefore be protected or certain plastic materials used.

● *Quicklime* (CaO). Used for neutralization, precipitation and carbonate removal.

Quicklime is used in water treatment in powder form. Its advantages over slaked lime area: lower cost, occupies less storage room for an equal quantity of calcium ions. But the caustic nature of the product means that precautions must be taken when storing and handling it.

Before use, it must be "slaked "by hydration in a mixing tank to obtain milk of lime. The reaction is exothermic. Its apparent density varies between 800 and 1 200 kg/m³. Quicklime must never contain less than 90 % calcium oxide; insoluble matter content (calcium carbonate, silica) must be less than 5 %.

The solubility of quicklime is given in the table on page 882. The use of lime in block or granular form requires complex apparatus.

● *Slaked lime* (Ca(OH)$_2$). Used for the same purposes as quicklime.

Flour of lime is used in water treatment in the form of powder obtained from the hydration of quicklime so that its affinity for water is chemically nullified. It is mainly composed of calcium hydroxide, magnesium hydroxide and impurities (calcium carbonates and silica). These powders are designated by their grain sizes, expressed as a screen number.

The corresponding screen numbers and particle sizes are as follows:

— 80 screen : particles less than 0.197 mm;
— 100 screen : particles less than 0.160 mm,
— 120 screen : particles less than 0.135 mm;
— 150 screen : particles less than 0.100 mm;
— 200 screen : particles less than 0.080 mm;
— 300 screen : particles less than 0.050 mm.

The apparent density of lime varies between 400 and 600 kg/m^3. Its solubility in water decreases with temperature (see table on page 882).

● *Caustic soda* (NaOH). Used for neutralization and for regeneration of anion exchangers.

Caustic soda is used in solid (blocks, flakes) or liquid form (lyes at various concentrations).

Soda flakes have an apparent density of 800 kg/m^3 and contain an average of 98 % of pure product. Caustic soda must be handled with care, and when making a solution a considerable amount of heat is generated: 1.066 kJ/kg of NaOH (0.457 BTU/lb) for a final concentration of 5 %.

Commercial soda lyes are defined by their NaOH content; they are readily crystallised when the temperature falls.

It is therefore advisable to heat the storage tanks or premises to prevent temperatures of crystallization being reached.

% soda concentration	Storage in metal tanks	
	Temp. < 50 °C	Temp. ⩾ 50 °C
⩽ 50 %	Bare steel, carbon content: 0.17 %	At 50 °C stainless steel is still suitable
> 50 %	Nickel	Nickel

For reagents not mentioned in section 1 of this Chapter refer to the alphabetical index.

2. STORAGE OF REAGENTS

The method of storing reagents used in water treatment varies according to whether they are powders, solutions, gases or liquefied gases.

2.1. Storage of reagents in powder form

The easiest way of storing reagent powders is in bags on a special storage floor. Because of the handling involved this type of storage is only suitable for small plants. At medium and large plants, these products are stored in containers, *hoppers or silos* which vary in capacity with the size of the plant and the length of time for which the plant has to work without fresh supplies; these silos are loaded in bulk from lorries or railway trucks.

The silos are made of metal, reinforced concrete or polyester reinforced with glass fibre and are round with conical bottoms; they are filled mechanically or, more often, by air pressure from a supply tanker the contents of which are fluidized and pumped by air pressure so that the reagent runs like a liquid into the silo. Under a variant of this method of pressure feed, the reagent is injected through a rotary-vane feeder into a low-pressure air circuit. In both cases, the air pressure is less than 1 bar and the rate of flow is about 500 m³/h (18 000 ft³/h). The reagent is carried to the top of the silo, and separated from the air, from which the dust is removed before it is released into the atmosphere.

A belt or bucket conveyor can of course always be used to fill these silos mechanically. But this type of equipment is not very suitable for handling dust-generating products like lime or activated carbon.

Some reagents tend to become compact and form arched voids when stored, so that extraction becomes difficult if not impossible. This difficulty can generally be overcome by two methods. First, the lower part of the silo has a number of judiciously-placed inflatable bags which are pressurized in turn to release the reagent from the walls and prevent the formation of voids. Secondly, the contents of the silo are fluidized by passing dry compressed air over the base; the fluidized product then flows without difficulty.

When there is less reagent and it can be stored in a metal hopper, the formation of arched voids is avoided by installing an intermittent mechanical vibrator, of suitable power for the size of hopper, on the conical or pyramidal part of the hopper.

The air which carries or fluidizes the reagent must have all dust extracted before being returned to the atmosphere. This is usually done by means of cloth filters at the top of the silos in a chamber which can be depressurized by a fan.

RANGE FOR USE OF MATERIALS
FOR STORAGE OF REAGENTS

Materials and type of lining / Reagents	Unlined concrete container	Steel tank			Plastic tank	
		Unlined	Lined		Epoxy, PVC, polythene, HD, polypropylene	Polyester, vinylester
			Epoxy, hypalon, polythene	Ebonite		
Sulphuric acid						
$H_2SO_4 \leqslant 60\%$			×			
$H_2SO_4 \leqslant 20\%$						×
$H_2SO_4 \geqslant 92\%$		×				
concentrated, normal temperature				×		
all concentrations					×	
Hydrochloric acid			×	×	×	
Soda NaOH < 50%, t < 50 °C		×				
Sodium silicate	×	×				
Aluminium sulphate			×	×	×	×
Ferric chloride			×	×	×	×
Potassium permanganate			×	×	×	×
Sodium hypochlorite			×	×	×	×
Sodium bicarbonate					×	×
Sodium chlorite					×	×
Polyelectrolytes					×	×

The **level of the reagent** in the silos can be checked in various ways. For example, a **freely-suspended motor which drives a blade** that can rotate in the reagent can be used. The presence of the reagent sets up an opposing torque which causes the housing of the drive motor to rotate; this rotation is detected by an electrical contact. In the absence of reagent, the housing returns to its normal position. **Capacitive systems,** which detect the permittivity difference of a dielectric, formed either by the product or by the air, can also be used to determine the minimum level in the silo. This level can also be detected by a **membrane** which is distorted by the weight of the product stored and operates an electric switch.

More complex systems are also used to give a continuous display of how much reagent is in the silo, either by means of **strain gauges** or by piezo-electric devices.

Other systems are based on **ultrasonic** waves and **gamma rays.**

Reagents stored in silos are extracted by means of **rotary-vane feeders, augers, slat-type or vibrator extractors, air slides or automatic valves.** A small-capacity storage hopper can be emptied manually through a simple slide-mounted damper-type flap.

Lastly, powder reagents can also be stored in sealed **containers** which are filled by the makers. These containers can be made of steel or synthetic rubber. This method of supply and of storage is especially suitable for small and medium-sized plants.

Fig. 369. — Reagent storage tanks and dosing pumps. Output of treated water: 11 700 m³/h (74 mgd). La Roche Water Works, NANTES, France.

2.2. Storage of liquid reagents

At small plants, liquid reagents are usually supplied and stored in carboys, jerricans or drums. At larger plants, they are delivered by road or rail tankers and transferred by gravity, air compressors or pumps to storage tanks which may need to be protected against the corrosive action of the reagent.

According to the nature of the reagent, storage tanks are made of:
— steel or concrete, with or without lining;
— some plastic material.

They normally stand above a lined concrete leakproof pit, with a capacity at least equal to that of the tank.

Storage tanks have some form of level recorder, ranging from a float and arm system with a pointer moving over a graduated scale, to the devices described on page 571 which also transmit the reading for remote display.

2.3. Storage of gaseous reagents

Gaseous reagents used in water treatment (chlorine, sulphur dioxide, ammonia gas) are compressed and stored in the liquid state in steel cylinders or tanks. These storage tanks are housed in special premises built and fitted as required by law.

In the case of *chlorine*, French laws are very strict; a circular of 1972 contains instructions for fixed storage vessels. A circular of 1977 lays down rules for mobile storage vessels (containers filled on the premises of the manufacturer).

These instructions stipulate the distance which must separate the storage site and residential areas, and also specify methods of neutralization according to storage capacity and the form in which the chlorine is fed. i.e. as gas or in liquid form.

If a neutralizing plant is required where chlorine is being stored, in order to deal with chlorine leaks, the chlorinated air to be neutralized is extracted by a fan and then expelled into the bottom of a tower in which a backflow of neutralizing solution (soda lyes alone or together with sodium hyposulphite) trickles across contact rings.

Any chlorine leak is so dangerous that all storage premises must be equipped with reliable leak detectors.

A leak of chlorine gas from a storage tank can only continue if enough heat penetrates into the tank from outside through the wall of the tank to maintain the heat of vaporization of the liquid chlorine at that rate of flow. Furthermore, as the pressurized gas expands when leaking out, the temperature drops and this tends to reduce the leak.

For these reasons, the storage tank must not be sprayed or immersed in water, except in the case of small chlorine cylinders where there is no neutralizing plant for chlorine leaks, and the only solution, if the cylinder valve is jammed, is to immerse the cylinder in a neutralizing solution (soda lyes alone or together with sodium hyposulphite).

3. PREPARATION OF SOLUTIONS AND SUSPENSIONS

In some cases, such as the regeneration of ion-exchange resins or the preparation of activated silica, the reagents (sulphuric or hydrochloric acid, soda lyes, sodium silicate) are sometimes dosed direct in concentrated form and then diluted to the required concentration. In other cases, the reagents are used in the form of dilute solutions or suspensions prepared from concentrated solutions or solid products.

● *Preparation of dilute solutions from concentrated solutions.*

The concentrated solution (fig. 370) stored in the tank (1) is transferred to a measuring tank (2) by means of pump (3). This measuring tank may have either a level-detecting device which stops the pump automatically when the required level is reached, or a variable-position overflow (4) which returns the liquid to the storage tank. The concentrated solution is then poured into tank (5) where it is diluted to the required concentration. Dilution may be assisted by means of a stirring device (6).

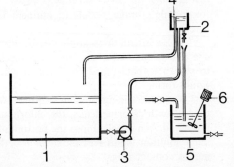

Fig. 370. — Preparation of dilute solutions from concentrated solutions.

● *Preparation of dilute solutions from solid products.*

When the reagent is supplied in the form of sparingly-soluble large grains, the apparatus shown in fig. 371 is used. The reagent is placed in a perforated basket immersed in tank (1) and dissolves gradually in the water. A stirrer (2) is used to produce a uniform solution after the reagent has dissolved. At large treatment plants, mechanical stirring is sometimes replaced by stirring with compressed air. The dissolving process can be accelerated by means of a pump which returns the solution from the bottom of the tank to the basket. At small treatment plants, and if it is easily soluble, the reagent is generally fed direct into the tank from bags.

The system shown in fig. 372 is also used. The reagent is taken from the storage hopper (1) by a rotary-vane feeder (2) and dropped into the preparation tank (3).

● *Preparation of suspensions.*

The equipment used to prepare suspensions is much the same as the types already described. The reagent is held in suspension by a slow stirrer in the preparation tank provided with anti-vortex partitions. This method is used to prepare **milk-of-lime** from slaked lime.

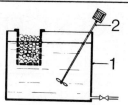

Fig. 371. — *Preparation of dilute solutions from immersed solid products.*

The time needed to slake quicklime and put it into suspension depends on the concentration required. For example, it takes 1 ½ min to 2 min to slake quicklime at 25 % concentration and 15 to 20 min at 10 to 15 % concentration. This explains why it is better to slake lime at 20-25 % concentration (at a mixing temperature of about 70 °C) and then to dilute it to the required concentration.

● *Mixing tank for preparing solutions and suspensions.*

The mixing tank shown in fig. 373 is used to dissolve reagents such as aluminium sulphate, sodium bicarbonate and polyelectrolytes and to put difficult products such as powdered activated carbon and lime into suspension.

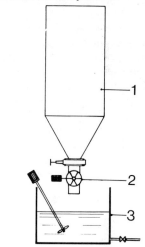

Fig. 372. — *Preparation of dilute solutions from solid products stored dry.*

The product is fed through the channel (1), drops into the vessel (2), and is sucked to the bottom by the turbine (3). The drop in pressure caused by suction prevents dust from forming as the reagent falls (especially in the case of activated carbon). The suspension is recycled with a large quantity of water, to ensure a thorough mixture and sufficient speed in the tank to avoid the formation of a deposit. The solution or suspension is distributed by gravity, a feed pump or a pump immersed in the tank.

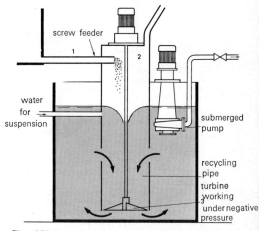

Fig. 373. —

4. REAGENT FEEDING

4.1. Distribution in liquid form

4.1.1. DOSING PUMP FEEDING

Dosing pumps are reciprocating displacement pumps; flow can be adjusted by modifying cylinder capacity or speed (working periods).

They can have a plunger piston, a mechanically-operated diaphragm, or a hydraulically-operated diaphragm. They are defined by their flow-rate, their maximum operating pressure, their accuracy, the nature of their hydraulic system (piston-operated or diaphragm feeder), and the nature of the materials of the feeder. When several of these pumps are interconnected mechanically in a multiple assembly, they can pump several liquids in proportionate amounts.

● *Piston-operated dosing pump:* this type is very accurate but precautions must be taken when it is used for abrasive or particularly corrosive products (sodium silicate, ferric chloride).

Depending on the type of pump (diameter of the piston, characteristic curve and working periods), the flow-rate can vary from some tens of millilitres to as high as several thousands of litres per hour per metering head.

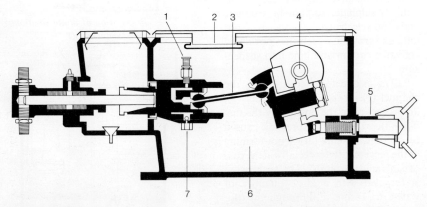

1. *Lubrication circuit safety valve.*
2. *Flexible membrane.*
3. *Pressure-lubricated drive mechanism.*
4. *Worm drive shaft.*

5. *Micrometric adjustment of stroke.*
6. *Lubricating oil.*
7. *Magnetic strainer.*

Fig. 374. — MILROYAL piston-operated dosing pump.

● *Dosing pump with coupled diaphragm :* this is less accurate than the pump described above, and its flow-rate seldom exceeds 20 l/h.

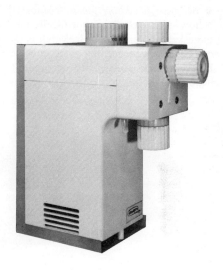

Fig. 375. — Pump with coupled diaphragm types M 18, M 28 and M 38.

● *Dosing pump with hydraulically-operated diaphragm:* this type is very accurate, but slightly less so than the piston-operated pump; it is used for corrosive, toxic, abrasive, polluted or viscous liquids.   It can have a single or double diaphragm. The flow-rate of this type of dosing pump can reach 2 500 litres/hour per metering head at high pressures.

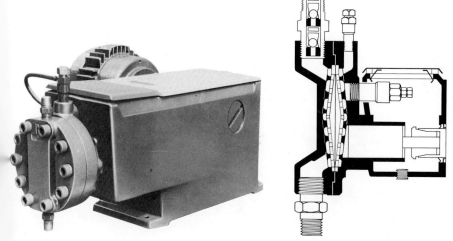

Fig. 376. — Dosing pump with diaphragm MILROYAL type.

● *Bellows-type dosing pump:* because the materials used in the manufacture of bellows-type feeders are chemically inert and because of the absence of any seal or gasket (giving absolute leak-tightness), this variant of the piston-operated pump can be used in the feeding of solvents, liquefied gases or highly corrosive products. Its accuracy is good and its maximum flow-rate can reach 1 500 l/h per metering head.   Its delivery pressure is rather low (generally less than 5 bars).

Fig. 377. — *MILROYAL pump with polythene bellows-type doser.*

● *Mutilple assembly:* in certain cases, a single motor can be used with advantage to drive several pumps (2 to 6).   With this arrangement a certain ratio between the flows of the different pump barrels can be established.   This is of great value, particularly in the manufacture of activated silica when an accurate ratio must be maintained between the flows of the two products to be distributed.

Fig. 378. — *MILROYAL diaphragm-type five-head dosing pump.*

● *Dosing unit:* dosing pumps can be fitted on to storage or preparation tanks, which may have blade-type mixers or level-gauges, thereby forming a monobloc dosing unit that also includes the electrical control cabinet.

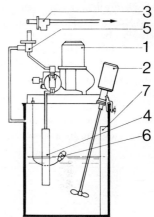

Figure 379 shows an installation of this type. The dosing pump 1 is located above the tank and draws through a stilling pipe 4 which has a diameter double that of the suction pipe. A level detector 6 ensures that the pump cannot run dry when the level is low.

Fig. 379. —

4.1.2. GRAVITY-FEED DOSING

● *DRC reagent feeders:* DRC feeders are particularly suitable for small and medium solution flows. The principle is extremely simple: a surface collector is lowered into a tank at a given speed.

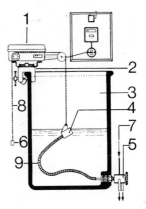

Figure 380 shows the basis layout of a DRC 1 feeder assembly. This feeder works like an electric clock. The time base is provided by a synchronous motor and an escapement mechanism controls the descent of the surface collector at a speed which can be regulated by changing the number of teeth in the escapement.

On the DRC 4 feeder, the electric motor is replaced by a clockwork winding system; this model can therefore be used to provide a constant flow of solution when no electricity is available.

Fig. 380. — Diagram showing the operation of type DRC 1 reagent feeders.

1. *DRC 1 distributor.*
2. *Distributor mounting.*
3. *Solution tank.*
4. *Surface collector scoop.*
5. *Three-way valve.*
6. *Counterweight.*
7. *Air inlet.*
8. *Stainless steel chain.*
9. *Flexible tubing.*

● *Orifice-type feeders:* orifice-type feeders can be used with both solutions and suspensions. They are designed for either constant flow or proportional to the flow to be treated.

C 110 feeder: a pump with a delivery greater than the flow to be distributed forces the solution or suspension through pipe 1 into a compartment 2 where the level is held constant by weir 3, over which the solution or suspension returns to the preparation tank via pipe 4 (fig. 381).

The amount of reagent used passes through the perforated stilling partition 5 and is fed in through an adjustable-section calibrated orifice 6 which is protected from clogging by a screen 7.

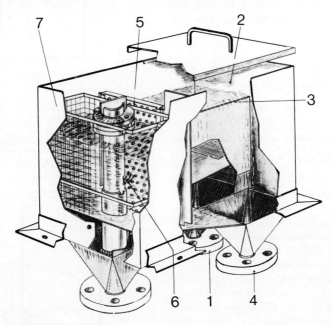

Fig. 381. — Diagram of a model C 110 feeder.

P 106 feeder: the P 106 feeder which is based on the same principle as the C 110 is illustrated in fig. 382. It delivers the reagent in proportion to the flow of water to be treated. In this case, the reagent passing through the calibrated orifice is not delivered direct to the point of use but runs down into an oscillating hopper 11 actuated by a hydropneumatic jack 9. In accordance with the pulses delivered by a proportional regulator such as the Chronocontact 402 (see page 565) the reagent is delivered either to the point of use in the plant via pipe 3 or ir returned to the preparation tank via pipe 4.

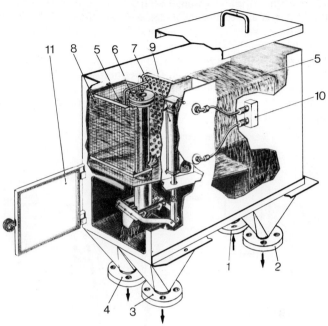

1. *Intake of liquid to be dosed.*
2. *Overflow return to preparation tank.*
3. *Dosed liquid outlet to point of use.*
4. *Liquid return to preparation tank.*
5. *Intake compartment of liquid to be dosed.*
6. *Adjustable-orifice feeder.*
7. *Stilling grating.*
8. *Protective screen.*
9. *Hydro-pneumatic jack.*
10. *Actuating fluid inlet.*
11. *Oscillating hopper.*

Fig. 382. — Diagram of P 106 feeder.

4.1.3. DISPLACEMENT FEED

● *Displacement feeder:* suitable for slightly-soluble reagents such as alum, fused sodium carbonate, certain polyphosphates, etc.

This system of measurement and feed is not particularly accurate, as the concentration of the solution can vary according to the quantity of reagents left in the tank. It cannot be employed for very soluble reagents such as aluminium sulphate.

● *Saturator:* the saturator is an instrument used to prepare and feed a saturated lime solution known as lime-water.

● *Static saturator:* with this equipment a saturated solution can be obtained by passing water through a lime bed so that the water is in contact long enough to become saturated.

Milk of lime is prepared intermittently. It is fed by gravity or pumping into the lower part of the saturator through a pipe (5), after the level in the saturator has first been lowered and carbonate sludge and impurities have been discharged through the drainage valve (3).   The apparatus is generally reloaded with milk of lime every 24 hours.

The water to be saturated is introduced into the lower part of the apparatus (1) and its flow is regulated by a rotameter; the saturated water is drawn off at (2).

In the form of saturated lime water, the saturator can provide from 1 to 1.2 kg of CaO per hour and per square metre of useful top surface area (1.3 to 1.6 kg/(h.m^2) of Ca(OH)$_2$).

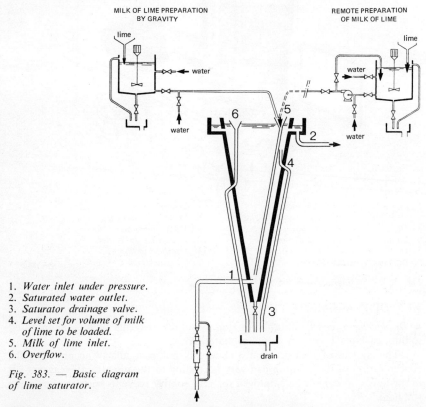

MILK OF LIME PREPARATION
BY GRAVITY

REMOTE PREPARATION
OF MILK OF LIME

1. *Water inlet under pressure.*
2. *Saturated water outlet.*
3. *Saturator drainage valve.*
4. *Level set for volume of milk of lime to be loaded.*
5. *Milk of lime inlet.*
6. *Overflow.*

Fig. 383. — Basic diagram of lime saturator.

● *Turbine saturator:* with this equipment a greater flow of lime-water can be obtained for the same top surface area than with the static saturator, i.e. 2.5 to 3 kg /(h.m^2) of CaO or 3.2 to 4 kg/(h.m^2) of Ca(OH)$_2$.

Milk of lime is prepared continuously or intermittently, but has to be distributed continuously.   It is introduced by gravity or pumping into the sludge-recycling draft-tube through the pipe (3).   Recycling is effected by a blade-type

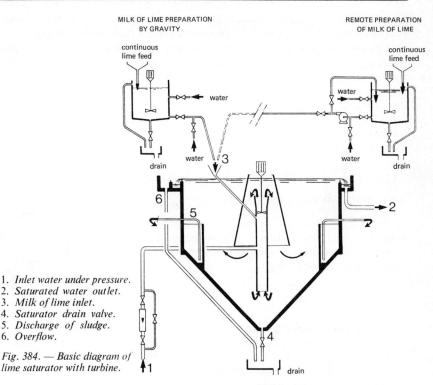

MILK OF LIME PREPARATION
BY GRAVITY

REMOTE PREPARATION
OF MILK OF LIME

continuous
lime feed

continuous
lime feed

water

water

water

water

drain

drain

1. *Inlet water under pressure.*
2. *Saturated water outlet.*
3. *Milk of lime inlet.*
4. *Saturator drain valve.*
5. *Discharge of sludge.*
6. *Overflow.*

Fig. 384. — Basic diagram of lime saturator with turbine.

drain

mixer located towards the top of the draft-tube which also receives the water to be saturated, and where water, milk of lime and carbonate sludge are thoroughly mixed together. The equipment operates as a sludge-blanket settling-tank, the level of the sludge blanket is regulated by the weirs of the concentrators which receive the sludge formed.

Sludge is usually discharged through pipes (5) and in exceptional circumstance (heavy sludge) through the drain-valve (4). The saturated water is drawn off at (2). When reagents like ferric chloride are used, even greater flow-rates can be obtained —as high as 5 to 8 kg of CaO per hour and per square metre of useful top surface area of the appliance.

Lime saturators can also be used to dissolve lime before saturation.

4.2. Measurement and distribution of reagents in dry powder form

Reagents can be measured and distributed in dry powder form by either volumetric or gravimetric appliances.

A feeder must be chosen carefully: account should be taken of the accuracy required, the type of product to be distributed and the required flow range.

It is very important to remember that manufacturers generally give the flow range of their feeders in terms of volume rather than mass, and that many factors, such as the extent to which the product is compacted, hamper the conversion between volume and mass. Added to this inaccuracy of the conversion is the inaccuracy of the feeder itself.

Confusion sometimes arises between the accuracy of distribution and the constancy of the feeder. A feeder with an accuracy of $\pm$ 3 % of the value distributed over the whole scale can be classified as an excellent apparatus.

4.2.1. VOLUMETRIC FEEDERS

● *Rotary vane:* this type of feeding is used in plants where great accuracy is not necessary. The vane can be controlled by a timed-contact or a variable-speed motor. Deliveries range from 50 to 1,000 l/h.

Fig. 385. — Rotary vane.

● *Revolving disc feeder:* the base of the feed hopper, which may not have a vibrator according to the reagent to be distributed, is fitted with a disc which rotates at constant speed. An adjustable angle plough above this disc deflects the reagent at a variable rate. The reagent runs at constant flow either into a vortex chamber under the chute where it is swirled round by water or into a tank, in some cases fitted with a stirrer, where a suspension or solution is formed.

This type of feeder is more accurate than the rotary vane. It is designed to distribute (at a rate of 10 to 1,000 l/h) standard reagents such as aluminium sulphate, lime, calcium or sodium carbonate, etc.

Flow is varied by means of a milled head which adjusts the cutting angle of the plough; the motor is set at a constant speed and operates continuously or intermittently. Alternatively, a variable-speed motor may be used.

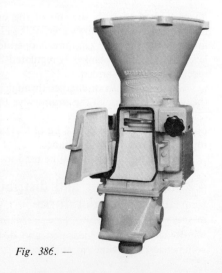

Fig. 386. —

● *Auger-type feeder:* with a feeder of this type, most powdered reagents, including those in very fine powder form, can be dosed with reasonable accuracy, provided the capacity of the hopper and the method of feed are both clearly defined.

The instrument consists of a feed hopper and a dosing auger fitted with a scraper arm which pulls the reagent to be fed through a calibrated pipe. The reagent has already been made uniform by a horizontal-shaft blade-type stirrer, which also prevents the formation of a dead zone at the entrance to the dosing auger.

Flow is varied either by directly adjusting the mechanical speed variator which controls the speed of rotation of the dosing auger, or by using a variable-speed motor, the speed of which is controlled by the flow of water to be treated. In the latter case, the variator motor control can also be adjusted at the same time by linking the speed ratio to the required hourly dose of reagent.

The feed hopper may be fitted with a vibrator or an oscillating system, the frequency or amplitude of which can be adjusted.

The flow-rate of the auger-type feeder variers from a few litres to a few cubic metres per hour.

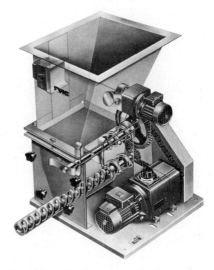

Fig. 387. — *Displacement feeder for reagent in powder from.* Type DP 350.

● *Standard plant for storage and volumetric feeding:* fig. 388 shows a plant for storing, feeding and distributing activated carbon in powder form. The product is discharged pneumatically from the lorry and conveyed to the top of the appropriate storage silo. The air and activated carbon are separated by means of a cyclone. The pumping air is then filtered and moistened so that it can be discharged to the outside atmosphere.

During distribution, dry air is forced in at intervals to fluidize the tapered part of the silo in order to ensure that the reagent flows freely and that the apparent density is kept as constant as possible. The reagent is distributed by an auger type feeder, supplied by a rotary vane governed by the high and low levels of the feed hopper. The activated carbon then falls into the recycling nozzle of a mixing tank operating at a pressure slightly below normal; a dosing pump or an immersed pump delivers the suspension to the point of injection.

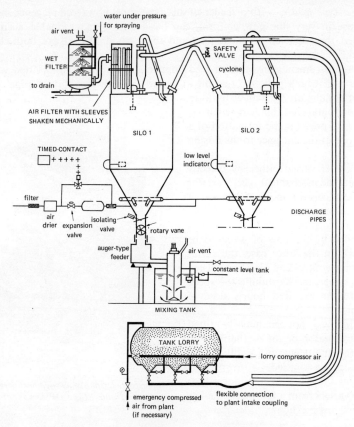

Fig. 388. — Typical storage and displacement feeding plant.

4.2.2. GRAVIMETRIC FEEDERS

The accuracy of dry feeders for reagents in powder form can be still further increased by using gravimetric feeders in which the powder to be distributed is weighed continuously. Any difference between the measured value and the required set value triggers action on the distributor tending to cancel the difference. Flow from such feeders is therefore unaffected by variations in the density of the powder distributed and can be adjusted with great accuracy, generally better than the precision normally required for water purification, for which the use of volumetric feeders is as a rule sufficient.

● *Loss-of-weight feeders (DEGREMONT model):* This feeder is designed to distribute an exact hourly mass flow of reagent by successive fillings of a weighing hopper which empties at intervals into a mixing tank. This type of feeder is very widely used and is suitable for the whole flow range of water treatment reagents. Fig. 389 shows an activated carbon dosing unit of this type. A control cabinet (8) emits a fixed number of pulses initiating filling and emptying cycles of the hopper (4). Figures for the flow of water to be treated and the dose of reagent are fed into a computer (9) which multiplies them, and by means of the control cabinet, fixes the weights of reagent to be distributed. Filling the weighing hopper (4) is a very short operation, whereas emptying is a long process, in order that the concentration in the mixing tank (7) is as constant as possible.

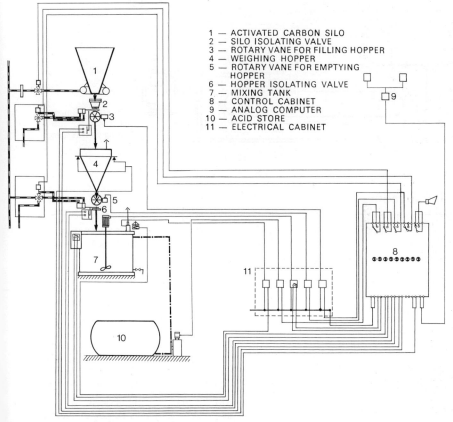

1 — ACTIVATED CARBON SILO
2 — SILO ISOLATING VALVE
3 — ROTARY VANE FOR FILLING HOPPER
4 — WEIGHING HOPPER
5 — ROTARY VANE FOR EMPTYING HOPPER
6 — HOPPER ISOLATING VALVE
7 — MIXING TANK
8 — CONTROL CABINET
9 — ANALOG COMPUTER
10 — ACID STORE
11 — ELECTRICAL CABINET

Fig. 389. — Principle of drawing off and dosing activated carbon in powder form.

● *Belt feeder:* this type of feeder consists of a belt fed by a hopper; its speed generally controlled by the flow of water to be treated, is variable. A blade of adjustable height regulates the thickness of the reagent layer, according to rate of treatment. A weighing device under the belt checks whether the weight of reagent corresponds to the product: flow of water to be treated multiplied by reagent dosing rate displayed. If there is a deviation, the height of the layer is automatically corrected.

While this feeder, in principle, is perfectly suitable for certain reagents, for others, such as activated carbon, it must be ruled out because it is not sealed tight. In general it is difficult to maintain.

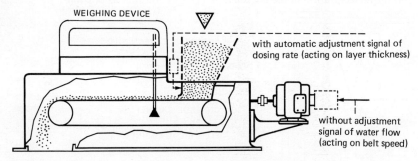

Fig. 390. — Belt feeder with or without weight adjustement and control.

4.3. Feeding of reagents in gaseous form

Appliances for feeding the various gases used as reagents in water purification are basically the same for all types of reagent. As chlorine gas is the most widely used, the following sections all relate to chlorine feeders.

4.3.1. CHLORINE SUPPLY TO DOSING EQUIPMENT

Chlorine is supplied by the chemical industry in liquid form in metal containers. The pressure within the containers varies with temperature, as shown in the graph on page 891.

The containers may be cylinders fitted at the top with a valve delivering chlorine gas at a pressure determined by the temperature of the liquid chlorine. As gas is drawn off, some of the liquid chlorine vaporizes, thus reducing the temperature of the remaining liquid. This drop in temperature must be offset by heat from outside and artificial heating of the cylinder may become necessary if large quantities of chlorine have to be drawn off. Such heating is rather dangerous because the temperature must not be allowed to rise too much. Consequently, a somewhat different procedure is followed when large volumes of chlorine are required.

Use is then made of horizontal cylinders, often called tanks and referred to in French legislation as "mobile containers" as distinct from "fixed tanks".

The type shown in fig. 392 is fitted with two valves so that either chlorine gas or liquid chlorine can be drawn off. The part of the cylinder containing gas is used when the rate of flow is not sufficient to require heating; on the other hand, when a very large volume of chlorine is needed it is taken from the part containing liquid chlorine, which is passed through a very narrow section immersed either in a water bath or in a bath heated artificially by thermostat and acting as an evaporator.

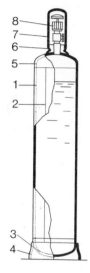

1. *Rolled sheet cylinder body.*
2. *Cylinder welding line.*
3. *Lower dished end.*
4. *Base.*
5. *Upper dished end.*
6. *Welded boss.*
7. *Valve cover.*
8. *Valve.*

Fig. 391. — Section through welded sheet cylinder.

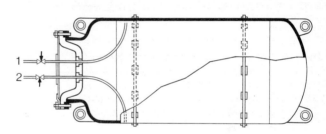

Outlet valves: 1. Gas. 2. Liquid.

Fig. 392. — Horizontal cylinder.

● *Standard plant layout:* Fig. 393 shows a plant for storing and distributing chlorine in liquid form and feeding it in gaseous form.

The tanks (1) are connected by the pressure-balancing circuit (3) in the gaseous state and by the distribution circuit (2) in the liquid state. If the isolating valves on the tanks and the evaporator close, an expansion cylinder on the circuit prevents excess pressure building up (expanding liquid chlorine may burst the system).

The evaporator (5) supplies via the resistor (6) enough heat to vaporize the liquid chlorine. Equilibrium is reached when the volume of evaporated liquid chlorine is equal to the volume of liquid chlorine entering the evaporator. The level then stabilizes in the evaporator cylinder.

The chlorine gas leaving the evaporator passes through a chlorine trap filter (7) and an expansion valve at operating pressure (10), before being distributed by a chlorinator (11).

In the event of excess pressure building up in the chlorine gas circuit between the evaporator and the chlorinator, a rupture disc (8) and a safety valve (9) will release the pressure into the neutralizing tower for chlorine leaks (14).

If there is a chlorine leak in the room housing the chlorine tanks or in the chlorinator room, a chlorine leak detector sets off an alarm and automatically starts the fan (18) which sucks the chlorinated air from the rooms affected and also starts the pump that delivers the neutralizing solution (soda, sodium hyposulphite). The chlorinated air passes through the contact rings (17) against the flow of the neutralizing solution, which is sprayed into the top part of the tower through perforated pipes (16).

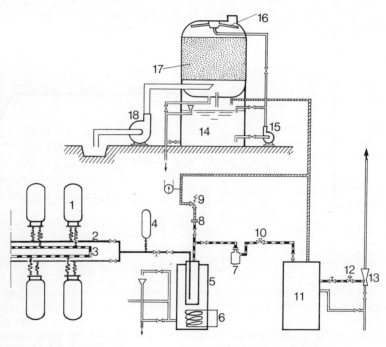

1. *Chlorine tank.*
2. *Liquid chlorine circuit.*
3. *Chlorine gas circuit.*
4. *Expansion cylinder.*
5. *Evaporator.*
6. *Heater element.*
7. *Chlorine trap filter.*
8. *Rupture disc.*
9. *Safety valve.*

10. *Pressure reducing valve.*
11. *Chlorinator.*
12. *Automatic isolating valve.*
13. *Suction ejector.*
14. *Neutralization tower.*
15. *Neutralizing solution pump.*
16. *Spray pipes.*
17. *Contact rings.*
18. *Chlorinated air fan.*

Fig. 393. — Storage and distribution of liquid-phase chlorine.

4.3.2. CHLORINATORS OPERATING UNDER A VACUUM

The use of vacuum equipment has been developed for the dual purpose of arresting the chlorine flow automatically if no water is available to dissolve the gas, and of preventing chlorine escaping into the atmosphere if a leak develops.

The vacuum is produced by an hydraulic ejector which also dissolves the chlorine in the water.

● *The compact chlorinator:* this type of appliance, which is particularly suitable for small chlorine flows, may be fitted to the top of the chlorine cylinder itself. However, it is sometimes attached to the cylinder by a short pipe.

Connections are always reduced to a minimum, thus avoiding the risk of leakages. On the other hand, if the chlorinator is fitted directly to the cylinder, special precautions must be taken when handling since the relatively fragile chlorinator has to be removed every time the cylinder is changed.

The principle of this appliance is shown in the form of a diagram in fig. 394. The vacuum produced by the hydraulic ejector (7) is regulated by the diaphragm-type of pressure-reducing regulator (4) as a function of the flow of chlorine required, this flow itself being controlled by the control knob (5) and monitored by the flowmeter (6).

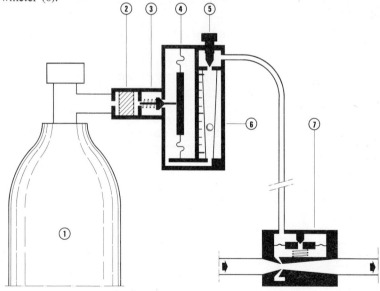

1. *Chlorine cylinder.*
2. *Filter.*
3. *Chlorine intake valve.*
4. *Regulating diaphragm.*

5. *Control knob.*
6. *Flowmeter.*
7. *Hydraulic ejector.*

Fig. 394. — *Diagram showing the principle of a compact chlorinator.*

In some appliances, the regulator consists of two diaphragm-type pressure-reducing valves working in series.

According to the vacuum created, the regulator diaphragm moves and displaces a needle constituting the chlorine intake valve (3).

When the appliance stops, a non-return valve incorporated in the ejector prevents the water rising in the chlorinator. In the event of accidental rupture or leakage in the pipe connecting the ejector to the dosing equipment, the vacuum above the diaphragm of the pressure-reducing regulator (4) is broken and the chlorine intake valve (3) closes.

Improvements have been made to the compact chlorinator and have been incorporated in most of the models; these are, in particular:

— the possibility of remote installation of the "chlorine flowmeter and control knob" assembly. This assembly can be installed in a control room while the chlorine cylinder and the remaining parts of the chlorinator (chlorine intake valve and pressure-reducing regulator) are installed in a room housing the plant or outside;

— the possibility of automatic changeover (by inversion) from one chlorine cylinder to another when the cylinder in service becomes empty;

— indicators showing "cylinder empty" or "chlorine low".

This type of "compact" chlorinator is suitable for feed rates of between 5 and 3 000 or 4 000 g/h of chlorine.

● *High-output chlorinators:* High-output chlorinators operate under a modulated vacuum.

The diagram below shows the basic layout of this chlorinator: the vacuum created by a hydraulic ejector (1) is regulated by a vacuum control valve (2) which ensures that a constant vacuum is maintained. The aspirated chlorine passes through an expansion valve (3), a flow-meter (4) and a flow control valve (5). With this type of chlorinator, the flow is adjusted by the valve (5) which can be governed by an electrical or a pneumatic device, so that full remote control is possible.

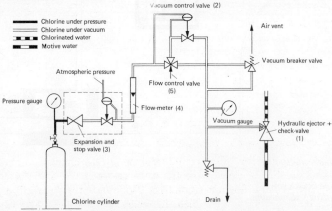

Fig. 395. — High-output chlorinator.

4.4. Monitoring the rate of reagent feed by the flow of water and the value of a physico-chemical parameter

● *Monitoring the rate of feed by the flow of water.*

Two systems can be used. The first and simplest operates as follows:

A flow meter in the supply pipe of the plant emits a pulse each time a set volume of water has passed through. These pulses, which are therefore emitted at a rate varying with flow (and the number or which is proportional over a given time to the flow), are received by a E 402 Chronocontact (fig. 396) which converts each pulse into a signal of constant adjustable duration. The reagent feeder operates automatically for the duration of the signal.

The actual operating process is a follows: on a dosing pump, the motor is started; on a P. 106 feeder the hopper tilts in the feed direction; and in the case of a Chlorinator, a valve on the expanded chlorine circuit or the ejector water circuit opens.

The second more complicated and therefore more expensive method involves continuous measurement of the flow of water to be treated and of the reagent or a parameter linked by a known law to that flow, combined with adjustment of the reagent feeder to maintain a constant ratio between the two flows.

Thus, in the case of a feed pump, a variable-speed drive motor is used and the parameter proportional to flow is the motor's speed of rotation measured by a tachometric dynamo. An electronic regulator then acts to adjust motor speed so that the ratio between the signals delivered by the flow meter and the dynamo remains constant

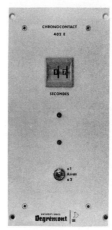

Fig. 396. — Chronocontact E 402.

when the former varies due to a variation of the flow of water to be treated.

In the case of a Chlorinator, the ratio is held constant by means of a servo-motor which automatically operates the regulating valve.

● *Monitoring the rate of feed by a physico-chemical parameter.*

The basic principles of this control are set out in the chapter on regulation. In most cases the flow of water to be treated is also variable, and this involves double control first by the flow and secondly by the value of the chemical parameter which is to be kept constant. When the feeder is itself fitted with two flow-regulating devices, one fairly simple method is to adjust one of the devices to the flow to be treated and then adjust the second, taking into account the difference between the set and measured values of the chemical parameter.

For example, if the flow of lime from a revolving disc feeder is to be governed by the pH of a varying flow of water to be treated, the first flow can be monitored by the second either by varying automatically with a Chronocontact the length of the interval between two feeds, or by adjusting the speed of the disc, which in that case is driven by a variable-speed motor. The difference between the pH of the water and its required value is used at the same time to correct the flow of lime by automatic adjustment of the position of the plough.

In the same way, in the case of a feed pump the flow of which can be regulated while running, reagent flow can be controlled by the flow to be treated by means of a Chronocontact and the effect of this proportional regulation can then be corrected by automatic adjustment of the flow-regulating device.

19

MEASUREMENT, MONITORING, CONTROL AND AUTOMATION

1. AUTOMATIC MEASUREMENT AND MONITORING PROCESSES IN WATER TREATMENT

1.1. General

The growing complexity of water treatment plants (where the techniques employed are moving closer and closer to those of chemical engineering) and the need to economize steadily-declining water resources have led to the continuous and automatic monitoring of treatment processes; these processes involve the automatic measurement of a number of parameters that can be grouped under the two main headings of common parameters and parameters which are specific to water.

The **COMMON PARAMETERS** which have to be measured are concerned mainly with flows, levels of liquids or solids, pressures and temperatures.

In a plant treating water for either human consumption or industrial use, knowledge of the **water intake flow-rate** is essential as a basis for the proper running of the plant; the **water outlet flow-rate** shows, by subtraction, the quantity of service water lost and hence forms the basis for systematic studies aimed at cutting such losses to a minimum.

During treatment a number of reagents are added to the water and these, by their physico-chemical action on the matter present in the form of dispersions or solutions, make the water fit for its intended purpose.

Reagent flows are measured so that they can be adapted to the flow of the water to be treated. Reagents are normally stored either in liquid form in tanks or in powder form in silos; automatic measurement of the levels held in stock shows continuously how long the plant can run without fresh supplies and provides a basis for planning future deliveries. **Measurements of level** are also taken on the various treated water tanks included in treatment plants.

Many **pressure measurements** are also taken; they include pump delivery pressures as well as pressures in pressure vessels, filters, air-water pressure tanks degassing units, etc. The proper running of a filter plant involves continuous knowledge of the degree of clogging present in the filter beds; this is generally calculated by measuring differential pressure or simply pressure in the case of operation of a constant-level open filter.

Measurement of the temperature of the water to be treated provides vital information for optimizing plant performance, because it is known that the speed of the chemical reactions, flocculation time and the degree of microbial activity all depend on the temperature of the water to be treated.

The more we know about a water at any particular moment the better it can be used, and this involves the continuous **automatic measurement** of a number of its **SPECIFIC PARAMETERS.** Such automatic measurement frees the specialist from the need to make tiresome routine analyses and considerably reduces the risk of error by cutting human intervention to a minimum. The main parameters measured are the turbidity of the raw or treated water, its resistivity, its pH value, the concentration of certain dissolved substances that must be kept within given limites, the optimal treatment rate for flocculation and the quantity of sludge produced by flocculation or biological treatment.

The continuous measurement of **turbidity** provides information concerning the degree of physical pollution of the water to be treated at any given moment and the quality of water delivered for human consumption or discharged at the outlet of a sewage treatment plant. This measurement is, however, not always enough to show the quality of the water or to give a full picture of the quantity of dissolved solids; in many cases the concentration of certain specific substances has also to be measured.

The continuous measurement of **resistivity** or inversely, of **conductivity,** indicates the salinity of a water when, for example, it is a high-resistivity demineralized water containing virtually only sodium chloride as an electrolyte. The figure is essential for the high-purity water used in high-pressure and high-vaporization-rate steam generators or in the manufacture of semi-conductors. Since every electrolyte has its own conductivity, the measurement of the resistivity of natural water can only be used to measure salinity if the relative proportions of the various dissolved salts remain constant.

Continuous measurement of the pH value of a water, mainly associated with measurement of its hardness or alkalinity, will indicate whether the treated water will cause either scaling or corrosion. It may also be used to determine automatically the amount of corrosion inhibitor which must be added to the water.

Continuous measurement of the oxygen in solution in the water provides a means, where effluent treatment is concerned, of monitoring the maintenance of aerobic conditions in biological treatment plants and, in the case of surface waters, of monitoring the conditions necessary to aquatic life.

The natural concern to discharge no harmful products into rivers, which are the usual outlets for sewage treatment plants, and not to deliver water polluted with these products for human consumption, leads to **automatic measurement of the concentration** of such substances in the water. At present, for example, concentrations of phenol, chrome, cyanide, detergent, etc., are commonly measured. The damage which can be caused at nuclear power stations by the presence of certain dissolved substances, such as silica, in the water fed to boilers and heat exchangers, and the need to maintain a given phosphate content in the circuits in certain circumstances, means that the concentration of these substances must be automatically and continuously recorded. Similarly, scaling can be avoided by measuring alkalinity and hardness to ensure that these parameters are kept within certain limits.

With the physico-chemical properties of a water measured automatically, as described above, it would be most beneficial if its bacteriological qualities could be quantified automatically. Since this is not possible, automatic measurements are taken to check that the water contains a minimum residual quantity of a disinfecting agent, e.g. chlorine, its derivatives, bromine or ozone.

The instruments concerned are not always used exclusively for automatic measurement; they are sometimes involved in the actual treatment process as monitoring devices. For example, a pH-meter is used to control the dose of an acidifying or neutralizing reagent; the instrument which measures the concentration of disinfecting reagent controls the apparatus which feeds in the reagent concerned, and so on. DEGREMONT have developed an apparatus which calculates automatically the **quantity of a coagulant** reagent which must be added to the water to obtain optimum coagulation and flocculation. The technical value of this device is that it ensures the production of water of constant, optimum quality; from the economic standpoint, it ensures that consumption of reagent never goes above the required level.

For the treatment of sewage by the activated sludge process, it is essential to know the quantity of sludge in an aeration tank and to keep it within the limits required to obtain the best results. This is achieved automatically with an instrument which **measures the percentage of sludge by volume** (Sedimometer).

Continuous measurement of the chemical oxygen demand (COD) makes it possible to evaluate the pollution load of chemical origin contained in the intake to the treatment plant. It is also a means of checking the efficiency of any treatment designed to remove this type of pollution.

1.2. Measurement and automatic monitoring of common parameters

1.2.2. Liquid flows are measured with conventional equipment such as turbine or piston-type meters, float-type flowmeters, and negative pressure devices connected to transmitters which may be either electrical or pneumatic. When the liquid, the flow of which is to be measured, contains a large amount of suspended solids liable to cause blockage, or when a high degree of accuracy is required over a wide range of flows, an **electro-magnetic flowmeter** is used. The operating principle of this type of meter is as follows. :

Water flowing through an insulating tube, subjected to a magnetic field of known constant strength at right angles to the line of flow, behaves like an electrical conductor moving through a magnetic field. An e.m.f. is generated in the water and a potential difference U is established between two electrodes immersed in the water at right angles to the magnetic field. This potential difference is given by the formula:

$$U = k.D.V.H$$

where D, V and H are respectively the diameter of the pipe, the velocity of the water flow and the strength of the magnetic field.

In practice, the magnetic field is usually generated by an electromagnet powered by an alternating-current supply to prevent polarization of the electrodes. The signal is fed to a high-input-impedance amplifier by the electrodes; as a result, the measurement is virtually unaffected by the resistivity of the liquid provided the latter is not too high.

The advantages of this flowmeter are that is does not create any head loss and that it is independent of suspended particles in the water. A further point is that, since the potential difference registered at the electrode terminals is proportional to the velocity of the fluid, the flow reading is linear and does not call for any ancillary equipment like the square root calculator needed for negative pressure flowmeters.

Ultrasonic flowmeters are based on the following principle: if two probes capable of emitting and receiving ultrasonic waves are immersed in a moving liquid in such a way that each is able to receive the wave signals sent out by the other, it will be found that there is a difference in the time of propagation depending on whether the sound wave is travelling upstream or downstream (differential velocity).

If v = the velocity of the liquid
   d = the distance between the probes
   c = the velocity of sound in the medium
   θ = the angle between the velocity vector and the orientation of the probes

then we can write the difference in the propagation time as follows:

$$\Delta T = \frac{2 \, v \, d \, \cos \theta}{c^2}$$

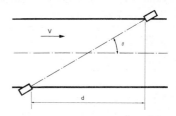

Fig. 396. — Diagram showing the working principle of an ultrasonic flowmeter.

This equipment also gives a linear reading of the flow rate and is capable of measuring flow rates of clean or slightly polluted liquids ranging from a few litres to several hundred cubic metres per hour.

1.2.2. **Liquid levels are measured** by numerous methods of which a few are described below.

Levels at atmospheric pressure can be measured by air purge methods.

In apparatus of this type, the flow of air is regulated by the valve R (fig. 397) so that it escapes in bubbles through the stand pipe submerged by a distance H below the surface of the water in tank C. Assuming the head loss in pipe T to be insignificant, the air pressure shown on the manometer M is equal to the height H of the the liquid above the pipe outlet. This manometer can be fitted with a transmitter to give a remote reading.

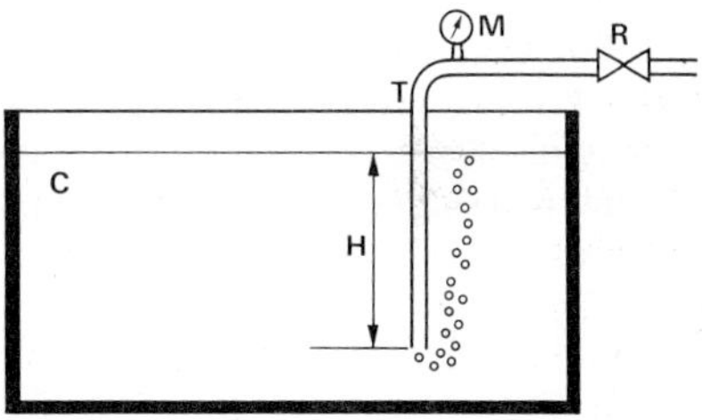

The present tendency is towards more universal systems which can be used to measure the levels of both liquids and powder solids. These include **diaphragm and capacitive systems.**

Fig. 397. — Measurement of liquid levels by the air purge system.

Diaphragm instruments work on the principle that a diaphragm is deformed in proportion to the load applied to it. Deformation of the diaphragm can be measured electrically by transmitting it to the slider of a potentiometer, the core of a differential transformer or to strain gauges. The electrical signal received is proportional to the force applied to the diaphragm. When a diaphragm is placed at the bottom of a tank containing a liquid or solid of known density, the level can then be determined.

Capacitive systems are also very valuable because there are no moving parts.

A capacitor, normally fed with an intermediate-frequency supply, is formed by two electrodes E_1 and E_2 in a vessel (fig. 398); one of these electrodes is plastic covered. Any variation in the depth h of the liquid alters the capacitance of the capacitor and hence, its impedance, so that, for a constant supply voltage, an electrical current results whose value represents the depth of the liquid or solid.

An **ultrasonic device for measuring levels** has no physical contact with the product; this type is therefore especially suitable when the product is corrosive. Because of its price, this type of equipment has so far been used only to resolve particularly difficult problems. Its basic operating principle is as follows:

An ultrasonic signal emitted periodically by a transmitter-receiver is reflected back from the free surface of the product of which the level is to be measured. The elapsed time be-

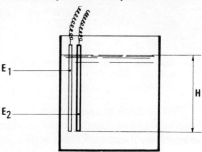

Fig. 398. — *Measurement of a level by the capacitance system.*

tween transmission of the signal and reception of the echo is proportional to the distance between the transmitter and the surface of the product, and the level is derived from this time lag.

1.2.3. **Measurements of head losses** in filters are equivalent to measurements of pressure.

For example, a differential manometer can be subjected to the pressures existing above and below the filter bed and the difference represents the head-loss. This is displayed on the dial of a pressure gauge or is transmitted to a remote display panel by means of a potentiometer, a differential transformer or strain gauges.

1.2.4. **The temperature of the water** is usually measured with a resistance-bulb-thermometer so that it can be transmitted for remote readings. Use is made of the fact that the resistivity of metals increases with temperature; the electrical current flowing through such a resistance supplied with a constant voltage will therefore be inversely proportional to the temperature of the surrounding medium.

1.3. Automatic measurement and monitoring of specific parameters

Various standard methods of analysis are applied automatically in the apparatus used for measuring and monitoring the specific parameters of the water. These methods comprise: nephelometric analysis (measurement of turbidity), measurement of resistivity (measurement of salinity), potentiometric analysis (measurement of pH), amperometric analysis (measurement of oxidizing agent content—i.e. chlorine, ozone, bromine), photocolorimetric analysis and titrimetric analysis (measurement of the concentration of certain substances dissolved in the water). Other types of apparatus, based on novel methods, are used to measure the concentration of sludge in the water or to determine the optimum dose of coagulant required for flocculation treatment.

The various types of apparatus can be classified under two main headings, namely automatic physical apparatus and automatic chemical apparatus, the latter carrying out one or more chemical reactions before taking a measurement.

1.3.1. AUTOMATIC PHYSICAL METHODS

A. Turbidity meters.

DEGREMONT make three types of turbidity meters for use according to the type of water to be checked; in all three cases, turbidity is determined by measuring the amount of light diffused by the Tyndall effect due to the presence of particles in the water.

The type EP TURBIDIMETRE (fig. 399) is a highly-sensitive instrument with a measurement threshold below 0.05 Jackson units; it is used to measure the turbidity of clean water. It comprises a constant-level tank (1) where the water can be deaerated; the tank (1) delivers a continuous constant flow to a measuring tank (2) equipped with two photo-resistive cells (3) immersed directly in the water. These cells are mounted at right angles to a light beam (4) focussed by a lens, one face of which is also immersed in the water under test.

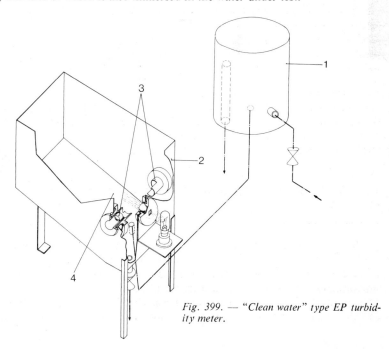

Fig. 399. — *"Clean water" type EP turbidity meter.*

The type **ES TURBIDIMETER**
(fig. 400) has no optical surface which
can be clouded by contact with the
water under analysis and is therefore
particularly suitable for analysing
very turbid water. It is based on the
principle of measuring turbidity at the
surface. A constant flow from an
overflow tank enters a second tank at
the bottom and leaves by overflowing.

A light beam touches the surface
at an angle such that a photo-resistive
cell located above the water and approximately at right angles to the zone of
reflection is not affected by either the
direct or the reflected beam. The light
diffused by the particles in suspension
in the water varies the resistance of the
cell in proportion to the number of
particles, thus providing a means of
measuring the turbidity of the water.

This type of turbidity meter has   *Fig. 400. — Type ES turbidity meter.*
been improved for use with liquids con-
taining large amounts of suspended solids which are liable to block water pipes.
Thus, the **DEGREMONT** type **ES TURBIDIMETER** has a device which
automatically drains and cleans the water pipes at intervals.

During cleaning, which can be set for various intervals, the measuring bridge
of the apparatus is of course not active.

These various types of apparatus can be fitted with a range of interchangeable
scales, so that a reading can be taken in the zone of greatest accuracy.

B. The "Coagulometer".

This apparatus (see figure 401) has been designed by **DEGREMONT** to
allow determination of the quantity of coagulating agent which has to be added
to a given water so as to bring about the optimum degree of flocculation.

The equipment works on the following principle: the tank in which the analysis is performed is fitted with two parallel, flat palladium electrodes, to which
a continuous electric field of approximately 10^3 V/m can be applied. A flat
beam of light, the thickness of which is determined by a slit some tenths of a millimetre wide, is made to pass through the liquid in a plane parallel to the electrodes
and in immediate proximity to the negative electrode. A measuring cell located
outside the tank on the opposite side to the light source converts the quantity
of light which it receives into an electrical signal.

If, when water containing colloids but no coagulating agent has been poured into the tank, a continuous electrical field is applied to the electrodes, the generally electronegative colloids will migrate towards the positive electrode, thereby clarifying the zone through which the beam of light passes.

When the same electrical field is applied to samples of water containing increasing quantities of coagulant, the migration velocity of the colloids will diminish until it reaches zero when their electrical charge has been neutralized. In these conditions, no significant change will be observed in the quantity of light transmitted.

Fig. 401. — Coagulometer.

Fig. 401 shows the portable version of the DEGREMONT coagulometer, into which the sample and reagent are poured by hand.

C. Resistivity Meters.

Pure water offers a certain amount of resistance to the passage of an electric current the magnitude of which, in the absence of dissolved salts, depends on the level of dissociation of the water molecules, which itself varies with temperature. Any dissolved salt lends the water a degree of conductivity which is measured in microsiemens/cm (μS/cm) and is specific to itself while varying as a function of its concentration. When a water contains several salts at the same time it is practically impossible to establish a relationship between its resistivity (the reciprocal of its conductivity) and its salinity. Only in the case of demineralized water, which generally contains only sodium salts in the ionized state, is it possible to establish a link between resistivity and salinity. By applying the same principle, it is possible to monitor automatically the concentration of an acid or basic solution used for ion exchanger regeneration.

The principle employed is simple, involving merely the measurement of the electric current flowing through the terminals of two electrodes of known geometry which are immersed in the water and subjected to a constant difference of potential. In practice, in order to avoid the phenomenon of polarization, an alternating potential difference is applied to the electrodes, the frequency of which has to be increased in proportion to the concentration of acids, salts or bases in solution.

Since the resistivity of water depends on the degree of dissociation of the molecules in solution, the most sophisticated apparatus are equipped to measure the water temperature by means of thermistors and to correct the resistivity measurement, giving a reading related to a temperature of 20 $^\circ$C. The more pure the water, the more complex do the devices needed for correcting the measured quantity become, because the relationship of resistivity to temperature then ceases to be linear.

When the liquid to be measured contains suspended solids, as in the case of river water containing sludge, this can accumulate on the measuring electrodes forming a resistant sheath which vitiates the measurement.

Frequent cleaning, even as often as once a day, may become necessary, despite which there is no guarantee that accurate readings will be obtained. To overcome this problem, the measurement can be based not on the potential at the electrodes but at two auxiliary electrodes immersed directly into the liquid.

In this case, the probe comprises 4 electrodes. Two large electrodes provide a constant, self-regulating current (any deposited coating merely resulting in an increase in the potential gradient between the current-carrying electrode and its outer surface). In these circumstances, the e.m.f. between two points a constant distance apart in the liquid will depend solely on the resistivity of the medium.

Fig. 402. — Multi-directional resistivity meter.

If two electrodes coupled to a high-impedance amplifier are placed at these points, any deposit will have virtually no effect on the reading.

By using conductivity meters of this kind it is possible to cut down by a factor of 10 the frequency with which the probes need to be cleaned.

By eliminating in large measure the polarization of the electrodes, these units provide an attractive means of performing measurement in solutions of low resistivity. Their use has made it possible, by measurement of resistivity, to monitor acid and basic solutions with concentration levels up to 20 %.

Fig. 402 shows a DEGREMONT multi-channel resistivity meter which allows continuous measurement by automatic switching on 6 channels. Each channel, with 2 interchangeable scales, is fitted with a circulation-type

Fig. 403. — Compact resistivity meter.

probe integral with an automatic temperature compensator and an adjustable threshold warning system.

The reliability and stability of modern electronic components has made it possible to cut down considerably the amount of adjustement.and maintenance required by the main measuring amplifiers used in water treatment.

These advances have led to the production of a type of compact unit in which signal-transmitting and indicating equipment is housed in the same cabinet suitable for direct attachment to the display panels of control rooms and operating posts.

The DEGREMONT resistivity meter shown in Fig. 403 belongs to this category of compact equipment. It can be used for measurements within the entire range from 10 Ω.cm to 1 M Ω.cm.

D. *pH meters.*

On the industrial scale, pH is always measured by electrometric methods, using two electrodes —a reference electrode and a measuring electrode; the reference electrode is immersed in a solution containing a fixed concentration of hydrogen ions. A partition, which allows the electric current to pass, separates the reference solution from the solution in which the measuring electrode is immersed to determine the pH value. A voltage, which is a linear function of the concentration of hydrogen ions in the solution, is then established at the electrode terminals. The pH value can be read by connecting these terminals to a voltmeter.

In practice, the electrodes are arranged to form a probe.

There are several types of **measuring electrode**; including the very accurate hydrogen electrode and the quinhydrone electrode (both of which are used in the laboratory only), the antimony electrode, which is a very robust industrial type, and the glass electrode which is of universal application.

The glass electrode consists of a small thin-wall bulb containing a silver electrode and a buffer solution.

The **reference electrode** generally uses calomel (HgCl) or silver chloride. The first is more commonly used and generally takes the form of a tube filled with a solution of potassium chloride above a quantity of mercury and calomel; the potassium chloride solution diffuses slowly into the liquid through a permeable membrane.

pH is therefore determined by measuring the e.m.f. of a cell in which one electrode—the glass electrode— has a very high internal resistance (several megohms); the e.m.f. attains several millivolts and it must be measured with no current to avoid polarization.

Two types of pH meters are used industrially. In one of them, the measurement is made potentiometrically; the e.m.f. of the cell is opposed by an equal e.m.f. of opposite sign and of known value, which inhibits any current passing through the circuit. In the second type of instrument, the electrodes are connected to a linear amplifier which has a very high input impedance and its output is connected to a voltmeter graduated in pH units.

The electrodes may be of the immersion or the circulation type and may be equipped with a mechanical cleaning device. Since the resistance of a glass electrode varies with temperature, pH meters using this type of electrode are usually compensated.

Fig. 404 shows the DEGREMONT pH meter on which the range of measurement can be expanded to one unit of pH plus or minus. This equipment belongs to the same series of compact units as the resistivity meter shown in Fig. 403

Measurement of the oxidation-reduction (redox) potential (E_H) also involves measuring an e.m.f., the measuring electrode normally being made of polished platinum. This value is determined in a nitrogen atmosphere.

Fig. 404. — pH meter.

When connected to **selective electrodes**, pH meters can be used to **determine the concentration** of certain ions such as sodium and chloride ions. The activity of sodium ions is measured potentiometrically using a glass electrode, while a silver-silver chloride electrode is used to measure the chloride ions.

E. Amperometers.

In water treatment amperometers are used industrially for continuous measurement of the concentration of oxidizing agents, employing a simplified method of amperometric analysis. The measuring cell, which receives a constant flow of the water to be analysed, has an inert but polarizable cathode, made for example, of platinum, and an anode which may be of copper, cadmium, silver, etc. In the absence of any oxidizing agent, the celle so formed is polarized and only a very small current can pass; its depolarization, and hence the current, is substantially proportional to the concentration of oxidizing agent reduced on the cathode. This method is used to measure concentrations of chlorine, bromine, ozone and oxygen in water. If the substance measured is liable to combine to form compounds, such as chloramines if the case of chlorine, automatic adjustement of the pH value is necessary to find the sum of free and combined chlorine. The measuring cells generally have a water-stirring device to stimulate migration of the depolarizer towards the cathode and thereby clean the electrodes continuously.

This device consists either of glass beads which are impelled against the electrodes by the flow of water or a motor-driven brush.

The disadvantage of such systems is that they measure the total effect of oxidizing agents and can only be used effectively when there is a variable concentration of one substance in solution. Under these conditions, the effect of any other substances present in constant concentration can be cancelled by adjusting the zero of the apparatus.

Under the heading of amperometers we may also consider the equipment used for the polarographic **measurement of the oxygen concentration.** The analyzer shown in Fig. 406 is an example. The probe of this unit

Fig. 405. — Amperazur.

comprises a gold cathode and a silver anode which are immersed in an electrolyte of potassium chloride gel isolated by a Teflon membrane from the liquid under examination and which are subjected to a constant potential difference. The dissolved oxygen diffuses through the membrane causing an oxidation-reduction reaction:

$$O_2 + 2\ H_2O + 4\ e^- \longrightarrow 4\ OH^-$$

which follows the electrolysis reaction:

$$4\ Ag + 4\ Cl^- \longrightarrow 4\ AgCl + 4\ e^-$$

The strength of the current is proportional to the partial pressure of the oxygen in the liquid phase.

F. Equipment for measuring the percentage of sludge (Sedimometer)

In effluent treatment, it is highly important to know the percentage of sludge contained in aeration tanks, as this information makes it

Fig. 406. — Oxygen meter.

possible to optimize sludge extraction rates. The DEGREMONT equipment illustrated in Fig. 407 provides for the continuous measurement of this percentage and for automatic adjustment of the sludge extraction.

In essence, this comprises a test tube on either side of which are arranged, parallel to a generatrix, a row of phototransistors and a bar-type lamp.

At regular intervals, a sample of sludge is placed in the test tube and allowed to settle out for a fixed time. At the end of this period, the lamp is switched on and those phototransistors which are not screened by the sludge are exposed to the light, whereupon they generate a signal proportional to their number. The equipment is, of course, fitted with a device for cleaning the test tube automatically after every measurement.

Fig. 407. — Sedimometer.

The unit is provided with two, mini and maxi, contacts which can be used to control sludge extraction.

G. Density meters.

The hydrometric determination of the mass of suspended solids in the activated sludge is a delicate operation which must be accurate to within one part in ten thousand (1 in 10^4). The actual value determined may not be the density of the suspension but the excess mass of the suspended matter in relation to the voluminal mass of water at the given temperature or, more precisely, of the suspensoid (which is in fact a solution of several mineral salts and gases in the water). This excess mass is virtually independent of variations in temperature or of the dissolved salt content of the suspensoid, whereas the density as compared with pure water at 4 °C varies considerably with these factors (fig. 408).

The measurement is taken by comparing the depths of two columns of liquid, consisting respectively of the sludge to be measured and the suspensoid, and exerting the same pressure on a common hydrostatic surface.

Two measuring tubes are branched at different levels into a vertical graduated tube, in which the sludge under test is circulating slowly; the suspended solids are removed from the measuring tubes by sedimentation, leaving the suspensoid.

The excess mass of the sludge is represented by a slight difference between the levels of the liquid in the two measuring tubes. This difference is equal to the excess mass multiplied by a factor which depends on the useful length of the graduated tubes (distance between the measuring tube inlets).

The difference in levels is measured by an ultrasonic level detector comprising

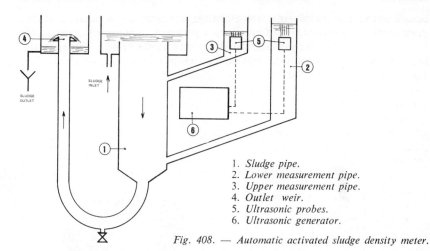

1. Sludge pipe.
2. Lower measurement pipe.
3. Upper measurement pipe.
4. Outlet weir.
5. Ultrasonic probes.
6. Ultrasonic generator.

Fig. 408. — *Automatic activated sludge density meter.*

basically two emitting and receiving probes which are immersed at a depth of a few centimetres below the liquid levels in the measuring tubes.

Each of these probes sends out ultrasonic signals which are reflected back from the surface of the liquid and the time taken by the signal to travel this distance is measured. The difference between the levels in the measuring tubes is proportional to the difference in propagation time of the ultrasonic signals.

On page 928 will be found the description of an item of equipment (Ponsarimeter) which can be used for rapid on-the-spot determination of the mass of suspended solids contained in activated sludge.

1.3.2. AUTOMATIC CHEMICAL METHODS

Automatic photo-colorimeters: with this apparatus, the preliminary operations for photocolorimetric quantitative analysis of certain substances dissolved in the water are carried out automatically. By means of these operations the element (the concentration of which is to be measured) is converted, by using suitable reagents, into a coloured compound, which is usually complex; the more intense the colour, the greater the concentration. The measurement consists of measuring the light intensity transmitted through the solution; according to Lambert-Beer's exponential law, this is inversely proportional to the degree of concentration. This method is now used for the completely automatic measurement of the concentrations of **silica, phosphate, phenols, detergents, iron** and **hydrazine** and to measure the **hardness of soft water** to within one tenth of one milliequivalent per litre, etc.

This apparatus is now so reliable that it can be used to control an automated water treatment process. For example, the DEGREMONT automatic photo-colorimeter illustrated in fig. 409 is currently used for continuous monitoring of the silica content of water leaving an ion-exchange demineralization plant.   It stops a demineralizing sequence and, if necessary, switches it to automatic regeneration if the amount of silica in the water exceeds the set warning value.   This apparatus has a number of automatic control devices which ensure maximum reliability. For example, it cannot give a wrong measurement if a fault develops in the measuring bridge, in the supply of water or in the feed of reagents required to form the coloured compound.

In the multi-channel version of this apparatus, each measuring line is scanned in succession.   Each channel is independent and can have its own measuring scale.

Automatic titrators measure a chemical parameter by titration.   There is a pre-set analytical sequence: a standard solution of a suitable reagent is gradually added to a sample of water of known volume; the reagent reacts twith the element (whose concentration is to be measured) to form a known compound.   The reaction ends at a certain pH value which is detected either with a standard pH meter or by a change in the colour of the water due to the presence of a previously-added indicator.

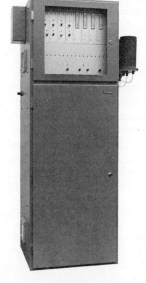

An example of a unit in which the end point of the reaction is detected by measurement of the pH is the DEGREMONT natrium hydroxide titrator (Shell Co. licence).

Detectors working on the principle of coulour-change are of two types:

— The **set point titrator** which auto-matically checks, without quantifying, that the value of a given chemical parameter does not exceed or fall below a pre-set value.   With this type of instrument a fixed amount of a standard titrating solution is added at regular intervals to a known volume of water containing a coloured indicator.

Fig. 409. — Photo-colorimeter.

A photo-electrical device checks whether a colour change takes place or not. This type of apparatus can be used, for example, to keep an automatic check on the running of a softener (check of hardness), of a carboxyl resin exchanger (check of TAC) or of a lime purifier (check of TAC or TA).

— The **measuring titrator** works on the same principle, and the quantity of titrating reagent added up to the end of the reaction is converted into the concentration of the element under test by an appropriate transducer.

Like the photo-colorimeters, DEGREMONT titrators are also made in a multi-channel version and have the same safety devices.

1.3.3. AUTOMATIC ANALYSING STATION

Monitoring the quality of surface waters calls for the measurement of parameters such as the pH value, resistivity, dissolved oxygen, temperature and turbidity. Information about the ammonia concentration is also useful.

The DEGREMONT automatic analysing station shown in fig. 410 performs the measurement of these parameters on a continuous basis. It has been specially developed for monitoring water intakes in rivers. The ammonia analyser, to the left of the picture, is an automatic colorimeter of the type illustrated in fig. 409. The other sensors are housed in one cabinet at the centre. The right-hand compartment accommodates all the electronic gear, the power supply, sequential automation system, measuring amplifiers, threshold comparators and multi-channel indicating and recording instruments.

The turbidity meter employed has been developed from the DEGREMONT ER type unit and is equipped with an automatic sequential cleaning mechanism. The resistivity meter is of the 4 electrode variety and is virtually unaffected by

Fig. 410. — DEGRÉMONT automatic analysing station.

fouling. Periodic cleaning of electrodes used to measure the pH value and the dissolved oxygen is effected by pressurized water spraying.

1.3.4. ICHTHYO-TESTING

Biological tests performed on living organisms complement the physico-chemical measurements carried out by automatic analysing stations of the kind described above. The behaviour of certain aquatic organisms, and particularly of fish, contained in a tank provided with an alarm system is used as a comprehensive method of monitoring the biological quality of surface water.

The DEGREMONT ichthyo-testing unit (from Greek ichthys = a fish) illustrated in fig. 411 comprises basically four long, narrow channels into which the water under supervision is made to flow in series at a constant velocity. A fish, preferably a rainbow trout which has a natural tendency to swim against the current, is placed into each of these channels.

Downstream from each channel is located a photo-optical barrier, immediately upstream of which are placed two electrodes.

When no pollution is present, the fish maintains its position in the upstream part of the channel. If toxic materials are introduced, the physical condition of the fish deteriorates, it is less able to combat the current and is borne towards the photo-optical barrier.

When a fish impinges on the barrier, impulses of a few volts amplitude are applied to the electrodes at intervals of about one second.

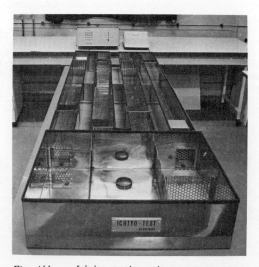

Fig. 411. — Ichthyo-testing unit.

Three possibilities then arise:

— either the trout has impinged on the barrier by accident and swims back to the upstream portion of the channel after receiving the first impulse;

— or the trout is tired and leaves the barrier only after having received 4 or 5 impulses;

— or it is dead and obstructs the barrier until the arrival of the operator.

All this information is recorded by a printer and alarms signalling "1 fish dead", "3 fish ill" and "3 fish dead" can be remotely transmitted.

2. AUTOMATIC CONTROL

2.1. General principles

Automatic control is achieved by means of a "controller" which establishes a given relationship between two physical quantities x and y, which are referred to as **coupled** if the effect of x on y is matched by a negative **feedback** of y on x, tending to cancel the effect of the action. x is the **controlled variable** and y the **regulated value.** If a value x_0, known as the set desired value, is imposed on x, any deviation between x_0 and x will have an action on y such that the value of y will alter to cancel this deviation. This forms a **closed loop** which will automatically return x to its set value x_0 whenever any external influence tends to change the value of x.

An automatic control unit comprises;

— a sensor to measure the controlled condition;

— a controller, which compares the value x with its set value x_0 and orders the action on the regulated value;

— a device for varying the regulated value.

To avoid influencing the measurement, the controller generally receives from an outside source the power required to couple the two values, and this energy is modulated by the signal from the measuring instrument. Consequently, if the controlled condition is too tightly coupled the system may be rendered unstable by sustained oscillations, a phenomenon referred to as **hunting.**

If the state of the system can be defined by a function which does not include the time factor, the state is **described** as **continuous.** When any parameter of the function changes in value, the function does not at once take on its new equilibrium value, and there is an interval during which this function varies with time, either exponentially or in accordance with a damped sine wave. During this period the state is said to be **transient.**

2.2. The four modes of control

2.2.1. The simplest form of **control** is the **on-off mode.** Applied, for example, to controlling the level h in a tank (controlled condition) from which a variable flow is drawn off according to consumption, this method of control involves a valve on the water intake (fig. 412), which is closed, and therefore cuts off the flow (regulated value) when the level exceeds the set value h by $\Delta\,h_1$ and which is open when, conversely, it is $\Delta\,h_2$ below the set value.   With this very simple form of control, the level cannot as a rule be kept sufficiently stable, except when the tank has a very large capacity in comparison with the flow.   In this case, performance is improved by relating response time to the deviation between the measured value and the set value, allowing a dead zone on either side of the set value where no change in the regulated value occurs.

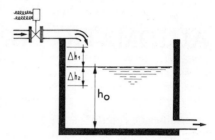

Fig. 412. —

2.2.2. Using **proportional control,** the extent of the action on the regulated value is proportional to the deviation between the value of the controlled variable at time t and its set value.   It can be shown mathematically that when a permanent disturbance occurs, the simple proportional mode is incapable of returning the controlled variable to its set value; a difference remains which can only be eliminated by adding integral mode to the proportional mode.   Control by proportional mode establishes a linear relationship between the controlled value and the regulated value.

The projection of the characteristic straight line for the controller on the abscissae representing the values of the controlled contidion (values of the regulated value on the ordinate axis) is known as the **proportional band width.**   It is generally expressed as a percentage of the complete range of the measuring unit. The width of this band is modified by varying the slope angle of the characteristic straight line of the controller.   The deviation between the set value for the controlled condition and the value recorded after a corrected disturbance, narrows with the proportional band but, if the band is too narrow, hunting occurs and the system becomes unstable.   This problem is overcome by adding integral mode to the control process.

2.2.3. This **method of control by the integral mode** consists of making the intensity of the action on the regulated value proportional to the integral in time of the deviations between the instantaneous value of the controlled variable and its set value.

This method of control can be expressed in the following mathematical form:

$$\frac{dy}{dt} = - kx$$

where y and x are respectively the relative variations of the regulated and controlled values, and the proportionality factor k is known as the integral control factor. Used in conjunction with control by proportional mode, the integral mode progressively eliminates the deviation left by purely proportional action between the value of the controlled variable and its set value. It involves, in practice, an automatic shift of the proportional mode band. Control by proportional and integral mode therefore gradually returns the controlled variable to its set value over a period of time which varies inversely with the integral action rate. Since the time factor is not involved in the proportional mode, and as the integral mode only takes effect progressively, it will be appreciated that a method of control confined to these two terms is incapable of correcting the effect of a sudden, violent disturbance. The necessary correction proportional to the rate of change of the disturbance is achieved automatically by using the derivative mode, which is also called differential control.

2.2.4. Derivative mode control consists of making the intensity of the action on the regulated variable proportional to the derivative (as a function of time) of the deviation between the instantaneous value of the controlled variable and its set value. It can be expressed in the following mathematical form :

$$y = - k \frac{dx}{dt}$$

In practice, this means narrowing the proportional-mode band-width at the moment of the disturbance, or in other words, increasing the sensitivity of the control and then returning it to its initial value, in proportion to the rate at which the effect of the disturbance disappears.

The curves in fig. 413 show, in simplified form, and for each of the three modes of control, how the value of the controlled variable varies with time after a violent disturbance.

The proportional mode (curve 1) leaves some offset between the desired value and the value of the controlled variable.

The proportional + integral mode (curve 2) slowly returns the controlled variable to its set value.

The proportional + integral + derivative mode (curve 3) enables the deviation caused by the disturbance to be cancelled more rapidly.

These curves indicate only the idealized form of behaviour, and return to equilibrium is always accompanied by major or minor oscillations dependent upon the relative influence of the various modes.

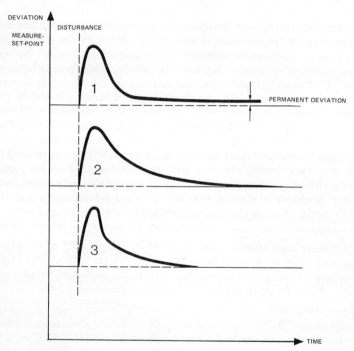

Fig. 413. — The three modes of regulation.

The sole purpose of these three modes of control used together is therefore to hold some quantity liable to variation automatically at a fixed, pre-set value. The problem to be resolved is often more complex when a correlation has to be established between the variable value of a given magnitude and some other magnitude, or when a **programme controller** is used to establish a relationship which changes the value of a physical magnitude with respect to time.

In the case of certain industrial processes involving a large number of often complex adjustments, the present tendency is to centralize all control functions in a digital computer which then works on a time-sharing basis. In accordance with a pre-set programme, it scans the various measuring points in turn and issues orders to the control systems. The computer is of course accompanied by various types of peripheral hardware such as devices for memorizing signals between the issue of two successive orders, digital-analogue converters enabling the signals given by the computer to be assimilated by certain control units, etc.

2.3. Choice of control method

2.3.1. CHOICE OF CONTROL MODE

All problems relating to control must be accurately analysed to determine the most suitable method to use.

The first step is to select the regulated variable (y) which is linked through the control system to the controlled variable (x), so that the latter can be stabilized at its set value (x_0) after any modification of the value (x) caused by any outside action.

When this regulated value has been selected, the next step is to examine the total system represented by the plant, the monitoring system, the control system, and the correcting system. The plant itself may have a varying susceptibility to changes in the variable regulated value and may show a varying degree of inertia. The system used to measure the controlled variable may introduce some delay into the measuring process, and this is added to the lag due to the correcting action of the controller itself and to some hysteresis of the correcting system. It will be appreciated that if this lag is too long, the corrective action may take place at the wrong time, i.e. after the deviation trend has reversed itself.

The method of control selected will therefore depend on the nature of the plant and on the required accuracy of control.

The **proportional mode** does not, as a rule, give very precise control. Its degree of accuracy is, however, very often adequate for the control of a level, for example, for which a relatively narrow proportional-mode band can be used, by a trade-off for continuous slight hunting. In other cases, such as regulation of filter water level, hunting can have an adverse effect on the quality of the filtered water, and it is better to use a fairly wide proportional band and accept a few centimetres difference of level between a recently-washed filter and a clogged filter.

The degree of accuracy obtained by the **proportional plus integral mode** is independent of the width of the proportional band. Since the integral mode acts as though the proportional mode coefficient were increased exponentially with time, it will be appreciated that this method of control can lead to erratic operation of high-inertia plants.

Control by proportional + integral + derivative mode, which links the amount of corrective action to the time variation of deviations of the controlled variable, allows the automatic control of systems which are essentially unstable and liable to sudden and relatively large disturbances. For example, the operation of a thermal deaerator can be controlled automatically, as regards the level in the tank,

by the simple proportional mode on the intake regulating valve, provided the
tank is big enough in relation to the flow of water. As regards pressure, the
control, which must be extremely accurate, will be achieved by the proportional
+ integral + derivative mode, if one of the disturbing factors such as flow of
water, temperature of water and vapour pressure is liable to sudden variation;
if not, the proportional + integral mode will be sufficient.

These four modes of control taken in isolation are not always sufficient
to solve certain problems connected with the treatment of water, and **multiple
control loops** are to be used. This happens when the value of the controlled
variable depends on the values of several parameters which can vary differently
and each has a major effect on that value. If any of the standard methods of
control is used in such a case, the plant becomes unstable and hunting occurs.
For example, the free chlorine content C, which must be held constant, of a water
treated under variable flow, is governed by three parameters: the flow Q of the
water to be treated, the chlorine demand D and the flow q from the chlorine
dosing device, which is used as the regulated variable. The problem has to be
resolved by making q proportional to Q as a first step, without waiting for C to
vary from its set value, and then to correct the deviations of C as a second step.
In the case of a high-inertia plant, the problem can be solved simply by on-off
control. In this case, if the correcting mechanism is a valve, the latter is opened
at time intervals proportional to Q for a constant time, thus automatically establishing proportionality between Q and q; the constant time is readjusted automatically according to the most appropriate law by reference to deviations of C
from its set value.

Since the assembly comprising the measuring unit, the system to be controlled, the control unit and the controller form a closed loop, it is sometimes necessary
to use **feedback controllers,** which are themselves closed loops, in the sense that
a fraction of the output signal from the regulator governed by the rate of feedback
is re-injected at the intake, so that the controller receives a signal corresponding
to the deviation between the signals representing the controlled variable and the
feedback variable. This arrangement, which is used more especially to control
filter levels, increases both reliability and stability.

2.3.2. CHOICE BETWEEN ELECTRIC AND PNEUMATIC CONTROLLERS

When the best method of control for the particular case has been decided,
a choice has to be made between a pneumatic and an electric control system.
Pneumatic controllers all have as their basic element the standard flapper/nozzle
system combined with an amplifying relay which governs the control unit. Pneumatic controllers, which are very easy to service and repair, are often adopted
for economic reasons, particularly when a source of control air is available, or for
safety reasons. However, when they have to be installed some distance from

the unit to be controlled, the time constant which they introduce into the control loop may rule them out. This delay can be reduced on the measuring side by using two electro-pneumatic converters enabling the pneumatic signal to be converted into an electrical signal at the output and to be reconverted from electrical to pneumatic at the controller input. The introduction of two converters into the chain naturally reduces the accuracy of control, and it does not eliminate the delay caused by the pneumatic control line of the control unit. When the line is so long that the delay is too great for corrective action, an electrical controller must be used. In that case, if the measuring sensor for the controlled variable gives the measurement in the form of a pneumatic signal, an electro-pneumatic converter is installed at the point of measurement. The control unit can then be actuated either electrically or, if this is impossible, or for reasons of cost, pneumatically through an electro-pneumatic converter in line with the control units and with an amplifying relay if necessary.

These few pages can do no more than offer general ideas on the subject of automatic control which, in the case of water treatment, is a matter for experts with a very thorough knowledge of the techniques involved.

3. AUTOMATION

3.1. General remarks

Water treatment processes usually involve a large number of repetitive stages. Automation of these stages has three main aims:

— to simplify plant operation;

— to avoid human errors;

— to ensure economic operation by cutting down working time.

3.2. Different forms of automatic operation

A water treatment process, such as filter washing or regenerating a deionization chain, can be automated to a degree which varies according to the economic factors involved and the risk of human error which increases with the complexity of the process.

There are three categories of automatic operation:

— remote manual sequential control;

— manually-started automatic sequential control;

— automatically-started automatic sequential control.

In the two latter categories, the reliability of the equipment can be increased by using "logical sequence" systems.

The purpose of any form of sequential automation is to link the individual operations which make up a sequence. In the case of water treatment, these operations consist mainly of closing or opening electrically- or pneumatically-actuated valves and the starting or stopping of pumps.

3.2.1. In the case of **remote manual sequential control,** transition from one sequence to the next is left to the discretion of the operator, as is also the start-up of the operation. This system may take the form of a multi-channel pneumatic distributor. Here, the operator checks the result of one sequence before passing on to the next.

3.2.2. **Manually started automatic sequential control** systems enable us to take plant automation one stage further. With this kind of automation, although the operator remains responsible for starting up the whole range of operations, transition from one sequence to the next is effected automatically. In these circumstances, the distributor mentioned above can be fitted with a pneumatic motor.

Another system employs a range of solenoid valves and pneumatic relays. With this system, the sequence of operations can be interrupted at any time by the operator and continued with remote control. The operator, who must be present to start the automatic sequence by hand, can check that operations are proceeding normally and remedy any fault in the automatic system.

3.2.3. In the case of **automatically-started automatic control,** the whole process can be completed without an operator. The normal running of the plant is then "linked" to the value of one or more governing factors, which start the automatic process when they are at the set values. The process can therefore be completed without an operator being present, which means that maximum reliability is essential over the whole automatic system. Current examples of fully-automated systems are filter plants in which washing of filters is monitored by readings from head-loss sensors and ion-exchange demineralizing plants which are automatically switched to regeneration by special sensors which check the quality of the water produced (resistivity, silica content).

With such automatic systems, maximum overall reliability is obtained by using a logical sequence process which checks automatically, for each sequence,

Fig. 414. — ECOPETROL. Barrancabermeja refinery, Colombia. Mimic diagram.

that the orders given have been received and carried out and, if not, stops the operations and warns the operator. In this case, the controlled elements, such as valves or pumps, must be fitted with devices to automatically check their situation, which can only be open or closed for the valves and "on" or "off" for the pumps. In that case, the automatic control can either give a general fault warning or can have a system for identifying and displaying the faulty element. This idea is of course most attractive as a means of increasing the reliability of the automated system, but it should be recognized that the price of the system will be considerably higher if logical sequences are introduced. In practice, the water engineer and the automation expert have to work together to find a solution meeting the joint requirements of both reliability and economy. Frequently, for example, a simple flow recorder, judiciously placed, can be used to detect an overall fault on a sequence. Similarly, a timing mechanism, combined with a monitoring resistivity meter, can indicate a probable defect if the expected result is not forthcoming after a certain time.

3.3. The programmable automatic unit

All modern industrial plants incorporate automated equipment which monitors and controls both machines and processes while at the same time improving their performance.

The choice of this equipment covers a wide range of possibilities extending from electromechanical units via static logic control devices to systems employing mini-computers.

The programmable automatic unit is one such system. It is an entirely static piece of equipment capable of performing the same functions as units with wired logic, whether these be of the electromechanical or static relay type.

It differs essentially from the latter in that it can be programmed, that is to say that all the functions and sequences necessary for monitoring the processes have no physical existence and can be altered at will without modifying the equipment. The programme is stored in a memory.

The automatic programmable unit comprises:

● input and output modules, the number of which varies according to the installation although it can usually be increased to a maximum of several hundred;

● a central unit, the function of which is to control the inputs and outputs and to execute the programme which has been stored in its memory whose capacity, related to the complexity of the programme, can usually be extended;

● the d.c. supply necessary to power the unit. The voltages are controlled in such a way that they are unaffected by fluctuations in the mains supply.

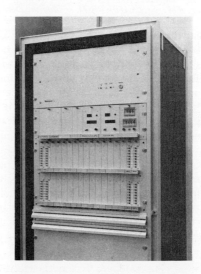

Fig. 415. — Programmable automatic unit.

A characteristic feature of the programmable automatic unit is its use of an extremely simple programming language made up of electrical or logical symbols. The majority of the functions necessary to an automation system are preprogrammed (logic, timing, counting, shift etc...).

In the field of water treatment, the programmable automatic unit is very suitable for controlling the washing of filters or the regeneration of deionizing chains.

The fact is that, in the majority of these applications, the number of necessary logical functions is fairly limited. A large number of sequences are solely a function of time. It is against this background that Degremont have developed a system in which both the duration of the sequences and the input/output control can be programmed.

3.4. Industrial data processing

While it is true that a large proportion of the automated equipment used in water treatment calls for no more than the kind of logical system which has been described previously in this chapter, some advantage may also attach to the use of process computers (the current prices of which are in some instances relatively low).

Computers are as useful for plant management and supervision as they are for automation itself. They provide the means for:

● centralized data gathering and monitoring of plant performance;

● direct calculation of the parameters affecting plant operation;

● the possibility of printing out selected historical data for the purposes of checking or inspection after the event;

● a maintenance programme.

Where automation is concerned, computers offer the advantages of:

● greater flexibility in the compilation of automated programmes;

● optimization of processes;

● integration of several automatic systems controlling a number of processes.

These advantages are particularly telling in large plants which require a certain level of computing facilities. The potentialities of the new mini-computers and programmable controllers mean that it is now possible to find the right solution for any type of plant.

Besides the computer(s), the peripheral items of hardware include printers, console, display screen, desk for communicating with the computer etc. as well as the links between the memory, the central unit, the inputs and outputs and so forth. The software is the translation into computer language of the input programme, the calculations, the read-out and possibly commands.

In every case, the potential advantages of a computer must be considered in the light of the reliability and accuracy of the sensing devices used (particularly where effluent treatment is concerned).

If the computer is to be used in real time for the control of the treatment processes themselves, the advantages to be derived may well be limited when the reaction involved in the processes is a slow one, as in the case of biological purification.

Where large-scale automated systems are handled by computer, the present-day tendency is moving towards a hierarchical method of digital control in which those items of plant or parameters which have to be monitored are equipped with control devices which are capable of functioning independently of the central unit. It follows from this that due attention must be given to the engineering of the interfaces.

At all events, the computer is an essential tool for the technical manager. It is capable of integrating a very wide range of data and must not be under-estimated. It also makes feasible the installation of modern equipment for the monitoring of circuits, plant items and operating parameters (in particular, by the use of colour television).

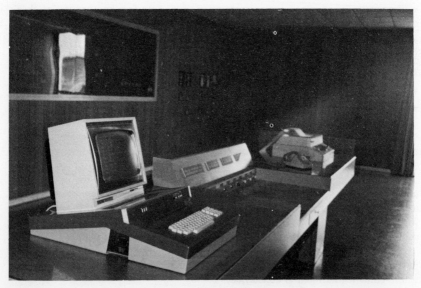

Fig. 416. — General control desk for the Water Pollution Control Works of the BORDEAUX Metropolitan Area, France.

Part Three

Treatment Methods According to the Nature and the Final Use of the Water

Veüe de la Cascade de Chantilli.

1. CHOICE BETWEEN SOURCES OF SUPPLY

The first problem to be resolved by authorities responsible for supplying the public with drinking water is to choose between the various possible sources—usually underground water taken from varying depths, surface water from rivers or lakes and possibly sea water.

This choice does not always arise, because in some regions there are no underground supplies in sufficient quantity or of the required quality, either on the spot or within a reasonable distance. Other areas have good local underground resources, and the choice is then automatic.

Other areas near to the coast have neither underground nor surface water, and sea water virtually has to be used, to avoid long fresh water mains or a costly system of supply by tanker.

When a choice exists, the main factors to be considered are the following:

1.1. Quality of the available water

Underground water is synonymous with "naturally pure" water and surface water is water sooner or later polluted.

Underground water, with its constant temperature, is admittedly usually more attractive than surface water which is liable to pollution by the waste created by the modern world. And in making the choice, account must be taken not only of the factors which are known when the decision is reached, but also of circumstances which may arise in the near future.

Unfortunately, underground water rarely retains the same degree of purity throughout the year; as in the case of surface water, the effect of periods of heavy

---

N.B. For information on drinking-water standards, see the Chapter 34 on legislation p. 1124

rainfall must be studied, because underground water can then become very turbid or can even be contaminated by the water table of a neighbouring river. Underground water can also be polluted from the surface by pesticides and animal or human sewage. While such pollution is normally less frequent and more limited in underground water than in surface water, accidents have a more prolonged effect on the former, because percolation is slow and the lost balance takes a long time to restore. Lastly, pollution of underground water is a slow process and may not be detected in time if the water is not analysed frequently and systematically.

Any water supply must therefore be judged on its properties as a whole.

In the case of drinking water, **temperature,** together with limpidity and taste, is an important criterion. As a result, water drawn from deep in the ground, which is virtually constant in temperature and mostly remains cool throughout the year, is quite rightly much appreciated; on the other hand, the temperature of surface water in temperate climates frequently varies from 1 to 22 °C according to the time of year. Temperature is also an essential factor in assessing the calcium carbonate balance of the water and the treatment which may be necessary; as the temperature of the water drops, chemical flocculation and clarification processes, if required, take longer and become more difficult immediately the temperature falls below 10 °C. On the other hand, a high temperature may stimulate a substantial growth of plankton or lead to over-saturation with dissolved gases which are liable to produce "white water" or to interfere seriously with the operation of the treatment plant.

More than for the other properties of water, it is important to know the rate of change as well as the extent of temperature variation.

The suspended solids content, which is judged only partly from the standpoint of **turbidity,** is one of the most revealing properties.

Drinking water should not contain any settleable suspended solids.

The World Health Organisation (WHO) allows a rating of 5 turbidity units and considers one of 25 excessive. In France, and in Europe in general, every effort is made to supply water with a turbidity rating of 5 drops of mastic or less, i.e. between 0.1 and 0.5 Nephelometric Turbidity Units (N.T.U.). The current French standard, however, accepts a turbidity limit of 15 drops of mastic.

Good drinking water must have no **colour,** and suppliers endeavour to ensure that a limit of 5 platinum-cobalt units is not exceeded (the French standard in force in 1977 accepts 20 units).

Like surface water, underground water is sometimes coloured. Even if clear when it emerges from the ground, it can become yellow, red or even black as soon as it takes in oxygen.

Smell and taste must also be taken into account when judging the quality of a water supply; drinking water must have neither, and if the available supply does not meet this requirement permanently, treatment, which will be costly if it is to be continuously effective, will be essential.

Furthermore, the available supply will contain **minerals,** which are generally found in greater quantity in underground water than in surface water. These salts are no more than a nuisance in the case of excess calcium, but they can be harmful when the water contains an excess of certain ions such as lead, arsenic, fluorine, etc.

The mineral content of water is assessed by reference to a number of chemical properties and contents, and more particularly:

— resistivity and pH value,
— TH, TCa and TMg,
— TAC and TA,   } see Chapter 27
— free CO_2,
— ammoniacal, nitrous and nitric nitrogen,

— Sulphates and chlorides, measured separately or together by SAF (salts of strong acids), determine the salinity of a drinking water supply. An excessive concentration is not only harmful to human beings (although contents well above the recommended 250 mg/l can be tolerated after a time), but also sets up corrosion in metal pipes.

In the latter case, therefore, the greater the salinity of the water the higher its TAC must be.

Deep-lying water has the highest salinity, and a low-content surface water is then particularly attractive.
— Iron and manganese are not harmful in themselves but must be eliminated. The French standards allow a combined Fe + Mn maximum of 0.3 mg/l, with a limit of 0.1 mg/l for manganese, but serious problems can arise with much lower concentrations; "black water" has been found in the low-flow-rate branches of a system carrying water containing much less than 0.05 mg/l of manganese and the same applies to iron; it is therefore essential to know the exact concentration of these metals in the water (underground water often contains large amounts) and to weigh the decision carefully before using ferrous or ferric salts as coagulants.

Enough **dissolved oxygen** must be present to make the water pleasant to drink and to allow it to be distributed without corroding metal mains. The fact that water from underground does not as a rule contain any dissolved oxygen does not mean that it is polluted; the oxygen present when the water originally entered the ground may have been lost during the auto-purification process. In such cases, the water must be aerated before it can be used.

The organic matter content of water likely to be used for human consumption must also be carefully examined. Underground water normally contains little organic matter, except in association with a source of pollution, and this, together with temperature, is its main advantage as compared with surface water. It has frequently been stated that the oxidizability of drinking water should not exceed 3 mg/l of oxygen measured in an alkaline medium; this is not a standard of any kind and cannot be regarded as a usual upper limit in regions where organic matter is chiefly of vegetable origin and harmless. Wine and tea, which contain much more, could not be condemned on grounds of excessive organic matter content.

Toxic products must be removed during traetment. Both mineral and organic micropollutants must be reduced to their minimum acceptable level. European standards currently being drafted include maximum values which must not be exceeded (cf. Chapter 34, Legislation).

Lastly, the biological properties of water are of special importance.

A close inspection must be made for **plankton,** which can cause serious trouble both at treatment plants and in mains and water towers. Plankton is found more especially in surface water where phytoplankton (of vegetable origin) is found side by side with zooplankton (of animal origin). In spring and summer, phytoplankton grows at a phenomenal rate and can clog badly-designed treatment plants, give the water an unpleasant taste and spread into the distribution mains, where it can be the indirect cause of damage to metal pipes. Zooplankton is even less acceptable. When some species begin to infest a water system, they are very difficult to eliminate and are even more unpleasant to find suspended in the water.

From the **bacteriological** standpoint, most water is disinfected before it enters the water mains; this is essential in order to destroy the pathogenic bacteria and germs present in virtually all surface water. It is also safer to disinfect underground water even though it is bacteriologically pure in most cases. Surface water is more likely to contain viruses and therefore must be more vigorously disinfected than underground water, though the latter is not, however, completely safe from such contamination.

To summarize the good and bad qualities of water which may be used for human consumption, it may be stated that underground water taken from various depths is safer than surface water as regards pollution, but that it has a number of other bad features which often tend to be overlooked.

Indeed, the defects of surface water are generally much more obvious than those of underground water, which may in fact be much more difficult to treat (cf. in particular "Removal of manganese and iron", page 624). The table on page 603 summarizes the main differences between the two types of water; to this we may add that some underground water is just as likely to be contaminated by bacteria and mineral or organic micropollutants if insufficiently protected, but is viewed with less mistrust than surface water. Because of its clarity on emergence, underground water may give a false impression of safety, and it should always be subjected before use to an examination as thorough as that for surface water.

If underground water has to be treated and if a potential surface supply is not too badly polluted, the difference between the two sources does not mean that one or the other should be automatically chosen, because there are no separate definitions of purity for surface and underground water respectively : and overall judgment has to be made in each particular case.

MAIN DIFFERENCE BETWEEN SURFACE AND UNDERGROUND WATER

Characteristics considered	Surface water	Underground water
Temperature	Varies with season	Relatively constant
Turbidity, suspended solids	Level variable, sometimes high	Low or nil
Mineral content	Varies with soil, rainfall, effluents etc.	Largely constant, generally appreciably higher than in surface water from the same area.
Divalent iron and manganese (in solution)	Usually none, except at the bottom of lakes or ponds in the process of eutrophication	Usually present
Aggressive carbon dioxide	Usually none	Often present in large quantities
Dissolved oxygen	Often near saturation level	Usually none at all
Ammonia	Found only in polluted water	Often found, without systematically indicating pollution
Hydrogen sulphide	None	Often present
Silica	Moderate proportions	Level often high
Nitrates	Level generally low	Level sometimes high; risk of methemoglobinaemia
Living organisms	Bacteria (some pathogenic), viruses, plankton	Ferrobacteria frequently found

1.2. Quantity of water required and regularity of flow

In choosing between various sources of water, account must be taken of the quantity required and the quantity available locally or at a reasonable distance. This concept of quantity must be considered together with the concept of reserves in the case of deep water strata that are not extensive and which have already been largely drawn upon; in the case of surface water, it must be compared with minimum flow at lowest level, because quality is liable to deteriorate seriously during the very dry season.

A constant supply is obviously of prime importance. While this can be assured in the case of surface water, where flows at different times of the year can easily be gauged, this is not always so for deep water supplies, which call for accurate studies supported by prolonged tests carried out by experts using the most modern investigation techniques.

1.3. Cost of finding, transporting, treating and distributing water

The cost of finding water has to be given special consideration in the case of deep underground supplies, particularly when demand is heavy.

In some cases, the cost of transporting water can also be one of the determining factors in choosing between alternative sources of supply. Underground water piped from a great distance is only an economically viable proposition if it is good enough not to need treatment, if there is no shortage in the area from which it is taken, and if the cost of transport is not prohibitive.

2. IMPURITIES FOUND IN WATER TO BE USED FOR HUMAN CONSUMPTION

There are basically four sources of general pollution affecting water; three are regularly encountered:

— **Waste water of animal origin** (manure heaps, cattle sheds, etc.) **or of human origin.** Animal water most often affects the quality of wells, nearby springs or small water tables close to the surface. Domestic waste water pollutes rivers, either by direct discharge or by discharge of the non-degraded fraction of effluent from purification plants. The pollutants in such waste are suspended matter, detergents, organic matter, phosphates, bacteria and sometimes viruses.

— **Industrial process waters or liquid effluents** vary so much that they may contain all known pollutants, including radioactive material, and sometimes mineral or organic carcinogens, in proportions which vary with the preceding treatment.

— **Run-off water** contains agricultural pollutants such as fertilizers, pesticides, detergents, etc.

— **Accidental contamination** due to concentrated discharge of a pollutant liable to affect surface water or even deep-lying water.

The many pollutants and micro-pollutants liable to be found in water for human consumption can be classified under three headings: mineral pollutants, organic pollutants and viral particles.

Suspended solids of organic or mineral origin (plastic bags, grit, clay, etc.) must also be taken into account, and a suitable form of preliminary traetment (bar screening, grit removal) selected to eliminate it.

2.1. Mineral pollutants and micropollutants

These include:

— undesirable or toxic substances: heavy metals, fluorine, arsenic, etc;

The effects of some of these are discussed in Chapter 27, page 921, under the heading "Toxic substances dissolved in water". National and international standards lay down maximum tolerable limits for some of these substances (Chapter 34, page 1136).

— substances such as iron, manganese, zinc and copper, which affect principally the organoleptic properties of the water. Copper, which the human organism requires in small quantities, becomes toxic if large doses are accumulated;

— phosphorus and its compounds, which are responsible for the growth of algae and the eutrophication of lakes.

— radioactive substances (for the record).

2.2. Organic pollutants and micropollutants

These are very numerous and can be classified as phenols, hydrocarbons, detergents, pesticides.

2.2.1. PHENOLS AND DERIVATIVES

Phenols and their derivatives are the mark of industrial pollution. Their worst effect is that, in the presence of chlorine, very small quantities of these products, depending on other organic matter in the water, leave a taste of chlorphenol. Normally there is no taste if the pure phenol content is kept down to 1 µg/l, but there is sometimes a slight taste of chlorphenol with a content of 0.1 to 0.01 µg/l.

The biodegradability of phenol derivatives varies with their composition. The maximum quantities found in river water leave a taste but would not be toxic (270 µg/l in the Rhine, 35 µg/l in the Meuse and 40 µg/l in the Seine).

2.2.2. HYDROCARBONS

The hydrocarbons capable of polluting surface or underground water supplies come mainly from oil refinery waste, industrial effluents of various kinds, gasworks effluent, fumes, etc.

Such waste may contain paraffin, kerosene, petrol, Diesel oil, fuel oil, other oils and lubricants.

Biodegradability is slow. Accidental pollution is short-lived at the intake of a river purification plant but can last a long time in the case of underground water (up to several years because of the soil's power of retention). This is why underground water supplies have to be strictly protected against the risk of hydrocarbon contamination.

● *Harmful and toxic effects:*

— formation of a film which interferes with the re-oxygenation and natural purification of surface water;

— interference with the operation of drinking-water treatment plants; flocculation and sedimentation are affected and the hydrocarbon is liable to remain in the filter material for a long time;

— the taste and smell threshold varies very widely according to the product involved (from 0.5 µg/l for petrol to 1 mg/l for oils and lubricants).

— Toxicity : there is a danger of toxicity in drinking water at concentrations above those at which taste and smell appear. Skin troubles have been caused by fuel oil additives. There may be a more serious risk with cyclic hydrocarbons which are suspected as possible carcinogens (3-4 benzopyrene for example), and the risk is increased by the presence of other compounds (surfactants).

2.2.3. DETERGENTS

Detergents are synthetic surface-active compounds which enter the water with municipal and industrial effluents.

Commercial products contain active compounds in the form of surfactants and aids.

● *Surfactants*, with a structure that modifies the physical properties of surfaces by lowering surface tension and gives them cleaning power.

There are various types:

— *anionic surfactants:* for a long time the most commonly used substances were "hard", slightly biodegradable, branched-chain products, such as the sulphonated alkylbenzenes (SAB), which have been mainly responsible for the problems created by the presence of detergents in water. According to country they accounted for 80 to 90 % of total production until the health authorities decided that they must be replaced by at least 80 % biodegradable, linear-chain detergents. (Law of 25th September 1970 in France) L.A.S. is the most frequently used of these.

The concentration of anionic surfactants can be measured easily by methylene blue analysis; their biodegradability over a period can then be followed without difficulty;

— *non-ionic surfactants* (those now used have an alkylphenol base). These are being used on an increasing scale, but problems have still to be solved as regards dosing;

— *cationic surfactants*, consisting of quaternary ammonium salts, are little used and are reserved for special uses linked with their biostatic properties.

● *Aids:*

These include:

— aids proper such as polyphosphates, carbonates, silicates;

— sequestrating and complex-forming agents (polyphosphates);

— reinforcing agents to improve the action of the active constituents (amino-oxides, carboxymethylcellulose, alkanolamides);

— additives: bleaching agents, perborates, optical bluers, dyes, perfumes;

— mineral salt fillers, to improve the appearance of products;

— enzymes, which should be regarded as pre-adjuvants and help to hydrolyse certain types of fouling.

● *Concentrations found in water:*

Before the introduction of biodegradable products, the concentration of anionic detergents in river water varied from 0.05 to 6 mg/l. The figure has dropped since.

Concentrations of non-ionic detergents are difficult to express because of the many methods of analysis and their limits of accuracy.

● *Harmful effects:*

The harmful effects caused by the presence of detergents in water are:

— formation of foam, which hinders natural or artificial purification, concentrates impurities and is liable to spread bacteria and viruses; anionic detergent concentrations of 0.3 mg/l and over are sufficient to produce a stable foam;

— formation of a barrier film on the surface, which slows down the transfer and dissolution of oxygen in the water, even when there is no foam;

— a soapy taste, at concentrations well above the foam point;

— higher phosphate content due to the combination of polyphosphates with surfactants, leading to the eutrophication of lakes and the growth of plankton in rivers; in some countries, a large proportion of polyphosphates is replaced by NTA (nitrilo-triacetic acid);

— a gradual increase in the boron content of surface and underground water supplies, due to the large quantities of sodium perborate used in detergents.

Detergents do not kill bacteria, algae, fish and other forms of river life, so long as the concentration does not exceed 3 mg/l.

Lastly, the enzymes recently added to detergents have no ill effects either on the receiving water or at purification plants.

● *Influence of biodegradable detergents.*

The introduction of detergents which are at least 80 % biodegradable has led to a very marked improvement, at least in the case of anionic detergents, which can be successfully checked.

The products resulting from the biological breakdown of these linear detergents are only very slightly toxic.

In some cases, the fraction which is not broken down is more toxic to fish than detergents which are not biodegradable.

Non-ionic detergents still raise problems because they stimulate the formation of foam by anionic detergents and then stabilize it. The non-ionic products now used are resistant to biological breakdown, particularly in cold weather.

Research now in progress suggests the possibility of new anionic and non-ionic detergents which will be almost completely biodegradable (linear-chain alcohols), and which will also have improved additives.

In general, it may be said that detergents are not harmful in themselves and that their indirect harmful effects will be greatly reduced when the formation of foam can be restricted by means of completely biodegradable products. All additives will, however, still have to be closely checked to ensure that no risk of toxicity is involved.

There is no French standard fixing a maximum detergent content for drinking water; it is, however, recommended that water should contain less than 0.2 mg/l of anionic surfactants. The European standard in course of preparation proposes a maximum permissible concentration of 0.1 mg/l.

2.2.4. PESTICIDES AND PLANT HEALTH PRODUCTS

Pesticides are products used to control organisms which are either harmful to health or attack materials and animal and vegetable sources of food.

They are themselves harmful to health and may, if allowed to accumulate in plant or animal cells, prove detrimental to the environment in general.

Pesticides include not only plant-health products (insecticides, fungicides, herbicides) but also certain products of industrial origin such as the polychlorinated biphenyls.

● *Classification.* They can be divided into five classes:

— *organo-chlorinated compounds,* some which are prohibited in a number of countries. The category includes: stable compounds, such as DDT, BHC, aldrin, among the insecticides, and the herbicides 2.4D, 2.4DT, MCPA, MCPP, as well as the polychlorinated bi- and triphenyls; less stable compounds such as Simazin;

— *organo-phosphorus compounds,* which are less stable and are thereby tending to replace the chlorinated products; in some cases incomplete oxidation produces oxon type derivates which are more toxic than the original compound (parathion);

— *organo-nitrogenous compounds,* relatively unstable compounds such as Simazin; stable compounds such as DNOC;

— *organo-metallic compounds,* such as derivatives of urea, of thiourcil and of nitrated or thionitrated tiazenes, used as weed killers. The carbamates and dithiocarbamates are used as fungicides;

— *mineral substances,* such as sulphur, copper sulphate and lead and calcium arsenate; the first two are still used fairly frequently for the treatment of plant diseases.

● *Origin of pollution and transfer factors.*

Pollution is caused by run-off water when there is heavy rain and by infiltration. Parathion has been found in water tables 60 m below the surface. In some cases, pesticides are absorbed by the soil and then transferred with it by erosion to sources of water. The factors affecting the transfer of pesticides to water are their solubility, their resistance to physical and biochemical breakdown, the nature of the soil and the volume and intensity of rainfall.

● *Harmful and toxic effects of pesticides in water.*

— *Organoleptic effects.* Pesticides can leave a smell or taste with thresholds varying from 0.1 to 1 000 $\mu g/l$ according to the product concerned. For example, 1 $\mu g/l$ of HCH is enough to leave a taste in the water, while the taste threshold for DDT is 1 mg/l.

— *Effect on aquatic fauna.* Pesticides have a direct effect in the form of slow or acute poisoning, and an indirect effect (represented by the disappearance of plankton), namely a drop in oxygen content and changes in pH value and CO_2

content. The organo-chlorinated insecticides are much more toxic to fish than the organo-phosphorus types, while weed killers are much less poisonous than insecticides in general.

— *Effect on human beings.* A distinction is made between acute and chronic poisoning. The acute forms are not caused by the water, but the chronic forms can be, because the effect of pesticides is cumulative; adipose tissue accumulates mainly organo-chlorinated pesticides, while the liver and kidneys are sensitive to DDT. In general, the organo-phosphorus pesticides are much more toxic to human beings and mammals than the organo-chlorinated types, with the exception of malathion (which is a slightly toxic phosphorus compound) and endrin (which is a highly toxic chlorinated compound not permitted in France).

2.3. Biological pollutants and micropollutants

2.3.1. MICROORGANISM AND VIRUSES

This form of pollution is caused by microorganisms and viruses in the different types of waste. They can affect both surface and underground supplies and are discussed in Chapter 28.

2.3.2. SECRETIONS OF MICROFAUNA AND MICROFLORA

Many organisms (algae and actinomycetes in particular) may develop in river water, especially if polluted by organic matter or substances causing eutrophication, in reservoirs and even in distribution systems. Their metabolites, released into the natural environment during their lifetime or after their death, cause certain problems:

● *Taste and smell:* the most common forms of taste are those of mud, earth and mildew, caused by actinomycetes and certain cyanophyceae which secrete various compounds, primarily geosmin. Highly unpleasant smells can also be attributed to the main categories of algae: chlorophyta, chrysophyta (chrysophyceae, diatoms), cyanophyta. The range of smells may vary widely with the type and concentration of species present, from aromatic (similar to the smell of certain fruits, flowers, or vegetables), to fish, grass, mud, rot, corked wine, etc.

● *Toxic substances:* some of the cyanophyceae produce products which are toxic to higher animals. These substances are usually intracellular, and do not represent an immediate danger provided the algae are carefully removed during treatment, but nothing is known of their long-term effects (accumulation in the organism) if they remain in the water in trace form after the algae have died and decomposed.

● *Appearance of water:* colour and/or turbidity caused by the secretions of microflora and excretions of microfauna.

2.4. Impurities from reagents used in water treatment

It is important that in the treatment of drinking water the reagents used should be relied on not to introduce any impurities which are likely to persist in the water after treatment.

When, for instance, caustic soda is used to correct the pH of filtered water before distribution through the system, care must be taken to ensure that it is mercury-free.

Regulations are currently being prepared which will define the maximum admissible impurity content for each reagent.

3. PRINCIPLES OF DRINKING WATER TREATMENT

The decision as to whether or not water for drinking needs prior treatment depends on the source of the water and a comparison of its physical, chemical and bacteriological properties with the required properties for drinking water, allowing for the possibility of micropollution.

The extent of the treatment will vary with the nature and number of faults. Several processes may be necessary, and these must be combined in the best way to correct any faults with the facilities available at the local purification plant

Another target must be to keep initial outlay and running costs to a minimum. In so doing, account must be taken of the possibilities now offered by automation and of the overriding need for reliable performance, so that the water fed into the distribution system is of the best quality in general and free from any risk of temporary deterioration.

The main processes used to purify drinking water are the following:

3.1. General processes

3.1.1. TREATMENT AT WATER INTAKES OR AT PUMPING STATIONS

In the case of underground water, the first problem is a catchment or pumping system which will extract the least possible earth or sand with the water; it is essential that limits of the protected area are clearly defined.

In the case of surface water, the intake must be designed to handle the different types of coarse material which may be present in the water. A well-designed intake is the first stage of the treatment process.

● *Designing a water intake*:

In the case of a **lake with a virtually constant water level,** the intake must be set at such a depth that the quantity of suspended solids, colloidal matter, iron, manganese and plankton in the water is as low as possible throughout the year. The temperature of the water must also be as low as possible.

For example, if the lake is fairly deep, it is generally best to have the intake from 30 to 35 m (say 100 ft) below the surface; at that depth the effect of light is so small that the plankton content will be limited, particularly at periods of great proliferation. The intake must, however, be at least 7 m (23 ft) above the bed of the lake; otherwise the water will be greatly affected by the movement of deposited particles and by bottom currents.

Allowances must also be made for the "stratum turn-over" of the lake water under the influence of temperature variations.

The same phenomena must be allowed for when locating an intake in an **impounding reservoir with variable water level;** the usual solution is an intake tower enabling water to be drawn off at different depths according to season.

One essential precaution, before a reservoir intended to provide drinking water is filled, is the destruction and removal of the plant cover from the flooded area, in order to prevent *eutrophication* of the impounded water (see page 965).

An intake **on a river** must be protected from matter such as earth, sand, leaves, reeds, grass, discarded packaging materials, particularly plastic, floating objects and films of moss or hydrocarbons carried along by the current. There is no ideal model, and the type used will vary with the nature of the entrained matter, variations in flow, the navigability of the river and the degree of accessibility of its banks. It may be a bottom intake, a side intake, a siphon intake and so on. Each case requires a special study.

According to the nature of the water supply, the first possible treatment is coarse straining to remove large particles which would interfere with subsequent stages of the purification process.

● *General pretreatment*:

General pretreatment may comprise (see Chapter 4):

— **a coarse bar screen** with bars set 8 to 10 cm (3 to 4 in) apart,

— **finer screening**, through bars set 25 to 40 mm (1 to 1.6 in) apart.

Coarse screening may be omitted if the suspended solids are not likely to damage the fine screen and its automatic bar cleaning system.

A cleaning system is essential if large quantities of foreign solids are entrained with the water. It is not enough if the water carries grass or leaves that might pass edgewise through the bars; it then becomes necessary to add:

— a rotating drum **screen** (when the water level does not vary greatly) or plate-type screens, when the water level varies. The mesh of such screens is usually from 1 to 5 mm. Cleaning must be automatic, and is generally controlled by the loss of head. In many cases, failure to install such a screen for reasons of economy causes serious trouble at purification plants, particularly when a pumping system is included

— **a grit removal system** which may be located before or after the screening system; it will vary in design according to the type of intake. Grit removal is essential when the water must be carried through a long pipe or channel, when it has to be pumped or when the subsequent treatment plant is liable to be seriously affected by substantial amounts of sand;

— **micro-straining** if the amount of plankton is limited of if there is no subsequent clarification. As micro-straining has only a limited effect, this treatment is rarely suitable for modern plants (see "Microstraining", Chapter 9, p. 247).

— **removal of surface oil;**

— **de-silting** (see page 123).

Fig. 417. — Water intake at a dam supplying raw water to a treatment plant designed for an output of 2 000 m³/h (12.7 mgd).

● *Pretreatment with chlorine:*

Pretreatment with chlorine or one of its compounds (Eau de Javel (bleach), hypochlorite, chlorine dioxide) may be necessary to protect mains carrying raw water.

When water with a high plankton or organic-matter content must be piped over a long distance to the main treatment plant, treatment with an oxidant is essential because the accumulation of plankton on the pipe walls will probably quickly reduce flow. The same treatment is necessary for short mains if fresh-water molluscs are present (Dreissena polymorpha).

Ferruginous or sulphate-reducing bacteria can attack the iron in metal pipes; this raises the iron content of the water, particularly when treatment is stopped; these problems can be overcome by adding chlorine.

3.1.2. RAW WATER STORAGE

Storage of raw water is advantageous in cases of prolonged drought (reduced river flow, with simultaneous deterioration of quality) or of accidental pollution. In the latter case intake from the river can simply be stopped and the stored water used instead.

While the water is stored, certain of its characteristics may be improved (reduction of suspended solids, of ammonia by nitrification and of bacterial flora).

Raw water storage has, however, some drawbacks. Where geographic and weather conditions favour plankton growth, algae and fungi may sometimes proliferate, and their metabolites may impart an unpleasant flavour to the water, which may prove very expensive to remove. In addition, the system requires a large area of land, which is costly in an urban environment. The reservoir may also require regular cleaning.

3.1.3. PRECHLORINATION

Prechlorination before clarification is usually advisable as a means of improving the quality of the water, in terms of filtrability and clarity. The chlorine has an oxidizing effect on the various substances present in the water:

— on the Fe (II) and Mn (II) ions;
— on the ammonia, forming chloramines or destroying them when the break-point is past (see Chapter 27, page 954). If the ammonia level is very high this treatment may be unsuitable, as it involves large amounts of residual oxidant and high operating costs;
— on the nitrites which become nitrates;
— on the oxidizable organic matter;
— on the microorganisms (bacteria, algae, plankton) which are liable to develop in the purification plant and to set up anaerobic fermentation, for example.

Pre-chlorination does not always reduce the colour of the water (except when such colour is due to humic matter).

The treatment can be simple chlorination, break-point chlorination or even superchlorination.

It is always best to pre-chlorinate to a **chlorine content slightly above the break-point** ,whenever this is possible and does not involve too heavy a dose of chlorine. This destroys all pathogenic germs and removes the maximum quantity of bacteria, harmless germs, plankton and chloramines; this also gives the lowest taste threshold.

These advantages are particularly appreciable at certain times of the year: at others, when the quality of the water is better, doses of chlorine smaller than the break-point level can be used.

The contact time is basically determined by the aim of the treatment. Chlorine acts very quickly and very vigorously during the first few minutes after it has been added; but it takes a long time to establish a stable residual chlorine content.

The time factor is very important in the case of disinfection, but less for pre-chlorination, where the main purpose is very quickly achieved even when powdered carbon is used in the clarifier.

Lastly, in certain cases where water is drawn from a surface source which may contain viruses, chlorination beyond the break-point may be necessary, with prolonged contact.

After prechlorination the chlorine is sometimes removed by sulphur dioxide (SO_2) or sodium thiosulphate.

3.1.4. AERATION (see Chapter 13, page 369).

The water may have to be aerated:

● *if it contains excess gases:*

— hydrogen sulphide (H_2S), which gives a foul taste and is easily removed by atmospheric aeration;

— oxygen, when the water is supersaturated and the release of oxygen is liable to cause trouble in the clarifiers (where the floc tends to rise to the surface) and in the filters, which are liable to false clogging caused by the release of gas within the filter bed;

— carbon dioxide, (CO_2) which makes water aggressive; it can be removed by aeration at atmospheric pressure. The degree of aeration varies with the mineral content of the water; in some cases it may be necessary to remove only part or the carbon dioxide; the remainder helps to raise the mineral content by reacting on the neutralizing agents;

● *if the water is short of oxygen;* the effect of aeration is then:

— to oxidize the ferrous and manganous ions;

— to nitrify the ammonia under certain conditions;

— to increase the oxygen content and thus to make the water more palatable; the addition of oxygen to water which is rich in ammonia or sulphates in some cases also facilitates control of anaerobiosis and prevents corrosion of metal pipes.

3.1.5. CLARIFICATION

The degree of clarification varies with the turbidity and colour of the water and its suspended, colloidal and organic solids content. According to these factors it may take the form of:
— complete coagulation, flocculation, clarification and filtration;
— partial coagulation, micro-flocculation and filtration.

As already mentioned in Chapter 3, page 61, the addition of a coagulant lowers the negative electric potential of the particles in the water. One method is to add a big enough dose to cancel all or almost all of this potential; this gives

Fig. 418. — The 10 Aquazur filters (each with a surface area of 80 m² (860 sq ft)) and their control desks. Total output: 4 800 m³/h (30.4 mgd). Water works of SAINT-ÉTIENNE (France).

complete coagulation of the colloids, leading to optimum clarification after flocculation, settling and filtration.

It is also possible to add only a small quantity of coagulant, this gives **partial coagulation** of the colloids, with the formation of very fine floc (micro-flocculation) which can be separated by filtration, with or without the addition of coagulation aids. The minimum levels for suspended solids, colour and organic matter are not reached by this method, but it is normally adequate if the raw water is not too heavily polluted.

A. Clarification by complete coagulation, flocculation, settling and filtration.

This procedure is reserved for water with one or more of the following characteristics:
— suspended solids in excess of 20 to 40 g/m³, during all or part of the year;
— colour in excess of 30 mg/l on the Pt-Co scale (other treatments, which can be used when the colour of the water is its only defect, are discussed later);
— a high organic matter content which must be reduced to a minimum;
— a concentration of heavy metals in excess of the maximum recommended limit;
— a high plankton content, even if only temporary; this treatment, combined with pre-chlorination, is in fact the only way of removing 95 to 99 % of the plankton, the remainder being removed by filtration. As described in Chapter 9, p. 248, microstraining is quite incapable of achieving this result.

The clarification process can be varied according to the amount of suspended solids in the water.

● *Clarification of very turbid water.*

When the water is liable to contain more than 2 000 to 3 000 g/m³ of suspended solids over a long period, the following will be necessary:
— either: single-stage settling and clarifying in a scraper-type flocculator/clarifier. This process is only possible if the maximum content is not too high and does not produce so much sludge that it is liable to choke the clarifier. The generally-accepted upward flow rate varies between 1 and 1.5 m/h (0.4 to 0.6 gpm/sq ft)
— or: two-stage settling (de-silter and a finishing clarifier). This process is suitable for water with a very high clay content.

In order to be effective, the first clarifier must not be considered as a grit removal unit. If the water contains large amounts of sand, particles with a diameter of 0.1 to 0.2 mm must be removed at a prior stage; otherwise the scraper of the desilter is liable to jamming or damage.

If the suspended solids content is high, the plant will have to be over-dimensioned if the amount in the effluent is to be reduced sufficiently without using reagents. In the case of a smaller clarifier, straightforward static operation is adequate during periods of average turbidity. At peak periods (over 5 000 to 1 000 g/m³), the addition of a smaller dose of coagulant and/or flocculation aid than that determined by the jar test will ensure the required degree of pre-treatment at a relatively high upward flow rate.

The actual rate is fixed basically by the nature and amount of the matter to be removed, by the volume of sludge produced and by the amount of coagulant added.

In times of spate, it may also be necessary to add a neutralizing agent to correct the pH of the water; at other times there is no advantage in adding such a reagent.

The second clarifier then receives water which varies in quality within the permitted limits, and after complete coagulation and flocculation, delivers clarified water which is always if good quality.

In the interests of the consumer's health, filtration is regarded as a finishing process which can be still further improved by polishing treatments.

On this basis, the addition of reagents in two stages at peak periods does not result in overdosage of reagent as compared with the single stage process. The reagent feed unit must, of course, be very carefully designed to handle peaks which can vary from year to year. Detailed knowledge of the water supply and a wide safety margin of reagent feed capacity are therefore necessary.

● *Clarification of moderately-turbid water* (20 to 2 000-3 000 g/m³) :

Complete coagulation combined with flocculation and single-stage clarification is normally sufficient in this case. The process can be carried out either in a flocculator followed by a static clarifier, or in a sludge-blanket flocculator/clarifier of the scraper type or with sludge recirculation.

To obtain the best results, it is first necessary to fix the right dose of coagulating agent and to adjust flocculation pH, which may have to be corrected by adding a neutralizing agent.

The use of a flocculating aid is almost always advisable, because it not only increases the rate of clarification but also ensures a better clarified water at all times. For an identical water quality, experience shows that there is as a rule no additive which allows the dose of coagulant to be reduced.

If a lower quality of water is accepted, without, of course, going outside the official standards, the dose of coagulant can be reduced whether or not a coagulation aid is used.

In countries where the use of synthetic adjuvants in drinking water is permitted, attempts have been made to reduce the dose of coagulant by adding cation adjuvants; laboratory tests are carried out before the adjuvant and dosage rate are selected.

Lastly, for these various reagents to have the maximum effect, flocculation must take place in a concentrated-floc medium; this explains the value of the flocculator/clarifier in which the concentration of flocculated sludge (and not of suspended solids) is highest.

Another possibility is a conventional flocculator followed by a short-retention-time clarifier which provides water within the required standards but not of the very best quality.

The process consisting of complete coagulation, flocculation, single-stage clarification and filtration is by far the most widely used; it is compatible with simultaneous use of additional treatments such as the removal of iron and manganese and final "polishing". It has the further advantage that it can be used as such during part of the year and can be switched to partial coagulation followed by with filtration for the rest of the time; the clarifier is then by-passed and is used as a single contact tank before filtration to allow pre-chlorination if necessary.

Lastly, in the case of surface waters which have a high organic matter content associated with significant quantities of iron, it may be necessary to carry out two coagulation-flocculation-clarification processes at different pH values. By this method it has been found possible to reduce organic matter by 75 to 80 % after filtration in cases where straightforward clarification could do no better than 50 % because of the presence of complexes due to various pollutants. These are, of course, exceptional cases.

Where the characteristics of the water and of the precipitates formed by flocculation are suitable, the latter may be separated out by flotation, a process which produces a highly concentrated sludge and sometimes economizes on reagent consumption.

B. Clarification by partial coagulation, flocculation and filtration.

When water is not permanently heavily polluted (less than 20 to 40 g/m³ S.S. as a rule), is only slightly coloured (less than 30 units), and contains only small quantities of organic matter, iron and manganese, it can be treated by partial coagulation followed by filtration. Addition of the coagulant normally requires some time of contact before the water reaches the filters, but the filtration aid is added at the filter intake. The dose of coagulant determines the final turbidity after filtration, while the adjuvant serves to slow down the rate at which the very fine, slightly cohering floc enters the filter bed. This type of treatment is reserved for a fairly limited category of slightly polluted waters; otherwise the final quality would not be acceptable because less organic matter is removed than by complete coagulation.

The use of this type of treatment is also limited by the maximum retention capacity of the filter media; this capacity can be increased by using a very thick layer or beds with materials of different grain size. (See Chapter 9, page 264).

Lastly, the limit can be further extended by using two filter stages, but this method is not as efficient as clarification by complete coagulation, flocculation, settling and filtration. When a slightly-polluted water is strongly coloured or contains a certain amount of iron or manganese, partial coagulation and filtration can be used if it is preceded by ozonization; with this method, the ozone acts as an oxidizing agent and not as a coagulant; experience shows that its addition does not modify the Zeta potential of the water. The use of a dual-media filter is particularly suited to this treatment.

3.1.6. DISINFECTION

● *Purpose of disinfection:*

— It is possible to inject a disinfecting agent which will at all times ensure a water supply completely free from putrid bacteria and pathogenic germs, in accordance with official standards and tests based on *Escherichia coli*, faecal streptococci and the sulphite-reducing Clostridium.

A minimum contact time of 20 to 30 minutes, with a recommended time of 1 to 2 hours, and a residual chlorine or chlorine dioxide content of 0.05 to 0.2 mg/l, is generally sufficient. Both the contact time and the residual content have to be varied according to the amount of ammonia in the water, the nature of the disinfecting agent and whether the water has been pre-chlorinated.

By using ozone, the contact time can be cut to about 5 minutes for the same residual content as above and with the dose unaffected by the amount of ammonia.

— The aim may also be to carry disinfection further than the official standards and to eliminate all harmless germs as well; for this purpose, the water usually has to be given a much stronger dose of disinfecting agent which leaves a substantial residual free content.

In the case of chlorine and its compounds a dose in excess of breakpoint has to be used with a contact time of at least 1 hour. The amount of residual free chlorine depends on the nature of the water and of the germs to be eliminated.

Experience with ozone shows that the dose used must leave a residual content of at least 0.35 mg/l for something like 4 minutes. It will be seen that this rule is the same as that for the destruction of viruses.

— Lastly, superchlorination may be utilized to eliminate certain parasites such as the vector of bilharzia. Super-ozonization with a residual content around 0.9 mg/l may also be used.

● *Choice of disinfecting agent.*

The choice of agent is determined by technical considerations (normal or over-standard disinfection, taste problems) and economic factors.

— **Chlorine,** Eau de Javel (bleach) or calcium hypochlorite can be used if the water to be disinfected contains no organic matter or chemical pollutants liable to form compounds which will give a foul taste to the water. This risk is reduced to a minimum when disinfection is carried slightly beyond the breakpoint, provided the amount of residual free chlorine at the plant outlet is not too high; it if is, the taste of chlorine must be removed by partial destruction with hyposulphite, or better still, sulphur dioxide.

Chlorine or hypochlorites can also be used before a final treatment with granular activated carbon, which removes the taste-carrying organic matter and destroys the excess chlorine by catalysis. After filtration through activated carbon, a very small additional amount of chlorine can be injected to maintain a level of residual chlorine in the system, without risk of leaving a taste, provided the lining material or earlier deposits in the mains do not themselves give a taste. Chloramines or chlorine dioxide can be used for this additional chlorination.

The action of the chlorine is strongly influenced by the pH; the higher the pH the bigger the dose of residual chlorine required to be equally effective with the same contact time. This factor must be taken into account when neutralization is necessary to raise the pH of the filtered water.

— The **chloramines** limit and often remove the taste which can develop if chlorine is used on its own. They can be effective in preventing a taste of chlorphenol, but this is not always the case.

— **Chlorine dioxide** (ClO_2) systematically prevents the formation of chlorphenol but has no effect on a number of other tastes, such as those of earth or slime. It should therefore be used only when there is a possibility of chlorphenol tastes and no others.

To avoid the presence of too much sodium chlorite in the water, the dose of dioxide must be limited, and the amount of chlorine involved in its formation must exceed the stoichiometric quantity because the reaction is reversible.

— **Ultra-violet rays** are an effective disinfecting agent and also kill large numbers of viruses, provided they are applied to a thin stream of water and at sufficient power; and provided also that the lamps are changed immediately when they lose an appreciable fraction of their power. The water must be clear, colourless and free from turbidity; it must contain no iron, organic colloids or planktonic microorganisms likely to form deposits on the pipes, with a consequent considerable reduction in radiation.

If these conditions are fulfilled, all active or sporulated living cells exposed to the UV rays die or at least are no longer capable of reproducing or acting on the surrounding medium.

To guarantee reliability, the unit must be large, well-monitored and well-maintained, and operated with water of constant quality throughout the year.

— **Ozone** is the ideal disinfecting agent; if certainly costs more than chlorine or its compounds, but it is much more effective to well beyond the actual disinfection stage. Ozone acts by oxidation, through the addition of an atom of oxygen, and by ozonolysis, which enables it to act on double bonds by fixing the complete ozone molecule on the double bond atoms (acting on proteins, enzymes, etc.). All these properties make it effective against viruses, taste, colour and certain micropollutants.

The required dose of ozone varies very widely according to the quality of the previous treatment. For example, in the case of an initially highly-polluted but perfectly-treated surface water, some 1 to 1.2 g of ozone per cubic metre are needed to give a residual content of 0.4 g/m^3; after a poorly-designed or badly operated treatment (poor prechlorination, insufficient coagulant, etc.) 1.7 g of ozone per cubic metre are needed to give the same residual content. The required dose is therefore greatly influenced by the quality of previous treatment equipment and the way it is used.

Unlike chlorine (which is greatly affected by the pH of the medium), the pH has little influence on the amount of ozone required for disinfection.

When a high residual content is required, a number of precautions must be taken; first, provision must be made for the effective destruction of the undissolved ozone escaping from the air vents of the contact towers (see Chapter 15, page 423); allowance must also be made for the proximity of the first consumers of water treated in this way; although ozone is an unstable gas, traces are still found in the water after more than an hour if the residual free ozone content is 0.4 g/m^3; this means that when the treated water is kept in the tank for only a short time there is a danger of corrosion in the homes of consumers living near to the purification plant. The excess ozone in the water for distribution should then be neutralized.

It frequently happens that the treated water remains in the storage reservoir for quite a time and that the first consumers are some distance from the treatment plant. Residual ozone content is then nil. As the water was perfectly disinfected to begin with, consumers are then in exactly the same situation as if the water treated with chlorine contained traces only.

However, the possibility of plankton growing on the walls of mains must be borne in mind.

It sometimes happens that a few phyto- or zooplankton survive in the treated water or in the mains, and feed on the organic matter which forms a mucilaginous coating on the pipe walls. This plankton, which is not exposed to any residual disinfecting agent, is liable to proliferate and again leave the water with a taste.

Ozonization is therefore followed by the addition of a very small dose of a persistent residual disinfecting agent to prevent such proliferation. Chlorine, or preferably chlorine dioxide, can be used for this purpose without leaving a taste, because the ozone has previously oxidized the organic matter which is the potential cause. This residual disinfecting agent can be added continuously in small quantities, or intermittently in larger quantities, so that traces will still be present at the farthest point of the system. The best method for each type of system must be decided according to the circumstances.

3.1.7. ADSORPTION

Where highly polluted surface water has to be made potable, micropollutants, tastes and odours are increasingly being removed by adsorption on activated carbon.

The place of ozone or chlorine disinfection and activated carbon adsorption processes in a treatment system must be individually studied in each case (for adsorption, see Chapter 3, page 86, and Chapter 11, and for micropollutant control pages 646 to 651).

3.1.8. TREATMENT OF SLUDGE

The sludge produced during the treatment of water for human consumption is extracted or drained from the clarification unit, when one is used, or washed out from the filters.

The suspended matter in this sludge comprises the matter present in the water before it is treated, i.e. plankton, flocculated mineral and organic matter, metallic hydroxides (iron, manganese), and matter added during treatment, i.e. metallic hydroxides from the coagulating agent and powdered activated carbon when it is used. When all the sludge comes from the filters only, the concentration is from 200 to 1 000 g/m^3 and higher than the content normally allowed for discharge into rivers.

When the treatment includes clarification as well as one or two filtration stages, the sludge has to be collected for treatment, either by recirculation of all the filter wash-water to the head of the plant by means of a small-capacity pump, or by clarifying this wash water separately. The clarified water is then recovered for recirculation or discharge, and the settled sludge is combined with the sludge extracted from the main clarifiers.

This sludge, which generally has a dry matter content of between 2 and 15 g/l, then has to be treated. In some cases, where the raw water is very heavily polluted, the concentration can be very much higher, reaching more than 100 g/l; this is, however, an exception and occurs when rivers are in spate, when waste can be discharged directly into them and the pollution is essentially mineral.

Assuming that the original sludge contains from 2 to 15 g/l of dry matter, various methods of treatment are possible:

— concentration and spreading on drying beds;

— concentration and freezing;

— concentration and passage through filter presses, vacuum filters (with or without pre-coat filters), filterbelt presses or centrifuges;

— recovery of the coagulating agent by acidification and treatment of the neutralized sludge on vacuum filters, filter presses or centrifuges.

These various forms of treatment are dealt with in Chapter 17.

Sludge from the purification of drinking water often contains a considerable amount of organic matter and a significant proportion of metallic hydroxides. The sludge concentration unit must therefore be of ample size, and conditioning will be necessary to give a clear effluent and a sludge that can be shovelled after dewatering and to prevent clogging of any filter fabrics that may be used.

The most suitable form of treatment for sludge is decided in each case by assessing all the technical and economic factors involved.

3.2. Specific forms of treatment to remove and correct the constituents of natural water

3.2.1. REMOVAL OF IRON AND MANGANESE

Iron and manganese have to be removed from drinking water for various reasons:

— they cause corrosion and pipe blockages (either directly, by precipitating and forming deposits, or indirectly by providing favourable conditions for the growth of specific bacteria);
— they affect the appearance of the water;
— they impart a metallic taste;
— they cause laundering difficulties.

They are equally undesirable in many process waters, including those used in the dairy, paper and textile industries (see Chapter 23).

In surface water iron and manganese are usually present in oxidized form, as precipitates; they can usually be removed by conventional clarification.

In underground water which is deprived of oxygen they are present in reduced form (oxidation level $+ 2$), and in solution. Such cases require special treatment, which is examined below.

A. Natural state of iron and manganese: problems of analysis.

Ferric iron other than in complex form is found as a precipitate; the **dissolved forms of iron,** which require the types of treatment considered in this chapter, may include:

● *Iron* (II), either as Fe^{2+} or as hydrated ions: $FeOH^+$ to $Fe(OH)_3^-$. In water with an appreciable TAC value, the Fe^{2+} ion is primarily found as a hydrogen carbonate (or bicarbonate), and its solubility will, according to the laws governing chemical equilibrium, be as follows:

$$(Fe^{2+}) = \frac{K'_{FeCO3}}{K'_2} \cdot \frac{[H^+]}{[HCO_3^-]} \neq \frac{[H^+]}{[HCO_3^-]}$$

with

$$[Fe^{2+}] \cdot [CO_3^{2-}] = K'FeCO_3$$

and

$$\frac{[H^+] \cdot [CO_3^{2-}]}{[HCO_3^-]} = K'_2$$

In the presence of H_2S, solubility is lower (because of the low value of the solubility product of the ferrous sulphide which precipitates).

● *Complexes*, formed with Fe^{2+} or Fe^{3+}:

— mineral: silicates, phosphates or polyphosphates, sulphates, cyanides etc.
— organic: a genuine complex forming process may be involved (chelation or peptization), particularly with humic, fulvic or tannic acids, etc.

To design an iron-removal process it is not enough, therefore, to know the total iron content; the different forms in which the element is likely to be found must be known as well. The various forms of iron in water may be summarized as follows:

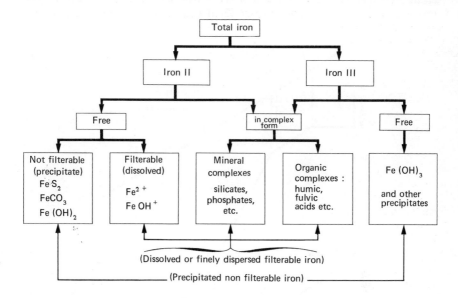

In practice, determination of total iron, total filterable iron, total Fe^{2+}, and filterable Fe^{2+} provides sufficient information. Removal of the dissolved forms will cause difficulty primarily in the presence of complexed iron. If it is impossible to make a detailed analysis on site, the presence of complexes, and hence of treatment difficulties, may be assumed if the total content of iron in solution exceeds the theoretical solubility value based on pH and alkalinity.

Manganese, like iron, may be present in various forms: bicarbonates, mineral and organic complexes, etc. The same analytical investigations should be carried out as in the case of iron.

Before any problem of iron or manganese removal can be tackled, most of the characteristics of the water must be determined on the spot, after sufficient water has been drawn from the well to ensure that the sample is representative.

The form in which the iron is present in the water is primarily dependent on pH and oxidation-reduction potential; figure 419 shows that a dissolved form of iron (such as Fe^{2+} or $FeOH^+$) can be changed to a precipitated form ($FeCO_3$, $Fe(OH)_2$ or $Fe(OH)_3$ by raising the potential (oxidation) or the pH or both together. Manganese obeys similar laws. The various treatment methods listed below are, in general, based on these principles.

B. Oxidation and filtration.

This is the most widely used technique, particularly on ground water. A number of supplementary types of treatment may be added: pH correction, chemical oxidation, clarification, etc. Water from a deep source which is deprived of oxygen must always be aerated, even if a chemical oxidant is also used.

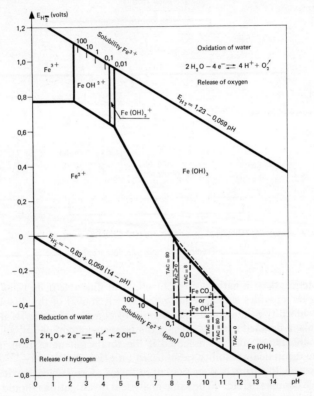

Fig. 419. — "pH-potential" diagram for iron (zones of ion and precipitate presence) according to HEM, 1961.

Prechlorination is sometimes advisable: it accelerates the oxidation of iron and allows ammonia to be removed chemically. It may also be harmful, if the break-point can not be reached; in such cases it stops certain forms of biological iron-removal and nitrification, without exerting a strong enough oxidizing power. Preliminary tests must therefore be carried out before any prechlorination treatment is attempted.

The most widely-used systems of application are reviewed below.

B1. *Iron removal without clarification* (aeration/filtration)

● *Principle:* this process is applied to raw water with a maximum iron content of 5 mg/l, and with no other unfavourable characteristics (manganese, colour, turbidity, humic acids); a low level of ammonia and moderate aggressive carbon dioxide content may be tolerated. In some cases water with an iron content of up to 10 mg/l may be treated in the same way.

The first stage of the iron-removal treatment is based on oxidation of the iron (II) by oxygen from the air. The aeration process may take place (see Chapter 13) at atmospheric pressure in installations operating by gravity, or under pressure, in which case compressed air is blown into oxidation towers filled with contact material (usually volcanic lava).

An advantage of the latter type of installation is that it can be operated at the system's delivery pressure, without pumping. On the other hand, aeration at atmospheric pressure often provides a cheap means of removing aggressive carbon dioxide, which, if present in large quantities, requires expensive neutralization treatment.

The rate at which iron (II) is oxidized by the oxygen depends on several factors, including in particular temperature, pH, and content of dissolved oxygen and iron. The reaction may be expressed as follows:

$$4\,Fe^{2+} + O_2 + 8\,OH^- + 2H_2O \longrightarrow \underline{4\,Fe(OH)_3} \atop \downarrow \tag{1}$$

and expressed kinetically according to STUMM & LEE's formula:

$$-\frac{d(Fe^{2+})}{dt} = k \cdot (Fe^{2+}) \cdot p_{o2} \cdot (OH^-)^2 \tag{2}$$

in which the constant k is a function of the temperature and buffering power of the raw water.

Reaction (1) shows that 0.14 mg oxygen is required to oxidize 1 mg of iron. Equation (2) shows that the reaction will be the faster the higher the pH and the nearer the water to the state of oxygen saturation.

The oxidation time calculated in the laboratory on a synthetic water may be considerably shorter in most installations, as result of the catalytic effect of:

— existing deposits;
— certain anions in the water (particularly silicates and phosphates);
— metallic catalysts introduced into the water during treatment; for instance, traces of copper sulphate may exert a strong influence over the oxidation of iron and manganese by oxygen or chemical oxidants.

Some biological processes, which are considered at a later stage, have a similar effect. The presence of humic acids, however, slows down the oxidation of iron.

According to the methods used, the precipitate formed may contain larger or smaller proportions of ferrous carbonate, which is more crystalline than ferric hydroxide. This feature explains the considerable differences recorded in the operation of certain plants; the effective size of the filter media may range from 0.5 to 1.7 mm, and the filtration rate from 5 to 20 $m^3/(m^2.h)$, i.e. 2 to 8 gpm/sq ft or even more.

The weight of iron retained per unit of filter surface area is also highly variable, for the same reasons; it may range from 200 to 2,500 g Fe per m_3 of sand according to the case. Generally speaking, dual-media filters (anthracite plus sand) are particularly well-suited to iron removal.

Some substances, such as humic acids, silicates, phosphates or polyphosphates, act as inhibitors of the precipitation and filtration of ferric hydrate. These effects can be controlled by additional treatment: oxidation (potassium permanganate, ozone), coagulation (aluminium sulphate) or flocculation (alginate) according to the case.

● *Type of installation:* the installations most commonly found are those operating **under pressure,** as illustrated in figure 420, which comprise:

— An *oxidation tower*, containing a bed of very pure, porous volcanic lava which divides the water and provides a large surface area for oxidation by contact with the air.

— A *filter* which is washable by a backflow of water and air scour. The filter may be combined with the oxidation tower.

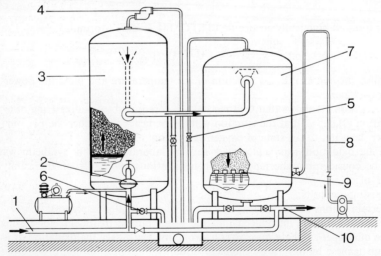

1. *Raw water inlet.*	6. *Drain valve.*
2. *Offset baffle-type mixer.*	7. *Sand filter.*
3. *Oxidation tower.*	8. *Scour air.*
4. *Air vent.*	9. *Floor with nozzles.*
5. *Air discharge valve.*	10. *Iron-free water outlet.*

Fig. 420. — Pressurized iron-removal plant.

Gravity units include aeration at atmospheric pressure (see above, page 369) followed by passage through open or closed filters (in the latter case, with or without pumping).

If the raw water also contains appreciable amounts of ammonia, the installation may be designed for simultaneous nitrification and iron removal: the aeration process then takes place in a basin filled with volcanic lava, on which the nitrifying bacteria develop. The volume of this nitrifying basin depends on the flow rate of raw water and the quantity of ammonia to be removed.

B2. Iron removal with clarification.

A clarification stage needs to be inserted between the aeration and filtration units in the following cases:

— high level of iron in the raw water, resulting in a undue amount of precipitate;
— presence of colour, turbidity, humic acids, complex forming agents, etc., which require the addition of a coagulant (aluminium sulphate or ferric chloride) at a dosage rate which is higher than about 10 g/m^3 of commercial product.

The sludge contact clarification processes described on page 172 et seq. are particularly suited to the treatment of such water. An aeration stage must precede clarification if the raw water is deprived of oxygen.

As a single plant is often supplied from several wells, the treatment must sometimes be adjusted to produce the most economical combination. In the installation illustrated in figure 421, for instance, two waters from different sources are treated. One contains a high proportion of iron, which is precipitated by lime in a Pulsator clarifier the second, with little iron, is mixed with the first after aeration by means of a sprinkler pipe fitted above the filters.

B3. Removal of manganese.

When a raw water contains manganese, iron is usually also present but the iron removal processes described above are generally incapable of efficient removal of manganese. Precipitation in the form of hydroxide, or oxidation by oxygen, would be feasible only at too alkaline a pH (at least 9 to 9.5); oxidation by chlorine is sometimes possible, but only in the presence of a large excess of chlorine which then has to be neutralized.

Sufficiently rapid oxidation can, however, be produced by chlorine dioxide, potassium permanganate or ozone, which raise the manganese (II) to an oxidation level of $+ 4$ and precipitate it in the form of manganese dioxide:

$$Mn^{2+} + 2ClO_2 + 2H_2O \longrightarrow \underline{MnO_2} + 2\ O_2 + 2\ Cl^- + 4\ H^+$$

$$3\ Mn^{2+} + 2\ MnO_4^- + 2\ H_2O \longrightarrow 5\ \underline{MnO_2} + 4\ H^+$$

$$Mn^{2+} + O_3 + H_2O \longrightarrow \underline{MnO_2} + O_2 + 2\ H^+$$

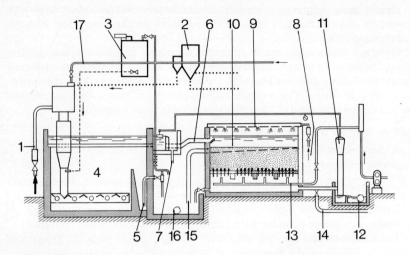

1. *Inlet for raw water rich in iron.*
2. *Lime dosing and injection.*
3. *Coagulant dosing and injection.*
4. *Pulsator clarifier.*
5. *Sludge concentration and extraction.*
6. *Clarified water from which iron has been partially removed.*
7. *Overflow.*
8. *Inlet for raw water containing only small quantities of iron.*

9. *Sprinkler pipe.*
10. *Open sand filter.*
11. *Filter regulation siphon.*
12. *Outlet for iron-free filtrate.*
13. *Filter-washing air scour.*
14. *Filter wash-water.*
15. *Outlet for sludge from washing process.*
16. *Drain.*
17. *Water main under pressure.*

Fig. 421. — Plant for the removal of iron from two waters with different iron contents.

According to these reactions, the theoretical quantities of oxidant required are (for 1 mg/l Mn (II):

Chlorine dioxide (ClO_2)	: 2.5 mg/l
Potassium permanganate ($KMnO_4$)	: 1.9 mg/l
Ozone (O_3)	: 0.87 mg/l

The actual dose may in fact differ very considerably from this, primarily as a function of pH, but also of contact time, existing deposits, organic matter content, etc. In practice it will range from 1 to 6 times the manganese content in the case of potassium permanganate, 1 ½ to 10 times in the case of chlorine dioxide and 1 ½ to 5 times in that of ozone.

The exact treatment rate can only be determined experimentally. A considerable saving in chemical oxidant can often be made by applying it **after aeration,** which will already have acted on the readily oxidizable substances, including iron (II) and hydrogen sulphide.

Technologically speaking, these energetic oxidation processes for the removal of manganese can be incorporated into the same gravity or pressurized units described above for iron removal purposes, with or without a clarification stage; it is merely necessary to add an additional reagent. Figures 422 and 423 give examples of gravity plants, with concrete filters, in which iron removal can be combined with manganese removal, coagulation — flocculation and pH adjustment. The exact point of reagent injection can sometimes only be determined during operation.

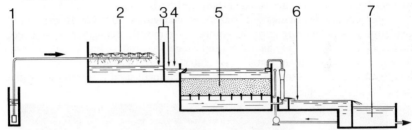

1. *Remote pumping of raw water, and pre-chlorination as necessary.*
2. *Aeration, elimination of CO_2 by spraying.*
3. *Correction of pH if necessary (to be carried out before or after total oxidation, according to circumstances; the exact point of injection will therefore be determined during operation).*
4. *Coagulant and/or flocculant if necessary. Complementary oxidation with $KMnO_4$ if the water contains manganese.*
5. *Sand (or dual-media) filter operating by gravity.*
6. *Chlorine disinfection.*
7. *Treated water storage tank.*

N.B. In the case of water containing manganese, aeration (2) and complementary oxidation (4) may be replaced by pretreatment with ozone.

Fig. 422. — Iron and manganese removal in an open plant—using aeration and filtration.

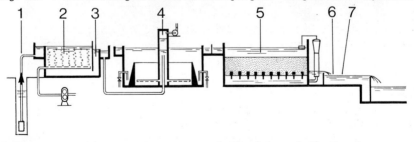

1. *Raw water pumping.*
2. *Aeration.*
3. *Injection of coagulant and adjuvant as necessary. If necessary, adjustment of pH and complementary oxidation with $KMnO_4$ or ClO_2.*
4. *Flocculation—clarification—iron removal.*
5. *Filtration.*
6. *pH correction, as necessary.*
7. *Disinfection.*

Fig. 423. — Iron and manganese removal by aeration, clarification and filtration.

In certain cases special requirements may have to be met for manganese removal:

— use of ozone: a water/ozone contact tower usually replaces the conventional aeration process referred to above but a pre-aeration stage may be advantageous, using air from the contact tower, which still contains usable traces of ozone (see fig. 447). This treatment may result in the formation of permanganate in the presence of excess ozone; to prevent a slight pink coloration of the treated water, it is worth filtering the oxidized water through anthracite (in a single or dual media filter) or through activated carbon, rather than sand;

— water containing manganese only: the manganese dioxide formed in the oxidation process may require, for thorough removal, a coagulation — flocculation stage and a dual media filter (anthracite + sand). But excellent results may also be obtained on sand, if the filter material is previously coated with a layer of manganese dioxide. The type of treatment used with zeolites or green sand can then be applied (see page 634).

Fig. 424. — Iron removal and neutralization plant at the Haguenau-Oberhoffen barracks (France). Output: 60 m³/h (264 gpm).

C. Treatment combined with carbonate removal.

Carbonate removal with lime, which results in a high pH, produces favourable conditions for iron and manganese removal. At pH 8.2 almost all the ferrous carbonate precipitates, as does the ferrous hydroxide at pH 10.5 (see fig. 425). In the presence of a high redox potential, the iron (II) in solution may be precipitated in the form of $Fe(OH)_3$:

$$Fe^{2+} + 3H_2O \longrightarrow \underline{Fe(OH)_3} + 3\,H^+ + e^-$$
$$\downarrow$$

In the case of manganese, the pH at which precipitation occurs is in the region of 9.2 in the case of carbonate and 11.5 in that of hydroxide.

Partial carbonate removal, at a pH in the region of 8, may thus result in complete removal of iron. In some cases, such as catalytic carbonate removal plants (Gyrazur, page 156), satisfactory manganese removal takes place at the same pH, whereas theoretically the process should be associated with total carbonate removal at pH 9.5 or 10.

The 1,000 m³/h installation built at Ratingen in the Federal Republic of Germany operates on this principle (see fig. 426): it simultaneously provides for removal of carbonates, iron, and mangagese, and for nitrification.

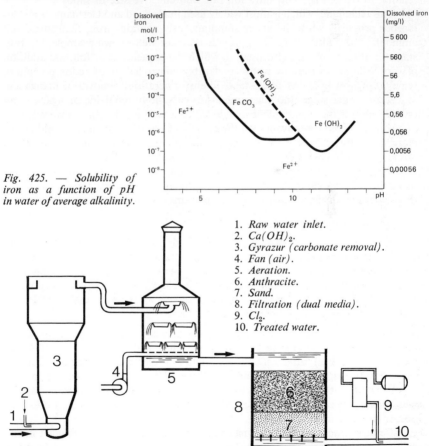

Fig. 425. — Solubility of iron as a function of pH in water of average alkalinity.

1. *Raw water inlet.*
2. $Ca(OH)_2$.
3. *Gyrazur (carbonate removal).*
4. *Fan (air).*
5. *Aeration.*
6. *Anthracite.*
7. *Sand.*
8. *Filtration (dual media).*
9. Cl_2.
10. *Treated water.*

Fig. 426. — Diagram of plant at RATINGEN (West Germany) designed for partial carbonate removal, aeration and filtration.

D. Biological treatment.

The metabolism of certain autotrophic microorganisms is based on the oxidation of iron and manganese; if conditions are favourable, reactions are very rapid and the two elements are completely removed. As ammoniacal nitrogen is essential to bacterial feeding, the presence of ammonia in the raw water will provide the right conditions for this type or treatment (simultaneous nitrification will often occur).

This principle does not differ essentially from that of the oxidation-filtration processes considered above, but the design of the units has certain special features, relating to oxygen content, filtration rate and effective size of sand.

This technique, which is most widely used in Holland and Germany, is advantageous primarily when the water simultaneously contains iron, manganese and ammonia. **Dry filters** (page 299) provide the most well-known example, but **two-stage filtration** plants are also used, in which the sequence is as follows: addition of very little air to the raw water (preferably controlled by its redox potential), removal of iron in a first filter stage, intensive aeration, removal of manganese in a second filter stage (the microorganisms capable of oxidizing manganese can only develop if the iron content is reduced to a very low level). However, it is impossible to count in advance on the growth of the microorganisms, and plants designed on this principle are sometimes difficult to control. The system should not be selected unless preliminary tests have proved it reliable.

The boundary line between purely chemical oxidation and biological treatment cannot be clearly defined, and many plants of the first type in fact owe their efficiency to the growth of microorganisms.

E. Filtration using special media.

● *Ion exchangers:* these would be considered only for the removal of small quantities of iron and manganese. There is a risk of rapid clogging, and they also remove other cations, such as Ca, Mg, etc. They are not generally used to deal with specific iron and manganese problems, particularly as they would not be cheap and would only rarely be accepted by the various Public Health Departments for the treatment of drinking water.

● *MnO_2 coated materials:* zeolites or green sands are sometimes used in the United States; they must first of all be enriched with MnO_2 (which they can fix by ion exchange), after which they function as electron exchangers. Manganese dioxide oxidizes iron (II) and manganese (II) (which are then precipitated and retained by the filter media), and is itself reduced to the sesquioxide Mn_2O_3. The dioxide can then be regenerated, in a continuous or a batch process, by potassium permanganate.

Certain products such as pyrolusite, CZ, etc., may also be included in this category, but the technique can be used only in small plants with a low iron and manganese content in water which is free of organic matter, and is uncommon in Europe.

If the raw water has been previously aerated, MnO_2 coated materials may also have a catalytic effect on the oxidation of iron and manganese by oxygen; in this case the manganese dioxide is unchanged and no regeneration is required.

● *Activated carbon:* iron and manganese are completely retained only if a powerful oxidant, such as ozone or permanganate, is previously applied. The process is expensive, but is warranted if unpleasant tastes also have to be removed.

● *Alkaline products:* carbonates or calcined dolomite have occasionally been used (Neutralite, Akdolit, Magno-dol, etc.). These methods of treatment offer little advantage unless it is necessary to neutralize aggressive carbon dioxide while increasing the alkalinity of the water; the dual function of the filter media (iron removal and neutralization) may also give rise to operating difficulties.

F. Powdered magnesia and diatomaceous earth.

In this process the powdered magnesia is introduced into the raw water after aeration, and the water is then filtered through diatomaceous earth. The magnesia appears to act in three ways:
— adsorption of ferrous ions
— precipitation of $Fe(OH)_2$ and $FeCO_3$;
— catalysis of iron oxidation by oxygen (primary function).
No major industrial developments have been based on this difficult technique.

* *

To summarize, iron and manganese removal treatment methods are often awkward to manage, and the same problem may be dealt with in several different ways. The essential thing is that the problem must be clearly defined: full knowledge of the raw water is required, and preliminary tests on a pilot plant are even more valuable in this case than they are in surface water treatment.

3.2.2. NEUTRALIZATION—REMINERALIZATION

Neutralization will be necessary if natural or treated water is not at carbonic equilibrium and contains excess carbon dioxide.

The water is then neutralized (see Chapter 14) by aeration or by the addition of alkaline reagents (lime, soda, sodium carbonate) or by filtration through alkaline earth products (marble, Neutralite, Akdolite, Magno, etc.). All these treatments, except aeration, increase the mineral content of the water. If this mineral content is insufficient to prevent the water from corroding metal pipes, and even if it is at carbonic equilibrium, further mineralization must be provided (see Chapter 14).

Aeration usually takes place at the start of the process, whereas chemical neutralization by an alkaline reagent may take place either at the end, after treatment, or partly at the start (e.g. pH adjustment of flocculation stage), with the rest at the end, or (more rarely) entirely at the start.

Remineralization usually occurs at the end of the treatment (in the case of very soft water or water which has been desalinated by reverse osmosis or some other method).

3.2.3. CARBONATE REMOVAL — SOFTENING

Water which is too hard, usually because of excess calcium, may be softened either by means of ion-exchange resins or by using lime to remove the carbonates.

● *Softening by means of resins* (see Chapter 10, page 321) : this method can be used only with clear or clarified water.

Only cation resins of the sulphuric type are allowed for the treatment of drinking water; anion resins are not officially approved in France.

The cation resins exchange their sodium ions for the calcium ions in the water. The alkali content and the sulphate and chloride contents of the water are not affected; the only change is that calcium salts are replaced by sodium salts.

The water so obtained has nil hardness and is unpleasant to drink; it is better to maintain a certain level of residual hardness (TH 8 to 15 French degrees) by softening only a fraction of the flow, which is then mixed with the rest of the water.

The value of this method of softening is that it leaves no solid waste and can be carried out under pressure. High-volume flows can easily be treated by means of continuous ion exchangers using the E.C.I.-Degrémont process.

● *Softening by removing carbonates with lime:* when the water is very hard and total alkali content (TAC) is high, it can be softened by using lime to remove the carbonates (see Chapter 6).
The process can be carried out:
— catalytically in a Gyrazur if simultaneous clarification is not necessary and the magnesia content is low,
— by clarification in all other cases.

Ferric chloride is the clarifying reagent to be used. At the pH at which treatment takes place, aluminium sulphate would solubilize the alumina, which might subsequently reflocculate.

Here again, and for the same reasons as before, it is necessary when treating drinking water:
— either to remove all the carbonates from part of the supply only and to mix it with the rest, which may also have to be completely clarified,
— or to remove part of the carbonates from the water while it is being clarified. In the latter case, the upward flow rate in the clarifiers will be lower the smaller the quantities of carbonates removed. Furthermore, the pH may be too high and will then have to be corrected.

In the case of water of high permanent hardness, carbonate removal with lime can be combined with sodium carbonate softening; in this case, after the minimum TAC has been reached, the calcium from the sulphates and chlorides is precipitated, and softening is continued by lowering the TH. So that the water will be pleasant to drink, some TAC must be restored to it by mixing with water from which the carbonate has not been removed.

3.2.4. REMOVAL OF SULPHATES AND CHLORIDES

As the use of ion-exchange resins to remove sulphates and chlorides from drinking water is not allowed, other processes have to be used, which, although involving a higher initial outlay than the resins, have the advantage of a much lower operating cost.

Large volumes of sea water or brackish water are usually desalinated by flash distillation. In the case of small and medium volumes, membrane desalination is the simplest solution; two main processes are available.

● *Electrodialysis* (see Chapter 12): with this process the water to be treated is divided into two fractions—a concentrated saline solution and a partially demineralized water—by means of membranes and an electric current. This process is only suitable for slightly brackish water, because the capacity of the membrane falls as the salinity of the water to be treated increases. It is only economical for water with a salinity of less than 3 g/l, which is reduced to 1.5 mg/l. The amount of electricity used increases as the salinity of the water decreases, because the demineralized water has higher resistivity. As a result, this method is of little use for water with a mineral content of less than 0.5 g/l.

The electrodialysis process is appropriate for daily flows up to several thousand cubic metres.

● *Reverse osmosis:* the principle is the opposite of electrodialysis and pure water is forced through the membranes at a pressure higher than osmotic pressure.

There is another basic difference: the salinity of water containing less than 500 mg/l can be substantially reduced (regardless of the original degree of salinity) for a very small expenditure of energy. This process is equally suitable for very saline water, such as sea water, and for slightly saline water of which the quality has to be improved.

3.2.5. FLUORIDATION AND FLUORIDE REMOVAL

The term "fluoridation" is preferable to "fluorination" because the element is used in the form of the fluoride, unlike chlorine which is added as such. By analogy, the term "fluoride removal" (defluoridation) will also be used.

It is generally considered that a small quantity of fluorine in drinking water (0.4 to 1 mg/l) promotes the formation of dental enamel and protects the teeth against caries.

On the other hand, too much fluorine leads to destruction of the enamel and causes a number of endemic conditions referred to collectively as "fluoroses", these include dental malformation, stained enamel, decalcification, mineralization of tendons, digestive and nervous disorders, etc.

The conditions occur in different people at very different fluorine contents. Water containing over 1 to 1.5 mg/l, expressed as fluorine, must be rejected or treated.

According to circumstances, therefore, it may be necessary either to add this element artificially or to remove it.

A. Fluoridation.

This treatment, which has the approval of the W.H.O., is practised mainly in the United States; a few isolated cases are also quoted in Europe, Australia, South America, etc. It is not yet becoming general, however, because it is not free from risk and has aroused some opposition. The following products may be used:

— sodium hexafluorosilicate: Na_2SiF_6 (the most common);
— hexafluorosilicic acid: H_2SiF_6;
— sodium fluoride: NaF;
— and more rarely ammonium hexafluorosilicate: $(NH_2)_2SiF_6$; hydrofluoric acid: HF; magnesium hexafluorosilicate: $MgSiF_6$; and calcium fluoride CaF_2.

The choice will be made according to the total quantity to be distributed and local economic conditions.

The treatment should be adjusted (taking into account fluctuations in the original fluorine content, if any, of the raw water) to give a fluorine content of between 0.4 and 1 mg/l, depending on the climate of the country concerned.

Every precaution should be taken to protect the personnel of the plant, and to prevent accidental overdosing.

B. Removal of fluorides.

Some kinds of natural water contain up to 10 mg/l of fluorine. The aim is to reduce this figure to about 1 mg/l (the permissible content falls as the mean annual temperature rises). The following processes may be used: '

B1. Tricalcium phosphate: the affinity of fluorine for this substance has long been known, because natural phosphates, such as apatites and phosphorites, always contain a substantial amount of fluorine (2 to 5 %), and the same applies to bones. It is believed that in apatite, for which the formula is $3 Ca_3(PO_4)_2$, $CaCO_3$, the carbonate ion is replaced by fluorine to give insoluble fluor-apatite. Hydroxyapatite also acts as an ion exchanger (the $CaCO_3$ being replaced by $Ca(OH)_2$) and can be indefinitely regenerated with soda, which re-converts the fluorapatite into hydroxyapatite.

In practice, either natural products, usually extrated from cattle bones—

bone ash (animal black) or powdered bone—or synthetic apatite, which can be produced in the water by a carefully controlled mix of lime and phosphoric acid, are used for this purpose.

They can be introduced into the water in the form of a fine powder, which also makes it possible to combine their action with that of aluminium sulphate (see below). In general, however, it is better to prepare them in the form of filter media, which then require regular regeneration.

B2. Use of alumina: aluminium sulphate can be used, but a very large dose is required, ranging from 150 to 1,000 g/m³ according to the case. Water treated in this way may contain a large amount of dissolved aluminium, and will then have to be reflocculated when the pH is adjusted.

Activated alumina has already been successfully used as a filter material; it can be regenerated by aluminium sulphate or by caustic soda and sulphuric acid. Its retention capacity may vary very considerably with the initial fluorine content of the raw water and the operating conditions, from 0.3 to 4.5 g F^- ion per litre of product.

B3. Softening the water with lime: this method can be used provided the water has a sufficiently high magnesium content, for it is the magnesia which adsorbs the fluorine. Otherwise the water must be considerably enriched with magnesium by the addition of magnesium sulphate or the use of dolomitic lime.

It is estimated that about 50 mg/l of magnesium are needed to remove 1 mg/l of fluorine.

B4. Filtration through activated carbon: this type of treatment (with regeneration by caustic soda and carbon dioxide) can be used only in a highly acid environment (pH not more than 3), and if followed by recarbonatation, which in most cases makes it impracticable.

B5. Other processes: if fluoride ions have to be removed simultaneously with excessive salt content from the water, reverse osmosis may provide the right answer. Electrochemical processes are also available, using aluminium anodes, although no major industrial application of such systems is yet known.

In all cases, preliminary tests (on the spot if possible) and an economic survey will be needed to determine the most suitable form of treatment.

*
* *

The application of the above techniques may result in processes which include either a clarification or a filtration stage, and some industrial plants, primarily of the latter type, have been built. The most commonly used filter media are products based on tricalcium phosphate (animal black or synthetic products), or activated alumina.

3.2.6. REMOVAL OF NITROGEN AND ITS COMPOUNDS

● The *ammonium* ion (NH_4^+), which is undesirable in drinking water, is not always easy to oxidize. If the quantity is small, chlorine can be used to convert it into chloramine. But as the required dose of chlorine is about 10 times the ammonia content, expressed as nitrogen, this process is ruled out when the content is high.

Chlorine dioxide and ozone do not act on the ammonium ion.

The ammonium ion can be *nitrified* through *dry filters* in the presence of iron and manganese, and also by aeration in contact towers containing pozzolana, in which nitrifying bacteria develop when a small amount of phosphate nutrient is added.

Preaeration at the head of a clarifier with no prechlorination stage (chlorine poisons the nitrifying bacteria), or ahead of low-rate filters, leads to partial nitrification, which can be supplemented by reinjecting re-aerated water into the ground and then collecting it again after a fairly prolonged percolation.

● **Nitrites,** which cannot be tolerated in drinking water, are easily transformed into nitrates by means of an oxidant (chlorine or ozone).

● When **nitrates** exceed the level laid down in official standards for drinking water some form of treatment must be provided to remove them; certain resins may be used, as may a biological *denitrification* treatment (see Chapter 24, page 102). When the mineral content is high enough to require treatment by reverse osmosis, nitrates are removed at the same time. The disinfection of water in which the nitrate concentration shows a tendency to increase must be particularly carefully monitored.

3.2.7. DESTRUCTION OF ALGAE AND PLANKTON

A. Elimination of algae

The growth of algae is promoted chiefly by the effect of sunlight on open tanks, clarifiers, reservoirs, swimming pools, etc. Free carbon dioxide is essential for their growth.

Algae can be destroyed or their proliferation restricted by means of copper sulphate (2 to 3 g/m^3) or by copper/chlorine disinfection. Unfortunately, copper is poisonous to fish, and there is no method of destroying algae without harming the fish in the water.

The proliferation of algae can be restricted by using lime to reduce the amount of free carbon dioxide in the water.

In pipes, the development of colourless algae is prevented by treating the water with lime or by copper/chlorine sterilization.

Intermittent massive doses of chlorine are the usual method of protecting condenser circuits.

Lastly, the proliferation of algae in settling tanks and open filters can be checked by adding a few grammes of powdered activated carbon per cubic metre. This acts by forming a barrier to sunlight entering the water, but chlorine must be added at the same time so that any microscopic algae already present in the raw water may coagulate; algicides (permitted in France for industrial water

treatment only) are now available commercially, including in particular salts derived from quaternary ammonium, sodium pentachlorophenate and certain organo-
sulphur derivatives.

B. Elimination of plankton

Plankton (see page 985) is the comprehensive name given to all the small
animal organisms (protozoa, worms, crustacea, insect larvae and so on) and
vegetable organisms (algae) which live in suspension in water. In size they vary
widely from one micron to several millimetres. They can be eliminated from
water by the following methods:

B 1. Microstraining (see Chapter 9) : the water can be filtered through a microstrainer, particularly if it contains algae only, is not excessively turbid or coloured, and does not contain a great deal of organic matter, etc.

The screen mesh must be suited to the species to be eliminated; it is usually
between 10 and 40 microns.

Microstraining is particularly effective for certain types of green algae (Pediastrum, Scenedesmus, Coelastrum, Cosmarium, etc.), of diatoms (Asterionella,
Melosira, etc.) and of cyanophyta (Anabaena, Microcystis, etc.); 90 to 95 %
elimination should be possible with these types; for phytoplankton as a whole,
the figure generally varies from 50 to 80 % over the year.

In the case of zooplankton, microstraining eliminates chiefly rotifers, ento-
mostracea, crustaceans (copepoda, cladocera) and nematodes.

In all cases, however, the reproductive elements (spores or eggs) are not
sufficiently eliminated, and microscreening has to be followed by chlorination
in order to be effective.

The microscreen for plankton removal should be located:
— either at the head of a slow filter unit, to take some of the load off the filter
at times of rapid plankton growth;
— or in the purification plant for lake or resevoir water, when only a limited
amount of plankton has to be eliminated.

B 2. Oxidation: all the disinfecting oxidizing agents (chlorine, chlorine dioxide,
ozone) destroy plankton, provided a sufficient residual content is maintained for
a time (which varies with the species). But some mechanical means of removing
the dead plankton is then required.

B 3. Complete treatment: if all plankton is to be eliminated, **pre-chlorination**
(to kill the organisms or destroy their activity in the sand of filter beds) must be
combined with **coagulation.**
— either coagulation through a sand filter if the water does not contain too
much plankton and suspended matter, and the frequency of washing depending
on the nature, size and quantity of the various forms of plankton. Rate of filtration
is linked with permitted frequency of washing and the automatic operation of
the filters. When water contains plankton, care must be taken in selecting the
rate of filtration;

— or coagulation adjusted to cancel the Zeta potential, followed by clarification preferably in an equipment of the Pulsator type with a concentrated sludge blanket and filtration through sand, all preceded by pre-chlorination to break-point. In all cases, plankton elimination rates of 98 to 99 % have been recorded for clarified water and of almost 100 % for filtered water.

With this last method, rapid filtration is possible, thus keeping the filters in continuous operation whatever the quantity of plankton.

3.2.8. ELIMINATION OF ORGANIC MATTER

The best method of removing organic matter, defined by the permanganate value test, is always the most difficult problem to resolve, and it is essential to choose the most suitable form of treatment. Only 10 to 30 % of the amount in the raw water is eliminated by chemical oxidation (even with ozone) or partial coagulation through filters. Good coagulation, flocculation and clarification eliminate from 40 to 70 %, sometimes with the addition of powdered activated carbon at the clarification stage.

Most of the remaining organic matter can be removed either by filtration through a pre-coat filter (with a coating of powdered **activated carbon** on plates or candles), or by filtration through granulated activated carbon.

This process removes from 75 to 95 % of the remaining organic matter. In the case of pre-coat filtration through activated carbon, which is renewed for each cycle, the rate of elimination is constant provided the nature of the water is also constant. A cycle usually lasts about a week with a continuous feed of powdered carbon (see page 341).

Fig. 427. — Eight superimposed filters filled with granular activated carbon used to treat river water. Ouput: 1 500 m³/h (95 mgd). LÜBECK (Federal Republic of Germany).

If only organic matter is to be eliminated, the life of the granulated carbon before saturation is shorter than for the elimination of detergents, pesticides and tastes. The time varies from 3 to 9 months according to the way the carbon is used, to the utilization rate expressed in volume of water per volume of carbon per hour and to the minimum permitted percentage reduction of organic matter.

The choice between these two methods depends on the simultaneous presence of pollutants and micropollutants, on their nature and on the cost of elimination.

Although the cycle through pre-coat filters is short, the use of powdered activated carbon is becoming expensive. Filtration through granulated carbon is more economical, although filtration is coarser. Similarly, if adsorption with prolonged contact is necessary, filtration through a very deep granular carbon bed is more suitable; it should not be overlooked, however, that the effect of the short contact time with the pre-coat is partly offset by the fineness of the powdered carbon which makes it more effective.

The final decision will be based on two factors—the quality obtained and the cost of the operation.

3.2.9. DEODORIZATION OF WATER

The "deodorization" of drinking water means the elimination of both the unpleasant tastes and the odours which are two outward signs of the same phenomenon.

A. Unpleasant taste in raw water.

Unpleasant tastes and odours are usually due to the presence of very small quantities of secretions given off by microscopic algae, and chiefly by actinomycetes (Streptomycae, Nocardia, Micromonospora, etc.), which develop on surface waters and on the beds of lakes and rivers under certain conditions of temperature and chemical composition. This process is often associated with the degree of organic pollution of the water, process agricultural activities (run-off of rainwater after manure has been spread, sugar beet season, etc.), and the seasons.

Actinomycetes and certain Cyanophyceae leave a muddy, earthy or mouldy taste; the substances responsible have recently been identified as geosmine and 2-methylisoborneol.

Water containing algae has the characteristic taste and odour of the predominant species; it is then said to taste mouldy, of grass, of geraniums, of beans, fishy and so on. These are, however, much less common than a "muddy" taste.

Some malodorous products may also be formed by the decomposition of vegetable or organic matter in the soil, certain kinds of fish spawn, industrial waste, etc. In the last-named case, however, the factory responsible can be identified by moving upstream and taking samples.

Apart from the qualitative aspect, every unpleasant taste can be expressed as a quantity by diluting the water under test with a pleasant-tasting reference water of similar mineral content, until the bad taste disappears; **the taste threshold** is then the inverse of the dilution required to obtain this result.

A properly-treated water should therefore have a threshold of 1, a value expressing the absence of unpleasant taste. Some countries measure by smell; in France, it is considered better to assess taste as if one were the consumer, although a measurement of smell is often more sensitive.

Tastes and odours can be eliminated;

● *by aeration* which eliminates *hydrogen sulphide* in particular;

● *by means of a powerful oxidizing agent.*

— *Ozone* is very effective. Bad taste may, however, be caused by several substances at the same time; a two-stage treatment is then possible, for example with powdered activated carbon (with simultaneous flocculation and clarification) followed by final treatment with ozone after filtration; in practice, both products are sometimes needed to remove all odours.

The action of ozone can vary with the temperature; below 5 °C it sometimes has only a limited action on the constituents which cause the taste.

This is why it is often useful to combine ozone with activated carbon when the water to be treated varies in temperature.

— *Chlorination* to beyond break-point followed by the removal of excess chlorine (page 402 and 342).

— *Chlorine* dioxide (page 405).

— A dose of potassium permanganate together, if necessary, with activated carbon.

— Other methods including treatment of the water with hydrogen peroxide and filtration through manganese dioxide.

● *by activated carbon.*

This product can be used alone in either powder or granular form. Provided that only a limited quantity of powdered activated carbon is required (15 to 20 g/m³) its use in a clarifier is more economical than that of granular carbon as a filter material. If powdered carbon has to be used in large doses from time to time, or if it does not eliminate all taste, good results can be obtained by combining a smaller dose in a clarifier with final ozonization, which also disinfects the water. The value of this combination, while capable of removing all unpleasant taste (this is the commonest case) lies in its great flexibility; when the weather is cold very little or no powdered carbon is required; as taste usually becomes more marked when the temperature rises, the dose of ozone is raised to the maximum capacity of the ozone production unit and the dose of carbon is then increased to deal with the highest taste levels. This cuts expenditure on powdered carbon and makes best use of the outlay on ozone treatment, which neither powdered nor granular carbon can rival as a bactericidal agent.

The cost of producing 3 g of ozone (depreciation included) is much the same as the cost of 10 g of powdered activated carbon; hence the value of using ozone as the basic treatment, when it is already used as a disinfecting agent, and of supplementing it with powdered activated carbon.

Finally, if the combination of ozone and powdered activated carbon does not remove all taste, if the average amount of carbon required exceeds 20 g/m³, or if no ozone is available, it is better to filter through granulated activated carbon, because the initial outlay is written off in a few years by the sums which would otherwise have had to be spent on large quantities of powdered carbon.

B. Unpleasant taste resulting from treatment.

The addition or substitution compounds formed when chlorine or ozone is used may leave an unpleasant taste. In particular, the presence of even small traces of phenol in association with chlorine produces chlorphenol which has a medicinal taste.

A taste is also produced by the combination of chlorine with a number of nitrogenous substances and by the formation of nitrogen trichloride NCl_3 which smells of geraniums.

The special odour of nitrogen trichloride can be identified by comparing the water in question with a sample prepared as follows: pour 250 ml of distilled water into a 500 ml flask and add a few crystals of ammonium chloride, followed by just enough chlorine water so that a small fraction of the sample immediately turns yellow when treated with orthotolidine.

In these conditions, nitrogen trichloride always forms in less than 15 minutes, because the liquid is acidified by the chlorine water. The flasks used to detect these odours should be kept stoppered and after use should be washed with running water to avoid any risk of explosion of the NCl_3.

Nitrogen trichloride forms more quickly with ammoniacal substances than with albuminoids, for which the reaction can last more than 2 hours. This explains why water which has no smell when it leaves the purification plant can have an odour when it reaches the consumer.

In all cases, disinfection to break-point (page 402) leaves the least taste. Superchlorination followed by total dechlorination eliminates all nitrogen trichloride and most of any chlorphenol present, but the removal of chlorine by chemical methods often leaves a taste of medicine, which is completely eliminated when activated carbon is used.

Chlorine dioxide (page 405) is effective in removing chlorphenols but not nitrogen trichloride.

All tastes and odours due to chlorination can be eliminated by using activated carbon (dealt with in the previous section).

C. Unpleasant taste developing in water mains.

The taste of chlorphenol caused by the reaction of chlorine with the coal tar in water mains normally disappears fairly quickly.

This problem has been eliminated by the general use of pipes lined with substances based on petroleum tar or bitumen. But other tastes are produced by bacteria and moulds which are very difficult to destroy. As a rule, however, such organisms only develop in water containing little oxygen and a lot of ammonia. Vigorous aeration of the water therefore checks their growth, as does the retention in the water of traces of antiseptic, chlorine or chloramines.

3.3.0. ELIMINATION OF MICROPOLLUTANTS

The methods of treatment discussed above eliminate most of the common pollutants and micropollutants. Consideration will now be given to types which can only be destroyed by special treatment; they include phenols, hydrocarbons, detergents and pesticides.

A. Phenols and phenol compounds.

Mechanical treatments and coagulation have no effect, and slow filtration does not eliminate these products completely.

Chlorine dioxide is a first means of removing the taste of chlorphenol. If the water contains variable or substantial amounts of phenol, excess chlorine dioxide must be added for reasons of safety even at the risk of adding too much sodium chlorite to the water. In such cases, ozone or activated carbon must be used.

● *Action of ozone:* ozone destroys phenol and phenol compounds provided the dose is properly matched to previous treatments, to the pH of the water, to the nature of the compounds involved and to the required final concentration.

pH is an important element in fixing the dose of ozone, which doubles when the pH drops from 12 to 7.

At the pH normally found at drinking-water purification plants (7 to 8.5 final), the amount of ozone consumed to remove each gramme of phenol is at a maximum. The dose of ozone varies according to whether the water contains pure phenol—which is rare—or diphenols, triphenols, cresols or naphthols, and according to whether these substances are associated with other compounds such as thiocyanates, sulphides, etc. The correct dose, which is usually around four times the compound content expressed as pure phenol, can only be determined by testing.

Prolonged contact is not necessary to oxidize phenols. When ozone is used, their destruction can be combined with complete disinfection of the water to be treated, provided phenol content is not high.

● *Action of activated carbon:* even after prechlorination, powdered activated carbon will reduce the amount of phenols in the water; the extent of the reduction varies with the type of phenol, the dose and the type of carbon and the concentration of the medium.

When all phenols have to be eliminated, they have to be adsorbed by filtration through granular activated carbon.

In general, phenols are easily adsorbed by the various kinds of granular carbon, even if the carbon is already exhausted for the removal of organic matter.

● *Combination of ozone and activated carbon:* a combination of ozone and activated carbon is normally only considered when there is a risk of sudden heavy increases in phenol content. In these circumstances, ozone on its own may not be sufficient, and there is no point in increasing the size of the ozone unit unduly when the problem can be resolved simply by adding an appropriate quantity of powdered activated carbon into the clarifier.

A combination of ozone and granular activated carbon should only be used when large quantities of phenol are present for a relatively long time.

B. Hydrocarbons.

Apart from the accidental discharges which stop a plant temporarily, any hydrocarbon film is generally eliminated at the water intake, and the traces are removed by coagulation-flocculation, clarification and filtration through sand.

If the hydrocarbons have left a taste in the water, this can be removed satisfactorily by adding a small quantity of powdered activated carbon to the clarifier (5 g per gramme of hydrocarbon). All remaining traces can be eliminated by filtration through granular activated carbon.

Activated carbon is, in general, the most suitable material for eliminating saturated chain hydrocarbons, the large molecules of which are only slightly soluble and cannot easily be destroyed by ozone.

Ozone destroys all polycyclic hydrocarbons such as 3-4 benzpyrene, which are suspected of being carcinogenic in the presence of other coumpounds, such as detergents.

C. Detergents.

A standard form af treatment comprising coagulation-flocculation, clarification and filtration generally removes little or no detergent from the water. Similarly, the addition of a pre-chlorination stage does not break down the detergents themselves. Foaming, ozone or activated carbon must be used.

● *Elimination by foaming:* the foaming method can be used when the detergent content is high and needs to be brought below the foam threshold (about 0.3 to 0.4 g/m^3).

Foaming consists of blowing large volumes of air into water at a point a little below the surface; it must be followed by additional treatment of the concentrated foam, either by evaporation or by adding activated carbon. It is wise to check the phosphate content of the water after it has been treated and to reduce it if necessary.

● *Action of ozone:* ozone decomposes a large part of those detergents that are not biodegradable: the dose can be high when the detergent content to be destroyed is considerable.

There is a first stage when the quantity of ozone to be used varies logarithmi-
cally with the drop in detergent content (see fig. 428).

Where C is the concentration of detergent obtained by adding a dose x of
ozone to water with concentration C_0, the value of x is governed by the formula:

$$\frac{C}{C_0} = e^{-ax}$$

where the coefficient a varies according to the nature of the detergents and the
other compounds traces of which are present in the water.

The dose required to reduce the quantity of non-biodegradable anionic
detergents by 50 % varies from 1.5 to 3 g of ozone per cubic metre of water.

When elimination has reached a given point (70 to 95 % according to the

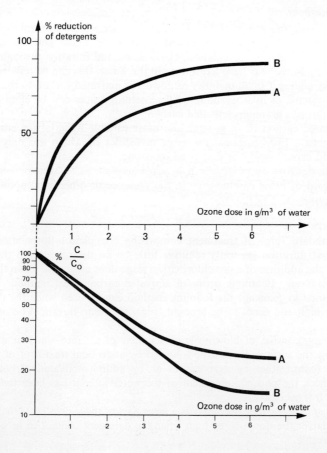

Fig. 428.

case), there is a residual detergent content which cannot be destroyed. This residue is, however, always less than what might be considered excessive because of possible indirect effects.

Prolonged contact has no effect on the amount of ozone required, and the same applies to treatment in one or several stages.

The addition of chlorine before ozonization does not reduce the amount of ozone required.

● *Action of activated carbon:* activated carbon eliminates hard detergents well by adsorption. It retains detergents much better than phenols (2 to 5 times according to the type of carbon) at the same final concentration in equilibrium. The usual method is to inject powdered activated carbon at the clarification stage where the pH is not important. A 50 % reduction in detergent content can be expected with doses of 12.5 to 25 g/m³ of water, according to the nature of the detergent and the type of clarifier used (static settling tanks or clarifier with a sludge blanket rich in carbon).

While a reduction of over 90 % is theoretically possible with powdered activated carbon, this figure, which is rarely necessary, would in practice mean using uneconomically high doses (40 to 80 g/m³).

If the detergent content is very high, granular activated carbon must be used.

● *Combined action of ozone and activated carbon:* in this case also a brief combination of powdered activated carbon and ozone is valuable when the detergent content reaches occasional peaks, or when a very low residual content is required (the threshold of detection is around 0.01 mg/l).

This method is only appropriate if large amounts of powdered activated carbon are not needed for long periods at a time; otherwise it is more economical to use granulated activated carbon on its own.

In practice, activated carbon and ozone should not be used together when the water contains detergents only, but rather when other pollutants and micropollutants have to be removed at the same time.

D. Pesticides.

The conventional treatment by flocculation/clarification and filtration is ineffective for pesticides in general, with the exception of DDT which is partially eliminated.

Slow filtration has a limited effect with some pesticides. Since this method does not remove all phenols or all detergents it cannot be considered as a modern polishing process.

● *Action of chlorine, chlorine dioxide and potassium permanganate:* chlorine and the peroxides have no effect on certain chlorine-base pesticides. Permanganate destroys some of them partially (e.g. lindane). On the other hand, certain weed-killers (organic compounds) are easily broken down by chlorine, chlorine dioxide and permanganate. These oxidizing agents cannot, however, be generally regarded as sufficiently effective against all pesticides found in surface waters.

● *Action of ozone:* some pesticides are little affected by ozone (dieldrin, HCH, etc.) while other organo-chlorinated formulations such as aldrin, in the amounts normally found in surface water, are destroyed by ozone doses of 1 to 3 g/m³.

● *Action of activated carbon:* activated carbon, in the form of either powder or granules, is the most effective means of eliminating pesticides. Neither the pH nor the temperature seems to have any marked effect on the rate of elimination.
— Using powdered activated carbon, with a maximum dose of 20 g/m³, the taste left by the majority of standard pesticides is eliminated. 5 g/m³ reduce taste effectively in certain cases (DDT), while 10 g/m³ eliminate 99 % of aldrin and dieldrin.
— Granulated activated carbon reduces the pesticide content to traces.

E. Heavy metals.

● *Coagulants:* coagulation by aluminium sulphate alone very successfully removes silver, lead and copper, and reduces the vanadium and mercury content by about 50 %, but reduces the level of zinc by only 10 %. The treatment has no effect on the levels of nickel, cobalt, manganese and chromium.

● *Powdered activated carbon:* the dosage rates normally used (20 g/m³) have very little effect in removing heavy metals. Much higher treatment rates would certainly have to be adopted.

● *Sand filters:* while coagulation using aluminium sulphate brings about an efficient reduction, the use of a sand filter reduces the levels of silver, mercury and copper virtually to zero, although the content of manganese, chromium, cadmium, vanadium and cobalt hardly varies. The levels of zinc and nickel are reduced, particularly in the presence of chlorine.

● *Granular activated carbon filters:* an adequate reduction in the content of undesirable or toxic ions can be obtained by filtration through granular activated carbon. Silver and mercury are completely removed, and the content of lead, copper, etc., is reduced below the level recommended as a guide in current regulations.

● *Chlorination:* chlorination, used in conjunction with coagulation, filtration through sand and filtration through granular activated carbon, improves heavy metal removal, particularly if the dose of chlorine used is slightly above breakpoint.

F. Conclusions.

The preceding analysis shows that conventional methods of treatment and slow filtration are of limited effect with micropollutants in general.

The most effective materials are activated carbon, with or without ozone.

● Activated carbon on its own is not sufficient to remove all pollution (some substances pass through it without being retained). The choice between powdered and granulated carbon, and the brand to be used, are determined by the nature of the micropollutants, the required quality of water and the possibility of regenerating the carbon.

● Ozone used alone has effects unattainable with carbon, particularly as a disinfecting agent.

A combination of the two is currently the best polishing treatment; the actual combination must be carefully worked out in each case by reference to the technical and economic factors involved.

When a purification plant is equipped with both an ozone disinfection unit and activated-carbon filters, there is no general rule that governs where the filters should be located in relation to the disinfection unit.

There may be a tendency to site the activated carbon filters after the ozonization unit in order to retain any oxidation products resulting from the action of the ozone and to prolong the life of the carbon; from the health point of view, however, it is better to locate the ozonization process at the end of the treatment sequence, particularly when viruses have to be destroyed. This dilemma can be resolved by locating activated carbon filters between two ozone treatment units.

4. PURIFICATION WORKS

4.1. Combined units and standard plant

4.1.1. SEMI-PORTABLE GSF PLANTS

Small hamlets, farms, long-term building sites, isolated communities and communities accidentally deprived of water can be temporarily supplied by combined units which can be quickly installed and put into service and removed when necessary.

DEGREMONT GSF plants meet these requirements; they are complete purification plants working under pressure, and are designed to supply drinking water from raw water with an average amount of pollution and a maximum turbidity of 500 mg/l. At higher concentrations, it is advisable to install a static pre-clarifier between the water intake and the pumps supplying the GSF; it can consist in most cases of a simple pit of sufficient size to precipitate the heaviest matter so that the raw water is within the design limits of the GSF.

These units have an electric raw-water pump which feeds a CIRCULATOR clarifier working under pressure followed by sand filters, and returns the treated water to a tank at least 10 m above ground. In this way, the filters can be washed by return flow and the treated water can be distributed by gravity.

Three reagents are added by dosing pumps:
— a coagulant before clarification;
— a neutralizing or pH correcting agent before clarification or after filtration;

— a disinfecting agent after filtration.

These reagents are prepared in plastic tanks.

The main features of the GSF are listed in the table below:

Type	Maximum output		Clarifier diameter		Filters			Minimum tank capacity	
	m³/h	gpm	mm	in	number	diameter		m³	gal
						mm	in		
GSF 0	1.5	6.6	1 200	48	2	450	18	5	1 320
GSF 1	3	13.2	1 600	64	2	650	25	9	2 400
GSF 2	6	26.4	2 300	90	2	800	32	12	3 150
GSF 3	9	39.6	2 800	110	3	950	38	18	4 800

Fig. 429. — Semi-portable GSF plant, type O and activated carbon filter.   Output 1.5 m³/h (6.6 gpm).   CNES, Ouagadougou, Upper Volta.

4.1.2. BIDONDO UNITS

These plants, which have a greater output than the GSF, can supply drinking water for a village or a larger community from raw water with an average amount of pollution.   They work under pressure and consist basically of one or more horizontal sludge-recirculation clarifiers, back-washed sand filters and three feed pumps to add reagents (coagulant, neutralizing agent or pH correcting agent and disinfecting agent).

The treated water is pumped to a tank at least 10 m above ground; the back-wash water is supplied from that same tank.

All the electrical and mechanical equipment, and also the reagent feed units, are housed in a building as shown in fig. 430.

The main features of the Bidondo plants are shown in the table below:

Type	Output		Clarifier		Filter
	m³/h	gpm			
VB 10	10	44	1 — diam 2 m (80″)	L = 5 m (16′)	1 — diam. 1.6 m (64″)
VB 15	15	66	1 — diam. 2.5 m (100″)	L = 5 m (16′)	1 — diam. 2.0 m (80″)
VB 20	20	88	2 — diam. 2 m (80″)	L = 5 m (16′)	2 — diam. 1.6 m (64″)
VB 30 A	30	132	3 — diam. 2 m (80″)	L = 5 m (16′)	3 — diam. 1.6 m (64″)
VB 30 B	30	132	2 — diam. 2.5 m (100″)	L = 5 m (16′)	2 — diam. 2.0 m (80″)

4.1.3. AQUAZUR PLANTS

The purification principles incorporated in the design of the BIDONDO plants are applied on a larger scale in the AQUAZUR plant.

The essential difference lies in the use of an open concrete clarifier; the plant cannot therefore work under pressure.

The filtrate is discharged into a storage tank placed in line with the clarifier, and is subsequently pumped away to distribution reservoirs. This storage tank also holds the reserve of wash-water for the filters. The reagents are administered by a dosing pump (fig. 431).

The cylindrical clarifier with conical bottom is of the sludge circulation type (Circulator).

Filter cleaning may be of two types:

— backwash only on CR plants;

— compressed air scour and backwash on CS plants.

The method selected will depend on the nature of the raw water.

It is preferable for the filtered water tank to be of the same diameter as the clarifier, so that the same formwork can be used for the reinforced-concrete constructional work. Many standardized versions of the AQUAZUR plant are available for flows ranging from 10 to 100 m³/h (44 to 440 gpm), and capable of handling the most common qualities of water, although preliminary treatment is necessary when the raw water is very polluted.

As these installations all comprise standard basic units, the output of a station which has become too small can be increased by merely adding further units and without any alteration to the existing plant, except possibly to the pumping arrangements.

The mechanical equipment of the plant, such as the raw water and treated water pumping sets, the wash-water pump, air blower and reagent feeders, are

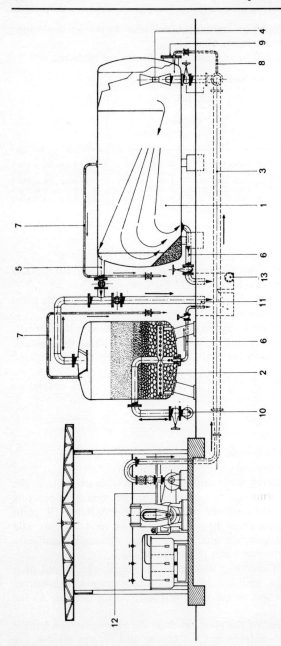

1. Clarifier.
2. Filter.
3. Raw water supply.
4. Ejector.
5. Clarified water outlet.

6. Drain.
7. Air bleed valve.
8. Tank bottom flushing feed pipe.
9. Manhole.
10. Filtered water pipe.

    Filter wash-water outlet.

11. Sludge outlet discharge.
12. Reagent supply.
13. To drain.

    Filter wash-water inlet.

Fig. 430. — *Bidondo plant and its associated clarifier.*

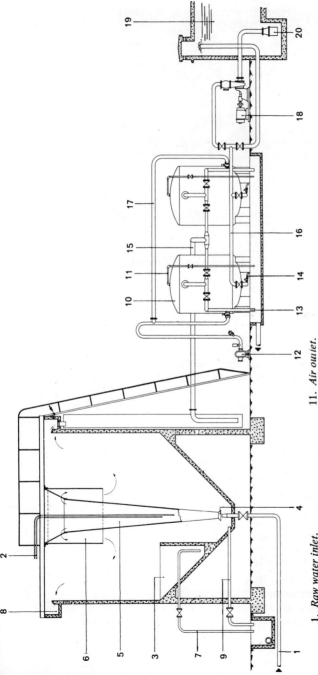

1. Raw water inlet.
2. Reagent injector.
3. Sludge concentrator.
4. Nozzle.
5. Mixing chamber.
6. Flocculation chamber.
7. Sludge outlet.
8. Clarified water channel.
9. Clarifier drain.
10. Filters.
11. Air outlet.
12. Scour air blower.
13. Wash-water outlet.
14. Air vent and drain.
15. Clarified water inlet.
16. Filtered water pipe.
17. Blown air header.
18. Filtered-water and wash-water lift pump.
19. Filtered-water storage tank.
20. Filtered-water suction unit.

Fig. 431. — Functional diagram of an AQUAZUR plant with CIRCULATOR clarifier.

Fig. 432. — Type CR 60 Aquazur plant, LE BRULÉ, Réunion. Output: 40 m³/h (176 gpm).

al housed in a building which can also contain the transformer or the diesel generators.

4.2. Medium-sized and large plants

4.2.1. PRINCIPLES

The design of medium-sized and large purification plants is based on the same treatment principles; these are dictated largely by the properties of the raw water and the required quality of the treated water; the actual application of these principles varies with the size of the plant, local conditions, availability and supply of reagents, etc.

For example:

— The depth of cylindro-conical or Circulator clarifiers not fitted with scrapers is limited by economic factors. When handling large flows, this equipment is kept to a substantially-constant depth, is sloped only gently at the bottom and has a scraper bridge to collect the sludge towards the centre.

— Reagent storage facilities naturally depend on the flow to be treated and the reserves to be held.

— Vertical metal pressure filters are generally limited to a size suitable for easy transport, but large diameter versions are possible if they can be assembled on the site.

— In open concrete filters scour air is supplied by pipes when the surface area is small, and through concrete ducts when the surface area is large.

— The degree of automatic control depends primarily on the size of the valves to be operated, on the staff available, on the complexity of the plant, etc.

Many different treatment sequences are needed to remedy the variety of defects to be found in water. They are distinguished principally according to:
— the physical properties of the water: suspended solids content, colour, lack of oxygen or excess of carbon dioxide;
— the chemical properties: presence of iron, manganese, and various pollutants which may or may not impart a taste to the water;
— the bacteriological properties: bacteria, plankton;
— the need to pump the water one or more times.
The commonest of the many possible sequences are as follows:

4.2.2. CLARIFICATION OF POLLUTED SURFACE WATER OR UNDER-GROUND WATER

A. *Slightly-turbid, slightly-coloured water containing little organic matter:*

● This is the case with underground water which is normally of good quality but becomes turbid at certain periods of the year (fig. 433).

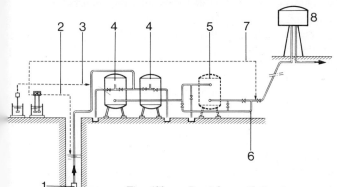

1. *Raw water pump.*
2. *Pre-chlorination (when necessary).*
3. *Injection of coagulant.*
4. *Pressure filtration.*
5. *Neutralization (when necessary).*
6. *Wash-water by-pass.*
7. *Injection of disinfection product.*
8. *Water tower.*

Fig. 433. — *Partial coagulation in pressure filters, with neutralization when necessary.*

Fig. 434. — *Aquazur plant, TAIAMA, Sierra Leone. Output: 50 m³/h (220 gpm).*

● This is also the case with water dammed in reservoirs, with properties that vary only slightly during the year, containing only small quantities of suspended matter and very little colour (fig. 436).

Pre-ozonization may be added if the water is highly coloured but only slightly turbid.

B. *Water of average turbidity, with colour, organic matter content, etc.:*

Water of this type generally requires complete treatment comprising coagulation/flocculation, clarification and filtration.

B 1. Case of limited pollution and limited micropollution: diagram 437 provides for complete treatment, including disinfection with chlorine dioxide or, preferably, ozone; this is done at the Limoges plant, an additional interesting feature of which is remineralization of the filtered water so that it is completely non-corrosive to metals.

Diagram 437 is also applicable in the case of clarification of water with a high plankton content; pre-chlorination (if possible break-point pre-chlorination) must always precede coagulation and clarification.

If the nature of the water and the type of sludge are suitable, a flotation process may be used, as in the Guingamp case (fig. 435).

In this case the reagents are injected into a mixing chamber, followed by the flocculator, flotation stage and filters.

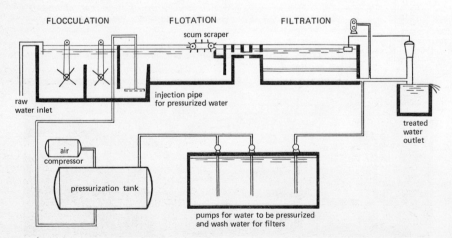

Fig. 435. — Clarification by flotation-filtration. GUINGAMP type plant, France.

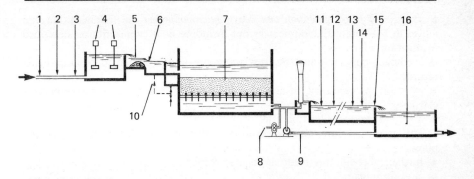

1. *Pre-chlorination.*
2. *Lime when necessary.*
3. *Coagulant.*
4. *High-speed mixers.*
5. *Raw water flow-meter.*
6. *Injection of activated silica.*
7. *Aquazur "V" filters with siphon control.*
8. *Scour air.*
9. *Wash-water.*

10. *Recovery of surface water (when required).*
11. *pH correction using lime.*
12. *Chlorination.*
13. *Neutralization of excess chlorine using SO_2.*
14. *NH_3 when necessary.*
15. *Treated water flow meter.*
16. *Treated water storage tank.*

Fig. 436. — Partial coagulation in open filters as in the case of the Majadahonda plant in MADRID. Phase 1: 3.5 m^3/s (80 mgd).

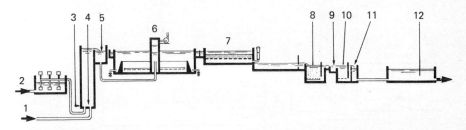

1. *Raw water fed by gravity.*
2. *Raw water supplied by lift pumps.*
3. *Pre-chlorination.*
4. *Injection of lime; coagulant and permanganate when necessary.*
5. *Injection of activated silica.*
6. *Pulsator clarifier.*

7. *Aquazur "T" filter with siphon control.*
8. *Ozonization tank.*
9. *Injection of lime-water.*
10. *CO_2 injection.*
11. *Chlorination when necessary.*
12. *Treated-water storage tank.*

Fig. 437. — Complete treatment cycle, including ozonization, for polluted water. As in the LIMOGES treatment plant. Output: 3 250 m^3/h (20.6 mgd).

B 2. Case of heavy pollution and heavy micropollution: the following may be
added to the complete treatment layout including final ozonization as described
in diagram 437.

● a contact tank in which the raw water previously treated to break-point is
returned;

● injection, as necessary, of activated carbon into Pulsator type sludge-blanket
clarifiers;

● a second stage of filtration after ozonization.

Diagram 438 shows this system as applied to the second stage of the Morsang
plant.

● further thorough treatment stages may comprise:

— a water intake;

— a holding tank;

— oxidation beds designed to treat the water fed to the holding tank or fed
direct to the clarifier;

— clarification;

— oxidation with ozone after the first sand filtration;

— filtration through granulated activated carbon;

— final disinfection with ozone;

— final chlorination before storage and distribution.

The diagram in fig. 440 representing the plant at **BLANKAART** (Belgium)
shows a complete treatment sequence which, of course, provides for the bypassing of any process which may not be required continuously.

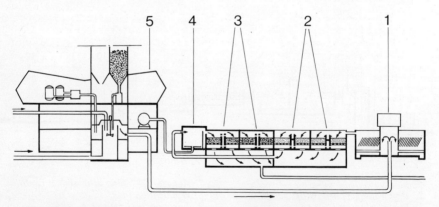

1. *Clarification and flocculation.*
2. *Sand filtration.*
3. *Filtration through activated carbon.*

4. *Disinfection with ozone.*
5. *Control building.*

*Fig. 438. — Flocculation and clarification in a Superpulsator. Sand filtration—Dis-
infection with ozone. Activated carbon filtration. System used at the MORSANG plant,
France (2nd stage). Output: 150,000 m³/h (39.6 mgd).*

Special reference should be made to the two-stage ozonization process separated by an activated carbon filter; this is the most effective method of using ozone and activated carbon to destroy pollutants and micropollutants.

Fig. 439. — Treatment plant at BOU-REGREG (Morocco). 3 Pulsator clarifiers (45 m × 19 m; 147 ft × 62 ft). 10 Aquazur filters each with a surface area of 140 m² (1 500 sq ft). Output: 10 800 m³/h (47.5 mgd).

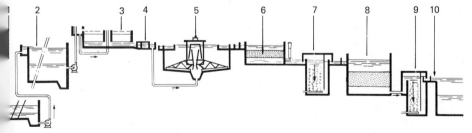

1. *Water intake and pump.*
2. *Contact and reserve tank.*
3. *Oxidation bed.*
4. *Parshall flume.*
5. *Circulator clarifier.*
   Carbonate removal if appropriate.

6. *Filtration over sand.*
7. *Oxidation using ozone.*
8. *Filtration over granular activated carbon.*
9. *Final disinfection using ozone.*
10. *Final chlorination.*

Fig. 440. — Treatment plant consisting of a contact tank, oxidation bed, flocculation, clarification, filtration over sand, ozonization, filtration over granular activated carbon, final disinfection with ozone, chlorination (BLANKAART, Belgium).

C. Heavily polluted water.

One or other of the following systems is used according to the amount and nature of the suspended matter in the water:

C 1. Single-stage clarification: diagram 442 represents a plant for the treatment of water containing a large amount of suspended solids; this is the case with the plant at Baghdad, which is equipped with scraper-type clarifiers (clarifloccula-tor).

The clarifiers are of the scraper type with an internally-located flocculator of the paddle type. The flocculator can also be external.

Fig. 441. — *Santillana drinking-water treatment plant supplying the city of MADRID, Spain. Output: 14 400 m³/h (91.3 mgd).*

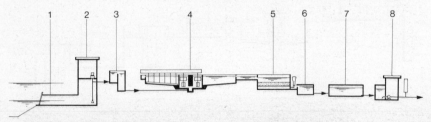

1. *Raw water intake.*
2. *Raw water pump unit.*
3. *Prechlorination and reagent injection.*
4. *Flocculation-clarification in 10 scraper type clariflocculators.*

5. *Aquazur V filters.*
6. *Disinfection.*
7. *Treated water storage tank.*
8. *Lift pump unit.*

Fig. 442. — *Single stage clarification and filtration, as used for the BAGHDAD, Iraq, treatment plant. Output: 15 800 m³/h (100 mgd).*

C 2. Two-stage clarification: when water contains large quantities of suspended solids it often contains sand; in such cases, the double clarification/filtration process must be preceded by grit removal.

The layout shown in diagram 443 is virtually that of the Cadarache plant which was built when the suspended solids in the water from the River Durance reached peaks of 10 g/l.

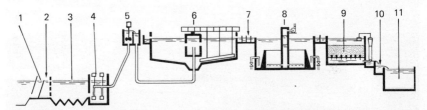

1. *Raw water intake.*
2. *Prechlorination.*
3. *Grit removal.*
4. *Raw water pump unit.*
5. *Primary injection of coagulant.*
6. *Preliminary clarification.*

7. *Secondary injection of coagulant, and additives if necessary.*
8. *Secondary clarification in a Pulsator.*
9. *Filtration.*
10. *Disinfection.*
11. *Treated-water storage tank.*

Fig. 443. — Grit removal, two stage clarification and filtration as at Cadarache treatment plant.

Fig. 444. — Treatment plant at FLORENCE (Italy). Output: 16,200 m³/h (103 mgd).

*Fig. 445. — Waterworks of CAEN, France, treating water from the Orne River.   Output:
2 500 m³/h (16 mgd).*

*Fig. 446. — Charles J. des Baillets Water Works, MONTRÉAL, Canada. Output :
1,136,000 m³/d (300 mgd).*

4.2.3. TREATMENT OF SLIGHTLY-POLLUTED SURFACE AND UNDERGROUND WATER

In most cases iron, manganese and colour have to be eliminated from this type of water, which also requires neutralization and disinfection.

● *Treatment under pressure:* the layout shown (fig. 420) represents the classic method of aeration of underground water, oxidation of the ferrous iron and retention of the ferric precipitates by filtration through sand. This method applies provided that the iron content is low.

Additional facilities can include pre-chlorination (to destroy bacteria), neutralization—either by alkaline reagent or final filtration through a neutralizing product—removal of manganese by means or a special aluminium product or even by filtration through granular activated carbon to eliminate any unpleasant taste.

● *Open-tank treatment:* open-tank treatment is used when large amounts of excess carbon dioxide have to be removed, a large flow has to be treated or the water needs to be clarified, ozonized, etc.

Figs 422, 423 and 447 show the three most common layouts:

— Fig. 422: sprinkling and filtration, useful for removing limited amounts of iron and manganese.

— Fig. 423: aeration, clarification and filtration, useful when large amounts of iron and manganese have to be removed, or when these elements are found together with other forms of pollution that require clarification.

— Fig. 447: sprinkling, ozonization and filtration, useful when the water contains only iron, manganese and colour, with little suspended or organic solids.

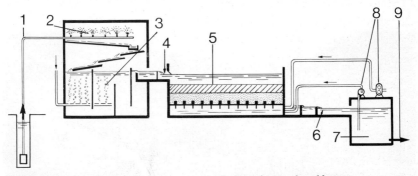

1. *Raw water pump unit.*
2. *Sprinklers.*
3. *Ozonization and manganese removal.*
4. *Filtration aid.*
5. *Dual media filter.*

6. *Disinfection by chlorine.*
7. *Wash-water tank.*
8. *Wash-water pump and air blower.*
9. *To treated-water storage tank.*

Fig. 447. — Elimination of CO_2, manganese removal using ozone, and filtration.

At Chalon-sur-Saône a dual media filter (sand and anthracite) is used which is particularly well suited to iron removal.

4.2.4. TREATMENT FOR BOTTLED BEVERAGES

Among the various special methods of treatment, the process for aerated beverages is interesting and fairly widely used. Fig. 448 shows the layout of plant which basically comprises the removal of carbonates by clarification in the presence of a high dose of chlorine, filtration through sand, storage of filtered water for filter-washing purpose, and a lift pump to convey the water to a subsequent stage of filtration through granular activated carbon and, sometimes, additional filtration to retain any fine particles of activated carbon.

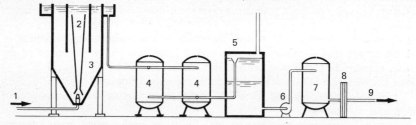

1. *Raw water inlet—prechlorination.*
2. *Injection of coagulant and lime.*
3. *Carbonate removal in a Circulator.*
4. *Filtration over sand.*
5. *Filtered-water storage and wash-water supply tank.*
6. *Intermediate lift pump.*
7. *Filtration over activated carbon Chlorine removal.*
8. *"Safety" filtration if applicable.*
9. *Discharge under pressure to process plant.*

Fig. 448. — A typical plant to treat water used in the manufacture of aerated beverages.

Fig. 448 A. — Iron removal unit consisting of 2 oxidation towers and 12 pressure filters. Output: 500 m³/h (3.2 mgd). Pumping plant at TOURCOING (France).

Fig. 449. — LA AYURA treatment plant for the city of Medellin (Colombia). Output: 5 000 m³/h (31.7 mgd).

Fig. 450 A. — HAMPTON LOADE drinking-water treatment plant (U.K.). Output: 3 120 m³/h (19.8 mgd).

Fig. 450 B. — La Roche Water Works. City of NANTES, France. Output: 11 700 m³/h (74.2 mgd).

Fig. 450 C. — San Martin Palermo II drinking-water treatment plant: City of BUENOS AIRES, Argentina.   Output: 36 000 m³/h (228 mgd).

21 SWIMMING-POOL WATER TREATMENT

1. PRINCIPLES

Swimming-pool water is almost always regenerated in a closed circuit, that s to say, the water gradually polluted by the bathers is returned to the pool after suitable treatment. Such regeneration is, in fact, much more economical than continuously renewing the water, thereby cooling the system and necessitating a costly supplementary heating process. Moreover, regeneration means a great saving in terms of water, and this factor is important in certain cases.

However, it is essential to introduce a certain amount of clean water each day in order to compensate for the water lost for various reasons, and to reduce the concentration of organic, non-organic and ammoniacal compounds in the water, which would continue to increase if the pool water were not progressively renewed.

This chapter is concerned particularly with public swimming pools, but similar standards may also be applied to "private communities' pools" and indeed, with a few modifications, to "lidos" which are increasingly included among the facilities offered by recreational areas; these "lidos" are much larger than conventional pools, usually shallower, and seek to re-create artificially the natural environment, bordered, for example, by sandy beaches, etc.

Treatment plants for swimming-pool water must be designed so as to ensure perfect sanitary conditions, but must also adhere closely to the regulations which vary from country to country. By way of example, we shall examine French regulations which may be referred to for countries where there are no special rules on these matters.

2. FRENCH REGULATIONS

The building and running of public swimming pools are governed by regulations which have recently been revised, replacing the decree of 13 June 1969. This new legislation comprises a series of decrees on safety and hygiene, within specified limits of application.

These rules lay down in particular the bacteriological and physico-chemical standards for swimming-pool water, the general arrangements and circulation of bathers in the swimming pool enclosure and deal with a certain number of technical questions concerning water treatment. The technical aspects are described below.

2.1. Recirculation rate and renewal of water

The recirculation rate is calculated as follows:
— 0.4 m³/h per m² of water surface when the maximum number of bathers at a given time is less than 1 person per 2 m² of water surface. But the turnover must under no circumstances exceed 4 hours;
— 0.5 m³/h per m² of water surface when the maximum number of bathers at a given time is greater than or equal to 1 person per 2 m² of water surface and les than or equal to 1 person per m². The turnover must under no circumstances exceed 4 hours;
— 0.6 m³/h per m² of water surface when the maximum number of bathers at a given time is greater than 1 person per m² of water surface or when the depth of the pool is no more than 1.30 m. The turnover must under no circumstances exceed 4 hours.

In open air pools with a surface area of 2,000 m² or more (this is usually the case with "lidos") a flow rate is to be adopted which will turn over the pool water in 4 hours.

With paddling pools, the flow rate adopted will renew the water in 45 minutes. For underwater swimming tanks and diving pools, the water should be turned over in 10 hours.

For open-air swimming baths in combined pools with both an open-air pool and an indoor pool which continues to operate in summer, the recirculation rate must not be less than 0.6 m³/h per m² of water surface; the rate for the indoor pools in this combined type of swimming bath must not be lower than 0.3 m³/h per m² of water surface.

In winter, when the open-air pool is closed, the water in the indoor pools must have a turnover in accordance with the general standards given earlier.

When the maximum number of bathers per day is more than double the maximum number at a given time (although triple the number is not considered permissible) these figures should be increased by a ratio of:

$$\frac{\text{maximum number per day}}{\text{twice the maximum number at a given time}}$$

However, water recirculation may be interrupted for a short time when the pool is open to the public, for the sole purpose of cleaning the filters.

In order to renew the water, each day a mandatory amount of clean water must be introduced, on the basis of 0.05 m³ for each person frequenting the pools during that day. Taken over the month, the amount of clean water introduced must be at least equal to the volume of the water in the pools.

In the case of pools supplied by sea water, the daily renewal water should be at least in part fresh water in order to compensate for the increased salinity caused by evaporation.

The clean water must be introduced by equipment which avoids contact between the water circulating in the pool and the water in the supply system.

It is essential to meter the renewal rate of the water (daily introduction of clean water) and the recirculation rate of the pool water.

In addition, swimming pools must be totally drained every three months for cleaning purposes and in order to completely renew the water.

Fig. 451. — Municipal swimming-pool at CARPENTRAS, France. Regeneration flow rate: 780 m³/h (3 400 gpm).

2.2. Water recirculation

An important innovation brought in by the new regulations concerns the direction in which the water circulates between the pool and the treatment plant: at least half the recirculation water (or if desired, all of it) must be drawn from the surface of the pool, and the treated water must be reintroduced at the bottom.

Under the system called **inverse hydraulics** the treated water is introduced into the pool by a central channel or a system of apertures covering the bottom of the pool.   The polluted water is drawn off entirely at the surface by means of gutters along two sides or all the way round.

If only part of the recirculation water (at least half) is taken from the surface, the system is called **mixed hydraulics**: the water is drained off partly at the surface and partly at the bottom.

These systems ensure that the maximum amount of pollution in the swimming bath, obviously concentrated at the very surface, is properly removed by drawing it off at this level.

Even before the previous regulations, which induced operators to remove water from the bottom, DEGREMONT had constructed pools in which the water circulated as described above, polluted water being drawn from the surface. The new system whereby part or all of the water is taken from the surface means that great care must be taken in the design and construction of collecting gutters installed at the surface along the sides of the pool.

2.3. Quality of water in the pools

The standards of quality required for swimming pool water are as follows: it must
— be transparent and limpid (turbidity less than 5 drops of mastic);
— have a pH between 7.2 and 7.7-7.8 if disinfected with chlorine, or between 7.5 and 8.2 if disinfected with bromine;
— contain less than 4 mg/l (expressed as oxygen released by hot potassium permanganate in an alkaline medium) of organic matter above that contained in the supply water;
— not contain ammonia or nitrous salts;
— contain less than 200 mg/l of chloride (expressed as chlorine) above the quantity in the supply water, and, in general;
— not contain toxic or undesirable substances.

In addition, of course, these standards of quality lay down a maximum number of bacteria characterizing pollution (less than 20 coliforms per 100 ml, total absence of E. Coli, less than 5 faecal streptococci, less than 10 staphylococci, total absence of pathogenic staphylococci; and there must not be more than 100 aerobic bacteria, revivable at 37 º C, per ml).

The water must not only be disinfected, but also slightly disinfectant, without acting as an irritant to the mucous membrane.

The water in indoor swimming pools must be heated to between 25 and 27 º C and the recommended temperature for open-air pools is 23 ºC.

An approved laboratory must monitor the quality of the pool water, carry out appropriate analyses and submit its findings each month to the Regional Health Authority.

Sets are available which include the necessary apparatus and reagents for carrying out all the standard checks on physico-chemical quality of swimming pool water (see page 904).

3. TREATMENT PROCESSES FOR CLOSED CIRCUIT SWIMMING POOL WATER

All of the pools' water must pass through the plant before being reintroduced separately into each pool. The treatment consists of two distinct stages: filtration and disinfection. The plant is arranged to provide the following sequence: prefiltration, pumping, filtration, heating (if necessary), disinfection.

As mentioned above, it is now compulsory in France to draw off at least 50 % of the recirculation water from the surface of the pool, but it is simpler and more economical to draw off all of this water from the surface and to reintroduce all the treated water at the bottom of the pool.

3.1. Prefiltration—Pumping

The water collected from the surface by gutters along the sides of the pool must be conveyed to a specially installed tank, which supplies the suction pumps recirculating the water. One or more prefilters must be installed immediately upstream of the pumps to protect the mechanism against sundry waste products which may be present in the pool water. The prefilter consists of a simple screen in a removable basket, which can readily be inspected and cleaned.

The number of pumps and prefilters depends on the magnitude of the recirculation rates and the possible combinations of these in order to adapt to the varying numbers of bathers in the pools. The specifications and installation of the pumps must be chosen carefully, taking into consideration the conditions in each pool.

3.2. Filtration

Filtration may be carried out by the following types of filters:

3.2.1. SAND FILTERS

Pressure-type sand filters are those most frequently used. Because of the low turbidity of the water high filtration rates are possible (15 to 40 m/h: 6 to 16 gpm/sq ft) and fine-grain sand is used. Filters designed for backwash by water alone are therefore particularly suitable for this application.

They comprise two types: standard filters operating at 15 to 20 m/h (6 to 8 gpm per sq ft) and high-rate filters (Hydrazur type) designed to operate satisfactorily at 40 m/h (16 gpm per sq ft). By the use of these filters, significant cost reductions can be made on the concrete work for the treatment plant.

Countercurrent washing of the filters is carried out at a rate of 30 to 40 m/h (12 to 16 (gpm/sq ft), regardless of the filtration rate. The amount of wash-water used in a period lasting several cycles is substantially the same regardless of the filtration rate.

It may, however, be advantageous to use pressure filters designed for water backwash and air scour; in this case washing is quicker and requires only a small volume of water and a small-diameter pipe discharging to the sewer. The instant-aneous rate of flow of wash-water per m² of filter is, in fact, about one-half that of filters washed by water alone.

It is important to realize that sand filtration without any added reagent to coagulate the water does not produce perfectly clear water. But this can easily be achieved by first adding a very small dose of aluminium sulphate to coagulate the suspended colloidal solids, which are then retained on the upper part of the filter, whereas they will pass completely through the filter bed if this precaution is not taken.

At the same time, it should be pointed that if no coagulant is used, an increased quantity of chlorine (or whatever disinfectant is used) is consumed because the chlorine will combine with the solids which have not been retained on the filter on the absence of coagulation. Increased consumption of disinfectant is harmful because it increases the concentration of chloride, and at the same time, produces chlorinated organic compounds, which are always undesirable.

The coagulant is injected by a dosing pump, in the same way as the sodium carbonate or acid which may be required to adjust the pH within the range recommended in the standards.

Fig. 452. — Swimming-pool water filtration unit consisting of 2 Hydrazur filters 2 200 mm (87 in) in diameter. Output: 300 m³/h (1 320 gpm).

3.2.2. DIATOMITE FILTERS

Diatomite[1] filters produce a very clear water provided that the filtration rate is limited to about 4 m/h (1.6 gpm per sq ft). Beyond this rate, the quality of the water depends on the method of applying the precoat, the type of diatomaceous substance used, and in particular, on the number of bathers: if this number is large, there is considerable risk in using rapid filtration through diatomites.

Faster filtration rates can be used, provided that less clear water is considered as acceptable and a washing frequency closely linked with pollution levels is allowed for.

The main drawback of diatomite filtration is that clogging rapidly occurs as soon as microscopic algae (plankton) appear in the swimming pool.

The fact that the diatomites must be replaced frequently when operating conditions are difficult (large number of bathers, presence of plankton, etc.) increases the running costs of this system to a considerable extent.

3.3. Disinfection

Disinfection is an extremely important stage in the treatment. Its purpose is both hygienic and aesthetic. It prevents the transmission of disease by contagion between the bathers, and halts the development of microscopic algae which cloud the water and give it a green tint.

Diseases transmissible by badly or inadequately treated swimming pool water are numerous, but the most common are:
— conjunctivitis, caused by a virus;
— sinusitis, tonsillitis and otitis due to streptococci and staphylococci propagated by nasal mucus.
— certain forms of enteritis due to the same microbes or to certain forms of virus ingested with the water;
— certain skin diseases (eczemas) for which Koch bacillus is sometimes responsible;
— cases of meningoencephalitis, some of which have been fatal, have been attributed to the Naegleria gruberi amoeba (this is destroyed by chlorine provided it is genuinely free, and by ozone);
— also certain skin diseases (epidermophytosis due to a fungus which attaches itself to the skin between the toes, plantar warts due to the papilloma virus) may be contracted by the bathers when walking on the areas around the pool: thus frequent cleaning and disinfection of the latter are essential.

To avoid contagion, the water must have a clearly-defined residual disinfectant power.

There are three common disinfection processes:

---

1. The term "diatomite", used in the French swimming pool regulations, is a synonym for "diatom", which will be found elsewhere in the present work.

3.3.1. CHLORINE AND DERIVATIVES

Either gas chlorine is used, distributed by a chlorinator from a cylinder of liquefied chlorine, or a solution of sodium hypochlorite (bleach), distributed by a dosing pump; the level of free chlorine in the pool water must be at least 1 mg/l but must not exceed 1.7 mg/l; the level of total chlorine must not exceed the level of free chlorine by more than 0.6 mg/l.

It should be noted that chlorine is the most commonly used disinfectant and that it is not unpleasant even when used in large doses in a water with a suitable pH — between 7.2 and 7.8.

Within this pH range, it has maximum effect and causes minimum inconvenience (irritation) if it is applied at the dose corresponding to the break point; this may sometimes mean relatively large doses.   See Chapter 15 for information about the different forms of chlorine in water.

3.3.2. BROMINE

It is applied so that the average level of total bromine in the pool water is at least 0.8 mg/l.   Under no circumstances must a level of 2 mg/l be reached.

The advantage of bromine is that it gives the treated water a less irritant character, but its action is controversial: some studies claim that the level required for proper disinfection is such that bromine becomes an irritant in the same way as chlorine.

When the water is disinfected with bromine, the pH must be higher than with chlorine; it is advisable to keep the pH between 7.8 and 8.2.

3.3.3. OZONE

Being the most powerful disinfectant known, ozone is particularly recommended for difficult cases (presence of amoeba, etc.).   It gives the water an attractive blue colour.   Ozone does not lead to the fermentation of products liable to irritate the mucous membrane and does not give the water a taste or odour, provided that the water contains no free ozone at the swimming pool inlet.   Swimming pool water treated only with ozone is thus more liable to later contamination by bathers through the lack of residual disinfectant properties in the water.   Therefore, after ozonization, it is essential to provide supplementary disinfection, adding for example a small dose of chlorine or bleach.

Ozone-enriched air is produced on the spot by technically sophisticated equipment.   Ozone production requires electricity only, approximately 20 Wh/m³ of recycled water.   Its action on the water being disinfected takes place in a special contact column and must be completed before the water is reintroduced into the swimming pool.

The minimum residual level of ozone after the ozone-enriched air has been in contact with the polluted water for at least 4 minutes must be 0.4 mg/l.   So as not to inconvenience the bathers, it is essential to eliminate the residual ozone completely before reintroducing the water into the pools.

In addition, after the ozone is eliminated, since it has no residual disinfectant power, chlorine or bromine must be added in sufficient doses to ensure the minimum levels indicated earlier. It is noticeable that the dose required is significantly smaller than that which would have been necessary without prior ozonization, and the irritant effect of chlorine or bromine is much less than would have been the case without ozonization.

Disinfection with ozone involves relatively high capital expenditure, but operating costs are reasonable.

Fig. 453. — Municipal swimming-pool at BORDEAUX, France. Regeneration flow rate: 815 m³/h (3 600 gpm).

3.3.4. OTHER PROCESSES

Other disinfectants are sometimes used, such as chlorine dioxide, iodine, chloramines, quaternary ammonium derivatives, silver produced by electrolysis, chlorocyanides. These products, although eminently suitable for other uses of water, are problematical when applied to swimming pools, for which they are seldom recommended. Before using any of them it is important to check whether they are approved by the regulations of the country in question. Certain countries (other than France) authorize the use of a solution of calcium hypochlorite instead of sodium hypochlorite.

There are disinfectant products on the market, often sold in capsule-form,

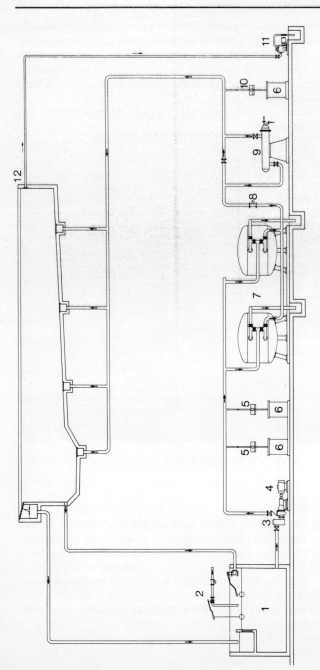

1. *Recovery tank.*
2. *Make-up water inlet*
3. *Prefilter.*
4. *Circulating pump.*

5. *Reagent dosing pumps.*
6. *Reagent storage tanks.*
7. *Filtration unit.*
8. *Treated water meter.*

9. *Heat exchanger.*
10. *Disinfecting closing pump.*
11. *Pump for pool sweeper.*
12. *Sweeper suction coupling.*

Fig. 454. — Principle of regeneration of swimming-pool water in a closed circuit.

which are intended to supplement, where necessary, the disinfectant power of swiming pool water; here again, it is necessary to check whether these products are approved.

3.3.5. DESTRUCTION OF ALGAE

There should be no proliferation of algae in pool water supplied by a water treatment plant if it is well designed and correctly operated. If, however, algae are detected, copper salts may be applied to destroy them; they are used either alone or in conjunction with chlorine, which may then be reduced in quantity.

A solution of copper sulphate, with about 10 % of commercial powder product, is prepared and introduced directly along the walls of the pool when the pool is not in use. The dose is 250 g of powder (or even 1 kg if the pool is very green) per 50 m³ of pool capacity. It is necessary to sweep the pool after several hours.

Copper sulphate may also be added to the recycle circuit in quantities of 1 to 2 g per m³, using a feeder or dosing pump operating intermittently to prevent the algae becoming inured to it.

Alternatively, large doses of chlorine may be intermittently applied (in the order of 20 g/m³ where the pH is acidic) when the pool is not in use.

3.4. Special cases

3.4.1. PH CORRECTION

It is often necessary to adjust the pH of the water in the recycle circuit in order to obtain the best conditions for using the disinfectant selected (see, page 401). When correcting the pH, a dosing pump is used to introduce an alkaline salt (sodium carbonate) or hydrochloric acid diluted in 20 times its volume of water.

3.4.2. REMOVAL OF IRON AND MANGANESE

The swimming pool water supply may sometimes contain a certain amount of iron or manganese. It is essential to eliminate these substances, otherwise they will cause reddish or blackish deposits to appear on the walls of the pool. See page 624 et seq. (Chapter 20) for details concerning the elimination of iron and manganese from natural waters.

3.4.3. SEA-WATER SWIMMING POOLS

Some swimming pools are supplied by sea water instead of fresh water. The principles for treating the water are identical, but precautions of a technological nature must be taken to protect the equipment against corrosion. Also, part of the water must be replaced by fresh water in order to prevent increasing concentration of salt.

3.5. Cleaning the pool

During the night, suspended matter is deposited on the bottom of the pool. It must be removed before the bathers arrive in order to prevent the further suspension of the substances. This is done with a pool sweep operated in the same way as a domestic vacuum cleaner, though the air is replaced by the pool water itself.

The simplest and most effective method consists of installing along the pool walls rapid couplings to which the floating flexible hose from the suction head can be connected. The couplings are connected to a pipe running around the the edge of the pool, and this pipe is itself connected to the suction inlet of a fixed pump fitted with a prefilter. This electric pump must be quite separate from the recirculation units which are incapable of creating a sufficient vacuum for easy suction of impurities.

A small, monobloc motor pump unit may also be used, mounted on a trolley, and fitted with a prefilter, to which the flexible suction hose is attached. This unit is moved along the edge of the pool in order to cover the whole surface area of the pool to be swept. When the drive motor of this mobile motor pump is electric, its voltage must match the regulations for each country. In France, the current may be at low voltage (220/230 V) or very low voltage (24/48 V). Electrical connection is made via a socket positioned outside the pool protection area, preferably in a room next to the area surrounding the pool and inaccessible to the bathers.

Petrol engines are sometimes permitted in outdoor pools.

Swimming pool sweeps are available with an immersed pump, supplied with current of low voltage or very low voltage. The water is drained from the pool by means of a floating hose, or returned to the pool after it has passed through a nylon bag or cartridge filter which forms part of the sweep.

Fig. 455. — Municipal swimming pool at TARBES, France. Regeneration flow-rate: 800 m³/h (3 500 gpm).

TREATMENT OF BOILER AND COOLING SYSTEM WATER

1. BOILER-WATER

1.1. Quality requirements for boiler-water

1.1.1. BOILER-WATER CIRCUIT

For all types of boiler, the water circuit can be very simply summarized as follows:

The boiler receives feed water, which consists in varying proportions of recovered condensed water (known as "returned water") and fresh water which has been purified in varying degrees and is known as "make-up water".

This water is converted into steam which escapes from the boiler to the outside. It might be supposed that this steam is made up of pure water molecules, but, in fact, it very often contains liquid droplets (priming) and gases (in particular, carbon dioxide generated by the decomposition of carbonates) and, at high pressures, it carries volatilized salts such as chlorides and silica by genuine "steam-stripping".

The water remaining in liquid form at the bottom of the boiler picks up all the foreign matter from the vaporized water (except the substances primed in the steam as described above).

The impurities would therefore concentrate increasingly in the remaining liquid if they were not systematically "blown down" by discharging some of the water from the boiler to the drains.

This sequence of events must be borne constantly in mind in order to arrive

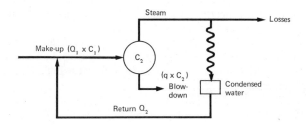

Fig. 456.

at a correct interpretation of the rules for purifying boiler water and drawing off the steam and waste.

Assuming that the boiler operates at a continuous rating and for simplicity, that the salinity entrained by the steam is negligible, the mineral content will be in stable equilibrium when the weight of salts discharged to the drains is equal to the weight of salts brought in by the make-up water (because the condensed water is considered to be pure). This gives the equilibrium state:

$$Q_1 \times C_1 = q \times C_2$$

where:

Q_1 = flow of make-up water concentration C_1
q = discharge flow
C_2 = concentration in the boiler,
and:

$$C_2 = C_1 \times \frac{Q_1}{q}$$

The salts content of the make-up water is multiplied in the boiler by the ratio of make-up flow to discharge flow.

If all the steam is lost and if the purification problem is stated not in cubic metres (or gallons) of make-up water per hour but in metric tons (or pounds) of steam T to be generated per hour, it must be remembered that $Q_1 = T + q$ and that the concentration factor will then be expressed by:

$$\frac{C_2}{C_1} = \frac{T + q}{q}$$

In practice, the permissible percentage of blow-down at a plant is strictly limited by running costs and initial outlay. Given the continual efforts to save energy, the tendency is to reduce this percentage to an ever smaller figure. The actual method of purification is determined by these calculations and by the standards of purity required either in the liquid phase or in the gas (steam) phase.

● *Nuclear generators and forced-circulation boilers:* certain nuclear generators used in modern reactors have no chamber, nor any provision for drawing off water, so that all impurities dissolved in the feed water are found on the steam-generating surfaces or in the steam. Therefore, the above method of calculation does not apply, and it is generally essential to limit the extraneous matter contained in the water to levels that can be tolerated in the steam.

The above also applies to all conventional boilers of the forced-circulation type and to those boilers used in the chemical industry known as "recovery boilers" which have no chamber or any provision for drawing off water.

1.1.2. DIFFICULTIES CAUSED BY IMPURITIES IN THE WATER

The principal difficulties caused by water in boiler or turbine operation are as follows:

A. Scaling, due to the deposition of crystalline precipitates on the walls of the boiler. This interferes with heat transfer and may cause hot spots, leading to local overheating.

Scaling is mainly due to the presence in the water of calcium carbonates or sulphates, which are less soluble hot than cold, or to too high a concentration of silica in relation to the alkalinity of the water in the boiler.

B. Priming, which is the carry-over of varying amounts of droplets of water in the steam. This lowers the energy efficiency of the steam and leads to the deposit of salt crystals on the superheaters and in the turbines

Priming is related to the viscosity of the water and its tendency to foam. These properties are governed by alkalinity, the presence of certain organic substances and by total salinity. The degree of priming also depends on the design of the boiler and its steaming rate.

C. Carry-over in the steam of volatile minerals at boiling point, the most harmful of which is silica. These minerals are deposited on turbine blades and cause serious running faults.

Carry-over increases with pressure and, therefore, with temperature. The quantity, of course, depends on the amount of harmful substances such as silica in the chamber.

D. Corrosion of widely varying origin and nature due to the action of dissolved oxygen, to corrosion currents set up between different metal surfaces, or to the iron being directly attacked by the water.

Corrosion is dealt with mainly:
— by eliminating all oxygen;
— by protecting the sheet metal with a coating of magnetite or phosphate;
— by adjusting the pH.

As we shall see below, water treatment techniques utilize the full range of corrective methods:
— external treatment of make-up water and condensates;
— physical or chemical elimination of oxygen;
— conditioning of boiler-water and condensate.

Before turning to corrective methods, it is first necessary to consider the quantity of the various harmful substances which can be allowed in the boiler water without risk of damage to the boiler or turbine.

Starting from these figures, and allowing for the amount which can be blown down, the next point to consider is the permitted concentration in the make-up water.

1.1.3. STANDARDS FOR WATER FOR USE IN STEAM GENERATING PLANT

Earlier editions of the Handbook included tables based both on the working pressure and the type of boiler.

Because of the growing tendency to increase the rate of heat transfer through the heated surfaces of modern boilers, this method of classification is often deceptive, and only relatively wide brackets can be given as to maximum levels of alkalis, salts, silica, phosphates, etc., in relation to working pressure; the actual maximum levels must of necessity be obtained from the boiler manufacturer, who will base them on the characteristics of the boiler in question.

A point constantly debated is the maximum level of NaOH, which decreases as the steaming rate per m^2 of tube increases, but which in low- and medium-pressure boilers can be raised if anti-priming conditioning is applied.

The Table below gives maximum recommended levels up to pressures of 100 bars for medium steaming rates and for volumes of water in the chambers sufficient to properly control the extraction rates.

REQUIRED STANDARDS
FOR BOILER-WATER

Working pressure		< 15 bars	15-30 bars	30-45 bars	45-75 bars	75-100 bars
pH at 25 °C	maximum	11.5-12	11.5-12	11-11.5	11	10.8
	minimum	10.5-11	10.5-11	10.4-11	10.3-10.8	10.0-10.5
TAC in French degrees	maximum	60-140	40-100	25-45	15-30	4-15
	minimum	20-25	10-20	10-20	5-7	—
Total salinity g/l	maximum	2-4	2-3	1.5-2	0.5-1	0.1-0.5
NaOH mg/l	maximum	250-700	200-500	120-250	50-150	15-50
SiO$_2$ mg/l	maximum	100-300	50-150	25-60	10-30	2-5
$\dfrac{\text{SiO}_2 \text{ in mg/l}}{\text{TAC (French degrees)}}$	maximum	1.5-2.5	1-2	1-1.5	1	1
Na$_3$PO$_4$ mg/l	minimum	50	50	50	20-40	10-20
	maximum	—	—	120-250	50-150	20-50
$\dfrac{\text{Na}_3\text{PO}_4 \text{ in mg/l}}{\text{NaOH in mg/l}}$	minimum	—	—	—	1	1

The foregoing table assumes that the feed water is always brought to zero hardness. Above 100 bars (1 450 psi) the 'Complexon' test for alkaline earth reaction should be negative.

Better knowledge of the sequestration and vaporization of dissolved salts has now led to a gradual lowering of the silica limits for very-high-pressure boilers. For the latest boilers installed by Electricité de France, the silica content is limited to 300 microgrammes/l, and any addition of colloidal silica is specially controlled by millipore filtration.

Again, in the case of very-heavy-duty boilers and, even more so, of hyper-critical boilers, the dissolved salts content must be reduced as far as possible, and this has led to the complete elimination of phosphate treatment.

That being so, there are few figures available for salinity and TAC limits for very-high-pressure boilers and very-heavy-duty boilers.

In fact, treatment of make-up water now has to include demineralization in an ion exchanger; the required production of water with a very low silica content is only possible if the total salinity is reduced to a fraction of a milligramme per litre. In these conditions, the TAC and the total salinity of the boiler water are determined essentially by conditioning and not by treatment.

Restrictions are also imposed on other elements such as iron, copper and oxygen, elements which are responsible for many corrosion phenomena. In 1969, Babcock and Wilcox (USA) published the following table in respect of feedwater for standard boilers:

	Under 40 bars (600 psi)	40 to 70 bars (600 to 1 000 psi)	70 to 140 bars (1 000 to 2 000 psi)
Iron (mg/l)	0.1	0.05	0.01
Copper (mg/l)	0.05	0.05	0.005
Oxygen (mg/l)	0.007	0.007	0.007

The standards for certain nuclear generators of whatever pressure are the strictest of the figures quoted above.

1.1.4. DIFFICULTIES CAUSED BY IMPURITIES IN THE CONDENSATES

Before the advent of heavy-duty boilers, forced-circulation generators and nuclear generators, the impurities introduced into the feed water by the condensates were considered, barring a few special cases, to be negligible and it was thought sufficient to carry out a proper purification of the make-up water.

This is no longer true now that purity standards are such that there is a danger of corrosion products (iron or copper oxides), dissolved salts from condenser leaks, or accidental pollution (hydrocarbons) of the steam used in heating circuits, exceeding the permitted levels in the feed water.

We shall examine later the main methods of treating condensates, so that the above impurities can be eliminated.

It is also true that condensates can cause corrosion in the circuits which carry them, due to the presence of carbon dioxide or substances produced by priming. In this case they must undergo suitable conditioning.

1.2. Purification and conditioning

The methods chosen for purifying make-up water and conditioning the boiler-water are designed to avoid these difficulties and satisfy the boiler-maker's requirements. The conditioning technique is linked to the purification process and vice versa. As a general rule, the more thorough the purification process, the less conditioning treatment is required.

1.2.1. CARBONATE REMOVAL AND SOFTENING PROCESSES

As already noted, water which has not been properly softened is no longer acceptable for boilers.

Make-up water must therefore always be treated by an ion-exchange process to bring TH as near as possible to zero.

At very low pressures, straightforward softening is still used sometimes (cation exchanger regenerated with sodium chloride), but at high pressures the only process used is demineralization. At intermediate pressures, the removal of carbonates (and if necessary of silica) is combined with softening by various methods.

The main processes now used comprise:

— the cold lime process for removing carbonates, with optional silicate removal by means of iron (III) chloride and aluminate, followed by softening (see p. 147);

— the hot lime and magnesia process for carbonate removal (95 to 110 °C) followed by softening (see p. 157);

— removal of carbonates through a carboxyl cation exchanger followed by softening and physical elimination of CO_2;

— removal of carbonates by a mixed exchange process through a hydrogen ion exchanger and softener; this is known as the "H-Na process" (little used today because of the danger of acid water in the boiler if the mixing system goes wrong)

To help in choosing between these methods, the results to be expected from each of them are summarized in the Table below.

All these processes must be followed by physical (deaeration, page 375) or chemical (page 395) elimination of oxygen, and by conditioning treatment.

● *Conditioning of water after carbonate removal and softening:* the above processes may possibly allow a very small quantity of calcium to enter the make-up water. It is therefore advisable to apply conditioning treatment using de-scaling products or dispersants (see page 392 : scale-inhibitors). It is always desirable to use an oxygen-reducing agent (see page 395); its nature and particularly the amount added will vary depending on whether any means of thermal deaeration is available. In order to obtain a satisfactory and constant steam purity with these partial purification techniques, it is often advisable to use an anti-priming product.

RESULTS TO BE EXPECTED FROM CARBONATE REMOVAL AND SOFTENING PROCESSES

	TA	TAC (average)	Total salinity	pH	SiO_2 in mg^2/l
Cold purification with lime and $FeCl_3$ + softening	0.5^o to 2°	2° to 4° [3]	ST_1 [1] - TAC_1 [2] + 3° to 6°	8.5 to 10	unchanged
Previous process + sodium aluminate	do.	do.	do.	8.5 to 10	2 to 5
Hot purification with lime and magnesia + softening	1° to 1.5^o	2° to 2.5^o	$ST_1 - TAC_1$ + 2° to 2.5^o	8.5 to 10	1 to 2
Carboxyl + softening with intermediate elimination of CO_2 (or H-Na process)	0°	1° to 3° [3] without pH correction 2° to 5° [3] with pH correction	$ST_1 - TAC_1$ + 1° to 3° $ST_1 - TAC_1$ + 2° to 5° with pH correction	6 to 7 without correction 7.5 to 8.5 with correction	unchanged

1. ST_1 : Salinity of raw water (expressed in French degrees)
2. TAC_1 : TAC of raw rater (expressed in French degrees).

3. If minimum TAC values do not give the required $\dfrac{SIO_2}{TAC}$ ratio, the necessary corrections should be made to obtain a suitable ratio between the silica content and the alkalinity in the boiler-water.

In the boiler all bicarbonates and some carbonates are dissociated, giving off CO_2. As soon as the water begins to condense, CO_2 dissolves in it and gives it a marked corrosive character. This difficulty is overcome by using volatile amines, of the neutralizing (see page 387) or of the film-forming type (see page 396).

● *Plant layouts for carbonate removal and softening:* some examples of plant layouts for boiler water treatment are given below.

These plants incorporate appliances such as filters, reagent feed units, ion exchangers and so on, which are described elsewhere. The following diagrams illustrate some suitable basic layouts.

Fig. 457 A. — Cold-process carbonate removal and softening through a stratified-bed carboxylic-sulphonic exchanger.

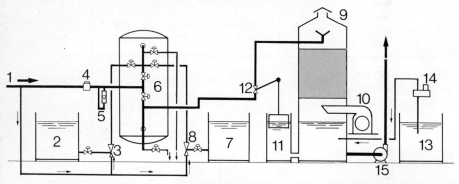

1. *Raw water intake.*
2. *Brine tank.*
3. *Brine injector.*
4. *Meter.*
5. *Flowmeter.*
6. *Stratified-bed exchanger.*

7. *Acid tank.*
8. *Acid injector.*
9. *CO_2 extractor.*
10. *Fan.*
11. *Level-monitoring tank.*

12. *Level-regulating valve.*
13. *Caustic soda tank.*
14. *Feed pump for pH correction.*
15. *Lift pump for treated water.*

1. *Lime storage silo.*
2. *Fluidization air.*
3. *Lime feeder.*
4. *Lime dilution tank.*

5. *Milk-of-lime pump.*
6. *Coagulation aid feeder.*
7. *CIRCULATOR clarifier.*
8. *Sand filter washed by water and scour air.*
9. *Softener.*
10. *Salt.*
11. *Salt storage.*
12. *Brine gauging tank.*
13. *Outlet for decarbonated and softened water.*

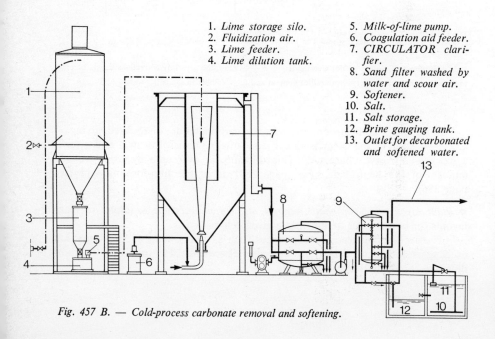

Fig. 457 B. — Cold-process carbonate removal and softening.

1.2.2. TOTAL DEIONIZATION OF BOILER MAKE-UP WATER

When the processes already described fail to achieve the required quality, boiler make-up water must be totally deionized.

As already noted in the chapter on ion exchangers, water obtained by total deionization contains a very low dry extract, and the degree of purification depends on the combination of the ion exchanger groupings. In order to keep to the required levels, depending on the pressure and type of boiler, of total salinity (expressed as the specific resistivity of the treated water) and of silica in the make-up water, one of three sequences is followed:

1. Deionization by a single run through cation and anion exchangers with co-current regeneration. According to the type of water and the regeneration rate, resistivities from 100,000 ohms.cm to 1 megohm.cm and levels of silica from 50 to 200 microgrammes per litre will be obtained.

2. Deionization by a single run through cation and anion exchangers with counter-current regeneration, (possibly through beds positioned one above the other). The resistivities obtained are between 500,000 ohms.cm and 5 megohms.cm while silica levels are between 20 and 100 microgrammes per litre.

3. Deionization involving a second stage of cation and anion exchangers (separate beds or mixed beds). In this case, resistivities are between 1 and 20 megohms.cm and silica levels between 10 and 100 microgrammes per litre.

Fig. 458. — Layout of a total deionization installation.

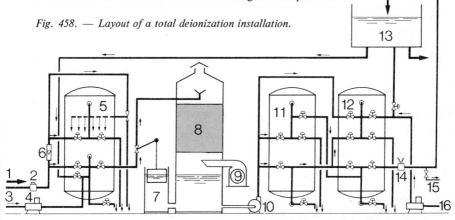

1. *Raw water intake.*
2. *Meter.*
3. *Acid intake.*
4. *Dosing pump.*
5. *Cation resin regenerated by back-flow.*
6. *Flowmeter.*
7. *Level-monitoring tank.*
8. *CO$_2$ removal.*
9. *Fan.*
10. *Lift pump.*

11. *Weakly basic anion resin.*
12. *Strongly basic anion resin.*
13. *Deionized water tank.*
14. *Resistivity meter electrode.*
15. *Sample to silica meter.*
13. *Deionized water tank.*
14. *Resistivity meter electrode.*
15. *Sample to silica meter.*
16. *Caustic soda dosing pump.*

All these figures are given as a guide only, and may not be feasible when dealing with water which is highly polluted or has not had sufficient pre-treatment.

The diagram fig. 458 illustrates a plant with a single cation exchanger and a combination of weakly and strongly basic exchangers.

Typically, a deionization plant for treating feed-water for high-pressure boilers may include:

— preliminary clarification or carbonate removal;

— a primary grouping of ion exchangers which must include a strong cation and a strong anion and may include a carboxyl exchanger and a weak anion;

— a secondary grouping comprising a cation and an anion or a mixed bed;

— conditioning (e.g. ammonia, phosphate, hydrazine);

— automatic control of regeneration and often of reagent preparation.

The following diagram (fig. 459) shows a total deionization plant for a high-pressure boiler, starting with underground water and including a complete primary sequence and a cation-anion finishing sequence.

● *Thermal deaeration and conditioning of deionized water:* conditioning of the deionized make-up water is greatly simplified if, in accordance with current practice, the mixture of make-up water and condensed water is subjected to intensive thermal deaeration (described in detail in Chapter 13).

It is then sufficient to add, either to the make-up water, or directly into the boiler, very small quantities of sodium phosphate, volatile amine and hydrazine.

In the case of high-pressure and heavy-duty boilers, the tendency is to reduce, or even eliminate completely, the sodium phosphate, which may cause local deposits by "sequestration".

Conditioning in this case (ammonia or morpholine + hydrazine) is called "volatile conditioning".

Depending on the general characteristics of the condenser-heater circuit of the boiler, deaeration can be carried out at temperatures around 105 °C or at much higher temperatures.

1.2.3. TREATMENT OF CONDENSATE

Depending on circumstances, the treatment of condensed water entails solving the following three problems:

— elimination of corrosion products from the turbine-condenser circuit;

— elimination of ions re-entering in the raw water as a result of leaks in the condensers;

— in the case of the petrochemical industry, elimination of oil from the condensates.

There are a number of possible solutions, all combining filtration and deionization. Therefore, filters and ion exchangers, alone or in combination, comprise the essential elements:

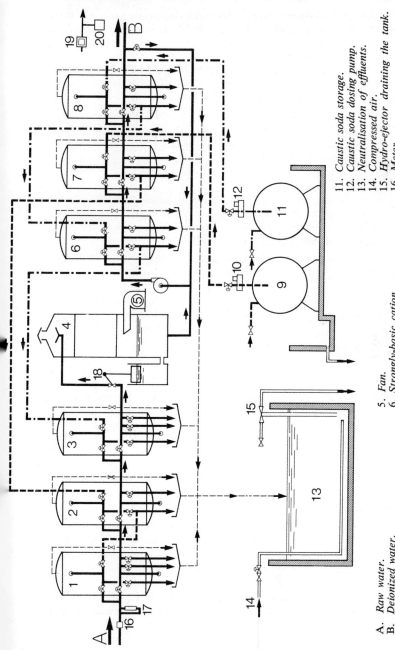

A. Raw water.
B. Deionized water.
1. Carboxylic cation.
2. Primary sulphonic cation.
3. Weakly-basic anion.
4. CO₂ removal.

5. Fan.
6. Strongly-basic cation.
7. Finishing sulphonic cation.
8. Strongly-basic finishing cation.
9. Acid storage.
10. Acid dosing pump.

11. Caustic soda storage.
12. Caustic soda dosing pump.
13. Neutralisation of effluents.
14. Compressed air.
15. Hydro-ejector draining the tank.
16. Meter.
17. Flow-rate indicator.
18. Level regulator.
19. Resistivity meter.
20. Silica meter.

Fig. 459. — *Typical lay-out of a total deionization installation for high-pressure boiler-feed water.*

A. Filtration through finely-divided (fibrous or granular) materials used in a layer a few millimetres thick on **pre-coat filters** (see page 250). In practice, use is made of:

● cellulose fibres, 40 to 100 microns long. These filters retain all suspended impurities, specifically metal oxides down to an average size of some 0.5 micron. The filter material begins to release soluble impurities from 60 °C upwards and decomposes above 85 °C;

● powdered, non-polar, synthetic resins, which have the advantage over cellulose of being capable of adsorbing colloids and withstanding temperatures of 100 °C. When treating condensates free from hydrocarbons, filtration rates with these materials usually vary between 5 and 10 m/h (2 to 4 gpm per sq ft);

● diatoms: with oily water, it is preferable, instead of cellulose or resin, to use high-porosity diatoms, which in addition to their filtering properties, have a specific adsorbent effect. The optimum temperature for oil removal, depending on the type of oils to be retained, is between 50 and 80 °C.

The filtration cycle can be prolonged by adding to the pre-coat at regular intervals during treatment, a quantity of material either different from or identical to the initial coat.

In the case of condensed water containing aromatic hydrocarbons, which are soluble in water to varying degrees, the coat must be topped up with activated carbon rather than diatoms, and the temperature must not exceed 50 °C.

Wherever oily water is concerned, filtration rates should be adapted to each problem, and will in any case be lower than for condensates containing only corrosion products.

B. Deionization with very-high rate filtration (80 to 120 m/h: 32 to 48 gpm per sq ft) through a mixed cation-anion bed.

This type of equipment is intended to fix the iron, copper, nickel and silica ions and to retain the salts carried into the condensate by the accidental entry of raw water into the condenser.

These mixed beds also act as filters, with an efficiency varying from 50 to 90 % according to particle size and working conditions. Practically no colloids are retained.

In order to withstand the head-loss caused by the combination of the high rate and clogging, ion exchangers have to be highly resistant; macropore resins are most often used. Special testing devices have been developed by DEGREMONT to check the behaviour of ion exchangers, so that correct types of resin can be recommended, and in addition, deliveries can be monitored.

In most cases, the resins are washed and regenerated in a system outside the treatment columns; the resins are transferred to the system by water. This ensures that no acid or soda enters the boiler accidentally when the resin is regenerated.

However, the extremely rigorous standards recently imposed by some manufacturers of pressurized water reactors (PWR) with regard to the sodium content of the circuit water have drawn attention to the significance of small quantities of sodium passing into the water issuing from a mixed bed where separation of the cation and anion exchangers is not complete prior to regeneration (i.e. Na ions are present in the regenerated cation exchanger where a part of this remains mixed with the anion exchanger).

There are various efficient methods of reducing this phenomenon to the extent that is no longer constitutes a problem.

C. Very-high-rate deionization with a cation exchanger followed by a mixed bed: When condensed water is conditioned with considerable quantities of volatile anions (ammonia, morpholine or cyclohexylamine) and only negligible amounts of salinity due to raw water leaks are returned to the system, the mixed bed is thrown off balance by the fact that there are many more cations to be retained than anions, other than the OH anion.

It may then be an advantage to use a cation exchanger regenerated with acid before the mixed bed stage; this will drastically reduce the NH_4 and amine ions as well as greatly prolonging the life cycle of the mixed bed.

The cation exchanger also has the effect of filtering out the corrosion products thereby limiting the mixed bed to the deionizing function.

D. Filtration and deionization combined in a single unit, by using very finely-powdered mixed resins (10 to 50 microns) in pre-coat filters as described in para. A above. Rates and pre-coat thicknesses are similar to the above. (See fig. 460).

This appliance serves both purposes for a lower capital outlay, but running costs are much higher because powdered resins are expensive to produce, and the resins often have to be replaced even though they have not had much use (either as a result of physical clogging by suspended impurities, or because of the presence of exceptional salinity caused by a leak in the condensers).

The maximum temperature for this process depends on the thermal resistance of the resins used and on the possible need to eliminate silica. If this is necessary, the limit is about 40 to 50 °C.

It is worth pointing out that this filtering process gives exceptional results.

E. Filtration through magnetic filters: by using easy-to-clean high-flow-rate magnetic filters it would be possible to filter certain metallic oxides from the condensate, at higher temperatures and without the expensive pre-coat materials that are inconvenient to use.

Present-day knowledge of real requirements and of the results obtained with these methods does not suggest that any one should be preferred to the other two. Mixed beds cannot on their own give the full filtration requirement in all cases, but, they do on the other hand provide valuable protection in the event of a leak in the condensers. The combined systems (pre-coat filters/mixed beds or cation exchangers/mixed beds) are costly but provide effective treatment in all cases.

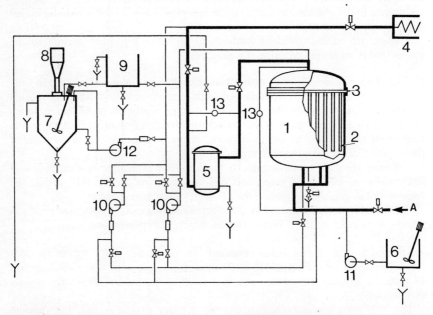

A. *Condensate inlet.*
1. *Pre-coat filter.*
2. *Filter candles.*
3. *Candle carrier.*
4. *Water.*
5. *Resin trap.*
6. *Feed-tank for pre-coat maintenance.*

7. *Pre-coat preparation tank.*
8. *Microresin feeder.*
9. *Equilibrium tank.*
10. *Recirculation pump.*
11. *Feed pump for pre-coat maintenance.*
12. *Pre-coat injection pump.*
13. *Differential pressure gauge.*

Fig. 460. — Filtration and deionization combined.

2. COOLING CIRCUITS

2.1. Types of cooling systems

Appliances that have to be cooled vary very extensively; the main types are as follows:
— condensers and heat exchangers;
— oil, air, gas and liquid coolers;
— motors and compressors;
— blast furnaces, steel furnaces, rolling mills, etc.;
— chemical reactors.

The cold water enters these appliances and is heated by contact with the hot walls. At the outlet, there are three possible conditions:
— the hot water is discharged into a river or drain; this is an **open circuit**, fig. 461;
— it is cooled by contact with a secondary fluid (air or water) and is returned to the appliance which has to be cooled without coming into contact with the atmosphere; this is a **closed circuit** (fig. 462);
— it is cooled by partial evaporation in a cooling tower and then returned to the appliances; this is a **semi-open circuit with recirculation through an atmospheric cooler** (fig. 463).

Cooling water which is used once only in an open circuit is expensive to treat with chemicals.

At most it is possible to correct the pH value slightly (page 698) or to add a precipitation inhibitor (page 699) in order to limit the formation of scale, or to add chlorine or hypochlorite in order to reduce the growth of algae and bacteria.

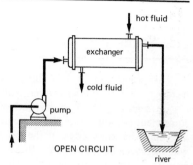

Fig. 461. —

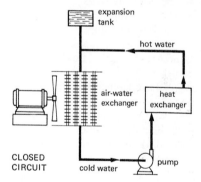

Fig. 462. —

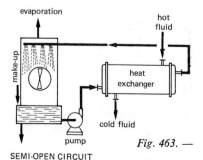

Fig. 463. —

In closed circuits, however, very little make-up water is needed and there is no contact with the atmosphere. Complete and virtually permanent protection can be obtained with the corrosion inhibitors described on page 700.

Because of the size of appliances and the complexity of the water problems involved, attention will be mainly concentrated on the protection of semi-open circuits.

2.2. Semi-open circuits

Air conditioning and humidifying systems can be included under this heading, at least so far as water problems are concerned.

In selecting the best method of protection for such a circuit, the type of equipment (which varies from manufacturer to manufacturer) is often less important than the knowledge of the operating parameters of the circuit. Attention will therefore be focussed on these factors:

$V(m^3)$: volume in circulation. This is the total amount of water in the system, i.e. in:
— the hot-water tank;
— the cold-water tank;
— the heat exchangers;
— the connecting pipes, etc.

$Q(m^3/h)$: circulating flow, i.e. the flow of hot water delivered to the atmospheric cooler.

$\Delta T(^oC)$: difference in temperature between the water entering and leaving the cooler.

T max.(^oC): temperature of the film of water ("skin" temperature) in contact with the hottest wall of the circuit.

$W(kcal/h)$ power of the atmospheric cooler (or cooling tower); this is the product of the two preceding figures, i.e.:

$$W = 1\,000\,Q\,\Delta\,T$$

$E(m^3/h)$: evaporation flow, i.e. the quantity of water evaporated to cool the main flow.

This evaporation flow consists of pure water which does not entrain any dissolved salts.

It is assumed with sufficient accuracy for water problems that one litre of evaporated water corresponds to 500 kcal; this gives the following equation:

$$E = \frac{Q\Delta T}{500} \tag{1}$$

$E_v(m^3/h)$: vesicular entrainment flow. This is the water entrained in the flow of air in the form of liquid droplets. Vesicular entrainment therefore consists of water with the same composition as the water circulating in the system.

Manufacturers are progressively reducing the amount of water lost by vesicular entrainment. A few years ago the figure was about 1 %, it is now around 0.1 % with the possibility of being reduced to 0.01 %. The following equation is used as an average:

$$E_v = \frac{Q}{1,000} \qquad (2)$$

P(m³/h): deconcentration blow-down flow. Evaporation increases the concentration of dissolved salts in the circulating water. To prevent the concentration from becoming so high that there is a risk of deposits forming, some of the water must be drawn off and discharged to the drains.

D(m³/h): total deconcentration flow:

$$D = E_v + P \qquad (3)$$

A(m³/h): make-up flow.

This water makes up all water lost from the system, chiefly by evaporation and total deconcentration:

$$A = E + D \qquad (4)$$

t(h): residence time

$$t = \frac{V}{D} \ln 2 \sim 0.7 \frac{V}{D}$$

C: concentration ratio.

This is the ratio between the concentration of dissolved salts in the circulating water and in the make-up water respectively.

This figure is normally determined by measuring the chlorides, which are easy to analyse and, because of their solubility, are the most stable in the system.

If the water in the circuit contains 12 Fr. degrees of Cl^- and the make-up water only 3⁰, the concentration ratio is:

$$C = \frac{12}{3} = 4$$

Some other relationships between the foregoing values may be useful:

Taking s to express salinity, i.e. the quantity of dissolved salts in the make-up water in, for example, grammes per cubic metre, salinity in the system is $s \times C$.

It then becomes possible to calculate the balance of dissolved salts entering and leaving the cooling circuit as follows:

— quantity of dissolved salts entering : $s + A$,
— quantity of dissolved salts leaving the cooling circuit: $D \times s \times C$.

As these two quantities shlould be equal, the following equatio is obtained:

$$A = D \times C \text{ or } C = \frac{A}{D} \qquad (5)$$

By applying equation (4), the following relationship is easily calculated:

$$D = \frac{E}{C - 1} \qquad (6)$$

The key values required to work out the appropriate means of protecting a circuit are the circulating flow, the temperature difference in the cooler (or the power of the cooler) and the maximum "skin" temperature.

If there are no other special operating requirements, the remaining parameters can be determined either by calculation or by analysing the water.

2.3. Scale and corrosion

Page 27 should be consulted for scale and page 21 for chemical corrosion. The deposits formed by salts which are difficult to dissolve (scale) should not be confused with iron oxide deposits which vary from orange-red to black in colour. The latter show that the cooling water is corrosive.

Scale at least partially prevents contact between the metal and the water and thereby prevents corrosion. When the formation of scale is prevented, corrosion is often found.

The water can be treated so that it is in equilibrium between these two extremes; this is a first protective process known as:

2.3.1. THE EQUILIBRIUM PROCESS

On the basis of certain data, acid (page 386) or alkaline reagents are added to adjust the pH, the TH and the TAC of the cooling water—but not of the make-up water, which means that the concentration ratio is taken into account—so as to remain half-way between scale and corrosion.

The Langelier index (see page 29)

$$I_L = pH - pH_s$$

($pH_s = pH$ of saturation or equilibrium) can be used to determine whether a water is scale-forming ($I_L > O$) or aggressive ($I_L < O$).

The two concepts of aggressivity to calcium carbonate and corrosivity especially to steel are not completely identical; a non-aggressive water may be corrosive.

J. W. RYZNAR (see page 32) suggested an empirical index:

$$I_R = 2 PH_s - pH,$$

which experience has shown to be more representative of the corrosive or scale-forming nature of cooling water.

I_R less than 6, scale-forming water;

I_R over 6 but less than 7, water close to equilibrium;

I_R over 7, corrosive water *.

The equilibrium process may appear attractive because it is simple, but it has serious limitations, first because the water comes into contact with the air

---

* J. W. Ryznar—A new index for determining the amount of calcium carbonate scale formed by a water. JAWWA, Vol. 36, April 1944, pp. 472-473.

in the atmospheric cooler and secondly because the temperature of the water in the system varies.

The pH value required for the water to be in a state of CaO/CO_2 equilibrium is not always compatible with the pH established in the atmospheric cooler, because the CO_2 is shared between the water and the atmosphere.

The graph in fig. 464 shows how the mean pH varies in these conditions with the TAC of the water in the system.

If the pH of the water in the system is kept above the values read in the graph, by adding an alkaline reagent, a very high level of consumption may result.

There is an equilibrium value for the water at each temperature. In the case of a cooling system, the water should remain stable over a certain temperature range, but this is normally impossible. Either the water is in equilibrium at the lower temperature and scale forms at hot spots, or it is brought into equilibrium

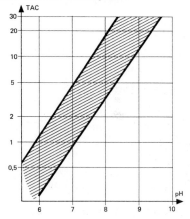

Fig. 464. — pH of the water circulating in cooling circuits.

for the hot spots and becomes corrosive at the lower temperatures. This is especially true since the high temperature to be taken into account is not that of the water returning to the atmospheric cooler, nor even that of the water leaving the hottest exchanger, but the temperature of the film of water as measured at the wall in contact with the hottest fluid to be cooled, i.e. the maximum "skin" temperature.

2.3.2. SCALE-INHIBITING PROCESSES (see page 392)

This is an advance on the previous method because chemicals are added to delay the precipitation of calcium carbonate, thus protecting the hot spots in the circuit. The pH, the TH and the TAC are then determined so that the water is in equilibrium or even scale-forming at low temperatures, while the retarding agent prevents any formation of scale. The materials used, mostly in combination, are firstly polyphosphates and phosphonates in particular (see page 394), which act at molecular level, and secondly dispersants (see page 395) which act at microscopic level.

Compared with the equilibrium technique and thanks to the increased stability of the available materials, these processes:
— widen the range of temperatures over which the water can be regarded as being in equilibrium;
— substantially improve the protection of hot spots against the formation of scale;

— allow a higher concentration ratio, thus reducing the quantity of water used.

However, a careful watch must be kept for steel corrosion where protection against scaling is total.

2.3.3. CORROSION INHIBITING PROCESSES

Corrosion inhibitors are reviewed on page 396.

These processes have very definite advantages over those already discussed:
— the cooling water is in a stable zone; its chemical properties are not exposed to the risks of a more or less precarious equilibrium;
— protection against corrosion is excellent; depending on the process, the thickness of metal lost can be cut down to 30 microns per year, or even less;
— there is no risk of scale formation because of the pH of the water;
— very high concentration ratios are possible.

In many cases, sufficient deconcentration is obtained by vesicular entrainment alone; no blow-down is then required.

The quantity of make-up water needed is very considerably reduced, thus leading to very substantial operational savings under the headings of water and reagents.

The required pH can be obtained by acid vaccination (see page 386).

Sulphuric acid is preferable to hydrochloric acid because it is easier to handle and costs less.

In these circumstances, the concentration ratio must often be limited to avoid the risk of calcium sulphate precipitation. It is then advisable:
— either to run a part of the make-up water flow through a carboxyl ion exchanger which simultaneously eliminates TAC and calcium and uses only slightly more acid than the amount needed for vaccination;
— or to purify, in by-pass, a small part of the water in circulation, using lime and sodium carbonate (see Chapter 6).

The circuit can then be used at maximum concentration ratio, without blow-down; consumption of water and reagents is substantially reduced.

2.3.4. WASTE DISPOSAL

Very often, the blow-down cannot be completely eliminated, and the problem of disposal then arises.

The disposal of water containing chromate is prohibited, usually above levels of 0.1 g/m^3 of CrO_4. But there are relatively inexpensive and well-tried industrial methods, for destroying and eliminating chromates; zinc can be removed at the same time. The chromate-zinc process therefore makes it possible to obtain a deconcentration blow-down which is free from the constituents of the corrosion inhibitor.

This is not the case with polyphosphates, and still less with the organic derivatives of phosphorus, the removal of which poses technical and economic problems which still await a satisfactory solution.

Given these circumstances, it is for example, difficult to adhere to the maximum level, sometimes required, of 2 g/m^3 of P_2O_5.

2.4. Fouling

This heading covers all solid deposits, other than scale and corrosion oxides, which are liable to occur in a cooling system.

Such deposits not only tend to complicate operation but, in the same way as scale, they also cause corrosion phenomena as a result of differential aeration, and these are sometimes aggravated by a pitting bacterial corrosion.

There are four possible sources of such fouling:

— the make-up water;
— the air from the atmosphere;
— the manufacturing process: fluids and appliances which are cooled;
— biological growths in the circuit.

● *Fouling due to the make-up water:* the make-up water may contain suspended solids which can be removed by suitable treatment to prevent their being deposited at points on the circuit where the flow is slowest.

Coming between suspended matter and matter in solution, there is what is commonly referred to as colloidal matter. This is unstable and may be transformed into an adhesive and adsorbent gel by a slight rise in temperature or by concentration. It is very often responsible for the fouling of condensers or exchangers with a gluey deposit which varies in colour according to the kind of solid matter adsorbed. Most of the least stable and therefore the most dangerous part of such matter can be eliminated by a physico-chemical process.

● *Pollution by air from the atmosphere:* an atmospheric cooler is an air scrubber; all matter entrained by the cooling air is transferred to the circulating water.

Both suspended solids and colloidal matter are found. When the concentration ratio is low (C less than 3), deconcentration blow-down is normally enough to entrain such matter.

Immediately upon the concentration ratio going above 3, a **fraction drawn** from the main cooling flow has **to be filtered** and possibly coagulated; this fraction can range from 3 to 15 %, according to the degree of pollution, the residence time and the sensitivity to fouling of the equipment which has to be cooled.

Organic dispersants can delay the formation of colloidal deposits. They were particularly valuable when a high proportion was blown down, so that the water in circulation was renewed fairly quickly.   In addition, the fouling potential remaining in the system was a constant threat.   Coagulation and filtration of a diverted fraction is more thorough and reliable, and costs no more.

Soluble alkaline matter (lime, ammonium salts) and acid matter (CO_2, SO_2, SO_3) can also enter the water from the air.   These permanently change the chemical composition of the water in circulation, and the change often differs according to the direction of the wind.   So far as possible, cooling towers should not be sited down the prevailing wind from a chimney or a lime kiln for example. These circumstances should be known when planning the conditioning of a circuit, if only to provide the necessary equipment for stricter control (pH for example).

● *Pollution caused by manufacturing process:* as it passes through the different production machines at a factory, the cooling water can collect all kinds of matter, including, for example, rolling-mill scale, liquid or gaseous hydrocarbons in the oil industry, and solid, liquid or gaseous chemicals.

Such matter is liable to disturb the operation of the system.   Remedies may be based on the guidance given in previous sections, but it is better to study each case on its merits.

● *Biological growths:* a cooling circuit is an ideal medium for the growth of living organisms, because it provides air, heat and light.

There is no need to emphasize the fact that algae proliferate in coolers and tanks open to the air, particularly in summer.   The common types of bacteria can form gelatinous masses in pipes and heat exchangers; they adsorb suspended matter and form a physical obstacle to the water flow.   In addition, they create local conditions favourable to the growth of ferruginous and sulphate-reducing bacteria.

Chlorine, in gaseous form or as hypochlorite, is often used to inhibit the growth of such organisms.   However, the doses normally used only limit growth and do not destroy the resistant biological forms which these organisms can adopt in conditions temporarily unfavourable to their survival.

It is best to introduce the chlorine product in massive doses and at intervals; the actual dose and the frequency of application depend on the particular case and the time of year.

The use of chlorinated materials is not recommended in a scale-inhibiting system: the higher the pH, the less effective they are, and in addition, phosphonates will react with them.   A suitable biocide should therefore be used (see p. 399).

Periodic short and drastic biocide treatment (once to three times a year) is also advisable, particularly at the start of the growing season, between February and June, in order to destroy the resistant forms.

Besides the question of cost-effectiveness, it is necessary to take into account:
— the compatibility of the biocides with the corrosion-inhibitors or dispersants;
— the effects on the receiving medium of the deconcentration blow-down.

Each day brings confirmation of the efficiency of by-pass filtration in dealing with suspended matter and colloids; this is quite understandable; but this effectiveness is also seen to extend to biological phenomena, which is not so obvious.

Lack of care at the start-up a new cooling system may lead to fouling which will be difficult to eliminate; whereas by-pass filtration, properly used, will from the start bring about a considerable and lasting reduction in the biological activity of the system.

2.5. How to design a cooling circuit

The chemical composition of the water is not of course the only factor to be considered when designing a cooling circuit, but it is involved in several ways. That is why there are many sound reasons for consulting a specialist on water treatment and conditioning right from the start:

— reliable new methods may have been developed which it would be a mistake not to adopt for a new factory;
— a choice may sometimes be possible between water from two or more sources;
— it may be possible to integrate treatment of the circuit make-up water with the general treatment of water for the factory;
— the quality of the water available on the site may lead to the choice of an equipment which may be easier to protect than another type; when the water in contact with the surfaces to be protected is renewed too rarely, it is virtually impossible to protect the equipment by adding chemicals to the water;
— the type of protection selected affects the choice of structural materials for coolers, pumps, connecting pipes, etc.

In the case of both an existing and a proposed system, accurate knowledge of:
— the equipment to be cooled,
— operating conditions,
— the composition of the water,
is essential in order to ensure the proper choice of the most suitable conditioning process.

In making this choice, the degree of protection (and therefore the reliability of service) must be set against the total running costs which include:
— amortization,
— maintenance,
— operating and supervisory staff,
— cost of make-up water,
— cost of conditioning reagents.

TREATMENT OF INDUSTRIAL PROCESS WATER

Water, solutions and water suspensions play an important part in the wide range of industrial processes. Each of these processes has its own individual requirements as regards quality of water. A detailed study of all these requirements would lie outside the scope of the present work. However, we shall be dealing with a number of guidelines and common factors.

First of all, we shall discuss, in a general way, the function of make-up water and the subjects of reuse and recirculation of water, which are gaining in importance. Then we shall describe the uses and treatment of water in the main industries, beginning with the processes undergone by the common metals. Among these uses are included hydrometallurgy and brine treatment since they both make use of techniques commonly used in water treatment.

The fight against pollution resulting from the use of industrial process water will be dealt with in Chapter 25.

1. GENERAL PROBLEMS

It has been shown that drinking-water treatment methods have a common purpose and involve a relatively small number of processes, though some of these processes are becoming ever more sophisticated; industry, however, requires many different qualities of water and the necessary treatment processes vary widely and sometimes have little in common. Many and varied techniques involving physical, chemical and sometimes thermal or even biological purification processes, must be used.

Industry often needs a very large and immediate supply; in their turn, industrial firms also often cause physical or chemical pollution of the water they use, and for this the firm concerned is responsible to the community at large. Planners developing new industrial projects or the operator of process water supplies must therefore make every effort to provide proper and adequate water management within the factory complex, and full consideration must be given to the various purposes for which water will be used, and to the planning of the different internal networks, including pretreatment processes common to all, and recirculation where appropriate.

Only then can actual methods of treatment be usefully considered.

1.1. Basic functions of water in industry

Water in industry is generally used for the following basic purposes:

● **generation of energy by production of steam** at conventional or nuclear power stations; these processes require the very best qualities of water;

● **transfer of heat** for the condensation of steam and the cooling of fluids or appliances. Considerable volumes are often required and average water qualities are therefore acceptable;

● **transport of raw materials or waste products:**
— beet (in the sugar industry),
— rolling-mill scale,
— coal in mining washers,
— fibres in paper mills;

● **mechanical action** (steel descaling at a four-high mill, a continuous strip mill, etc., at pressures of 60 to 150 bars—840 to 2 100 psi);

● **manufacture of products** (paper mills, textile factories, food industry, etc.) for which strict and specific standards are often laid down;

● **transport of ions** in hydrometallurgy and electroplating;

● **rinsing of components or products** (metal finishing treatments, semi-conductors, agricultural and food industry, etc.) which according to circumstances can require drinking water quality, completely sterile water or even total deionization;

● **quenching of white-hot products** (slag, coke, cinders);

● **scrubbing of gases:** this rapidly expanding process is used in the metallurgical and chemical industrie;

● **preparation of baths** for various purposes (electrophoresis, soluble oils, etc.);

● **air conditioning,** for certain textile and other applications;

● **maintaining pressure** in oil fields by injecting water which has been treated very carefully and conditioned againts corrosion, fouling and scaling.

1.2. Typical quantities of water used by industry

The actual quantities of water vary widely according to the industry, from a few hundred gallons per day to several hundred gallons per second.

Thermal power stations, steel mills and paper mills are the biggest consumers (particularly for cooling purposes) and the following figures may be quoted as a guide:
— 30 m³/s (680 mgd) for condenser cooling water at a conventional 700 mW power station with a temperature rise of 7 °C;
— 40 m³/s (910 mgd) at a thermonuclear power station of 900 mW capacity with a temperature rise of 1 to 12 °C;

— 200 m³ (52 800 gal) are needed to make 1 metric ton of steel, and this means 12 000 m³/h (76 mgd) for descaling and cooling at a strip mill with a yearly output of 4 millions metric tons.

— 50 to 300 m³ (13 000 to 80 000 gal) to make 1 metric ton of paper, which means 10 000 m³/h (63 mgd) for a newsprint machine producing 1 000 metric tons daily. These figures represent instantaneous flows through the systems.

In many industries, the volume of water can be limited to make-up water by recirculation.

1.3. Industrial process water deterioration

In the course of every industrial process the water used undergoes changes which may be physical, chemical or biological. The resultant pollution is discussed in Chapter 25.

In designing his water system, therefore, the planner or operator must bear in mind the possibility of the following sources of deterioration and must adapt his layout accordingly:

— temperature rise;
— dissolving of gases (CO_2, SO_2, HF), dust (CaO) or chemical products, or conversely the removal of gases (CO_2, H_2S);
— suspension of various sorts of dust;
— precipitation (by heating) of salts which are only slightly soluble ($CaCO_3$, $CaSO_4$, etc.);
— deposit of salts by vaporization of the water.

1.4. Make-up waters

1.4.1. ORIGIN, VARIABILITY

At government instigation, industry is tending towards the use of mainly surface water drawn from rivers and lakes, or even rainwater. The properties of such water vary according to:

— temperature, with the time of year;
— salinity, with the nature of the soil (crystalline, siliceous or limestone) and with the amount of waste discharged upstream;
— organic pollution, with the nature of the soil and the waste;
— physical pollution by suspended solids, with rainfall and the melting of snow.

The correct choice of treatment processes calls for detailed knowledge of how pollution varies in nature and extent over a period.

Underground water, on the other hand, has the advantage of fairly constant temperature and salinity. It sometimes contains varying amounts of iron, organic matter levels often linked with increasing pollution, and too little dissolved oxygen to protect metals from corrosion.

1.4.2. PRELIMINARY TREATMENT

Regardless of the final use of such water and any subsequent treatment, it is often advisable to carry out general treatment close to the intake or well. The purpose is to protect the distribution system itself and at the same time to provide initial or sufficient treatment for some of the main uses of the water.

● **In the case of surface water,** general protection must first be provided against clogging and deposits.

— **The obstruction** or clogging of apertures and pipes by foreign matter is, in fact, an elementary hazard which can be avoided by screening or straining through a mesh suitable for the system to be protected.

The protection used is either a bar screen, in which the gap between the bars can be as narrow as 2 mm, or a drum or belt filter, with a mesh of over 250 microns. This type of protection is often sufficient for condenser cooling water at thermal power stations.

— **The fouling** or deposit of a slime including predominantly organic matter and metallic hydroxides is caused by sedimentation or 'attachment' when the water slows down or is flocculated by a rise in temperature. According to the requirements of the equipment and the amount of pollution in the water, a 250-micron filter may be used on an open system, or microstraining down to 50 microns may be necessary in certain specific cases.

In some cases, rapid filtration through siliceous sand may be necessary after screening and will eliminate suspended matter down to a few microns.

Where there are large amounts of suspended matter, grit removal and/or some degree of settling should be provided.

● **In the case of ground water** the main risks are abrasion by sand or corrosion.

— **Abrasion,** which occurs when the water contains considerable quantities of grit, causes wear on the moving parts and on the glands of pumps and other appliances.

The pumps must therefore be suitably designed, and the protection, which concerns only the parts of the system downstream of the pumps, will take the form of very rapid filtration through sand, straining under pressure or the use of hydrocyclones, if the grit is of the right grain size.

— **Corrosion** frequently occurs on systems carrying underground water and leads to the formation of tuberculiform concretions, which must not be confused with scale.

This corrosivity is often caused by the lack of oxygen.

The best method, therefore, of preventing corrosion is by oxygenation and filtration, processes that have the dual advantage of removing the grit and any iron present, and of feeding into the water the minimum amount of oxygen needed for the system to protect itself.

1.5. Conditioning of process water

Conditioning reagents (cf. Chapter 14) to prevent formation of scale or corrosion and fouling, are commonly used in clean water systems, especially if closed. Many precautions are needed in the case of dirty water, particularly when recycling water used in washing or various manufacturing processes. Additional factors must be taken into account, such as:

— the presence of varying amounts of salts,
— the existence of sludge capable of adsorbing the sequestering substances which will therefore be consumed in excessive quantities.
— if colloids are predominant, their precipitation is prevented by the use of dispersants.

The addition of agents to destroy algae and microorganisms is more difficult and the use of chlorine is sometimes impossible.

1.6. Reuse and recirculation

In view of the growing demand for high-grade water for industry and of the need to economize and reduce the quantities drawn and discharged, methodical use is now being made of the available supplies principally by reuse and recirculation.

● **Reuse** or use in series means using water in an open system for two successive but different purposes, which may be separated in some cases by an intermediate lifting or treatment stage. The second purpose is usually not as demanding as the first, and therefore a poorer quality of water may be used. The commonest example is the use of water first to cool exchangers or condensers and then for washing or rinsing. In another interesting case, waste water from toilets and laboratories is recovered, biologically purified and neutralized and then used, after tertiary treatment, as make-up water in open cooling systems. Special care is taken to keep under control physical characteristics of the water such as temperature, amount of suspended solids or any factor which may promote bacterial growth.

● **Recirculation,** on the other hand, means the indefinite reuse of the same water for the same purpose, with make-up water used only to replace unavoidable losses, deliberate drainoff or evaporation.

The recirculation ratio of water can thus be very high and the concentration of inorganic or organic salts or the gradual accumulation of suspended matter can quickly reach troublesome levels and require continuous purification.

Measures should therefore be taken to limit the following:
— alkaline earth sulphates and carbonates, in order to prevent their precipitation;

— all soluble mineral salts, so as not to increase the conductivity of the water and cause an unacceptable amount of corrosion;

— degradable organic matter, ammonium salts and phosphates, so as not to promote the growth of aerobic or anaerobic bacteria;

— detergents, to prevent frothing and other troublesome phenomena;

— settleable and even non-settleable suspended solids, in order to prevent the clogging or fouling of plant;

— and, of course, heat, in order to avoid setting up intermediate cooling, or discharging excessively hot water into a river.

1.6.1. RECIRCULATION RATIO

Depending on whether evaporation takes place during recirculation, the recirculation ratio of water can be expressed in two ways:

● **The concentration ratio:** $C = \dfrac{a}{p}$

where: C is the ratio of the quantity of water supplied to the quantity of water eliminated in liquid form, p,

a is equal to the quantity evaporated plus the quantity settled, p.

In cooling systems equipped with atmospheric coolers, provided the surrounding air is pure, C represents roughly the increased salinity of the water in the system, S, compared with that of the make-up water, s.

$$C = \frac{S}{s} = \frac{a}{p}$$

In cooling systems for condensers and exchangers, C generally varies from 1.5 to 6, but in extreme cases, values between 20 and 40 have been obtained. Since carbonates can easily bee liminated during purification of the make-up water, it is the sulphates which generally constitute the main limiting factor.

When scrubbing waste gases, concentration by evaporation may be completed by dissolving a number of gases and salts (SO_2, SO_3, NH_4, Cl^-, SO_4^{2-}, etc.). In this event, the concentration ratio no longer expresses the increased salinity, which may be much higher for certain compounds mentioned above or, conversely, lower for a few precipitable or adsorbable compounds.

● **The recirculation ratio (R):** if there is no evaporation or if, in practice, this is negligible, R is the ratio of the circulating flow of water (Q) to the flow of make-up water.

$$R = \frac{Q}{a}$$

When designing recirculation plant in industry, considerable attention should be given to uncontrollable conditions, which limit the recirculation ratio, and especially to the increasing temperature, a problem which is often overlooked. Account should also be taken of the presence of sulphates resulting from the use of metal coagulants in intermediate purification.

The purpose of treatment of all or part of the recirculated flow is to limit the harmful accumulations mentioned above. Depending on the nature of the compounds to be eliminated, any of the following processes may be used:
— total deionization by ion exchange or reverse osmosis, the latter being mostly used in electroplating;
— settling-clarification, to eliminate gas scrubbing dust or particles removed by the degradation of various materials;
— filtration through granular material, to eliminate oxide dust and various crystalline precipitates.

Only part of the water in the system (5 to 50 %) is treated when the pollution is of minor proportions and a partial deconcentration is required: in the by-passed fraction of the recirculated flow, alkalinity and hardness are precipitated and atmospheric dust entrained in cooling water is filtered out.

Inorganic coagulants must not be used for the above purification processes; use instead various polyelectrolytes (Prosedim, see page 143). Purification of the recirculated water is often backed up by anticorrosion or descaling **conditioning** and elimination of living organisms.

1.6.2. REUSE OF WASTE WATER

A. Reuse of purified domestic sewage in circuits including an atmospheric cooling system.

This practice has heen adopted in several countries, and consequently some experience has been acquired. The sewage must undergo biological purification, followed by **tertiary treatment,** the object of which is to:
— precipitate the dissolved phosphates
— eliminate waste products in suspension and colloids.

It is known that, apart from bacteria, the presence of ammoniacal nitrogen and phosphorus affects the behaviour of waste water in cooling towers where the rise in temperature and aeration create ideal conditions for biological growth. Phosphorus in particular has proved to be a major limiting factor as regards the growth of algae. Ammonia (10 to 40 mg/l) also plays a critical role in recirculation with atmospheric cooling. When the ammonia cannot be eliminated economically during initial purification of the waste water, it can be removed by nitrification in the coolers. Since this reaction causes acidification, treatment of the water in the system must:
— maintain sufficient alkalinity in the make-up water;
— use intermittently or continuously biocidal agents which do not cause discharge of toxic waste;

— deconcentrate biological sludge produced in the system by increasing the by-passed flow of recirculated water treated by filtration.

B. *Reuse of industrial waste water.*

Effluents discharged by industry, predominantly organic and purified in the first instance by biological means, are sometimes used a second time.

The above processes should be considered, but account must also be taken of the presence of salts (Cl and SO_4).

An increased ammonia content may impede the process of nitrification, making operation easier.

1.7. Organisation of systems

It is clear from this brief discussion of industrial water problems that, without necessarily being an expert on water treatment, anyone planning to build a workshop or factory must be at least aware of the complexity of the issues involved, and must be prepared to call in qualified technicians at a sufficiently early stage.

The sections which follow, therefore, give not only a brief description of how purification processes are adapted to industry but also a review of the various processes which may concern the individual reader.

Fig. 465. — In the foreground, two 20 m diameter (65 feet) CIRCULATORS for the removal of carbonates from process water and water for general purposes. Output: 3 800 m³/h (24 mgd).

2. METALLURGICAL INDUSTRIES

Industries which extract and work metals make use of complex and very varied chemical processes. However, there are two common factors which should be emphasized with regard to the use of water. They are gas scrubbing and hydrometallurgy, which we shall deal with first.

2.1. Gas scrubbing

The gases given off during the roasting or reduction of ores contain dust and a number of harmful gaseous substances. Before these gases are discharged or reused, they must be scrubbed in purifiers fed with water in a closed circuit. The following phenomena are observed with all metals:
— dissolving of gases such as H_2S, SO_2, SO_2, CO, HF, HCN and NH_3;
— suspending of salts of Ca, Mg, K, Na, which are then partially dissolved by the acids corresponding to the above gases, by the action of their relatively high partial pressure;
— dissolving of the metals;
— breaking of the solubility equilibrium of the salts formed, when the water is subjected to atmospheric pressure;
— precipitation of the metal hydroxides by release of the equilibrium CO_2 through the atmospheric coolers.

All these reactions directly or indirectly cause scale-forming precipitations, acidification or a considerable increase in alkalinity.

Once the nature of the gases and the ore is known (basicity index, percentage of alkalis), general rules for conditioning the water and adjusting its pH and alkalinity can be applied.

In addition, treatment generally includes settling of the recirculated water n one or two stages, and, if necessary, detoxification of the blowdown water.

2.2. Hydrometallurgy

This is the production of metals by a water process; it can be applied to a large number of metals, often extracted from ores where they are present in small amounts (e.g. copper, uranium, nickel, cobalt, zinc, aluminium, lead, titanium, precious metals, etc.). The main stages are as follows:
— **preparation** of the ore;
— **leaching** (acid, alkaline or neutral) using techniques of percolation, mixing or bio-leaching;
— **separation** of solid from liquid by settling, filtering, etc.;
— **purification** and/or **concentration**;

— **extraction of the metal:** depending on which process is used, the metal will be obtained in the solid state as pure metal or one of its compounds.

There is a great variety in the design of systems. They are even more complex where polymetallic ores are being processed.

The processes and appliances used for all solid/liquid separations and for a good deal of the stages in liquor purification are largely the same as for industrial process water: flotation units, scraper-type settling tanks, hydrocyclones, degritting units, centrifuges, filter presses, vacuum filters, ion exchange resins. But in hydrometallurgy, one of the first problems to be encountered is *technological*, since pH values and heat nearly always entrain difficulties. The choice of constituent materials for the appliances is therefore very important in this field, and maintaining the plant is often very expensive.

However, recent progress in various areas of water treatment now permits the quality of the liquor to be improved before purification or extraction:

— **Suspended solids:** after clarification, the liquors still frequently contain 100 to 200 mg/l of suspended solids, sometimes several g/l. These residual colloids are just as troublesome in the direct extraction of the metal as in purification by organic solvent or resins. Many users want to reduce them to less than 10 to 20 mg/l; this can be achieved by filtration (preferably through sand with pressure filters) possibly preceded by flocculation-clarification in a sludge contact clarifier (Pulsator clarifier, Circulator or Turbocirculator, depending on the circumstances) if the liquor is high in suspended solids.

These sludge contact clarifiers can also be used to reduce the content of **colloidal silica** (flocculation by polyelectrolyte and industrial gelatine or by other organic flocculants).

— **Calcium sulphate:** limestone and dolomite ores treated with sulphuric acid give rise to liquors supersaturated with $CaSO_4$, causing formation of scale and precipitation. These problems can also be overcome by the use of sludge contact clarifiers in the presence of the previously formed precipitates. This treatment may be complemented by the use of sequestering agents.

— **Organic solids:** in liquid-liquid extraction the residual solvent impedes precipitation of the metal, especially in the case of electrolysis. Approximately 80 to 90 % can be eliminated by *coalescence through sand*, a process developed by **DEGREMONT**, which has the additional advantage of removing 60 to 70 % of suspended solids present in the liquid phase, without any risk of irreversible clogging. This treatment can be complemented by filtration through activated carbon in granular form.

Figure 466 shows how different systems can be improved by putting to good effect experience gained in the water treatment industry. The example taken here is electrolysis following extraction by a solvent.

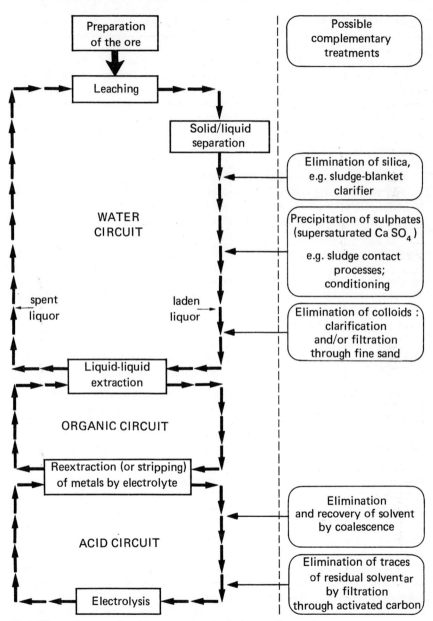

Fig. 466. — An example of improvement of a hydrometallurgical process.

2.3. Iron and steel

The iron and steel industry, which uses vast quantities of water, causes mainly physical pollution. For many years, therefore, the industry has installed systematic closed circuits and has used treatment plants adapted to the specific working conditions experienced in the following types of circuit. The characteristic effluents produced are described in chapter 25.

2.3.1. COKING PLANT

— Cooling system for the indirect condensation of gases.
— Wet dust-removal system for flue-gases in the discharge of coke from the oven (DEGREMONT clarifiers and specil filters);
— Wet dust-removal system for gases in the charging of preheated coal.

2.3.2. BLAST FURNACES

— *Cooling of blast furmace components*, such as the tuyères, tymps and air-boxes. This can be achieved by circulating water in CaO/CO_2 equilibrium or conditioned with chromates. In open circuits, very rapid filtration of part of the circuit water is necessary to keep the water at a minimum standard of cleanliness.

— *Quenching of slag:* slag particles, which are abrasive and of low density, are suspended in large batch flows, and are retained by clarifiers or by tanks with filtering bottoms. The water is very hot and contains sulphides and cyanides.

— *Scrubbing of blast-furnace gases:* water from the gas scrubbers contains suspended matter or fine ash varying in amount with the degree of physical purification of the gases. Although until a few years ago a scraper-type circular or longitudinal settling tank was sufficient to discharge water that was merely clarified, chemical purification or even flocculation is nowadays essential so that the water can either be discharged or recirculated without causing fouling on the frequently delicate equipment of the gas scrubbers. This treatment involves the use of flocculants or dispersants, of acid or lime. The water feeding the scrubbers and in particular the venturis, may also have to be filtered.

2.3.3. DIRECT REDUCTION

This new process now under development uses large quantities of water
— In scrubbing and cooling gases released by the ore reduction process or in cooling sponge iron. The volumes of water used (12 to 20 m^3 per tonne of sponge) are higher than in blast furnaces and, because of the high temperature of the gases, the water is very hot when it leaves the scrubbers (50 to 60 °C). The

amount of dust collected is of the order of 4 to 15 kg per tonne of sponge, depending on which process is used (Korf, Midrex, Purofer, H.Y.L., etc.).

— In the cooling of machinery (compressors and oil coolers) and the direct cooling of gases. A considerable quantity of deionized water is required to produce steam in the reforming of the reducing gas.

2.3.4. CONVERTERS

The predominant use of LD, OLP, LWS or BOS processes requires large-scale water systems:

— Water is used for cooling the hood and the lance (steam is sometimes used for cooling),

— gas-scrubbing water undergoes clarification followed by descaling conditioning. Depending on the steelworking techniques, the water, which is very alkaline, contains large amounts of calcium or sodium; calcium may cause prolific formation of scale.

As converters work intermittently, the waste water varies widely in temperature, pH, chemical pollution (SO_2, CaO) and physical pollution (iron oxide and slag dust). Pretreatment in a degritting unit or hydrocyclone is often required prior to the main purification process, which is basically the same as for the scrubbing of blast-furnace gases, with adaptations for the different kinds of pollution.

Fig. 467. — Clarification and filtration of rolling-mill process water.   Output: 1 500 m³/h (9.5 mgd).   National Steelworks, Portugal.

2.3.5. CONTINUOUS CASTING

There are normally 3 types of system:
— for cooling the ingot mould in a closed system with deionized water make-up;
— for cooling the machine in a conventional system with careful inhibitor conditioning;
— for spraying the machine and the bloom. The waste water contains scale and flame-cutting slag. Entrainment of these substances in the water means that it must be settled and filtered according to techniques adopted in hot rolling. It should be borne in mind that sometimes considerable pollution is caused by the presence of hydraulic fluids (oils, phosphate ester, glycol).

As converters work intermittently, the waste varies widely in temperature, pH, chemical pollution (SO_2, CaO) and physical pollution (iron oxide and slag dust.) Pretreatment in a degritting unit or hydrocyclone is often required prior to the main purification process, which is basically the same as for the scrubbing of blast-furnace gases, with adaptations for the different kinds of pollution.

2.3.6. HOT ROLLING MILLS

These include a great variety of plants, from the strip mill, which can circulate as much as 60,000 m³/h (380 mgd), blooming/slabbing mills, heavy plate mills, merchant bar mills, to billet mills, rod mills and wire-drawing mills of the "mini-mills". In addition to the conventional processes, we should note:
— automatic scarfing, which produces large quantities of granular slag;
— cooling of the blooms in pools, in tunnels or by sprays, using large quantities of water ,without heavy pollution.

Separate cooling systems supply the motors, the oil tanks and the heating furnaces.

Descaling systems are designed for the prior removal of various oxides (wurtzite, haematite) by washing with water. The water collects fine flakes and scale, in varying quantity and of varying fineness, according to the kind of mill and the hardness of the steel. The treatment must include settling before pumping, either in appropriately designed scale pits or in circular grit traps which are wrongly named "hydrocyclones".

According to the concentration and grain size of the scale remaining in the water after this first operation, the water is then clarified and filtered, or filtered only.

As this water rarely causes clogging, high-rate, deep bed, horizontal or vertical filters are used (single-layer type FV2B or two-layer type UHR).

The amount of suspended solids at the inlet to the scale pit is 0.5. to 4 % of the metal passing through the mill.

2.3.7. COLD ROLLING MILLS

The manufacture of thin rolled and galvanized steel products involves metal pretreatment processes such as degreasing and pickling. The latter process uses sulphuric acid and, with increasing frequency, hydrochloric acid. As a result of the controlled attack on the coating of oxides, ferrous iron in the form of sulphate or chloride is dissolved in the pickling tank and the rinsing water.

In both cases, the waste water, rich in acid and iron, has to be successively neutralized in several stages with quicklime, oxidized by aeration and clarified. In most cases, a large thickening unit is required for treating the large quantities of sludge produced. The amount of reagent to be added is also considerable. Effluents from degreasing circuits, which contain too much grease and soap for the pickling sludge to be filtered, should be treated separately.

Deionization plant in a make-up arrangement or in a closed circuit is required in the supply of final rinsing water to galvanising and tinning works. Water of a very pure and soft quality is also required for the preparation of the vast soluble oil baths.

2.3.8. IN FOUNDRIES

The water used to scrub cupola gases must also be purified. According to whether a hot-blast or cold-blast cupola is used, the treatment unit consists of a clarifier with or without prior flocculation.

2.4. Copper production

● **The metal can be obtained** by either dry or wet processes, according to the nature of the ores. Hydrometallurgical processes (see page 713), which usually involve leaching with sulphuric acid and electrolysis, are used on an increasing scale, because they treat low-grade ores and flotation residues. At the same time, greater use is now being made of ion exchange resins and particularly liquid-liquid extraction (or extraction by a solvent) in the treatment of solutions before precipitation of the metal, so that, depending on the case:
— the liquor can either be purified to obtain a commercial product, or
— the depleted leach liquors can be enriched.

The improvements set out in the diagram page 715 are especially useful in copper hydrometallurgy. Figure 468 shows a sand filtration plant in a copper mine in Zambia.

● **Copper wire bar is rolled** to make sections, cables and wires.

Roughing involves washing the surface of the metal with water; valuable copper oxides are left in suspension in the water, and these are usually worth recovering by straightforward clarification, by cyclone extraction, or by mechanical or sand filtration according to the grain size of the oxides. During wire-drawing operations, the wire is often dipped in soluble oil baths, which are prepared to strict standards, and may need maintenance treatment.

Fig. 468. — Sand filtration of a copper sulphate solution.   Output: 3,300 m³/h (21 mgd).
Chingola mines, Zambia.

The production of some types of electric cables, and the preparation of surfaces which require subsequent finishing, involves prior pickling with sulphuric acid.   The pickling process gives rise to the simultaneous formation of metallic copper in suspension and to copper sulphate in solution.   The pickling bath, enriched with copper sulphate, can be regenerated by electrolysis, during which process some of the copper is recovered on the cathodes.   The copper sulphate entrained by the rinsing water can also be recovered after the water has been concentrated in a cation exchanger from which the acid eluate is recirculated to the electrolysis unit.

2.5. Nickel production

● **Extraction by hydrometallurgy** mostly involves low-grade silicate ores (garnierites, laterites).   For a long time, the only processes used were ammonia and ammonium carbonate leaching.   Now, other methods are used: leaching by sulphuric acid (hot and pressurized), and chloridation.

● **The matte is also refined by** hydrometallurgy, using two main leaching processes —sulphuric acid or chloride— in both of which the metal is extracted by electrolysis.   The chloride process can also include solvent and/or ion exchange resin purification, as well as removal of sulphates with barium chloride.   In this case too, marked improvement can be achieved, with supplementary treatment such as sand filtration, coalescence and/or adsorption by activated carbon in the case of solvent extraction.

2.6. Zinc production

● **The blende is enriched** by flotation, a process that produces large volumes of waste water containing deads and zinc salts in suspension or in solution. Since these salts are toxic and cannot be discharged into normal receiving waters, all the zinc must be precipitated by alkaline treatment in clarifiers.

● **The blast-furnace gas scrubbing water** is clarified by slow sedimentation in scraper-type settling tanks, with sludge recirculation in certain cases. The harmful effects of certain secondary reactions are limited by means of special devices.

● **In hydrometallurgy** sulphide ores firstly undergo roasting, then sulphuric acid leaching, solid-liquid separation, precipitation of other metals (Cu, Ni, Cd) by zinc dust, and electrolysis in which cathodic zinc is collected (and H_2SO_4 is recirculated). Solvent extraction processes are also possible. Enrichment of low-grade ores can be achieved by treating them with ammonia or caustic soda.

2.7. Aluminium production

This involves the extraction of alumina from bauxite, followed by electrolysis of the alumina in the presence of a cryolite flux.

● **The alumina is extracted** by the Bayer alkali process, and this involves the clarification at 95 °C of an aqueous solution of sodium aluminate, which contains iron and silica oxides in suspension (red sludge).

The treatment comprises a number of circular scraper-type settling tanks in which the liquor is clarified by fractions. Settling of the red sludge is greatly enhanced by using a specific organic aid. Decomposition of the aluminate by hydrolysis and cooling to 50 °C leads to the precipitation of alumina which is then roasted.

● **Electrolysis of the alumina** gives rise to the release of large volumes of gas containing varying amounts of SO_2, of fluorides or hydrofluoric acid according to whether the furnace is of the prebaked or continuous anode type. These gases have to be scrubbed to prevent pollution of the atmosphere. The acid or alkaline wash-water must itself be neutralized and can be returned to the scrubbers to reduce water losses. The Circulator clarifier is ideal for this type of treatment.

Fig. 469. — Crown Zellerbach paper mills at St Francisville, Louisiana, U.S.A. : Two Accelator. 75 000 m³/d (20 mgd).

● **Aluminium smelting and the rolling of finished and semi-finished products** involve the use of water for washing, cooling and degreasing. In particular, the use of soluble oils means that methods must be devised for the purification and continuous maintenance of the oils and for the emulsion break down and purification of deconcentration blowdown from the dirty oil baths.

2.8. Uranium production

● **Uranium is extracted** by leaching the ore with sulphuric acid or an alkaline detergent. Extraction entails at this stage:
— separation of the deads from the uranium-bearing liquid by clarification and filtration;
— chemical treatment using one of the following methods:
    ● concentration through ion exchange resins;
    ● solvent extraction (tributylphosphate or liquid amines);
    ● precipitation or uranium in the form of ammonium, calcium or magnesium uranate, from the resin-regenerating eluate or by separation of the enriched solvent and the dead liquid.

With acid treatment, the acid leach liquid, after separation from the uranium, must be neutralized with lime and the calcium sulphate precipitated in a scraper-type settling tank.

With solvent extraction, the solvent can be recovered by coalescence.

● **The treatment of irradiated uranium fuel elements** from nuclear reactors makes use of filtration, deionization and conditioning techniques in the storage pools and in the process of de-cladding fuel rods extracted from the reactors.

3. THE PAPER INDUSTRY

The standard qualities to be adhered to in the manufacture of paper and paper pulp are laid down for each type of product (fine quality paper, bleached or unbleached Kraft papers, various types of pulps) by the American organization TAPPI (Technical Association of the Pulp and Paper Industry).

	Maximum concentration in process water				
	fine quality papers	Kraft papers		mechanical pulp papers	soda and sulphite pulp
		bleached	unbleached		
Turbidity (in SiO_2)	10　mg/l	40　mg/l	100　mg/l	50　mg/l	25　mg/l
Colour (in platinum units)	5　—	25　—	100　—	30　—	5　—
Total hardness, in French degrees	10°	10°	20°	20°	10°
Calcium hardness = TH in French degrees	5°				5°
Magnesium hardness (in Ca) in French degrees					5°
Alkalinity in methyl orange (in Ca) = TAC in French degrees	7.5°	7.5°	15°	15°	7.5°
Iron (Fe)	0.1　mg/l	0.2　mg/l	1.0　mg/l	0.3　mg/l	0.1　mg/l
Manganese (Mn)	0.05　—	0.1　—	0.5　—	0.1　—	0.05　—
Residual chlorine (Cl_2)	2.0　—				
Soluble silica (SiO_2)	20　—	50　—	100　—	50　—	20　—
Total dissolved solids (salinity)	200　—	300　—	500　—	500　—	250　—
Free carbon dioxide (CO_2)	10　—	10　—	10　—	10　—	10　—
Chlorides (Cl)		200　—	200　—	75　—	75　—

References : TAPPI standards : E 600 s-48, E 601 s-53, E 602 s-48, E 603 s-49.

According to these standards, raw river water must be clarified by clarification or flotation, often after coagulation. Carbonates are sometimes removed from the water by lime treatment in a sludge contact clarifier, and it usually undergoes a final rapid filtration treatment.

These industries (particularly pulp mills) use large quantities of steam. The treatment of boiler feed water is, therefore, of considerable importance.

4. THE FOOD AND AGRICULTURAL INDUSTRIES

4.1. Sugar factories and sugar refineries

4.1.1. BEET SUGAR FACTORIES

Water treatment techniques are used at the following successive stages of manufacture:

● *Washing the beet:* recirculation of the polluted water through settling tanks, of the scraper type, to recover the water after separation of the sludge (anti-algae treatment, flocculation with polyelectrolytes).

● *Diffusion of chips (preparation of raw juice):* treatment through cation exchangers of condensed ammonia water recirculated to the head of the diffusion battery; disinfection with chlorine.

● *Treatment of thin juices:*
— Deliming of second carbonatation juices (protecting evaporation units against scale). The cation resin can be regenerated conventionally with sodium chloride or by a newer technique, using low-grade mother-liquor from the delimed juice preventing regeneration effluents rich in mineral salts).

Fig. 470. — Deliming of sugar juices.   Output : 3 × 125 m³/h (3 × 0.8 mgd).

— Deionization of the juices by double cation and anion exchange (to reduce level of molasses).
— Removal of colour from juices through activated carbon or adsorbent resins.

● *Concentration of juices, crystallization of sugar:*
— conditioning of juices in evaporators
— treatment of boiler make-up water where recirculated condensate is insufficient (essential in sugar factories comprising a syrup plant or a distillery).

● *Treatment of low-grade products* (mother liquor or molasses) in order to reduce the level of molasses-sugar:
— Treatment of mother liquors through cation resin, regenerated with magnesium chloride (Quentin process). Replacing the Na and K ions by Mg ions reduces the sugar remaining in the molasses.
— Deionization of mother liquor or molasses by ion exchange, with possible complementary treatment to obtain liquid sugar.

Note that with these processes it is often necessary to treat the regeneration effluents before they are discharged. This treatment can be made more economical by concentrating the effluents and by marketing the products recovered (ammonia and potassium salts, amino acids).

4.1.2. CANE SUGAR MILLS

In the manufacture of cane sugar, industrial clarification methods are used to purify raw juices:
— accelerated clarification of the juices defecated with lime;
— flotation of the purified juices in order to separate the very fine cellulose fibres ("folles bagasses").

4.1.3. SUGAR REFINERIES (BEET AND CANE)

Colour removal methods through activated carbon or adsorbent resins are used in the treatment of remelt syrups or in the thorough removal of colour from liquid sugar.

4.2. Dairy industries

Water treatment techniques can be used in:
— the purification of boiler feedwater and cooling systems;
— the clarification and disinfection of process water;
— dechlorination of disinfected water for washing butter;
— removal of iron from ferruginous water used in manufacturing;
— deionization or neutralization of milk or serums (manufacture of powdered milk);
— treatment of condensate from milk concentration;
— purification of water used in reconstituting milk from powdered milk.

4.3. Breweries, distilleries and aerated drink plants

The main water treatment processes used comprise:
— carbonate removal from water for beer brewing;
— softening and/or carbonate removal from water used to produce lemonade and soda water;
— partial carbonate removal followed by disinfection by superchlorination and deodorization (by activated carbon) of water used in the preparation of beverages such as Coca-Cola and Pepsi-Cola, etc.;
— carbonate removal or softening of water used for washing bottles in all these industries and for washing wine and spirit bottles. In the case of very salty raw water, these processes may be insufficient and should be supplemented by deionization by ion exchange or reverse osmosis;
— deionization of water used to dilute spirits.

Fig. 471. — Water treatment plant for the Pepsi-Cola factory at BARCELONA, Spain. Carbonate removal.

4.4. Vegetable canneries

Partial softening of process water is generally essential. A reduction in total salinity may also be necessary.

5. OTHER INDUSTRIES

5.1. The textile industry

In this industry large volumes of water are required at the various stages of production, and call for the following treatments:
● softening or deionization of water used when making yarn, particularly in the case of artificial fabrics;
● softening, frequently preceded by carbonate removal, of the water used in the bleaching and dyeing of fibres;
● treatment of boiler-feed water, often involving large volumes of make-up water.
● deionization of the water used to condition the air in spinning or weaving shops (reverse osmosis, ion exchange).

5.2. Chemical and pharmaceutical industries

The problems arising in these industries are too numerous to list. Apart from the general problems relating to water systems which involve all the techniques described in Part 2, chemical engineering itself often uses exactly the same equipment as that employed in water treatment.

Examples of the use of ion exchangers to solve many problems include:
— purification of glycerines, gelatins, citric acid, phosphoric acid from breakdown of bone, formol, methanol etc.;
— concentration or separation of dilute valuable metals;
— separation and concentration of ionized organic products: amino acids, antibiotics, etc.

In all cases, the technical knowledge acquired in the treatment of water can be transferred directly to the particular field of chemical engineering.

5.3. Manufacture of electronic components

The manufacture of semiconductors uses considerable quantities of water. This water must be extremely pure, physically, chemically and bacteriologically, since the slightest mineral or organic deposit on the surfaces of these components causes irremediable damage.

This industry therefore has plants which prepares "ultra-pure" water. Varying combinations of all the modern processes are used: coagulation, clarification, filtration, chlorination, activated carbon treatment, deionization in several stages, reverse osmosis, final ultraviolet disinfection, safety filtration through membranes.

Primary treatment of river water must be particularly thorough, using all methods to reduce to a minimum the mineral or organic colloids and the dissolved organic matter.

5.4. Metal finishing

Water treatment techniques are involved in the chemical or electrochemical processing of metal surfaces for:
— preparation of deionized water from raw water or recirculated rinsing water;
— continuous regeneration treatments of concentrated baths (ion exchange, ultrafiltration, reverse osmosis);
— recovery of the metals contained in the effluents;

The detoxification of effluents, essential in modern industry, is dealt with in page 849.

Fig. 472. — Clarification plant including two Pulsator of 10 mgd each. Gulf Chemicals Co, Houston, Texas, U.S.A.

5.5. Extraction of crude oil, refining and petrochemicals

These industries use large quantities of steam, a large part of which is used up, another part being recoverable in the form of condensates which are polluted to varying degrees. The treatment of boiler water (make-up and condensates) is therefore of great importance. A number of processes also require softened or deionized water, or water from which the carbonates have been removed. Protecting the general circuits against corrosion by appropriate conditioning is another important aspect.

5.6. Miscellaneous industries using deionized water

Demineralization by ion exchange or reverse osmosis has replaced practically everywhere the old distillation processes, in a variety of applications, such as:
— supplies of pure water for industrial laboratories;
— deionization of water for storage battery production and maintenance;
— deionization of water for mirror silvering baths;
— supplies of water to photographic workshops;
— diluting spirits and washing containers used in perfumery, etc.

In the case of flows exceeding a few hundred gallons per day, the more or less advanced deionization units described in Chapter 10 are used; for smaller flows interchangeable cartridges which contain a mixture of cations and anions and can be regenerated, are often used.

6. BRINE PURIFICATION

A. In the production of purified salt, by recrystallization of raw rock salt:

— clarification and deaeration of raw brine;
— precipitation of magnesium by lime and of calcium by sodium carbonate in reactors of the Circulator or Turbocirculator type,
— partial removal of sulphates with lime or calcium chloride, recirculating the crystallization mother liquor, in clarifiers of the Pulsator type with high sludge concentration.

B. Make-up water in electrolytic systems manufacturing caustic soda and chlorine:

— total removal of sulphates using barium carbonate or barium chloride;
— precipitation of magnesium and calcium using caustic soda and sodium carbonate.

The above methods of treatment may involve intermediate deaeration or pH adjustment. They are frequently followed by sand filtration (for example, through filters of the FV2B type).

C. *In continuous purification of electrolytic metal shaping brine:*

The object here is to limit the increase of metal hydroxides content by treating a bypassed part of the brine flow in a centrifuge or through a precoat filter of the Cannon type.

Fig. 473. — Mitsubishi fertilizer plant at BASRAH, Iraq; 3 deionization lines for feed water treatment. Output: 60 m³/h (264 gpm).

Fig. 473 a. — Process water clarification and carbonate removal. Output: 500 m³/h (3.2 mgd). Usinor Steelworks, Mardyck, France.

DOMESTIC SEWAGE TREATMENT

1. NATURE OF DOMESTIC SEWAGE

1.1. Composition

The impurities present in domestic sewage consist of inorganic and organic matter entrained by the liquid flow in the form of suspended solids (settleable, floating and colloidal) or, to a varying extent, matter dissolved in the water.

In addition to such matter, it contains micro-organisms liable to decompose organic matter and cause putrid fermentation.

One of the main characteristics of a domestic sewage is its biodegradability (susceptibility to biological purification); this depends on the existence of a balanced food supply for the bacteria (nitrogen and phosphorus).

It is desirable for sewage to reach the treatment plant in a sufficiently "fresh" state. Malodorous sewage is toxic to the treatment process and should undergo preliminary aeration or prechlorination before primary settling.

1.2. Assessment of domestic sewage

The pollution of domestic sewage is assessed according to its flow, its concentration of suspended solids and its biochemical oxygen demand. Measurement of the chemical oxygen demand (COD) permits assessment of pollution of industrial origin.

It is accepted that in any community there will be a fixed and very specific quantity of pollution (basis of the inhabitant-equivalent) per person for a given country or region, according to the water supply conditions, the standard of living, and type of connection to the sewage system.

A. Flow:

In France, except under special circumstances, the following daily volumes (per capita) are accepted as standard:
— less than 10 000 consumers: 150 l (40 gal);
— from 10 000 to 50 000 consumers: 200 l (53 gal);
— more than 50 000 consumers: 250-500 l (66-132 gal);
— the flow varies during the day and one or more peaks are noted. By taking Q_j as the 24-hour flow, the following can be defined:

Average hourly flow in daytime $Q_d = \dfrac{Q_j}{14}$.

Average hourly flow over 24 hours $Q_m = \dfrac{Q_j}{24}$.

Q_d may vary between $\dfrac{Q_j}{12}$ and $\dfrac{Q_j}{18}$ depending on the conditions and volume of the industrial effluent.

The dry-weather peak can be calculated by the formula:

$Q_p = Q_m \left(1 \cdot 5 \times \dfrac{2 \cdot 5}{\sqrt{Q_m}} \right)$ where Q_p and Q_m are expressed in litres per second up to a peak coefficient not exceeding 3. The wet-weather peak (in a combined system) is generally agreed to be between 3 and 5 times the average flow Q_m.
— Generally the dry-weather peak occurs once a day. The shorter the system, and the smaller the population served, the greater the relative size of the peak.
— If the raw sewage contains an appreciable proportion of industrial waste water (from slaughterhouses, dairies, etc.) the variations in pollution may be far more abrupt and extensive than for a sewage effluent alone. This must be taken into account when designing the treatment plant.

B. BOD₅ loadings.

In France, the BOD_5 loading contributed by raw sewage, per head per day, is normally estimated as follows:
— separate system .... 60-70 g
— combined system .... 70-80 g

The lower figures apply to populations of up to 5 000 and the higher figures to populations exceeding 20,000.

As the standard of living rises, it is found that the pollution load per capita and also the volume of sewage increases (indeed, the sewage flow increases faster than the loading, so that the sewage tends to increase in quantity but becomes less concentrated).

The BOD_5 concentration varies according to time of day. The hourly pollution peak may be as much as ten times the mean hourly pollution.

C. Suspended solids (SS) loading.

Suspended solids loadings contributed by raw sewage, per head per day, are generally estimated as follows:

— separate system .... 70 g, including 70 % volatile matter
— combined system .... 70 g, including 66 % volatile matter
     These figures are for suspended solids after screening and grit removal, and
do not include material collected during preliminary treatment, which is roughly
as follows:
— screening: 2-5 dm³ of screenings per person per year in the case of screens
with bars approximately 35-50 mm apart;
          5-10 dm³ of screenings per person per year in the case of screens
with bars approximately 15-25 mm apart.
The water content of these screenings is about 70-80 % after natural drainage.
— grit removal: volume of grit per head per year about 5 dm³ (for densely populated areas) to 12 dm³ (for less densely populated areas);
— the volume of matter retained by the bar screens increases with the standard
of living of the user population, mainly because of the increase in the proportion
of fibrous matter;
— when household refuse is ground in the sink and discharged to the sewer (a
practice prohibited in France), this matter is not retained by the screens and the
load entering the treatment plant may then be substantially increased (BOD_5 and
SS are practically doubled in some towns in the United States).

D. Examples of loadings in various countries:

Country	Flow		BOD_5 g/head day	SS g/head day	Sewer system
	l/head day	gal/head day			
Italy	150-350	40-92	70	80	combined
			60	70	separate
Canada, USA	400-500	105-132	80-100	100-120	
Japan 1970	300-500	79-132	44	40	
Japan 1990	300-500	79-132	64-84	58-76	
Switzerland	500	132	75	100	

E. Biodegradability (see page 100).
     In the presence of organic matter that is only very slowly degradable, or of
chemical reducing agents or biological inhibitors, there will be an increase in the
ratios $\dfrac{DCO}{BOD_{21}}$ (theoretical) and $\dfrac{COD}{BOD_5}$ (usual), indicating the presence of industrial
pollution.
     There is a risk of effluents of this type retaining an abnormally-high COD
after treatment.
     The origin of the BOD_5 loading of the organic matter can also give an indi-
cation of the presence of industrial wastes.

In municipal effluents where there is no industry, the BOD_5 may be divided approximately into:

    66 % for the BOD_5 of suspended or colloidal matter;
    34 % for the BOD_5 of dissolved matter.

In general, the 'soluble' proportion of the BOD_5 increases when the percentage of industrial waste in the effluent rises.

F. Influence of industrial effluents.

The proportion of industrial effluent in sewage is constantly increasing; it is often more economic to treat the mixture as a whole than to treat the two types of effluent separately, but where toxic substances or biological inhibitors are present in the industrial effluent, prior treatment at the factory is essential.

The conditions governing the discharge of effluents into sewers are laid down in regulations.

In France implementation of the legislation is the responsibility of the 'Agences de Bassin' (River Basin authorities).

The pollution of industrial effluents may be expressed by the 'coefficient of population equivalent', the unit representing the BOD_5 loading of the daily quantity of domestic waste water per head.

This idea of establishing an equivalent by the BOD_5 loading is convenient, but inadequate. It would also be necessary to estimate the population-equivalent as a function of suspended solids, and use whichever of the two should be predominant in the systems concerned.

The importance and the effects of the presence of an industrial effluent in sewage effluent can be estimated by comparing the qualities of this effluent with those of municipal effluent free of any industrial waste. This comparison may be based on the following criteria:

— **Presence of nitrogen:** the total nitrogen (TKN) content amounts to about 15 to 20 % of the BOD_5 in domestic effluents considered alone. A higher content may be a sign of industrial pollution.

— **Effect of pH:** the pH of sewage alone can generally be said to be 'neutral', and is from 7 to 7.5 approximately. A different pH indicates industrial pollution. Biological purification is possible between pH 6.5 and 8.5.

— **Oxidation-reduction (Redox) potential (E_H); oxidation reduction power (rH):** unmixed domestic sewage that is sufficiently fresh has an oxidation reduction potential of about $+ 100$ mV, corresponding to an rH of about 17 to 21 for a pH around 7. A potential of $+ 40$ mV (rH = 15 at pH 7) or a negative potential indicates a reducing medium (septic waste water, putrid fermentation, connection of septic tanks, presence of chemical reducing agents). A potential exceeding $+ 300$ mV (rH = 24 at pH 7) reflects an abnormal oxidizing medium.

— **Toxicity and inhibition:** the presence of heavy metals Cu^{2+}, Cr^{6+}, Cd^{2+} ..., even in small quantities (0.1 mg/l), may suppress the action of bacteria.

Sulphides in concentrations of 25 mg/l completely inhibit biological growth in a non-acclimatized activated sludge. After a few days' adaptation, the tolerance increases to 100 mg/l.

A large number of products are toxic and their discharge to sewers and especially to natural receiving waters is prohibited by law (e.g. cyanides, cyclic hydroxyl compounds, etc.).

Some pharmaceutical products may also be harmful to bacterial life (e.g. antibiotics).

— **Nutritional equilibrium:** a nitrogen and phosphorus deficiency is often found in industrial effluents. In some cases, it may be necessary to add nutrients to restore the ratios:

$$\frac{BOD_5}{N} \simeq 20 \text{ and } \frac{BOD_5}{P} \simeq 100$$

necessary for biological purification.

Unbalance may cause the biological treatment system to malfunction, with bulking of the sludge and low purification efficiency.

— **Salinity:** high salinity may reduce efficiency; abrupt changes are much more harmful than slow variations.

— **Temperature:** Temperature variations also affect the purification processes.

G. Introduction of effluent emptied from cesspools.

In built-up areas, some dwellings remain connected to fixed cesspools or septic tanks for some time before they are connected to the main sewer. The effluent emptied from these tanks generally contains much grit, sometimes pebbles etc., and disturbs the purification processes.
Its composition is as follows:
— BOD_5 : about 4 000 to 10 000 mg/l
— COD : about 6 000 to 16 000 mg/l
— SS : about 5 000 to 17 000 mg/l
— NH_4 : about 1 500 to 5 000 mg/l
These effluents must be pretreated before being added to the raw sewage reaching the purification station:
— screening and grit removal;
— dilution with good mixing in the raw sewage in a maximum proportion of 1 % by volume, immediately before it enters the primary settling tank (preaeration is sometimes necessary).

1.3. Objectives

The purpose of treatment is to obtain a purified effluent in which pollution is limited to such an extent that is does not cause any harm to the flora and fauna of the receiving waters.

In all countries, health regulations lay down the pollution limit tolerated in effluent discharged into rivers (see section dealing with legislation, chapter 34).

The discharge standards of various countries are set out in the table on page 1144. These are general standards constituting the basis for discharge studies; particular local conditions may induce the competent authorities in some cases to stipulate more severe conditions: low dilution, protection of drinking water supplies or recreational waters, etc.

The purification criteria must be determined for each case on the basis of the self-purification capacity of the receiving medium and on the total pollution of all types of discharge taken together. The maximum pollution level of a river depends on its classification (in France (*a*) high-pollution rivers used for the discharge of waste water with a dissolved-oxygen content $O_2 \leqslant 4$ mg/l, (*b*) rivers stocked with cyprinidae with $6 \geqslant O_2 > 4$ mg/l, and (*c*) rivers stocked with salmonidae with $O_2 \geqslant 7$ mg/l).

The purified effluent still contains mineral impurities (nitrogenous or phosphated) and non-biodegradable organic impurities (detergents, pesticides, etc.). Tertiary treatment is sometimes necessary (see page 798).

1.4. The main process lines and their purification efficiencies

Treatment of sewage may involve the following, either separately or in combination:

● physical treatment facilities :
— preliminary treatment (bar screening, grit removal, etc.) (see page 111);
— primary settling (primary treatment) (see page 164) for removal of settleable and floating suspended solids;

● physicochemical treatment facilities: nonsettleable suspended solids can be separated in these facilities by coagulation, giving more thorough separation than in the case of settling alone; coagulation also eliminates certain heavy metals and phosphates;

● biological treatment facilities (see page 204), for more complete elimination of organic pollution by bacterical action;

● sludge treatment and drying facilities (for primary sludge deposited in primary settling tanks and excess sludge from biological treatment) (see page 427).

This basic configuration may be varied in some medium-sized treatment plants: primary settling may be omitted, or sewage and sludge may undergo simultaneous aerobic biological treatment (extended aeration, etc.).

Physical treatment may sometimes be sufficient on its own, where only partial purification is required.

On the other hand, if a very high degree of polishing or the elimination of

residual non-biodegradable matter is necessary, tertiary treatment must be provided, for example:
— further reduction of BOD and SS;
— removal of phosphates;
— denitrification;
— elimination of surfactants;
— chlorination.

In some cases, when the required degree of purification comes midway between biological purification and ordinary primary treatment, physico-chemical treatment of the water by flocculation (coagulants or polyelectrolytes) and clarification can be used. Physico-chemical processes are also suitable for advanced degrees of purification, for example in stations with a seasonal peak (spas, tourist centres), However, the sludge disposal conditions and operating costs must be determined with great care.

To avoid over-sizing the treatment plant it may be desirable, especially in combined systems, to limit in turn the maximum flow entering:
— biological treatment unit;
— primary treatment plant;
— and if applicable, the pretreatment unit.

When heavy rainfall increases the sewage flow, the river into which the effluent is discharged often also carries a larger flow, and consequently its natural purifying capacity is greater.

Purification efficiency of the different processes.

Purification efficiency depends on many factors, in particular:
— regularity of the flow and the effluent loading to be treated;
— proportion of industrial waste;
— concentration of the raw sewage;
— temperature of the sewage.
The efficiency of elimination, as a percentage, can be estimated as follows:

● **Primary settling** (physical treatment)
For municipal sewage alone :
— elimination of BOD by about 35 %;
— elimination of settleable solids up to 90 %.
The presence of industrial pollution may bring the elimination efficiency figure down to as low as 10 %.

● **Settling after flocculation** (physicochemical treatment)
— BOD elimination in the region of 70 %;
— Total suspended solids elimination up to 90 %.

● **Biological purification:** By biological purification it is possible to obtain a high BOD reduction (more than 95 %) in conventional installations with a low sludge loading (less than 0.3 kg BOD_5 per kg SS in the aeration tanks).

LIMITS OF PERFORMANCE OF DOMESTIC SEWAGE TREATMENT PROCESS LINES
(SS = 300 mg/l; BOD = 250 mg/l; COD = 450 mg/l)

	SS mg/l	BOD_5 mg/l	COD mg/l	N(TKN)	P	SALINITY	INHIBITION Parasite eggs	INHIBITION Bacteria
= Primary settling.............	130-170	150-200	260-340	— 15 %	— 15 %	—	effective	slight or nil
2 = 1 + chemical flocculation (poly-electrolytes)	50-70	110-160	200-250	— 2 %	— 20 %	—	total	slight
3 = 1 + biological treatment......	15	12	50	— 9 % (with nitrification)	— 35 %	—	effective	slight
4 = 3 + sand filtration...........	1	8	40	— 95 % (with nitrification)	— 40 %	—	total	slight
5 = 3 + chlorination	15	10	45	— 90 % (with nitrification)	— 35 %	—	effective	substantial to subtotal
6 = 3 + 1 lagooning + chlorination.	15	8	40	— 90 % (with nitrification)	— 35 %	—	total	subtotal
7 = 1 + chemical precipitation (Fe^{2+}) + biological precipitation (Al^3)	10	10	40	— 90 % (with nitrification)	— 90 %	—	total	slight
8 = 3 + chemical precipitation (Fe^{2+}) (Al^{3+})	15	10	40	— 90 % (with nitrification)	— 90 %	—	total	substantial
9 = 3 + biological denitrification.	15	8	50	— 95 %	— 30 %	—	effective	slight
10 = 4 + activited carbon	11	10	10	— 90 % (with nitrification)	— 40 %	—	total	substantial
11 = 1 + Ca^{2+} precipitation....... + stripping................ + restoration of carbonates. + sand filtration	40-60	100-150	200-250	— 75 %	— 90 %	—	total	subtotal
12 = 3 + chemical precipitation... + sand filtration.......... + activated carbon .... + chlorination	~ 1	~ 1	< 10	— 95 %	— 95 %	—	total	total
13 = 11 + deionization............	~ 1	~ 1	< 10	— 99 %	— 99 %	< 100 % (ion exch.) < 90 % (reverse osmosis)	total	total

An increase in this efficiency involves an increase in power consumption; this is especially marked if nitrification is included.

It should also be noted that biological purification includes aeration and clarification plant together forming an indivisible treatment unit.

The quality of the clarification phase is as important in determining efficiency as that of the aeration phase.

● **Tertiary treatment** (see page 798): This is a blanket term covering various treatments usually used to supplement biological (or secondary) purification. Chief among these are filtration, lagooning, chlorination, chemical postprecipitation, stripping, adsorption on activated carbon, deionization, etc.

The efficiencies given by the different types of processes are set out in the table page 738.

2. BIOLOGICAL TREATMENT

2.1. General

Aerobic biological treatment of sewage makes use of systems causing the growth of bacteria which act physically and physiochemically to retain organic pollution, on which they feed. This growth can be achieved by placing the bacteria in suspension in the sewage (activated sludge) or on fixed films (biological filters); these processes are discussed in Chapter 8.

To ensure that the operation of biological treatment systems is not upset by heavy or bulky matter, they are preceded by suitable preliminary treatment facilities. In small plants, primary settling may be dispensed with; in this case, preliminary treatment—which must include thorough grease removal—is absolutely essential.

2.2. Extended aeration type treatment plants

In these systems, mainly used in small plants, sewage is treated at a sludge loading low enough for simultaneous sludge stabilization to be possible.

This process has the advantage of operational simplicity, but on the other hand it requires tanks of relatively large volume. It is used in several different systems, the most important of which are described below.

2.2.1. TYPE M.A. PLANT

The M.A. is a compact facility consisting of an oxidation tank with one or more surface aerators and a final clarifier with sludge recirculation by a suction type scraper bridge.

● **Oxidation tank:** the raw sewage is stirred for "integral mixing" with the return sludge and the pre-existing liquor.   Transfer of the activated sludge into the next

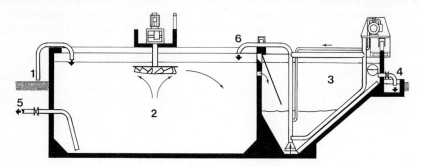

1. *Intake of raw sewage after coarse screening and grit removal.*
2. *Oxygenation.*
3. *Clarification.*
4. *Treated water outlet.*
5. *Sludge extraction.*
6. *Return sludge.*

Fig. 474. — M.A. plant.   Diagrammatic representation.

Fig. 475. — M.A. plant for a population-equivalent of 1 500.

compartment (the final clarifier) is controlled by a distribution device. The sewage is agitated and oxygenated by an ACTIROTOR surface aerator mounted on a concrete platform. The hydraulics of the system, based on the dimensional proportions of the tank and the specific power of the aerator, are calculated to give a homogeneous sludge suspension and to avoid dead areas which might impede the purifying function of the facility.

● **Final clarifier:** the sludge deposited in this tank is scraped together on the bottom, from which it is collected by air lifts. The scraper and air lifts are mounted on a reciprocating bridge whose reversals of direction are controlled by a clock. This automatic equipment:
— recirculates the floc without damaging it, and
— collects the entire sludge without risk of its rising to the surface.

Sewage enters the final clarifier by way of a submerged duct which prevents any agitation at the point where the activated sludge is introduced. At the outlet, the clarified water is discharged through a channel or a submerged pipe. The discharge weir is protected by a scum baffle for interception of floating matter, which is trapped by a collection hopper on each transit of the bridge.

The design of the M.A. eliminates the disadvantages of uncontrolled denitrification, such as the rising of patches of sludge to the surface. It gives a high recycle rate and keeps the concentration in the aerator at a level compatible with the low sludge loading used. Finally, it avoids the risk of obstruction associated with low-flow recirculation pumps.

● **Extension and variant:** at high altitudes, air blowers may be used instead of mechanical aerators. The facility can be extended by the addition of a stabilization tank, giving the configuration known as S.A.M. (see page 748).

2.2.2. MINIBLOC A.P. PLANT

The Minibloc A.P. combines the various compartments required for the treatment process in a single factory-made metal tank of rectangular cross-section; these compartments are for: preliminary screening, aeration, final clarification, pumping of the treated sewage (optional).

Blown air aeration is used; the same air also operates the air lifts, which are housed in convergent ducts and are used to recirculate sludge from the final clarifier to the aeration compartment.

A scum baffle on the top of the clarifier compartment retains floating matter, which an air lift recirculates periodically to the aeration compartment.

The MINIBLOC A.P. is available in five models with capacities for populations of 100, 200, 300, 400 and 500 respectively. The basic parameters for calculation are the same as those of a conventional extended aeration system.
— aeration: sludge loading 0.07 kg BOD /(kg SS.d);
— final clarification: upward flow rate: 0.7 m/h for peak flow.

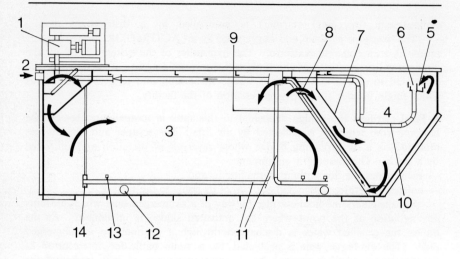

1. *Electric blower unit.*
2. *Raw sewage inlet.*
3. *Extended aeration.*
4. *Clarification.*
5. *Treated water collection channel.*
6. *Scum collection channel.*
7. *Scum baffle.*

8. *Transfer orifice.*
9. *Sludge recirculation.*
10. *Pumping of scum.*
11. *Air supply.*
12. *Anti-flotation orifice.*
13. *Vibrair diffuser.*
14. *Drain plug for air feed-pipe.*

Fig. 476. — Diagrammatic representation of Minibloc A.P. unit.

2.2.3. TYPE M.V. PLANT

This system is identical in design to the Minibloc A.P. It is intended for populations of 600 to 2 000 and is made of reinforced concrete.

2.2.4. TYPE U.I. PLANT

Activated sludge from extended aeration can withstand relatively long periods of anoxia. The type U.I. plant, in which two tanks operate alternately, exploits this feature, so that its operation is particularly simple and reliable, making it very suitable for small and medium-sized installations.

The U.I. consists basically of two identical tanks 1^1 and 1^2 each equipped with an Actirotor surface aerator. The raw sewage (5) enters a distributing tank with an overflow weir. Two pneumatic valves 4^1 and 4^2 can direct the flow to be treated either to tank 1^1 or to tank 1^2. When valve 4^1 is open, valve 4^2 is closed and vice versa. The two tanks 1^1 and 1^2 communicate either by a weir or by a pipe (8). Two valves 3^1 and 3^2 can cut off the flow at the outlet from tanks 1^1 and 1^2, each of which have a split collecting pipe (2^1 and 2^2 and 6). From the bottom of the tanks, two pipes (9) discharge the excess sludge.

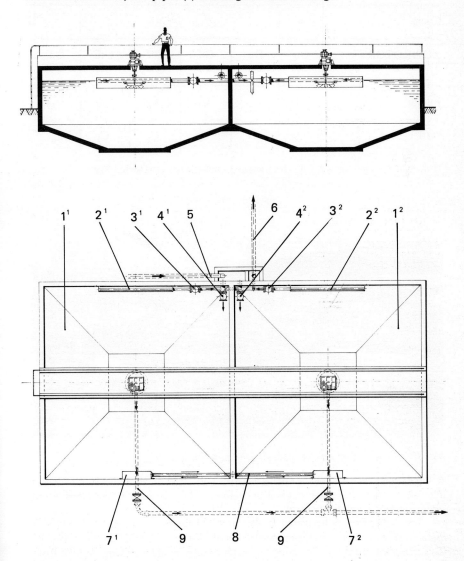

Fig. 477. — Typical U.I. plant.

● **How it works:** this is shown in the following table:

Time	t_0	t_1	t_2	t_3	t_4	t_5
Cycle	0		I			II
ACTIROTOR C_1	Off	On	Off	Off	Off	On
C_2	Off	Off	Off	On	Off	Off
Inlet to	C_2	C_1	C_1	C_2	C_2	C_1
Outlet from	C_1	C_2	C_2	C_1	C_1	C_2

The time intervals are approximately : t_1—t_0 and t_5—t_4, 6 hours; t_2—t_1 and t_4—t_3, 1 hour,

Fig. 478. — U.I. plant serving a population-equivalent of 2 500.

● **Operation:** the ACTIROTOR is started and stopped and the motor-driven inlet and outlet valves are opened and closed from an automation console not requiring any manual attention.

The treatment by activated sludge is very simple to operate. The date of extraction of the excess sludge (and with this there is great latitude) is determined by the fortnightly check on the content of settled sludge made by taking samples from the two tanks during the aeration period.

Excess sludge is removed by opening the discharge valve on the tank during the settling period. The degree of concentration of the sludge removed from that tank may reach 20 g/l.

● **Technical data:** the volume of the tanks, based on 0.18 to 0.30 kg BOD_5/m^3 d (10 to 17 lb/1 000 ft³ d), corresponds to a retention period that generally exceeds 24 hours and largely compensates for the variations due to the alternating method of operation.

● **Nitrogen elimination** (see page 107) : the succession of aerobic phases which nitrify the effluent and anoxic phases in which denitrification takes place gives high nitrogen elimination efficiency, which may reach 80 % and more and depends very closely on temperature.

The type U.I. system is capable of satisfying the requirements of Level 6 of the French discharge standards (i.e., the most severe level), in which a maximum concentration of 7 mg/l is stipulated for total Kjeldahl nitrogen.

● **Extension and variant:** U.I. systems are particularly suitable for installations intended to be subsequently enlarged. They can easily be converted into type S.A. units (see page 746); the capacity of the system can be doubled by merely adding a final clarifier.

In a variant of the U.I. system, blown air diffusion replaces ACTIROTOR aeration.

2.2.5. TYPE A.O.S. PLANT

In this plant, an aeration tank equipped either with one or more surface aerators or with blown air diffusion facilities is combined with a separate final clarifier which may or may not have a scraper system.

Activated sludge is recirculated by pumping or by gravity return to the raw effluent pumping tank.

Like the M.A., this unit can be extended at a later date and used at medium loading; it is then supplemented by a stabilization tank, in which case it comes to ressemble the type S.A. system (see pages 746-748). This system will often be suitable for populations of over 1 500.

2.2.6. TYPE A.C. PLANT

This is a compact facility comprising a circular central aeration tank surrounded by an annular final clarifier without a scraper.
The sewage is oxygenated by an ACTIROTOR surface aerator, whose pumping effect, enhanced by a draft tube, also provides the necessary recycling.
The A.C. unit is suitable for populations in the range 500 to 4 000.

Fig. 479. — A.C. plant serving a population-equivalent of 2 000.

2.2.7. OXIDATION DITCHES, CARROUSELS

In a channel (generally shallow) forming a closed circuit, one or more horizontal brush aerators oxygenate the liquor and sweep it along with a rotating motion. There are several possible arrangements:
— one tank for batch operation;
— two tanks operating alternately;
— one continuously-operating tank, in which case the channel is supplemented by a final clarifier with activated sludge recirculation. This is sometimes designed in the form of a circular channel round the final clarifier.

The CARROUSEL works on a similar principle, but the brushes are replaced by vertical-shaft aerators.

There are many other systems of extended aeration in the world; they are all more or less closely related to those described above.

2.3. Plants using medium load purification followed by aerobic stabilization

2.3.1. S.A. COMBINATIONS

● **Principle and description:** S.A. units treat the raw sewage without primary settling and also stabilize the excess sludge aerobically in separate units equipped with ACTIROTOR mechanical aerators. This configuration is suitable for populations up to over 50,000.

Apart from preliminary screening, grit and grease removal facilities and pumping where necessary, the S.A. installation comprises:
— an aeration tank;
— a final clarifier with bottom and surface scraping;
— a sludge stabilization tank.

In most cases, especially in small units, the aeration and stabilization tanks are positioned side by side. The S.A. then consists of only two structures: aeration/stabilization, and final settling.

After pretreatment, the raw water enters the aeration tank A_1, fitted with one or more ACTIROTOR.

From the aeration tank outlet, the mixed liquid is taken to the final tank D. The treated water Qt is discharged into the receiving water.

The sludge settling on the bottom of the final tank is returned to the aeration tank to maintain therein an adequate concentration of activated sludge. This is the return sludge qr. The sludge can be recycled by a pump P or returned by gravity to the head of the station. This operation is automatically controlled.

The excess sludge is pumped to the stabilization tank A_2 by the sludge pump P or by a separate pump system or by gravity. This operation can be automatically controlled by a DEGREMONT Sedimometer (see page 579). The supernatant water is returned to the head of the station.

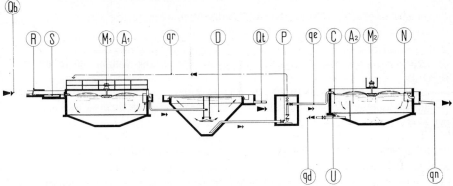

Fig. 480. — Diagrammatic representation of a S.A. unit.

Qb *Raw sewage.*	**D** *Final tank.*	**M₂** *Motor and reduction gear.*
R *Screening.*	**Qt** *Treated sewage.*	**N** *Surface draw-off.*
S *Grit removal.*	**P** *Sludge pump.*	**qd** ⎰ *Removal of stabilized*
M₁ *Motor and reduction gear.*	**qe** *To sludge stabilization tank.*	**U** ⎱ *sludge.*
A₁ *Aeration tank.*	**C** *Sludge concentrator.*	**qn** *Extraction of supernatant*
qr *Return sludge.*	**A₂** *Sludge stabilization tank.*	*liquor.*

Fig. 481. — S.A. plant serving a population-equivalent of 15 000.

● **Operation:** the cycle of main operations is controlled automatically from a general control console.

The ACTIROTOR surface aerators can operate continuously or at intervals, according to the variations in the mean daily flow and the degree of pollution.

Operation of a type S.A. unit entails mainly:

— supervision of the mechanical operation of the aerators;

— adjustment of the activated sludge return rate and the aerator operating cycle.

This control is based on simple measurements of the sludge percentage and oxygen content in the tanks:

— transfer of excess sludge to the stabilization tank every second day;
— periodic removal of stabilized sludge to drying beds.

● **Characteristics:** the S.A. unit is designed on the following basic data:
— aeration volume: 1 to 2 kg BOD_5 per day per m^3 of tank volume (60 to 120 lb/ft^3 d);
— stabilization volume: 25 to 50 l/head (1 to 2 ft^3);
— surface loading of final tank: 10 to 20 $m^3/(m^2$. h) at peak times (4 to 8 gpm per sq ft).

The characteristics are determined by the rate of removal of stabilized sludge, the frequency and size of peak flows, and peak pollution loadings.

● **Variant with primary settling:** in some cases, especially to make provision for extensions, it may be useful for the installation to be preceded by primary settling, to which the excess sludge is returned. The fresh sludge is stabilized at a concentration within the tank limited to 25 g/l.

2.3.2. TYPE S.A.M. PLANT

In this system, aeration and final settling are combined in a compact structure identical in form with that of the M.A. system; the stabilization tank, however, remains separate. A S.A.M. system sometimes results from extension of a M.A. plant (see page 740).

2.3.3. MINIBLOC AC PLANT

The MINIBLOC A.C. is a single-structure system in which the combined aeration and final settling tank is adjacent to the sludge stabilization tank.

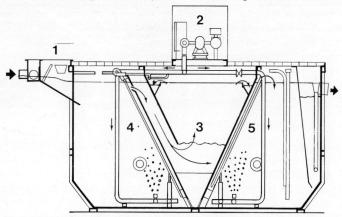

Fig. 482. — Diagrammatic representation of MINIBLOC A.C.

1. *Raw sewage inlet. Coarse screening.*   4. *Aeration.*
2. *Blower.*   5. *Aerobic stabilization.*
3. *Clarification.*

The assembly is factory-built of steel and is conveyed fully assembled to the site. The MINIBLOC A.C. comprises:
— an aeration compartment with air injection by Vibrair diffusers;
— a final clarification compartment with excess sludge recycling and transfer air lifts;
— an aerobic stabilization compartment with Vibrair air diffusers and a sludge concentrator;
— a blown air production and closed loop control system.
The basic calculation parameters are as follows:
— aeration: BOD loading of 1 kg BOD/d per m^3 tank volume;
— final clarification: rate of 0.30 m/h at average loading (0.12 gpm/sq ft);
— stabilization : same volume as aeration.
The MINIBLOC A.C. is standardized in five types: populations of 70, 150, 200, 270 and 350.

2.4. Medium- or high-load plants with anaerobic digestion or direct sludge treatment

2.4.1. GENERAL DESIGN

This method of treatment is suitable for medium- and large-sized plants.

A. The typical configuration of a plant of this type is:
— preliminary screening (generally mechanical);
— grit removal, if necessary (aerated grit remover);
— grease removal, if necessary (often combined with grit removal);
— lifting by pumps or Archimedean lift screws;
— primary settling (total or only partial primary settling of peak flow), in a circular or rectangular settling tank with scraper;
— aeration tanks with blown air or surface aerators, often constructed in several channels to give different hydraulic working configurations (see page 757);
— final settling in circular or rectangular tanks either with scrapers or, more commonly, with suction facilities;
— activated sludge recycling by pumps or Archimedean lift screws, excess sludge being returned to the head of the plant;
— pumping of fresh sludge with or whitout thickening;
— one- or two-stage anaerobic digestion with separate gasholder;
— mechanical sludge dewatering (or natural dewatering where the climate so allows);
— thermal drying (if necessary);
— sludge incineration (if necessary) at the plant or together with household refuse in a nearby incineration plant.
The effectiveness of anaerobic digestion is increased when the sludge is fresh

and concentrated.   The simplest method of concentration is to mix the excess activated sludge with the raw influent before primary settling.   This settling also produces a 25 to 35 % reduction in the BOD_5, giving a saving in the capacity of the aeration tanks and in power costs. (p. 746).

This type of station is very common in Western Europe, and is now meeting competition from plants using separate aerobic sludge stabilization (described in the preceding section).   However, it has useful advantages: sludge drying is facilitated by a greater reduction in weight and volume and by the large storage capacity of the digester.   Almost complete destruction of pathogenic germs is also achieved.

These plants are "made to measure", and only some of the structures and units are standardized.

Fig. 483. — Conventional type of purification plant using activated sludge.

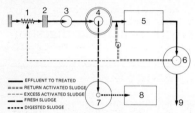

1. *Lift screw.*
2. *Screen.*
3. *Grit and grease removal.*
4. *Primary settling tank.*
5. *Aeration tank.*
6. *Final clarifier.*
7. *Digester.*
8. *Drying beds.*
9. *Effluent to river.*

Fig. 484. — Purification plant without digester, with Oxyrapid and chemical sludge stabilization.

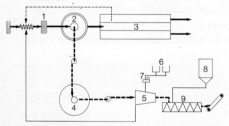

1. *Screen.*
2. *Primary settling tank.*
3. *Oxyrapid unit.*
4. *Thickener.*
5. *Centrifuge.*
6. *Reagent tanks.*
7. *Dosing pump.*
8. *Lime silo.*
9. *Incinerator.*

B. Variants:

— Biological treatment: a valuable variant here is to place the aeration and final settling tanks side by side in a single structure of the Oxyrapid type including sludge recycling by air lifts (see page 216);

— sludge treatment: digestion facilities can be dispensed with here in order to limit capital costs (in this case, incidentally, the entire calorific value of the sludge can be utilized where the sludge is incinerated).

Sludge dewatering (which must always be mechanical), must then be preceded by chemical or thermal stabilization.

It is desirable to install a fresh sludge thickener with provision for lime treatment, to serve as a buffer. Another possibility is separate thickening of the primary sludge in a conventional thickener with scraper and of the excess biological sludge by flotation.

On the other hand, the volume of sludge to be treated is higher and incineration has to be provided on the spot to limit nuisance; the "buffer" effect of the digester capacity no longer exists, and operation may become critical if a fault occurs in the sludge treatment plant.

— Sludge treatment: another possibility here is direct composting of the fresh sludge with biological support (screened household refuse, grape stalks, etc.).
C. *Additional facilities:*

Tertiary treatment such as phosphate removal, denitrification or polishing is sometimes used to complement a purification plant (see page 798).

2.4.2. MONITORING AND OPERATION

A. *The following daily checking and control operations are required in the operation of sewage treatment facilities:*

a) Monitoring of preliminary treatment facilities, disposal of screenings and grit.

b) Extraction of fresh sludge from the primary settling tank.

c) Adjustement of aeration facilities in accordance with the dissolved oxygen content of the activated sludge; this is generally measured continuously by an electrode type oxygen analyser. It is becoming more and more common for this analyser to adjust the aeration facilities automatically, maintaining the dissolved oxygen content within a set range (e.g. between 1 and 2 mg oxygen per litre). Depending on the aeration system in use, an adjustement is applied to air production (see page 754) or to the surface aerators (by on/off or fast/ slow syncopation, or less commonly, by variation of the turbine submersion depth or speed).

d) Monitoring of activated sludge in the aeration tank:
— firstly, by measurement of the percentage sludge in a test tube (see page 948). This tedious operation can be automated using a Sedimometer (see page 579).
— where appropriate, by measurement of sludge density, either manually (Ponsarimeter, see page 928), or automatically (using an adapted densimeter or turbidity meter). This measurement may be essential where the activated sludge quality is not constant, in which case measurement of the sludge percentage becomes meaningless. The weight of dry solids in the sludge can be estimated quickly by density measurement. (The density can be determined more accurately in the laboratory, but the result is delayed for several hours).

e) Extraction of excess sludge: sludge extractions may be initiated either manually or automatically by electric cycle/duration timers. Closed-loop control of sludge extractions by means of a Sedimometer or densimeter is also possible (see above).

f) Adjustement of activated sludge recycle rate.

B. Grease and floating matter removed at the preliminary treatment stage and during primary settling, generally stored in collection pits, is eliminated periodically.

C. The operation of sludge digestion facilities also calls for certain daily operations, viz.:

— introduction of fresh sludge into the digester;
— constant-temperature control of the digester heating;
— digester mixing cycles: in single-stage digestion the mixing cycles are generally associated with introductions of fresh sludge and may be automatically controlled by electric cycle/duration timers. In two-stage digestion, the primary digester is mixed continuously;
— monitoring of pH, volatile acids and alkalinity;
— flow measurement of the gas produced;
— gas quality control (Co_2 and/or CH_4 content);
— quality monitoring of the overflow recycled to the head of the plant.

D. Operation of mechanical sludge dewatering facilities may require more frequent supervision, in particular to ensure satisfactory sludge conditioning and optimum operation of the equipment used. The latter must therefore be designed for a daily period of operation equivalent to the time when the facilities are manned.

E. Finally, if its function of reducing energy consumption and simplifying operation is to be performed effectively, it should be noted that the **automation and closed loop control** equipment, and in particular the sensors, must be maintained in perfect working order and frequently checked.

F. The following is a model for a **log-book** to be used in the daily operation of a medium-size treatment plant comprising activated sludge process and digestion.

This log-book can be amended to suit various individual cases and may be supplemented (to include artificial sludge drying or incineration, etc.). Other documents may be more specifically devoted to the maintenance of mechanical equipment (equipment operating times, frequency of lubrication and overhaul, periodic checks, etc.).

In medium-size or large plants, which are generally operated by more highly skilled staff than small units, the main emphasis should be placed on the reduction of unskilled staff. Again, in medium-size plants, one-shift operation should be aimed at. Any equipment liable to faults should then be equipped with alarm facilities to alert the person responsible, who will be accommodated nearby.

Safety and alarm facilities (e.g., pH monitors, accidental level detectors, etc.) and control units for bypassing and shutting down certain equipment may be supplied with electricity by a special circuit using automatically rechargeable storage batteries.

PURIFICATION PLANT AT:																		DAILY LOG			
D	**RAW EFFLUENT**					**AERATION**												**Week ended :**				

RAW EFFLUENT columns: Tot. E | Daily av. l/s | Aver. flow l/s | Peak hour h | Total screenings m³/d | Total grit m³/d

AERATION columns: O₂ a.m. Head/Outlet | O₂ p.m. Head/Outlet | % 30 min | 60 min | Factor | Mohlman index | mg/l | Vol. air m³/d m³/h | Treated flow m³/d m³/h | Return flow m³/d m³/h

Rows: 1, 2, 3, 4, 5, 6, 7, Av.

REMARKS

D	**BOD₅ mg/l**			**S.S. mg/l**			**pH**			**DIGESTION**												**GAS**	**SLUDGE**					

DIGESTION sub-columns: pH (PD/FD) | TAC (PD/FD) | Volatile acid (PD/FD) | Temperature (Inside PD/FD, Inlet heat exch., Exch. outlet)

GAS: Flow m³/d

SLUDGE: Raw (m³/d, %SS) | Digested (pH, m³/d, %SS)

BOD₅/S.S./pH sub-columns: RS | APS | TE

Rows: 1, 2, 3, 4, 5, 6, 7, Av.

REMARKS

PD = Primary digester
FD = Final digester
RS = Raw sewage
APS = After Primary Settling
TE = Treated effluent

2.4.3. LARGE PLANTS

A. *Common arrangements.*

Owing to their size, large sewage treatment plants call for particular attention. In these plants, it is desirable to plan:
— automatic levelling-off of peak flows after pretreatment and primary settling;
— the completion of each treatment stage in several units operating in parallel to facilitate maintenance in particular.

The plant is generally equipped with a transformer station, the electricity supply being medium voltage (3 000 to 5 000 V) or high voltage (15 000 to 30 000 V). Some installations may contain an HV-MV transformer station, an MV internal mains and local MV-LV transformers. Capacitor banks are often useful to increase the power factor.

The size of the installations and the adverse effects on the natural environment of any stoppage justify:

— The use of high-performance electrical and mechanical switchgear and equipment fitted with all the desirable safety devices (unlike small stations where ease of operation and maintenance is the primary consideration).

— The employment of skilled labour for operating purposes (to or three-shift working, or at the very least a night watchman making rounds), maintenance of equipment, laboratory (analyses and checks), etc. The running of the installation depends on periodic laboratory checks and is recorded on written record sheets: daily and monthly operating tables, quarterly balance sheets, maintenance vouchers and records of equipment operating times, machine inspection sheets, etc. Statistical documents are prepared to show the changes in the pollution load, to help in the forecasting of daily, weekly and seasonal peaks, and in studying the possibility of extending the structure.

— A group of ancillary buildings, together or separate, comprising:

● an administrative and social building;

● a laboratory equipped with the necessary apparatus for taking samples and making checks and analyses;

● an electrical and mechanical engineering shop for plant maintenance and repair;

● a store for products, consumables and spares;

● garages;

● housing for the staff.

B. Aeration facilities

● Either blown air or surface aerators are used for aeration (see page 222).

If the plant is close to a residential area, special precautions must be taken to avoid equipment noise propagation. The sound level measured in the immediate vicinity of residential buildings should preferably not exceed:

— 45-55 dB (A) during the day;

— 35-45 dB (A) at night.

Plants should always be surrounded by one or more lines of evergreen trees (not deciduous varieties, as leaves might then fall into the open tanks).

Blowers are housed in soundproofed buildings or are located underground. A carefully-soundproofed building provides internal attenuation of 10 to 30 dB, and 20 to 30 dB between outside and inside, depending on the quality of the building and the frequency bands concerned.

Certain machines generate a sound intensity of about 90 to 110 dB (A);

Fig. 485. — Water pollution control works of TUNIS, Tunisia. Population equiva-lent 400,000.

good soundproofing can reduce the noise near the blower building to about 55 to 65 dB (A).

The air intake on the outside (and sometimes the air outlet) must be equipped with a silencer (attenuation about 15 to 30 dB). The heat dissipated by the machines can be removed by separate ventilation or by circulating the air drawn in by the blowers through the building before it is filtered.

It should be remembered that soundproofing is expensive and that the choice of apparently economical blowers which are too noisy is liable ultimately to involve an increase in total capital cost.

The noise of surface aerators, at a level of about 90 dB (A) close to the units, and the aerosols produced by water spray from them, are more difficult to control. The aeration facilities could be roofed over, in which case good ventilation with soundproofing will be required.

● With a blown air system, the blown air is generally produced by equipment housed in a separate building and may be of two types:
— displacement blowers of the Roots type,
— centrifugal blowers with one or more stages.

The range of delivery pressures used in aeration tanks is generally between 300 and 500 mbars (4 to 7 psi).

Roots displacement blowers with double rotating pistons give an air output more or less proportional to the speed of revolution with a virtually constant motor torque. The total mean efficiency is about 60 to 70 % for 400 mbars (5 psi) and for a speed variation between 1 and 2.

High-powered centrifugal blowers have an efficiency of the same order, but the delivered flow varies with the back-pressure. This may be a drawback when the liquid-level variations in the areation tanks are high in relation to the depth. It may be necessary to provide an automatic control maintaining a constant flow, by modifying the aerodynamic supply conditions (adjustable vanes in the air stream usptream or downstream of the blower, controlled by an electric or pneumatic servomotor governed by a set control point).

Roots blowers can be used up to flows of about 6 000 m³/h (3 300 ft³/min). Above that figure, it is preferable to install the centrifugal type of blower.

Variation of output, which should be kept within a ratio of 1 to 2 to maintain high efficiency, is ensured:

● with a Roots blower, by varying the rotational speed;

● with centrifugal blowers, generally by regulation of the vanes in the air stream.

● with three-stage blowers, by varying the speed of revolution of the third stage.

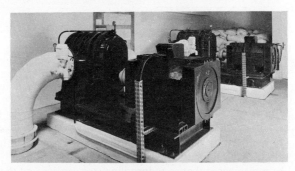

Fig. 486. — Blower room, TOURS, France.

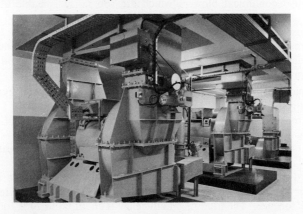

Fig. 487. — Centrifugal blower room, BORDEAUX, France.

Air filtration is necessary in the case of "fine bubble" diffusion, to avoid the accumulation of dust in the diffusers.

It is necessary to reduce the dust content in the filtered air to about 15 mg per 1 000 m³ air.

Filtration can be effected in two phases:
— by a prefilter with an efficiency of 80 to 90 % by weight;
— by a finishing filter, e.g. of the dry-media glass-fibre pocket filter type having a very high efficiency.

The overall filtration head loss can be about 20 to 50 mm on a water gauge.

Finally, dust control treatment is required for concrete surfaces which might be contacted by the filtered air (concrete pipes and ducts, rooms behind the air filters, etc.).

Air pollution expressed by weight of dust
Rural and suburban regions = 0.03 to 0.5 mg/m³ air
Urban regions   = 0.07 to 1.0
Industrial areas   = 0.25 to 4.0

C. Flow control.

a) Large plants treating large quantities of sewage effluents pose a number of problems regarding flow control; examples are:
— automatic and even distribution of flows to similar structures operating in parallel (with elimination of differences resulting from head losses in supply channels, which vary with the flow);
— automatic levelling-off of peak flows on the basis of set maximum flow control points at different points on the hydraulic circuit, e.g. at the plant intake (limitation of raw sewage intake flow), upstream of the primary settling tanks, upstream of the biological purification tanks, with direct discharge of excess flows to the receiving water;
— the multiplicity of hydraulic circuits in the aeration tanks to obtain optimum biological action (operation in parallel, in series, step feed, etc.). or equalization of the dissolved oxygen content throughout the liquid mass.
b) These problems can be solved by the use of **partialized regulating siphons** operating under negative pressure.

● This control system makes use of:
— regulating devices consisting of reinforced concrete or steel plate siphons with no moving parts or submerged mechanical equipment;
— automatic valves controlling the admission of partialization air to the siphons;
— pneumatic valve actuators.

● When not primed (communicating with the atmosphere), the siphons also act like closed valves, preventing all flow.

Fig. 488. — Partialized siphon. LYONS,   Fig. 489. — Siphon regulating sewage flow
France.   in Oxyrapid units: Paris-Achères plant,
                                           France.

— priming requires an initial vacuum and this can be done in several ways (pressure-reducing fan, vacuum pump, pneumatic air-ejector, hydraulic air-extractor,
etc.);

— when it is merely a matter of "on-off" control, it is possible to use a pressurized on-off siphon consisting of a simple metal bell covering a submerged weir:
when the bell is open to the air, flow is natural, but when it is under air pressure
the liquid level inside is pushed down below the weir and the flow is interrupted.

● Control problems of a widely varying nature can be solved by "partialization"-
type siphons with closed loop control, e.g.:

— control of an upstream or downstream water level;

— control of fixed or programmed flow regardless of the upstream or downstream levels;

— operation as head compensator, etc. (see page 301).

D. Measurements and closed loop control facilities.

Measuring, monitoring and control instruments must be designed to provide
a happy medium between entirely manual operation with local controls only and
an automatic installation with remote control and monitoring only from a control
room.   The installation will tend towards one or other of these extremes, depending on:

— the degree of skill of the operating personnel, which may be few in number
and highly specialized (when automation is preferable) or more numerous and
less skilled (when manual operation is preferable);

— power costs (some automatic controls facilitate optimization of power consumption) and costs of consumables (control of reagent dosage, etc.);

— available repair and maintenance facilities for highly-technical plant in the
region or country concerned.

a) Measuring and monitoring instruments

— The following table shows the indicating (I), recording (R) or summating (S) measuring instruments that are highly desirable (1), desirable (2) and optional or only required in highly technical plant (3).

Various small pressure and temperature indicating instruments are not listed.

Installation of the desirable or optional instruments provides permanent information on the points concerned, but the same measurements can be made periodically by the operating personnel with portable equipment.

Depending on the design of the installations, the indicating, recording and summating instruments can be grouped together or installed separately (e.g. indicating instrument fitted locally with recording and summation in the control room). Generally the control room is equipped with a display panel which, depending on its size, may contain:

— pilot lamps showing whether machinery is running or not;
— pilot lamps indicating electrical faults;
— pilot lamps showing equipment in service, ready for operation (programmed equipment);
— other accessories such as valve-position indicators.

The display panel may also contain a group of fault indicators (e.g. liquid levels too high or too low in certain channels or pump tanks, pressure drops in control fluids, water or compressed air, position of the gas holder bell too high or too low, etc.).

The recorders and summating instruments are sometimes mounted on the display panel and included in the schematic representation of the installation.

The same control room often contains a general desk for the main electrical control gear (pushbuttons, switches, selectors) and monitoring equipment (voltmeters, ammeters, etc.).

b) Closed loop control instruments (see Chapter 19)

A number of controls can be automated, in particular:

b1) Closed loop control of the aeration facilities in accordance with the oxygen demand in the aeration tanks:

● A sensor measuring the dissolved oxygen content of the aerated liquor is generally used for this purpose. A single oxygen probe is sufficient if integral mixing is used in the tank; several probes may be called for if the dissolved oxygen content is not uniform throughout the tank, in which case the system will be controlled by the probe indicating the lowest value. The range chosen is in the region of 0.3 to 0.5 mg oxygen per litre on either side of the set point— generally 1.5 to 2 mg/l, or higher (more than 3 mg/l) if the biological conditions so require.

Main measurement and monitoring instruments

Point concerned	Principle of measurement (in general)	Instrument I	R	S	Notes
Effluent flows					
— Raw efffuent	Weir or Parshall flume	1	2	2	total
— Settled effluent	Weir or Parshall flume	2	3	3	total or per settling tank
— Excess effluent extracted	Weir or Parshall flume	3	3	3	total
— Effluent passed for biological treatment	Weir or Parshall flume	2	3	3	per treatment line
— Purified efifluent	Weir	1	1	1	total
Sludge flows					
— Fresh sludge	Electro-magnetic flowmeters	1	2	1	total and where there is no thickener
— Return sludge	Weir, Parshall or venturi	2	3	3	total or per line
— Excess sludge	Weir	2	3	3	total or per line
— Thickened sludge	Electro-magnetic flowmeters	1	2	1	per thickener
—· Digested sludge	Electro-magnetic flowmeters	1	1	1	total
Air flows					
— General	Orifice or venturi	1	1	1	total
— By blower	Orifice or venturi	2	3	3	
— By treatment line	Orifice or venturi	1	3	3	
Sludge density					
— Fresh sludge	Gamma ray absorption	2	2	—	where there is no thicknener
— Thickened sludge	Gamma ray absorption	2	2	—	
— Activated sludge	Degrémont density meter	3	3	—	
— Return sludge	Degrémont density meter	3	3	—	
— Digested sludge	Gamma ray absorption	2	3	—	
Aeration tanks					
— Dissolved oxygen	Electrode analyser	1	1	—	one or more points per tank
— Percentage of sludge	Degrémont Sedimometer	2	3	—	per tank
Treated water					
— Turbidity	Degrémont H.R. turbidity meter	3	3	—	

● The following are controlled:

— the Roots blowers: by variation of rational speed, either continuously or in steps;

— centrifugal blowers: by variation of the vane angle or rotational speed (see page 756);

— surface aerators: by on/off or slow/fast syncopation, or possibly by progressive variation of submersion depth or speed.

Where a single blown air production plant supplies a number of aeration tanks, closed loop control of the oxygen content of each tank also requires variation of the settings of the automatic total air flow distribution valves, thus appreciably complicating the control system.

b2) Closed loop control of the hydraulic configuration of the aeration tank in accordance with dissolved oxygen content:

● This system is used only in step feed aeration tanks; it is similar to the first system described in that the oxygen measurements in the tank are carried out by the same probes, which can also control the distribution of the treatment

Main measurement and monitoring instruments

Point concerned	Principle of measurement (in general)	Instrument I	R	S	Notes
Digested gas					
— Gas flows produced	Orifice-venturi-meter	1	2	1	global or per digester
— Gas flows consumed	Orifice-venturi-meter	2	3	2	global or per unit
— CO_2 content	Analyser	3	3	—	
— CH_4 content	Analyser	3	3	—	
Digestion					
— pH of digested sludge	pH meter	3	3	—	in primary digester
— Internal temperatures	Thermometric probe	1	1	—	in primary digester
— Hot sludge temperatures	Thermometric probe	1	2	—	exchanger outlet
— Hot water temperatures	Thermometric probe	1	2	—	exchanger inlet
Sludge drying (filtration or centrifuging)					
— Flow of feed sludge	Electro-magnetic flowmeters	1	2	1	(if this flow is different from the flow of digested or thickened sludge)
— Concentration of feed sludge	Gamma ray absorption	3	3	3	
— Flows (or hourly weights) of flocculants	Various	2	3	2	
— Hourly weight of sludge cake	Weighing conveyor belt	2	3	2	
— Turbidity of liquid phase	Special turbidity meter	3	3	3	
Incineration					
— Hourly weight of sludge entering incineration	Weighing conveyor belt	1	2	1	(if this weight is different from the hourly weight of cake)
— Air flows	Orifice or venturi	1	1		
— Flow of make-up fuel-oil	Various	1	1	1	
— Temperature at various levels in furnace	Pyrometric probe	1	1		
— Smoke temperature	Pyrometric probe	1	1		
— O_2 in smoke	Analyser	2	3		(essential if furnace control is based on O_2 content)

flow among the different channels of the tank to give uniform oxygen content throughout the liquid mass. If the difference between the dissolved oxygen contents measured by two probes exceeds the set point (e.g. 0.5 mg/l), the distribution of the influent flow between the different admission points is varied in stages in accordance with a programmed cycle until the deviation in the oxygen measurements again falls within the set range.

● This configuration allows the full potential of the previous system described to be realized, as air consumption is then minimum.

b3) Closed loop control of excess sludge extraction:

● This problem is solved by the methods described in Chapter 8 (page 242).
● All the possible biological purification control systems are shown in the OBC diagram (figure 490) which involves the use, *inter alia*, of a Sedimometer (see page 579) which can be used in a closed loop system where the activated sludge is sufficiently constant in quality, or a continuous densimeter of suitable design (see page 580).

● Some light absorption photometers, in which photoelectric cell clogging and aging are measured separately, can provide measurements sufficiently well correlated with the dry solids concentration of activated sludge, but must be calibrated individually for each application, and the calibration must be corrected if the Mohlman index or the $\dfrac{\text{volatile matter}}{\text{inorganic matter}}$ ratio of the dry solids varies.

b4) Sludge drying:

Linking of the dosage of conditioning reagents to the weight of suspended solids or sludge flow.

b5) Sludge incineration:

Linking the furnace temperature to the sludge flow, to the fuel-oil flow and to the combustion-air flow.

c) Programming

A number of operations can be programmed in accordance with a specific time schedule (use of time switches, cam programmers, cyclic dosing apparatus, etc.). For example:

● Cycle and duration of operation of certain plant, grit removal travelling bridges, grit extractors, scraper bridges, etc.

● Programming of fresh sludge extraction:

— either cycle/duration programming,

— or cycle programming, in which case the extraction time is limited by the minimum specific gravity of the sludge removed (measured by a gamma ray absorption densimeter).

● Programming of return sludge flows according to the daily load curve for the plant.

● Programming of excess sludge extraction (see page 242).

● Programming of supply, mixing and heating operations in the digester.

● Programming of the sludge filtration cycle on filter presses.

d) Use of computers for plant monitoring and management:

● Data processing techniques are used in large plants, the following tasks being performed by computer:

— acquisition and monitoring of analog measurements, the various sensors being connected to the computer;

— operator input of external data—e.g., entry of laboratory measurements (BOD_5 of the raw sewage, settled sewage, and purified effluent, percentage volatile matter in the sludge, etc.);

— basic data processing, with calculation of the characteristic operational parameters of the plant (sludge weight, pollution weight, sludge loading, purification efficiency, energy balance, etc.);

— keeping of logbook, either periodically or on demand, the following being printed out:

— measurements of the sensors;

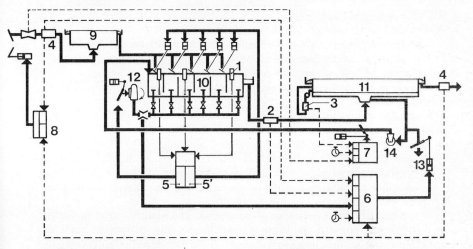

1. O_2 *analysers.*
2. *Sedimometer or densimeter.*
3. *Sludge blanket detector.*
4. *Pollution meter.*
5. O_2 *regulator.*
5'. *Distribution regulator for sewage leaving the primary settling tanks.*
6. *Excess sludge regulator.*

7. *Recycling regulator.*
8. *Pollution regulator.*
9. *Primary settling tank.*
10. *Aeration tank.*
11. *Final tank.*
12. *Blower.*
13. *Excess sludge extraction.*
14. *Recycling pump.*

Fig. 490. — *Diagram illustrating the OBC (Oxygen/Sludge/Load) principle of control.*

Fig. 491. — *Automated controlling units for the biological treatment process. Pierre Bénite plant, Lyons, France.*

— operator-entered values;
— calculated values;
— alarms (when a given parameter exceeds a threshold), showing the points of occurrence and elimitation of the fault;
— display of several simultaneous measurements;
— issue of instructions for closed loop control purposes and normal automation facilities.
● Since the computer prints out the measurements, the number of chart recorders can be minimized.
● The computer can also be used for:
— management of the spare parts store;
— programming of electromechanical maintenance of equipment;
— identification and accounting of the operating and shutdown times of equipment;
— recording of faults (printout of occurrence and disappearance of fault).
● If sufficiently powerful hardware is available, an attempt may be made to achieve process optimization for treatment of the maximum pollution compatible with:
— a given operating budget;
— the technical potential of the plant;
— observance of the specified purification standards.

After an initial phase of operation based on the above principles, a mathematical model can be constructed and used as an operator guide, issuing process control instructions.

If the mathematical model and operator guide prove to be sufficiently reliable, it may be possible to proceed to the optimization stage.

E. Power autonomy.

In a large installation with sludge digestion facilities not followed by incineration, the use of high-performance equipment and the stability of gas production will allow the plant to be partially or totally self-sufficient in energy.

The sale of gas to an outside user is also possible, but is not economic because of the usual requirements regarding scrubbing, purification, compression, etc.

It is far more advantageous to use the gas to meet the station's own power requirements, and in this case the heat given off by the heat engines can be recovered.

There are two possible systems:
● An electricity power station supplying current to the whole installation including the blower motors.
● A combined power station producing simultaneously:
— compressed air (blowers driven by diesel engines or gas turbines);
— some extra electricity.

Dual-fuel (or diesel-gas) diesel engines

Generally calorific power is converted into mechanical power (to drive a blower, alternator or both in series) by dual-fuel engines. An advantage of this type of engine is that it is virtually unaffected by variations in the gas flow and the calorific value. Consequently very reliable operation is ensured. Ignition is arranged by fuel-oil injection with a normal regular consumption of 6 to 7 % of total calorific consumption.

— Average characteristics of diesel-gas engines:

Gas consumption :

at full load   9 100 to   9 700 kJ/kWh (6 400-6 800 BTU/hp h)
at 3/4 load   9 700 to 10 200 kJ/kWh (6 800-7 200 BTU/hp h)
at 1/2 load 10 200 to 10 800 kJ/kWh (7 200-7 600 BTU /hp h)

Total heat recovery:

3 400 to 4 000 kJ/kWh (2 400-2 800 BTU/hp h) (from cooling water, lubricating oil, exhaust gas);

Efficiency of engine :

60 to 70 % (allowing for above-mentioned heat recovery).

— Energy balance.

In normal operation, if all the gas produced by digestion is used for power production by dual-fuel engines, the calorific power recovered from these engines is quite adequate to meet the heating requirements of the digesters in all seasons. However, if one or more dual-fuel engines should stop, the heat balance may become deficient (especially in winter) since the amount of heat recovered drops. Consequently stand-by boilers must be provided; it is desirable to have a boiler capacity sufficient to supply the heat required for digestion, but without a stand-by boiler.

The gas produced can normally take care of:

— the aeration system requirements;
— the electricity requirements of the auxiliaries;
— the heat requirements for digestion.

There may be a surplus of electric power of up to 20 % of the initia total gas energy, usable for ancillary equipment (e.g. sewage-pump stations).

The surplus can sometimes be kept in gas form for various uses (sludge heating and drying, incineration).

2.4.4. COMPARISON OF THE MAIN SLUDGE TREATMENT SYSTEMS

For a sanitary engineer responsible for the construction of a sewage treatment plant, selection of the best sludge treatment system is one of the most difficult tasks and the most important. It is also a task falling increasingly to those responsible for the design and operation of surface-water treatment plants.

It is unrealistic to adopt a general-purpose system and it is essential to find the system best suited to the conditions governing the construction and operation of the station concerned. An attempt is made pages 767 and 768 to define the main advantages and drawbacks of the sludge treatment systems most frequently used for municipal sewage effluents.

Where incineration of sludge and household refuse are combined, the sludge must normally undergo prior mechanical dewatering to ensure correct thermal balance. Dewatering also makes for greater flexibility in the operation of the entire plant.

2.4.5. SLUDGE HANDLING

Reliable, clean and functional handling of material is normally an essential prerequisite for high-quality sludge treatment facilities.

The method of handling must conform to the nature of the sludge. Centrifugation gives a slurry-like sludge which can be totally enclosed during transport as it can be pumped through pipes.

With filter presses, intermediate storage is almost always necessary. Very-steep-slope silos are equipped with powerful scrapers and extractors (the latter usually of the screw type).

Sludge cakes may require prior breaking-up before transport; cakes are transported on conveyor belts or Redler conveyors (trough conveyors with endless chains).

Where the sludge is incinerated, two important points must be borne in mind: the advantage of intermediate storage, and the need to provide a regular charging flow to the furnace.

Fig. 492. — Overall view of the digested sludge treatment complex of Achères IV, Paris, France. Daily output: 240 tonnes of dry solids.

Treatment process line	Advantages	Disadvantages	Normal range of applications
1. Separate sludge stabilization (or extended aeration) + drying beds	a 11 - Lowest capital cost a 12 - Moderate operating cost a 13 - Simplicity of process	d 11 - Very large land area required d 12 - Limited reduction of volatile solids d 13 - Labour requirements for collection of dried sludge d 14 - Operation greatly affected by weather conditions	Population less than 15,000
2. Digestion + drying beds	a 21 - Moderate capital cost a 22 - Low operating cost a 23 - Excellent volatile matter reduction if heating provided. Under same conditions, no risk of unpleasant smells for landfill or open-air storage a 24 - Very substantial elimination of pathogenic bacteria a 25 - Large available storage volume	d 21 - Large land area occupied d 22 - Same as d 13 d 23 - Strict monitoring of digestion required if there is a high proportion of industrial effluents liable to contain toxic substances (Cr, heavy cations) d 24 - *Same as* d 14	Population 30,000 to 100,000
3. Sludge stabilization + filterbelt press	a 31 - Sludge can be disposed of regardless of plant operating conditions and weather a 32 - Compact plant a 33 - Moderate capital cost a 34 - Immediate operation and very easy supervision	d 31 - Chemical conditioning required; high operating cost for consumable products d 32 - *Same as* d 12	Population 10,000 to 100,000
4. Sludge stabilization + centrifuging	a 41 - *Same as* a 31 a 42 - Very compact plant; enclosed treatment of sludge a 43 - Even the most difficult sludge can be dewatered a 44 - Immediate use; easy supervision	d 41 - *Same as* d 31 d 42 - *Same as* d 12 d 43 - Standby machine or beds required (during overhauls and maintenance) d 44 - Preliminary sludge treatment (grit removal) essential	Mixed or industrial effluents with highly colloidal sludge Population 10,000 to 100, 000
5. Sludge stabilization + filter presses	a 51 - *Same as* a 31 a 52 - Dryness of sludge cake may reach 35- 40% DS a 53 - Additional chemical stabilization of sludge	d 51 - Fairly high capital cost d 52 - Need for chemical conditioning; fairly high operating cost d 53 - Increased quantity of DS produced	Population 50,000 to 150,000
6. Vacuum filtration of fresh sludge	a 61 - *Same as* a 31 a 62 - Moderate capital cost a 63 - Rapid use; very easy supervision a 64 - Partial chemical stabilization of sludge possible a 65 - Treatment plant very compact	d 61 - *Same as* d 31, except for very fibrous sludge d 62 - No reduction of volatile solids d 63 - Relatively high energy consumption.	Mainly for cellulosic and/or relatively inorganic sludge

Treatment process line	Advantages	Disadvantages	Normal range of applications
7. Fresh sludge filtration by filterbelt press	a 71 - *Same as* a 31 a 72 - Very moderate capital cost a 73 - *Same as* a 63 a 74 - *Same as* a 65	d 71 - *Same as* d 61 d 72 - *Same as* d 62 d 73 - No sludge stabilization	Mainly for relatively inorganic sludge
8. Centrifuging of fresh sludge + incineration	a 81 - Minimum weight of residues a 82 - Final product totally innocuous a 83 - Operational flexibility and possible elimination of dried sludge collection a 84 - Enclosed sludge treatment	d 81 - Very high operating costs for consumable products (conditioners and fuel oil) d 82 - High capital cost d 83 - *Same as* d 43 d 84 - *Same as* d 44 d 85 - Sludge not easily removed during furnace overhauls	Population 50,000 to 300,000
9. Digestion + filter presses	a 91 - Dryness of sludge cake up to 40-45 % a 92 - *Same as* a 23 a 93 - *Same as* a 24 a 94 - *Same as* a 25 a 95 - Gas recovery for energy purposes possible a 96 - Highly flexible operation	d 91 - High capital cost (very high with generating station) d 92 - *Same as* d 52	Population 100,000 to 1,000,000
10. Filter presses for fresh sludge + incineration	a 101 - Sludge cake normally self-burning a 102 - Very small quantity of residues a 103 - *Same as* a 82	d 101 - High capital cost d 102 - Sludge cakes must be collected before incineration	Population over 200,000
11. Digestion + heat treatment + filter presses	a 111 - Minimum weight of residues without incineration; cake dryness 45-50 % for all sludges a 112 - Very low consumption of consumables a 113 - *Same as* a 23 a 114 - *Same as* a 24 a 115 - *Same as* a 25 a 116 - Sterile sludge produced	d 111 - Very high capital cost d 112 - Odour protection desirable d 113 - Overload of biological treatment	Population over 100,000
12. Heat treatment of fresh sludge + filter presses incineration.	a 121 - Minimum consumption of consumables with incineration (if with heat recovery) a 122 - *Same as* a 101 a 123 - *Same as* a 81 a 124 - *Same as* 82	d 121 - Very high capital cost d 122 - *Same as* d 112 d 123 - *Same as* d 113	Population over 300,000

2.4.6. ENVIRONMENTAL NUISANCE CONTROL

Modern man is now very aware of his environment and insists that it be protected; for this reason the possibility of environmental nuisances due to sewage treatment plants must increasingly be combated. The risk of such nuisances is exacerbated by the increasing dearth of sites remote from inhabited areas, which are always to be preferred, so that additional protection involving considerable extra expenditure is needed. The protective facilities must be designed and constructed by very experienced technicians to ensure that the most suitable techniques for the relevant purposes are chosen.

A. Noise.

Noise is caused mainly by rotating machines, and in particular by air compressors. The strictest requirements can be met by the construction of totally enclosed buildings, as far as possible of underground construction, and the use of sound-proofing materials. With very large items of equipement (e.g. lift screws or large surface aerators), only the drive motors and transmissions can be insulated. Submerged pumps are, of course, silent.

The noise of water agitation is sometimes felt to be annoying. Such noise is very evident with surface aeration, but far less so with a blown air system. The use of submerged aeration pumps is very satisfactory in this connection, although energy consumption is greater.

B. Smells.

Smells are sometimes due to septic sewage arriving at the plant (caused by hot weather or long sewers), containing compounds such as hydrogen sulphide, mercaptans, etc. Strong oxygenation of the influent sewage with pure oxygen is a highly effective preventive measure.

Most commonly, however, unpleasant smells originate from facilities containing insufficiently aerated fermentable matter or from which such matter is extracted. For this reason preliminary treatment facilities tend increasingly to be covered (this also gives more operator comfort in winter). Fresh sludge pits or thickeners are also sometimes covered. This sludge may be limed or even chlorinated to control its fermentation. The golden rule for management is, of course, to eliminate fermentable matter as it arises. Thus, in fresh sludge treatment, the sludge must be dewatered as soon as possible after thickening. The following arrangements are recommended for the heat treatment of sludge: covering of cooked sludge settling tanks, and additional cooling of this sludge.

Where sludge is heat-dried, there is a considerable risk of unpleasant smells. These can be effectively controlled by raising the temperature of the off gas to about 750 °C. The off gas may also be scrubbed. Gas scrubbing techniques (single- or multiple stage using oxidizing, alkaline or acid products) may also be used for the deodorization of certain enclosures by controlled extraction of the gas mixture which is then fed to the scrubbing tower(s). Activated carbon adsorption columns give excellent elimination of sulphur compounds. Ozone, used either directly or dissolved in water, is also a powerful deodorant.

Total positive deodorization of an entire treatment plant is extremely expensive. It is carried out in the VEVEY and MONTREUX treatment plants in Switzerland, where the entire atmosphere is periodically renewed after scrubbing with ozonized water.

C. Visual intrusion

This kind of environmental nuisance is of a subjective nature (this is sometimes also true of the other nuisances discussed, but to a lesser extent). It can be mitigated by a style of architecture consistent with the site and the region, while remaining functional. The construction of underground facilities, although often expensive, is to be recommended in this connection. The reduction or elimination of incineration smoke plumes is sometimes demanded and is possible (see page 533). Well tended green areas improve the appearance of the plant and operator comfort, and the operator thus benefits from a more pleasant working environment.

Fig. 493. — Overall view of the sewage treatment plant, Bologna, Italy.

2.4.7. EXAMPLES OF LARGE PLANTS

To illustrate present-day sewage treatment techniques as used in FRANCE, some typical examples are described in the following pages. DEGREMONT alone was responsible for the plants at EVRY, QUIMPER, and BORDEAUX, and played the principal part at ACHERES.

The wide range of treatment flows (14,000 to 2,700,000 m³/day; 3.7 to 713 mgd) and the great variety of methods used for sludge treatment will be noted: digestion at ACHERES and BORDEAUX, no digestion at EVRY and QUIMPER; vacuum filtration at ACHERES; filtration under pressure after heat treatment at ACHERES and BORDEAUX; centrifugation and incineration by rotary-hearth furnace at EVRY; incineration by rotary kiln at QUIMPER.

Fig. 494. — General diagram illustrating a large plant using activated sludge (separate tanks), digestion and sludge filtration.

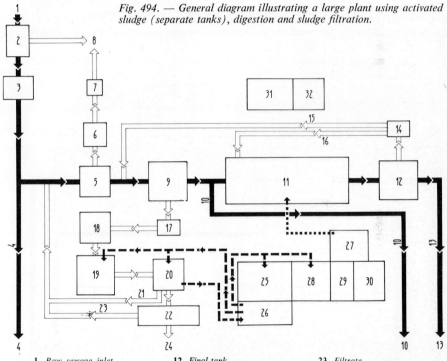

1. *Raw sewage inlet.*	**12.** *Final tank.*	**23.** *Filtrate.*
2. *Bar screening of raw sewage.*	**13.** *Treated effluent disposal.*	**24.** *Disposal of dried sludge.*
3. *Pumping of raw sewage.*	**14.** *Sludge pumping.*	**25.** *Digester heaters.*
4. *Raw sewage by-pass.*	**15.** *Excess sludge.*	**26.** *Digester mixing.*
5. *Grit removal.*	**16.** *Return sludge.*	**27.** *Blown air production.*
6. *Grit washing.*	**17.** *Pumping of fresh sludge.*	**28.** *Power generation.*
7. *Hopper.*	**18.** *Thickener.*	**29.** *Electrical transformer station.*
8. *Extraction of grit and refuse from bar screens.*	**19.** *Primary digester.*	**30.** *Control room.*
9. *Primary settling tank.*	**20.** *Secondary digester.*	**31.** *Administrative building.*
10. *Settled sewage by-pass.*	**21.** *Digester overflow.*	**32.** *Laboratory.*
11. *Aeration tank.*	**22.** *Mechanical sludge drying (vacuum filters).*	

Biological treatment involves aeration with fine bubbles, with separate or adjacent final clarifiers depending on circumstances.

Two flow sheets of large plants will be found pages 771 and 772, showing two different configurations for the treatment phases. Both include biological treatment by activated sludge.

The first system (figure 494) includes separate aeration and final clarification tanks, anaerobic sludge digestion, and vacuum filtration of sludge; the second (figure 495) makes use of combined tanks, excludes digestion, and includes heat treatment of fresh sludge followed by dewatering by filter presses and incineration.

Fig. 495. — General diagram illustrating a large plant using activated sludge (compact tanks), heat conditioning for the fresh sludge and sludge incineration.

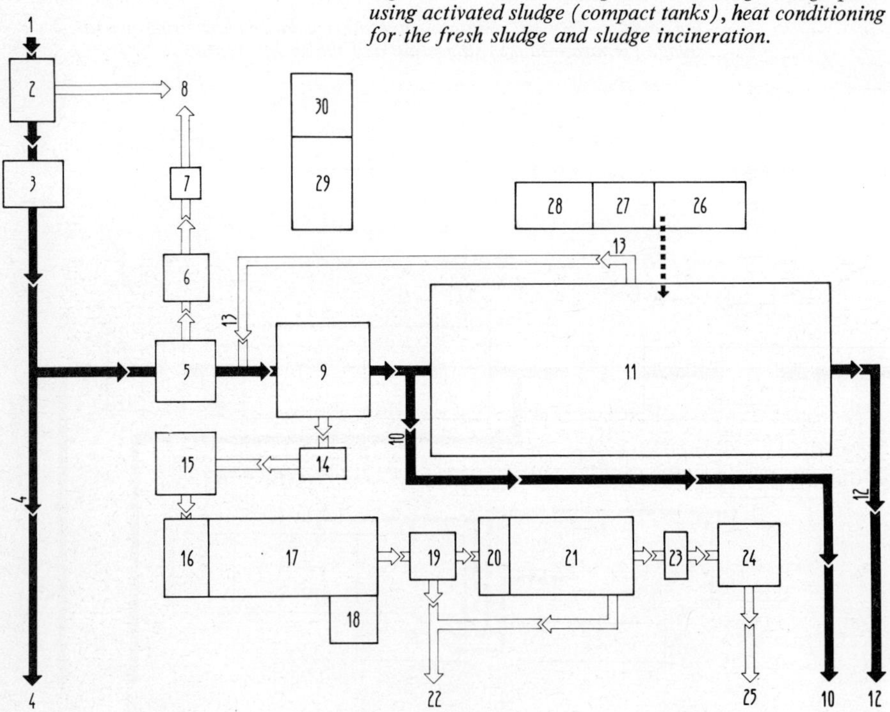

1. *Raw sewage inlet.*
2. *Bar screening of raw sewage.*
3. *Pumping of raw sewage.*
4. *Raw sewage by-pass.*
5. *Grit removal.*
6. *Grit washing.*
7. *Grit hopper.*
8. *Disposal of grit and of refuse from bar screens.*
9. *Primary settling tank.*
10. *Settled sewage by-pass.*
11. *Combined unit.*

12. *Treated effluent disposal.*
13. *Excess sludge.*
14. *Pumping of fresh sludge.*
15. *Thickener.*
16. *Pumping of thickened sludge.*
17. *Heat conditioning of sludge.*
18. *Conditioning heater.*
19. *Settling and storage of treated sludge.*
20. *Pumping of treated sludge.*
21. *Filtration.*

22. *Effluent from the sludge heating process.*
23. *Sludge cakes.*
24. *Incineration.*
25. *Ash storage and disposal.*
26. *Blown air production.*
27. *Electrical transformer station.*
28. *Control room.*
29. *Administrative building.*
30. *Laboratory.*

A. Evry treatment plant:

● Complete plant designed for a population equivalent of 420,000 in four equal-sized stages.

● Fisrt stage capacity 21,000 m³/day (5.5 mgd).

● The current plant includes, in particular:

— a circular primary settling tank with scraper, diameter 33 m (108 ft), liquid capacity about 3050 m³ (107,700 cuft);

— biological treatment in an Oxyrapid combined-tank facility (DEGREMONT patent) comprising: a 1970 m³ (69,500 cuft) aeration channel with air injection by fine-bubble porous diffusers, two final clarification tanks of 720 m² (7750 sq ft) total area,closed loop control of excess sludge extractions by a Sedimometer;

— a circular fresh sludge thickener, with scraper, diameter 14 m (45 ft), capacity 675 m³ (23,800 cuft), with liming, preceded by a cyclone-type grit remover for the sludge with grit extraction by Archimedean screw classifier;

— slude centrifugation by continuous decanter;

— sludge incineration in furnace with 7 rotary hearths, 5.75 m (18′6″) diameter, scaled for the first two stages of the plant, with off gas scrubbing.

Fig. 496. — Aerial view of treatment plant at Evry New Town, France.

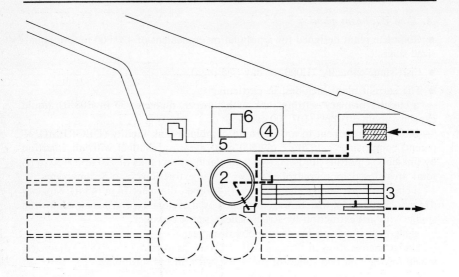

1. *Lifting. Pretreatment.*   4. *Thickener.*
2. *Primary settling tank.*   5. *Sludge centrificaiion.*
3. *Compact tank.*   6. *Sludge incineration.*

Fig. 497. — Treatment plant at Evry New Town.

B. Quimper treatment plant:

● Complete purification plant for a population of 334,000 with a substantial preponderance of food industry influents.

● To be constructed in two stages each designed for 10,000 kg (22,000 lb) BOD per day and 14,000 m³/day (3.7 mgd) raw sewage.

● The first completed stage comprises, in particular:

— one circular primary settling tank with scraper, diameter 27 m (89 ft), capacity 1 650 m³ (58,200 cuft);

— two step feed aeration tanks, each of 3 500 m³ (123,500 cuft), with fine-bubble air injection by porous discs;

— one circular final clarifier with scraper and suction facilities, 35 m (115 ft) diameter, capacity 2 400 m³ (84,700 cuft) with excess sludge extraction by Sedi-mometer;

— chlorination of the purified effluent;

— one earth tidal tank with flexible lining, capacity 8 000 m³ (282,400 cuft), emptying automatically as the tide goes out;

— one circular fresh sludge thickener, with scraper, diameter 11 m (36 ft), pre-ceded by a hydrocyclone type sludge grit remover with grit extraction by Archi-medean screw classifier; fine screening of sludge by mechanically cleaned screen with a bar spacing of 5 mm;

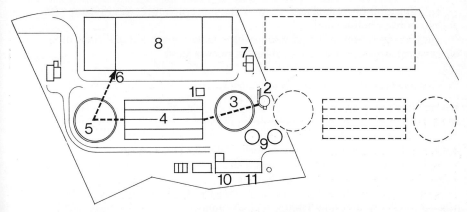

1. *Lifting unit.*
2. *Pretreatment.*
3. *Primary settling tank.*
4. *Aeration tank.*
5. *Final clarifier.*
6. *Metering of treated effluent.*

7. *Chlorination room.*
8. *Tidal tank.*
9. *Sludge thickeners.*
10. *Sludge centrifugation.*
11. *Sludge incineration.*

Fig. 498. — Treatment plant at Quimper, France.

Fig. 499. — Aerial view of the treatment plant, Quimper, France.

— centrifugation of fresh sludge in two continuous decanters (centrifuges) with polyelectrolyte conditioning;
— sludge incineration in a rotary kiln with countercurrent hot gas circulation; off gas treated by cyclone and scrubbed to remove dust.

C. Treatment plant for the Bordeaux district:

● Complete treatment plant designed for a population equivalent of 400,000.
● Facilities completed for a population equivalent of 300,000 and a dryweather flow of 90,000 m³/d (24 mgd).
● Peak flow admitted to physical treatment: 3 120 l/s. (49,500 gpm).
● Peak flow admitted to biological treatment: 2 600 l/s. (41,200 gpm).
● The facilities in use include in particular:
— sewage lifting by three 2,25 m (7'4") diameter Archimedean screws, lifting height 9 m (29'6"), unit flow 1 100 l/s (17,400 gpm);
— three longitudinal primary settling tanks with scrapers, width 20 m, unit capacity 3 780 m³ (133,400 cuft);
— biological treatment in six Oxyrapid type compact tanks each comprising: one aeration channel, capacity 1 150 m³ (40,600 cuft), with fine-bubble air injection by porous diffusers; two final clarification channels, total area 630 m² (6,780 sqft);
— two circular fresh sludge elutriator-thickeners, with scrapers, 19 m (62 ft) diameter, unit capacity 1 000 m³ (35,300 cuft), with automatic sludge extraction control by gamma-ray densimeter and flow monitoring by electromagnetic flow meter;
— one high-rate anaerobic digestion system (sized for the complete plant) with: two 24 m (79 ft) diameter primary digesters, unit capacity 5 000 m³ (176,500 cuft), heated by external heat exchanger and mixed by recycled gas injection; two 18 m (59 ft) diameter secondary digesters, unit capacity 2 500 m³ (88,200 cuft);
— two digested sludge thickeners, with scrapers, 16 m (52 ft) diameter, unit capacity 700 m³ (24,700 cuft);
— sludge heat treatment at approximately 200 °C sized for 18 tonnes/day of solid matter with: two lines each including a 120 m² (1,290 sqft) tube-type heat exchanger and a 12 m³ (420 cuft) direct-steam-injection reactor; settling of cooked sludge in two settling tanks with scrapers, 10 m (33 ft) diameter, unit capacity 240 m³ (8 400 cuft); storage of settled cooked sludge in two storage tanks with scrapers and mixing facilities, 10 m diameter (33 ft), unit capacity 240 m³ (8 500 cuft);
— sludge filtration by two filter presses, unit filtration area 276 m² (2 970 sqft);
— ancillary facilities for receiving 700 m³/day (18 500 gal/day) of drained matter, comprising: receiving channel, diluting water inlet, mechanical bar screen, grit remover with grit washing system, homogenization tank, and activated carbon internal air deodorization;
— control room: this features a computer with printer and alphanumeric display console, capable of performing the functions described on page 762.

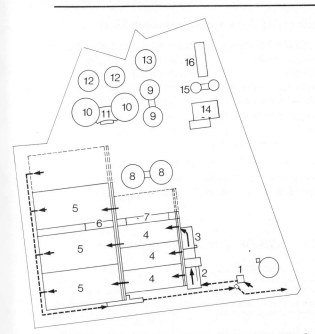

1. *End of sewer system.*
2. *Lifting unit.*
3. *Pretreatment.*
4. *Primary settling tanks.*
5. *Biological treatment.*
6. *Blower building.*
7. *Sludge lifting.*
8. *Elutriators.*
9. *Thickeners.*
10. *Primary digesters.*
11. *Heat exchanger.*
    Gas compression.
12. *Secondary digesters.*
13. *Gasholder.*
14. *Thermal treatment.*
15. *Settling tanks/storage tanks.*
16. *Filter presses.*

Fig. 500. — Flow diagram of the treatment plant at Bordeaux, France.

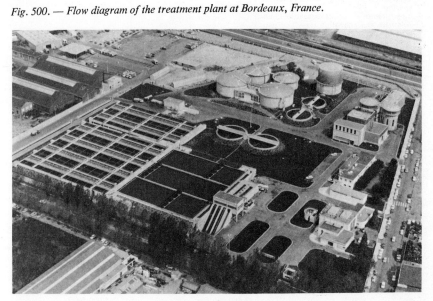

Fig. 501. — Aeral view of the water pollution control works, Bordeaux, France.

D. ACHERES purification plant (serving Paris and surrounding area):

The construction of the ACHERES station has been planned in five stages:

- ACHERES I   : completed about 1940 for   200 000 m³/d (53 mgd)
  later developed for   300 000 m³/d (79 mgd)
- ACHERES II   : constructed from 1962 to 1965   300 000 m³/d (79 mgd)
- ACHERES III : constructed from 1968 to 1972   900 000 m³/d (238 mgd)
- ACHERES IV   : constructed from 1974 to 1978   600 000 m³/d (158 mgd)
- ACHERES V   : future stage for   600 000 m³/d (158 mgd)

giving a total programme for 2 700 000 m³/d (713 mgd), 1978.

— Characteristics of the average raw effluent at ACHERES in 1976.
— BOD$_5$ 200 mg/l — Suspended solids 250 mg/l average figure for year;
— BOD$_5$ 250 mg/l — Suspended solids 300 mg/l — figure for month with greatest pollution.

MAIN TECHNICAL FEATURES OF THE FIRST FOUR STAGES

ACHÈRES I AND II	ACHÈRES III	ACHÈRES IV
Preliminary treatment	**Preliminary treatment**	**Preliminary treatment**
— End chamber for sewer cleaning sphere at the end of each main sewer, and reception tank — Mechanical screening, 25 mm (1 in) bar spacing — 12 grit removal and oil separation channels with grit extraction travelling gantries, and air production station for aeration of grit removal plant.	Extension of installations using the same processes	Extension of common installations (additional grit removal tanks)
Machine room	**Machine room**	**Machine room**
— Two Roots displacement blowers of 25,000 m³/h (14,600 ft³/min), driven by electric motor or gas engine — Two centrifugal blowers of 60,000 m³/h (35,000 ft³/min), driven by electric motors or dual-fuel engine — 3 centrifugal blowers of 80,000 m³/h (47,000 ft³/min), driven by 882 kW dual-fuel engines — 2 × 1 050 kVA alternators driven by 882 kW dual-fuel engine — 1 × 580 kWA alternator driven by 500 kW dual-fuel engine — 1 × 210 kVA alternator driven by 191 kW diesel engine (final standby)	— 3 centrifugal blower of 65,000 m³/h (38,000 ft³/min), driven by 1230 kW dual-fuel engines — 2 identical blowers driven by electric motors — 2 × 1 440 kVA alternators driven by 1 230 kW dual-fuel engine	— 4 blowers of 65,000 m³/h (38,000 ft³/min) at 500 mbar driven by 1 150 kW electric motors (in a separate underground room) — 1 blower driven by 1 230 kW dual-fuel engine — 1 alternator of 1 440 kVA driven by 1 230 kW dual-fuel engine (these 2 units housed in the Achères III machine room) (111,000 gpm)
Lifting at flood water periods	**Lifting at flood water periods**	**Lifting at flood water periods**
— 2 propeller pumps, unit discharge 8,000 m³/h (35,200 gpm) — 2 propeller pumps, unit discharge 12,500 m³/h (55,000 gpm)	— 3 propeller pumps, unit discharge 25,200 m³/h (111,000 gpm)	— 2 propeller pumps, unit discharge 25,200 m³/h (111,000 gpm)
Primary settling	**Primary settling**	**Primary settling**
8 circular settling tanks (4 of 35 m (115 ft) diameter and 4 of 50 m (164 ft) diameter), total capacity 37,500 m³ (1,320,000 ft³)	8 circular settling tanks with scrapers, 52 m diameter (171 ft), total capacity 70,000 m³ (2,460,000 ft³)	4 circular settling tanks with scrapers, 60 m (196 ft) diameter, total capacity 50,000 m³ (1,765,000 ft³)

ACHÈRES I et II	ACHÈRES III	ACHÈRES IV
Biological treatment Aeration tanks of 70,900 m³ (2,480,000 ft³) total capacity; "coarse-bubble" air injection	**Biological treatment** Aeration tanks of 55,000 m³ (1,930,000 ft³) total capacity, 75 % of which is in separate tanks and 25 % in Oxyrapid compact tanks; "fine-bubble" air injection through porous domes	**Biological treatment** Aeration tanks of approx. 51,000 m³ (1,790,000 ft³) total capacity in two sets each comprising 15 channels in series; "fine-bubble" air injection through porous domes
Secondary settling 8 circular settling tanks (4 scraper-type of 35 m (115 ft) diameter and 4 suction-type of 50 m (164 ft) diameter), total capacity 43,750 m³ (1,540,000 ft³). 2 rectangular suction-type settling tanks total capacity 10,000 m³ (350,000 ft³)	**Secondary settling** Total capacity 101,200 m³ (3,550,000 ft³), 82,500 m³ (2,900,000 ft³) in 9 circular suction type tanks of 50 m (164 ft) diameter and 18,700 m³ (650,000 ft³) in combined tanks; hydraulic flow control by "partialized" siphons	**Secondary settling** Total capacity 63,000 m³ (2,230,000 ft³) in 6 settling tanks of 60 m (197 ft) diameter with double peripheral collection weir
Primary digesters Number: 10 (6 of 3,000 m³ (105,000 ft³) capacity and 4 of 5 250 m³ (185,000 ft³) capacity) mixed and heated	**Primary digesters** 6 digesters of 8,125 m³ (285,000 ft³) with heating by external exchangers and mixing by gas injection	**Primary digesters** 3 digesters of 12,000 m³, (425,000 ft³) 33 m (110 ft) diameter, with heating by external exchangers and mixing by gas injection.
Secondary digesters Number: 3 (1 of 4,000 m³ (140,000 ft³) and 2 of 4,750 m³ (166,000 ft³)	**Secondary digesters** 4 digesters of 8,125 m³ (285,000 ft³) with heating by external exchangers and mixing by gas injection	**Secondary digesters** 2 digesters of 12,000 m³, (425,000 ft³) 33 m (110 ft) diameter, with heating by external exchangers and mixing by gas injection
Gasholders 3 of 4,800 m³ and (170,000 ft³) 2 of 4,400 m³ (155,000 ft³)	**Gasholders** 2 of 7,200 m³ (254,000 ft³) and one 15 m (49 ft) diameter sphere at 3.5. bar	**Gasholders** 1 hydraulic-seal gasholder of 10,000 m³ (35,400ft³) capacity and 1 storage sphere of 15 m (50 ft) diameter, 3 bar, capacity 1,750 m³ (62,000 ft³).
Tertiary digesters 2 of 3,600 m³ and (126,000 ft³) 4 of 6,000 m³ (210,000 ft³)	**Tertiary digesters** 4 each of 8,000 m³ (283,000 ft³)	**Thickening of digested sludge** 8 circular scraper-type thickeners of 60 m diameter, (197 ft), unit capacity 13,000 m³ (460,000 ft³).
Drying beds Total area 107,600 m² (26.3 acres) (including 90,000 m² (22 acres) with mechanized equipment)	**Drying beds** Total area 90,000 m² (22 acres) with mechanized equipment	
Drying of digested sludge by vacuum filters: — 2 filters of about 50 m² (540 ft³) each with chemical sludge conditioning	**Sludge drying with:** — storage and blending of digested sludge — heat treatment — storage and thickening of treated sludge — filtration through 8 filter presses each having 100 filter plates 1,20 m (4 ft) square. Total filtration area 2,200 m² (23,600 ft²)	**Drying of sludge with:** — storage and blending of digested sludge — 7 heat treatment lines — storage and thickening of treated sludge in closed tanks — filtration through 10 filter presses each having 140 filter plates 1.50 m (5 ft) square. Total filtration area 5 800 m² (62,400 ft²)

Fig. 502. — Aerial view of ACHERES IV, PARIS, sewage treatment and sludge digestion plant.   Output: 600 000 m³/d (158 mgd).

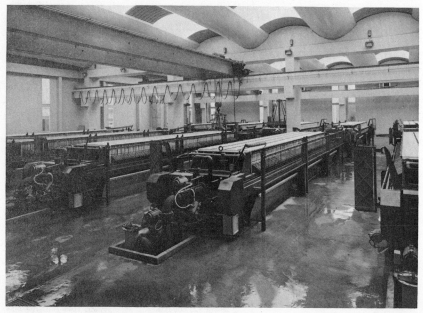

Fig. 503. — Partial view of the bay housing 10 filter presses at ACHERES IV (PARIS). Total filtering surface area: 5 800 m² (62,400 sq ft).

2.5. Other biological treatment systems

2.5.1. TREATMENT PLANTS USING PRIMARY SETTLING, CONVENTIONAL-MEDIA BIOLOGICAL FILTER, AND ANAEROBIC SLUDGE DIGESTION

The general layout is similar to that of activated sludge plants, the aeration tanks being, however, replaced by one or more high-rate biological filters.

The effluent quality can be improved by using two beds able to operate either in series or in alternating double filtration (ADF).

Conventional-media biological filters are now tending to be used less commonly in medium-size plants. This decline in popularity is due to progress with the activated sludge technique, which is more flexible and has a lower capital cost; however, energy consumption is higher. If the biological flora is poisoned, it is easier to drain an activated sludge tank than to change or clean the filter media of a biological filter.

A large number of plants were built in the past using a primary settling tank and a digester combined in a two-stage tank or Imhoff tank, followed by a biological filter and a separate final clarifier.

This configuration is now obsolete, as the operation of the Imhoff tank is often difficult to monitor.

One ingenious design achieved some degree of success some years ago: the MONOBLOC plant, in which the biological filter surmounts the final clarifier. However, this arrangement is now very seldom used, mainly for economic reasons and because it does not lend itself to successive extensions.

2.5.2. TREATMENT PLANTS USING PLASTIC-FILLED BIOLOGICAL FILTERS ASSOCIATED WITH ACTIVATED SLUDGE TREATMENT

In many medium-sized country towns, one or more agricultural and/or food-processing industries constitute important centres of activity. The pollution discharged by these plants is sometimes greaterthan the contributi on of domestic effluents. Mixing of domestic sewage and industrial effluents normally appears to be the simplest and most economical approach. However, the characteristics of industrial effluents are generally not consistent with the satisfactory operation of treatment plants unless their design is specifically adapted.

Biological filters with geometric plastic filling are particularly well suited to the treatment of agricultural and food-processing industry effluents, which have the following important features:

— they may contain energy-supplying carbon compounds which, while readily bioassimilable, are liable to trigger the growth of filamentous bacteria which could cause the bulking of the activated sludge. Final clarification is then seriously disturbed and the excess sludge becomes difficult to thicken and dewater;

— their concentration of soluble BOD_5 is high;

— their temperature is sometimes high, so that these effluents are more likely to undergo acid fermentation.

In plastic-filled biological filters, the most readily assimilable fraction of the pollution can be rapidly destroyed by a biological film which, even if composed of filamentous organisms, breaks up into readily settleable sheets. Hence 50-70 % of the pollution can easily be treated with low energy consumption.

For plants of not less than a certain size, the most economical way of achieving the 95-99 % purification efficiencies commonly specified for effluents of this type is to combine the high-rate plastic-filled biological filter with a finishing stage of activated sludge treatment.

In addition to the economic aspect, the plastic-filled biological filter + activated sludge combination has the following advantages:

— elimination of a high proportion of pollution with low energy consumption, as oxygenation takes place by natural draught through the filter;

— greater resistance to abrupt load variations, which are frequent with industrial effluents, than a conventional activated sludge plant;

— greater resistance to nutritional unbalance of the incoming effluent;

— breakdown of rapidly assimilable carbon compounds without clogging the filter media, so that the activated sludge can then be used at medium loading and avoiding the risk of bulking.

● *Features of plastic-filled biological filter + activated sludge plants:*

— *Preliminary treatment:* this is an important but often neglected point. Industrial effluents often contain certain matter which must be eliminated before admission to the biological filter if serious operational difficulties are not to be experienced (clogging of effluent distribution facilities on the filter surface, in particular).

This matter includes, for example, fat in dairy industry effluents, draff in brewery effluents, particles of meat, fat, straw and viscera, etc., in abattoir effluents, and hair and fleshing wastes in tannery effluents.

As a rule, primary settling does not appear warranted (unless, of course, toxic compounds are to be eliminated by precipitation), because the risk of clogging of the biological filter by fine suspended solids is low.

— *Intermediate settling tank:* the use of a high surface loading intermediate settling tank is recommended to prevent fragments of biological film from the plastic-filled filter from being carried into the activated sludge tank. These fragments of zoogloea contain highly active bacteria and also most of the particulate pollution trapped in the filter, even if it is not yet entirely metabolized. The oxygen demand at the activated sludge stage is considerably reduced by retaining all this sludge in a settling tank.

— *Sludge treatment:* the excess sludge extracted from the final clarifier is mixed with the sludge from the intermediate settling tank; the combined sludge after thickening can be dewatered either direct or after aerobic stabilization.

2.5.3. TREATMENT PLANTS USING PURE OXYGEN

The flexibility and reliability of activated sludge plants can be enhanced by the use of pure oxygen.

"All-oxygen" plants are well adapted to the treatment of highly concentrated industrial effluents with very high degradable pollution levels or containing reducing or inhibiting compounds which can only be accepted by a large mass of high-activity purifying bacteria.

Special design is then required for the treatment plant: the entire area of the aeration tanks and possibly also the stabilization tanks is covered with a gas-tight roofing, and facilities may also be provided to prevent excessive acidification of the activated sludge by carbon dioxide originating from intense bacterial metabolism. Acidification control can be ensured by air stripping of the free excess CO_2 or by adding a conventional air treatment stage to the oxygen treatment.

In addition, suitable control is required to limit oxygen losses caused by inevitable draw offs of the tank atmosphere and to maintain optimum dissolved oxygen content in the activated sludge liquor.

Special precautions must be taken to avoid any accumulation of inflammable volatile substances in the internal atmosphere of the tanks.

Special arrangements are required for the final clarification of highly concentrated activated sludge. Where very large quantities of sludge form, the aeration tank must itself be fitted with extraction facilities to relieve the final clarifier.

Where sewage and industrial effluents of variable quality are mixed a treatment plant designed for variable-mode operation may be considered: the aeration tank then has a number of compartments fed either with oxygen or with air.

An example of such a system is the LA GRANDE SYNTHE installation near DUNKIRK, which is the first oxygen-based plant built in Europe. It is designed to treat the effluent of a population equivalent of 45,000, with industrial waste of miscellaneous origins. (see fig. 504).

● *Oxygen doping*

Where treatment concerns mainly domestic sewage or relatively innocuous industrial effluents of constant quality, the use of pure oxygen or oxygen-enriched air in moderate size plants is worth while only for the purpose of increasing the oxygenation capacity of the system without thereby necessitating excessive modifications to the design and dimensioning of the plant, e.g., where an existing

Fig. 504. — GRANDE SYNTHE treatment plant at DUNKIRK, France. Biological oxygen reactor and final clarifier.

installation has to treat effluents from an industrial establishment newly connected to the sewerage system.

However, superoxygenation or oxygen doping is of particular value in plants designed to cope with variable loads. This is a worthwhile alternative to physico-chemical treatments for variable-population resorts in popular tourist areas.

Oxygen can be temporarily introduced in this case either direct to the aeration tanks by high-efficiency dissolving facilities located between the normal aerators or by intensively recirculating the liquor through a superoxygenation compartment outside the tanks (see fig. 505).

This provision of additional oxygen is particularly effective in treating highly septic domestic sewage arising when the sewers are flushed out at the time of a massive influx of holiday-makers.

Uncontrolled denitrification in the final clarifiers can be limited by main-taining a high dissolved oxygen level at peak times.

Pure oxygen injection into the sewers under pressure or into the sewage arriving at the lifting facilities allows rapid oxidation of any reducing sulphur compounds and increases the redox potential, thus limiting foul smells. This use of oxygen is advantageous in old or very long systems and in particular in holiday areas and hot climates.

Fig. 505. — Treatment plant at FOUESNANT, France.   Superoxygenation compartment.

2.5.4. NATURAL AEROBIC LAGOONING; AERATED LAGOONING

The addition of a plastic-filled biological filter preceding an activated sludge plant, and temporary oxygen introduction to aeration tanks, make for enhanced reliability and operational flexibility by intensification of processes; in some cases, however, a similar result can be achieved by the opposite approach, using an extensive process.

Natural aerobic lagooning is performed in shallow tanks (0.8 to 1.20 m deep; 2' 7" to 4") into which light can penetrate and favour the development of green algae, whose process of photosynthesis produces oxygen which enables the aerobic purifying bacteria to grow.

This simple process requires considerable space as reaction times are very long.   To ensure optimum aerobic conditions for lagooning without the creation of smells and insect proliferation (flies and, in particular, mosquitoes), the incoming effluent must undergo thorough preliminary treatment, if not prior settling.   This prevents rapid clogging of the tanks.

Depending on location, the treatment capacity of this process ranges from 25 to 50 kg of BOD_5 per hectare per day (22 to 44 lb per acre per day).

The purified sewage still contains large quantities of algae which are hard to eliminate; hence it is not unusual for the treated effluent to still contain 80-120 mg/l of suspended solids.

In **aerated lagooning,** effluents containing only small quantities of suspended solids can be purified in very large, deeper tanks (2-3 m deep; 6' 6" — 9' 10").

An equilibrium arises between the input of biodegradable pollution and the mass of bacteria growing in the tank from this pollution.   The settled activated sludge is not recycled from a final clarifier in this case.

The aeration facilities are dimensioned in accordance with oxygen requirements. In view of the large volume used, the specific powers applied are low (2-5 W/m³); a part of the pollution and of the bacteria produced settles on the tank bottoms where they form a deposit in which anaerobic fermentation occurs.

The water temperature in the lagoon has an enormous effect on the performance of the highly dispersed bacterial cultures which develop in it, and this must be allowed for in the sizing of the tanks. It is generally accepted that the reduction of a pollution level Lo to its final value Lf is described by the equation $\frac{Lf}{Lo} = \frac{1}{1 + Kt}$, in which t is the retention time in days and K depends on the temperature T according to the relation $K_T = K_{20} \times 1.07^{T-20}$.

For an influent of the domestic sewage type, the purification quality does not exceed Level 3 of the French standards (see page 1141) in spite of retention times of 10 days and longer.

Even with a final clarification zone, the suspended solids content of the purified effluent remains high, often exceeding 50 mg/l.

On the other hand, the large volumes used constitute an important buffer capacity which, when strengthened by internal recirculation, increases the dilution effect of the raw sewage in the stored mass.

From the economic point of view, aerated lagooning is comparable with traditional processes only if land is very cheap, the volume of earthworks required is limited and if the sides and bottom require limited protection. Finally, the risk of groundwater pollution by seepage into the ground must not be ignored.

Aerated lagooning is more valuable when used for tertiary purification in holiday areas. A high proportion of pathogenic bacteria can be eliminated cheaply by prolonged storage of a well-purified effluent kept under aerobic conditions, especially where the lagoon water temperature exceeds 18 °C. If the retention time is between 1 1/2 and 3 days, the BOD_5 of the effluent (after normal biological treatment) can be reduced to less than 10 mg/l. No permanent reduction of residual nitrogen and phosphorus has been observed under these conditions. To conform to the strictest discharge standards (Levels 5 and 6 of the French standards), such a tertiary treatment must be preceded by biological purification on a sufficient scale to nitrify the effluent.

2.6. VARIABLE-POPULATION PLANTS

Some plants serving seaside or winter sports resorts are subjected to considerable seasonal variations in load which may occur more or less abruptly. The load may vary by a factor of 2 in a town where tourism is important but not the only activity, but much higher factors are possible in some cases. In extreme cases —e.g., camp-sites— the facilities are closed out of season.

Only the case of activated sludge treatment will be considered here. Biological filters do not perform well where such variations occur, as problems of load and, in particular, of minimum hydraulic loading arise. The possibility of physicochemical treatment will be discussed later (see page 788).

A. Variability of a conventional plant: considerable load variations are possible in medium-loading plants with separate sludge stabilization: aeration facilities designed to cope with a BOD loading of 1.5 kg/day m³ will still operate correctly if this loading falls to 0.3 provided that the sludge is then stabilized; the possible range of variation is thus 1.5. The stabilization tank can be used for sludge storage, thus constituting a buffer capacity enabling the dewatering facilities to operate intermittently.

B. Conversion of multipurpose tanks: considerable flexibility can be achieved by varying the number of treatment lines or by using different modes of operation ranging from medium load to extended aeration. A good example is a plant on the French Atlantic coast in which the load varies from 7 000 kg/day in summer to 2 800 kg/day in winter:
— in summer, the plant operates in two stages: aeration in a 4 500 m³ tank at a BOD loading of 1.56 kg/(m³.d); stabilization in a tank identical with the aeration tank;
— in winter, extended aeration is carried out in both tanks at a BOD loading of 0.31 kg/(m³.d).

C. Other methods:

● *Use of activated carbon:* seasonal variations may sometimes be very abrupt —e.g., at a winter sports resort over Christmas. The growth of the biomass does not keep pace with the increase in load, so that the sludge loading increases substantially, thereby impairing the quality of the effluent and of the activated sludge. These abrupt variations can be tolerated by the addition of a quantity of powdered activated carbon; it acts not only by adsorption but also by facilitating the growth of the biomass, for which it serves as a support.

● Increasing the plant capacity by oxygen doping (see page 783).

● Relieving the load on the biological treatment facilities by increasing settling efficiency by chemical flocculation (see page 790).

3. PHYSICOCHEMICAL TREATMENT

3.1. General

This term is most commonly applied to raw sewage treatment comprising flocculation or precipitation followed by liquid-solid separation by settling or flotation.   Separation by filtration or screening is used in some techniques.

These processes can be used by themselves, in which case only partial purification is possible.   They may also be installed before a biological treatment facility to reduce the pollution reaching the latter, either continuously —e.g., for highly polluted mixed domestic sewage and industrial effluents— or temporarily, as in plants treating the effluent from a town or community of highly variable population.

Although these processes have been known for many years, they had developed little.   For a waste water of essentially domestic origin and without substantial variations in pollution, they are less effective in treating organic pollution than biological treatments; their operating costs are generally higher and they produce more sludge.   However, renewed interest in these techniques has been shown with the growth of pollution control in holiday areas.   Their main advantage lies in their virtually immediate response to any substantial variation in load; no other process allows intermittent operation of treatment plants serving complexes of second (week-end) homes, camp-sites, hotels or certain winter sports or seaside resorts with very small permanent populations.

Where sewage is discharged to the sea, the general trend today —at least in France— is to confine purification to thorough elimination of suspended solids and moderate (65-75 %) reduction of organic pollution.   A clear and easily disinfected effluent can then be discharged not too far from the shore.   Physicochemical treatments come into their own in such a case, as the relevant plants can be of compact design and small size and their visual intrusion can be limited by accommodating them in closed buildings.

The general trend for large urban areas, in which per capita water consumption is increasing, to produce less concentrated effluents also favours partial purification processes.   The development of new inorganic and organic flocculation agents and flocculating aids allows better adaptation of reagents to effluents of widely varying quality, so that many problems which would have been insoluble 10 years ago can now be solved, while running costs are at the same time reduced.

Fig. 506. — Aerial view of the plant at CANNES, France, comprising pretreatment, flocculation, settling and sludge treatment. Serving a population of 153 000 in winter and 225 000 in summer.

Finally, physicochemical treatments give excellent results in phosphate removal where metal salts or lime are added.

Experience with many plants of all sizes has shown that the nature and doses of the reagents to be used vary considerably from one effluent to another. Some general rules may nevertheless be formulated:

● **Mechanical flocculation** without added reagents improves contact between particles and hence the efficiency of static settling.

● **Polyelectrolytes** used alone facilitate flocculation of part of the colloidal matter. Anionic products must be used for dilute, fresh sewage and cationic products for concentrated and septic effluents. These reagents give elimination efficiencies by settling of 60 % for BOD_5 and 75 % for suspended solids; cationic products could also be used for flotation.

Synthetic polyelectrolytes are highly specific and the performance of an installation in which they are used cannot easily be predicted without prior tests.

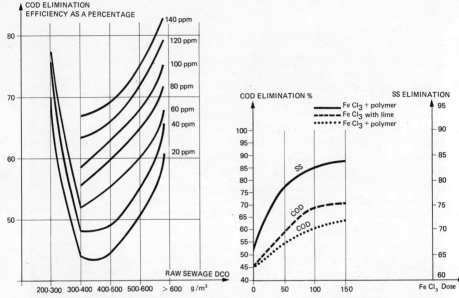

Fig. 507. — Evolution of COD elimination efficiency in terms of the concentration of raw sewage and the quantity of ferric chloride injected (CANNES plant).

Fig. 508. — Influence of the dose of ferric chloride on flocculation-settling efficiency.

● **Iron and aluminium salts** can be used with polyelectrolytes or with lime; the former combination is favourable to flotation and the latter to settling.   For a given effluent, the suspended solids, COD and BOD_5 elimination efficiency curve can be plotted for different reagent doses or flocculation aid doses (fig. 507). Efficiencies of 75 % for BOD_5 and over 90 % for suspended solids may be expected. Note that, as in the case of surface waters, the flocculation agent doses to be used are affected by temperature.

● **Lime treatment** at high pH (not less than 11.5) with moderate addition of iron salts gives equivalent results to the previous method while at the same time providing a considerable degree of disinfection.   However, water treated in this way must normally be neutralized before discharge or transfer to a subsequent stage of biological treatment, and this involves high consumption of acid reagents. Again, modification of the calcium carbonate -carbon dioxide equilibrium of the water due to the addition of high doses of lime may cause scale, giving deposits in equipment and scale in pipes.

3.2. Flocculation and physical separation

Flocculation of sewage follows similar rules to those applicable to surface water flocculation, except that the facilities used can be simplified because contact is facilitated by the large number of particles present.

The flocculated particules can be separated from the purified effluent by all available techniques. Straining seems to give random and relatively unpredictable results, but static settling gives good separation with heavy sludge. In cylindrico-conical settling facilities, the use of upward flow rates exceeding 1.5 m/h (0.6 gpm per sq ft) with polyelectrolytes used alone and 2 m/h (0.8 gpm/sq ft) with metal salts is hazardous. Settling is normally facilitated by the use of lime as an aid, but this does not appear to be necessary with certain waters which flocculate readily.

Where settling tanks with internal sludge recirculation (such as the TURBO-CIRCULATOR) are used, upward flow rates can be increased substantially and reagent doses reduced. Settling rates may be as much as 15-20 m/h (6-8 gpm/sq ft) at peak times if parallel-plate type clarifiers are used. Combination of a parallel-plate unit with a settling system using sludge recirculation (the RPS settling tank) or a sludge blanket (plate-type PULSATOR) gives a compact facility of high performance. However, in plate-type appliances with or without narrow plate spacing, it is essential for preliminary treatment to include fine bar screening and thorough grease removal.

Many settling facilities have the disadvantage of producing a thin sludge which must be concentrated. If this sludge contains iron or aluminium hydroxides, the concentration process is slow and a SS content exceeding 35-40 mg/l cannot be achieved unless further reagents are added.

From this point of view, dissolved-air flotation has the advantage of giving a concentrated sludge (SS greater than 50 g/l) without additional thickening. However, it appears limited to peak rates of 10 m/h and often requires the use of specific reagents (cationic polyelectrolytes) to facilitate the fixing of air micro-bubbles on the floc.

Finally, reference must be made to filter flocculation techniques which can sometimes be used for treating relatively unpolluted waters such as storm water —once the initial flood has passed— or for tertiary treatment.

Filtration after initial separation by settling or flotation is useful for final improvement of the quality of the treated effluent. This last treatment can be utilized to eliminate a fraction of the residual dissolved biodegradable pollution; the result is a combination of physicochemical and biological processes.

3.3. Combination of physicochemical and biological processes

Where physicochemical treatment is regarded as a temporary supplement or designed to have only moderate efficiency, a system combining physicochemical and biological treatment becomes virtually mandatory.

The simplest approach is to use an activated sludge treatment system preceded by a static settling tank including flocculation facilities either with added poly-electrolyte for more thorough reduction of colloidal pollution or without reagent in the form of simple mechanical flocculation. In any case, an activated sludge installation should not be preceded by excessively intensive flocculation and settling because the biological process will then act only on the dissolved pollution. The activated sludge produced is very rich in organic matter and difficult to settle, so that the installation, and in particular the final clarifier, must be designed on an unnecessarily large scale.

However, thorough physicochemical treatment is worth while in combination with biological filters. This approach is particularly well suited to existing plants which are overloaded. Another possibility is to add reagents at the biological treatment stage, as, for example, in the technique of simultaneous phosphate removal, in which ferric chloride is introduced at the head of the aeration tank.

If an abrupt increase in organic pollution is to be temporarily tolerated in an activated sludge tank, powdered activated carbon can be introduced and will fix medium-sized organic molecules by adsorption; the action of the bacterial flora is then confined to attacking the small molecules which can be adsorbed and are readily metabolized; the large particles are fixed in the floc and slowly metabolized subsequently.

Fig. 509. — Treatment plant at GUETHARY, France, for a population-equivalent of 13 000.

By means of simultaneous or consecutive physicochemical and biological treatments, the quality of the treated effluent can be maintained at a high level irrespective of loading. Where a medium standard of quality is acceptable at peak periods, the plant can be designed with biological treatment facilities limited to a certain population equivalent, any excess being received by a relief system of physicochemical treatment which will be more or less intensive depending on requirements. The advantage of this configuration is that the biological treatment can be operated at constant loading and independently of the physicochemical facilities.

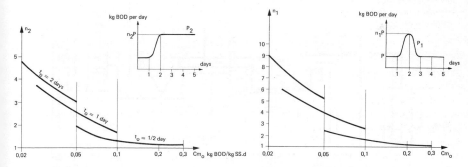

Fig. 510. — *Examples of curves indicative of the rate of variation in daily loading admissible in biological treatment.*

The FLOPAC process constitutes a highly effective and flexible combination of physicochemical and biological techniques. After flocculation/settling or flotation, the water is passed through a filter media of high specific area—i.e., with open macroporosity. Activated carbon in sometimes suggested for this purpose, but this product is expensive, brittle and liable to clogging. Its microporosity, which makes it adsorbent, is not well utilized in this case. Expanded minerals such as **biolite,** which fix aerobic bacteria in the infrastructures of their surface, are more suitable. After each periodic expansion wash, these bacteria rapidly restart the biological action, which can be controlled and adjusted by varying the amount of oxygen available for the aerobic organisms. The filtered water can be recycled to an oxygenation tank in which the necessary oxygen is introduced by air injection or surface aeration; the recycle rate can also be reduced, the dissolved oxygen concentration being concurrently increased by addition of pure oxygen; in this case the oxygenation tank and filters are closed and pressurized.

Out of season, when the raw sewage flow is much less than during the active period, primary treatment can be limited to mere static settling without addition of reagents, the settled effluent being fed to the biological filter and a very high recycle flow being maintained. Satisfactory results are thereby obtainable.

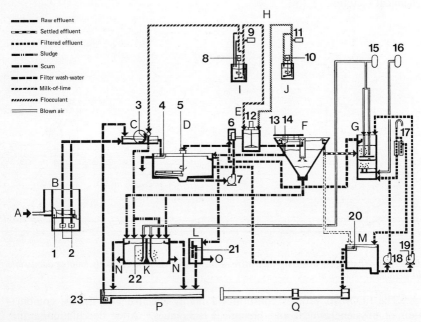

A *Raw sewage inlet.*	**1.** *Basket screen.*	**13.** *Scraper.*
B *Lifting unit.*	**2.** *Raw sewage lift pumps.*	**14.** *Scum recovery channel.*
C *Coarse screening.*	**3.** *Rotary screen.*	**15.** *Preaeration and stirring blower.*
D *Storage tank.*	**4.** *Storage tank overflow.*	
E *Flocculator.*	**5.** *Mixer.*	**16.** *Decompacting blower.*
F *Clarifier.*	**6.** *Constant-level tank.*	**17.** *Adjustable loop.*
G *Filtration tower.*	**7.** *Recirculation pump.*	**18.** *Filter wash-water pump.*
H *Reagents.*	**8.** *Paddle-type mixer for milk-of-lime tank.*	**19.** *Dilution pump.*
I *Milk-of-lime tank.*		**20.** *Treated effluent discharge overflow.*
J *Flocculant tank.*	**9.** *Milk-of-lime dosing pump.*	
K *Sludge tanks.*	**10.** *Paddle-type mixer for flocculant tank.*	**21.** *Scum tank overflow.*
L *Scum tanks.*		**22.** *Stirring air pipes.*
M *Wash-water storage tank.*	**11.** *Flocculant dosing pump.*	**23.** *Pump sump.*
N *Scum extraction.*	**12.** *Paddle-type mixer for flocculator.*	
P *Duct.*		
Q *Spreading system.*		

Fig. 511. — Treatment plant at LA BARRE DE MONTS, France.
FLOPAC SYSTEM comprising primary physicochemical treatment, with biological filtration as a secondary treatment.

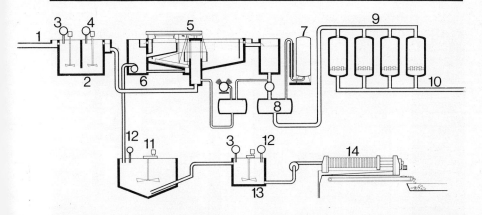

1. *Raw effluent.*
2. *Flocculation tank.*
3. *Reagents.*
4. *Flocculants.*
5. *Sediflotazur.*
6. *Sludge pit.*
7. *Pure oxygen.*

8. *Pressurization.*
9. *Filtration through Biolite.*
10. *Treated effluent.*
11. *Sludge storage tank.*
12. *Milk-of-lime.*
13. *Conditioning tank.*
14. *Filter press.*

Fig. 512. — FLOPAC physicochemical treatment with OXYAZUR process using pure oxygen. Waste water treatment plant at BRIANÇON, France.

3.4. Physicochemical treatment sludge

Where intensive flocculation is performed in order to "trap" material which will not normally settle, the amount of sludge collected in the settling tanks is necessarily increased. The use of organic polymers is highly advantageous because virtually nothing is added to the suspended solids in the raw sewage. On the other hand, all inorganic electrolytes lead to the formation of settleable hydroxide flocs ($Fe(OH)_3$, $Al(OH)_3$) and simple or complex phosphate and carbonate precipitates, which substantially increase the amount of sludge to be extracted from the settling tank. Furthermore, whereas lime when used in high doses gives a dense sludge, and particularly so if it simultaneously removes carbonate present in the effluent, iron and aluminium hydroxides settle in the form of a very loose and hydrophilic floc which cannot be readily thickened. The increase in the mass of sludge is accompanied in this case by an even greater increase in its volume.

Fig. 513. — Method of extracting cake from filter presses, CANNES, France.

Fig. 514. — Sludge incineration plant equipped with a fluidized bed furnace, CANNES, France.

Production and concentration of sludge
obtained by flocculation/settling at Le Lavandou.

Treatment		Quantity of sludge produced $(g/m^3$ treated effluent)	Concentration (g/l) of extracted sludge
Settling without flocculating agents		173	40
$FeCl_3$ + Lime	100 mg/l 250 mg/l	456	32.5
$Al_2(SO_4)_3$ + Lime	260 mg/l 280 mg/l	493	20.6
Lime	800 mg/l	1 100	60

Supplementary thickening is then essential, with or without addition of lime, according to the type of reagents previously used.

This is unnecessary with flotation.

Such sludge can be mechanically dewatered by centrifugation and filtration under continuous pressure (filterbelt press) or intermittent (pressure filter presses). Centrifugation and filterbelt pressing after polyelectrolyte conditioning give dry concentrations of 20-25 %, while filter presses give dry concentrations of 35-45 % depending on the type of conditioning chosen and the filtration pressure used.

Another possibility is dewatering in drying beds after intensive liming and additional polyelectrolyte conditioning.

4. TERTIARY TREATMENT

4.1. General

This process improves the characteristics of an effluent after biological or equivalent treatment.

Applications can be divided into the following classes, in increasing order of quality:
— agricultural requirements; irrigation;
— industrial cooling;
— preservation of the biotic equilibrium of the receiving medium;
— recycling in industry;
— replenishment of aquifers;
— pisciculture;
— domestic use, extending to human consumption.

The extent of the range of treatments applied will vary with the purpose for which the water is used.

The following tertiary treatment processes exist:
— polishing, which tends to further reduce the suspended solids and BOD_5 levels;
— phosphate removal to combat eutrophication of lakes;
— nitrification/denitrification, intended to eliminate all or part of the organic and ammoniacal nitrogen;
— elimination of nonbiodegradable COD and organic or inorganic toxic substances;
— colour and detergent elimination;
— disinfection and elimination of pathogens and parasites.

Tertiary treatment is just as applicable to the purification of industrial effluents as to that of domestic sewage, if not more so.

4.2. Polishing techniques

These techniques may be either biological or physical.

A widely used biological technique is the **polishing lagoon.** The effluent from the final clarifier passes through a shallow pond supplied with oxygen either naturally (by photosynthesis of algae) or artificially. Since suspended solids and BOD_5 levels are low, deposits are limited and the lagoons seldom require cleaning, even in the former case. Retention in a lagoon for 48 hours at 15 °C reduces the BOD_5 content of the water from 40 to 20 mg/l.

In natural lagoons, growth of algae considerably reduces the nitrogen and phosphorus content, although the degree of reduction varies with the season. However, the problem of eliminating the algae present in the discharged effluent does not yet appear to have been satisfactorily solved. In shallow lagoons, prolonged exposure to light has a considerable disinfectant effect.

Biological treatment can be supplemented by percolating through the ground.

The most widely used polishing process is **tertiary filtration,** which is basically a physical technique.

Direct sand filtration gives 60-80 % suspended solids elimination and 30-40 % elimination of organic carbon compound pollution. The lower the sludge loading of the preceding biological treatment, the higher the efficiency.

The effective size of the sands used ranges from 0.95 to 2 mm. With smaller sizes filtration cycles are too short and with larger grains purification efficiency is too low. The depth of sand varies from 1 to 1.5 m. The purification efficiency of a filter is affected by the filtration rate, which different authors give as between 5 and 30 m/h (2-12 gpm per sq ft); efficiency tends to decline rapidly above 20 m/h (8 gpm per sq ft). In most applications, a rate of 10 m/h is used, increasing to 20 m/h at peak times (4-8 gpm per sq ft).

For many years it was believed that waste waters had to be disinfected prior to tertiary filtration. In fact, BOD_5 elimination efficiency is improved by the bacterial growth which may take place in the filter, and it is preferable to exploit this if possible. If aerobic conditions are maintained in the filter, excessively

Fig. 515. — Treatment plant at La TREMBLADE, France.

fast clogging and too frequent break through can be avoided. An additional supply of oxygen by preaeration is not very effective in the presence of sand.

With porous materials such as Biolite, on the other hand, such oxygenation is very advantageous, and tertiary filtration then becomes a genuine form of **biological filtration,** retaining 70-90 % of suspended solids and 40-66 % of the BOD_5. Again, cycle times are much longer with Biolite than with sand: the former can fix 8-12 kg/m^2 of material before clogging, compared with 3-5 kg/m^2 with sand. The results of Biolite filtration compare favourably with those of filter flocculation, while reagents are not required (alumina 15-30 mg/l, anionic polyelectrolyte 0.5-1 mg/l). If only suspended solids are to be eliminated and if the purified effluent after secondary clarification contains no colloids or very fine floc, the sand or Biolite filters could be replaced by microstrainers.

4.3. Phosphate removal

Elimination of phosphates is very advantageous if the effluent is discharged into a lake or very slow-flowing watercourse, as eutrophication may be stimulated by the discharge of large quantities of assimilable phosphates with the purified effluent. These phosphates act as limiting factors for the development of algae and plankton. Consequently, countries such as Switzerland wishing to protect their lakes have stipulated a maximum limit of about 1 mg/l for total phosphates in waste water discharged into or close to lakes.

Phosphate removal in conventional biological treatment plants is incomplete. Furthermore, bacterial action favours the conversion of polyphosphates into directly assimilable orthophosphates. While the composition of raw waste waters is such that two thirds of the total phosphorus is present as polyphosphate and one third as orthophosphate, this ratio is reversed for biologically purified water.

Modern laundering products are the main source of polyphosphates, the amount of phosphates in the effluent therefore tends to increase with detergent consumption.

The trend is towards chemical elimination of phosphates usingr eagents (lime, iron and aluminium salts), which produce insoluble precipitates or complexes.

Two techniques are recommended: **simultaneous precipitation** by introduction of an iron or aluminium salt to the activated sludge, and **separate precipitation,** which constitutes a third stage of purification, with flocculation and settling or flotation. In the latter case, the quality of the purified effluent is further improved because suspended solids and the corresponding BOD_5 are also reduced.

Simultaneous precipitation has been successfully used in Switzerland on a large scale. Large quantities of reagents must be used, in the region of 1-1.5 mg of iron per mg of phosphate (expressed as PO_4^{3+}); elimination efficiency is then

Fig. 516. — MORGES, Switzerland. Phosphate removal by simultaneous precipitation.

80-90 %. The activated sludge becomes heavier, the Mohlman index falls, and purification efficiency seems not to be affected, although difficulties have been observed in the case of low-rate treatment. Anaerobic digestion of biological sludge containing precipitated phosphate does not seem to liberate an excess of PO_4^{3+} in recirculated overflows.

Separate precipitation calls for large quantities of flocculation aids in order to obtain correct clarification. The use of a sludge blanket clarifier improves the degree of phosphate removal. The amount of flocculation aid required could be reduced by flotation.

For lime treatment, the pH must be raised to about 11 and additional neutralization is required before discharge.

A new avenue of approach seems to be opening up with the observation that certain biological processes of denitrification permit a luxury uptake of phosphorus by bacteria. However, it is uncertain whether such a process by itself is sufficient to provide the very low phosphorus levels demanded by some countries' regulations, but it should reduce the emphasis laid on chemical phosphate removal.

4.4. Nitrogen elimination

The French discharge standards mention total nitrogen in three ou of the six levels (see page 1141). Nitrogen is included among the substances to be controlled in many countries, the levels specified for discharges often being very low. There are several reasons for these requirements:

— limitation of oxygen consumption in receiving waters, as the oxidation of 1 mg of ammoniacal nitrogen requires about 4.5 mg of oxygen;
— limitation of eutrophication of lakes and slow-flowing watercourses (in this case nitrogen elimination must be coupled with phosphate removal);
— facilitation of the use of surface waters for certain industrial or domestic applications in which the presence of nitrogen is harmful or prohibited.

The simplest method is to oxidize the nitrogen at the treatment plant itself to the nitric state (N/NO_3^-), in which it is regarded as being completely harmless. In fact, biological nitrification converts nitrogen from the ammoniacal state N/NH_3 to the nitric state by way of the intermediate nitrous state N/NO_2^-; the latter is somewhat toxic in water consumed by young children, and certain faults can halt the nitrification process at this stage. Again, the opposite process, known as assimilative reduction, may happen to take place in a watercourse. Finally, nitrification consumes oxygen, and hence energy, whereas additional dissimilative reduction, by converting nitrates into gaseous nitrogen, releases a part of the oxygen used for nitrification and consumes some of the pollution due to carbon compounds. The trend is therefore to aim for complete elimination of nitrogen. There are two possible approaches here :
— physicochemical elimination
— biological nitrification/denitrification.

A. Physicochemical nitrogen elimination.

The simplest approach is to raise the pH to a high level by adding lime so as to displace the NH_4^+ ions and to eliminate ammonia by stripping with air in contact towers. Although this process can be combined with phosphate removal, it is little used, for large structures with a huge amount of ventilation are required for large flows, entailing problems in winter and also risks of calcium carbonate precipitation in the towers.

There have been various attempts to utilize ion exchange techniques with resins or natural zeolite (clinoptinolite), electrodialysis, electrochemical precipitation as $Mg\ NH_4PO_4$, and chlorination, but none of these has proved to be sufficiently effective and economic to warrant the construction of an industrial-scale plant.

B. Biological nitrification/denitrification.

Current research is preferentially directed towards biological processes: conversion of all the ammonia into nitrate during secondary treatment and supplementary denitrification in an anoxic medium. Satisfactory nitrification can be obtained if biological purification is performed at a sludge loading below a certain limit, which varies with temperature and the pH of the effluent (e.g., $C_m \leqslant 0.25$ kg BOD/ (m^3.d) if $t^o \geqslant 18\ ^oC$ and pH $\geqslant 7.2$), provided that sufficient oxygen is supplied to oxidize the ammonia into nitrites and then nitrates. Where

Fig. 517. — Treatment plant serving the ROYAN district, France. The denitrification tank is shown in the foreground.

the load is relatively high, no NO_3^- ions are found to appear in the treated effluent, but the NH_4^+ concentration may sometimes fall slightly.

Conversion of these nitrates into gaseous nitrogen entails a supplementary denitrification stage, which may take a number of forms. While nitrification involves only highly specific autotrophic bacteria (Nitrobacter and Nitrosomonas), denitrification makes use of a wide range of heterotrophic microorganisms. However, in order for denitrification to occur, three conditions must be satisfied:
— good nitrification must first have been achieved;
— an energy source such as assimilable carbon must be available;
— anoxic conditions must obtain.

The different processes of denitrification can be classified in accordance with the origin of the carbon source:
● *Processes using an external carbon source:* the use of methanol is generally recommended; it is added to the activated sludge liquor discharging from the aeration tank, which is sized to ensure good nitrification. The reagent is added at the inlet to the denitrification tank, which is mixed slowly and kept free of oxygen. The mixture is then fed in the normal way to the final clarifier. In this case, 2.4-3 mg of methanol is required to reduce 1 mg of $N\text{-}NO_3^-$. Retention time in the denitrification tank depends on temperature and may be as much as several hours.

Much faster denitrification is achieved by biological **filtration on Biolite,** and in this case, if the nitrate content is not very high, denitrification can be effectively coupled with tertiary filtration, the dose of methanol being minimized.

Cheaper carbon sources might be found among the by-products of the agricultural and food industry.

● *Processes using endogenous respiration of activated sludge:* when undernourished, activated sludge passes completely to the stage of endogenous respiration. A number of cells are lysed and the nutrients thus solubilized are used as food. The energy liberated by endogenous respiration appears as secondary, but has

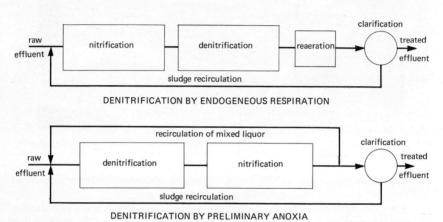

DENITRIFICATION BY ENDOGENEOUS RESPIRATION

DENITRIFICATION BY PRELIMINARY ANOXIA

Fig. 518 —

the advantage of costing nothing. The reaction rate depends on the sludge age and on temperature. The installation has the same layout as in the previous case, but the carbon supply is omitted and anoxic retention times prolonged. In an extended aeration tank, partial denitrification of this type can be achieved by intermittent operation of the aerators. Endogenous sludge respiration is also the reason for the uncontrolled denitrification occurring in certain final clarifiers in which sludge rises to the surface owing to the formation of nitrogen microbubbles. This is a parasitic phenomenon which should be avoided.

● *Processes using the carbon contained in the raw sewage:* the type U.I. plant (see page 742), calculated for a very low loading, eliminates nitrogen by nitrification/ denitrification with an overall efficiency of 70 % by virtue of the supply of raw sewage during the general shutdown between two cycles.

The raw sewage is fed to an initial tank kept under anoxic conditions, to which the activated sludge liquor from the downstream aeration tank is recycled at high flow (this tank is dimensioned so that total nitrification takes place). The result is rapid denitrification in contact with the carbon pollution of the raw sewage, which also allows 50-60 % of the oxygen necessary for nitrification to be recovered.

Fig. 519. — U.I. unit for a population equivalent of 1000: nitrification-denitrification.

This process is applicable to the treatment of sewage or industrial effluents in which nitrogenous pollution is limited to 300 mg/1 TKN; it is particularly effective and gives about 95 % nitrogen elimination. Furthermore, a luxury uptake of phosphorus also occurs during denitrification.

4.5. Elimination of non -biodegradable pollution

Adsorption on powdered or granulated *activated carbon* can be used to fix dissolved nonbiodegradable molecules. If the granulated product is used, almost all the biodegradable pollution must first be eliminated if proliferation of bacteria in the mass of the activated carbon bed is to be avoided. This treatment is incorporated in some American plants, but it has not been used on any scale in Europe.

The activated carbon is usually regenerated thermally or chemically.

The treatment essentially affects detergents and a number of organic molecules responsible for the coloration of various industrial effluents.

4.6. Disinfection

After biological and even tertiary treatment, the waste water may still have to be disinfected before discharge. This is necessary with certain waters that may be suspected of harbouring large quantities of pathogens, e.g., effluents from hospitals, sanatoriums, etc. Disinfection is also to be recommended if the treated effluent is to be reused by spraying with appliances liable to create aerosols.

Disinfection is pointless unless the water has first been properly purified and clarified. Chlorine disinfection is the most widely used method, and calls for a minimum contact time of 15 minutes if it is to be effective. To reduce the number of coliform bacteria by over 99.9 %, the doses required will generally be as follows:

● after chemical treatment .... 3-10 mg/l
● after biological treatment .... 2-10 mg/l
● after sand filtration or activated carbon treatment .... 2-5 mg/l

The better the quality of the preceding purification, the more effective the disinfectant action of chlorine will be. The presence of NH_4^+ reduces the bactericidal effect of chlorine by forming relatively inactive chloramines. Intensive nitrification/denitrification thus favours disinfection with chlorine. This disadvantage can be largely avoided by the use of chlorine dioxide.

To avoid the formation of organochlorinated compounds liable to be toxic to the fauna and flora of the receiving environment, the use of bromine or ultraviolet radiation has been suggested. The economy and effectiveness of these processes still remain to be proved in the purification of waste waters.

Ozone is both an active disinfectant, especially for viruses, and an oxidant of organic matter; for this reason it is likely to be used more in the relatively near future for the treatment of waste waters.

Fig. 520. — Treatment plant of LA TREMBLADE, France. Treated effluent-chlorine contact channels.

25 TREATMENT OF INDUSTRIAL EFFLUENTS

1. GENERAL

Whereas all domestic sewage contains inorganic and organic impurities whose nature and concentration are relatively constant from one instance to another, so that similar treatment systems can be used in all cases, the extreme diversity of industrial effluents calls for an individual investigation for each type of industry and entails the use of specific treatment processes.

Although industrial pollution has certain major factors in common with domestic sewage, the means of purification must normally be determined industry by industry.

In the enumeration of the principal industries covered by this chapter, first place is taken by the agricultural and foodstuffs industries, the pollution from which warrants mainly biological treatment similar to that used for domestic sewage.

Industrial effluent treatment plants must satisfy discharge standards concerning not only BOD_5, COD, and suspended solids content, but also a number of inorganic and organic compounds. In some countries different standards are laid down for the various sectors of industry (see Chapter 34, page 1146).

The choice of treatment system must always be based on:
— a knowledge of the various pollutants;
— characterization of the effluents;
— organization of sewers and separation of waste streams;
— choice of purification techniques from the different physicochemical and/or biological possibilities available.

The correct operation of the treatment plant will therefore depend on a detailed initial survey of the relevant polluting factory, as the omission of any constituent of the pollution may seriously impair the working of the plant.

1.1. Specific pollution factors

The principal types of pollutants are set out below, classified in accordance with the types of treatment to which they may be subjected.

● **Insoluble substances which can be separated physically with or without flocculation:**
— floating greasy matter (greases, aliphatic hydrocarbons, tars, organic oils, etc.);
— solids in suspension (sands, oxides, hydroxides, pigments, colloidal sulphur, latexes, fibres, etc.).

● **Organic substances separable by adsorption:**
— dyes, detergents, miscellaneous macromolecular compounds, phenolic compounds.

● **Substances separable by precipitation:**
— toxic and nontoxic metals, Fe, Cu, Zn, Ni, Be, Ti, Al, Pb, Hg, Cr, which can be precipitated within a certain pH range;
— sulphites, phosphates, sulphates, and fluorides, by addition of Ca^{2+}.

● **Substances which can be precipitated in the form of insoluble iron salts or which can be chelated:**
— sulphides, phosphates, cyanides, sulphocyanides.

● **Substances separable by degassing or stripping:**
— H_2S, NH_4, alcohols, phenols, sulphides.

● **Substances requiring a redox reaction:**
— cyanides, hexavalent chromium, sulphides, chlorine, nitrite.

● **Acids and bases:**
— hydrochloric, nitric, sulphuric and hydrofluoric acids;
— miscellaneous bases.

● **Substances which can be concentrated by ion exchange or reverse osmosis:**
— radionuclides such as I*, Mo*, Cs*;
— salts of strong acids and bases; ionized organic compounds (ion exchange) or non-ionized organic compounds (reverse osmosis).

● **Substances treatable by biological methods:**
— all biodegradable substances by definition, e.g., sugars, proteins, phenols; biological treatment is also applicable after acclimatization to organic compounds such as formaldehyde, aniline, and certain detergents.

The following points should be remembered:

1. The ratio of COD to BOD_5 in industrial effluents differs very substantially from that of domestic sewage. It changes during the stages of treatment, the final COD sometimes reaching a value more than five times that of the corresponding BOD.

2. The presence of very active toxic substances may conceal that of biodegradable substances and thus seriously falsify the measurement of BOD.

3. Basic information on the biological treatability of waste waters is given in Chapter 8, page 232.

1.2. Characterization of effluents

For the correct design of an effluent treatment plant, the following parameters must be carefully established:
— daily volumes;
— minimum and maximum hourly flows;
— *composition of the make up water used by the industrial plant*;
— continuous and intermittent manufacturing processes;
— intensity and timing of pollution peaks;
— possibility of separating sewer systems;
— possibilities of local or partial treatments or recycling;
— secondary incidences of pollution, even if slight or occasional, liable seriously to disturb the working of certain parts of the treatment facilities (glues, tars, fibres, oils, sands, etc.).

When a new factory is being designed, these parameters will be ascertained after analysis of the manufacturing processes and compared with data from existing factories.

Where treatment facilities are to be built for an existing plant, it is worth while to compare the quantities of pollutants revealed by continuous and systematic effluents analysis with the plant's consumption of different chemicals.

1.3. Separate treatments

It is often advantageous to isolate certain waste streams and subject them to specific treatments. This approach is essential whenever the effluent from a manufacturing unit displays one of the following features:
— very high COD or BOD_5 concentrations due to the presence of soluble compounds;
— average or high concentrations of H_2S, NH_4, or toxic substances.

Rather than dilute such effluents, it is often more economic to use one of the following processes:
— concentration to allow reuse of the product;
— destruction by direct pyrolysis of the liquid or its stripping vapour;
— liquid-liquid extraction.

Here are three examples of reduction of the pollution of the overall effluent of an industrial plant:
— regeneration of a wide variety of spent baths (electroplating, machining) by intermittent or continuous elimination of their dissolved or suspended impurities;
— chemical of salt or acid liquors whose molar concentration is greater than treatment the solubility threshold of the corresponding calcium salt, which can then be precipitated;
— treatment of soluble oils by chemical or thermal processes or by membrane separation.

1.4. Preliminary treatment

The conditions of preliminary treatment for the overall effluents of industrial plants, too, are more varied than in the case of domestic sewage.

Automatic **bar screening** operations are desirable in most industries and essential in some others (agricultural and foodstuffs, and paper mills).

Grit removal is effected only in a few special cases (rolling mills, sandpits, foundries, and rainwater).

Oil removal is often performed: hydrocarbons and oils sometimes originate from manufacturing processes, and almost always from lubrication or fuel storage circuits.

Equalization of the liquid flow and of the pollutant load is also often aimed at and can be achieved as follows:

— By the use of **buffer tanks** which store stormwater in the case of a combined system and where rain (always less in volume than in the case of municipal sewage) entrains and dilutes the pollutants. The function of these tanks is to obviate the need to scale the treatment system in accordance with occasional flow peaks.

— By the use of **homogenization tanks** in which the whole of the effluents produced by a unit or by the entire factory are stored for a few hours, or indeed for several days. It is essential for these tanks to be fitted with stirring facilities. Their function is to smooth out pollution peaks to avoid concentration overloads harmful to the regular operation of the purification system. These facilities also allow some degree of forecasting for operational purposes.

Prior neutralization, oxidation, and reduction operations are often carried out for the treatment of concentrated or toxic effluents. Automatic pH or redox potential controllers are used for these purposes.

1.5. Physicochemical treatment

Physicochemical purification may be an intermediate or a final stage in the treatment process as a whole depending on the individual situation. It has one or more functions:

— precipitation of toxic metals or salts;

— elimination of oils in emulsion and various substances in suspension;

— clarification with attendant reduction of colloidal BOD and the corresponding COD.

For this treatment it is essential to hold the pH within a relatively narrow range. Depending on the nature of the process (precipitation, crystallization, adsorption, or flocculation), physicochemical purification can be effected in combined reactors and settling units or clarifiers of very different types:

— flotation units such as the Flotazur or the Sediflotazur, for elimination of oils or fibres;

— reactors such as the Turbactor and reactor/settling facilities such as the Circulator, Densator, or Turbocirculator, for precipitation of calcium salts or hydroxides;
— sludge circulation type clarifier/reactors such as the Turbocirculator, the RPS settling unit, or the Accelator, in mixed situations;
— sludge blanket type clarifiers, such as the Pulsator and the Superpulsator for separation of a light floc or development of the adsorbent properties of the sludge blanket.

The choice of the appropriate facility from these units depends not only on the predominant technique to be used but also on other parameters specific to the relevant industry.

Depending on circumstances, this physicochemical purification may be preceded or followed by one the following processes:
— neutralization;
— oxidation or reduction;
— degassing or stripping.

Filtration is required only in the case of very strict discharge standards for suspended solids and total metals.

1.6. Biological treatment

The use of biological purification techniques depends on the biodegradability of the efffuents, and certain particular features of industrial waste strcams must be taken into account in the design of the appropriate facilities:
— effluents which have undergone prior physicochemical treatment—often for several purposes—contain few suspended solids;
— their nutrient composition is seldom balanced, and phosphorus and/or nitrogen correction will be necessary;
— an initial deficiency of microorganisms must be compensated by appropriate seeding with an acclimatization of specific organisms ;
— where biodegradable compounds are present, it may be necessary to keep their concentration relatively constant and to develop a specific flora;
— excessively high concentrations of inorganic salts, and in particular rapid variations in these concentrations, may disturb the progress of purification;
— nitrification-denitrification may be impeded by excessively high COD and ammonia concentrations (see page 107) and in certain pH ranges;
— particular attention must be devoted to the maintenance of relatively constant temperature zones. The temperature of certain effluents is favourable to the development of thermophilic bacteria.

The system may comprise the following stages:
— activated sludge (high- or medium-rate or, in most cases, with extended aeration);
— regular-configuration plastic-filled biological filters, as preliminary or polishing treatment;

— traditional biological filters;
— Biolite filters, as main or polishing treatment;
— aerated or mixed lagoons, as polishing treatment.

1.7. Elimination of non-biodegradable COD

Biological purification is the most rational method of reducing BOD_5 and the corresponding COD. Nevertheless, the application of increasingly strict regulations may call also for the elimination of nonbiodegradable COD, colour, and certain specific coumpounds.

This COD is due to organic compounds, which are generally dissolved and differ widely in type: solvents, aromatic hydrocarbons, nitrocompounds, sulphonated compounds, etc.

Common techniques for eliminating this COD are as follows:
— adsorption on thermally or chemically regenerated activated carbon (DEGREMONT process) or on miscellaneous adsorbents;
— ultrafiltration and reverse osmosis;
— ion exchange;
— miscellaneous oxidation techniques (using air, oxygen, ozone, or chlorine).

1.8. Industrial sludge

The observed specificity of industrial effluents is, of course, also found in the sludges arising, which are sometimes predominantly organic and sometimes —indeed frequently—predominantly inorganic.

In general, more sludge is produced by physicochemical treatments than by biological purification processes. Note, finally, that sludge from the clarification of industrial make-up waters sometimes predominates. All the treatment techniques specified for domestic sewage sludge are applicable here. Only some special points relevant to thickening and mechanical dewatering will be discussed in this section.

A. Sludge thickening.

Sludge is concentrated mainly by settling techniques, using highly variable surface loadings—ranging from 10 to 800 kg solid matter/(m^2.d) depending on the sludge composition. If large quantities of hydrocarbons are present, a second liquid phase may form in the thickener, thus impeding its operation. This sometimes causes thickeners used in refineries to perform no more than a storage function.

B. Sludge dewatering.

Thermal conditioning is not normally justified by the volume of organic sludge arising. Chemical conditioning, on the other hand, is highly developed, using synthetic polyelectrolytes and/or inorganic reagents, and—less frequently— inert fillers (such as kieselguhr, calcium carbonate, and wood flour).

The filtrability or centrifugability of sludge differs not only in accordance with its chemical composition but also with its manner of formation. Products may differ in filtrability by a factor of 10, and even one and the same product may exhibit a difference in filtrability by a factor of 3.

Prior tests are essential where the sludge is insufficiently known.

Vacuum filtration of certain hydroxides or flocculated oils requires the use of a precoat (wood flour, diatoms, etc.).

By-products of the factory can sometimes be used to improve the filtrability of a sludge.

The Pressdeg belt filter is particularly suitable for dewatering oxide and carbonate sludge or other grainy or fibrous sludges.

C. Final disposal of sludge.

This sludge will be disposed of in widely differing ways according to its nature:

● Relatively stable and nontoxic inorganic sludges (calcium carbonate, ferric hydroxide, calcium sulphate) can be spread on land (amelioration), discharged in the open, or used as inert fillers (e.g., for road construction).

● Unstable or toxic inorganic sludges (ferrous hydroxides, miscellaneous hydroxdes of metals, calcium fluoride) must be stored in controlled or stabilized dumps or, in some cases, incinerated; toxic sludges, especially those containing heavy metals, may only be stored in watertight dumps isolated from the groundwater.

● Organic sludge, which is in general liable to fermentation, must be stabilized before spreading or tipping, or else it must be destroyed by incineration.

● Oil-rich sludges which cannot be reclaimed must be incinerated. Where they are substantially autocombustible, they can facilitate the destruction of other, poorer sludges by incineration.

The best form of disposal is, of course, reintroduction of the sludge into the manufacturing process, but this is seldom possible. Wherever feasible, attempts will be made to find openings for the sludge, but its marketability is closely tied with the cost of transport.

2. AGRICULTURAL AND FOODSTUFFS INDUSTRIES

The characteristics common to all effluents from the **foodstuffs industry** are essentially organic and biodegradable pollution, and a general tendency to acidfication and rapid fermentation. All these effluents are treated primarily by biological methods, but the medium often becomes deficient in nitrogen and phosphorus.

2.1. Dairy products industries

The effluents from these industries differ in composition depending on their origin. Pasteurization and packaging shops for full milk products discharge only washing waters, consisting of very dilute milk. There may be very acid or very alkaline peaks due to the use of nitric acid or soda for cleaning pasteurizers and other equipment.

Cheese dairies and casein factories also produce serum rich in lactose but poor in proteins, while butter dairies produce buttermilk which is rich in lactose and proteins but poor in fats. Buttermilk and serum give rise to considerable pollution: BOD_5 of 60 000-70 000 mg/l for buttermilk and 30 000-40 000 mg/l for serum. They can be treated by controlled anaerobic digestion, but the operation of such plants is difficult and expensive. In practice these by-products are generally recovered, especially for use in cattle feed, and at present only the wash-waters are industrially purified.

The flows and compositions of the effluents vary greatly depending on manufacturing conditions: losses of milk, mixture of cooling waters, treatment processes for the milk itself. For illustration, the BOD_5 of full cream milk is about 100 000 mg/l.

The pollution discharged by the dairy or factory may be estimated approximately as follows:

Principal activity	*grams BOD_5 per 100 l milk treated*
Powdered milk	100-300
Butter and powdered milk without buttermilk recovery	370-630
Butter	100-300
Cheese	650-1 050
Liquid milk and multiproduct dairies	350-750

Grit and grease must often be removed from the effluents (always prior to a biological filter). The use of a buffer tank with stirring facilities is recommended to smooth out pH peaks.

The most suitable solution is biological treatment—especially suitable is the activated-sludge method with extended aeration—but the upward flow rate in clarification must be low. Extended aeration enables the volume of sludge produced to be considerably restricted and provides, owing to the large aeration volume, a considerable buffer capacity which makes it possible to cope with the particularly sharp pollution peaks encountered in this type of industry.

For large plants (over 800 kg (1 760 lb) BOD_5/d), the activated lsudge technique with extended aeration can be used by itself or preceded by a plastic-filled biological filter.   These filters are not only economical (low operating costs) but also have better resistance to abrupt variations in load and will break down rapidly assimilable carbonated substances (lactose).   This can protect the second treatment stage (activated sludge) from sludge bulking problems.   In the dairy industry, recirculation to the biological filter is almost always essential to maintain a minimum liquid flow (approximately 2 m^3/(m^2.h) (0.8 gpm per sq ft), variable according to the filter material).   A load of 2-5 kg BOD_5/(m^3.d) in the biological filter gives 50-70 % BOD_5 purification efficiency.   The sludge produced is, of course, not stabilized.

A concentration of more than 1 or 2 % of milk or serum in the effluent quickly leads to uncontrollable acid aerobic fermentations (lactic fermentation) which can completely stop biological activity.

Many large plants have already been established.

Spreading and sprinkling can also be used for disposing of these effluents. The flows must be limited to 20 to 40 m^3 of water per day and per hectare (1 800 to 3 600 gal/acre d), depending on the permeability of the soil.

Fig. 520. — Dairy waste water treatment at Isigny, France.   Daily output: 3 820 m³ (1 mgd).
BOD : 4 165 kg (9 160 lb) per day.

2.2. Vegetable and fruit canning factories

These industries are seasonal; they discharge washwater containing relatively little pollution and waters from "bleachers" which in fact represent very concentrated broths (BOD_5 of the order of 25 000 mg/l). The degree of pollution varies greatly depending on the methods used and the products treated. An indication is given by the following table:

Products	Total pollution discharged (g of BOD_5 per kg of canned product)
French beans	3 to 5
Butter beans and the like	5 to 7,5
Carrots	18 to 20
Green peas	15 to 18
Spinach	25 to 35
Tomato purée	6
Celery	2 to 9
Mushrooms	20
Fruit in syrup	7 to 12
Fruit juices	3 to 6

These effluents are generally rich in glucides. There is normally enough nitrogen in vegetable effluents (beans and peas), but their phosphorus content is often insufficient for biological purification.

Conversely, effluents from fruit canning plants are often deficient in nitrogen, while they contain enough phosphorus.

The treatment of these effluents must always include fine straining to retain fragments of vegetables, leaves and peelings. The substances thus collected must be composted or incinerated. Irrigation or sprinkling are theoretically feasible methods, but in actual fact are often performed under unsatisfactory conditions, as the surface areas available are not always adequate and the effluents are very fermentable.

Actually, taking into account the nature of the pollution, low rate biological purification appears to be the most suitable method of treatment for these industries.

2.3. Abattoirs and meat-packing factories

The wastes discharged vary depending on the method of evacuation of stercoral matter, the size of the tripe and offal-processing shops, and the nature of the animals slaughtered. On the basis of the weight of the dressed carcass produced, which can easily be ascertained from the abattoir, the following flows can be assumed:

Animals slaughtered	Tripe shop	Evacuation of stercoral matter	Effluent flow (per kg of carcass)
Pigs, small livestock			8 litres
Poultry			10-13 litres with recycling of conveying water 18-21 litres without recycling
Cattle multi-purpose abattoirs	small	dry	5 litres
		hydraulic	15 litres
	large	dry	13 litres
		hydraulic	27 litres

Fig. 521. — Abattoir waste water treatment (1 600 kg BOD per day; 3 520 lb).

The pollution load of these effluents depends on the rate of recovery of blood and also on the size of the tripe and offal-processing shop and the method of evacuation of stercoral matter. After an extremely detailed survey, the French Ministry of Agriculture has drawn up a graph which combines all these factors.

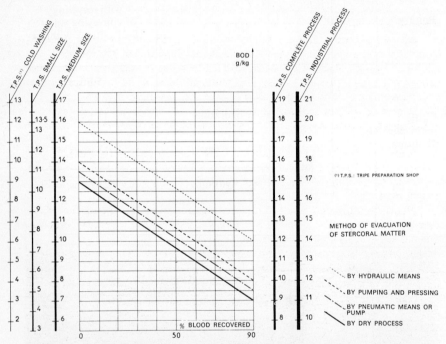

Fig. 522. — BOD expressed in g/kg of carcass according to the type of slaughter-house.

For the associated industries the following amounts should be allowed for:
— 3 to 5 kg of BOD_5/kg of product for salted meats;
— 14 to 16 kg of BOD_5/kg of product for canned meats.

Any discharge of abattoir effluent into the sewage system must obligatorily be preceded by a grit removal and degreasing pretreatment, screening and, if possible, fine straining, the effect of which is to reduce the content of pollutants by from 10 to 15 %. An aerated storage pond may be desirable if the weekly production variations of the abattoir are high.

Fatty matter and proteins contained in the effluent may be recovered by preliminary treatment using flocculation followed by flotation. The sludge arising can sometimes be used to feed animals. The BOD is reduced by 70-90 %.

Abattoir effluents are very suitable for biological treatment. Medium- to low-rate activated sludge treatment gives very good efficiency. Plastic-filled biological filters are also used, but these call for thorough preliminary treatment to obviate the risk of clogging by grease. Where plant size so allows, it is desirable to include an aerobic stabilization stage for the sludge from the biological treatment.

2.4. Breweries and fermentation industries

Brewery effluents originate in the cleaning of the brewing rooms, cooling troughs, fermenting and storage vats, and the washing of bottles and casks. In addition to these effluents (which are polluted by substances in suspension, nitrogenous substances, residues of beer and **yeast,** particles of draff, and kieselguhr), cooling waters containing very little pollution are sometimes encountered.

It is recognized that production of one hectolitre of beer entails a discharge of pollution of, on average, nearly 800 g of BOD_5 (8 lb of BOD per 120 gal of beer).

While the bottle-washing effluents are of low concentration (BOD_5 from 200 to 400 mg/l), those from the cleaning of fermentation vats or filters reach 3 000 mg/l of BOD_5, and those from the washing of storage tanks as much as 16 000 mg/l.

Nutrients (nitrogen and phosphorus) must sometimes be added where intensive revovery of by-products is practised. Preliminary treatment comprises fine bar screening (elimination of draff) followed by neutralization,

Fig. 523. — Biological and tertiary treatment for malt-house effluent recirculation. Output 100 m³/h (0.63 mgd).

Low-rate activated sludge treatment of these effluents is highly effective, giving over 95 % BOD reduction, but clarification rates must be low.

In large plants, it may be worth while to use two stages of treatment. The first, or roughing, stage gives rapid sugar elimination. A plastic-filled biological filter can be used for this purpose. The second stage comprises an activated sludge facility operating at low or medium rate. The biological filter gives up to 60 % BOD_5 reduction after clarification. The odour problems often caused by the use of biological filters for brewery effluents must be mentioned. The sludge arising from the biological filter settles very well but is extremely liable to fermentation.

The sludge produced must be treated after concentration by vacuum filtration or pressure filtration, with conventional conditioning using iron salts and lime, or by centrifugation with organic flocculants.

Industries using **fermentation** processes (amino acids, enzymes, antibiotics, yeasts, etc.) have concentrated effluents, often rich in nitrogen, with acid pH, and generally devoid of suspended solids (except in the case of accidents such as loss of cultures). The whole range of biological methods is applicable to these. The purifying bacteria are very adaptable to the various types of effluents. Thus waste products from the manufacture of antibiotics can be successfully treated with activated sludge.

2.5. Sugar refineries and distilleries

Polluted effluents discharged by **sugar refineries** are of varied origin. The main sources of pollution are:
— the muddy waters from the **washing of the beets;**
— the process waters (diffusion and pulp-press waters, mud conveying waters);
— the regeneration effluents from the sugar-juice demineralization plants.

The water used for washing and for fluming the beets is generally recycled. For this purpose, settler-thickeners are inserted in the circuit in order to reduce the proportion of suspended solids in the waters, as during the campaign these solids can very rapidly reach several grammes per litre.

The tare of the raw beets defines the concentration of suspended matter in the washing waters and largely determines the dimensions of the settling tank.

It is possible by settling to obtain sludge with a concentration which may exceed 300 g/l. It is advisable for the settling tank to be preceded by a screening device and a grit remover. Lime or polyelectrolyte is sometimes injected in order to improve settling. Traditionally this sludge is conveyed into large earth tanks which gradually fill up. The water which overflows from these tanks has a high pollution loading and is very fermentable.

In fact, during the campaign (which lasts about two to three months in Europe) the washwater also contains polluting substances from damaged beet and also from the humus.

Increases in BOD_5 of 70 mg/l per day have been observed, so that by the end of the campaign the pollution of these effluents can reach 3 000 to 5 000 mg/l of BOD_5.

Anaerobic fermentation of all the waste matter from a sugar refinery, stored in large tanks throughout the inter-campaign period, is common pratice.

Well operated, with sufficiently large tanks, checking of infiltrations and annual maintenance of the embankments, this method still remains the most economical, and certainly one of the safest. However, it requires very large areas of ground, and creates inevitable nuisances for the surrounding region. The use of aerated lagoons is possible to overcome these disadvantages.

The other biological treatment processes are also applicable.

Distilleries discharge very concentrated and very polluted exhausted molasses or residuary liquors (BOD_5 reaching 40 000 mg/l for the liquors and 5 000 to 10 000 mg/l for all the waste products combined). Residuary liquors from lees are heavily loaded with suspended solids and are treated by centrifugation with or without conditioning. Biological treatment of distillery effluents with plastic-filled biological filters in one or more stages appears to be particularly well suited to solving this problem. Molasses and residuary liquors are discharged at a temperature of 95 °C and at a pH between 4 and 5; after neutralization and homogenization, running them through a plastic-filled biological filter with a high recycle rate (20-100 times) must allow sufficient cooling for biological purification. Biological filtration may be followed by lagooning or by activated sludge treatment. Beet molasses may also be subjected to anaerobic digestion.

Experiments with a view to making use of the residuary liquors have shown that it is possible to develop in them **yeasts** of the Torula genus; one can thus reduce the pollution to around 1 000 to 1 500 mg/l of BOD_5.

Fig. 524. — Sugar-beet wash water clarification. Output: 1 200 m³/h (7.6 mgd).

2.6. Starch factories and potato processing industries

The effluents from these industries are very fermentable, as they contain starch proteins. Their amount is increasing owing to the ever-increasing sales of potato crisps, dehydrated mashed potatoes, etc.

The polluted effluents from the **potato industry** have two sources: firstly there is the water used for washing and conveying the potatoes, which contains earth, vegetable debris and fragments of potato, and secondly the water from the peeling machines, which contains peel and pulp. The quantities of residua pulp may be sufficient to warrant consideration of their recovery for animal feeds.

Such recovery, a rather difficult operation, can be effected by mere static settling in this circuit alone, after which the pulp which has settled can be centrifuged and then dehydrated by a drying drum. The overflow from the settling tank can be mixed with the washwater and the combined discharge treated with activated sludge by a low-rate or medium-rate process. The concentration of BOD_5 at the inlet to the biological treatment system can usually reach 500 to 1 200 mg/l.

The storage conditions and the nature of the stabilizing or antioxidant products used in manufacture may impair the settlability of the activated sludge. Very low settling rates must be applied. The use of a plastic-filled biological filter may be considered for an initial stage of BOD reduction.

Fig. 525. — Treatment of mashed potatoes manufacture effluent. Output: 500 m³/h (3.2 mgd).

In the case of starch-works, the pollution concentration in the waste products is higher owing to the considerable amounts contributed by the evaporator
condensates and the starch-washing waters. Normally the average pollution
amounts to 1 500 to 2 500 mg/l of BOD_5 and can still be dealt with by extensive
aerobic biological treatment.

2.7. Oil mills and soap factories

The effluents from these industries often have an extremely high or low pH,
depending upon the shops involved. Thus the fat-washing shops discharge
extremely acid effluents (pH between 1 and 2), while the saponification of fatty
acids in the soap works produces very alkaline effluents (pH in the region of 13).

It is therefore useful to mix these different effluents—a process which entails
the use of large and appropriately scaled neutralization tanks.

Degreasing of the waste waters is generally carried out by acid cracking,
sometimes combined with flocculation, followed by separation by flotation.

Generally, physicochemical degreasing and flocculation pretreatment reduces
the organic pollution by 50 to 70 %. Supplementary biological treatment is then
necessary. The sludge from the pretreatment ferments very readily and can
only be dried by filtration if fermentation is not too far advanced. The technical
methods used for treating these effluents must be particularly adapted and very
carefully designed in each case.

Fig. 526. — Biological treatment of oil-mill effluent.

2.8. Tanneries and leather industries

The quantities of water used by these industries are very large, amounting to 5 m³ per 100 kg (600 gal per 100 lb) of dried hides treated. The effluents are highly polluted and contain protein colloids, fats and tannins, fragments of flesh and hair, colouring matter and also toxic substances such as the **sulphides** from the unhairing shops, and above all, **chromium** from chemical tanning. The BOD_5 levels easily reach 700 to 900 mg/l. Prior screening is essential before any treatment process.

If all the effluent waters are mixed together, an alkaline effluent is obtained in which the chromium is precipitated in the trivalent state and is therefore mainly to be found in the sludge. There can be no question of digesting this sludge by the anaerobic method, because chromium is a toxic element in methane fermentation.

The sulphides gradually decompose by natural oxidation; if it is desired to speed up this process, use must be made of catalysts, namely cobalt or manganese salts. Acid stripping of the sulphurous waters can also be envisaged.

Sulphides can also be eliminated by precipitation using ferrous sulphate or alum; flocculation simultaneously occurs, reducing the BOD by about 70 %.

After storage and homogenization, followed by clarification and neutralization, biological treatment can be applied. This raises no special problems and can be carried out at fairly large BOD loadings. The only disadvantage is the danger of foaming, which should be prevented by sprinkling and the use of anti-foaming agents. The vegetable tannins are unaffected by the biological treatment and colour the purified waters light brown.

Effluents from the **manufacture of animal glues and gelatines** from debris such as hides, bones and fish residues are very heavily polluted. One must reckon with 5 kg of BOD_5 per 100 kg of glue.

Flocculation with lime alone has a considerable purifying effect. This can be followed by biological treatment.

2.9. Piggeries and stock raising effluents

This source of pollution, widely scattered up to the present, is tending to become more concentrated with the development of industrial methods of stock-breeding, especially in the case of pigs. The amount of effluent and the degree of pollution depends on the method of cleaning the sties, which can be either with water, or dry, or combined.

The liquid effluent amounts to 17 to 18 litres per day and per pig in the case of water-washing, and 11 to 13 litres per pig for "dry" cleaning. The peak coefficients are very high, ranging from 8 to 12. There is considerable organic pollution. From 150 to 200 g of BOD_5 per pig can be expected with water-washing and from 80 to 100 g of BOD_5 per pig with "dry" cleaning.

The manure can be separated from the water by a mechanical process.

The residue obtained can be mixed with quicklime or dolomite to produce a baggable fertilizer.

On small livestock farms, the liquid fraction can be used for spreading on land after deodorization. On large farms, biological purification is performed using extended aeration or, preferably, aerated lagooning.

. PULP AND PAPER INDUSTRIES

3.1. Paper and board mills

The effluents from these mills are characterized by:

— high flow: from 30 m³/tonne (1 200 gal per short ton) produced (newsprint, kraft, and boards) to 150-200 m³/tonne; 36 000-48 000 gal per short ton (fine coated papers, magazines, etc.);

— high "insoluble" pollution: fibres lost in the manufacturing cycle, fillers used to improve the paper quality ($CaCO_3$, kaolin, silica, starch, $Al(OH)_3$, coating products such as latex, etc.).

It is therefore logical to use physicochemical processes to purify these effluents, since this form of purification is normally intended to maximize recycling of materials and water.

Virtually complete elimination of suspended solids (SS) and colloids is possible by flocculation, for which alum will preferably be used for reasons of recycling and because it is readily available in this industry, followed by separation by settling or flotation (Turbocirculator or Sediflotazur).

The sludge extracted from the settling tank or flotation unit is reintroduced to the manufacturing process, or dewatered and disposed of (depending on the quality of the paper manufactured).

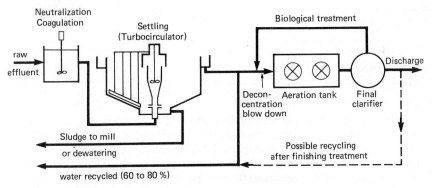

Fig. 527. — Exemple of a paper-mill effluent treatment plant.

The treated effluent may be partially recycled, the recycle rate depending on the permissible concentration factor, which in turn depends on the soluble pollution entering the circuit (salinity, BOD). In most of the cases studied, the water saving may be as much as 60-80 % without any evident disadvantage to the manufacturing process.

The deconcentration blowdown is treated biological without any special problems (see diagram 527, page 825).

The table below sets out the results obtained in three different paper mills which are representative of their respective categories.

PAPER MILLS—EXAMPLES OF TREATMENT RESULTS USING FLOCCULATION, SETTLING OR FLOTATION AND BIOLOGICAL TREATMENT

Products	Input			After flocculation/settling			After biological treatment			Remarks
	SS mg/l	CODas$_2$ [3] mg/l	BODas$_2$ [3] mg/l	SS mg/l	COD mg/l	BOD mg/l	SS mg/l	COD mg/l	BOD mg/l	
Packaging and board	300-3 000	1 550	780	30	1 230	700	30	100-150	30-40	Materials recycled: 100 % Water recycled :75 % Biological treatment: activated sludge with air
Bleached kraft based printings and writings	100 [1] 700	200 [1]	90 [1]	20 25	100 [1] 210 [2]	55 [1] 100 [2]	30	50	20	1. Open water circuit 2. 50 % water recycle Biological treatment: activated sludge with air
Newsprint and thermo-mechanical pulp	550	2 700	1 230	150	2 500	1 110	80	300	60	Large-scale recycling at mill level Biological treatment: biological filter + activated sludge

3. CODas$_2$, BODas$_2$: COD or BOD measured after 2 hours' settling.

3.2. Paper pulp mills

3.2.1. NATURE OF EFFLUENTS

Their composition depends on the manufacturing process and on the nature of the vegetable fibres used. There are four broad process classes, each of which produces unbleached—or, more frequently, bleached pulps. Bleaching, which allows more intensive attack of lignin residues (mainly by alkaline dissolving of chlorolignin), makes a substantial contribution to pollution, as the table on page 827 shows.

A. The kraft process, based on alkaline cooking (NaOH, Na$_2$S) of wood gives 40-50 % solution of the dry solids of the wood (efficiency 60-50 %), this organic matter reappears in the form of cooking liquors (black liquors) and then of pulp washwaters.

When a pulp of this kind is bleached, overall efficiency is reduced by about 10 %, and the bleaching process contributes considerably to the colour discharged by the effluents.

B. The bisulphite process is based on acid cooking of wood (solubilization of lignin by bisulphites); these pulps are almost always bleached, overall efficiency reaching 50 %.

C. "Semichemical" processes combine chemical and mechanical action: the best knowns is the NSSC (Neutral Sulphite Semi Chemical) process, which has an efficiency of 75 %.

D. Processes using "mechanical or thermomechanical disintegration" of wood: efficiency 90-95 %.

● Very substantial flow: between 50 m^3/tonne (mechanical pulp) and 400 m^3/tonne (bleached bisulphite pulp without recovery of black liquors).

● High insoluble pollution (fibres and fibrils, clays, sticks, etc.). Characterized by a content of "nonsettlable" matter which is not negligible (between 50 and 300 mg/l).

● High soluble pollution, variable within the following ranges:
— BOD$_5$: between 100 and 500 mg/l;
— COD: between 300 and 4 000 mg/l (50 % biodegradable at most);
— colour: considerable and refractory to biological processes.

Typical pollution values, per tonne of finished product, are given in the following table, which applies to mills which incinerate their cooking liquors.

	BOD kg/tonne	SS kg/tonne	Colour (Pt-Co) kg/tonne	Water consumption m^3/tonne
Kraft:				
● unbleached	25-35	20-30	10-50	30-120
● bleached	45-65	25-35	170-240	120-350
Bisulphite:				
● unbleached	50-160	20-55	10	100-300
● bleached	60-250	25-60	75	200-650
Bleached NSSC	60	20-40	40	250

3.2.2. PURIFICATION TREATMENT FOR PAPER PULP EFFLUENTS

A number of techniques are used for complete treatment of these effluents:
— suspended solids are normally eliminated by settling without the addition of reagent; owing to the considerable cohesion of the primary sludge, particular care must be devoted to the design of the sludge extraction system;
— about 90 % of the BOD_5 is normally eliminated by biological treatment—either aerated lagooning (4-10 days) or activated sludge (air or pure oxygen);

Fig. 528. — Aerated lagooning treatment of paper-mills effluent. Örebro paper-mills, Frövifors, Sweden.

Fig. 529. — Treatment of manufacturing process effluent. Output: 1 100 m^3/h (7 mgd). Zanders paper-mills, Western Germany.

— colour removal and nonbiodegradable COD elimination call for specific treatment: adsorption on resin or activated carbon, ultrafiltration of the most intensely coloured effluents (alkaline extractions), or chemical precipitation applied to the total flow.

4. TEXTILE INDUSTRIES

The basic feature of these industries is their enormous diversity, expressed in:
— the fibres treated (natural, artificial, or synthetic);
— the dyeing processes (full-width, padder, jig, kier, roller printing, screens, etc.);
— the products used (dyes in particular).

This diversity obviously has repercussions on the nature and quantity of pollution discharged.

4.1. Nature of discharges

The textile industry involves two main types of activity:
— mechanical activity (spinning, weaving, etc.), which causes only slight pollution (except for wool carding and scouring shops);
— textile finishing-bleaching, dyeing, printing and finish-treatment.

The second form of activity, which generally gives rise to substantial pollution, is characterized by:
— dilution of the pollution, frequently due to the use of massive quantities of rinse water;
— intense coloration, depending on the dye used and its application technique.

The following types of dyes may be used:
— dyes discharged in solution: "acid", "basic", "mordant", "reactive", and other dyes;
— dyes discharged in solid form: "vat", "oxidation", "disperse", "naphthol" (soluble at certain pH values); "pigment", "sulphur", and other dyes.

The other pollutants comprise:
— organic acids (usually biodegradable) and inorganic acids;
— alkalis (caustic soda, carbonate);
— oxidants, from bleaching —carried out with oxygenated water, hypochlorite solution, chlorite or perborate— or from bichromates used as developers for certain dyes;
— reducing agents (sodium hydrosulphite and sulphite);
— adjuvants (wetting agents and detergents);
— mercerizing and finishing products (starch, alginate, enzymes, carbomethyl-cellulose, etc.);
— emulsifiers: alginates and white spirits used in the preparation of printing pastes.

Fig. 530. — Treatment of dye-works effluent. Output: 2 500 m³/d (0.64 mgd).

4.2. Order of magnitude of pollution loads

A. The overall effluent of textile finishing (in the case of cotton and artificial and synthetic fibres) typically has the following values:

— Flow .... 80-400 m³/tonne of fibre
— pH .... 3-12 (usually basic)
— COD .... 200-1 200 mg/l
— BOD_5 .... 60-400 mg/l
— $\dfrac{COD}{BOD_5}$ 2.5-6
— SS .... 30-100 mg/l (mainly down, fluff, and fibres).

The effluent may also contain hexavalent chromium (up to 2-3 mg/l) and sulphides (up to 100 mg/l), which are toxic and which must be taken into account if biological treatment is used.

B. Cleaning and degreasing of wool with detergents:

— Flow .... 8-35 m³/tonne of wool
— pH 9-10
— BOD_{5as2}[1] .... 2 000 mg/l
— SS .... 20 000 mg/l
— Grease .... 5 000 mg/l

---

1. BOD_5 measured after 2 hours' settling.

4.3. Purification techniques

In most cases the initial action taken will be aimed at maximizing water saving. Facilities required in all cases after this will be fine bar screening (for down and fibres), possibly oil, grease, and grit removal stages, and a homogenization tank (essential).

Thereafter, a choice must be made from three types of processes, all of which are applicable and effective:

A. Physicochemical treatment:

Comprising neutralization, coagulation, flocculation (with metallic salts), and settling.

Elimination efficiencies vary in individual cases between 35 % and 70 % for COD (but are low for BOD_5: 10-30 %). Depending on the type of dye and the proportion of insoluble dyes used, 50-95 % colour elimination will be obtained.

This treatment is essential whenever greases or toxic products (Cr^{6+}, S^{2-}) are present and for solvent-medium printing shops; in the case of the latter, up to 90 % COD elimination is possible.

The disadvantage of physicochemical treatment is the production of relatively hydrophilic, unmarketable sludge which must be dewatered (e.g., by filter press or by centrifugation).

In wool carding and scouring shops, lanolin can be recovered after refining if degreasing (by solvent or centrifugation) is carried out.

B. Biological treatment.

The most suitable techniques are activated sludge treatment and aerated lagooning.

This biological treatment (after elimination of toxic substances) is the most effective. BOD_5 elimination efficiencies generally exceed 80 % for a sludge loading of 0.5 kg BOD/(kg VS.d) and may even exceed 90 % for a loading of 0.2 kg/(kg VS.d).

However, colour removal is slight, as the dyes are not biodegradable.

If a combination of physicochemical and biological treatment is used, it is in most cases possible to restore to the receiving waters an effluent from which over 85 % of the colour has been removed and whose residual BOD_5 is less than 40 mg/l, a value which can be substantially reduced still further by finishing treatment on Biolite (submerged biological filters —see page 793).

C. Finishing treatment.

To eliminate residual colour and COD, the best technique remains adsorption on activated carbon. Where this process is used as a tertiary phase, after physicochemical and biological treatment, an effluent from which all colour has been removed can be obtained, so that the possibility of recycling the treated water to the manufacturing process can be considered. In some cases, where the BOD_5 of the effluent is very low, the use of activated carbon immediately following flocculation and settling could be contemplated.

5. PETROLEUM INDUSTRY

The main polluting agents in the specific discharges of the petroleum industry are hydrocarbons, but there is also a wide variety of other components, viz.:
— organic compounds (phenols, sulphonic acids, alcohols, etc.);
— sulphurous compounds (sulphides, mercaptans, thiosulphates, etc.);
— sodium salts;
— suspended solids (sands, clays, coke pitch, catalysts).

Since hydrocarbons are the common denominator of the pollutants, treatment will in general commence with oil removal in two phases: an initial phase of gravity separation followed by a final phase of flotation; if biological treatment is applied, it must be reserved for effluents from which most of the oil has been removed and which do not contain sulphides or heavy metals.

5.1. Main sources of pollution

Hydrocarbon pollution of the natural environment originates chiefly from accidental or deliberate discharges during transport and, in particular, when the finished products are used; however, this section will be confined to a discussion of the known discharge sources —which are therefore controllable— encountered in the following three phases of the petroleum industry:
— production;
— transport;
— refining.

5.1.1. PRODUCTION

The following arise:

● *Formation waters* (i.e., those present in the oil-bearing rock), entrained with the crude and separated from it during the latter's demulsification and heating.

The temperature of these waters is 50-90° C. They are saline and constitute mechanical emulsions containing 0.5-2 g/l of hydrocarbons. They can be destabilized essentially by coalescence (DEGREMONT process); the new international regulations on discharges at sea cannot normally be satisfied by gravity separation in a parallel-plate unit or by mechanical flotation.

● *Drilling sludges:* When wells are drilled, large volumes of sludges rich in bentonite are extracted and subsequently undergo static settling. The overflow of the tanks contains a stable emulsion of hydrocarbons in water which must be broken and separated by flotation ·before discharge.

5.1.2. SEA TRANSPORT

● *Ballast waters at terminals:* depending on the size of the tankers, these represent from 33 to 25 % of the useful capacity of these vessels, and it is estimated that the quantity of hydrocarbons contained amounts to about 0.4 % of the product carried, corresponding to a potential concentration of hydrocarbons in the water in the region of 1 %.

These waters, which are usually highly saline, are at sea water temperature. After the customary buffer storage, they contain only small quantities of suspended solids and hydrocarbons, which can be eliminated by a simple dissolved-air flotation treatment after flocculation with a cationic adjuvant (such as Prosedim).

● *Tanker cleaning waters:* powerful water jets are used in the cleaning of oil tankers before maintenance, a thick encrustation of viscous slops (bituminous or asphaltic substances) being thereby stripped off. Massive doses of detergents (0.1-3 g/l) or/and solvents are sometimes used.

The cleaning waters, which are heated where necessary to remove heavy substances, consist of high-stability chemical emulsions, which must be broken under difficult conditions, involving the consumption of large quantities of coagulants.

5.1.3. REFINING

Different types of effluents are encountered, depending on the organization of the sewerage systems:

● process waters, comprising:
— desalter effluents (2-10 % of the flow of crude), which are relatively hot and saline;
— distillation condensates (1-1.5 % of the flow of crude), which usually undergo preliminary stripping treatment;

Fig. 531. — Treatment plant for ballast and process water in an European oil refinery.

— catalytic cracking condensates (2-6 % of the flow of crude cracked), which are heavily polluted with ammonia, sulphide, and phenol, and must be steam-stripped before the main purification process;
— steam-cracking condensates, much less polluted;

● oily waters, comprising: rainwater from paved areas, washwaters from circuits and pumping facilities, spray waters and tank bottom draws.

These effluents, whose flows are highly variable, are polluted mainly by hydrocarbons and suspended solids; they are usually cold.

● concentrated effluents: spent caustic soda solutions, generally calling for highly specific treatments.

5.2. Treatment processes

5.2.1. PRELIMINARY OIL REMOVAL

Several different types of facilities are used for preliminary oil removal from effluents entering the treatment plant. The choice of facility depends on the effluent characteristics —whether they contain heavy or bituminous substances, waxes, and paraffins, a high level of suspended solids, etc. :
— longitudinal settling tanks to **API** standards, whith must be mechanically scraped;
— the **DEGRÉMONT** circular oil removal unit, with preliminary separation of light products in a unit protected from the atmosphere (see fig. 532);
— parallel-plate settling tanks; as the plates gradually become clogged, the effectiveness of these units declines, so that they must be cleaned periodically or reserved for the treatment of hot effluents or effluents containing little sediment;
— mechanical flotation with dispersed air.

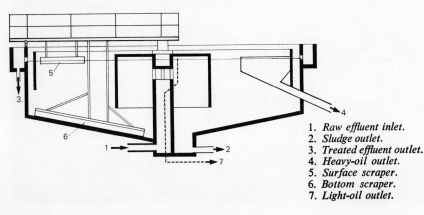

1. *Raw effluent inlet.*
2. *Sludge outlet.*
3. *Treated effluent outlet.*
4. *Heavy-oil outlet.*
5. *Surface scraper.*
6. *Bottom scraper.*
7. *Light-oil outlet.*

Fig. 532. — Diagram of a circular oil-removal tank.

5.2.2. OIL REMOVAL FROM REFINERY EFFLUENTS

Depending on the effluents to be treated, the above facilities allow them to be discharged with hydrocarbon contents between 25-100 mg/l —sometimes more. Secondary oil removal is necessary to reduce these levels to the range 5-20 mg/l, compatible with the discharge standards or the requirements of subsequent biological treatment phases. Several processes can be used for this purpose:

— Dissolved-air flotation (Sediflotor, Sediflotazur), after addition of poly-electrolyte. Precipitation of sulphides and complete clarification of the effluent will also be obtained if a metallic coagulant is added at the same time.

— Coalescent filtration under pressure, with the addition of a cationic polyelec-trolyte (Prosedim CS 53). Special arrangements (Degrémont patent) must be made for treating the filter washwaters to break the emulsion formed.

5.2.3. DESULPHURIZATION

There are various possibilities, depending on the initial concentration:
— steam stripping of concentrated effluents such as FCC condensates;
— slow air oxidation, with or without catalysts (spent caustic soda solutions);

Fig. 533 A. — Oily water treatment. Sand filtration: 7 filters 4.3 m (14 ft) in diameter. Total output: 1 500 m³/h (9.5 mgd). S.I.B.P. oil refinery at Antwerp, Belgium.

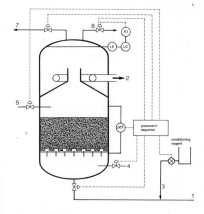

Fig. 533 B. — Degremont coalescence de-oiler.
1. *Raw water inlet.*
2. *Treated water outlet*
3. *Conditioning*
4. *Scour gas inlet*
5. *Washing effluents*
6. *Washing water inlet*
7. *Vent*
8. *Oil removal*

— displacement of H_2S by acidification, cold stripping with air of H_2S, which is then burnt (effluents containing 100-500 mg/l of S^{2-});
— precipitation by iron salts (effluents containing less than 25 mg/l of S^{2-}).

5.2.4. BIOLOGICAL PURIFICATION

Several methods exist, the choice of which depends on the quantity and nature of the BOD_5 to be eliminated:
— aeration/oxidation, to satisfy the immediate oxygen demand due to the presence of inorganic reducing agents;
— medium-rate activated sludge or extended aeration;
— traditional biological filters, or preferably biological filters filled with geometrically shaped plastic media; these require much more effective protection against influent hydrocarbons than the equipment used in the previous techniques;
— trickling through Biolite. This patented process is indicated for the treatment of refinery effluents whose BOD_5 is less than 50-60 mg/l after dual stage oil removal.

5.2.5. TERTIARY TREATMENT

This may take the form of filtration, trickling through Biolite or activated carbon, ozonization, flocculation, and sludge blanket settling.

Low levels of phenols can be reduced by trickling through Biolite.

5.2.6. SLUDGE TREATMENT

The sludge obtained after flotation without an inorganic coagulant is very often recycled in the manufacturing process. If not, it can be mechanically dewatered by continuous or intermittent centrifugation (sludge flotation and sludge from the bottom of API settling tanks) or by filter press (biological sludge).

The sludge is then destroyed by incineration in a fluidized bed furnace or by pyrolysis.

5.3 Oil removal from formation waters discharged during production

These very hot, saline waters contain 200 to 5000 mg/l of hydrocarbons. They cannot be treated by flotation and require oil removal in a single stage in order to keep the plant compact.

The DEGREMONT coalescence de-oiler answers this specific problem.

The process consists of percolating the effluent upwards through an oleophilic granular mass. An oil film forms continuously at the surface and the oils in emulsion in the effluent coalesce on this film. The film is itself torn away from the top of the filter bed and breaks up into globules a few millimetres in diameter which rapidly separate from the water by reason of their own, much higher, flotation speed. A very fine degree of separation may be achieved by this process. An automatic device washes the filter with air and water.

6. SYNTHETIC CHEMICALS INDUSTRY

The organic synthetic chemicals industry is highly complex and varied.

Starting from simple products obtained by distilling coal or oil —e.g., ethylene, propylene, benzene, toluene, phenol, formaldehyde, etc.— it produces monomers, which are in turn used as intermediates for making plastics and plasticizers, surfactants, paints, varnishes, solvents, adhesives, resins, latexes, cosmetics, pharmaceuticals, and other products.

The elementary operations, which are often discontinuous, are numerous: oxidation, hydrogenation, dehydrogenation, alkylation, sulphonation, amination,, and polymerization. They make use of organic and metallic catalysts (Zn, Cu, Co, Th, Cr, Ti), sometimes take place in the presence of high saline concentrations (calcium, sodium and ammonium salts), and employ sulphur compounds which may give rise to the production of sulphides, sulphites, or mercaptans, as well as other compounds.

These operations are followed by separation and miscellaneous types of purification, such as precipitation, washing, centrifugation, etc., usually performed in an aqueous medium, whereby the finished product is separated and excess primary reagents and unwanted intermediate compounds are discharged to the sewer. These operations, together with equipment washwaters, are normally the most concentrated sources of effluents, but they are not the only ones —polluted water also rises from washing floors, scrubbing gases, and even rainfall on storage areas.

In view of the complexity and variability of these effluents, the general recommendations given on page 809 are particularly important here: a prior survey workshop by workshop, segregation of excessively concentrated discharges, regularization of concentrated flows, etc.

No one process can solve all these problems alone. A complete sequence of treatment is therefore required, in general combining both physicochemical and biological techniques. This is illustrated by the examples given below, which are drawn from DEGRÉMONT's experience based on industrial trials and plants actually constructed.

6.1. Complete treatment systems

6.1.1. PHYSICOCHEMICAL PRELIMINARY TREATMENT

Liquid-liquid or liquid-solid separation treatment, either direct or after coagulation/flocculation, is essential wherever:

— oil or insoluble solvent levels exceed a few tens of mg/l;

— suspended solids or colloid levels are sufficient for the pollution to be substantially reduced (minimum 20-30 %) at low cost;

— inorganic toxic substances such as sulphides, fluorides, and heavy metals can thereby be eliminated.

In a **paint and varnish** factory, a treatment plant with oil removal, homogenization (8 hours' retention time), flocculation at high pH (ferric chloride plus $Ca(OH)_2$ plus polyelectrolyte), followed by dissolved-air flotation, reduces suspended solids from 300 to 15 and COD from 2 900 to 1 650 mg/l. This method also eliminates certain toxic metals (Cr, Pb, Zn), and ensures problem-free subsequent biological treatment.

In a **pharmaceutical and phytosanitary product** manufacturing and packaging unit, where colloid, bactericide and pesticide residues from certain formulations must be eliminated, adsorption treatment (by addition of powdered activated carbon) allows subsequent biological treatment of manufacturing process effluents after mixing with the sewage from the plant's sanitary facilities.

6.1.2. SECONDARY BIOLOGICAL PURIFICATION

Biological purification is the process of choice for treating large flows of effluents containing complex pollutants, a high proportion of which often consists of light solvents (alcohols, acetic acid, acetones) which are readily biodegradable. There are few organic compounds which cannot be attacked by bacteria if the medium so permits and if the discharges are sufficiently constant in time for effective bacterial strains to adapt and survive.

Fig. 534. — Petrochemistry effluent treatment.   Average output: 650 m^3/h (4.12 mgd).

There are, however, exceptions. In general, unsaturated aliphatic compounds are oxidized more readily by biological means than aromatic compounds. Organic chemicals can be classified by their biodegradability: some, such as epichlorohydrin, are toxic, while others, such as isopropyl ether, diethanolamine, the polyethylene glycols, morpholine, etc., are highly resistant, although certain strains of bacteria can be adapted to them.

Similarly, activated sludges have been successfully acclimatized to the use of aniline, although this is considered to be toxic. In the presence of phenol and formaldehyde, aniline even has a favourable effect on biological purification.

Another example: a plant producing **solvents** and synthetic rubbers (polyurethane, polypropylene) was equipped with activated sludge purification facilities sized for a flow of 15 000 m^3/d. The effluent contains diisobutyl, isobutene, butyl alcohol, isopropyl alcohol, paraffin, furfurol, alkylphenols, styrene, and butadiene, representing a COD of 20-30 tonnes/d and a BOD_5 of 10-15 tonnes/d; with 8-10 hour retention times in the aeration tank, 85 % COD and 95 % BOD_5 reduction are possible.

The pilot study preceding the design and construction of the treatment plant had shown that in this case the use of pure oxygen not only added nothing to the quality of the treatment but also was economically unwarranted. The opposite conclusion was reached in a parallel study by the same workers on a similar effluent from a nearby plant in which, however, the discharges featured high calcic salinity and high levels of reducing sulphur.

Biological treatment using activated sludge was employed successfully to treat the discharges from a plant making **phenol-formaldehyde, urea-formaldehyde, and epoxy resins and epoxy hardeners.** The effluents contain typically 400-500 mg/l phenols, 300-400 mg/l formaldehyde, and —at times— 1 000-1 500 mg/l aniline. Continuous efficiencies obtained are as much as 97.5-99 % for phenols, 96.6-99.8 % for formaldehyde, and 95-100 % for aniline.

In a plant manufacturing **ion exchange resins** by copolymerization of styrene and divinylbenzene, sulphonation or amination, and chloromethylation, discharging about 2 000 m^3/d of effluents, activated sludge treatment facilities give BOD_5 values of less than 30 mg/l BOD_5 with loadings in the region of 1 $kg/(m^3.d)$. The residual COD, on the other hand, remains high (250-400 mg/l).

Further treatment is necessary to give better purified discharges: tertiary filtration through porous material (Biolite) to keep the suspended solids content below 30 mg/l, and adsorption on activated carbon to reduce the COD below 200 mg/l.

In a plant producing **synthetic polymers, glyoxal, ethylsulphonate, gallic acid,** etc., a complex treatment system was installed comprising preliminary neutralization, homogenization, colour removal with lime at pH 12 in a Turbocirculator, activated sludge purification preceded by denitrification under anoxic conditions, and final clarification. This treatment reduces the BOD_5 from 1 800 to 30 mg/l, the COD from 3 000 to less than 400 mg/l, total nitrogen from 200 to 30 mg/l TKN, and nitrates from 200 to less than 50 mg/l.

6.1.3. SECONDARY PHYSICOCHEMICAL PURIFICATION

Where the pollution is not degradable, or is toxic, physicochemical purification is used —generally, adsorption on activated carbon, the only effective way of eliminating dissolved organic compounds.

The effluents of a production unit manufacturing **nitrated intermediates for the pharmaceutical and dyestuffs industries** containing aromatic nitrocompounds (e.g., dinitrotoluene and dinitrobenzene) and chlorinated solvents (e.g. methyl chloride and chlorobenzene) are treated, after homogenization and sand filtration, by trickling through activated carbon in an acid medium (pH 3) and final neutralization at pH 7 before discharge.

In this way, over 85 % of organic carbon is eliminated, the level of the latter in the initial effluent having been 800-1 000 g/l TOC. The BOD of such effluents is obviously not measurable.

Fig. 535. — Effluent treatment plant for a chemical manufacturing works. Output: 400 m³/h (2.5 mgd). Hoechst French Co.

6.2. Manufacture of polymers and fibres

Some manufacturing processes for important products such as polymers and fibres are applied in self-contained plants. Where the process used is precisely known, the relevant characteristic effluents can then be subjected to the appropriate specific treatments. The following information is therefore given for the sake of illustration only.

A. Nylon: Nylon-6, and sometimes also nylon-12, is made by polymerization of caprolactam, which may itself be synthesized from cyclohexanone and hydroxylamine sulphate. The effluents contain high levels of ammonium sulphate, although this compound is partially recovered by crystallization. They have high BOD values, due to the presence of soluble and highly biodegradable compounds (low COD/BOD_5 ratio) in spite of the presence of epichlorohydrin and a marked tendency to cause bulking of the activated sludge.

In the production of nylon-11 from aminoundecanoic acid, soluble pollution is less significant than in the above process, but the insoluble compounds require more intensive prior physicochemical treatment.

B. Latexes: production of organic polymers by the aqueous process (styrene, butadiene, and acrylonitrile copolymers) causes two types of pollution:
— settlable solids, predominantly organic (coagulated latex), which are directly separable by flotation;
— emulsified or colloidal substances, which must be coagulated before flotation. The final biological purification stage will itself treat a relatively low residual BOD_5.

The synthesis of ethylene-propylene copolymers also gives rise to discharges of metals (Ti and Al) from catalysts.

C. Butyl rubber: copolymerization of isobutylene gives rise to predominantly alkaline discharges containing different soluble organic compounds, and zinc stearate in emulsion.

D. Polystyrene: the cold polymerization and H_3PO_4 dispersion process gives effluents containing high levels of phosphoric acid and smaller quantities of hydrochloric acid. These effluents also contain a suspension of fine polystyrene particles and generally have a very low BOD_5. They are treated by neutralization with lime, flocculation, and settling.

E. Polyethylene: there is a great diversity of processes and grades (high- and low-density, high- and low-pressure). With high-pressure polymerization under anhydrous conditions, the pollution consists of polymer pellets and large quantities of oils from the compressors. After separation by flotation, the residual BOD_5 is usually negligible.

F. Polypropylene production: the effluents contain substantial pollution due to the presence of suspended solids (polypropylene dusts, and aluminium and titanium hydroxides), pigments used as additives, and non-ionic surfactants used for fibre rinsing. The purification facilities usually include a physicochemical stage and a biological stage designed for high detergent elimination and to take account of the sensitivity of the activated sludge to load variations and its liability to bulking.

G. Styrene: the ethylbenzene produced by alkylation of benzene is washed with caustic soda and purified before conversion.

The effluents generally have a strongly alkaline reaction and contain considerable quantities of suspended solids (aluminium hydroxide) or emulsified matter (oils, tars, styrene, and ethylbenzene). After flocculation and flotation, residual BOD_5 pollution is slight.

H. Rayon: in the manufacturing process, cellulose is dissolved in the form of viscose, which is precipitated in acid spinning baths containing copper and zinc salts.

Zinc losses in the rinse waters may amount to 3-6 kg per tonne of rayon. The zinc can be recovered by single- or two-stage neutralization of these waters and acidification of the zinc hydroxide sludge arising. The sludge has been separated, in particular, in the Sediflotor unit.

7. METALLURGICAL AND ASSOCIATED INDUSTRIES

7.1. Iron and steel industry

The main effluents encountered in this industry, which recycles most of its water (see Chapter 23), are ammoniated waters from coking plants, blowdowns from blast furnace and oxygen converter gas scrubbing circuits, and cold-rolling effluents.

A. Ammoniated coking plant effluents: these consist of the coal water (4 % formation water and 8 % moisture content if the coal is not preheated) and gas scrubbing effluents originating mainly from softened make-up water.

The pollution of these effluents consists mainly of several g/l of phenols and ammonia, and secondarily of sulphides, cyanides, and sulphocyanides. These effluents generally contain high chloride concentrations.

Treatment comprises the following phases:

— removal of tar from the ammoniated effluents by settling and/or filtration;

— stripping of volatile ammonia, and then of fixed ammonia after displacement with caustic soda, or (more economically) with lime. DEGRÉMONT is developing a preliminary treatment allowing the use of lime while avoiding the difficulties associated with scale deposition;

— buffer storage of the stripped effluents and cooling to 35 °C;

— biological purification, usually with activated sludge in one or two stages, possibly with lagooning;

— tertiary clarification treatment with possible adsorption of residual COD on activated carbon.

B. Blowdowns of blast furnace gas scrubbing circuits: these contain ammonia, cyanides, sulphocyanides, and traces of phenols and metals (Zn, Pb).

They are treated by precipitation of metals by alkalinization, chelation of cyanides by addition of ferrous salts or their oxidation by peroxygenated compounds, and elimination of ammonia by biological nitrification or possibly by stripping.

C. Blowdowns of oxygen converter gas scrubbing circuits: these may contain fluorides, which must be precipitated by the addition of lime where the effluents contain sodium bicarbonate (as in the case of caustic soda desulphurization of steel).

D. Cold-rolling effluents: these comprise mainly:

— acid rinse waters following pickling with sulphuric, hydrochloric, or nitric acid, which are treated by neutralization, air oxidation, and precipitation of ferric iron;

— chromic passivation effluents, which are sometimes combined with the previous effluents or are deionized by ion exchange for recycling;

— electrolytic degreasing effluents, which have high COD values and can be treated in open circuit by coagulation and flotation;

— four- or five-stand train lubrication effluents (oils and greases) from emulsions which must be deconcentrated and treated by flocculation and flotation;

— soluble cooling oils from three-stand mills, which must be maintained and periodically drained and destroyed by acid breaking. The lower, water phase

Fig. 536. — Treatment of rolling mills effluent and process make-up water. Solmer steelworks, France.

is treated by flocculation, flotation, or settling, or by ultrafiltration (see Chapter 25, page 856).

— other passivation, bonderization, or zinc removal effluents, as well as effluents from miscellaneous concentrated baths which can be regenerated or treated (see metal finishing, Chapter 25, page 845).

7.2. Aluminium metallurgy

Aluminium extraction and metallurgy involve a number of consecutive operations each of which gives rise to specific discharges, which are usually treated separately:

— sodium aluminate liquors from bauxite leaching. These concentrated liquors (290 g/l Na_2O) are hot (95 °C) and contain 100-180 g/l dry solids. After successive exhaustions, they can be treated in scraped settling/thickening facilities;

— cryolite electrolysis gas scrubbing effluents from processes using Söderberg continuous anodes or prebaked anodes. These effluents, which are usually acid, contain sulphites and fluorides; their suspended solids are carbon, alumina, and cryolite;

— effluents from prebaked anode manufacture. These are acid and contain dissolved aluminium, fluorides, and tars; they have a high COD;

— alkaline effluents from lute crushing, which contain fluorides and cyanides;

— cooling waters from aluminium pours in ingot moulds containing small quantities of grease and kaolin;

— acid chlorination effluents, which contain aluminium chloride.

All acid effluents can be treated by neutralization, precipitation and coagulation, and settling, applied in clarifiers of the Circulator and Turbocurcilator type. Sludge concentrations and settling velocities differ considerably according to the materials comprising the sludge: Al_2O_3, $Al(OH)_3$, CaF_2, C, etc.

Alkaline scrubbing of electrolysis gases, which is less common, makes use of a more complex effluent purification technique, in which cryolite is recovered.

Surface finishing shops for manufactured products discharge effluents whose pollutant composition is less complex:

— alkaline degreasing effluents;

— satin finishing effluents;

— anodization effluents containing sulphuric acid and aluminium;

— possible chromic passivation effluents.

While degreasing effluents, which contain high levels of aluminium, are often treated by open-circuit precipitation, satin finishing and chromic passivation effluents are recycled after total deionization.

All processes tend to reduce acid and caustic soda consumption and to increase the concentration of the final sludge.

7.3. Metal finishing industries

7.3.1. METAL FINISHING TECHNOLOGY; POLLUTION CAUSED

Surface finishing is applied mainly to metal parts, but also to certain synthetic materials, to protect them against corrosive attack, to modify their physical surface characteristics, or to improve their external appearance for decorative purposes.

Finishing processes employing aqueous solutions will be discussed below. Surface preparation and the finishing processes themselves involve successive immersion of the parts in several baths, in which chemical or electrolytic reactions take place. The parts carry with them a significant quantity of liquid on leaving each bath owing to their shape and to surface tension effects. They must therefore be thoroughly rinsed before proceeding to the next treatment stage.

Metal finishing effluents therefore fall into two classes (see fig. 537, overall organization of a metal finishing shop):
— spent treatment baths, with high pollutant concentrations;
— dilute rinse waters.

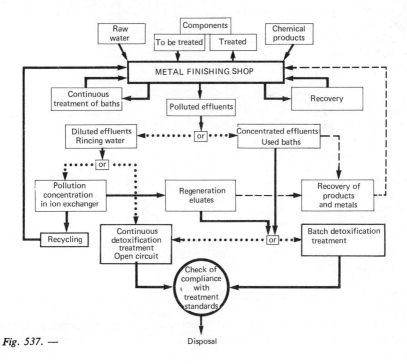

Fig. 537. —

TABLE OF SURFACE TREATMENTS AND POLLUTIONS

Column groups:

- **PRETREATMENTS**
  - DEGREASING: Chemical, Electrolytic, Alkaline (on Al)
  - PICKLING: Acid
  - PREPARATION OF PLASTICS: Sensitizing, Activation, Activation
  - NEUTRALIZATION: Alkaline, Acid
- **MODIFICATION OF SURFACES**
  - PHOSPHATIZATION: Amorphous, Crystalline
  - ANODIC OXIDATION
  - SEALING
  - BRILLIANCY TREATMENT (Electrolytic and chemical)
  - PASSIVATION
  - ELECTROLYTIC POLISHING
- **DEPOSITION OF META[L]**
  - Acid
  - COPPER-PLATING: Alkaline, With pyrophosphate, With fluoborate
  - NICKEL-PLATING: With sulphate, Cold, With fluoborate, With sulphamate, With pyrophosphate
  - CHROMIUM-PLATING: Acid, With fluosilicate, Black
  - GALVANIZING: Alkaline, With sodium zingate, With fluoborate, Acid

NATURE OF POLLUTION sub-columns: Toxic, Changes pH, Suspended matter, General regulations

MAIN POLLUTANTS ENCOUNTERED IN IONIC FORM (row labels):

C^{III}, Sn, Cd, Cu, Al, Zn, Ni, Fe^{II}, Fe^{III}, Ag, Au, Pt, Rh, Pd, Pb, In, Sb, As, Ba, Na, K, NH_4, $C2^-$, PO_4^{3-}, HPO_4^{2-}, $P_2O_7^{...}$, S^{--}, HSO_3, SO_3^{2-}, $NH_2SO_3^-$, SO_4^{2-}, NO_2^-, NO_3^-, F^-, BF_4^-, $H_2BO_3^-$, CN^{--}, $HCrO_4^-$, CO_3^{--}, AsO_4H_2, SiO_3^-, SiF_6^{2-}, $C_2H_4O_6^{2-}$, CH_3COO^-

Main user industries: Motor vehicles, Elec. domest. equipt., Aircraft, Metal furniture, Cycles, Decoration, Metal architecture, Electricity, Electronics

Column groups (top headers)

ATING | DEPOSITION OF METALS BY CHEMICAL METHODS | REMOVAL OF METAL PLATING | ETCHING | TABLE OF SURFACE TREATMENTS AND POLLUTIONS

Column headings (left to right):

- BRASS-PLATING
- SILVER-PLATING — Alkaline / With fluoborate
- GILDING
- PLATINUM-PLATING
- RHODIUM-PLATING
- LEAD-PLATING
- IRON-PLATING
- DEPOSITION OF INDIUM
- DEPOSITION OF ARSENIC
- DEPOSITION OF ANTIMONY
- COPPER-PLATING
- NICKEL-PLATING
- TINNING
- SILVER-PLATING
- GILDING
- PALLADIUM-PLATING
- REMOVAL OF NICKEL
- REMOVAL OF COPPER (Chemical, from iron)
- REMOVAL OF ZINC (Electrolytic)
- Chemical
- REMOVAL OF CHROMIUM (Electrolytic / Chemical)
- REMOVAL OF TIN
- REMOVAL OF LEAD
- REMOVAL OF SILVER (From brass / From iron)
- ELECTROCHEMICAL
- CHEMICAL (Alkaline / Acid)

METHODS OF TREATMENT	Results obtainable using current techniques (mg/l)	Swiss Standards		
Reduction / Oxidation / Neutralization / Precipitation Settling		Discharge to water course	Discharge to sewer	
	1	2	2	Cr^{III}
		2		Sn
	1	1	1	Cd
	1	1	1	Cu
	5	10		Al
	1	2	2	Zn
	1	2	2	Ni
		1	1	Fe^{II}
	1	''	''	Fe^{III}
		0,1	0,1	Ag
				Au
				Pt
				Rh
				Pd
	2	1	1	Pb
				In
				Sb
		1	1	As
		10		Ba
				Na
				K
				NH$_4$
				Cl$^-$
		2		PO$_4^{3-}$
				HPO$_4^{2-}$
				P$_2$O$_7^{4-}$
		0,1		S^{2-}
				HSO$_3^-$
		1 à 10	10	NH$_4$SO$_3^-$
				SO$_3^{2-}$
	1	1	10	NO$_2^-$
				NO$_3^-$
	15	10	10	F$^-$
				BF$_4^-$
				H$_3$BO$_3$
	0,2	0,1 à 0,5	0,5	CN$^-$
	0,1	Cr 0,1	Cr 0,1	HCrO$_4^-$
				CO$_3^{2-}$
				AsO$_4$H$_2^-$
				SiO$_3^{2-}$
				SiF$_6^{2-}$
				C$_2$H$_4$O$_8^{2-}$
				CH$_3$COO$^-$

The specific pollution arising includes:

— organic substances originating mainly from degreasing operations (wetting agents and chelating agents, which are now being used more and more commonly in modern baths);

— suspended solids (oxides, hydroxides, soaps, etc);

— and, in particular, dissolved and ionized inorganic compounds.

The table on pages 846-847 sets out the main types of baths and their dissolved constituents. Plainly, the baths listed are not all of equal importance, and although the enumeration of reagents used includes all possible constituents, some are predominant while others are encountered only occasionally.

All bath components reappear in the rinse waters, which may also contain metallic ions from chemical attack of the treated parts.

In France, the Ministry of the Environment, in combination with specialist metal finishing and effluent detoxification firms, has conducted a great deal of work on the monitoring and control of pollution due to these effluents and baths, culminating in a Ministerial Circular which sets out precise discharge rules and standards for the various pollutants, together with recommendations for the facilities to be provided at workshops and treatment plants in order to minimize the technical and financial demands resulting from their application.

7.3.2. POLLUTION PREVENTION AND PRODUCT RECLAMATION

To achieve the detoxification standard laid down as efficiently as possible, the pollution control engineer must approach each purification project from the twin angles of **prevention** and **reclamation.**

First of all he must thoroughly familiarize himself with the detailed configuration of the surface finishing shop and its activities, to enable him to apply these two concepts correctly.

The next step is aimed at reducing pollutant emissions from the shop itself by the application of different methods:

— reduction of bath-to-bath carry-over, by modifying the geometry of fixtures and workpiece mountings, optimizing drip times, especially in drum processes, installation of sprinkler rinsing facilities, etc.;

— direct action on the manufacturing processes themselves if necessary, perhaps by replacing polluting baths by ones giving the same results but which are chemically less toxic or less hazardous—e.g., substitution of low-cyanide or in some cases acid or alkaline baths without cyanide or organic chelating agents for high-cyanide deposition baths.

These initial efforts should be continued at shop level with a view to reclaiming, and if possible marketing, certain products commonly lost.

The first aim will concern rationalization of water use in the workshops: it can be shown that, for a given operation and equal rinsing and manufacturing quality, there are rinsing structures which allow substantial savings of water consumption.

The possibility of modifying processing cycles by interposing structures of the following types will therefore be considered (other possibilities also exist):
— installation of static or reclamation rinsing facilities;
— installation of double or triple cascade rinsing facilities.

With regard to reclamation of materials used in the work of the shop, it may in some cases be possible to install facilities for their recovery and utilization:
— recovery of nickel and copper salts, or the salts of other metals, by reverse osmosis;
— recovery of chromium salts from pickling bath acid by ion exchange.

7.3.3. PURIFICATION OF RESIDUAL POLLUTION

When the above fundamental action has been taken, the balance of the initial pollution must be treated in the form of spent baths which cannot be reclaimed or reused and of rinse waters.

A. Classification of forms of pollution and effluent treatment.

This pollution can be subdivided into four main categories:
— toxic pollutants such as cyanides, hexavalent chromium, and fluorides;
— pollutants which modify pH, i.e., substances with acid or basic functions;
— pollutants whose presence increases the suspended solids content, e.g., hydroxides, carbonates, and phosphates;
— pollutants covered by special regulations, e.g., because they affect the COD (sulphides and ferrous salts).

Treatment techniques also break down into four main groups:
— oxidation treatment for cyanides, bivalent iron, sulphites, and nitrites;
— reduction treatment for hexavalent chromium;
— pH adjustement treatment;
— precipitation and settling treatment, possibly followed by filtration.

B. Basic reactions of oxidation and reduction treatments.

Owing to the importance of cyanide and chromate removal, particular emphasis is laid below on the relevant reactions.

● **Example of oxidation. Cyanide treatment:** high-toxicity cyanides are converted into virtually nontoxic cyanates by the action of strong oxidants such as sodium hypochlorite, chlorine gas, or peroxosulphuric acid (Caro's acid) in an alkaline medium.

The overall reactions used are:

— with sodium hypochlorite:

$$NaCN + NaClO \longrightarrow NaOCN + NaCl$$

— with chlorine gas:

$$NaCN + Cl_2 + 2NaOH \longrightarrow NaOCN + 2NaCl + H_2O$$

— with Caro's acid:

$$NaCN + H_2SO_5 \longrightarrow NaOCN + H_2SO_4$$

The first two reactions are practically instantaneous at pH levels above 12, but the reaction rate falls rapidly as the pH is reduced (the critical threshold being pH 10.5). Whatever the pH, the first compound formed is cyanogen chloride CNCl, which is just as dangerous as hydrocyanic acid:

$$NaCN + NaClO + H_2O \longrightarrow CNCl + 2NaOH$$

From pH 10.5, however, cyanogen chloride is hydrolysed as soon as it forms, by the reaction:

$$CNCl + 2NaOH \longrightarrow NaCl + NaCN + H_2O$$

With Caro's acid, the observed reaction rate is sufficient when the pH exceeds 9.5.

Note that in all cases the transition from cyanate to nitrogen by the reaction:

$$2NaOCN + 3Cl_2 + 6NaOH \longrightarrow 2NaHCO_3 + N_2^\nearrow + 6NaCl + H_2O$$

takes place at the same pH as that from cyanide to cyanate, but requires three times the amount of reagent and a reaction time of 5 to 90 minutes. Such an expenditure is not normally warranted owing to the very low toxicity of cyanates.

Table 1 shows the quantities of reagents required to oxidize to cyanate 1 gramme of CN^- present in an effluent which has first been adjusted to the optimum pH for reaction.

Table 1. Oxidation of free cyanides

Reagents	Commercial NaClO [1] ml	Cl_2 in g — NaOH in g	Commercial H_2SO_5 [2] in ml — NaOH in g
Stoichiometry for 1 g of CN	18.2	2.75 — 3.10	22 — 3.10
Commercial practice for 1 g of CN	21 excess 20 % [3]	3 — 3.5 } excess 10 % [3]	24 — 3.5 } excess 10 % [3]

1. NaClO, commercial solution of 47-50 chlorometric degrees, or 150 g/l active chlorine.
2. H_2SO_5, commercial solution of 200 g/l.
3. Normal excess for cyanide concentration of less than 100 mg/l (as in the case of rinse waters).

● **Example of reduction: treatment of hexavalent chromiums:** the object is to reduce toxic hexavalent chromium to trivalent chromium, which is less toxic and can be precipitated in the form of hydroxide.

This reduction takes place in an acid medium by the action of sodium bisulphite or ferrous sulphate.

Reactions:

— *Sodium bisulphite:*

$$H_2Cr_2O_7 + 3NaHSO_3 + 3H_2SO_4 \longrightarrow Cr_2(SO_4)_3 + 3NaHSO_4 + 4H_2O$$

— *Ferrous sulphate:*

$$H_2Cr_2O_7 + 6Fe(SO_4) + 6H_2SO_4 \longrightarrow Cr_2(SO_4)_3 + 3Fe_2(SO_4)_3 + 7H_2O$$

The first of these reactions is practically instantaneous at pH values below 2.5, but the reaction rate falls rapidly as the pH increases (critical threshold pH 3.5).

Reduction with ferrous iron is more tolerant and can be effected at a pH of less than 6.

Table 2 sets out the reagent consumptions required to reduce 1 g of Cr^{6+}.

Table 2. Example of reduction: treatment of hexavalent chromiums

Reagents	$NaHSO_3$, d : 1.33 [1] (ml) —— H_2SO_4 (g)	$FeSO_4 . 7H_2O$ (g) —— H_2SO_4 (g)
Stoichiometry for 1 g of Cr^{6+}	5.7 — 0.95	16 — 1.90
Commercial practice for 1 g of Cr^{6+}	6.5 excess 15 % [2] — 1.0	20 excess 25 % [2] — 2

1. $NaHSO_3$, commercial solution d: 1.33 at 530 g/l.
2. Normal excesses for rinse water treatment.

C. Treatment systems.

There are two main treatment configurations; a system may sometimes include elements of both.

● **Open-circuit** treatment, in which the water is discarded.

It is impossible to enumerate all possible plant layouts here. However, they all involve a number of elementary functions (illustrated in diagram 540), using oxidation, reduction, and neutralization reactions, followed by reactions in which the various toxic substances and metallic hydroxides are precipitated.

Note that the open-circuit technique often uses high-rate reactor/ mixers (Turbactor), closed, pressurized vessels, which give shorter detoxification and neutralization reaction times than conventional tank or pit type systems, so that more compact installations can therefore be constructed (diagram 541 and figure 539).

The TURBACTOR (fig. 538) is a powerful recirculation type mixer/ reactor, in which the basic components of the "core" of the Circulator (Chapter 7, page 174) are combined in a vertical, cylindrical vessel. The scaling (which is matched to the various reaction types) of the injection pipe and the convergent/diffuser assembly facilitates both the development and the active recirculation of the nuclei in formation.

1. *Raw effluent.*
2. *Injection nozzle.*
3. *Rapid circulation zone.*
4. *Vegative pressure diffuser.*
5. *Control rH meter.*
6. *Control pH meter.*
7. *Reagents.*
8. *Treated effluent.*

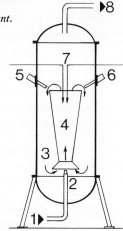

Fig. 538. — Rapid reactor "Turbactor".

Fig. 539. — Turbactor detoxification unit.   Output: 50 m³/h (220 gpm).

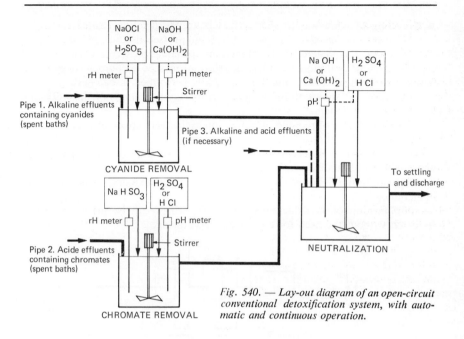

Fig. 540. — *Lay-out diagram of an open-circuit conventional detoxification system, with automatic and continuous operation.*

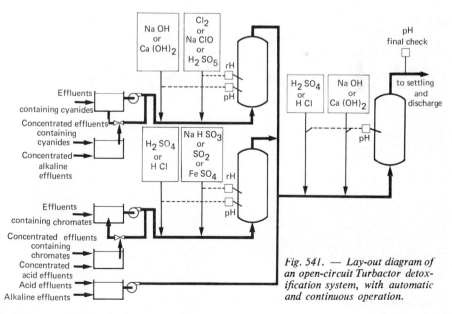

Fig. 541. — *Lay-out diagram of an open-circuit Turbactor detoxification system, with automatic and continuous operation.*

● Recycling of rinse waters after ion exchange purification **(closed circuit).**

Diagram 542 illustrates the use of this technique (see also figs. 543 and 544).

Note that this system has two functions: reclamation (reuse of water) and concentration of pollution. It must therefore always be accompanied by elimination treatment of toxic substances (generally performed in batches).

Diagram 537 shows all the various possible configurations of the workshop and purification system.

The advantages of recycling through ion exchangers are emphasized:

— substantial reduction of water consumption;

— very high purity water obtained at low cost, so that rinsing quality, and hence workpiece quality, can be improved;

— concentration of pollution and possibility of reclaiming some expensive compounds;

— simplification of detoxification operations, which can be performed batch by batch without the need for continuous monitoring.

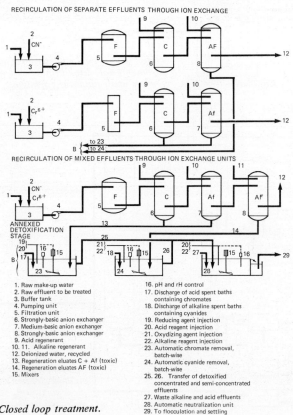

1. Raw make-up water
2. Raw effluent to be treated
3. Buffer tank
4. Pumping unit
5. Filtration unit
6. Strongly-basic anion exchanger
7. Medium-basic anion exchanger
8. Strongly-basic anion exchanger
9. Acid regenerant
10. 11. Alkaline regenerant
12. Deionized water, recycled
13. Regeneration eluates C + Af (toxic)
14. Regeneration eluates AF (toxic)
15. Mixers
16. pH and rH control
17. Discharge of acid spent baths containing chromates
18. Discharge of alkaline spent baths containing cyanides
19. Reducing agent injection
20. Acid reagent injection
21. Oxydizing agent injection
22. Alkaline reagent injection
23. Automatic chromate removal, batch-wise
24. Automatic cyanide removal, batch-wise
25. 26. Transfer of detoxified concentrated and semi-concentrated effluents
27. Waste alkaline and acid effluents
28. Automatic neutralization unit
29. To flocculation and settling

Fig. 542. — Closed loop treatment.

Fig. 543. — Treatment of acid and alkaline effluents containing chromates. Automatic recirculation through ion exchange units. Output: 2 × 20 m³/h (2 × 88 gpm).

Fig. 544. — Treatment of effluents containing chromates and cyanides. Recirculation through ion exchange units. Output: 2 × 25 m³/h (2 × 110 gpm).

D. Centralized detoxification plants.

The idea of sub-letting the detoxification of pollutants from metal finishing shops has led to the development of centralized detoxification plants.

Recent economic studies have shown that this approach is only economic (compared with an on-site plant) if the quantities to be processed and transported are not too great and if the distance from the detoxification centre to the shop is short.

In practice, rinse waters can only be treated on site. Centralized treatment of ion exchange resin regeneration effluents is conceivable in the case of nearby small shops, but the cost is prohibitive for large installations.

Fig. 545. — Neutralization and settling. Output: 80 m³/h (350 gpm). Michelin tyre works MONTCEAU-LES-MINES, France.

7.4. Purification of used soluble oils

Ther term "soluble oils" is generally used to cover all fluids employed for cooling and lubrication in machine tools (in which they are known as cutting oils), presses, dies, rolling mills, etc., in the metal processing industries.

At least three product groups must be distinguished:

● **Emulsified oils,** whose main constituent is a petroleum fraction kept in the form of a stable emulsion by an emulsifier (an anionic or nonionic surfactant).

● **Synthetic oils or true solutions,** which no longer contain petroleum products but instead synthetic materials with similar if not superior lubrication characteristics: polyglycols, hydrocarbon polyoxides, etc.

● **Semisynthetic oils,** which are blends of the two previous types in various proportions.

In addition, all formulations include "antioxidant", "bactericidal", "anti-foam", "ultra high pressure", and other additives.

Before considering the treatment of these used fluids, it is necessary to know the nature of the products or product mixtures confronting one, and it must be remembered that use in machines brings about major changes in the composition of the oils, which, after all, are the reason for discarding them:
— incorporation of product from the manufacturing process (swarf, filings, metallic oxides, grinding paste, **machine oils and greases,** etc.);
— oxidation and polymerization of the least stable products ("gums");
— bacterial attack, which may be more or less intensive depending on the efficacy and stability of the bactericides used.

The table page 858 sets out the principal types of treatment which can be used.

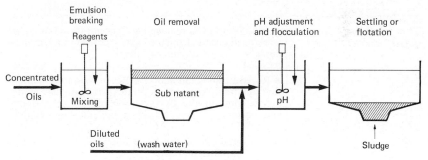

Fig. 546. — Chemical oil-breaking and flocculation.

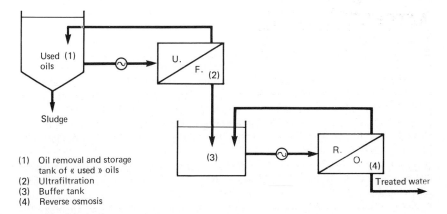

(1) Oil removal and storage tank of « used » oils
(2) Ultrafiltration
(3) Buffer tank
(4) Reverse osmosis

Fig. 547. — Complete treatment using membranes.

SOLUBLE OIL TREATMENT METHODS

Fluids to be treated	Treatments	Residues
Concentrated or semisynthetic emulsions (20-80 g of oil per litre)	● Chemical breaking (cold or hot) followed by flocculation of the water phase (see diagram 1) ● Ultrafiltration	● Oil (90 %) + flocculation sludge (10-30 % of the initial emulsion volume) ● Oil (40-50 %) (autocombustible)
Dilute emulsions (< 1 g of oil per litre)	● Chemical flocculation (possibility of reusing iron salts from pickling baths, etc.)	● Hydroxide sludge containing 50-100 g/l SS, to be dewatered
True solutions (COD: 15-50 g/l)	● Ultrafiltration to remove working impurities from the fluids so that they can be recycled ● Reverse osmosis in addition to ultrafiltration to eliminate most of the COD and bactericides (see diagram 547) ● Biodegradation in addition to ultrafiltration after dilution (necessary to avoid toxicity of bactericides)	● Concentrate (45-50 % organic matter) (autocombustible) ● Reverse osmosis concentrate (8-20 % organic matter depending on the osmotic pressure of the solution) ● Excess activated sludge

8. MISCELLANEOUS INDUSTRIES

8.1. Fossil fuel power plants

Conventional power plants (coal-, lignite-, or fuel-oil-fired) feature periodic discharges of highly concentrated effluents in small quantities, which can therefore be treated intermitently in batches.   The following different waste streams arise:
● *Boiler tube nest cleaning effluents*, containing citric acid as well as sodium fluoride or sulphuric or hydrochloric acid.   In a 250 MW plant, each discharge may contain 1 000-1 500 kg of iron oxides.
● *Effluents from annual cleaning of the air heaters*, containing mainly oxidizing alkaline sludge (ammonium persulphate) and cold hydrofluoric acid in air emulsion.   Discharges from fuel-oil-fired power plants contain copper, magnetite, vanadium, and substantial oily deposits.   After lime neutralization, supersaturation with calcium sulphate must be avoided.

● *Ion exchange resin regeneration effluents*, which contain large quantities of sulphates or sodium chloride, which in general need only to be neutralized.

● *Rainwater from the fuel oil storage area.* Depending on the discharge standards, this may be simply deoiled in an oil removal unit, or it may undergo complete flocculation/flotation.

● The recently introduced technique of *flue gas desulphurization* may constitute a considerable source of pollutants, varying with the process used:
— simultaneous adsorption of SO_3 and SO_2 in an excess of ammonia, with recovery of bisulphite reduced to sulphur, by the Claus reaction;
— adsorption by NH_3 of SO_3 only in the form of ammonium sulphate, whose ammonia is displaced by lime. The treatment involves precipitation of calcium sulphate from the liquor used in closed circuit, and can be performed in a Circulator; the sludge is then dewatered by filter press or filter belt press.

In lignite-fired power plants, the wet process of treating large quantities of fly ash calls for the installation of scraper type settling tanks.

8.2. Nuclear industry

8.2.1. ORIGIN OF DISCHARGES

Radioactive discharges are produced mainly by nuclear power plants, nuclear fuel processing establishments, and nuclear research centres.

Modern pressurized water (PWR) or boiling water (BWR) nuclear power plants include various types of treatment for circuits and radioactive effluents, e.g.:
— purification of reactor water in a by-pass circuit (clean-up);
— condensate treatment;
— decontamination of general service waters (laundry, floor, washing, etc.);
— treatment of storage pond waters;
— treatment of regeneration effluents.

PWR plants also have deionization systems associated with the primary circuit, for monitoring the latters's boron and lithium content.

8.2.2. TREATMENT

Radioactive effluents undergo a variety of treatments depending on their activity level and their content of solid matter and dissolved substances.

Low-salinity effluents containing little suspended solids are treated with precoat filters or ion exchange resins (clean-up, condensates, and storage pond waters).

Waste streams liable to contain a heavy load of suspended solids or to be highly saline (condensate, floor washing, laundry, and miscellaneous drainage treatment effluents) should preferably be treated by flocculation/adsorption. In addition to suspended solids, a high proportion of soluble radioactive matter can be eliminated by combining this flocculation/absorption treatment with coprecipitation in the presence of carriers such as calcium carbonate, calcium phosphate, tannate, copper or nickel ferrocyanide, etc.; the residual soluble substances are then fixed on ion exchange resins.

Soluble and colloidal radioactive substances can be concentrated by reverse osmosis, giving decontamination factors of up to about 10.

With pond waters, the main aim is to keep them clear: most treatment systems include fine, cartridge, or precoat filtration stations, the precoat in the latter case consisting of inert material or ion exchange resins.

In addition to decontamination of the water, all types of treatment must eliminate radioactive matter in as concentrated a form as possible:
— solid residues (flocculation sludge, saturated ion exchange resins), which can be stabilized and then stored;
— liquid residues; these normally undergo final concentration in an evaporator.

To obtain high-concentration effluents, each process should preferably be exploited to the limit of its performance; for this reason it is often worthwhile to combine two or more techniques, e.g., precoat filtration and deionization on ion exchange resins.

Fig. 548. — 970 MW nuclear plant, Gösgen, Switzerland. Make-up water treatment for the cooling system. Output: 3,000 m³/h (19 mgd).

8.3. Fertilizer and phosphoric acid manufacture

Many fertilizer manufacturing processes discharge specific effluents, common to all of which, however, is the presence of either ammonia, nitrates, and urea, or of phosphates, fluorides, and silica. The most frequently encountered effluents are as follows:

● *Superphosphate production effluents* from processes using sulphuric acid attack (plain superphosphate) followed by phosphoric acid attack (triple superphosphate) of natural tribasic calcium phosphates. The effluents to be treated are acid and contain H_2SO_4 and SiF_6H_2 (silicon hexafluoride), which breaks down into CaF_2 and SiO_2 during lime neutralization. This silica precipitates in the form of a bulky gel.

● *Phosphoric acid manufacturing effluents* from the process in which phosphate rock is intensively attacked by sulphuric acid. These effluents consist of an acid slurry of calcium sulphate anhydrite (phosphogypsum), lime neutralization of which precipitates various phosphates, calcium fluoride, and a silica gel.

● *Anhydrous phosphoplaster drying effluents*, the product being made by dehydration of the above residual phosphogypsum.

Depending on whether acid or alkaline washing is used, these effluents contain HF, H_2SO_4, $CaSO_4$ or NaF, $Si(OH)_4$ and $Si(F_6Na_2)$. Their reaction to treatment is similar to that of the gas scrubbing effluents from cryolite electrolysis.

For these the reactions consume large quantities of lime and generate abundant sludge which cannot be readily settled and dewatered.

● *Condensates from ammonia synthesis processes*, with high concentrations of ammonium carbonate and bicarbonate. They can be treated by stripping followed by ion exchange on carboxylic resin after cooling. They are then recycled to the boilers within limits set by their content (which is considerable) of organic compounds (methanol, methylamine, etc.).

● *Ammonium nitrate manufacturing effluents*, comprising:
— distilled condensates of nitric vapours (2-3 g/l of $NH_4 NO_3$ and 150-250 mg/l NH_2);
— workshop washwaters, which have a much higher concentration of nitrates.

Depending on their concentration, these effluents can be treated by reverse osmosis or distillation.

● *Urea manufacturing effluents*, consisting of condensates containing 8-16 g/l urea and 0.5-1 g/l NH_3. They can be treated by alkaline hydrolysis and/or stripping, with high caustic soda and steam consumption. The treatment is completed with biological purification.

8.4. Effluents from other industries

Industry	Origin	Characteristics
Glassmaking and mirrors	— Glass frosting, decorating, and shaping — Sawing, polishing, finishing	— High hydrofluoric acidity, ammonium bifluoride — Corundum, pumice, and emery powders; cerium oxide; garnet
Glass fibres	— Fibre manufacture and oiling	— High SS, BOD and COD due to dextrin, gelatins, silicones and miscellaneous acetates; phenolic resins
Abrasives on supports	— Preparation of adhesive mixtures	— Phenol-formadldehyde and urea-formaldehyde resins; gelatins; starch; epoxy and solvent resins — High COD; average BO BOD and SS
Sand pits	— Screening and washing	— Coloured waters; organic and clayey solids
Aircraft washing and maintenance	— Cleaning of paints — Engine maintenance	— High concentrations of BOD and detergents; silicates; phosphates; chromic acid; oils; greases; phenols; etc.
Cement works		— SS (clinker dusts) — High alkalinity — Liable to set hard
Cosmetics	Packaging	— High levels of grease and detergents — Dissolved, readily biodegradable COD
Paint spray booths	Gas scrubbing and aerosol capture	— High alkalinity and SS — Dissolved COD due to solvents

Industry	Origin	Characteristic
Ceramics	Centrifugation of pastes Floor washing	— High SS level — Possible crystallization reactions
Fibreboard		— Fibres — High dissolved COD from binders
Chlorine industry Mercury cells	Washing of gases produced Cleaning Brine losses	— Mainly mercury pollution (metallic and ionic) — Possibility of recycling mercury
Adhesives		— Glues and gums; high COD — Liable to set hard
Tyres	Casing manufacture	— Iron and copper — Lubricating oils — Soluble oils — Pickling effluents
	Rubber synthesis Tyre manufacture	— Soaps and metallic salts, solvents — Hydraulic oils — Greases
Explosives and explosive powders	Manufacture of nitrocompounds	— High sulphuric and nitric acidity; coloration and non-biodegradable COD — treatment by incineration or activated carbon
	Granulation and impregnation of powders Loading and unloading shops	— Ethyl acetate; nitrocellulose; nitroglycerine, Na_2SO_4, glues, plasticizers — relatively biodegradable effluents; — phosphorous, perchlorates — primary physicochemical treatment

Fig. 549. — Treatment of process wastec water.   Kronenbourg Brewery, OBERNAI, France.
Output: 14 600 m³/d (3.85 mgd).

Part Four

Water:
General
Information

26 THE CHEMISTRY OF WATER AND REAGENTS

1. CHEMISTRY

Substances found in nature usually consist of mixtures of pure substances which can be separated by physical means and by the application of low energy ratings. A **pure substance** is one with identical physical constants whatever the size of the specimen taken.

The molecule is the smallest part of a pure substance having all the physical and chemical properties of the said pure substance.

A molecule can be dissociated by chemical means, but the energy needed is high. The following can be obtained:

— identical particles **(atoms)**, in the case of an **element**;
— different particles, for a chemical compound.

The atom is composed of a group of elementary particles, and its diameter is of the order of 0.1 to 0.5 nanometre (nm).

The main elementary particles are **the electron, the proton** and **the neutron.**

The electron is a particle having a negative electrical charge $e^- = 1.6 \times 10^{-19}$ coulombs. Its mass when at rest is 9.107×10^{-28} g. The electron possesses an angular moment (spin) caused by its rotation around its axis. This causes a dipolar magnetic moment responsible for the properties of the majority of para- and ferro-magnetic substances.

The proton is a positively charged particle ($e^+ = 1.6 \times 10^{-19}$ coulombs) with a mass 1 836 times that of the electron.

The neutron is an electrically-neutral particle with a mass almost equal to the mass of the proton.

An **atom** consists of:

— A **nucleus** containing Z protons and N neutrons. Its electrical charge is Ze^+ and its size is 10^{-5} nm (approx.).

The term **nucleon** is used either for a proton or a neutron. The nucleus, which is the heavy part of the atom, is chemically stable.

— A cloud of Z **electrons**, electrical charge: Ze^-.

In an electrical-neutral atom, the electrons attracted by the nucleus are moving rapidly around the nucleus in a space of a fraction of a nanometre.

● **Atomic number — Isotopes**

The **atomic number** is the number of protons in the nucleus. The **mass number** is the sum of $Z + N$ (number of nucleons in the nucleus). The **atomic mass** is equal to the mass number multiplied by the mass of the nucleon.

Two atoms differing only in the number of neutrons are called **isotopes.**

Isotopic atoms have the same chemical properties because they have the same number of protons. Only their chemical kinematics are different.

Isotopes are distinguished by writing the approximate atomic mass above and to the left of the atom symbol:

Example: ^1H light hydrogen

^2H heavy hydrogen (deuterium)

^{35}Cl and ^{37}Cl.

● **Atomic mass of elements — The mole**

The mass of an atom is expressed as the sum of the masses of the nucleons; this approximation is sufficient in conventional chemistry. The numerical value of atomic masses is very small and inconvenient to use, and so the mass of a quantity of atoms is used conventionally. The nuclide ^{12}C was chosen as the basis for atomic mass relationship at the 14th General Weights and Measures Conference, held in 1971.

The mass of a mole of an element is used instead of its atomic mass. A *mole* is defined as the amount of substance in a system containing the same number of elementary entities as there are atoms in 12 g of carbon ^{12}C (symbol mol). When the mole is used, the elementary entities must be specified; they may be atoms, molecules, ions, electrons, other particles, or specified groups of such particles.

The mole, as the unit of amount of substance —just as the kilogram is the unit of mass— has been one of the basic units of the SI system since 1917. The number of atoms contained in one mole of an element is 6.023×10^{23} (Avogadro's number).

The term mole replaces the terms gram-atom, gram-molecule and gram-ion; it is specified that the mass of one mole of carbon ^{12}C atoms is exactly 12 g, that the mass of one mole of chlorine atoms is 35.453 g, and that the amount of substance contained in 1 kg of H_2O is 55.533 mol.

The molar mass of a substance is defined as the mass of this substance divided by the amount of substance contained in it; it is expressed in grams per mole (g/mol).

$$* \atop {* \ *}$$

The properties of chemical elements are not haphazard but depend on the electron structure of the atom: two different atoms with identical peripheral

	Ia	IIa	IIIb	IVb	Vb	VIb	VIIb	VIII			Ib	IIb	IIIa	IVa	Va	VIa	VIIa	O
1	1 **H** 1.0080																	2 **He** 4.00260
2	3 **Li** 6.941	4 **Be** 9.01218											5 **B** 10.81	6 **C** 12.011	7 **N** 14.0067	8 **O** 15.9994	9 **F** 18.9984	10 **Ne** 20.179
3	11 **Na** 22.9898	12 **Mg** 24.305											13 **Al** 26.9815	14 **Si** 28.086	15 **P** 30.9738	16 **S** 32.06	17 **Cl** 35.453	18 **Ar** 39.948
4	19 **K** 39.102	20 **Ca** 40.08	21 **Sc** 44.9559	22 **Ti** 47.90	23 **V** 50.9414	24 **Cr** 51.996	25 **Mn** 54.9380	26 **Fe** 55.847	27 **Co** 58.9332	28 **Ni** 58.71	29 **Cu** 63.546	30 **Zn** 65.37	31 **Ga** 69.72	32 **Ge** 72.59	33 **As** 74.9216	34 **Se** 78.96	35 **Br** 79.904	36 **Kr** 83.80
5	37 **Rb** 85.4678	38 **Sr** 87.62	39 **Y** 88.9059	40 **Zr** 91.22	41 **Nb** 92.9064	42 **Mo** 95.94	43 **Tc** 98.9062	44 **Ru** 101.07	45 **Rh** 102.9055	46 **Pd** 106.4	47 **Ag** 107.868	48 **Cd** 112.40	49 **In** 114.82	50 **Sn** 118.69	51 **Sb** 121.75	52 **Te** 127.60	53 **I** 126.9045	54 **Xe** 131.30
6	55 **Cs** 132.9055	56 **Ba** 137.34	57 **La** 138.9055	72 **Hf** 178.49	73 **Ta** 180.9479	74 **W** 183.85	75 **Re** 186.2	76 **Os** 190.2	77 **Ir** 192.22	78 **Pt** 195.09	79 **Au** 196.9665	80 **Hg** 200.59	81 **Tl** 204.37	82 **Pb** 207.2	83 **Bi** 208.9806	84 **Po** (209)	85 **At** (210)	86 **Rn** (222)
7	87 **Fr** (223)	88 **Ra** 226.0254	89 **Ac** (227)															

58 **Ce** 140.12	59 **Pr** 140.9077	60 **Nd** 144.24	61 **Pm** (145)	62 **Sm** 150.4	63 **Eu** 151.96	64 **Gd** 157.25	65 **Tb** 158.9254	66 **Dy** 162.50	67 **Ho** 164.9303	68 **Er** 167.26	69 **Tm** 168.9342	70 **Yb** 173.04	71 **Lu** 174.97
90 **Th** 232.0381	91 **Pa** 231.0359	92 **U** 238.029	93 **Np** 237.0482	94 **Pu** (244)	95 **Am** (243)	96 **Cm** (247)	97 **Bk** (247)	98 **Cf** (251)	99 **Es** (254)	100 **Fm** (253)	101 **Md** (256)	102 **No** (254)	103 **Lr** (256)

Atomic number
SYMBOL
Atomic mass

Periodic table of the elements.

layers have similar chemical properties (the alkali metal group, the halogen group, etc.).   These similarities form the basis of the periodic table of the elements (Mendeleyev's table — see page 869).

● **Gram equivalent**

The gram equivalent or equivalent weight is equal to the mass of a unit of amount of substance (one mole) of a substance divided by the number of charges of the same sign carried by the ions released by an elementary particle of that substance in aqueous solution.

For instance, a molecule of orthophosphoric acid (H_3PO_4) releases three positive charges and three negative charges.   One gram equivalent of H_3PO_4 is therefore equal to one third of the mass of one mole of H_3PO_4.

The gram equivalent is no longer included in the latest publications of the International Standards Organization (ISO 1973) or of the International Union of Pure and Applied-Chemistry (IUPAC 1975).   It is, however, retained in this book because of its value in routine water treatment analyses and determinations.

● **Formulae for compounds**

Compounds are represented by formulae using the symbols for their constituent elements.   Where the number of atoms of the elements making up the compound differs, whole-number indices are assigned to their symbols to represent the numerical proportions of the different atoms.   These figures are written as subscripts to the right of the symbols; e.g., H_2O: two hydrogen atoms and one oxygen atom.

A formula showing how particular groups of atoms are combined in the molecule is often used for organic compounds.   For example, the formula CH_3COOH is used for acetic acid to show that a CH_3 group is attached to a carbon atom to which are also attached one oxygen atom and an OH group.

● **Rules for writing formulae**

The complete rules of inorganic chemistry nomenclature laid down by the International Union of Pure and Applied Chemistry were published inter alia by the Société Chimique de France (special bulletin dated February 1975).

Some of the basic rules are given below:

2.15

In formulae, the electropositive component (cation) must always be written first, e.g., KCl, $CaSO_4$.   If the compound contains more than one electropositive or electronegative component, the sequence within each class should be the same as the alphabetical order of the symbols.

Acids are treated as hydrogen salts, e.g., H_2SO_4 and H_2PtCl_6; see Section 6.2 and 6.32.3 for the position of the hydrogen.

2.16.1

With binary compounds between non-metallic elements, the component to be written first is that shown first in the following list: B, Si, C, Sb, As, P, N, H, Te, Se, S, At, I, Br, Cl, O, F.

Example: NH_3, H_2S, SO_2, ClO_2, OF_2.

● **Molar mass of the main salts**

Compound		Formula	Molar mass g/mol
Aluminium	sulphate	$Al_2 (SO_4)_3, 18 H_2O$	666.4
Ammonium	nitrate	$NH_4 NO_3$	80.0
—	nitrite	$NH_4 NO_2$	64.0
—	sulphate	$(NH_4)_2 SO_4$	132.1
Barium	hydroxide (baryta)	$Ba (OH)_2, 8 H_2O$	315.5
—	carbonate	$Ba CO_3$	197.4
—	chloride	$Ba Cl_2, 2 H_2O$	244.3
Calcium	carbonate	$Ca CO_3$	100.1
—	hydrogen carbonate	$Ca (HCO_3)_2$	162.1
—	chloride	$Ca Cl_2, 6 H_2O$	219.1
—	sulphate	$Ca SO_4, 2 H_2O$	172.2
Copper	sulphate	$Cu SO_4, 5 H_2O$	249.7
Iron	chloride (ferric)	$Fe Cl_3, 6 H_2O$	270.3
—	sulphate (ferrous)	$Fe SO_4, 7 H_2O$	278.0
—	— (ferric)	$Fe_2 (SO_4)_3, 9 H_2O$	562.0
Lead	carbonate	$Pb CO_3$	267.2
—	sulphate	$Pb SO_4$	303.2
Magnesium	carbonate	$Mg CO_3$	84.3
—	chloride	$Mg Cl_2, 6 H_2O$	203.3
—	sulphate	$Mg SO_4, 7 H_2O$	246.5
Manganese	hydroxide (manganous)	$Mn (OH)_2$	89.0
—	carbonate	$Mn CO_3$	115.0
Potassium	nitrate	$K NO_3$	101.1
—	permanganate	$K MnO_4$	158.0
—	phosphate	$K_3 PO_4$	212.3
Silver	chloride	$AgCl$	143.3
Sodium	aluminate	$Na_2 AlO_4$	137.0
Sodium	hydrogen carbonate	$NaHCO_3$	84.0
—	carbonate	$Na_2 CO_3$	106.0
—	—	$Na_2 CO_3, 10 H_2O$	286.1
—	chloride	$Na Cl$	58.4
—	disodium hydrogen orthophosphate	$Na_2HPO_4, 12 H_2O$	358.1
—	trisodium orthophosphate	$Na_3 PO_4, 12 H_2O$	380.1

The molar mass of other compounds also used for water treatment is given on page 894.

6.2 — Salts containing acid hydrogen atoms
The names of these salts are formed by preceding the name of the anion with
the word hydrogen to show that substitutable hydrogen is present in the salt.
These salts cannot, of course, be called acid salts.

Examples: $NaHCO_3$ sodium hydrogen carbonate
        LiH_2PO_4 lithium dihydrogen phosphate
        KHS   potassium hydrogen sulphide

6.3 — Double+and triple salts

6.31 — Cations:
All cations must predece anions in formulae.

6.32.1
Cations other than hydrogen must be written in alphabetical order; the order
may differ as between formulae and names.
Example: $KNaCO_3$: sodium potassium carbonate.

6.32.3 — Acid hydrogen:
Hydrogen is written last among the cations when Rule 6.2 is not applied.
Example: $NaNH_4HPO_4.4H_4O$: sodium ammonium hydrogen phosphate with
four molecules of water.

6.33 — Anions:
Anions must be written in alphabetical order; the order may differ as between
formulae and names.

● **Properties of electrolytes in solution — Dissociation**
    A knowledge of chemical equilibria in an aqueous medium is one of the
fundamentals of the science of water treatment.
    An inorganic compound dissolved in water is dissociated to a greater or
esser extent, with the formation of negatively charged ions (cations). The
dissolved substance is called an electrolyte and facilitates the flow of electric
current:

$$AB \rightleftharpoons A^+ + B^-$$

    Where a single solution contains a number of electrolytes, each is dissociated
to a certain extent and the ions formed may combine with one another to form
new compounds. For example, if two compounds AB and CD are dissolved,
the solution will be found to contain molecules AB, CD, AD and CB in equilibrium with ions A^+, B^-, C^+ and D^-. The equilibrium may change if insoluble
compounds, complexes or gases form (Le Chatelier's law). For example, if
compound AD is insoluble, the equilibrium is almost entirely displaced to the
right, in accordance with the reaction:

$$AB + CD \rightleftharpoons AD + CB$$
$$CB \rightleftharpoons C^+ + B^-$$

● Common commercial form of the principal reagents

	Commercial name	Pure compound formula	Density	Degrees Baumé	Concentration % by weight or g/l
Acids	Hydrochloric acid............	HCl	1.18	22	35.4% HCl
	Nitric acid.................	HNO₃	1.35-1.36	38	56-58 % HNO₃
	Phosphoric acid.............	H₃PO₄	1.45	45	45 % P₂O₅
	Phosphoric acid.............	H₃PO₄	1.72	60	65 % P₂O₅
	Sulphuric acid..............	H₂SO₄	1.83	65.5	92.3% H₂SO₄
	Sulphuric acid..............	H₂SO₄	1.84		98 % H₂SO₄
Bases	Lime, calcium oxide.........	CaO	quicklime (rock or powder)		80-95 % CaO
	Lime, calcium hydroxide......	Ca(OH)₂	sieved, powder form		85-97 % Ca(OH)₂
	Caustic soda................	NaOH	flake-pebble		98 % NaOH
	Caustic soda — 35% solution ..		1.38	40	485 g/l NaOH
	Caustic soda — 41% solution ..		1.45	45	590 g/l NaOH
	Caustic soda — 47% solution ..		1.50	48.5	715 g/l NaOH
Clarifying reagents	Ferric chloride..............	FeCl₃, 6H₂O	crystallized		60 % FeCl₃
	Ferric chloride..............	FeCl₃	crystallized		99 % FeCl₃
	Ferric chloride, 39-41% solution	FeCl₃	1.41-1.45	41-45	596 g/l FeCl₃
	Ferric chlorosulphate........	FeClSO₄	1.6		594 g/l FeCl₃
	Basic aluminium chloride (WAC)...................	Alₙ(OH)ₘCl₃ₙ₋ₘ	1.2		10 % Al₂O₃
	Aluminium sulphate.........	Al₂(SO₄)₃, 18 H₂O	crystallized		17.2% Al₂O₃
	Aluminium sulphate, 7.5-8.5% solution............		1.3	34	8.3% Al₂O₃
	Sodium silicate.............	nSiO₂, Na₂O	crystallized		60 % SiO₂
	Sodium silicate, 25% solution..		1.33	36	330 g/l SiO₂
	Sodium silicate, 28% solution..		1.37	39	380 g/l SiO₂
Carbonate removal, silica removal	Sodium aluminate ....⎰	1.4 Na₂O, Al₂O₃, nH₂O	1.48	46	21 % Al₂O₃
	Sodium aluminate ....⎱		crystallized		40 % Al₂O₃
	Sodium carbonate...........	Na₂CO₃	powder		99 % Na₂CO₃
	Sodium carbonate...........	Na₂CO₃, 10 H₂O	crystals		36 % Na₂CO₃
	Magnesium chloride.........	MgCl₂, 6 H₂O	crystallized		19 % MgO
	Magnesia	MgO	powder		98 % MgO
Disinfection	Sodium chlorite.............	NaClO₂	crystallized		80 % NaClO₂
	Sodium chlorite, 25% solution.		1.15	20	375 g/l active Cl₂
					300 g/l NaClO₂
	Sodium chlorite, 7,5% solution.		1.05	7	112 g/l active Cl₂
					90 g/l NaClO₂
	Sodium hypochlorite 47-50 chlorometric degrees....	NaOCl		30	149-159 g/l active Cl₂ approximately
	Calcium hypochlorite........	Ca(OCl)₂	crystallized		600 g/kg active Cl₂
Conditioning	Sodium bisulphite...........	NaHSO₃	1.32	35	23-24 % SO₂
	Sodium metabisulphite.......	Na₂S₂O₅	crystallized		60-62 % SO₂
	Sodium phosphate, dibasic....	Na₂HPO₄, 12 H₂O	crystallized		20 % P₂O₅
	Sodium phosphate, tribasic....	Na₃PO₄, 12 H₂O	crystallized		20 % P₂O₅
	Sodium phosphate, tribasic, anhydrous	Na₃PO₄	crystallized		40 % P₂O₅
	Sodium sulphite.............	Na₂SO₃, 7 H₂O	crystallized		23-24 % SO₂
	Sodium sulphite, anhydrous...	Na₂SO₃	crystallized		48 % SO₂
Miscellaneous	Aluminium chloride.........	AlCl₃, 6 H₂O	crystallized		20 % Al₂O₃
	Sea salt....................	NaCl	crystallized		97 % NaCl

● **Law of mass action**

In the case of a chemical reaction at equilibrium:

$$mA + nB \overset{1}{\underset{2}{\rightleftarrows}} m'C + n'D$$

The reaction velocity in direction 1 is given by the relationship:

$$V_1 = k_1 (A)^m (B)^n$$

The velocity in direction 2 is: $V_2 = k_2 (C)^{m'} (D)^{n'}$

At equilibrium $V_1 = V_2$, so that $k_1 (A)^m (B)^n = k_2 (C)^{m'} (D)^{n'}$

$$\frac{(A)^m (B)^n}{(C)^{m'} (D)^{n'}} = \frac{k_2}{k_1} = k$$

This is the law of mass action. K is called the thermodynamic dissociation constant; (A), (B), (C) and (D) represent the activities of the compounds in solution.

● **Titrated solutions**

To facilitate calculation in volumetric chemical determinations, the "normal solution" has been chosen as the unit of concentration.

A **normal solution** is one containing one gram equivalent of the relevant substance per litre.

Multiples and submultiples of the normal solution are also used (2N, N/10, N/25, N/50, N/100, etc., solutions).

In general, when a volume V_1 of an electrolyte of normality N_1 is acted upon by another electrolyte of normality N_2, the volume V_2 is determined from the relation:

$$N_1 V_1 = N_2 V_2$$

● **Units of concentration**

The unit often used in practice is the **milliequivalent per litre (meq/l),** which is obtained by dissolving a quantity of the electrolyte equal to one thousandth of its gram equivalent in one litre of water. This is the concentration of an N/1 000 solution.

The **French degree,** which corresponds to the concentration of an N/5 000 solution, is used in practical water treatment.

Example: A solution of a calcium salt at 25 French degrees,

i.e., $\dfrac{25}{5} = 5$ meq/l contains $\dfrac{40 \times 5}{2 \times 1\,000} = 0.1$ per litre of calcium Ca (of molar mass 40 g/mol and valency 2).

● **Measurement of pH** (see definition on page 12)

a) Electrical measurement: the pH is measured by means of a glass electrode/ reference electrode pair immersed in the liquid to be analysed; the potential difference of the resulting cell depends on the pH of the solution. The potential difference is displayed on high input impedance electronic millivoltmeters graduated direct in pH units. These instruments must be calibrated regularly with reference solutions; the most commonly used reference solutions are the following buffer solutions:

	pH
— HCl, 0.1 N(H_3O^+/H_2O buffer)...............................	1
— HCl, 0.01 N...	2
— Chloroacetic acid, 1N....................................	2.9
— Acetic acid, 1N...	4.8
— Saturated $NaHCO_3$ + saturated CO_2..........................	7.4
— (NH_4Cl + NH_3), 1N....................................	9.2
— NaOH, 0.01 N (H_2O/OH buffer)............................	12
— NaOH, 0.1 N...	13

The pH of the buffer solutions can be adjusted by varying the relative proportions of the acid and the base; it can always be calculated from the relationship:

$$pH = pK + \log \frac{\text{base concentration}}{\text{acid concentration}}$$

b) Use of coloured indicators: the pH can be determined by coloured indicators whose colour can be compared with that of a set of standard colours of known pH values. There are also "universal" papers and indicators consisting of mixtures of indicators, whereby the pH can be measured with an accuracy of within one unit.

Note: Addition of the coloured indicator modifies the pH in only slightly mineralized waters; the electrical method is preferable in such cases.

Principal coloured indicators

Name	Solution use conc. in	Colour acid — alkaline	pH at colour change
Thymol blue..............	1 °/$_{00}$, alcohol ethyl	red-yellow	1.2 — 2.8
Methyl orange or helianthin .	1 °/$_{00}$ water	red-orange-yellow	3.1 — 4.4
Bromophenol blue	1 °/$_{00}$, water	yellow-blue	3.0 — 4.6
Bromocresol green	1 °/$_{00}$, alcohol	yellow-blue	3.8 — 5.4
Methyl red	2 °/$_{00}$, alcohol	red-yellow	4.2 — 6.2
Chlorophenol red	1 °/$_{00}$, alcohol	yellow-red	4.8 — 6.4
Bromothymol blue	1 °/$_{00}$, alcohol	yellow-blue	6.0 — 7.6
Phenol red	1 °/$_{00}$, alcohol	yellow-red	6.4 — 8.0
Cresol red	1 °/$_{00}$, alcohol	red-yellow/brown	7.2 — 8.8
Thymol blue..............	1 °/$_{00}$, alcohol	yellow-blue	8.0 — 9.6
Phenolphthalein............	1 °/$_{00}$, alcohol	colourless-carmine	8.2 — 9.4
Thymolphthalein	1 °/$_{00}$, alcohol	colourless-blue	9.3 — 10.5

● **Measurement of redox potential** (see definition on page 95)

The redox potential is measured with an electrode pair connected to an electronic millivoltmeter. One of the electrodes is usually made of non-corrodible material (platinum or gold) and the other is a reference electrode (normally KCl-saturated calomel).

The redox potential is expressed in volts.

The measured potential E_{Hg} (positive or negative relative to the calomel electrode) must be compared with the potential of the hydrogen electrode E_H; it should be remenbered that the former is positive (A = + 0.248 V at 20 ºC).

The rH (oxydo-reduction capacity) is calculated from an equation derived from Nernst's equation:

$$rH = \frac{E_H}{0.029} + 2\,pH = \frac{A \pm E_{Hg}}{0.029} + 2\,pH$$

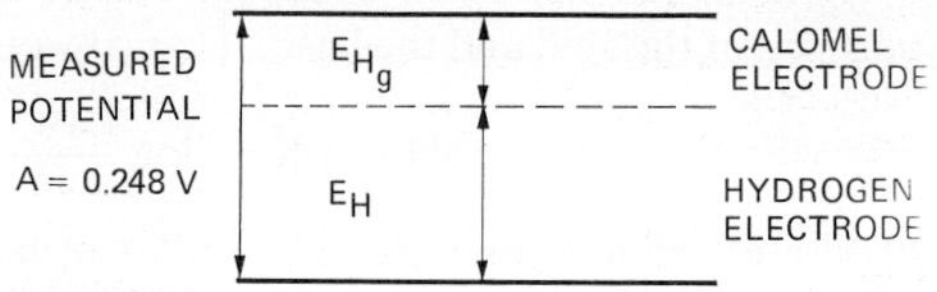

Fig. 550. — Measurement of the redox potential.

● **Measurement of resistivity or conductivity**

The resistivity of a solution is measured by a cell consisting of two electrodes made of non-corrodible material, connected to an indicator via an amplifier. Polarization is minimized by using an alternating signal of appropriate frequency.

Conductimetry cells have a constant which allows for the cell geometry. They must be recalibrated periodically with N/50 or N/100 potassium chloride solutions.

	Resistivity in Ω cm	
Temperature ºC	N/50 KCl	N/100 KCl
15	446	872
16	436	852
17	426	834
18	417	817
19	408	800
20	400	782
21	392	766
22	384	751
23	376	736
24	369	721
25	362	708

$$\text{Cell constant } K = \frac{\text{theoretical resistivity}}{\text{measured resistivity}}$$

Resistivity is normally measured in ohm.cm; conductivity, the reciprocal of resistivity, is expressed in siemens/metre (the most common submultiple being the microsiemens/cm).

pk Table [1]
Constants for standard acid-base combinations in water at 25 °C

Acid name	Acid formula	Base formula	pK
Oxonium ion	H_3O^+	H_2O	− 1.74
Pyrophosphoric acid	$H_4P_2O_7$	$H_3P_2O_7^-$	0.85
Oxalic acid	$H_2C_2O_4$	$HC_2O_4^-$	1.2
Dichloroacetic acid	Cl_2CHCO_2H	$Cl_2CH_2CO_2$	1.3
Phosphorous acid	H_3PO_3	H_2PO_3	1.5 — 1.8
Hypophosphorous acid	H_3PO_2	$H_2PO_2^-$	1.7
Sulphurous acid	H_2SO_3	HSO_3^-	1.8
Hydrogen sulphate ion	HSO_4^-	SO_4^-	1.9
Trihydrogen pyrophosphate ion	$H_3P_2O_7$	$H_2P_2O_7^{2-}$	2.0
Orthophosphoric acid	H_3PO_4	$H_2PO_4^-$	2.2
Arsenic acid	H_3AsO_4	$H_2AsO_4^-$	2.3
Monochloroacetic acid	$ClCH_2CO_2H$	$ClCH_2CO_2^-$	2.9
Hydrofluoric acid	HF	F^-	3.2
Nitrous acid	HNO_2	NO_2^-	3.4
Formic acid	HCO_2H	HCO_2^-	3.7
Cyanic acid	$HCNO$	CNO^-	3.8
Hydrogen oxalate ion	$HC_2O_4^-$	$C_2O_4^{2-}$	4.1
Benzoic acid	$C_6H_5CO_2H$	$C_6H_5CO_2^-$	4.2
Anilinium ion	$C_6H_5NH_3^+$	$C_6H_5NH_2$	4.6
Hydrazoic acid	NH_3	N_3^-	4.7
Acetic acid	CH_3CO_2H	$CH_3CO_2^-$	4.8
Aluminium ion	Al^{2+} aq.	$AlOH^{2+}$ aq.	4.9
Hexamethylenetetramine ion	$(CH_2)_6 N_4H^+$	$(CH_2)_6 N_4$	4.9
Pyridinium ion	$C_5H_5NH^+$	C_5H_5N	5.15
Hydrogen sulphite ion	HSO_2^-	SO_3^{2-}	7.1
Hydroxylamine ion	NH_3OH^+	NH_2OH	6.0
Carbonic acid	CO_2 aq.	HCO_3^-	6.4
Dihydrogen pyrophosphate ion	$H_2P_2O_7^{2-}$	$H_2P_2O_7^{3-}$	6.7
Dihydrogen phosphite ion	$H_2PO_3^-$	$HP_2O_3^{2-}$	6.2 — 6.6
Dihydrogen arsenate ion	H_2AsO^-	$HAsO_4^{2-}$	4.4 — 7.1
Hydrogen dichromate ion	$HCrO_4^-$	CrO_4^{2-}	6.4
Hydrogen sulphide	H_2S	HS^-	7.1
Dihydrogen (ortho) phosphate ion	$H_2PO_4^-$	HPO_4^{2-}	7.2
Hypochlorous acid	$HClO$	ClO^-	7.3
Monohydrogen pyrophosphate ion	$HP_2O_7^{3-}$	$P_2O_7^{4-}$	8.5
Hydrocyanic acid	HCN	CN^-	9.1
Hydrogen arsenate ion	$HAsO_4^{2-}$	AsO_4^{3-}	9.2 — 12.0
Arsenious acid	$HAsO_2$	AsO_2	9.2
Ammonium ion	NH_4^+	NH_3	9.2
Orthoboric acid	HBO_2	BO_2^-	9.3
Hydrogen carbonate ion	HCO_3^-	CO_3^{2-}	10.2
Monohydrogen phosphate ion	HPO_4^{2-}	PO_4^{3-}	11.9 — 12.6
Hypoiodous acid	HIO	IO^-	11.0 — 12.2
Calcium ion	Ca^{2+} aq.	$CaOH^+$	12.6
Hydrogen sulphide ion	HS^-	S^{2-}	14.9
Water	H_2O	OH^-	15.7
Hydroxyl ion	HO^-	O^{2-}	24.0

1. See definition of pK: page 11.

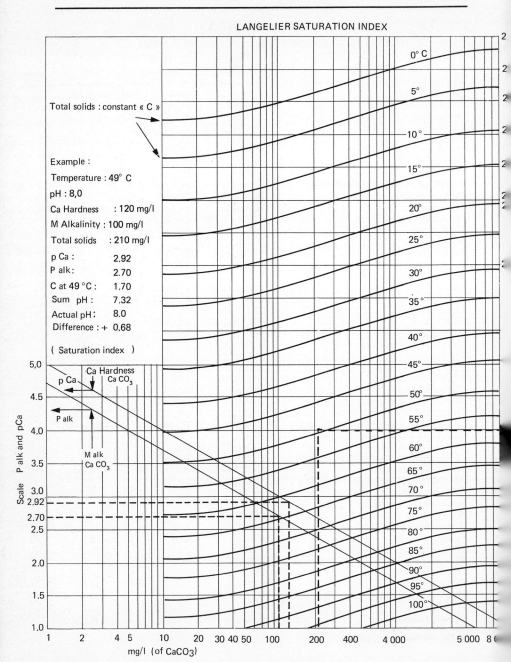

LANGELIER SATURATION INDEX

Total solids : constant « C »

Example :
Temperature : 49° C
pH : 8,0
Ca Hardness   : 120 mg/l
M Alkalinity : 100 mg/l
Total solids   : 210 mg/l

p Ca :   2.92
P alk :   2.70
C at 49 °C :   1.70
Sum pH :   7.32
Actual pH :   8.0
Difference : + 0,68

(Saturation index)

Fig. 551. — Langelier Diagram. From Water Conditioning for Industry, by Sheppard T. Powell. McGraw-Hill Book Co., 1954.

● **Viscosity of various liquids and gases at atmospheric pressure**

Description	Temperature (°C)	Kinematic viscosity v m²/s
Water	0	$1.8 \cdot 10^{-6}$
—	20	$1.0 \cdot 10^{-6}$
NaCl (Saturated brine)	0	$2.5 \cdot 10^{-6}$
—	10	$1.8 \cdot 10^{-6}$
Acetic acid at 100 %	20	$1.2 \cdot 10^{-6}$
Nitric acid at 95 %	0	$1.5 \cdot 10^{-6}$
—	10	$1.2 \cdot 10^{-6}$
Sulphuric acid, 66° Baumé	— 10	$4.4 \cdot 10^{-5}$
—	0	$2.6 \cdot 10^{-5}$
—	15	$1.7 \cdot 10^{-5}$
—	25	$1.3 \cdot 10^{-5}$
—	50	$6 \cdot 10^{-6}$
Hydrochloric acid, 20-21° Baumé	— 10	$2.6 \cdot 10^{-6}$
—	0	$2.2 \cdot 10^{-6}$
—	10	$2.0 \cdot 10^{-6}$
—	20	$1.7 \cdot 10^{-6}$
Soda lye, 49 %	15	$7.9 \cdot 10^{-5}$
—	20	$5.4 \cdot 10^{-5}$
—	25	$3.6 \cdot 10^{-5}$
Soda lye, 41 %	15	$4.5 \cdot 10^{-5}$
—	20	$3.4 \cdot 10^{-5}$
—	25	$1.7 \cdot 10^{-5}$
Sodium silicate, 38-40° Baumé	0	$5.5 \cdot 10^{-4}$
—	5	$2.9 \cdot 10^{-4}$
—	10	$2.05 \cdot 10^{-4}$
—	20	$1.13 \cdot 10^{-4}$
Aluminium sulphate, 34° Baumé	8	$1.38 \cdot 10^{-5}$
Ferric chloride, 45° Baumé	— 15	$2.5 \cdot 10^{-5}$
—	0	$1.0 \cdot 10^{-5}$
—	20	$3.0 \cdot 10^{-6}$
Air	0	$1.33 \cdot 10^{-5}$
—	20	$1.51 \cdot 10^{-5}$
—	100	$2.31 \cdot 10^{-5}$
Oxygen	20	$1.51 \cdot 10^{-5}$
Chlorine	20	$4.5 \cdot 10^{-6}$
Saturated steam	100 (1 bar)	$2.2 \cdot 10^{-5}$
	180 (10 bars)	$3.5 \cdot 10^{-6}$
	250 (40 bars)	$1.5 \cdot 10^{-6}$

NOTA: Some flocculation aids have a high viscosity: this should be ascertained from the supplier.
— 1 m²/s $= 10^4$ stokes.
— For calculation of the Reynolds number, see page 1052.

2. CHARACTERISTIC CONSTANTS OF SOLUTIONS

● **Relation between the density and the concentration of solutions of acids and bases (Gram of pure product per litre of solution at 15 °C)**

Baumé degrees	Density	H_2SO_4	HCl	HNO_3	NaOH	Ammonia Density	Ammonia NH_3
0	1.000	1.2	2	1.6	0.8	0.998	4.5
0.7	1.005	8.4	12	10.7	5	0.996	10
1.4	1.010	15.7	22	20.0	10	0.994	13.6
2.1	1.015	23	32	28	14	0.9915	19.8
2.8	1.020	31	42	38	19	0.990	22.9
3.5	1.025	39	53	47	23	0.9875	29.6
4.2	1.030	46	64	56	28	0.986	32.5
5.6	1.040	62	85	75	38	0.983	39.3
6.9	1.050	77	107	94	47	0.982	42.2
8.2	1.060	93	129	113	57	0.979	49
9.4	1.070	109	152	132	67	0.978	51.8
10.7	1.080	125	174	151	78	0.974	61.4
11.9	1.090	142	197	170	88	0.970	70.9
13.1	1.100	158	220	190	99	0.966	80.5
15.5	1.120	191	267	228	121	0.962	89.9
17.7	1.140	223	315	267	143	0.958	100.3
19.9	1.160	257	366	307	167	0.954	110.7
22.0	1.180	292	418	347	191	0.950	121
24.0	1.200	328	469	388	216	0.946	131.3
26.0	1.220	364		431	241	0.942	141.7
27.9	1.240	400		474	267	0.938	152.1
29.8	1.260	436		520	295	0.934	162.7
31.6	1.280	472		568	323	0.930	173.4
33.3	1.300	509		617	352	0.926	184.2
35.0	1.320	548		668	382	0.923	188
36.6	1.340	586		711	412	0.922	195.7
38.2	1.360	624		780	445	0.918	205.6
39.7	1.380	663		843	478	0.914	216.3
41.2	1.400	701		911	512	0.910	225.4
42.7	1.420	739		986	548	0.906	238.3
44.1	1.440	778		1 070	584	0.902	249.4
45.5	1.460	818		1 163	623	0.898	260.5
46.8	1.480	858		1 270	662	0.894	271.5
48.1	1.500	897		1 405	703	0.890	282.6
48.8	1.510	916		1 474	723		
49.4	1.520	936		1 508	744		
50.0	1.530	956			766		

● **Relation between the density and the concentration of solutions of salts (and of milk of lime)**
(Gram of pure product per litre of solution at 15 °C)

Baumé degrees	Density	Aluminium sulphate $Al_2(SO_4)_3$ 18 H_2O	Ferric chloride	Ferrous sulphate	Anhydrous sodium carbonate	Common salt	Sodium hypochlorite Cl (approx.)	Milk of lime (approx.)
1	1.007	14	10.1	13.1	6.3	10.1	2.8	7.5
2	1.014	28	20	26.4	13.1	20.5	5.5	16.5
3	1.021	42	29	40.8	19.5	30.5	8	26
4	1.028	57	37	55.5	29	41	10.5	36
5	1.036	73	47	70.5	35.4	51	13.5	46
6	1.044	89	57	85.5	41.1	62	16	56
7	1.051	103	66	102	50.8	73	18.5	65
8	1.059	119	76	116.5	58.8	85	21	75
9	1.067	135	86	132	67.9	97	23	84
10	1.075	152	96	147	76.1	109	25	94
11	1.083	168	106	163	85.0	121	27.5	104
12	1.091	184	116	179	93.5	134	30	115
13	1.099	200	126	196	101.2	147	32	126
14	1.108	218	138	213	110.6	160	34	137
15	1.116	235	150	230	122	174	36	148
16	1.125	255	162	247	131	187	38	159
17	1.134	274	174	265	141.5	200	40	170
18	1.143	293	186	284	150.5	215		181
19	1.152	312	198	304	162.5	230		193
20	1.161	332	210	324		248		206
21	1.170	351	222	344		262		218
22	1.180	373	236	365		277		229
23	1.190	395	250	387		292		242
24	1.200	417	263	408		310		255
25	1.210	440	279	430				268
26	1.220	462	293	452				281
27	1.230	485	308	474				295
28	1.241	509	323	501				309
29	1.252	534	338					324
30	1.263	558	353					339
32	1.285	609	384					
34	1.308	663	416					
36	1.332	720	449					
38	1.357		483					
40	1.383		521					
42	1.411		561					
44	1.437		601					
45	1.453		626					
46	1.468		650					

● Table of reagent solubilities [2]
(Grams of substance of given formula per litre of water)

Substance	Formula [1]	0 °C	10 °C	20 °C	30 °C
Aluminium sulphate	$Al_2(SO_4)_3, 18\ H_2O$	636	659	688	728
Calcium chloride	$CaCl_2$	595	650	745	1 020
Calcium sulphate	$CaSO_4, 2\ H_2O$	2.22	2.44	2.58	2.65
Copper sulphate	$CuSO_4, 5\ H_2O$	233	264	297	340
Ferric chloride	$FeCl_3$	744	819	918	—
	$FeCl_3, 6\ H_2O$	852	927	1 026	—
Ferrous sulphate	$FeSO_4, 7\ H_2O$	282	331	391	455
Potassium permanganate	$KMnO_4$	28	44	64	90
Ammonium phosphate	$NH_4H_2PO_4$	184	219	261	
	$(NH_4)_2H\ PO_4$	364	386	408	
	$(NH_4)_2H\ PO_4, 2\ H_2O$	340	388		
Ammonium sulphate	$(NH_4)_2SO_4$	413	420	428	
Sodium carbonate	$Na_2CO_3, 10\ H_2O$	250	305	395	568
Sodium chloride	$NaCl$	357	358	360	363
Sodium fluoride	NaF	40	—	42.2	—
Sodium hydrogen carbonate	$NaHCO_3$	69	81.5	96	111
Monosodium phosphate	$NaH_2PO_4, 2\ H_2O$	615	735	888	1 010
Crystalline disodium phosphate	$Na_2HPO_4, 12\ H_2O$	233	252	293	424
Anhydrous trisodium phosphate	Na_3PO_4	15	41	110	200
	$Na_3PO_4, 12\ H_2O$	231	257	326	416
Sodium hydroxide	$NaOH$	420	515	1 090	1 190

1. When the commercial form of a reagent has a formula different from that given above, its solubility must be re-calculated.
2. These solubilities apply to substances dissolved in water in the absence of other salts.

● Solubility of lime

Temperature: °C	0	10	20	30	40	50	60	70	80	90	100
CaO g/l	1.40	1.33	1.25	1.16	1.06	0.97	0.88	0.80	0.71	0.64	0.5
$Ca(OH)_2$ g/l	1.85	1.76	1.65	1.53	1.41	1.28	1.16	1.06	0.94	0.85	0.7
Titration of the lime water in TAC degrees	250	238	223	207	190	173	157	143	127	115	104

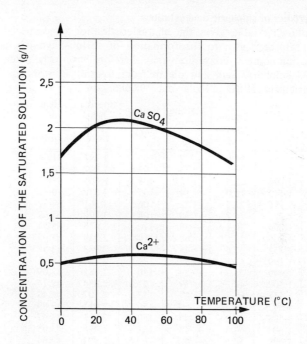

Fig. 552. — Solubility of calcium sulphate in pure water in relation to the temperature.

● **Composition of sea water**

Standard sea water defined by the Copenhagen hydrographic laboratory, Denmark.

	Cations			Anions	
	mg/l	meq/l		mg/l	meq/l
Na^+	11 035	479.8	Cl^-	19 841	559.6
Mg^{2+}	1 330	109.4	SO_4^{2-}	2 769	57.6
Ca^{2+}	418	20.9	HCO_3^-	146	2.5
K^+	397	10.2	Br^-	68	0.9
Sr^+	13.9	0.3	F^-	1.4	

Total salinity: 36.047 g/l
Total alkalinity: 119.8 mg/l

● Characteristics of sulphuric acid solutions

The following table gives the characteristics for concentrated solutions. Attention must be drawn to the formation of various hydrates with widely varying freezing points. Precautions may therefore have to be taken to prevent concentrated sulphuric acid and oleum from freezing in storage.

Sulphuric acid (H_2SO_4): Molar mass 98.086 g/mol.

Concentration % H_2SO_4	Volume mass kg/l 15 °C	Baumé degrees	Melting point °C	Boiling point °C	Specific heat kJ/kg 18 °C	Heat of solution kJ/mole	Formula
5	1.033	4.6	− 2	101	3.992	71.06	
10	1.068	9.1	− 5	102	3.857	70.22	
15	1.104	13.5	− 8	103.5	3.666	68.97	
20	1.142	18.0	− 14	105	3.532	70.22	
25	1.182	22.2	− 22	106.5	3.361	67.72	
30	1.222	26.2	− 36	108	3.200	65.62	
35	1.264	30.1	− 58	110	3.026	63.54	
40	1.306	33.6	− 68	114	2.830	61.03	
45	1.352	37.6	− 48	116.5	2.688	58.52	
50	1.399	41.1	− 37	124	2.533	56.01	
55	1.449	44.7	− 29	130	2.383	53.09	
60	1.502	48.2	− 29	141.5	2.236	49.74	
65	1.558	51.7	− 47	153.5	2.107	46.40	
68	1.592	53.7	− 50	159	2.015	43.89	
68.5	1.598	54.0	− 45	160.5	2.011	43.47	
69	1.604	54.3	− 43	162	1.998	43.05	
69.5	1.609	54.6	− 42	163.5	1.994	42.64	
70	1.615	54.9	− 41	165	1.985	42.22	
70.5	1.621	55.3	− 40	166.5	1.977	41.80	
71	1.627	55.6	− 40.5	168	1.969	41.38	
71.5	1.633	55.9	− 40	169.5	1.965	40.96	
72	1.639	56.3	− 39	171.5	1.956	40.55	
72.5	1.644	56.5	− 39	173	1.948	40.13	
73	1.650	56.8	− 39	175	1.939	39.71	
73.5	1.656	57.2	− 39.5	176.5	1.935	39.29	
74	1.662	57.5	− 40	178	1.931	38.87	
74.5	1.668	57.8	− 40	180	1.923	38.46	
75	1.674	58.1	− 41	182	1.873	38,04	
80	1.733	61.0	− 1	202	1.772	32.60	
85	1.784	63.4	+ 7	225	1.735	27.17	H_2SO_4
90	1.820	65.0	− 6	255	1.659	18.39	$+ H_2O$
91	1.825	65.2	− 11	268	1.597	16.72	
92	1.829	65.4	− 24	274.5	1.584	14.63	
93	1.833	65.6	− 38	281.5	1.513	12.54	$2H_2SO_4$
94	1.836	65.7	− 28	288.5	1.496	10.87	$+ H_2O$
95	1.839	65.8	− 19	295	1.484	8.78	
96	1.8406	65.9	− 11	307.5	1.450	7.11	
97	1.8414	65.94	− 5	317.5	1.434	5.02	
98	1.8411	65.92	+ 0	326	1.404	2.93	
99	1.8393	65.85	+ 6	333	1.409	1.25	
100	1.8357	65.72	+ 10.4	338	1.400	—	H_2SO_4
17 % SO_3	1.9000		− 11				$4H_2SO_4 + SO_3$
45 % SO_3	2.000		+ 35				$H_2SO_4 + SO_3$
60 % SO_3	2.020		0				$H_2SO_4 + 2SO_3$
100 % SO_3	1.980		+ 40				SO_3

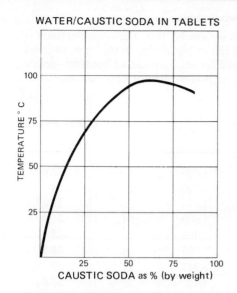

WATER/CAUSTIC SODA IN TABLETS

TEMPERATURE ° C

CAUSTIC SODA as % (by weight)

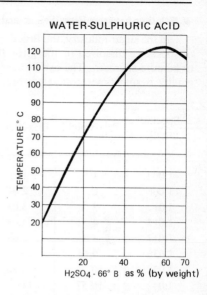

WATER-SULPHURIC ACID

TEMPERATURE ° C

H_2SO_4 - 66° B as % (by weight)

Fig. 553. —   Fig. 554. —
Exothermic reactions (maximum temperature reached during preparation of the mixture).

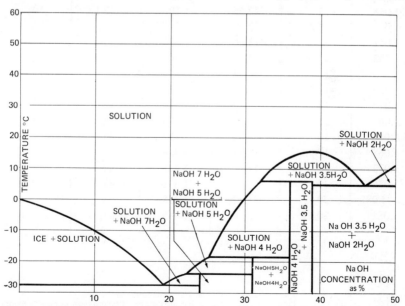

Fig. 555. — *Solubility diagram of caustic soda.*

● Relation between Brix-Weight and Brix-Volume

In the sugar industry, the Brix degree (or simply "Brix") indicates the quantity of substance dissolved in 100 g (or 100 ml) of liquid; the substance is assumed to be pure sucrose. The hydrometer used for measurement of apparent specific gravity is sometimes graduated directly in Brix-weights or Brix-volumes.

Specific gravity	Brix-weight	Brix-volume	Specific gravity	Brix-weight	Brix-volume
1.002	1	1	1.085	21	22.8
1.006	2	2.01	1,090	22	24
1.01	3	3.03	1.094	23	25.17
1.014	4	4.08	1.099	24	26.4
1.017	5	5.09	1.103	25	27.6
1.036	9.65	10	1.108	26	28.8
1.038	10	10.38	1.113	27	30
1.04	10.57	11	1.127	30	33.8
1.042	11	11.47	1.15	34.8	40
1.044	11.5	12	1.175	40	47.6
1.046	12	12.56	1.19	42.1	50
1.048	12.4	13	1.225	49	60
1.051	13	13.66	1.23	50	61.5
1.052	13.32	14	1.26	55.5	70
1.055	14	14.77	1.286	60	77.2
1.056	14.22	15	1.316	65	85.6
1.059	15	15.89	1.347	70	94.4
1.060	15.10	16	1.378	75	103.5
1.063	16	17,02	1.412	80	
1.068	17	18.15	1.418	81	
1.072	18	19.3	1.425	82	
1.076	19	20.45	1.431	83	
1.081	20	21.62	1.438	84	
1.082	20.33	22	1.445	85	

NOTE: The values in the table have been chosen to include the concentrations most commonly found in the treatment of sugar-containing liquids.

3. CHARACTERISTIC CONSTANTS OF GASES

● **Density and specific mass of gases**

	Density in relation to air	Masse per litre at 0 °C and 760 mm Hg in grammes
Air.................	1	1,29349
Oxygen (O_2)........	1.1052	1.4295
Nitrogen (N_2).......	0.967	1.2508
Hydrogen (H_2)......	0.06948	0.08987
Carbon dioxide (CO_2)	1.5287	1.978
Chlorine (Cl_2).......	2.491	3.222
Ammonia (NH_3)....	0.5971	0.772
Sulphur dioxide (SO_2)	2.263	2.927
Hydrogen sulphide (H_2S)	1.1895	1·539

If M_0 is the mass per litre at 0 °C, the mass per litre at t °C, at the same pressure, is:

$$M_t = \frac{M_0}{1 + 0.00367\ t}$$

If M'_o is the mass per litre at 760 mm of mercury, the mass per litre at pressure P is:

$$M_p = \frac{P}{760}\ M'_o$$

● **Solubility of CO_2, O_2 and O_3 in water (milligrams of gas per litre of water at atmospheric pressure, under an atmosphere of the pure gas)**

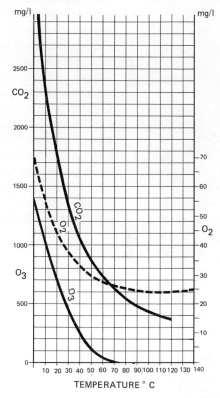

Fig. 556. —

● Solubility of gases in water

(Litres of gas at NTP per litre of water under an atmosphere of the pure gas at a pressure of 1 bar)

Température °C	Gas									
	Air	O_2	N_2	H_2	CO_2	H_2S	Cl_2	NH_3	SO_2	O_3
0	0.0373	0.0489	0.0235	0.0215	1.713	4.621	4.61	1 135	75.00	0.640
5	0.0330	0.0429	0.0208	0.0204	1.424	3.935	3.75	1 005	62.97	0.571
10	0.0293	0.038	0.0186	0.0196	1.194	3.362	3.095	881	52.52	0.502
15	0.0265	0.0342	0.0168	0.0188	1.019	2.913	2.635	778	43.45	0.432
20	0.0242	0.0310	0.0154	0.0182	0.878	2.554	2.260	681	36.31	0.331
25	0.0223	0.0283	0.0143	0.0175	0.759	2.257	1.985	595	30.50	0.273
30	0.0208	0.0261	0.0134	0.0170	0.665	2.014	1.769	521	25.87	0.207
35	0.0195	0.0244	0.0125	0.0167	0.592	1.811	1.570	460	22.00	0.151
40	0.0184	0.0231	0.0118	0.0164	0.533	1.642	1.414	395	18.91	0.103
50	0.0168	0.0209	0.0109	0.0161	0.437	1.376	1.204	294	15,02	0.045
60	0.0157	0.0195	0.0102	0.0160	0.365	1.176	1.006	198	11.09	
70	0.0150	0.0183	0.0097	0.0160	0.319	1.010	0.848		8.91	
80	0.0146	0.0176	0.0096	0.0160	0.275	0.906	0.672		7.27	
90	0.0144	0.0172	0.0095	0.0160	0.246	0.835	0.380		6.16	
100	0.0144	0.0170	0.0095	0.0160	0.220	0.800				
110		0.0168			0.204					
120		0.0169			0.194					
130		0.0170								
140		0.0172								

● **Solubility in water of atmospheric gases (g/m³) at atmospheric pressure**

A. *Solubility of nitrogen.*
B. *Solubility of oxygen.*
a. *Partial pressure of nitrogen.*
b. *Partial pressure of oxygen.*
c. *Partial pressure of water vapour.*
d. *Partial pressure of air.*

Fig. 557. —

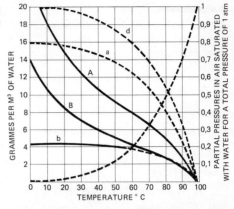

● **Absolute humidity of atmospheric air at saturation point plotted against the dew point.**

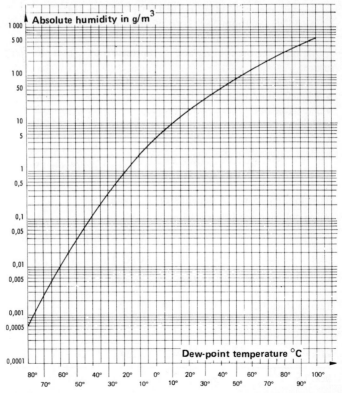

Fig. 558. —

● **Ammonia.**

Vapour pressure of liquid ammonia

Temperature °C	— 31	0	10	20	30	40	50
Pressure bars	1	4,2	6,1	8,5	11,6	15,7	19,6

Legislaltion requires containers to be rates at 20 bars and tested to 30 bars.

● **Chlorine.**

Characteristics

In its normal state chlorine is a greenish-yellow gas with the following physical constants:
— Density in relation to air: 2.491;
— Relative atomic mass: Cl = 35.46;
— Specific mass: 3.214 g/litre at O °C, 760 mm Hg;
— At 15 °C, 760 mm Hg, 1 kg of chlorine gives 314 litres of chlorine gas and 1 litre of liquid chlorine corresponds to 456 litres of gas.

It liquefies on cooling and compression at a pressure varying with temperature:

10 bars at 40 °C;   5 bars at 18 °C

● Liquefaction point (at 1 bar)................ — 34.1 °C
● Freezing point........................... — 102 °C
● Critical temperature...................... 144 °C
● Critical pressure......................... 77.1 bars
● Heat capacity of gas...................... 0.518 kJ/kg (0.124 kcal/kg) (from 15° to 100° C at 1 bar)
● Heat capacity of liquid.................... 0. 92 kJ/kg (0.22 kcal/kg)

Its solubility in water depends on temperature:

6.7 g chlorine per litre of water at 20 °C ⎫
9.6   —   —   10 °C ⎬ (in pure chlorine atmosphere)
11.8   —   —   5 °C ⎪
14.8   —   —   0 °C ⎭

● Latent heat of vaporization of chlorine:

Température °C	0	10	20	30	40	50	60
J/mole	17.64	17.14	16.59	16.01	15.47	14.88	14.30
kJ/kg	249.1	242	234.1	226.1	218.2	209.8	201.5
kcal/kg	59.6	58.9	56.1	54.1	52.2	50.2	49.2

Chlorine is an irritant, suffocating gas which is not corrosive when pure and dry. With even only small amounts of moisture, however, it becomes highly corrosive.

Chlorine is highly reactive with most elements an may give rise to explosive reactions with ammonia gas, hydrogen, etc.

● **Pressure-temperature curve for saturated chlorine gas.**

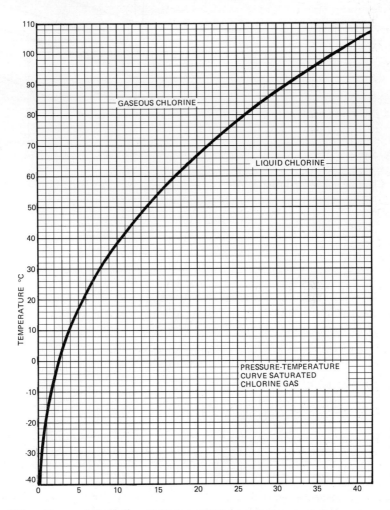

Fig. 559. — Pressure in kgf/cm². Multiply by 0.981 to obtain pressure in bars.

● **Variation of the density of liquid chlorine with pressure at different temperatures**

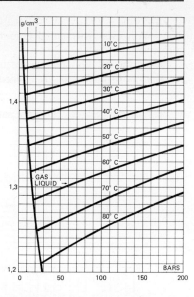

Fig. 560. —

● **Chlorine solubility in water.**

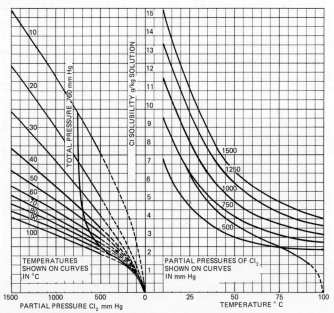

Fig. 561. — Multiply by 0.00133 to obtain pressure in bars.

27 METHODS OF ANALYSIS

1. GENERAL

1.1. The use of the milliequivalent and the "degree" in the chemistry of water

A. French definitions

In order to facilitate calculations, it is usual in chemistry to evaluate the results of analyses not in grammes per litre but in gramme-equivalents per litre (page 874).

For practical purposes, and in order to avoid decimals, the milli-equivalent is taken as the unit.

The abbreviation used for the gramme milli-equivalent (unit of mass) is "meq".

The abbreviation used for the gramme milli-equivalent per litre (unit of concentration) is "meq/l".

For example, as the atomic weight of chlorine is 35.5 g, if a given sample of water contains 2 g of chlorine per litre this result can be expressed by saying that it contains $\dfrac{2 \times 1}{35.5} = 0.056$ gramme-equivalents of chlorine per litre (or 56 meq/l).

In the case of a polyvalent element, the value of the unit of concentration (meq/l) will be the molecular weight divided by the valency and expressed in milligrammes per litre.

For calcium, for example, that being a bivalent element with a molecular weight of 40, a concentration of one meq/l corresponds to $\dfrac{40}{2} = 20$ mg/l.

The advantage of this notation is that it facilitates the immediate calculation of saline concentrations. If, in the previous example of a solution containing 56 gramme milli-equivalents of chlorine per litre, we are dealing with a pure calcium chloride solution, the concentration of $CaCl_2$ will be

$$0.056 \left(\frac{40}{2} + \frac{35.5}{1} \right) = 3.1 \ \text{g/l}$$

and the corresponding concentration of calcium will be:

$$0.056 \times \frac{40}{2} = 1.1 \ \text{g/l}$$

In the chemistry of water, it is often necessary to know not the actual dissolved salts in the sample but the balance-sheet in respect of anions and cations. This is indicated more clearly by the notation in equivalents.

For the analysis of alkaline earths, BOUTRON and BOUDET chose an arbitrary standard corresponding to 11.4 mg/l of calcium chloride, without reference to the atomic system.

The concept of the "degree" (page 874) is now preferred, with a correction of the Boutron and Boutet hardness unit so as to obtain a simple correlation with the milli-equivalent.

Defined in this way, the degree then becomes a unit of concentration which may be used, like the meq/l, to express the concentration of any soluble salt contained in ware and no longer merely as the degree of hardness due to lime and magnesium salts only.

● TABLE SHOWING THE CONCENTRATION OF SOLUTIONS
(Values of the different degrees)

	Formula	Molar mass	Value in mg/l of the various units			
			meq/l	Fr. °	Germ. °	CaCO₃ ppm
1. Calcium and magnesium salts and oxides causing hardness in water(degree of hardness)						
Calcium carbonate	$CaCO_3$	100	50	10.0	17.8	1.00
Calcium hydrogen carbonate (bicarbonate)	$Ca(HCO_3)_2$	162	81	16.2	28.9	1.62
Calcium sulphate	$CaSO_4$	136	68	13.6	24.3	1.36
Calcium chloride	$CaCl_2$	111	55.5	11.1	19.8	1.11
Calcium nitrate	$Ca(NO_3)_2$	164	82	16.4	29.3	1.64
Quicklime	CaO	56	28	5.6	10.0	0.56
Hydrated lime	$Ca(OH)_2$	74	37	7.4	13.2	0.74
Magnesium carbonate	$MgCO_3$	84	42	8.4	15.0	0.84
Magnesium hydrogen carbonate (bicarbonate)	$Mg(HCO_3)_2$	146	73	14.6	26.1	1.46
Magnesium sulphate	$MgSO_4$	120	60	12.0	21.4	1.20
Magnesium chloride	$MgCl_2$	95	47.5	9.5	17.0	0.95
Magnesium nitrate	$Mg(NO_3)_2$	148	74	14.8	26.4	1.48
Magnesia	MgO	40	20	4.0	7.1	0.40
	$Mg(OH)_2$	58	29	5.8	10.3	0.58
2 Anions						
Carbonate ion	CO_3	60	30	6.0	10.7	0.60
Hydrogen carbonate (bicarbonate) ion	HCO_3	61	61	12.2	21.8	1.22
Sulphate ion	SO_4	96	48	9.6	17.3	0.96
Sulphite ion	SO_3	80	40	8.0	14.3	0.80
Chloride ion	Cl	35.5	35.5	7.1	12.7	0.71
Nitrate ion	NO_3	62	62	12.4	22.1	1.24
Nitrite ion	NO_2	46	46	9.7	16.4	0.07
Orthophosphate ion	PO_4	95	31.66	6.32	11.25	0.63
Hydrogen silicate ion ⎫ expressed as Silicate ion ⎬ SiO_2	SiO_2	60	60	12.0	21.4	1.20

	Formula	Molar mass	Value in mg/l of the various units			
			meq/l	Fr.°	Germ. °	CaCO₃ ppm

	Formula	Molar mass	meq/l	Fr.°	Germ. °	CaCO₃ ppm
3. Acids						
Sulphuric acid	H_2SO_4	98	49	9.8	17.5	0.98
Hydrochloric acid	HCl	36.5	36.5	7.3	12.8	0.13
Nitric acid	HNO_3	63	63	12.6	22.5	1.25
Orthophosphoric acid	H_3PO_4	98	32 66	6.52	11.64	0.65
4. Cations and oxides						
Calcium	Ca	40	20	0.4	7.15	0.40
Magnesium	Mg	24.3	12.1	2.43	4.35	0.24
Sodium	Na	23	23	4.6	8.2	0.46
	Na_2O	62	31	6.2	11.1	0.62
Potassium	K	39	39	7.8	13.9	0.78
	K_2O	94	47.1	9.4	16.8	0.34
Bivalent iron	Fe	55.8	27.9	5.6	10.0	0.50
Trivalent iron	Fe	55.8	18.6	3.7	6.6	0.37
Aluminium	Al	27	9	1.8	3.2	0.18
	Al_2O_3	102	17	3.4	6.1	0.34
5. Bases						
Ammonium	NH_4	18	18	3.6	6.4	0.36
Sodium hydroxide	NaOH	40	40	8.0	14.3	0.80
Potassium hydroxide	KOH	56	56	11.2	20.0	1.12
Ammonia	NH_4OH	35	35	7.0	12.5	0.70
6. Various salts						
Sodium bicarbonate (hydrogen carbonate)	$NaHCO_3$	84	84	16.8	30	1.65
Sodium carbonate	Na_2CO_3	106	53	10.6	18.9	1.06
Sodium sulphate	Na_2SO_4	142	71	14.2	25.3	1.42
Sodium chloride	NaCl	58.5	58.5	11.7	20.9	1.17
Sodium orthophosphate	Na_3PO_4	164	54.7	10.9	19.5	1.09
Sodium silicate	Na_2SiO_3	122	61	12.2	21.8	1.22
Potassium carbonate	K_2CO_3	138	69	13.8	24.6	1.38
Potassium bicarbonate (hydrogen carbonate)	$KHCO_3$	100	100	20	35.7	2.00
Potassium sulphate	K_2SO_4	174	87	17.4	31.1	1.14
Potassium chloride	KCl	74.5	74.5	14.9	26.6	1.49
Potassium orthophosphate	K_3PO_4	212.3	70.8	14.1	25.2	1.41
Ferrous sulphate	$FeSO_4$	152	76	15.2	27.1	1.52
Ferric sulphate	$Fe_2(SO_4)_3$	400	66.6	13.3	23.8	1.33
Ferric chloride	$FeCl_3$	162.5	54.2	10.8	19.3	1.08
Aluminium sulphate	$Al_2(SO_4)_3$	342	57	11.4	20.3	1.14

B. Other units

The equivalent and the milli-equivalent have the advantage of being international.

On the other hand, for the measurement of hardness, the Germans use an arbitrary unit corresponding to 10 mg/l of lime expressed as CaO.

Furthermore, the English degree corresponds to a concentration of one grain of $CaCO_3$ per gallon of water, that is, 14.28 mg/l.

This unit is little used. It is usually replaced by ppm of $CaCO_3$, which corresponds to 0.02 milliequivalent per litre.

C. Equivalent hardness and alkalinity values

1 French degree   = 0.56 German degree
                  = 0.7 English degree   = 10 ppm CaCO₃

1 German degree = 1.786 French degree
                  = 1.25 English degree   = 17.86 ppm CaCO₃

1 English degree = 1.438 French degree
                  = 0.8 German degree   = 14.38 ppm CaCO₃

1 ppm CaCO₃   = 0.1 French degree
                  = 0.056 German degree = 0.07 English degree

The table on pages 894 and 895 shows the respective values of these different units for the principal elements found in water.

1.2. Determination of hardness and alkalinity of water
(see analysis methods No. 302, 303, 304, 305 and 306).

1.2.1. SIGNIFICANCE OF TH—THE TITRATION FOR HARDNESS

The titration for hardness (TH) indicates the total content of calcium and magnesium salts; these are the salts which make the water hard (scaling salts which impede both the cooking of vegetables and the formation of lather with soap).

● **Various hardness values:** The degree of hardness gives the quantity of alkaline earth salts (bicarbonates, sulphates, chlorides) etc. present in the water.

The following aspects are covered:

A. Total TH: this indicates the overall content of calcium and magnesium salts.

B. Calcic TH: this is an overall measure of the content of calcium salts.

C. Carbonate hardness: indicates the content of calcium and magnesium carbonates and the hydrogen carbonates of the same elements. It is equal to the TAC (see below) if the TH exceeds the TAC, or to the TH if the TAC exceeds the TH.

D. Non carbonate hardness: indicates the content of calcium and magnesium sulphate and chloride. It is equal to the difference between A and C.

1.2.2. INTERPRETATION OF ALKALINITY TITRATIONS (TA AND TAC)

The relative values of TA and TAC titrations indicate the quantities of alkali or alkaline-earth hydroxides, carbonates and hydrogen carbonates in the water.

Alkalinity is measured by means of a standard acid solution in the presence of either phenolphthalien (simple alkalinity titration TA or "p Wert") or methyl orange or helianthin (complete alkalinity titration TAC or "m Wert").

Phenolphthalein changes from red to colourless (fig. 562) at pH 8.3; reactions (1) and (2) are complete and reaction (3) starts at this pH value and CO₂ appears in the solution. The TA value therefore includes all the hydroxide content but only half the carbonate content.

Methyl orange changes from yellow to orange at pH 4.3 immediately there is a trace of free, strong acid and reaction (3) is complete. The TAC therefore indicates the hydrogen carbonate content.

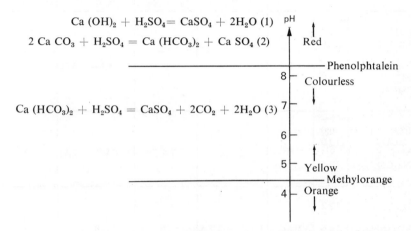

$$Ca(OH)_2 + H_2SO_4 = CaSO_4 + 2H_2O \quad (1)$$

$$2\,Ca\,CO_3 + H_2SO_4 = Ca(HCO_3)_2 + Ca\,SO_4 \quad (2)$$

$$Ca(HCO_3)_2 + H_2SO_4 = CaSO_4 + 2CO_2 + 2H_2O \quad (3)$$

The three main types of alkalinity in the water, namely hydroxides, carbonates and hydrogencarbonates, can therefore be ascertained from the results of the TA and TAC titrations.

The table on page 898 provides the basis for this calculation.

The various salts which may be present in the water are listed in the first column.

The quantity of each in mg/l is obtained by multiplying the corresponding figure in the second column by the TA or TAC values shown on page 898 in the subsequent columns.

1.2.3. GERMAN TESTS OF ALKALINITY

The German method of determining the alkalinity of water consists of measuring the number of ml of decinormal acid added to 100 ml of water in order to change the colour of the phenolphthalein or the methyl orange (helianthin).

The number of ml used is represented by:

with phenolphthalein = p
with methyl orange = m

Phenolphthalein alkalinity in German degrees P = 2.8 p
Methyl orange alkalinity in German degrees M = 2.8 m
Equivalent in French degrees:

$$TA = 5\,p \quad TAC = 5\,m$$

Dissolved salts	mg/l per degree	Respective values of TA and TAC titrations				
		if TA = O	if TA < $\dfrac{TAC}{2}$	if TA = $\dfrac{TAC}{2}$	if TA > $\dfrac{TAC}{2}$	if TA = TAC
(OH)	3.4					
CaO	5.6					
Ca(OH)$_2$	7.4					
MgO	4.0	0	0	0	2 TA — TAC	TAC
Mg(OH)$_2$	5.8					
Na(OH)	8					
CO$_3$	6.0					
CaCO$_3$	10.0					
MgCO$_3$	8.4	0	2 TA	TAC	2 (TAC — TA)	0
Na$_2$CO$_3$	10.6					
HCO$_3$	12.2					
Ca(HCO$_3$)$_2$	16.2					
Mg(HCO$_3$)$_2$	14.6	TAC	TAC — 2 TA	0	0	0
NaHCO$_3$	16.8					

1.3. List of French "AFNOR[1]" standards for water analysis

Number
Publication date
Nature:

NF	French standard	
ENR	registered	
HOM	officially approved	Title and method
EXP	experimental	
FD	documentation booklet	

NF T 90 000   1974 — ENR   Guide for the drafting of **analytical reports.**

NF T 90 001   1951 — FD   Sampling of **boiler waters** and **pressurized circuit waters.**

NF T 90 003   1958 — HOM   **Hardness** measurement with chelating reagent.

1. AFNOR: See page 905.

Number Publication date Nature:		Title and method
NF T 90 004	1968 — HOM	Spectrophotometric determination of **fluorine.**
NF T 90 005	1958 — HOM	**Magnesium** determination.
NF T 90 006	1950 — HOM	Colorimetric measurement of **pH.**
NF T 90 007	1950 — HOM	Determination of **silica.**
NF T 90 008	1953 — HOM	Electrometric measurement of **pH** with glass electrode.
NF T 90 009	1954 — HOM	Determination of **sulphate** ions.
NF T 90 010	1952 — HOM	Colorimetric determination of low levels of **free oxygen.**
NF T 90 011	1962 — HOM	Determination of **carbon dioxide.**
NF T 90 012	1975 — HOM	Determination of **nitrates.**
NF T 90 013	1975 — HOM	Determination of **nitrates.**
NF T 90 014	1952 — HOM	Determination of **chlorine** ions — volumetric measurement with silver nitrate.
NF T 90 015	1975 — HOM	Determination of **ammoniacal nitrogen.**
NF T 90 016	1958 — HOM	Gravimetric determination of **calcium.**
NF T 90 017	1959 — HOM	Colorimetric determination of **iron.**
NF T 90 018	1960 — HOM	Determination of **oxygen** given off by potassium permanganate.
NF T 90 019	1959 — FD	Determination of **sodium** ions.
NF T 90 020	1959 — FD	Gravimetric determination of **potassium** ions.
NF T 90 021	1962 — HOM	Photometric determination of **iodine.**
NF T 90 022	1966 — FD	Spectrophotometric determination of low levels of **copper.**

Number Publication date Nature:		Title and method
NF T 90 023	1963 — HOM	Spectrophotometric determination of **orthophosphates** and **polyphosphates.**
NF T 90 024	1963 — HOM	Spectrophotometric determination of **manganese.**
NF T 90 025	1967 — HOM	Spectrophotometric determination of **selenium.**
NF T 90 026	1975 — ENR	Determination of **arsenic.**
NF T 90 027	1967 — FD	Overall determination of **heavy metals.**
NF T 90 028	1969 — HOM	Colorimetric determination of **lead.**
NF T 90 029	1970 — HOM	Determination of dry **residues,** calcinated residue, and sulphate residue.
NF T 90 030	1973 — HOM	Determination of **clogging power.**
NF T 90 031	1973 — HOM	Determination of electrical **resistivity** (or conductivity).
NF T 90 032	1975 — ENR	Table of **oxygen solubility** in water.
NF T 90 033	1975 — ENR	Measurement of light diffusion index "**turbidity measurement**" .
NF T 90 034	1975 — HOM	**Colour** measurement by comparison with the Hazen scale.
NF T 90 035	1975 — HOM	**Taste** evaluation.
NF T 90 100	1972 — FD	Sampling—precautions when taking, preserving, and treating **samples.**
NF T 90 101	1971 — HOM	Determination of **chemicdal oxygen deman (COD).**
NF T 90 103	1975 — ENR	Determination of **biochemical oxygen demand (BOD).**

Number Publication date Nature:		Title and method
NF T 90 104	1972 — HOM	**Putrescibility** test.
NF T 90 105	1972 — EXP	Determination of **suspended solids.**
NF T 90 106	1973 — ENR	Determination of **dissolved oxygen.**
NF T 90 107	1975 — EXP	Determination of **total cyanides.**
NF T 90 108	1975 — EXP	Determination of **free cyanides.**
NF T 90 109	1976 — EXP	Determination of **phenol index.**
NF T 90 111	1975 — ENR	Evaluation of **dissolved salts content.**
NF T 90 112	1976 — EXP	Determination of ten metallic elements: **(Cr, Mn, Fe, Co, Ni, Ca, Zn, Ag, Cd, Pb)** by atomic absorption spectrometry.
NF T 90 113	1976 — EXP	Determination of **mercury** by atomic absorption spectrometry.
NF T 90 201	1973 — EXP	Aqueous effluents from **oil refineries—sampling.**
NF T 90 202	1973 — EXP	Aqueous effluents from **oil refineries**—determination of hexane-extractable **organic matter in suspension** in water.
NF T 90 203	1973 — EXP	Aqueous effluents from **oil refineries**—determination of **total hydrocarbons.**
NF T 90 204	1973 — EXP	Aqueous effluents from **oil refineries**—determination of **phenols.**
NF T 90 301	1974	Determination of **inhibition of mobility** of **Daphnia magna** (crustaceans, Cladocera).

2. TAKING SAMPLES FOR ANALYSIS

2.1. Drinking water

Analysis is meaningless unless sampling is done with sufficient care to ensure that the analysed water is truly representative of the water to be tested.

In particular, a sample should only be taken from well-water after prolonged pumping. In the case of waters which may vary in composition, several samples (if possible representing the extreme characteristics) should be taken.

This procedure should be adopted with river water, which should be examined both at times of low water and flood water, and also with spring water affected by rainfall.

Samples must be taken in perfectly clean bottles which have been rinsed several times in the water to be examined.

The sampling bottles must be filled as full as possible.

In principle, the bottle should be filled without splashing the water and bringing it into contact with the air. For this purpose, take a rubber tube connected to the tap from which the sample is drawn and lead it to the bottom of the bottle; allow the water to fill the bottle and overflow until several times the capacity of the bottle has passed through it. Stopper immediately.

To take depth samples use a weighted bottle fitted with a stopper with two tubes passing through it. One of these reaches to the bottom and the other is provided with a stopper which is removed by a wire attached to it when the bottle is at the desired depth. This makes it possible to remove the air from the bottle without bubbling it through the water sample.

When testing for iron and manganese, a special sample must be taken and acidified (1 ml of nitric acid per litre), to avoid precipitation during transport.

Samples for bacteriological tests are taken in sterilized bottles after sterilizing the bottle neck and the sampling tap by a flame. They must then be removed as quickly as possible to the analytical laboratory, preferably packed in ice.

Recommendations for the preservation of samples are set out in the table on page 904.

2.2. Waste water

The amount of pollution in waste water is measured on correctly-taken samples. As with drinking water, the various analyses are of little value unless the sample is truly representative of the conditions and quality actually found in practice.

Sampling is difficult because the raw waste varies both in composition and flow. The first step is to select a sampling point where there is a good mixture of the matter to be sampled, unaffected by earlier deposits.

Isolated samples should be taken whenever unusual or undesirable elements or concentrations are detected. For example:
— toxic waste such as cyanides (detected by smell), chromium or copper (detected by colour);
— oil and grease;
— concentrated organic waste form dairies, slaughter-houses, canning factories, tanneries, etc.

Average samples should be taken when the aim is to measure average quality over a period not exceeding 24 hours. It is often useful to know how pollution varies during the day, to assess the peak flows and to ascertain the extent of the day-time or night-time pollution.

Apparatus has been designed for taking average samples proportional to the flow of water, either by collecting a certain volume of water at regular intervals, with the volume of the sample fixed or varying with flow, or by drawing off uniform quantities according to a time schedule varying with the flow. The water is circulated through the pipes at a velocity of at least 60 cm/s (2 ft/sec) to prevent the formation of deposits.

Any of the following types of appliance can be used to take samples automatically:
— peristaltic pump,
— diaphragm pump,
— suction by negative pressure set up by creating a vacuum in the sampling vessel,
— sampling with a scoop moving to and fro or in circles.

Hand sampling retains its full value when new sources of supply are sought, as only simple apparatus is then needed.

The sample should be stored and transported in a lagged ice-box at around 4 °C.

The following points should be noted during sampling:
— variations in turbidity and colour,
— the presence of coarse solids carried by the current and, as far as possible, their nature (lumps of grease, rags, hair, textile fibres, etc.).
— surface iridescence caused by the presence of oil,
— odours (putrid, organic, etc.),
— variations in temperature.

A number of automatic sampling units which can be used to obtain representative samples of the relevant pollution are now available on the market.

Using this apparatus, samples proportional to the flow can be taken on an hourly basis. The equipment includes facilities for storing the sample taken at a temperature of 4 °C.

2.3. Preservation of samples for analysis

For the performance of analyses and to obtain representative results, it is essential for the samples not to change between sampling and analysis.   Various techniques for "fixing" these samples are suggested below.

Type of analysis	Techniques or products to be used	Maximum stable time
Acidity and alkalinity	Refrigeration at 4 °C	24 hours
BOD	Refrigeration at 4 °C	6 hours
Calcium	No recommendation	24 hours
COD	2 ml H_2SO_4 d = 1.84 per litre	7 days
Chlorides	No recommendation	7 days
Colour	Refrigeration at 4 °C	24 hours
Cyanides	NaOH up to pH 10	24 hours
Dissolved oxygen	To be determined on the spot	
Fluorides	No recommendation	7 days
Hardness	No recommendation	24 hours
Total metals	5 ml HNO_3 d = 1.33 per litre	several weeks
Dissolved metals	Filter and add 3 ml HNO_3 d = 1.33 per litre	several weeks
Ammoniacal nitrogen	40 ml $HgCl_2$ per litre or refrigeration at 4 °C	7 days
Total Kjeldahl nitrogen	ditto	
Nitrogen, nitrate nitrite	ditto	
Oil and grease	2 ml H_2SO_4 per litre or refrigeration at 4 °C	24 hours
Organic carbon	2 ml H_2SO_4 per litre pH = 2	7 days

3. RAPID FIELD ANALYSIS

For the preliminary study of any purification scheme, just as for the operation of any treatment plant, certain water quality tests must be carried out on the spot.

The pH, free CO_2 content and dissolved oxygen and ammonia content can change while the samples are being transported to the laboratory and it is always better to perform these tests on the spot.

Hydrocure kits[1] are specially designed for these tests and contain all the necessary equipment in compact form.

---

1. Hydrocure, P.O. Box 45, 94380 Bonneuil-surMarne, France.

There are several different models covering all normal requirements.

Wherever possible, simple colorimetric methods of analysis are employed, using a comparator with reference screens; these quickly give a result in milligrammes per litre.

In other cases, the method is volumetric, using special burettes which give a direct reading in French degrees.

Details are given below of the methods used in the various different cases.

For more accurate laboratory analysis, it is best to consult the methods standardized by AFNOR in France listed on pages 898 to 901 or the "Standard methods for the examination of water and sewage" (American Public Health Association—1015, Eighteenth Street NW Washington DC 20036.

3.1. Colorimetric analysis methods for water
(Using the HYDROCURE comparator)

These methods involve the use of a comparator consisting of a plastic case which holds test tubes containing the water to be analysed; appropriate reagents are added and develop a colour which varies in intensity with the concentration of the element under investigation.

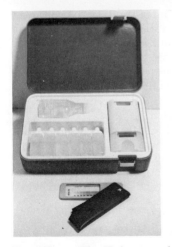

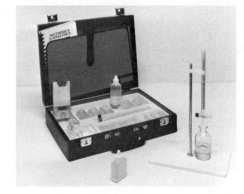

Fig. 563. — The Hydrocure colorimetric comparator.

Fig. 564. — Portable analysis kit for volumetric and colorimetric measurements.

2. The excerpts of the AFNOR standards contained in this book are reproduced with the permission of the "Association Française de Normalisation, Tour Europe, Cedex 7, 92080 Paris La Défense (France), to whom any order for the complete document is to be addressed. Only the latest edition of the relevant original Standard is to be deemed authentic.

A removable plate carrying a series of reference screens can be slipped into the comparator; a reading is obtained by comparison with the colour of the sample.   The plate should be set horizontally or vertically according to the concentration of the element.

N.B. : An instruction manual is included with each analysis kit.   The measuring test tube will normally be filled up to the mark B.

No. 201—Measurement of pH (3.7 to 11.8)

Fill the comparator measuring glass; add to this sample the number of drops of indicator stated on the bottle and find the reference screen which matches the colour obtained.   If the colour observed corresponds to one of the ends of the scale, repeat the test with the indicator of the nearest range.

No. 202-C—Determination of free chlorine (0.1 to 4 mg/l)

Fill the comparator measuring glass; add the dose of DPD reagent (diethyl-p-phenylene diamine) to this sample and find the reference screen which matches the colour obtained.

No. 202-D—Determination of total chlorine (free + combined)

Fill the comparator measuring glass; add the doses of potassium iodide and DPD reagent one after another and find the reference screen which matches the colour obtained.

No. 203—Determination of iron (0.06 to 5 mg/l)

Add a pinch of sodium hydrosulphite to 100 ml of the water to be analysed (pH between 6 and 8 and temperature between 15 and 25 °C); stir and wait 5 minutes.   Then add 0.5 ml dimethylglyoxime and 1 ml ammonia.   Stir.   Wait 15 minutes for the pink colour to develop.   Pour the solution into a comparator measuring glass and find the reference screeen (plate 0.3-6) which matches the colour obtained.   Observe the sample through the top of the test tube if the iron content is between 0.06 and 1 mg/l (plate 0.06-1).

No. 204-A—Determination of dissolved silica (3 to 50 mg/l)

Add a drop of phenolphthalein to a 25 ml sample of the water to be analysed; neutralize with sulphuric acid until any pink colour which has appeared is eliminated; add a measure [1] of ammonium molybdate and shake until dissolved.

Add 8 drops of sulphuric acid; leave to stand for 5 minutes and add 2 ml of stabilizer.   Pour the solution into a measuring glass and find the reference screen which matches the colour obtained.

No. 206-A—Phosphate determination (0.5 to 10 mg/l)

Pour 10 ml into a graduated flask and make up to 50 ml with distilled or deionized water.

Then add 5 ml of vanado-molybdic indicator.   Shake and leave the colour to develop for 5 minutes.

Then fill a measuring glass on the comparator up to the mark B and check which reference screen matches the colour, by looking through the upper part of the test-tube.

---

1. Small quantity of product supplied with the analysis kit.

If the water contains tannin (boiler water), it must first be boiled after adding potassium nitrate, and then left to cool and filtered through activated carbon.

No. 206-C—Measurement of polyphosphates in drinking and industrial process water.

The previous method can also be used to measure poly-phosphates: begin by taking 100 ml of water, add 0.5 ml of sulphuric acid and boil for an hour. Leave to cool. Make up to 100 ml with distilled water. Remove 10 ml and then proceed as above.

No. 208—Nitrous nitrogen determination (0.1 to 2 mg/l)

To 50 ml of the water to be analysed add 2 ml of Zambelli reagent. Stir and leave to stand for 10 minutes.

Then add 2 ml of ammonia solution. The liquid should turn yellow at once. Find the reference screen of the same shade.

No. 209—Determination of ammoniacal nitrogen (0.1 to 1 mg/l)

To 50 ml of the water to be analysed, add 2 ml of Rochelle salt ($C_4H_4O_6$ KNa, $4H_2O$) and stir. Then add 2 ml of Nessler reagent. Leave the colour to develop for 10 minutes.

Find the reference screen of the same shade while observing the sample from above the test-tube.

No. 210—Determination of manganese (0.05 to 2 mg/l)

Adjust the pH of the water to a value close to neutrality; eliminate any excess of oxidants (ozone, chlorine dioxide, chlorine, etc.). Rinse the two measuring glasses with this water.

Fill one of the glasses up to mark A and place in the comparator on the side opposite the "reagents" mark.

Fill the other glass up to mark A; add 2 or 3 grains of sodium periodate and stir until dissolved.

Add 1.5 ml of acetic acid, 3 ml of tetramethyldiaminodiphenylmethane (TDD) indicator; stir, and place the glass in the comparator.

Take the reading immediately, by finding the reference screen which matches the colour obtained, observing the sample through the top of the test tube.

No. 211—Determination of hexavalent chromium (0.03 to 1 mg/l)

Fill one of the measuring glasses up to mark A; add 0.5 ml sulphuric acid and one measure [1] of diphenylcarbazone. Mix and transfer to the other glass placing the latter in the comparator on the "reagents" side. Rinse the empty glass, fill it with water up to mark A and place in the comparator. Wait 10 minutes.

Find the reference screen of matching colour, observing the sample through the top of the test tube.

No. 212—Determination of cyanides (0 to 0.5 mg/l)

Fill one of the glasses; add 4 drops of buffer solution and 0.5 ml of chloramine T; stir and wait 1 minute; add a measure [1] of barbituric acid; stir and wait

1. See footnote page 906.

2 minutes; finally add 1 ml of pyridine. Mix by transferring to the other measuring glass and then place in the comparator on the "reagents" side.

Rinse the empty measuring glass, fill it with the sample of water to be analysed, and place in the comparator. Wait 30 minutes.

Find the reference screen which matches the colour obtained.

No. 213-A—Determination of hydrazine (0 to 1 mg/l)

Fill one of the measuring glasses up to mark A with the sample of water to be analysed and make up to mark B with para-diamethylaminobenzaldehyde (DAB) indicator. Mix by transferring to the other measuring glass and place in the comparator on the "reagents" side.

Rinse the empty glass, fill to mark B with the sample of water to be analysed, place in the comparator, and find the reference screen which matches the colour obtained.

No. 214-B—Ozone determination (0.5 to 3 mg/l)

Pour 10 drops of potassium iodide into a 40/100 graduated flask, fill the flask with ozonated water up to the 50 mark, add a measure of Vitex (an iodine indicator used in iodometry) and shake well.

Then fill a measuring glass up to mark A and check which reference screen matches the colour.

No. 220—Measurement of water colour

Fill a measuring glass with colourless or mineral spring water up to mark C and then place in the comparator equipped with its extension piece.

Fill another measuring glass with the water under test up to mark C and check which reference screen matches the colour, by looking at the sample through the upper part of the test-tube.

3.2. Volumetric analysis methods for water

Hydrocure kits contain special burettes which are graduated in 1/2 ml and are provided with an extra mark 1/2 ml above zero.

This mark is used only for measuring the degree of hardness with soap solution. In all other cases, the burette must be filled to the zero mark.

As the solutions used are at N/25 normality, the reading obtained from the burette gives the result directly in French degrees if 100 ml samples of water are tested.

By dividing this figure by 5 the result in milli-equivalents per litre is obtained.

If an ordinary burette graduated in millilitres is used instead of a special burette, the result in French degrees is obtained by multiplying the millilitre reading by 2.

An ordinary burette graduated in millilitres and tenths of a millilitre is also used for certain titrations, such as liability to oxidation by potassium permanganate, when a result in milligrammes per litre is required.

Fig. 565. — Laboratory analysis cabinet with automatic burettes.

No. 301 — Measurement of free CO_2

Introduce $V = 5$ ml of the $N/25$ alkaline solution and 4 drops of phenolphthalein into an Erlenmeyer flask graduated to 100 ml. Make up to 100 ml with the sample of water to be analysed, avoiding agitation or splashing (pour the water down the wall of the flask). Stir gently and wait 5 minutes: the solution should now be pink. If it is not, repeat with $V = 10$ ml of alkaline solution.

Titrate with $N/25$ acid until the colour disappears; let A be the number of degrees shown on the burette.

Carry out a blank test with the same volume of alkaline solution; let B be the number of degrees indicated on the burette.

The concentration of free CO_2, expressed in mg/l, is equal to:

$$\frac{B - A}{100 - V} \times 880$$

Add 1 mg/l to the result for every 10.9 French degrees of hardness.

No. 302 — Measurement of total hardness (TH)

● *Soap method—302-A*

Work on 100 ml of the liquid on which the free CO_2 has been measured.

Fill the special burette with soap solution up to the extra mark above the zero mark.

Drop by drop pour in the soap solution, shaking briskly after each addition. The reaction is complete when a stable lather is produced which persists for at least 5 minutes and re-forms after shaking.

The French degree of hardness is then read off directly from the burette.

If the reading exceeds 20 ⁰f work on only 50 or even 25 ml of water, dilute to 100 ml with distilled water and multiply the figure read off of the burette by 2 or 4 as the case may be.

● *Complexon method—302-B*

To 100 ml of the water to be analysed, add 2 ml of the buffer solution K 10 and 10 drops of Eriochrome T Black and hydroxylamin chlorhydrate indicator (heat to 45 °C to accelerate the reaction).

With the special burette, titrate with E.D.T.A. solution until the liquid turns from ruby red to pure blue (if possible, perform a control test with the same volume of distilled water).

The French degree of hardness is then read directly off the burette.

If the reading exceeds a value of 30 °f, use only 50 (or 25) ml and dilute with distilled water.

Then multiply the result by 2 (or 4) as the case may be.

No. 303—Measurement of calcium hardness

To a 100 ml water sample, add 2 ml of K 12 buffer solution and about 0.2 grammes of MRX indicator (heat to 45 °C to accelerate the reaction).

With the special burette, run in the special "complexon" solution until the solution turns from pink to mauve (if possible make a control experiment with the same volume of distilled water).

The degree of calcium hardness is read directly off the burette.

If the result is more than 30 French degrees, take only 50 (or 25) ml of the water sample and dilute it with distilled water.

Multiply the result by 2 (or by 4).

The MRX indicator can be replaced advantageously by calcein without changing the procedure.   Appearance of green fluorescence from the calcein indicates the end of titration.

No. 304—Measurement of hardness due to magnesium salts

The difference between total hardness TH and calcium hardness gives the magnesium hardness.

No. 305—Measurement of the degree of caustic alkalinity (TA)

Fill the special burette to the zero mark.

Add a few drops of phenolphthalein into 100 ml of the water to be analysed; then with the burette add N/25 acid, drop by drop, until the pink colour disappears.

The TA is then read off directly on the burette.

If the liquid remains colourless after a few drops of phenolphthalein have been added to the water sample, the TA is equal to zero.   This is normally the case for natural waters with a pH of less than 8.3.

No. 306—Measurement of total alkalinity (TAC)

Fill the special burette to the zero mark.

Add a few drops of methyl orange into 100 ml of the water to be analysed and then, with the special burette, add the N/25 acid solution, drop by drop, until the liquid turns orange.   Total alkalinity is then read off on the burette.

If the value is over 30 °f, use only 50 (or 25) ml of water and multiply the burette reading by 2 or 4 as the case may be.

When the water to be analysed has a caustic alkalinity value TA, measure this and then measure the TAC, as described above, **without again filling the burette with acid solution.**

No. 307—Measurement of solvent action on marble (agressivity)

Fill a flask with 125 ml of the water sample, to which are added 1 to 2 grammes of finely crushed marble, previously washed several times with distilled water, then with the water to be analysed.

The flask must be completely filled without entrapping an air bubble.

Shake slowly for 24 hours and measure the pH and the TAC.

The corrosive action of the water on marble, or its scale-forming capacity, is determined by comparing these values with the pH and TAC of the water before it comes into contact with the marble.

No. 308—Measurement of the free sodium hydroxide content of boiler water

Any determination of sodium hydroxide by reference to the TA and TAC values shown in the table on page 898 is inaccurate if the TA is much above $\dfrac{TAC}{2}$.

It is then better to proceed as follows:

To 100 ml of boiler water, add 5 ml of 20 % strontium chloride solution and then measure the TA, i.e. the alkalinity due to NaOH. If a TA value greater than 25 °f is recorded, dilute the sample and multiply by the dilution factor.

The concentration in terms of sodium hydroxide (NaOH) in mg/l = 8 TA.

TAC being the total alkalinity of the boiler water, the concentration of sodium carbonate NA_2CO_3 in mg/l = 10.6 (TAC-TA(NaOH)).

No. 309—Measurement of the strong free acid content (TAF)

Strong free acid content (TAF) is nil for any water with a pH above 4.5.

To 100 ml add two drops of methyl orange which turns pink (otherwise TAF = 0).

Add an N/25 sodium hydroxide solution until the colour turns to yellow and read the TAF value directly off the special burette.

No. 312—Measurement of oxidation by permanganate

Use a burette graduated in tenths of a millilitre and an N/80 solution of potassium permanganate.

1° In an alkaline medium (No. 312-B)

To 100 ml of the water to be analysed, add 1 ml of saturated sodium bicarbonate solution; then bring the sample to boiling point. Add 10 ml of N/80 solution of potassium permanganate and boil for exactly 10 minutes. Cool quickly, and add 2.5 ml of 50 % sulphuric acid and 10 ml of Mohr's salt solution (5 g/l, slightly sulphuric).

With the burette add the permanganate solution drop by drop until a pale pink colour is obtained. Let A be the total number of ml added.

Comparison of oxidizing and reducing solutions: to 100 ml of distilled water add 2.5 ml of 50 % sulphuric and 10 ml of Mohr's salt solution (5 g/l).

Add permanganate solution with the burette until a pale pink colour is obtained. Let B be the number of ml added.

The quantity of matter liable to be oxidized by permanganate in the water under analysis, expressed in milligrammes of oxygen per litre, is A—B.

2° In an acid medium (No. 312-G)

To 100 ml of the water under analysis, add 2.5 ml of 50 % sulphuric acid, then bring to boiling point. Add 10 ml of permanganate solution and boil for exactly 10 minutes.

Cool quickly, then add 10 ml of Mohr's salt solution (5 g/l, slightly sulphuric).

With the burette, add permanganate solution drop by drop until a pale pink shade appears. Let A be the total number of ml added.

As in the previous case, compare the oxidizing and reducing solutions. Let B be the number of ml.

The quantity of matter liable to be oxidized by permanganate in the water under analysis, expressed in milligrammes of oxygen per litre, is A—B.

Note

(*a*) Greater accuracy can be obtained by using 200 ml of water and doubling the amounts of the reagents; the result obtained has to be divided by 2.

(*b*) If liability to oxidation exceeds 2 mg/l, the water under analysis must be diluted with distilled water containing no organic matter: the quantity of standard permanganate solution must always be less than 20 % of the quantity added to the sample to be analysed.

(*c*) For a given water, liability to oxidation is usually less in an alkaline than in an acid medium.

No. 313—Determination of dissolved oxygen

See method 403, page 924.

No. 314-G—Determination of the chloride content

Fill the special burette with N/25 silver nitrate solution.

To 100 ml of the water under analysis add one drop of phenolphthalein; if the water turns red, add 10 % nitric acid.

Add 3 or 4 drops of 10 % potassium chromate; then, with the burette, add the silver nitrate solution drop by drop, and shake, until the solution turns brick red.

Perform the same experiment with distilled water and subtract the figure obtained from the chloride content read off from the burette. One French degree corresponds to 7.1 mg/l of Cl.

No. 315-G—Measurement of salts of strong acids SAF

(total concentration of salts of strong acids)

In a tube 2 to 3 cm in diameter, place 100 ml of strongly-acid cation exchange resin, supported on a glass-wool plug.

The lower part of the tube must be extended by a swan-neck glass tube raised to the level of the resin layer, so that the latter is always submerged.

Regenerate the resin by passing through it 100 ml of a solution of hydrochloric acid (15 ml of HCl of S.G. 1.18 in 100 ml of water).

Rinse with 500 ml of distilled water, followed by 200 ml of the water under analysis.

Flow 100 ml of the sample through the column at the rate of 1 1/2 litre per hour. Collect the filtrate and measure its strong acid content TAF.

The figure obtained represents the content (SAF) of salts of strong acids of the water examined.

The German method of determining free content of strong acids consists of measuring the number of ml of an N/10 sodium hydroxide solution added to 100 ml of water filtered as above, through a strongly-acid cation exchange resin, to change the colour of the methyl orange. The number of ml recorded is expressed by:—m Wert or by: negativer m Wert.

No. 322—Determination of hexavalent chromium content

Fill an ordinary burette graduated in ml with N/17.33 Mohr's salt solution.

To 100 ml of the water to be analysed:
— 10 ml of 6 N sulphuric acid;
— 5 ml of concentrated phosphoric acid;
— 6 to 8 drops of indicator (0.5 % barium diphenylamine sulphonate in distilled water).

Add the Mohr's salt solution until the violet colour which appears at the start of titration disappears.

1 ml of Mohr's salt solution corresponds to 1 mg of hexavalent chromium.

No. 324—Determination of hydrazine

Fill the burette to the zero mark with special potassium bromate solution.

To 100 ml of the solution to be analysed, add 3 to 4 drops of methyl orange and then hydrochloric acid until a pink colour appears. Add a further 10 ml of hydrochloric acid and then add the potassium bromate solution drop by drop until the methyl orange colour disappears. Stir vigorously at the end of the experiment as the colour change is sudden and irreversible.

If n is the number of ml potassium bromate necessary to obtain the colour change, the hydrazine content expressed in g/l is: $\dfrac{n \times 8}{100}$.

4. PHYSICAL MEASUREMENTS

pH
Redox potential $\Big\}$ see Chapter 19, pages 575 to 578.
Conductivity

No. 316—Measurement of turbidity by light diffusion. Calibration with formazine.

In this method, the intensity of the light scattered at 90º by the sample is compared with the result from standard solutions of formazine.

Photoelectric cell instruments measure the intensity of this scattered light; the manufacturers can supply stable standards expressed in formazine.

Preparation of the control suspension

Solution A: Weigh out 1 gramme of hydrazine sulphate and dissolve it in a small quantity of distilled water; then make up to 100 ml with distilled water.

Solution B: Weigh out 1 gramme of hexamethylene tetramine and dissolve by heating in a small quantity of distilled water; then cool and make up to 100 ml with distilled water.

Pour 5 ml of solution A and 5 ml of solution B into a 100 ml graduated flask. Mix well and leave for 48 hours at 20 °C. The fluid becomes opalescent.

Make up to 100 ml with distilled water; this gives a control suspension rated at 400 Jackson units.

Measurement of turbidity

A set of controls can be obtained by diluting the control suspension with perfectly clear water.

The units used are taken from ASTM standards; the following three units are equivalent:

— JTU ....Jackson Turbidity Units.
— FTU ....Formazine Turbidity Units.
— NTU ....Nephelometric Turbidity Units.

This type of apparatus can also be calibrated with other suspensions. Kieselguhr (special treatment for nephelometry) is often used, the results then being expressed in mg/l silica.

No. 316-A—Measurement of diffusion index (AFNOR NF T 30 033) [1]

In this method, the intensity of the light scattered by the suspended particles is measured; the measurement is effected at 90° to the axis of the incident light beam. The scattered intensity is compared with the incident intensity, the measurement being carried out at a wavelength of 546.1 nm (green line of mercury).

The equipment must be calibrated with chemically pure liquids whose diffusion index is precisely known (e.g., n = 1.63 for benzene).

The standards or the sample must be introduced into a perfectly clean measuring dish without scratches, mist, etc. Stir the sample and allow air bubbles to escape before measuring.

Fig. 566. — Hydrocure turbidity meter with photo-electric cell.

1. See footnote page 905.

No. 317—Platinum wire measurement of limit of visibility
(raw or clarified water; wastewater)

Place the water to be examined in a glass tube 3 cm in diameter and approximately 60 cm long, closed at one end by a black stopper carrying a loop of bright platinum wire 1 mm in diameter.

Measure the height of the liquid in the tube beyond the point at which the wire ceases to be visible.

No. 317-A—Measurement of visibility limit using the Secchi disc
(raw or clarified water, waste water)

Progressively immerse a 20 cm diameter white disc on the end of a pole and measure the depth beyond which it ceases to be visible.

No. 318 — Checking the turbidity of filtered water (DIENERT and GUILLERD method)
— Prepare a 1 $^o/_{oo}$ alcohol solution of vegetable mastic.
— A special lamp directs two light beams across two small vessels, containing a sample of the water under analysis and perfectly clear water respectively.

To the latter, add drops of the mastic solution until a degree of opalescence comparable to that produced by the light beam across the sample under examination is obtained.

The number of drops of mastic indicates the turbidity of the sample.

No. 320—Checking coloration by the platinum-cobalt method

Compare the water to be examined with platinum-cobalt solutions of different concentrations. The colour of the solutions is expressed by the concentration (mg/l) of platinum.

The unit of colour corresponds to 1 mg of platinum per litre.

For this comparison use Nessler tubes, or 40 cm^3 tubes for very slight degrees of coloration.

If the water contains matter in suspension, eliminate it by centrifuging.

If the colour is greater than 70, repeat the comparison after suitably diluting the sample.

The scale used is the following: 5, 10, 15, 20, 25, 30, 35, 40, 50, 70.

It is prepared from a 500 unit colour stock solution containing the following per litre of 10 % hydrochloric acid:
1.245 grammes of potassium chloroplatinate;
1.000 gramme of cobalt chloride ($6H_2O$).

No. 321—Determination of suspended solids (AFNOR standard NF T.90.105)[1].
Method of filtration through a glass fibre disc.

Wash a filter disc in distilled water, stove-dry it at 105 °C, cool in a desiccator, and weigh to within 0.1 mg (weight M_1).

Place the filter on its support and connect to the vacuum pump.

Filter a volume V such that the mass of material retained on the filter is at least 1 mg/cm^2. V must not be less than 100 ml.

---

1. See note page 905.

Rinse the sampling vessel in distilled water (10 ml) and run the wash waters through the filter.

Dewater and dry the filter at 105 °C, cool in the desiccator and weigh (weight M_2). The weight of suspended solids in mg/l is:

$$\frac{(M_2 - M_1)\ 1\ 000}{V}$$

5. PRESENTATION AND INTERPRETATION OF ANALYTICAL RESULTS

5.1. Presentation of results of analyses

A salt is a combination of a cation (metal radical) and an anion (acid radical). Chemical analysis can only distinguish between the anions and the various cations, and cannot, therefore, indicate exactly the nature and amounts of the dissolved salts, since each cation can combine with any anion and vice-versa. In most cases, it is impossible to indicate in analyses what salts result from the combination of anions with cations, except by making certain assumptions regarding the nature of these combinations. Analysis results reached in this manner are therefore often very far removed from reality.

Such assumptions always give a false picture since, as electrolyte solutions are more or less dissociated, in reality the water contains only ions.

It is moreover not important to know the exact nature of the dissolved salts. It is sufficient to know the respective proportions of the different anions and cations. The notation in equivalents or degrees thus readily indicates the different possible combinations between the anions and the cations and permits a check to be made of the accuracy of the analysis, as the sum of cations must be equal to the sum of anions (except in the case of excess acid or base).

Liability to oxidation by permanganate, and the content of iron, manganese, free CO_2, dissolved oxygen, silica, nitrogen and ammoniacal, nitrous and nitric nitrogen, content, are expressed in milligrammes per litre.

The cations and anions which combine to form the various salts dissolved in water are generally grouped in two tables, where they are expressed in mg/l and milli-equivalents per litre or in degrees.

Total salinity of the water is expressed by total cations and anions in mg/l.

Total cations expressed in milli-equivalents/litre (or in degrees) must be the same as total anions expressed in milli-equivalents/litre or in degrees.

5.2. Interpretation of analysis results

Any assessment of the results obtained depends entirely on the use to which the water will be put. There are no specific rules in this connection but certain basic factors make it possible to discuss the results of analyses.

- **Colour** (Methods Nos. 220 and 320)

  The true colour is due to the presence of dissolved or colloidal organic matter.

  There is no relation between the colour and the quantity of organic matter, which may be coloured or not.

  Water which is coloured is unpleasant when used for domestic purposes, and particularly for drinking, as it always gives rise to doubts regarding its potability.

  Certain industries require water to be completely free of any colour which might be transferred to the products manufactured and detract from their quality; examples are the paper, rayon, cellulose, starch and dyeing industries.

- **Odour and taste**

  These are generally due either to pollutants or to organisms living in the water, e.g., algae or fungi (metabolic or decomposition products).

  Certain chemical products, sometimes in minute quantities, also give off unpleasant odours. This is the case with phenols, which combine with chlorine to form very evil-smelling compounds.

  Unpleasant tastes are not considered to be important from the health point of view, but they are extremely objectionable in drinking water.

  All water samples, however, have their own particular taste due to the salts and gases dissolved in them.

- **Turbidity** (Methods Nos. 316, 317, 318)

  Together with measurement of suspended solids, this gives an indication of the level of colloidal matter of inorganic or organic origin.

  Some industries (such as paper-making) demand water clarified to a very advanced degree while others allow a measure of turbidity (cooling systems).

  Waters containing suspended matter which may eventually settle always cause difficulty due to the formation of deposits in pipes and tanks.

  Turbidity is related to transparency, and a knowledge of turbidity is an indispensable factor in water treatment.

- **Resistivity**

  Resistivity depends on the concentration of dissolved conducting salts. Its measurement, which is quite an easy matter, gives some idea as to the salt concentration in the water. A knowledge of this factor is particularly useful for the periodic checking of a given water, as an immediate indication of variations in composition is obtained.

  It is essential to specify the temperature at which the resistivity was measured.

● **pH** (Method No. 201)

pH values indicate whether the water is acid or alkaline.

They have no health significance, but represent a very important factor in determining the aggressive action of the water.

The regular measurement of the pH is an essential factor in water treatment practice.   In particular, it plays an important part in the efficiency of the coagulation process.

It is essential to specify the temperature at which the pH was measured.

● **Measurement of hardness** (Method No. 302)

The total hardness expresses the combined concentration of dissolved calcium and magnesium salts.

These salts form an insoluble compound with soap, which then loses part of its detergent power and can only with difficulty be made to lather.   Furthermore, calcium and magnesium salts prevent satisfactory cooking of vegetables.

For these reasons, water is said to be "hard" if it has a high TH and "soft" in the opposite case.

Under the influence of heat, calcium bicarbonate forms a precipitate of insoluble carbonate which is deposited as scale on the walls of pipes and appliances.

Water having a total hardness of up to 50 French degrees can be used for domestic purposes, but the most palatable value is between 8 and 15 degrees.

In the case of town water supplies, the presence of a certain degree of carbonate hardness is useful in order to form a layer on the inside of the pipes which are thus protected against corrosion.

The degree of magnesium hardness must also be known when studying water softening treatment.

● **Measurement of degree of alkalinity (TA and TAC)** (Methods Nos. 305 and 306).

The significance of these two determinations is explained on page 896.

A knowledge of these values is essential when considering a given type of water and particularly its aggressivity (solvent action on marble), or conversely its tendency towards the formation of scale, since these two phenomena depend on the equilibrium between free CO_2 and the bicarbonates.

Alkalinity measurements provide the data for studying and checking water-softening processes using chemical precipitation.

In the case of boiler water, a high degree of alkalinity is liable to cause troubles due to priming and caustic embrittlement.

From the health standpoint there is no limit to the permitted TAC value.

● **Measurement of salts of strong acids (SAF)** (Method No. 315)

Natural waters contain no free strong acids but only their salts, in particular the sulphates and chlorides of calcium, magnesium and sodium. The SAF expresses the total content of these salts of strong acids.

A knowledge of this value is necessary when planning ion-exchange deionization units because the salts are decomposed to free the corresponding strong acids when the water is passed through a strongly acid cation exchanger.

● **Measurement of free strong acids (TAF)** (Method No. 309)

This measurement expresses the total content of free strong acids. These acids are not found in natural waters, but may be present in waters treated with cation exchange resins (form H) and in certain industrial effluents; but a knowledge of the TAF of water which has passed through an acid cation exchanger is essential for the operation of a water treatment plant.

● **Iron** (Method No. 203)

The presence of iron in water often creates difficulties. Even in small concentrations, iron precipitates on contact with air and forms red flocs which produce turbidity and mark linen.

Certain bacteria live on iron and attach themselves to the walls of pipes, thus causing the latter to corrode with the formation of bulky and hard incrustations.

● **Free carbon dioxide** (Method No. 301)

It is essential to know the carbon dioxide content when investigating the stability of water and its power of causing either corrosion or scale in water pipes.

The presence of CO_2 noticeably improves the taste of drinking water, which is insipid without this substance.

● **Organic matter** (Method No. 312)

This term includes all substances capable of being oxidized by potassium permanganate at boiling point.

The results are expressed either in mg/l O_2 or in mg/l $KMnO_4$; it is compulsory to state the reference (1 mg/l O_2 = 3.95 mg/l $KMnO_4$).

The health significance of these substances is not clearly understood and it is not necessary dangerous to health to drink water which contains large quantities of organic matter (a cup of tea contains 2 000 mg/l of organic matter expressed as oxygen).

Some forms of organic matter give rise to colour and bad taste, as they favour the development of such organisms as algae, fungi, and bacteria, which attach themselves to pipe walls and secrete essences having an unpleasant smell.

They can also create malodorous compounds by reacting with the chlorine used for sterilization purposes.

Water rich in organic matter must always be suspected of bacteriological or chemical contamination (reducing agents).

● **Nitrogen**

The presence of large quantities of ammoniacal nitrogen generally indicates recent contamination by decomposing organic matter. Bacterial contamination, which should be checked by special analysis, must then the suspected.

While water is permeating through soil, and under the influence of certain bacteria, ammoniacal nitrogen passes into nitrites and then to nitrates.

Water which is poor in ammonia and rich in nitrates thus indicates that it has been effectively filtered and purified in the soil.

Certain deep well waters may, however, be rich in ammonia without necessarily being polluted.

Ammonia is favourable to the development of certain bacteria which in turn give rise to unpleasant taste.

The nitrate content of drinking waters must be limited owing to the risk of methaemoglobinaemia in infants.

● **Aggressivity** (Method No. 307) — **Scale formation** (see page 28)

A water is in calcium carbonate equilibrium if, when in contact with limestone where air is excluded, it does not dissolve the former and does not allow calcium carbonate to precipitate.

If, during measurements, the pH increases as well as the TAC, the water is aggressive. If, on the contrary, the TAC falls, the water has scale-forming qualities.

In general, the aim is water which will favour the formation of a protective layer on metal surfaces, as that is a means of preventing corrosion.

● **Ionic balance-sheet**

An approximate ionic balance-sheet can be established by calculating as remainders certain elements which are not measured directly.

Thus:

— magnesium is deduced from the TH — TCa value;

— bicarbonates, carbonates and OH ions are deduced from the TA and TAC values (page 896);

— sulphates are deduced from: SAF — (chlorides + nitrates);

— alkaline cations (Na, K, NH_4) are the remainder of: (sum of anions)—(TH+Fe)

● **Sodium**

Some waters have a very high content (brackish water) and for that reason have a disagreeable taste.

Sodium chloride has a very marked taste beyond a concentration of 500 mg/l.

● **Bacteriological examination**

Chemical analysis on its own is not sufficient to decide whether a water is potable or not. Information regarding organic matter, nitrogen, etc. can only serve as a guide towards the possibilities of pollution.

The potability of a water can only be determined by additional bacteriological analysis. Bacteriology is essentially a laboratory matter and involves specialists, and this handbook is not intended to include such methods (see, however, several remarks on page 973).

Water bacteriology is based on the search for germs of the bacterium coli type, and in particular Escherichia coli, and for certain other sporulating germs, which are not dangerous in themselves but indicate the presence of pollution by faeces. Water which contains these germs can become dangerous in the event of an epidemic.

5.3. Unwanted or toxic substances dissolved in water

It is difficult to lay down appropriate and maximum levels for substances commonly found in water, such as calcium, magnesium, iron, chlorides, sulphates, etc.

In very hot and dry countries where large quantities of water (up to 15 litres) may be consumed every day, the mineralization of the water may be very important. Magnesium sulphate, in particular, may cause intestinal troubles, which disappear after a time as the body becomes accustomed to this salt.

Maximum levels for certain substances can often be laid down on the basis of their toxicity. Health regulations (see the Chapter on "Legislation" (p. 1128) specify the "maximum permissible concentration" of the majority of unwanted or toxic organoleptic or physicochemical factors as well as microbiological factors.

5.4. Chemicals deliberately added to water

It may be necessary to discuss the toxicity or the utility of certain substances deliberately added to water.

Alum, ferric chloride, activated silica, and in general all inorganic products used in clarification are unobjectionable, but the level of residual reagent in the water is nevertheless limited by regulations.

Some organic flocculation aids and certain polyphosphates may be used in drinking water treatment; the approval of the competent authorities is required for their use.

Doubts have been expressed as to the harmlessness of chlorine used in disinfection; simultaneous research is in hand on both chlorine and ozone in order to determine their limits of use.

6. EXAMINATION OF WASTE WATERS Evaluation of pollution

This section sets out the principal methods of analysis used to characterize a waste water before and after purification.

The measurements to be carried out are as follows:
— pH;
— BOD_5 in mg/l O_2 dissolved in the water;
— COD in mg/l of O_2 with potassium dichromate;
— posssibly, permanganate oxidizability;

— methylene blue test;
— suspended solids;
— ammoniacal nitrogen;
— total Kjeldahl nitrogen (organic nitrogen + ammoniacal nitrogen);
— nitrous nitrogen;
— nitric nitrogen;
— toxicity of an effluent discharged into the natural environment;
— special determinations.

No. 401—Colorimetric measurement of pH
    See Chapter 26 page 906.

No. 402—Measurement of BOD$_5$ (Biochemical oxygen demand)

No. 402-A—Dilution method

For this operation, suitable dilutions are prepared by mixing the water to be examined with pure inoculated water (see 402-C), with which periodical checks are made to ensure that it does not itself absorb appreciable quantities of oxygen. Inoculation is not necessary in the case of domestic sewage. The bes results are obtained when oxygen loss during the test is between 35 and 60 % of the initial content. The dilutions used depend on the degree of pollution. They can be determined by measuring the COD, knowing that the BOD$_5$ will normally be less than the COD and that the ratio of COD to BOD$_5$ is most often between 1.5 and 3. Assuming that a water stabilized at 20 °C contains about 8 mg/l of oxygen, the equation can be written as follows:

$$\text{dilution factor} = \frac{\text{presumed BOD}_5}{4} \quad .$$

At least three different dilutions must be prepared to be sure of including the presumed value.

The dilutions are kept in the dark for five days at 20 °C. The dilution water used must remain at this temperature and be in a state of perfect equilibrium with the atmosphere; this is easily achieved by keeping the reserve of pure water to be used for measurement in the controlled-temperature stove or in the thermostatic bath used for incubating diluted samples.

● *Procedure*

Prepare 500 ml of the diluted sample (use a ground-glass-stoppered flask, holding exactly 500 ml when stoppered). After thorough mixing, draw off two aliquot parts; put one in the incubator and use the other to measure the immediate dissolved oxygen. Prepare a control sample of diluting water, which may or may not be inoculated with microbes as mentioned above. After 5 days, measure the remaining oxygen. The difference between the two results for each dilution, less the consumption of the control sample, and allowing for the dilution factor, gives the quantity of oxygen per litre necessary for the biological purification of the waste water; the result is known as the five-day biochemical oxygen demand or BOD$_5$. This is a conventional value, since the total demand is only obtained after a longer period of incubation.

● *Notes:*

1. After dilution, the pH of the sample should be between 6.5 and 8. If it is not, the effluent must first be neutralized.

2. The diluting water can be prepared from drinking water or water from a very pure source; it may also be prepared from distilled water or deionized water of equivalent purity enriched with mineral salts at the rate of 1 ml per litre of each of the following solutions:

— solution of phosphates: 8.5 grammes of potassium di-hydrogen-phosphate (KH_2PO_4), 21.8 grammes of potassium mono-hydrogen-phosphate (K_2HPO_4) and 44.6 grammes of sodium mono-hydrogen-phosphate (Na_2HPO_4, 12 H_2O) dissolved in 1 litre of water;

— magnesium sulphate solution containing 20 g/l of $MgSO_4$, 7 H_2O;

— calcium chloride solution containing 25 g/l of $CaCl_2$;

— ferric chloride solution containing 1.5 g/l of $FeCl_3$;

— ammonium chloride solution containing 2 g/l of NH_4Cl.

● *Special cases*

When the effluent from a complete purification plant in which nitrification proceeds very actively has to be analysed, the nitrifying organisms sometimes continue their action in the incubating sample. Additional oxygen is absorbed and the BOD value of the effluent is substantially increased.

The procedure for analysis has then to be modified as follows:

1. Acidify the sample so that its pH is between 2 and 3;

2. Allow this acidity to act for 15 minutes;

3. Neutralize until the pH is between 7 and 7.4;

4. Use a dilution water inoculated with 5 ml of freshly-settled sewage effluent per litre of dilution water. The BOD$_5$ so obtained should be corrected, if necessary, to allow for the small quantity of organic matter added by inoculation.

N. 402-B—Use of respirometers

Place the incubating sample in a sealed vessel and shake it in the presence of air. The exchange of gases is checked by elimination of the evolved CO_2 (by alkali absorption) and measuring the consumption. The record taken can be either the loss of pressure (WARBURG respirometer and derivations) or the quantity of oxygen supplied to restore the initial oxygen pressure as required (SIERP respirometer and derivations). By this method it is possible to draw the consumption/time curve. It does not represent the phenomena observed in a river unless the water analysed is first diluted in the same proportion as the discharged effluent.

No. 402-C—Measurement of BOD$_5$ in non-putrefying industrial effluents

Since the measurement of BOD$_5$ is biological, care will naturally be taken to ensure the presence of microorganisms capable of breaking down the pollutants. The necessary microbial fluid can be obtained by trickling the water under analysis through loam or through sludge collected from the river bank downstream of the discharge point.

No. 403—Measurement of dissolved oxygen

No. 403-A—Chemical methods

a) In the water

In the case of the measurement of the oxygen content for the determination of the BOD_5, the measurement is to be made directly in the 125 ml flask in which the sample has been stored for five days.

In all other cases use also flasks holding approximately 125 ml. Fill them carefully with the water under examination, making sure that no bubbles can enter.

Quickly add 1 ml of manganese sulphate and 1 ml of nitrided potassium iodide with 2 ml pipettes reaching to the bottom of the flask, while discharging the reagent only as far as the 1 ml mark.

Re-stopper without entrapping an air bubble and shake the bottle.

After 10 minutes, again remove the stopper and add 0.5 ml of pure sulphuric acid or 1 ml of 50 % acid.

Shake, and when all the precipitate previously formed has redissolved, take 100 ml of the liquid and titrate it with a solution of N/80 sodium thiosulphate in the presence of starch or, preferably, vitex.

The number of ml of sodium thiosulphate indicates the dissolved oxygen content of the sample in mg/l.

Reagents used:

— manganese sulphate solution (400 g/l of $MnSO_4$ in distilled water);
— potassium iodide (dissolve 700 g of potassium hydroxide, 150 g of potassium iodide and 10 g of sodium nitride in 1 litre of distilled water);
— sodium thiosulphate (N/80 solution);
— iodine (N/80 solution);
— starch 4 g/l plus 1.25 g/l of salicylic acid, or, preferably, powdered vitex.

b) In the activated sludge

The activity of the sludge must be stopped quickly. To do so, pour 10 ml of inhibiting solution into a 500 ml flask and then transfer by siphoning about 700 ml of water containing activated sludge, taking care to fill the flask without entrapping any air bubbles; stopper, still without bubbles, shake and then leave the sludge to settle.

Siphon off the clear water at the top and collect it in a 125 ml flask. Continue with the measurement as before.

Copper sulphamate inhibiting solution: dissolve 32 g of sulphamic acid in 475 ml of cold water and dissolve 50 g of copper sulphate ($CuSO_4$, 5 H_2O) in 500 ml of water. Mix the two solutions and add 25 ml of acetic acid.

No. 403-B—Electro-chemical methods

There are several types of apparatus with probes (polarography or amperometry) which can be used to determine directly either the oxygen content or the saturation percentage as compared with a water of known oxygen content.

No. 404—Liability to oxidation by potassium permanganate

a) Cold (in an acid medium)

This measurement gives an idea of the organic matter in the effluent. The method consists of allowing a certain volume of the water under analysis to remain in contact with an acid solution of potassium permanganate, the excess being then measured after four hours' reaction time.

- *Reagents used*
- — N/80 permanganate solution, 1 ml of which is equivalent to 0.1 mg oxygen;
- — Mohr's salt solution (ferrous iron and ammonium sulphate), 5 g/l;
- — sulphuric acid solution, 50 % by volume.

- *Procedure*

50 ml of the water to be analysed is placed in a long-necked flask and 5 ml of 50 % sulphuric acid is added, followed by 50 ml of permanganate. The flask is then left for 4 hours at laboratory temperature. The specimen is then decolorized by adding 10 ml of Mohr's salt solution. The excess is then back-titrated with the N/80 permanganate solution until a pink colour is obtained. As the relation between the Mohr's salt and the permanganate solution is known, as well as the volume of water used, it is easy to deduce the quantity of oxygen needed to oxidize the organic matter contained in 1 litre of the treated water.

It is essential that there should be a permanent excess of permanganate throughout the four hour's oxidation time. Care should also be taken to ensure that the water being analysed does not contain large quantities of nitrites.

b) Hot (in an acid medium)

Add 10 ml of 50 % sulphuric acid to 100 ml of the water under analysis and heat to boiling point. Add 20 ml of N/80 permanganate and continue to boil for exactly 10 minutes.

Cool quickly and add 20 ml of 5 g/l Mohr's salt. Titrate with the N/80 permanganate and deduct from this figure the figure obtained when applying the same procedure to distilled water.

Taking V as the volume of permanganate recorded, liability to oxidation is expressed by V in mg oxygen per litre.

If oxidation is greater than 3.5 mg/l, the experiment must be repeated with a dilute sample.

No. 404-B—Chemical oxygen demand (COD)

This measurement is now replacing the old potassium permanganate method of assessing the "organic-matter content of water". The oxidizing power of potassium dichromate is greater than that of permanganate; but a number of straight aliphatic chains are not easily oxidized and the same applies to certain nuclei.

● *Reagents used:*
— distilled water or water of equivalent purity;
— mercury sulphate crystals;
— sulphuric acid in which silver sulphate has been dissolved.
Dissolve silver sulphate crystals in the sulphuric acid (specific gravity 1.84 at 20 °C) at the rate of 6.6 g/litre.
— 0.25 N ferrous ammonium sulphate solution (Mohr's salt) obtained by dissolving 98 g $FeSO_4$, $(NH_4)_2SO_4$, 6 H_2O in water, adding 20 ml of H_2SO_4 and making up to 1 litre.
Determination of the exact strength of the solution of iron sulphate and ammonium sulphate:
Dilute to approximately 250 ml (using distilled water) exactly measured 25 ml of the dichromate solution. Add 75 ml of sulphuric acid (ρ 20 = 1.84 g/ml), cool to room temperature and titrate with the iron and ammonium sulphate solution in the presence of several drops of ferroine (see below) solution (let V_T be the titration quantity).
— 0.25 N potassium dichromate solution. In the water, dissolve 12.2588 g of $K_2Cr_2O_7$ which has previously been dried for two hours at 110 °C, then make up to 1 litre in a graduated flask.
— Ferroine solution: dissolve 1.485 g of 10 % phenanthroline and 0.695 g of ferrous sulphate ($FeSO_4$, 7 H_2O) in water. Make up to 100 ml.

● *Procedure*
    A flat-bottomed flask with a side funnel and teflon tap can be used.
    The oxidation flask (see Fig. 567), equipped with a bar magnet, has a reflux cooler mounted above it. The ground neck should be lubricated beforehand with sulphuric acid.

Important: grease must not be used under any circumstances.

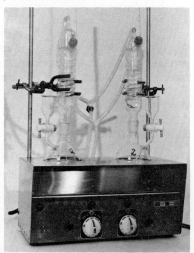

Fig. 567. — Hydrocure COD meter.

The stirring device, which may or may not be coupled with the heating device, is placed below the flask.

Through the funnel, pour in 50 ml of the effluent under examination or an aliquot made up to 50 ml with distilled water.

Start the stirrer.

Add 1 g mercury sulphate, then 5 ml sulphuric acid (containing silver sulphate) and dissolve. Make sure that no silver chloride precipitate persists—otherwise heat slightly until the precipitate dissolves. Add 25 ml (measured exactly) of the dichromate solution and then 70 ml sulphuric acid (with 6.6 g/l silver sulphate).

Close the tap and boil for 2 hours. The bar magnet, whether moving or not, assists steady boiling.

● *Measurement*

After boiling, leave to cool and make up to about 400 ml with distilled water; add a few drops of ferroine solution (see above) and reduce the excess dichromate with the ferrous iron and ammonium sulphate solution. The colour changes from green to purplish red; let V_1 be the volume used. At the same time, prepare a blank test using 50 ml of distilled water instead of the sample; let V_0 be the volume of ferrous iron and ammonium sulphate used.

COD expressed in mg/l of O_2 is given by the formula:

$$\frac{8\ 000\ (V_0 - V_1)\ t}{V}$$

where:

t   = strength of the ferrous iron and ammonium sulphate solution

V   = volume of the sample in ml of waste water

V_0, V_1 = volumes in ml of ferrous iron and ammonium sulphate.

This formula can be simplified to:

$$\frac{1\ 000\ (V_0 - V_1)}{V_T} \times \frac{50}{V}$$

V_T being the titration volume, if care has been taken to titrate exactly 25 ml with the ferrous iron and ammonium sulphate solution.

Note: This analysis method has been taken from French standard **AFNOR** NF T 90-101, September 1971 ([1]).

No. 405—Methylene blue test

The interpretation of this putrescibility test is very often disputed. It should be used with extreme caution because it can give false indications, particularly with industrial effluents containing reducing agents, and with weak effluents from purification plants (BOD_5 less than 15 mg/l) containing large quantities of oxidized salts which are liable to anaerobic reduction.

---

1. See note page 905.

Use a 100 ml flask with a ground-glass stopper. Into the flask put 1 ml o 0.05 % methylene blue and fill with the water to be analysed. Replace the stopper, taking care not to entrap an air bubble; heat the solution to 30 °C. Keep a watch on the flask for five days and note the moment at which the colour disappears (if such a change does occur) and the nature of the smell which emanates from the flask. The test is said to be positive if the colour disappears before the five days have elapsed, and if the odour from the flask is putrid.

No. 406—Measurement of suspended solids

Suspended solids present in waste water can be measured in several ways:

● *By centrifugation*

100 ml of well-shaken sewage effluent is placed in several tubes and centrifuged for 10 to 15 minutes at 3 000 rev/min (the normal speed for commercial laboratory centrifuges). The suspended solids thus become compacted at the bottom of the tubes by the effect of rotation, and the clear liquid, which is later used for determining the dissolved solids, is carefully decanted. The precipitate is then suspended in distilled water, centrifuged again and the supernatant liquid discharged to the drain.

When this second operation has been repeated twice more, the precipitate is washed by means of a jet from a wash bottle into a silica or platinium evaporation dish. It is first dried over a water-bath, then heated at 105 °C to constant weight (24 hours are quite sufficient) and then weighed (total suspended solids); it is then ashed at 600 °C and the volatile fraction (loss by combustion, sometimes taken to be the same as the organic matter) is obtained by the difference between total solids and the ashed residue.

In the case of river water or purified sewage effluents, the volumes to be handled are much larger and require a centrifuge fitted with large tubes or a rapidly rotating vertical separating bowl. Lastly, the centrifuging operation may be replaced by filtration or a 24-hour settling process; the results obtained from these latter methods are, however, somewhat less accurate.

● *By filtration*

A fixed quantity of water to be analysed is percolated through an ashless filter paper, previously dried at 105 °C, and weighed. Dry at 105 °C for about 2 hours until the weight is constant (total suspended matter); then ash at 600 °C (volatile matter or loss by combustion).

The ashless filters can be replaced by glass fibre filters (Millipore AP 20, Durieux D 28, Sartorius FN 501, etc.). These glass fibre filters may only be used for the determination of total suspended solids.

● *Densimetric method* (for activated sludge). *Use of the "Ponsarimeter"*

A rapid method of determining suspended solids, giving approximate results which are of value for the day-to-day operation of activated sludge treatment plants, is to weigh by densimetry a given volume of activated sludge liquor contained in a very light plastic bag and immersed in the settled (interstitial) water of the activated sludge. The areometer indicates the weight of suspended solids (p) in the relevant activated sludge sample in centigrammes.

In parallel with the weighing of the sludge, the sludge may also be subjected to a settling test in a 2.5 dm³ "enlarged test glass", thus giving the apparent volume (v) of the sludge deposited after 30 minutes; the two parameters required to define a quality index are then available:

$$Ip = \frac{v}{p} \text{ (in ml/g)}$$

This is as good a measure of the properties of an activated sludge as the Mohlman or Donaldson index (Chapter 3, page 71).

The combination of the sludge weighing system and the "enlarged test glass" constitutes the *Ponsarimeter*.

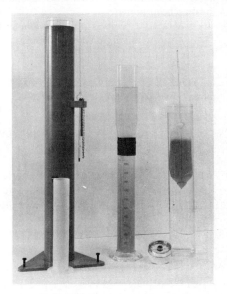

Fig. 568. — Ponsarimeter.

No. 407-A1—Determination of ammoniacal nitrogen (after French standard NF T 90 015) [1]—acidimetry

This method (acidimetry after distillation) is applicable to waters containing more than 4 mg/l of NH_4^+ (with a 50 ml sample). Amines and ammonia-hydrolysable substances interfere.

● *Reagents:*
— Sodium carbonate, 100 g/l solution;
— Boric acid, 10 g/l solution;
— Sulphuric acid, solution titrated to 0.1 N or 0.02 N;
— Tashiro indicator (methyl red + methylene blue).

● *Procedure*

Sampling:

Take a volume V_2 of sample containing between 0.2 and 20 mg of ammoniaca nitrogen expressed as NH_4.

---

1. See note page 905.

Determination:

Introduce the sample into the flask of the distillation apparatus. Add 20 ml of sodium carbonate solution and, if necessary, an antifoaming agent. Admit steam for at least 20 minutes, collecting the distillate in 5 ml of the above boric acid solution to which a few drops of indicator and enough water for bubbling has been added.

Check that there is no ammonia in the final distillate fractions.

Titrate with sulphuric acid, using:

— a 0.1 N titrated solution if the sample contains between 2 and 20 mg of ammoniacal nitrogen expressed as NH_4;

— a 0.02 N titrated solution if the sample contains between 0.2 and 2 mg ammoniacal nitrogen expressed as NH_4.

Blank test:

Carry out a blank test under the same conditions as the determination.

● **Expression of results**

Notation:

V_1 volume in millilitres of the sulphuric acid solution used for determination;
V_o volume in millilitres of the sulphuric acid solution used for the blank test;
T normality of the titrated sulphuric acid solution used (0.1 N or 0.02 N);
V_2 sample volume in millilitres.

The ammoniacal nitrogen content per litre is given by the expression:

$$\frac{(V_1 - V_o)\, T \times 1\,000 \times 18}{V_2} \text{ in mg/l } NH_4$$

$$\frac{(V_1 - V_o)\, T \times 1\,000 \times 14}{V_2} \text{ in mg/l } N$$

No. 407-A2—Determination of ammoniacal nitrogen—Colorimetry
(in accordance with French standard NF T 90 015)[1]

In determining ammoniacal nitrogen in waste waters, interference due to suspended solids and to alkaline earth ions liable to precipitate in an alkaline medium must be eliminated. Suspended solids are eliminated by clarification with alkaline zinc sulphate; alkaline earth ions are chelated by the addition of EDTA or Rochelle salt (potassium-sodium tartrate). NH_4^+ ions are determined by colorimetry using Nessler's reagent; this method is applicable for concentrations in the range 0.1 to 6 mg/l.

● *Reagents*

— Distilled water containing no ammonium ions.

— Zinc sulphate solution:

— $ZnSO_4$, 7 H_2O 100   g/l

— Caustic soda solution:

— NaOH 250   g/l

---

1. See note page 905.

— potassium-sodium tartrate:
  — Potassium-sodium tartrate.............................. 392 g
  — Water .... 748 ml
  — Potash lye d = 1.33 .... 5.5 ml
  — Soda lye d = 1.33 .... 4.5 ml
— Nessler's reagent: prepare:
  — Mercuric iodide....................................... 45.5 g
  — Potassium iodide..................................... 35 g
  — Water .... 15 ml
in a 1 litre flask. Mix the following separately:
  — Potash lye d = 1.33................................... 333 ml
  — Water .... 500 ml
and add to the mercuric iodide solution; make up to a litre.

NH_4^+ standard solution = 133 mg NH_4 Cl/l (2.5 meq/l).

● *Procedure*
— *Clarification:*
  Add 1 ml of zinc sulphate solution to a 100 ml sample. Stir, and then add 0.5 ml of soda solution (the pH must be close to 10.5). Centrifuge or filter through paper.
— *Calibration:*
  Introduce the following into a series of 50 ml graduated flasks:

— Standard NH_4^+ solution (ml)	1	2	3	4	5	6
— meq/l NH_4^+	0.05	0.1	0.15	0.2	0.25	0.3
— mg/l NH_4^+	0.9	1.8	2.7	3.6	4.5	5.4
— mg/l N	0.7	1.4	2.1	2.8	3.5	4.2

  Add 2 ml of potassium-sodium tartrate solution, stir, and add 2 ml of Nessler's reagent; stir and leave to stand for 10 minutes. Measure the intensity of the coloration at 420 nm. Plot the calibration curve.
— *Determination:*
  Add the tartrate solution and Nessler's reagent to a 50 ml sample containing less than 0.3 mg NH_4, proceeding as in the case of calibration. Measure on the photometer at 420 nm and read off the concentration from the calibration curve.

No. 408—Determination of total nitrogen (organic + ammoniacal)
  Kjeldahl's method converts organic nitrogen into NH_3 by digestion with sulphuric acid and potassium sulphate in the presence of mercuric sulphate. The temperature must remain below 380 °C. After dilution, the solution is alkalinized by caustic soda. It is then distilled, the ammonia being collected in a boric acid solution. The ammonium borate is then determined by a titrated sulphuric acid solution. Interference from organic substances is eliminated by treatment with a solution of sulphuric acid, mercuric sulphate, and potassium sulphate (50 ml/g suspended solids).

● *Reagents:*
— solution of sulphuric acid + mercuric sulphate + potassium sulphate.

Dissolve 267 g of K_2SO_4 in 1.3 l of water; add 400 ml of concentrated H_2SO_4, and then 4 g of $HgSO_4$, and dilute to 2 litres;
— thiosulphate solution: 25 g of $Na_2S_2O_3$ and 500 g of NaOH to 1 litre of water;
— phenolphthalein indicator: 0.5 %;
— mixed indicator: 2 volumes of 0.2 % methyl red in alcohol + 1 volume of 0.2 % methylene blue in alcohol;
— boric acid: 20 g/l. Add 10 ml of mixed indicator per litre;
— 0.02 N solution of sulphuric acid, 1 ml = 0.28 mg nitrogen N;
— concentrated sulphuric acid.
● *Procedure:*
— *Sample:*

Total N in mg/l	sample in ml
0 to 10	250
10 to 20	100
20 to 50	50
50 to 100	25

● *Determination:*

Introduce the sample into an 800 ml receiver. Add 50 ml of sulphate acid reagent + 50 ml per gramme of suspended solids present. Boil until the solution turns clear; cool and add 300 ml of water containing no NH_4^+. Alkalinize with the alkaline thiosulphate solution in the presence of phenolphtalein to pH 9.5. Distil in 50 ml of boric acid until 200 ml of distillate has been collected (temperature of lower condenser 29 °C). Titrate with the 0.02 N sulphuric acid solution until the indicator turns lavender (V_1). Carry out a blank determination with the reagents (V_2).

The total nitrogen content N is given by:

$$N \text{ mg/l} = \frac{(V_1 - V_2) \times 280}{\text{Sample volume (ml)}}$$

No. 409—Determination of nitric nitrogen (in accordance with French standard NF T 90 012)[1]

Nitrates react with sulphosalicylic acid to form a coloured compound; they are determined by spectrophotometric measurement at 415 nm. The method is applicable to levels exceeding 0.05 mg/l expressed as NO_3.
● *Reagents:*
— distilled water or water of equivalent purity;
— sulphuric acid d = 1.84;
— crystallizable acetic acid;
— ammonia d = 0.9;
— sodium salicylate, 10 g/l, to be prepared fresh each day;
— sodium azide, 5 g/l;
— silver sulphate, washed in water to eliminate nitrates;

1. See note page 905.

— standard solution, 5 mg/l NO_3;
— stock solution, 100 mg NO_3 per litre.

● *Procedure:*
Sampling:
— 25 ml if the sample contains less than 1 mg/l;
— 5 to 25 ml if the sample contains between 1 and 5 mg/l.
The analytical sample must contain less than 25 μg of NO_3.
Calibration (40 mm vessels):

Introduce into a set of borosilicate glass dishes 0, 1, 2, 3, 4 and 5 ml of reference solution with 5 mg/l NO_2, corresponding to 0, 5, 10, 15, 20 and 25 μg of NO_3.

Add 0.2 ml acetic acid to each dish. Wait 5 minutes and evaporate until dry on a boiling water bath. Then add 1 ml sodium salicylate solution, homogenize, and evaporate until dry. Allow to cool in a desiccator. Add 1 ml sulphuric acid, and wait 10 minutes. Add 10 ml water and 10 ml ammonia. Transfer the contents of each dish to a 25 ml graduated flask. Make up to 25 ml with the dish rinse water. Perform the colorimetric measurement at the absorption maximum -(about 415 nm) against pure water.

● *Determination:*
— *Checking the chlorides content:*

If the chlorides content is less than 100 mg/l, take a sample and develop the coloration directly. If it exceeds 100 mg/l, neutralize to pH 6 by addition of acetic acid. Add enough silver sulphate to precipitate the chlorides (4.4. mg Ag_2SO_4 per mg of Cl). Heat to 40 ºC, allow to stand for 15 minutes, and filter through a 0.45 μm membrane. Take a sample and develop the coloration.
— *Development of coloration and spectrophotometry:*

Check the pH; if it exceeds 8, neutralize with acetic acid. Place the analytical sample in a borosilicate glass dish. Add 0.5 ml of sodium azide solution and 0.2 ml acetic acid. Wait 5 minutes, and then evaporate until dry on a boiling water bath. Add 1 ml of sodium salicylate solution, homogenize, and evaporate until dry. Allow to cool in the desiccator. Continue in the same way as for plotting the calibration curve.
— *Compensation test:*

Organic matter may interfere; perform a compensation test by treating a fresh specimen as above without adding sodium salicylate.

● *Expression of results:*

The results read off the calibration curve and expressed in mg/l NO_3 can be converted into mg/l N.

Nitric nitrogen expressed as N = Nitric nitrogen expressed as $NO_3 \times 0.226$.

No. 410—Determintaion of nitrous nitrogen (in accordance with French standard NF T 90 013)[1]

1. See note page 905.

The intensity of the coloration (at 537 nm) of the diazotized compound formed by reaction of nitrites on sulphanilamide is measured.

● *Reagents:*

— Standard nitrite solution, 1 mg NO_2/l, prepared at the time of use from a solution containing 150 $\pm$ 0.1 mg/l $NaNO_2$.

— Diazotation reagent: Add 100 ml concentrated orthosphoric acid and then 40 g of sulphanilamide to 800 ml water. Allow to dissolve, and then add 2 g of N—(1-naphthyl) ethylenediamine dichloride. Stir until completely dissolved and make up to a litre. Keep in a dark-coloured bottle in the refrigerator.

● *Procedure:*

— *Calibration (10 mm tubes):*

Introduce the following into a set of 50 ml graduated flasks: 0, 1, 2, 5, 10, 20, 30, 40, and 50 ml of standard solution with 5 mg NO_2/l, corresponding respectively to 0, 0.1, 0.2, 0.5, 1, 2, 3, 4, and 5 mg/l NO_2. Make up to 50 ml and homogenize. Add 1 ml of diazotation reagent to each flask, and stir. Wait 10 minutes and carry out photometric measurement at the maximum of the absorption peak (close to 537 nm) relative to water. Plot the calibration curve.

● *Determination:*

Add 1 ml of diazotation reagent to the 50 ml analytical sample (containing ess than 1 mg NO_2/l), and proceed as with calibration. The nitrite concentration of the sample, expressed in milligrammes of NO_2/l, is derived from the calibration curve. The content of nitrous nitrogen, expressed in mg/l N, is obtained by multiplying the previous results by the factor 0.304.

No. 411-G—Determination of phosphates

Reduction of ammonium phosphomolybdate by stannous chloride forms a compound whose blue coloration is proportional to the phosphates content.

● *Reagents:*

— distilled water;

— concentrated sulphuric acid containing no phosphates;

— ammonium molybdate solution. Dissolve 50 g ammonium molybdate in 750 ml of 5 N sulphuric acid and make up to 1 litre with the same acid;

— stannous chloride solution. Dissolve 5 g of stannous chloride in 750 ml 10 % vol. hydrochloric acid and make up to 1 litre with the same acid. Keep in a dark-coloured bottle in the presence of tin swarf (1 g/100 ml). The solution will keep for 5 to 8 days on average;

— phenolphthalein indicator;

— 0.1 N potash solution;

— 100 g/l sulphamic acid solution;

— 100 mg/l standard orthophosphate solution.

● *Procedure:*

— *Calibration:*

Introduce the following into a set of 250 ml graduated flasks: 0, 1, 2, 3, 4, 5, 6, ... 18, 19, and 20 ml of 0.1 g/l standard solution, corresponding respectively to PO_4 contents of 0, 1, 2, 3, 4, 5, ... 20 mg/l on the calibration curve. Add 5 ml sulphuric acid, make up to 250 ml and homogenize. Introduce 25 ml of this solution into a 100 ml graduated flask. Add 1 ml sulphamic acid solution, 1 ml ammonium molybdate solution, and 1 ml stannous chloride. Make up to 100 ml and homogenize. Wait for exactly 10 minutes and carry out photometric measurement at 670 nm against pure water. Plot the calibration curve.

● *Determination:*

Introduce exactly 100 ml of sample (after filtration if necessary) into a 250 ml graduated flask. Add a few drops of phenolphthalein and neutralize so that a very pale pink colour remains. Then proceed as in the case of calibration, referring to the curve to obtain the content of PO_4^{3-} in mg/l.

● *Expression of results:*

The phosphates content can be expressed in mg/l of PO_4, P, or P_2O_5.
1 mg/l PO_4 = 0.326 mg/l P = 0.747 mg/l P_2O_5.

No. 412 G—Determination of polyphosphates

The polyphosphates are hydrolysed in an acid medium and total phosphates are determined by colorimetry. The difference (total phosphates—phosphates determined without acid hydrolysis) gives the polyphosphate content. Introduce a water sample (200 ml) into a COD flask. Add 2.5 ml nitric acid (d = 1.3). Boil for 2 hours and then cool. Take a sample of exactly 100 ml and introduce into a 250 ml flask. Neutralize to the colour change point of phenolphthalein. Then determine phosphates as above.

No. 413—Determination of phenol index (in accordance with French standard NF T 90 109)[1]

● *Principle:*

After separation by distillation phenols react in the presence of potassium ferricyanide with 4-aminoantipyrine to form a coloured compound.

● *Reagents:*
— sodium chloride;
— phosphoric acid, 85 % solution;
— copper sulphate, 100 g/l solution;
— buffer solution:
   — ammonium chloride . . . . 34 g ⎫
   — potassium-sodium tartrate. . . . . . 200 g ⎬ to 1 litre
   — ammonia (d = 0.92) . . . . 15 ml ⎭
— potassium ferricyanide, 2 % solution;
— 4-aminoantipyrine, 2 % solution (in dark-coloured bottle);
— phenol, 1 g/l standard solution.

Samples:

When the sample (500 ml) is taken, add 5 ml of copper sulphate solution and acidify to pH 4 with phosphoric acid.

---

1. See note age 905.

● *Procedure:*
Distillation:
Introduce 200 ml of sample into a 500 ml flask and acidify to pH 1.5 with H_3PO_4. If the sample contains sulphites, blow away with nitrogen for 15 minutes. Add 50 g sodium chloride and distil, keeping the water level approximately constant (adding distilled water through a lateral funnel).

● *Determination:*
Add 5 ml of buffer solution, followed by 1 ml of 4-aminoantipyrine solution and 2 ml of potassium ferricyanide solution, to an analytical sample containing less than 0.5 mg phenol equivalents. Wait 5 minutes and then carry out photometric measurement at 510 nm against a calibration curve plotted from pure phenol solutions (0 — 5 mg/l).

No. 414-A—Determination of total hydrocarbons (in accordance with French standard NF T 90 203) [1]

● *Principle:*
— extraction of hydrocarbons with carbon tetrachloride;
— separation of hydrocarbons from other organic compounds by chromatography in a column containing adsorbent;
— spectrophotometric determination at the wavelength corresponding to the absorption maximum (approx. 3 420 nm).

● *Reagents:*
— adsorbent: synthetic adsorbent silica $+$ magnesia;
— carbon tetrachloride, spectrophotometry grade;
— anhydrous sodium sulphate;
— 1 N solution of hydrochloric acid;
— sodium chloride.

● *Calibration:*
Synthetic control:
Make a 100 mg/l solution of the following mixture in CCl_4:
— 37.5 % isooctane ⎱
— 37.5 % hexadecane ⎰ volume percentages
— 25 %   benzene ⎱
Refinery standard:
Whenever possible, the standard should be prepared from slops recovered from the refinery separator and dried on anhydrous sodium sulphate.

● *Procedure:*
Add the following to a 1 litre glass bottle containing a weight m of effluent: 1 N hydrochloric acid until the pH reaches 5, followed by 5 g of sodium chloride (in the case of soft waters) and 50 ml carbon tetrachloride. Stir for 15 minutes (high-speed magnetic stirrer). Allow to settle for 10 minutes. Draw off 20 ml of the CCl_4 phase and filter through anhydrous sodium sulphate. Then trickle

---

1. See note page 905.

through a 1 cm diameter column containing 5 g of adsorbent. Determine the absorbency of the filtrate at the absorption maximum (about 3 420 nm) relative to pure carbon tetrachloride. Refer to the calibration curve; let C be the hydrocarbon concentration of the sample. The content of total hydrocarbons in the effluent, expressed in milligrammes per kilogramme, is:

$$T = C \times \frac{V}{m}$$

in which V is the volume of CCl_4 (usually 50 ml) introduced into the bottle.

No. 414-B—Determination of hexane-extractible organic matter in suspension in water (in accordance with French standard NF T 90 202)

● *Principle:*
— absorption and flocculation by aluminium hydroxide of the organic matter suspended in the water, including hydrocarbons;
— hexane extraction of flocculate;
— gravimetric determination after evaporation of hexane.

● *Reagents:*
— alum, 20 g/l solution;
— sodium carbonate, 20 g/l solution;
— hydrochloric acid, 1 N solution;
— anhydrous sodium sulphate;
— hexane.

● *Procedure:*
Flocculation of organic matter:
　　Add 10 ml alum solution to the glass sampling bottle containing a known weight of water (about 900 g), stir for 10 minutes, and adjust the pH to 6-7 with the hydrochloric acid solution. Leave to stand for 2 hours. Filter.
Extraction of hydrocarbons from the flocculate:
　　Spread out the filter at the bottom of a crystallizing dish and add 30 ml hexane. Rinse the bottle with 10 ml hexane and transfer to the crystallizing dish. Leave in contact for at least 1 hour. Then pass through another filter containing a few centigrammes of anhydrous sodium sulphate. Collect in a previously weighed anti-creep dish. Rinse the crystallizing dish and the first filter with 10 ml hexane and transfer to the weighed dish. Make up to 50 ml with hexane. Place 50 ml of pure hexane in a second dish.
N.B.: Dishes must have been previously weighed (weights M_1 and M_2).
Evaporation:
　　Place the two dishes in an oven under vacuum (50 mm Hg) at 30 °C for five hours. Transfer them to a desiccator and weigh after 2 hours (weights M'_1 and M'_2). The content of extractible organic matter is:

$$\frac{(M'_1 - M_1) - (M'_2 - M_2)}{\text{sample weight}} \times 1\ 000 \text{ mg/kg}$$

No. 415—Determination of "free" cyanides (in accordance with French standard NF T 90 108) [1]

● *Reagents:*
— Sodium hydroxide, 1 N solution;
— Sodium hydroxide, 0.2 N solution;
— Phosphates buffer (pH = 7.1):
   — 112 g $Na_2HPO_4 . 12 H_2O$ } to 1 litre;
   — 42.5 g KH_2PO_4 }
— Potassium cyanide, 1 g/l solution:
   — 2.5 g KCN to 1 litre of 0.2 N NaOH solution;
— Potassium cyanide, 4 mg/l solution [2];
— 1/50 acetic acid;
— Acetate buffer (pH = 6);
— Chloramine T, 10 g/l solution [2];
— Pyridine-pyrazolone reagent :
   4 g of 1-phenyl-3-methyl-5-pyrazolone } in 100 ml pyridine.
   + 0.08 g of bis-pyrazolone }

Microdiffusion:
   Introduce 2 ml of 1 N caustic soda into the central cavity of a microdiffuser. Introduce 2 ml of phosphate buffer and 5 ml of the analytical sample (containing less than 20 µg of CN^-) into the outer cavity.   Close the microdiffuser and place in darkness for 24 hours.

Calibration curve:
   Introduce 0, 0.5, 1, 2, 3, 4, and 5 ml of a 4 mg/l CN^- solution into a set of beakers.   Make up the volume to 5 ml with 0.2 N NaOH solution, then add 20 ml distilled water and 5 ml acetate buffer.   Adjust to pH 6.5 and transfer to a 50 ml flask.   Add 0.5 ml chloramine T, stir, and wait 2 minutes.   Finally add 1 ml pyridine-pyrazolone reagent, make up to 50 ml, stir, and allow to stand in darkness for 25 minutes.   Mesure in the photometer at 620 nm.

● *Determination:*
   After 24 hours' microdiffusion, draw off 1 ml of solution from the central cavity and place in a 50 ml beaker.   Add 25 ml of distilled water, followed by 5 ml of acetate buffer.   Adjust the pH to 6.5 and proceed as above.

No. 416—Toxicity of an effluent discharged to the natural environment
   This method is described in a French experimental standard (NF T 90 301), to which reference may usefully be made [1].
   The parameter determined is the short-term inhibition in the effluent of the mobility of Daphnia magna Straus (crustaceans, Cladocera), commonly known as daphnid.   The results are expressed in equitox units, defined as follows:an effluent contains one equitox/m³ if it gives rise in 24 hours to the immobilization of 50 % of a population of daphnids under the test conditions.

---

1. See note page 905.
2. To be prepared at the time of use.

The daphnids used for this test are selected by size (between 560 and 800 μm); a sample from a batch of such daphnids is subjected to a preliminary test with potassium dichromate as a toxicity standard (immobilization in 24 hours of 50 % of daphnids at a concentration $C_0 = 1.2$ mg/l of $K_2Cr_2O_7$, experimental value between 0.9 and 1.5 mg/l).

Then perform a series of dilutions of the effluent to be examined and note the concentration C_1 which immobilizes 50 % of the daphnids. A preliminary test is carried out for approximate determination of C_1 followed by a final test from which the actual result is determined. The number of equitox/m³ effluent is equal to the reciprocal of the concentration C_1 corrected in accordance with the value of C_0.

7. EXAMINATION OF FILTERING MATERIALS AND POWDERS

In the case of high-density filtering materials (such as sand) larger test quantities will be used than in the case of low-density materials such as anthracite, granulated activated carbon, etc.

No. 501—Particle size distribution of filter material

Weigh out 100 grammes of material (50 grammes in the case of granulated activated carbon) after drying for 4 hours at 120 °C.

Pass the material through the series of standard sieves (AFNOR No. X 11-501) [1] and note the weight retained on each sieve.

From these results, calculate the weight of material which has passed through each sieve (total material retained or passed by all the sieves of smaller mesh than the one under consideration) and express this as a percentage of the weight of material used for the analysis.

Draw the curve showing these percentages in terms of the mesh void of each sieve.

For preference, use semi-log paper.

Effective size:

This is the size corresponding to 10 % on the graph.

Coefficient of uniformity:

From the graph, read off the size corresponding to 60 %.

The coefficient of uniformity is the ratio: $\text{C.U.} = \dfrac{60 \% \text{ oversize}}{10 \% \text{ oversize}}$

The coefficient should be lower than 1.6, but up to 1.8 is acceptable.

Note: All standardized gauzes and perforated plates are set out in the tables

---

1. See note page 905.

on pages 941 to 943, together with their equivalents for the measuring sieves used in British, American, and other standards.

No. 502—Friability

The friability of a material is calculated by assessing the quantity which can be used after milling, i.e. with the same effective size as the original sample.

● *Procedure:*

The friability test is performed on 35 ml of accurately weighed material.

The material is placed in a metal cylinder with an internal diameter of 40 mm and a working height of 100 mm.   Secure the cylinder radially to a wheel 34 cm in diameter.   Rotate the wheel on a central spindle at 25 rev/min.

Place 18 steel balls 12 mm in diameter in the cylinder.

Two measurements are necessary to assess friability: first after 15 minutes (750 strokes or 375 revolutions) and then after 30 minutes (1 500 strokes or 750 revolutions).   The particle size distribution curve for the material is plotted after each experiment.

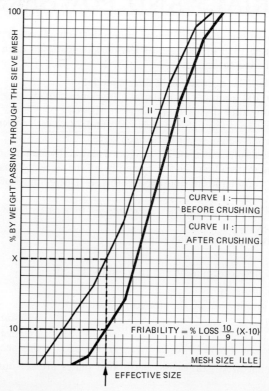

Fig. 569. — Friability study for filter media.

Measuring sieves

WIRE GAUZES				PERFORATED PLATES			
NOMINAL MESH SIZE		NOMINAL MESH SIZE		NOMINAL SIZE OF SQUARE HOLE		NOMINAL SIZE OF ROUND HOLE	
Main dimensions	Supplementary dimensions	Main dimensions	Supplementary dimensions	Main dimensions	Supplementary dimensions	Main dimensions	Supplementary dimensions
mm		μm		mm		mm	
125		800	900	125		125	
	112				112		112
100			710	100		100	
	90	630			90		90
80			560	80		80	
	71	500			71		71
63				63		63	
	56		450		56		56
50		400		50		50	
	45		355		45		45
40		315		40		40	
	35.5		280		53.5		35.5
31,5		250		31.5		31.5	
	28		224		28		28
25				25		25	
	22.4	200			22.4		22.4
20			180	20		20	
	18	160			18		18
16			140	16		16	
	14	125			14		14
12.5			112	12.5		12.5	
	11.2	100			11.2		11,2
10			90	10		10	
	9	80			9		9
8			71	8		8	
	7.10	63			7.10		7.10
6.3				6.3		6.3	
	5.6		56		5.6		5.6
5		50		5		5	
	4.5		45		4.5		4.5
4		40		4		4	
	3.55		36 *				3.55
3.15						3.15	
	2.8	32 *					2.8
2,5			28			2.5	
	2.24						2.24
2		25				2	
	1.8		22 *				1.8
1.6						1.6	
	1.4	20					1,4
1.25						1.25	
	1.12						1.12
1						1	

Figures underlined are given in ISO Recommendation R 565.

* These three values are rounded from the standard numbers 35.5, 31.5, and 22.4, as the number of significant digits in the latter has no real physical meaning.

MEASURING SIEVES
Table of wire gauze equivalents
STANDARDS IN FORCE

ISO Main (mm)	ISO Suppl. (mm)	FR/GER Main (mm)	FR/GER Suppl. (mm)	ASTM Nom. mesh (mm)	ASTM Desig. inches	ASTM Desig. N°	TYLER Main inches	TYLER Suppl. inches	TYLER Desig. mesh	TYLER Corresp. opening mm	UK Main (mm)	UK Suppl. (mm)
125		125		125	5							
			112									
				106	4.24							
	100	100		100	4							
90.0			90	90	3 1/2							
	80.0	80										
				75	3							
			71									
63.0		63		63	2 1/2							
			56									
				53	2.12							
	50.0	50		50	2							
45.0			45	45	1 3/4							
	40.0	40										
				37.5	1 1/2							
			35.5									
31.5		31.5		31.5	1 1/4							
			28									
				26.5	1.06		1.050	1.050		26.5		
	25.0	25		25.0	1							
22.4			22.4	22.4	7/8			0.883		22.4		
	20.0	20										
				19.0	3/4		0.742	0.742		19.0		
			18									
16.0		16		16.0	5/8			0.624		16.0	16.0	
			14									
				13.2	0.530		0.525	0.525		13.2		13.2
	12.5	12.5		12.5	1/2							
11.2			11.2	11.2	7/16			0.441		11.2	11.2	
	10.0	10										
				9.5	3/8		0.371	0.371		9.5		9.50
			9									
8.00		8		8.0	5/16			0.312	2 1/2	8.0	8.00	
			7.10									
				6.7	0.265		0.263	0.263	3	6.7		6.70
	6.30	6.3		6.3	1/4							
5.6			5.6	5.6		3 1/2		0.221	3 1/2	5.6	5.60	
			5									
				4.75		4	0.185	0.185	4	4.75		4.75
			4.5									
4.00		4		4.00		5		0.156	5	4.00	4.00	
			3.55									
				3.35		6	0.131	0.131	6	3.35		3.35
			3.15									
2.8			2.8	2.80		7		0.110	7	2.80	2.80	
			2.5									
				2.36		8	0.093	0.093	8	2.36		2.36
			2.24									
2.00		2		2.00		10		0.078	9	2.00	2.00	
			1.8									
				1.70		12	0.065	0.065	10	1.70		1.70
			1.6									

Column groups:
- **ISO** — ISO/R 565 1967, Nominal mesh size (Main / Supplementary, mm)
- **FRANCE / GERMANY** — NF X 11-501 (1970), DIN 4188 1969, Nominal mesh size (Main / Supplementary, mm)
- **U.S.A. and CANADA** — ASTM E 11-70 (1970), Nominal mesh size (mm), Designation (inches, N°)
- **U.S.A. TYLER** — Nominal mesh size (Main / Supplementary, inches), Designation (mesh), Corresponding opening (mm)
- **UNITED KINGDOM** — BS 410 1969, Nominal mesh size (Main / Supplementary, mm)

STANDARDS IN FORCE

ISO		FRANCE-GERMANY		U.S.A. and CANADA			U.S.A.				UNITED KINGDOM	
ISO/R 565 1967		NF X 11-501 (1970) DIN 4188 1969		ASTM E 11-70 (1970)			TYLER				BS 410 1969	
Nominal mesh size		Nominal mesh size		Nominal mesh size	Designation		Nominal mesh size		Designation	Corresponding opening	Nominal mesh size	
Main	Supplementary	Main	Supplementary				Main	Supplementary			Main	Supplementary
mm	mm	mm	mm	mm	inches	N°	inches	inches	mesh	mm	mm	mm
1.4			1.4	1.40		14		0.055	12	1.40	1.40	
		1.25										
			1.12	1.18		16	0.046	0.046	14	1.18		1.18
1.00		1		1.00		18		0.039	16	1.00	1.00	
μm		μm	μm	μm		N°	in 10⁻⁴	in 10⁻⁴	mesh	μm	μm	μm
			900									
				850		20	328	328	20	850		850
710		800	710	710		25		276	24	710	710	
		630		600		30	232	232	28	600		600
500		500	560	500		35		195	32	500	500	
			450	425		40	164	164	35	425		425
355		400	355	355		45		138	42	355	355	
		315		300		50	116	116	48	300		300
250		250	280	250		60		97	60	250	250	
			224									
				212		70	82	82	65	212		212
180		200	180	180		80		69	80	180	180	
		160		150		100	58	58	100	150		150
125		125	140	125		120		49	115	125	125	
			112									
				106		140	41	41	150	106		106
90		100	90	90		170		35	170	90	90	
		80		75		200	29	29	200	75		75
63		63	71	63		230		24	250	63	63	
			56									
				53		270	21	21	270	53		53
45		50	45	45		325		17	325	45	45	
		40		38		400	15	15	400	38		38
			36									
		32										
			28									
		25										
			22									
		20										

● *Calculation of friability:*

X represents, after crushing, the percentage of material smaller than the initial "effective size"; the fraction of bigger size is $(100 - X) \%$ and represents 90 % of the material usable after crushing.

The fraction: $\dfrac{100}{90} (100 - X)$ can therefore be used.

The loss is:

$$100 - \frac{100}{90} (100 - X) = \frac{100}{90} (90 - 100 + X)$$

The percentage loss is therefore $\dfrac{10}{9} (X - 10)$.

This loss measures the friability of the material.

● *Friability limits*

		15 minutes (750 strokes)	30 minutes (1 500 strokes)
Normal range of use	very good	6 to 10 %	15 to 20 %
	good	10 to 15 %	15 to 25 %
	poor	15 to 20 %	25 to 35 %
	reject	Over 20 %	Over 35 %

This quality scale is based on results obtained with the majority of everyday filter materials.

No. 503—Loss in acid

This is the loss of weight after contact for twenty-four hours with a 20 % hydrochloric acid solution.   The figure must be less than 2 %.

No. 504-A—Bulk density in air

Weigh 100 grammes of material and pour into a measuring cylinder.

Let V be the volume read on the cylinder.   The bulk density of the uncompacted material is:

$$\rho_a = \frac{100}{V} \ \text{g/ml}$$

Density can also be measured after the material has been tamped down in the cylinder.

No. 504-B—Bulk density after washing and draining the interstitial water

Take about 100 grammes of material and pour into a beaker.   Wet thoroughly with distilled water and remove all air from the grains by boiling and stirring for 5 minutes.

Cool and drain off the interstitial water. Weigh out 100 g of the damp material and place it in a graduated cylinder. Let V' be the reading for the volume in the cylinder. The density of the damp material is:

$$\rho_a' = \frac{100}{V'} \text{ g/ml}$$

It is useful to correct the volume V' by weighing the same volume of water in the cylinder used.

No. 504-C—True bulk density

● *Non-porous material:*

Weigh 50 g of material and introduce into a 250 ml cylinder containing 100 ml of water. Let V be the volume read on the cylinder. The true bulk density is:

$$\rho = \frac{50}{V - 100} \text{ g/ml}$$

The volumes read on the cylinder can be corrected by weighing the same volumes of water.

● *Porous material:*

Weigh 50 g of material and pour into a beaker. Wet thoroughly with distilled water and eliminate all air present in the grains by boiling and stirring for 5 minutes.

After cooling, drain off the interstitial water. Weigh the wetted material (weight P) and place it in a 250 ml cylinder containing 100 ml of water. If V is the volume read off the cylinder, the true bulk density is:

$$\rho = \frac{50}{V - P - 50} \text{ g/ml}$$

The volumes read off the cylinder can be corrected by weighing the same volumes of water.

No. 505—Water content

This concept applies both to materials in granular and in powder form (e.g. activated carbon in powder form).

Weigh exactly about 50 grammes of the filtering material (or 5 grammes of powder); let P_1 be this weight. Place the sample in a drying oven for 4 hours at 120 °C. After cooling in a desiccator weigh again and let P_2 be the weight recorded.

The percentage water content H is expressed by: $\dfrac{P_1 - P_2}{P_1} \times 100 =$ percentage of weight

No. 506—Grain size of activated carbon in powder form

Dry the carbon for 4 hours at 120 °C. Weigh accurately about 10 grammes of carbon and place it on the first sieve (0.149 mm). After damping, wash the carbon remaining on the sieve with water under pressure. Wash and check with an enamelled white bowl that no more carbon passes through the sieve. Then place the sieve in the drying oven for 4 hours at 120 °C to dry. Weigh the carbon remaining on the sieve. By subtracting this figure from the original weight, calculate the amount of carbon which has passed through the sieve. Express the ratio as a percentage.

Repeat the operation with the smaller mesh sieves (0.074, 0.053 and 0.044 mm).

No. 507—Adsorbing power of activated carbon

The adsorbing power of activated carbon can be expressed by the isothermic adsorption curve for the particular type of carbon and the given pollutant.

This isothermal curve establishes the relationship between the equilibrium concentration of water containing a given pollutant (phenol, detergent, iodine etc.), after contact with a certain quantity of carbon on the one hand, and the weight of pollutants retained by weight of carbon on the other. This curve is drawn in the laboratory and can be used to test the effect of a given type of activated carbon in static water. On the basis of these isothermal curves, a number of "indices" can be defined reflecting the given operating conditions (iodine, benzene, methylene blue, detergents, phenol, etc.).

● *Procedure:*

— take six 1-litre coloured glass flasks;

— fill with 750 ml of water containing the pollutant of which the elimination is to be studied;

— if a naturally-polluted water is under analysis, fill the bottles with it in the natural state;

— if a made-up water is to be analysed, fill with water containing substantially more pollutant than the dose to be studied later.

For example, if a water containing 10 mg/l of detergent is to be studied, fill with water containing 14 mg/l of detergent.

The concentration of pollutant must be exactly the same in each bottle.

— Grind the carbon under analysis to a powder in a mortar, and sift dry through a 0.40 mm mesh sieve.

Keep the particles which pass through the sieve.

— Dry the carbon for 4 hours at 120 °C in an evaporating dish.

— Add increasing quantities of carbon to the bottles containing the polluted water samples:

Bottle No.	1	2	3	4	5	6
Dose (mg/l)	0	10	20	30	40	50

After stirring for 1 hour, filter each sample through a 0.45 µm cellulose acetate membrane which has been previously weighed. Discard the first 100 millilitres and determine the pollutant remaining in the rest of the filtrate. For each amount of carbon retained on the membrane and exactly weighed, the equilibrium concentration of the pollutant in the water after contact will be obtained in this way.

Draw the isothermal curve on log-log paper. Plot the equilibrium concentration in mg/l on the x-axis. On the y-axis, plot the weight of the pollutant retained in milligrammes per gramme of activated carbon (mg/g).

No. 508—Dechlorinating power

The dechlorinating power of a carbon is expressed as the depth of carbon required to remove half the chlorine present.

Remove all air from the carbon by boiling it in distilled water. Place the damp carbon in an airtight resin tube until a column exactly 10 cm in height is obtained.

With a sodium hypochlorite solution prepare a chlorinated water solution containing 10 mg/l of active chlorine and with a pH of 7.5. Run this solution over the carbon at a rate of 20 metres per hour. After 30 minutes, titrate the chlorinated water accurately at the top of the column (that is, *a* mg/l) and at the bottom of the column (that is, *b* mg/l).

h being the depth of the layer in centimetres, calculate the length G at which half the chlorine is removed:

$$G = \frac{0.301 . h}{\log \frac{a}{b}}$$

No. 509—Carbon ash

— Take approximately 1 gramme of dry carbon, weigh it carefully, and place it in a crucible; let P_1 be the quantity so weighed.

— Ash the carbon sample at 625 °C ($\pm$ 25°). Check that ashing is complete. After cooling, weigh the ash P_2.

The ash content C is expressed by:

$$C = \frac{P_2}{P_1} \times 100 \text{ as a percentage}$$

8. PHYSICOCHEMICAL TESTS FOR DRINKING WATER TREATMENT

Chemical analysis establishes the characteristics of the water but does not indicate its behaviour during the various purification processes.

To determine such behaviour, individual tests must be carried out, and these are meaningless unless they are made on the spot or very shortly after the sample is taken.

Tests Nos. 701 to 712 relate to the treatment of drinking water and to certain industrial process waters.

No. 701—Study of natural settling

Use the laboratory flocculator in exactly the same way as for the flocculation-coagulation study, but do not use any reagent (see no. 704).

No. 702—Volume of sludge in simple clarification

Pour the water in a 1-litre measuring cylinder or better still into an Imhoff cone. Note the percentage of sludge deposited after 30, 60 and 120 minutes, etc.

No. 703—Study of the coagulation and flocculation of water

The purpose of this study is to ascertain the nature of the reagents and the necessary reagent doses to be used for the best treatment of a given water supply.

The reagents most frequently used are aluminium sulphate, ferric chloride, ferrous or ferric sulphate for clarification purposes and lime, sodium hydroxide or sodium carbonate to adjust the pH or to reduce hardness; other reagents, such as activated silica, alginates and polyectrolytes— the latter being used as aids to facilitate flocculation and sedimentation.

It is often advisable also to add oxidizing agents (chlorine and chlorine dioxide) or adsorbants, such as activated carbon, bentonite, etc. The first essential is an analysis of the water to be examined, with particular reference to its temperature, turbidity, colour, pH, alkalinity and organic matter content.

The amount of electrolyte needed can be determined either by electrophoresis or by flocculation tests.

● *1⁰ Electrophoresis :*

In this technique, the displacement of the colloids placed in an electric field is observed. Measurement can be performed either by microscopic observation (zetameter) or by means of a photoelectric detector (coagulometer).

a) *Zeta potential:* the instrument used for this measurement (zetameter) comprises a control box, an electrophoretic cell, a lighting device, a binocular microscope for examining particles near to 1 micron in size. The speed of movement is checked by means of a micrometric eye-piece and a stop-watch.

Movement of the particles rendered visible by the Tyndall effect is observed in the cell-connecting tube.

Measurements are taken first on the raw water, then with increasing doses of electrolyte, the speed of movement of the colloidal particles being measured in each case.

The zeta potential of the colloidal particles for each speed and temperature is given by graphs. The figures obtained can be used to plot a curve showing how this potential varies (in millivolts) with the amount of electrolyte. The result of the electrophoresis test is not affected by moving the water sample from the sampling site to the laboratory.

b) *Coagulometer:* this instrument (Degrémont patent) comprises a regulated direct current generator, a rectangular-section optical vessel with two electrodes a parallel light beam generator, a photoelectric cell, and an electronic amplification system.

The method involves measurement of the variation in optical density in the immediate vicinity of an electrode during the first few seconds after establishment of the electrophoretic current. Displacement of colloids releases a zone of lower colloidal concentration and hence of lower optical density close to the electrode of the same sign as the charge.

The treatment causing the colloids to discharge can thereby be determined; when this dose is reached, the colloids are no longer displaced when the electrophoretic current is set up and no variation in optical density is observed (fig. 401)

● *2° Flocculation tests:*

These tests must be carried out at a temperature close to the actual temperature of the water during its industrial treatment.

Begin by experimenting with different doses of a single reagent; if the result is unsatisfactory, start a fresh test by repeating the treatment which gave the best result during the first experiment and by trying another treatment at the same time.

It is important to use a flocculator in which the liquid in a series of beakers can be stirred simultaneously at a set speed (fig. 570).

For the results to be comparable, the speed of rotation must be exactly the same for each beaker; the optimum figure is around 40 rev/min for a blade 1×5 cm rotating in a 1-litre beaker.

The test is continued for 20 minutes.

Note the following details:

a) *Amount of reagents.*

b) *Appearance of the floc expressed by a mark:*
   0 — no floc;
   2 — floc scarcely visible, small spots;
   4 — small floc;
   6 — medium-size floc;
   8 — good floc;
   10 — very bulky floc.

c) *pH after flocculation:* the following data are to be added for the best results:
— percentage of sludge after 1/2 hour settling, method 702;
— rate of flocculant settling, method 704;
— sludge cohesion coefficient, method 705 or rate of zone settling, method 706;
— susceptibility to oxidation by permanganate;
— colour and turbidity of the filtered water;
— special measurements for specific treatments.

● *3° Examination of sedimentation:*

The electrophoresis and flocculation tests are not sufficient on their own to transfer the results to the industrial scale, as the main question is to ascertain the rate at which it will be possible to operate the clarifier. An examination of sedimentation is therefore necessary.

Two separate cases may arise:
— During a laboratory test it may be possible to obtain a relatively slight degree of flocculation, in such a way that if the flocculated water is allowed to stand, each particle will fall as if it were alone, some at high velocity and others more slowly.

The liquid thus clarifies gradually and a deposit forms at the bottom of the beaker. This can be called free or flocculant settling.
— In the other case, the flocculated liquid may be very cloudy and all the flocculated particles may settle together to leave a clear liquid at the top of the beaker above a layer of sludge. This can be called hindered or zone settling. In practice, this second type of sedimentation only occurs during the treatment of liquids containing a large amount of matter suitable for flocculation.

The measurements that have to be made vary slightly for these two different types of settling (see Methods no. 704 or 706 as applicable).

No. 704—Measurement of the flocculant-settling rate

Fig. 570. — Electrically operated Hydrocure flocculator, with speed control and timer.

The aim is to ascertain the quality of the clarified water if the flocculated water is taken directly into a clarifier subject to a given upward flow rate.

The "clarification ratio" is given by:

$$\frac{\text{turbidity of clarified water}}{\text{turbidity of raw   water}} = \frac{\text{depth of raw water a}}{\text{depth of clarified water b}}$$

and the "sedimentation ratio" by:

$$\frac{\text{turbidity of clarified water}}{\text{turbidity of flocculated water}} = \frac{\text{depth of flocculated water c}}{\text{depth of clarified water b}}$$

No. 705—Examination of sludge cohesion

If this same experiment is carried out (No. 704) by adding to the water an increasing quantity of sludge obtained from an earlier flocculation test, it is found that the sedimentation rate increases, which means that the clarification ratio for a given rate decreases.

This happens until the liquid has received a sufficient quantity of sludge to reach the point of zone settling. This phenomenon forms the basis for the industrial use of "sludge contact" clarifiers.

As direct experimentation into the enrichment of sludge during flocculation to ascertain the fastest sedimentation rate would be too lengthy an operation, recourse can be had to measuring the degree to which the sludge expands when t is subjected to a known, rising current of water.

It is then found that a layer of sludge subjected to a rising current of water expands and takes an apparent volume almost proportional to the velocity of the water, in accordance with a relationship expressing the cohesion of the sludge.

● *Measurement of the sludge cohesion:*

Take a 250 ml measuring cylinder and pour in the sludge collected in the various beakers during the flocculation test, each beaker having received the same amounts of reagent.

Leave to stand for 10 minutes. Then siphon off the excess sludge so that an apparent volume of approximately 50 ml is left in the measuring cylinder.

Then insert into the measuring cylinder a small funnel with an extension tube reaching to approximately 10 mm from the bottom of the cylinder. Then via this funnel, which must penetrate slightly into the top of the measuring cylinder to prevent the entrainment of air bubbles, pour in the water; for this purpose it is essential to use the water already clarified in the flocculation test so as not to bring about **any change in the pH or the temperature.** The water must be added gradually in small quantities; the excess water will run off by overflowing over the top of the measuring cylinder.

The addition of this water causes the sludge to expand and it is thus possible to determine the upward flow rates corresponding to different states of sludge expansion.

The time T in seconds corresponding to the introduction of 100 ml of water is measured for apparent sludge volumes V ml of: 100, 125, 150, 175 and 200 ml. The velocity v expressed in m/h is equal to $\dfrac{3.6\,A}{T}$ where A is the distance in mm on the measuring cylinder corresponding to 100 ml (distance between the 100 and 200 ml marks on a 250 ml measuring cylinder).

The results are shown on a graph with v plotted vertically and V horizontally.

The curve representing velocity as a function of sludge expansion is found to be a straight line:

$$v = K \left(\frac{V}{V_0} - 1 \right)$$

V: is the apparent volume of the expanded sludge,
v: is the upward flow rate in the measuring cylinder required to obtain volume V,
V_0: is the volume of compacted sludge corresponding to nil velocity and measured on the grapht.

The coefficient "K" expresses the degree of sludge cohesion and is known as the sludge cohesion coefficient. It depends on the temperature which must, therefore, be carefully recorded.

In the case of a well-constituted rapidly-clarifying sludge, the value of the coefficient K can reach 0.8 to 1.2.

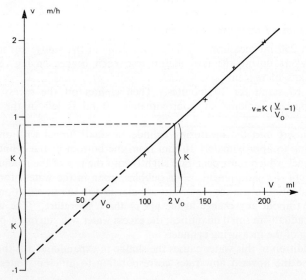

Fig. 571. — Measurement of the cohesion coefficient of a sludge sample.

Conversely, for a light, water-rich sludge, formed by a fragile flocculate, the value of the coefficient K can be as low as 0.3. The measurement of this coefficient is therefore of high value to ascertain the manner in which precipitates will behave in a "sludge contact" clarifier and in order to determine the influence of a flocculation aid.

No. 706—Measurement of the "zone settling" rate

When the flocculation test gives rise directly to "zone settling", it is of course useless and even harmful to enrich the liquid with sludge, and any idea of using a "sludge contact" clarifier should therefore be abandoned.

The aim is to take a direct measurement of the rate of contraction of the sludge mass obtained from the flocculation test; that is, a measurement of the rate at which it will be produced naturally in an industrial-scale clarifier.

The procedure is the same as for measuring the coefficient K, except that the concentration of sludge used is equal to that obtained by flocculating 1 litre of the water under analysis.

Fill a 250 ml measuring cylinder with flocculated liquid; allow the sludge to settle for 5 to 10 minutes until the floc re-forms; then, using the funnel, gradually add water in several lots until the expanding sludge returns to its initial apparent volume of 250 ml.

The rate thus measured indicates the upward flow rate that will be theoretically possible in an industrial clarifier.

At the end of this test, it is advisable to allow the sludge to re-settle spontaneously in the measuring cylinder and to note the apparent volumes of the sludge layer, related to the initial volume, for given periods of time (0 to 2 hours). This will give an indication of the quantities of sludge to be extracted and will thus facilitate the calculation and design of certain aspects of the clarifier such as the dimensions of the sludge pits, the shape of the scraper blades, etc.

The operations are summarized below:
— Distance on the measuring cylinder corresponding to 100 ml .. : A (mm)
— Volume of water added in 1 minute in order to keep the top of the sludge layer at the level of the liquid in the 250 ml measuring cylinder... : B (ml)
— Theoretical settling rate: V_s (m/h)

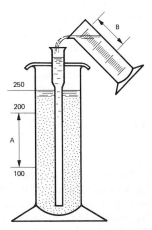

Fig. 572. — The rate Vs in m/h is equal to $\dfrac{0.6\ AB}{1\ 000}$

No. 707—Lime-softening test

In principle the coagulant will be ferric chloride but tests may also be made with aluminium sulphate, with or without activated silica.

In the latter case, the alumina and the silica (where appropriate) will be measured after filtration through paper (aluminium is easily soluble in an alkaline medium).

A first test should be made without the coagulant in order to determine the amount of lime, and at the same time 1 g/l of powdered $CaCO_3$ (50 μm) should be added. Stir for 5 minutes, decant, and filter.

Lime dose	100	125	150	175	200	225
	250	275	300	325	350	375

TA   of filtered water
TAC   —
TH   —

Then, using the amount of lime that gives either a TA exceeding by 1.2 degree half the TAC or the minimum TH, the flocculation test should be repeated but with increasing amounts of flocculant.

No. 708-A—Chlorine absorption test (break-point) of a water (normal method)

A set of bottles all with the same capacity and made of the same glass is used.

Introduce the same volume of sample water into each bottle together with increasing doses of chlorine from bottle to bottle.

After a contact time usually equivalent to the retention time of the water in the plant, at constant temperature and in the shade, the chlorine remaining in the water in each bottle is measured (it may sometimes be worth while to perform this test with different contact times: 1, 2, 5 ... 24 hours).

In most cases it is found that instead of increasing regularly with the initial dose introduced, the residual chlorine passes through a maximum (M), declines, passes through a minimum (m), and then increases again regularly. This minimum represents an initial dose of chlorine characterizing the "break point".

The curve for residual chlorine vs chlorine introduced is linear beyond the break point; if the horizontal and vertical scales are equal, its slope will be 45° (fig. 276, pp. 402 and 956).

In the zone OP in fig. 276, all the chlorine introduced is consumed by the oxidizable substances contained in the water (metals, reducing agents, etc.); the chlorine can be assumed to have no bactericidal action.

In zone PM, the chlorine introduced combines with ammonia and certain organic substances (which may or may not be azotized) contained in the water; the majority of the residual chlorine found is in the form of chloramines (depending on the pH and the Cl/NH_4 ratio, the relative proportions of monochloramine and dichloramine may vary).

In zone Mm, additional amounts of chlorine introduced break down chlorine compounds (in particular, chloramines); an increase in the proportion of free chlorine in the residual and total chlorine is observed, as well as a reduction of chloramines.

In the zone of introduced chlorine quantities exceeding point m, almost all the residual chlorine is free and the water can be regarded as possessing a certain bactericidal reserve.

No. 708-B—Rapid method of break-point determination

In this case only a single measurement is performed, a huge excess of chlorine being introduced to the raw water (A in fig. 276); after contact, the break point (Om = OA — Aa) can be determined approximately from measurement of the residual chlorine (Aa).

No. 710—Iron removal test

This test must be carried out on the spot immediately after the samples are taken.

Iron removal by oxidation with air is not always possible, especially with waters containing large quantities of organic matter. To allow for this, proceed as follows:
— aerate rapidly by transferring the water from one beaker to another 20 times;
— filter on blue-band Durieux paper;
— measure the quantity of residual iron and the variation of pH, dissolved oxygen, and carbon dioxide.

If the amount of residual iron is not less than 0.1 ppm, more thorough tests must be carried out, if possible in a pilot plant, using other oxidants and/or miscellaneous coagulants and flocculants (e. g., alginates).

No. 711—Deaeration test

It is sometimes useful to remove free CO_2 by trickling the water in the presence of air. To ascertain the efficiency of the operation, proceed as follows:
— using 2 one-litre beakers, pour the water from one to the other, allowing the water to fall through 20 cm (8 inches) and at a flow-rate of approximately 1 litre in 10 seconds;
— measure the free CO_2 and note the pH in relation to the number of pouring operations until the pH no longer varies appreciably.

No. 712-E—Fouling index

● *Principle:*

Determination of the fouling of a cellulose acetate membrane of 0.45 μm porosity after filtering the analytical sample for 15 minutes.

● *Apparatus:*
— 47 mm diameter filter support;
— cellulose acetate filter, 0.45 μm, diameter 47 mm;
— 0.5 bar manometer;
— needle valve for pressure regulation.

● *Procedure:*

Place the filter on its support, moisten, and adjust the ring seal.   Bleed air from the circuit and fix the support so that the membrane is orientated vertically. Adjust the pressure to 2.1 bar (30 psi), and, using a chronometer, measure the time t_0 for filtration of 500 ml of water (this time must exceed 10 seconds).   Repeat if the pressure varies by $\pm 5\%$ during measurement.

Leave the filter in position and in operation, correcting the pressure regularly if necessary.

After 15 minutes, measure on the chronometer the time t required to filter 500 ml, checking that the pressure remains constant at 2.1 bar.

Isolate the filter and remove the membrane, which should be kept for any further analyses to be carried out.

● *Calculation:*

The fouling power P is given by the relation:

$$P\% = 100 \left(1 - \frac{t_0}{t}\right)$$

If this percentage exceeds 80 % in 15 minutes, the test must be repeated with times of 10, 5, or even 3 minutes.

The fouling index I_C is calculated from P % and the time T between two measurements:

$$I_C = \frac{P\%}{T}$$

Example: at 2.1 bar $\left\{\begin{array}{l} t_0 = 28 \text{ seconds} \\ t = 44 \text{ seconds (after 15 min, i.e., T = 15)} \end{array}\right.$

$$P\% = 100 \left(1 - \frac{28}{44}\right) = 36.4\%$$

$$I_{C15} = \frac{36.4}{15} = 2.4$$

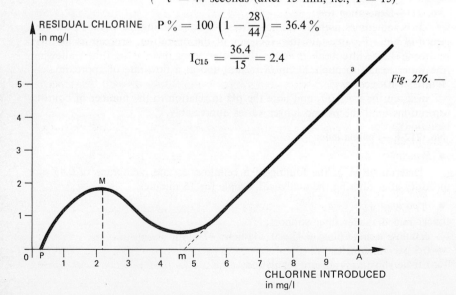

Fig. 276. —

9. EXAMINATION OF WASTE WATER SLUDGE

No. 801—Dry residue, or dry solids at 105 °C

The dry residue is also known as the dry extract or dryness of the sludge. It is measured by placing a sludge sample (25 to 100 ml depending on the sludge concentration) in a drying oven at 105 °C until constant weight is obtained.

If M_1 is the weight of the wet sample and M_2 the weight after drying:

$$\text{dry residue} = \frac{M_2 \times 100}{M_1} \%$$

No. 802—Dry residue at 175-185 °C

The water of crystallization content of the salts, and the levels of particularly volatile substances, as well as other parameters, can be evaluated by drying to 175-185 °C and to 105 °C and comparing both results.

No. 803—Solids suspended in liquid sludge

Suspended solids are determined either by method 803 (a) or by method 803 (b) below depending on the dryness of the sludge concerned. The value obtained differs from the dry residue because it does not include the content of dissolved substances present in the sludge. These methods differ from method 406, page 928, due to solid concentrations in sludge being higher than in waste water.

No. 803 (a)—Centrifugation method

Use a laboratory centrifuge with 100 ml graduated cups. Fill each cup with 80 ml of sludge and centrifuge at 4 000-5 000 rev/min for 10 minutes. After centrifugation, eliminate the supernatant liquid and carefully recover all the sludge residues, which are dried in a drying oven at 105 °C until the weight remains constant (in general, for at least 12 hours).

Note: If the sludge is first flocculated with a few mg/l of polyelectrolyte (flocculation in situ in the cup), homogenous residues which can be readily recovered without loss of material can be obtained.

Notation: M(g) = weight of dry residue obtained; V (ml) = volume of sludge centrifuged (V = 160 or 320 ml).

$$\text{Concentration of suspended solids} = \frac{M \times 1\,000}{V} \text{ g/l}$$

No. 803 (b)—Filtration method

This method is more applicable to relatively thin sludges (e.g., of 5 to 30g/l concentration):

Exactly weigh a paper filter (150 mm diameter extra-rapid ashless filter type, e. g., Durieux) and filter the sludge (25 to 100 ml depending on concentration) through a glass funnel. This may take a long time in some cases where the sludge is not readily filtrable. Then oven-dry the filter at 105 °C until the wieght remains constant.

$$\text{Suspended solids concentration} = \frac{M - F}{V} \times 1\,000 \text{ g/l}$$

M = dry weight of filter and cake
F = weight of filter alone
V = volume of sludge filtered

No. 804—Calcinated residue at 550 °C and volatile matter

The dry residue at 105 °C is heated for 2 hours to 550 °C in a muffle furnace which has been preheated and whose temperature is thermostatically controlled. Silica dishes with about 10 to 20 g of finely crushed dry sludge are generally used.

The content of volatile matter gasified at 550 °C must not be confused with the organic matter content; there are several reasons for this:
— a part of the inorganic matter and salts may decompose between 105° and 550 °C;
— part of the organic matter (in particular, certain organocalcic or organo-metallic complexes) may volatilize not at 550 °C but only towards 650-700 °C.

Nevertheless, for most sludges, determination of volatile matter is a rough evaluation of the organic matter content. The volatile matter level is generally expressed as a percentage of the dry solids (% of DS).

No. 805—Rapid method of determining the TAC and the volatile acids level in a liquid sludge

These determinations are important, for example, for satisfactory operation of the anaerobic digester.

Measure 25 ml of sludge as exactly as possible. Initially, centrifuge this sludge at 5 000 rev/min for 10 minutes. Collect the supernatant liquid in a 400 ml beaker. Dissolve the residue in 50 ml of distilled water, taking care not to lose any of the solid part.

Again centrifuge at 5 000 rev/min for 10 minutes and collect the supernatant liquid in the beaker. Repeat this residue washing operation once more.

The liquid collected contains in particular bicarbonates and soluble volatile acids.

● *Measurement of TAC:*

The liquid recovered in the beaker is stirred with a magnetic stirrer. The electrodes of a pH meter are immersed in the liquid and the initial pH is noted.

Using a 1/10 ml burette, add 0.1 N sulphuric acid until the pH = 4; hence V (ml).

$$\text{TAC} = \frac{V \times 0.1 \times 1\,000}{25} = V \times 4, \text{ in meq/l}$$

or:

$$\text{TAC} = V \times 4 \times 0.05, \text{ in g/l CaCO}_3$$

● *Measurement of volatile acidity:*

Again add 0.1 N sulphuric acid until the pH = 3.5. Now boil the liquid at pH 3.5 for exactly 3 minutes. Allow to cool. Apply the electrodes of the pH meter to the cooled liquid and, using a 1/10 burette, add 0.1 N caustic soda, stirring all the time, until the pH = 4 (volume V_2).

Continue to add NaOH until the pH = 7 (new volume = V_3).

$$\text{Volatile acidity} = \frac{(V_3 - V_2) \times 0.1 \times 1\,000}{25} = (V_3 - V_2) \times 4, \text{ in meq/l}$$

or:

Volatile acidity = $(V_3 - V_2) \times 4 \times 0.06$, in g/l CH_3COOH

No. 806—Test of vacuum filtrability with the Büchner funnel — measurement of specific resistance of a sludge under a vacuum of 0.5 bar (see fig. 573)

This determination serves for approximate evaluation of the capacity of an industrial vacuum filter. It can also be used to determine the optimum doses of reagents.

● *Apparatus:*
— a chronometer;
— a vacuum pump;
— a manometer;
— a filtration system comprising a 200-300 ml Büchner funnel attached to a 250 ml test tube.

The Büchner funnel (made of china or plastic) consists of two detachable parts with the following held between them:
— the filter support (perforated plate);
— the filter gauze (or filter paper);
— the rubber gasket.

● *Procedure:*

Fill the Büchner funnel with the sludge to be filtered (previously conditioned); 150 ml of sludge is generally sufficient to give a final cake 8-10 mm thick.

Set up the vacuum by connecting the test tube to the vacuum pump; the required negative pressure of 0.5 bar (about 40 cm Hg) should be set up quickly and must be kept constant throughout the test.

As soon as the required vacuum level has been attained, start the chronometer and note volume of filtrate already collected (volume V_0 corresponding to the time t = 0, which must be subtracted from the volumes obtained subsequently). Note the volumes of filtrate collected during the test for different filtration times —every 10, 15, 20, 30, or 60 seconds depending on the flow rate of the filtrate. The test is continued until the cake is dry (loss of vacuum due to cracking of the cake).

● *Calculation of specific filtration resistance:*

The filtrate volumes V_0, V_1, V_2, V_3, etc., corresponding to the times T_0, T_1, T_2, T_3, etc., have been noted.

Draw a graph of V_x against

$$\frac{T_x}{V_x - V_0}$$

These points are in principle in alignment (except at the beginning of filtration and during drying). The slope of the linear part of the resulting curve is equal to the coefficient a (see Chapter 17, page 481). The other terms of the expression for the resistance are known (see Chapter 17, page 482):

$$r_{0.5} = \frac{2\,a\,PS^2}{\eta\,C}$$

a: expressed in s/m^6
P: expressed in pascal (49×10^3 Pa)
S: expressed in m^2
η: expressed in Pa.s (at 20 ºC, η approximates to 1.1×10^{-3} Pa.s)
C: expressed in kg/m^3
r: expressed in m/kg

Note: C, the solid matter concentration, is an approximation to W (weight of solid matter deposited per unit volume of filtrate and determined by measurement of the dryness of the cake obtained at 105 ºC divided by the total filtrate volume).

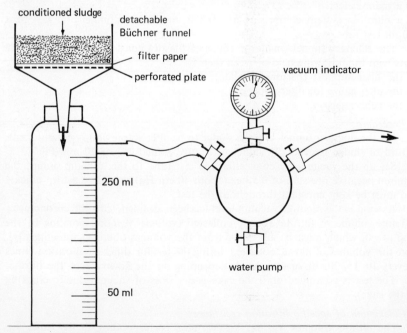

Fig. 573. — Filtrability test under a vacuum of 0.5 bar.

No. 807—Pressure filtrability test using the Pont-à-Mousson pressure cell (see fig. 574)

This instrument is used not only to determine the specific resistance but also to establish the coefficient of compressibility of filter cakes and their limiting dryness.

The principle is the same as for test 806.

● *Procedure:*

— take a sample (100 to 150 ml) of conditioned sludge;

— set up the measuring cell by placing first the filter gauze disc and then the paper filter on the perforated support;

— moisten the paper filter and apply slight excess pressure to seal the bottom of the cell and eliminate the excess of water retained by the filter;

— correctly position the test tube below the cell funnel;

— pour the sludge sample into the cell;

— leave to stand for 15 seconds before applying the pressure, to facilitate formation of a precoat;

Progressively apply the selected pressure (0.5 to 15 bar); it is inadvisable to use the piston for pressure below 2 bar;

— drain off the prefiltrate and note its volume V (10-20 % of the volume of sludge to be filtered);

— start the chronometer and note the volume V of filtrate as it varies with time.

Draw the curve $\dfrac{T_x}{V_x - V_0} = f(V_x)$.

The frequency of measurements will depend on the filtrate flow.

— See test 806 for calculation of resistance.

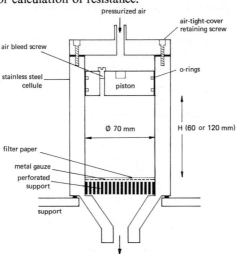

Fig. 574. — Pressure filtrability test using the PONT-A-MOUSSON pressure cell, under a pressure of 0.5 to 15 bar.

No. 808—Determination of compressibility coefficient

Measure the specific filtration resistance r at several pressures and draw the curve $\log r = f(\log P)$.

Make sure that it is linear and measure its slope, which is equal to the compressibility coefficient.

The same apparatus is used as in test 807.

For maximum precision, regular intervals of pressure should preferably be used. The following values are recommended:

$$P = 49 \text{ kPa}, 147 \text{ kPa}, 441 \text{ kPa}, \text{ and } 1\,323 \text{ kPa}$$

(or the maximum pressure permitted by the laboratory apparatus).

The compressibility coefficient is expressed in the form of a dimensionless number.

No. 809—Determination of limiting dryness

The apparatus is identical with that used for tests 808 and 807, the piston, however, in this case being essential to transmit to the sludge the entire pressure admitted to the cell and to prevent cracking of the cake. The procedure is the same as in test 807, but the following operations must also be carried out after placing the sludge sample in the cell:

— press down the piston, with the bleed screw removed, until it touches the sludge;

— fit and tighten the bleed screw;

— then follow the rest of the procedure set out for test 807 and continue measurement of the specific resistance until the vertical asymptotic branch of the curve $\frac{t}{V} = f(V)$ is obtained. In practice, filtration should be stopped when the slope of the tangent to the relative point is 5 times the slope of the initial straight-line portion. Determine the dryness of the cake at the pressure chosen. If H is the percentage moisture content of the cake, determined at the end of measurement of the specific resistance at pressure P by drying in an oven at 105 °C $\pm$ 2 °C to constant weight, the limiting dryness, expressed as a percentage, will be:

$$S_{L(P)} = 100 - H$$

 **BIOLOGY**

1. GENERAL REMARKS

Life originated in water; many vital phenomena still occur in the original marine environment as well as in lakes, rivers, ponds, reservoirs, etc.

In some respects, these phenomena play a role advantageous to man—the best example being the self-purification of a river down-stream of an effluent discharge. Various organisms are also able to multiply when favourable conditions are created for them by Man. Advantage is often taken of such useful phenomena in the course of biological treatment (see pages 203 to 244).

However, numerous forms of waterborne life are an inconvenience to man especially when pure water is required, and it is this aspect which will be principally dealt with here. Plankton, in both the animal and vegetable forms, can clog filtration media (sand or resins), invade clarifiers, and cause sludge to overflow. If water is not treated effectively, plankton can cause colonies to form in pipes and reservoirs; it can also act as a vector for pathogenic microorganisms; and the iron and sulphur bacteria take part in corrosion processes. Many diseases are waterborne, being caused either by bacteria, protozoa (such as the amoebae), or by worms (bilharzia, fluke worms, etc.). All these species, for which water serves as a natural habitat, or as a means of transport, must be eliminated.

Treatment procedures have been discussed elsewhere in this book, and this chapter will review the main types of organisms found in water, so that, when the occasion arises, they may be correctly classified and identified. First, however, definitions will be given of some of the principal terms used to describe waterborne life and its relationship with the environment. Some concepts of bacteriology will also be restated. At the end of the chapter will be found a description of the organisms present in activated sludge.

2. CONCEPTS OF ECOLOGY

2.1. Definitions

The science of the laws of life, or **bionomy,** may be divided into two branches: **ethology,** which deals in particular with the habits and behaviour of animals and **ecology,** the science concerned with the environmental conditions under which living creatures exists.

One of the many branches of this new science—it has been developing for less than fifty years—**synecology,** considers, more especially, the relationship of individuals belonging to the various species in a group with one another and with their environment.

Every environment, then, may be identified with a **biocenosis,** which is an association of living creatures, the composition and physiognomy of which are determined by the characteristics of the environment and by the life relationship of the creatures with one another; the quantity of the corresponding living matter, per unit volume or unit area, known as the **biomass,** may be determined quantitatively for a given environment in both space and time.

The habitat of a biomass, together with the prevailing conditions as a whole constitute the **biotope.**

The biocenosis and the biotope form two inseparably linked elements which react upon each other to form a more or less stable system, called **ecosystem.**

Most of the ecosystems in nature have formed in the course of prolonged evolution and they are the result of protracted processes of adaptation between the species and the environment.   As a result of autoregulation, they are, within certain limits, capable of resisting changes in the conditions of life or sudden changes in population density.

Biological systems are open systems involving constant interchange of matter and of energy between them and the environment.   These open systems tend towards a stable state by passing through a series of successive biocenoses until reaching a stable biocenosis which is in equilibrium with the environment and which is termed a **climax.**

An ecosystem is composed necessarily of a number of **trophic** levels, the successive stages of which lead to the formation of a **food chain.**

The typical food chain consists, firstly, of primary **producer** organisms, capable of producing and of accumulating potential energy, in chemical form, in the organic matter which they synthesize (for example, vegetable plankton) from inorganic materials.   These are **autotrophs.**

These are followed by the **heterotrophic** organisms, which may be subdivided into :

— **first order consumers** which eat the autotrophic producers (for example, animal plankton);

— **second order consumers** which are carnivores, living at the expense of herbivorous organism: it is possible, according to circumstances, to identify third, or higher, order consumers, which are predators or parasites or organisms feeding on carrion.

The organisms causing **decomposition** or bioreduction form the final link in the chain.   These are principally the **saprophytic** microorganisms (heterotrophic bacteria, yeasts or fungi).

Ecosystems can vary very greatly in size, from the macrosystems such as the sea, deserts, deciduous forests, to microsystems such as the stump of a tree, or the underside of a stone, in a given place.

In the aquatic sphere with which we are concerned, each lake, each river or reach of a river, forming a defined biotope, may be considered as a particular ecosystem and be studied as such.

In water treatment, artifical ecosystems are created such as: an activated sludge tank, a sludge digester, the biological membrane in a slow filter, etc.

2.2. The ecology of lakes: eutrophication

This term which, in principle, is used solely in connection with lakes, was used originally to describe a natural evolution phenomenon, which can be summarized as follows:
— a lake which is young and deep is **oligotrophic:** its water is blue and transparent; dissolved oxygen is present down to the bottom; the biomass is sparse;
— as ageing proceeds, the lake becomes enriched in organic matter, due to its primary photosynthetic generation (algae) and possibly due to contributions from external sources; it then becomes, successively, **mesotrophic,** then **eutrophic;** the following phenomena are then found: decrease in depth by progressive silting; discoloration of the water (green to brown); reduced transparency; oxygen depletion in the deeper levels; greater biomass, with the appearance of species indicative of eutrophication (especially of Cyanophyceae, or blue algae, of which the best known is Oscillatoria rubescens);
— the final stage is the pond, swamp, etc.

The transition from one type to the next takes a very long time, which can be measured in thousands of years. But this natural process has in some cases been accelerated to such an extent as to become apparent during a human life-span. This situation is the result of human activities, such as agriculture and the discharge of domestic sewage or industrial effluents, which carry organic matter and fertilizing elements (nitrogen and phosphorus, in particular) to the stagnant waters.

The consequences of this artificial eutrophication can be disastrous for the tourist trade and to fishing; furthermore, the cost of water treatment is considerably increased as a result of the equipment and reagents necessary for the elimination of the organisms themselves or of their metabolic products (see "problems due to plankton", page 610).

It is possible to combat eutrophication with:
— curative measures (oxygenation, destratification, chemical or biological methods);
— preventive measures: diversion of the effluents by the use of a ring waterside interceptor (e.g. lake of Annecy) or by a change in course; or tertiary treatment in purification plants (see page 798).

2.3. River ecology: biotic indices

The same problems do not arise in running water, to which the full definition of eutrophication cannot be applied. For a long time now degrees of **saprobity** have been used to describe the extent of pollution of a river; these are linked to the chemical properties of the water and to biological zones, which are defined by the presence of members of the various animal and vegetable families (Kolkwitz and Marson's system, 1909).

This system of saprobes is very much disputed at the present time; in France the preferred method is that of **biotic indices,** due to Verneaux and Tuffery (1967), adapted from the method developed in Great Britain by the Trent River Authority: analysis of the benthic macroinvertebrate population (Molluscs, Crustaceae, Worms, Insect larvae), from which an index of the river quality can be deduced, expressed on a scale increasing from 0 to 10.

Changes in biocenoses are particularly distinct downstream of a pollutant discharge (change in the type of dominant zoological group, reduction in the number of species, increase in the number of individuals in each species); using the method quoted above, peak pollution with organic matter is immediately reflected by a massive drop in the biotic index downstream, which increases again, further down, as a result of self-purification. Such methods give valuable information for the study of pollution or in the assessment of the efficacy of effluent treatment, etc.

2.4. Biological tests for pollution

Toxic substances discharged into a natural environment are a threat to living creatures. Certain tests have been developed to assess the toxicity of an effluent; these relate in general to a single organism chosen from the following 4 categories:
— bacteria (Pseudomonas, Mycobacterium, Escherichia coli...);
— algae (Chlamydomonas, Dunaliella, Selenastrum...);
— invertebrates (Crustaceae, above all, but also, Worms, Protozoa, etc...);
— fish (Trout, Minnow, Guppy, Carp, Brachydanio).

These tests can be either static or dynamic. Cabridenc & Lundahl developed a static test in France on one of the common Crustaceae, the water flea (Daphnia magna): determination of the inhibitory concentration which, in a period of 24 h, immobilizes 50 % of the Daphnia used in the test. The result may be expressed as the **equitox** (inverse of the 24 h-I.C. 50, see page 938). Other methods are under investigation.

Biological tests can also be used to detect occasional pollution in a river, upstream of a drinking water treatment plant, of a fish breeding farm, etc. A fish is normally used (most often, trout or carp); a variety of apparatus has been proposed for the provision of automatic and continuous control. The system developed by Leynaud, Barbier et al., is used in France: its principle was used as a basis for the DEGREMONT **Ichthyotest,** described on page 584. These dynamic tests can be integrated with the more complex warning stations, at which automated instruments simultaneously measure various physicochemical parameters of the water to be monitored.

3. PRINCIPLES OF CLASSIFICATION OF LIVING CREATURES

Before embarking upon an examination of the principal organisms which live in water and which can interfere with its treatment, it may be as well to recall the main classification lines or living creatures, in order to be able to assign them to their correct place without too much difficulty.

There are 3 kingdoms: **bacterial, animal and vegetable**; the first has its place at the lowest limit of the other two. The **viruses**, large molecules of nucleic acid (combined with proteins), may be considered to mark the borderline between life and inert matter; they form a world of their own, which is quite separate from the "cellular" life represented by the 3 kingdoms already mentioned.

A simplified table is given below for each kingdom, in which certain subdivisions only are mentioned (branch, class, order, family and genus) and in which solely those groups which live in aquatic environments are considered. For the purpose of illustration, at the level of order, family and genus, only a few isolated examples are quoted. Further details will be found in the subsections on bacteriology and the study of plankton.

3.1. The bacterial kingdom

This is composed of unicellular, microscopic organisms, either isolated or in colonies. Many characteristics distinguish it from the other 2 kingdoms; in particular, the bacteria are prokaryotes (having no nuclear membrane), a characteristic which they share with the Cyanophyceae (see later: prokaryotic Protista).

Nevertheless the Bacteria show resemblances, in some respects, to one or other of the two following kingdoms; in the classification of the bacteria given here (Prevot, 1961) there will be found 4 branches: alongside that of the true bacteria (Eubacteria), the 3 others show certain analogies with the fungi (Mycobacteria) or with the Algae (Algobacteria), or with the unicellular animals, or Protozoa (Protozoobacteria).

3.2. The vegetable kingdom

At the base of the present-day classification are to be found the Myxomycetae (lower fungi) and the Eumycetae (higher fungi). Then come the numerous branches of algae, after which, going through the Bryophyta (mosses) and the Pteridophyta (vascular cryptogams), the higher plants are reached.

THE BACTERIAL KINGDOM

Branch	Class	Order	Family	Genus (examples) *
Eubacteria	Asporulales	Micrococcales	Neisseriaceae Micrococcaceae	Neisseria **Streptococcus**, Staphylococcus
		Bacteriales	Pseudomonadaceae Enterobacteriaceae	**Pseudomonas**, Serratia **Escherichia**, Salmonella, Shigella
			Parvobacteriaceae Ristellaceae Protobacteriaceae	Pasteurella, Brucella Ristella **Nitrobacter, Nitrosomonas, Thiobacillus**
			Bacteriaceae	Bacterium, Lactobacillus
		Spirillales	Vibrionaceae Spirillaceae	**Vibrio**, Cellvibrio Spirillum
	Sporutales	Bacillales	Bacillaceae Innominaceae	**Bacillus** (B. subtilis) Innominatus
		Clostridiales	Endosporaceae Clostridiaceae	Endosporus **Clostridium** (Cl. perfringens)
		Plectridiales	Terminosporaceae Plectridiaceae	Terminosporus Plectridium
		Sporovibrionalcs	Sporovibrionaceae	**Sporovibrio,** (or **Desulfovibrio**)
Mycobacteria	Actinomycetales	Actinobacteriales	Sphaerophoraceae Actinomycetaceae Streptomycetaceae	Sphaerophorus **Actinomyces**, Nocardia **Streptomyces,** Micromonospora
		Mycobacteriales	Mycobacteriaceae	Mycobacterium
	Myxobacteriales	Myxococcales	Myxococcaceae	Myxococcus
		Angiobacteriales	Archangiaceae Sorangiaceae Polyangiaceae	Archangium Sorangium Polyangium, Chondromyces
		Asporangiales	Cytophagaceae	Cytophaga, Flexibacter
	Azotobacteriales	Azotobacteriales	Azotobacteriaceae	Azotobacter
Algobacteria	Siderobacteriales	Chlamydobacteriaceae	Chlamydobacteriaceae Crenothricaceae Siderocapsaceae	**Sphaerotilus, Leptothrix** **Crenothrix,** Clonothrix Siderocapsa, **Sideromonas,** Ferrobacillus
		Caulobacteriales	Caulobacteriaceae Gallionellaceae	Caulobacter **Gallionella**
	Thiobacteriales	Rhodothiobacteriales	Thiorhodaceae Thiobacteriaceae Athiorhodaceae	Thiocystis, Chromatium Thiobacterium, Thiospira Rhodopseudomonas
		Chlorobacteriales	Chlorobacteriaceae Chlorochromatiaceae	Chlorobium, Pelodyction Chlorochromatium
		Leucothiobacteriales or Beggiatoales	Beggiatoaceae Achromatiaceae	**Beggiatoa, Thiotrix** Achromatium
Protozoobacteria	Spirochaetales	Spirochaetales	Spirochaetaceae Treponemaceae	Spirochaeta Treponema, Leptospira

* The genera most frequently found in the literature on water treatment, in bold type.

THE VEGETABLE KINGDOM

Branch or Sub-branch	Class (examples)	Order (examples)	Family (examples	Genus (examples)
— *Thallophyta* (cellular cryptogams)				
— Myxomycetae	Myxogastromycetae	Physarales	Didymiaceae	Didymium
— Eumycetae	Phycomycetae Zygomycetae Ascomycetae Basidiomycetae	Leptomitales Mucorales Aspergillales Agaricales	Leptomitaceae Mucoraceae Aspergillaceae Amanitaceae	Leptomitus Mucor Aspergillus Amanita
— Chlorophyta (green algae)	Chlorophyceae	Volvocales Chlorococcales Ulothrichales Chaetophorales Conjugales	Chlamydomonadaceae Hydrodictyaceae Ulothrichaceae Chaetophoraceae Zygnemataceae	Chlamydomonas (fig. 580) Pediastrum (fig. 584) Hormidium (fig. 588) Draparnaldia (fig. 589) Spirogyra (fig. 589)
— Euglenophyta	Euglenophyceae	Euglenales	Euglenaceae	Euglena (fig. 597)
— Chrysophyta (yellow algae)	Bacillariophyceae or Diatoms	Centrales Pennales	Araphideae Monoraphideae Biraphideae	Cyclotella (fig. 599) Synedra (fig. 602) Rhoicosphenia (fig. 606) Navicula (fig. 605)
	Chrysophyceae	Ochromonadales	Synuraceae	Synura (fig. 609)
	Xanthophyceae or Heterokontes	Heterotrichales Heterosiphonales	Heterotrichaceae	Tribonema Vaucheria (fig. 611)
— Cyanophyta (blue algae)	Myxophyceae	Chroococcales Hormogonales	Chroococcaceae Oscillatoriaceae	Microcystis (fig. 613) Oscillatoria (fig. 614)
— Rhodophyta (red algae)	Rhodophyceae	Bangiales Florideae		Bangia Batrachospermum
— Pyrrhophyta	Cryptophyceae Dinophyceae (or Peridinians)	Cryptomonadales Peridiianles Dinocaspales		Cryptomonas Peridinium (fig. 616) Gleodinium
— Phaeophyta (brown algae)	Phaeophyceae	Fucales		Fucus
— *Bryophyta* (Muscineae)	Hepaticae Mosses	Jungermanniales Sphagnales		Riccia Sphagnum
— *Pteridophyta* (Vascula) cryptogams)	Filicineae			Ferns
— *Spermatophyta:* or *Phanerogams:*				
— Gymnosperms	Coniferales	Abietales	Pinaceae	Fir
— Angiosperms	Monocotyledons	Fluviales Glumiflores Cyperales Spadiciflores Pandanales	Potamogetonaceae Gramineae Juncaceae Cyperaceae Lemnaceae Typhaceae	Pond-weed Cereals Rush Sedge Duckweed, Wolffia Cat's tail
	Dicotyledons	Ranales Leguminosae Fagales	— Ceratophyllaceae — Nympheaceae Papilionaceae Fagaceae	Ceratophyllum Yellow and white water lilies Pea, bean Oak, beech

3.3. The animal kingdom

The animal kingdom begins with the single-celled animals (Protozoa) which include numerous parasitical forms (among the flagellates, the only types that include free-living forms are the Polymastiginae and the Diplomonadinae; the Sporozoa are all parasites).

Among the plankton and the fauna in activated sludge large numbers of protozoa are found; the signifiance of and the role played by these is very important.

The Spongiae are the transition from the Protozoa to the Metazoa, the primitive sponge taking the form of a colony of differentiated protozoa. Then, moving up towards the higher animals, important branches in relation to the plankton are met, particularly among the worms, the Vermidae (Rotifers), and the Arthropods (Crustaceans and insect larvae).

•

In actual fact, the system of classification proposed here cannot be regarded as universal, as there are others.

For example, Haeckel has proposed the regrouping of all the most primitive organisms into a separate kingdom, that of the **Protists,** divided into **prokaryotic** Protists (Bacteria and Cyanophyceae or blue algae) and **eukaryotic** Protists (algae, fungi, lichens or protozoons); this classification system is still used sometimes by some authors. The same applied to the vegetable kingdom, in which the large groups of algae may be considered to be either branches or sub-branches; in the animal kingdom, the Rotifera can form either a separate branch or a class of round worms (still called Aschelminthes in some countries); the Tardigrada may be classified either among the Annelida or among the Arthropoda, etc.

THE ANIMAL KINGDOM Sub-Kingdom I: **Protozoa**

Branch	Class (examples)	Order (examples)	Genus (examples)
Rhizopoda	Amoebians Foraminifera Radiolaria Heliozoa		Amoeba (fig. 619) Polystomella Actinophrys (fig. 621)
Flagellata	Zooflagellata	Herpetomonadinae Polymastiginae Diplomonadinae Hypermastiginae	Trypanosome Trichomonas Lamblia
Sporozoa	Telesporidiae Cnidosporidiae	Gregarinae Coccidae Hemosporidiae	Plasmodium
Protociliata		Opalinae	
Infusoria	Ciliata Tentaculifera (Acinetae)	Holotrichiae Heterotrichiae Peritrichiae Hypotrichiae	Paramecium, coleps (fig. 617) Stentor Vorticella (fig. 618) Euplotes, Aspidisca (fig. 634) Podophrya Acineta

Classification of living creatures

THE ANIMAL KINGDOM Sub-Kingdom II: **Metazoa**

Branch	Class (examples)	Sub-Class (examples)	Order (examples)	Genus (examples)
Sponges	Chalky sponges / Siliceous sponges	Demospongiae	Tetraxonidae	Spongilla
Coelenterata	Hydrozoa / Scyphozoa / Anthozoa	Hexacoralliariae	Hydrariae / Medusae / Madreporariae	Hydra / Craspedacusta / Corals
Echinoderms	Echinoidae			Sea-urchins
Platyhelminthae (flat worms)	Turbellariae / Trematodae / Cestodae	Neoophores / Eucestodae	Tricladae / Digeniae / Cyclophyllidiae	Planaria / Fluke-worm, bilharzia / Tapeworm
Nemertinae	Nemertae	Armatae	Hoplonemertae	Prostoma
Nematyhelminthae (round worms)	Gasterotrichae / Nematoda / Acanthocephala / Gordiaceae	Aphasmidariae / Phasmidariae	Chetonoidae / Trichinoidae / Tylenchoidae / Ascaridoidae / Gordiodae	Chaetonotus / Thread-worm / Anguillula(fig. 629) / Ascaris / Pomphorhyncus / Gordionus
Rotifera	Bdelloidae / Monogononta		Bdelloidae / Ploimidae / Flosculariaceae	Philodina (fig. 625) / Keratella(fig 622) / Hexarthra
Echiuria	Gephyrinae		Echiurinae	Bonellia
Annelida (ringed worms)	Polychetae / Clitellatae		Errantae / Oligochetae / Hirudinidae	Nereis / Tubifex, nais / Leech
Lophophoria	Bryozoa / Brachiopoda	Ectoprocta / Testicardinae	Phylactolemae	Cristatella
Molluscs	Amphineura / Scaphopoda / Gasteropoda / Lamellibrancha	Euthyneura	Pulmonae / Eulamellibrancha	Chiton / Planorba / Dreissena (mussel) (fig. 630)
	Cephalopoda	Dibranchiata	Octopodia / Decapodia	Octopus / Cuttle-fish
Arthropods	Crustaceae	Entomostraceae	Ostracoda / Copepoda / Cladoceriae	Cypris (fig. 628) / Cyclops (fig. 627) / Daphnia, Bosmina (fig. 626)
		Malacostraceae	Isopoda / Amphipoda / Decapoda	Asellus, wood-louse / Gammarus / Astacus (crayfish)
	Tardigradae / Myriapoda / Arachnidae / Insects	Diplopoda / Pterygota	Eutardigradae / Chilognathae / Acariae / Diptera	Macrobiotus / Galley-worm / Hydrachna / Chironoma
Stomochordata	Enteropneustae			
Chordata — Tunicata	Ascidians			
— Acraniae				Amphioxus
— Vertebrata	Cyclostomata / Chondrichthyae / Osteichthyae	Elasmobranchia / Neopterygii	Selachii / Teleostei	Lamprey / Shark / Trout
	Batrachiae		Urodelae / Anoura	Proteus / Frog
	Reptiles / Birds / Mammals	Placentaria	Cheloniae / Anatidae / Primates	Tortoise / Duck / Man

4. BACTERIOLOGY

The study of bacterial microorganisms is very complex, a subject on which, moreover, many excellent works may be consulted (see bibilography).

4.1. General characteristics of bacteria

Bacteria are generally identified following cultivation on special media, and, by observing the specific cultural characteristics which each species exhibits on or in the appropriate medium, it is possible to isolate and classify them.

Bacteria reproduce by binary fission (including nuclear division) provided the environment is favourable, but, under less favourable conditions, they may assume spore forms, from which, later, they can develop.

Certain bacteria are capable of proliferating in animals thereby causing diseases of varying severity.

Their virulence varies according to the environmental conditions. It may be reduced by artificial means (preparation of vaccines).

The immunity resulting from the injection of a vaccine, or from a natural attack by microorganisms of lowered virulence, is due to the production of antibodies which arise in response to the antigen derived from the invading pathogen.

The injection of blood serum from an animal that has undergone a mild attack of a disease is also a means of curing or prevention. This works through the action of serum antibodies which destroy or immobilize (coagulate) the corresponding pathogenic micro-organisms. The specific agglutination of cultures of pathogenic bacteria by the corresponding antibody is a means of identifying pathogens.

Some bacteria, normally non-pathogenic, may, in exceptional circumstances, become so under suitable conditions, or when, by accident, they happen to enter the bloodstream (e.g. Bacillus coli).

Living beings may carry pathogens in their systems without being troubled by them, either because they are naturally immune or because the virulence of the microbial species concerned has been temporarily reduced or because the particular bacterial strains are pathogenic only for some other type of host. Such self-immunized apparently non-infected persons are known as "carriers", who can transmit the specific disease in a virulent form without being affected themselves.

Certain viruses, the **bacteriophages,** possess the property of attacking and causing the disappeance of microbial colonies (the phenomenon called bacteriolysis). The specificity of their mode of action is variable: a given phage may affect a single strain, a single species or even several bacterial species (generally within the same group). The detection of these can be useful in the bacteriological examination of drinking water, since they can reveal previous contamination, the germs of which since disappeared.

4.2. Pathogenic bacteria found in water

Water can transmit a number of diseases of bacterial origin; among the germs responsible for these, are:

- **the typhoid fever bacilli:**
— Bacillus typhi or Eberth's bacillus: Salmonella typhosa;
— Paratyphoid bacilli A and B; Salmonella paratyphi and S. schottmülleri, respectively.

- **the dysentery bacilli, such as:**
— Shigella dysenteriae and Sh. paradysenteriae.
— Shigella flexneri or Flexner's bacillus (pseudo-dysenteric).
Gastrointestinal infections, which comprise salmonellosis and shigellosis, are not all well defined: there are many benign cases.

- **the cholera vibrio** (Vibrio cholerae or V. comma), discovered by Koch, which is in the form of small, curved, flagellate and motile rods.
V. proteus (cholera nostras et infantum) should be compared with the preceding organism.

- **Koch's bacillus** (Mycobacterium tuberculosis) is responsible for tuberculosis in all its forms. It is present in sanatorium effluent. An excess chlorine level of more than 1 mg/l, for 1 h, is needed for its destruction.

- **Proteus morganii,** which is associated with summer diarrhoea, especially in children.

- **Leptospira (or Spirochaeta) icterohaemorrhagiae,** the cause of haemorrhagic jaundice, which is very widespread throughout the world. Many carriers of the germ exist, notably the wild rat; sewermen sometimes contract this disease.

- **Escherichia Coli** is the cause of colibacillosis.

- **Proteus vulgaris** causes diarrhoea, intestinal catarrh (rather similar to typhoid fever) and a variety of diseases.

- **Bacillus pyocyaneus** (Pseudomonas aeruginosa), is frequently found in sewage.
The Pasteurelleae cause haemorrhagic septicemia in horses, pigs, sheep, cattle, rabbits, cats, dogs, etc.

- **Pasteurella tularensis** (Francisella tularensis) causes tularemia, a disease which is normally passed from person to person through the bites of blood-sucking insects, but which can also be spread by water.
Diseases due to viruses are treated separately, later.

4.3. Bacteriological analysis of Water

We shall not enter into the details of the bacteriological examination of water: works covering this subject in full detail will be found in the bibliography.

The methods to be used for the examination are generally laid down by the regulations in force in each country. In France, the methods laid down by the Conseil Supérieur d'Hygiène Publique, relating to drinking water, are given in detail in the Circulaire ministérielle of 21 January 1960. The legal requirements with regard to sampling methods, determinations to be carried out and their interpretation are defined in the Circulaire ministérielle of 15 March 1962.

4.4. The enzymes

The cell biocatalysts, or enzymes, only transform certain substrates and only catalyse certain reactions. They are specific. They consist, generally, of a prosthetic group or coenzyme and of a specific protein or apoenzyme. Depending upon the type of combination, enzymes are said to be induced or constitutive, such as:

— the oxidoreductases; hydrogen or electron carriers (cytochrome, flavine, etc.);
— the transferases; these transfer chemical groups;
— the hydrolases which cause the hydrolytic cleavage of certain bonds in the presence of water (maltase, amylase, etc.);
— the lyases which dissociate groups from molecules;
— the isomerases which bring about changes in internal structure;
— the ligases which, without the necessity for water, produce new linkages.

The oxidoreductases are extremely important as, due to their presence, bacteria can control certain exothermic reactions from which they draw the energy necessary for their metabolims. This is so in the case of the oxidation of ammonia or of nitrites (nitrification), the oxidation of sulphides and of sulphur, of iron and of manganese (removal of iron, removal of manganese, corrosion).

The activity of enzymes is thus essential, not only for the control of fermentations in the strict sense, which are caused essentially by the action of hydrolases and lyases, but also for the operation of the nitrogen and sulphur cycles and the process of bacterial corrosion.

4.5. Fermentations

Fermentations are chemical reactions produced in nature by certain microorganisms. These microorganisms, or ferments, act by secreting specific enzymes (see above). Consideration of these will be restricted to the main types of fermentation reactions.

● **Alcoholic fermentation:** this consists of the aerobic decomposition of sugars into alcohol and carbon dioxide. The organisms responsible for this are the fungi: yeasts (Saccharomyces) and moulds (Aspergillus, Penicillium, Mucor):

$$C_6H_{12}O_6 \longrightarrow 2\ C_2H_6O + 2\ CO_2$$

Small quantities of glycerin and succinic acid are always produced as by-products.

● **Acetic fermentation:** this is the conversion of alcohol into acetic acid by the action of Mycoderma aceti (Acetobacter aceti) under aerobic conditions:

$$C_2H_6O + O_2 \longrightarrow C_2H_4O_2 + H_2O$$

● **Lactic fermentation:** this is the conversion of certain sugars (e.g. lactose) into lactic acid by the action of a bacterium (Bacillus lacticus) in aerobic conditions:

$$C_{12}H_{22}O_{11} + H_2O \longrightarrow 2\ C_6H_{12}O_6 = 4\ C_3H_6O_3$$

Excess acidity (1 % of lactic acid) stops the fermentation.

● **Butyric fermentation:** certain bacterial species, such as Clostridium pastorianum, decompose ternary organic matter such as cellulose as well as the sugars and starchy materials, with the production of butyric acid $C_4H_8O_2$. This is an anaerobic fermentation and hydrogen and carbon dioxide are produced at the same time.

$$C_6H_{12}O_6 \longrightarrow 2\ CO_2 + 2\ H_2 + C_4H_8O_2$$

Alcohol is also oxidized to butyric acid by Terminosporum kluyveri:

$$C_2H_6O + C_2H_4O_2 \longrightarrow C_4H_8O_2 + H_2O$$

● **Methane fermentation:** this type is anaerobic and accompanies butyric fermentation. It is due to certain species of Clostridium and gives rise to the production of methane, one of the components of marsh gas.

$$C_6H_{12}O_6 = 3\ CH_4 + 3\ CO_2$$

● **Decomposition of cellulose:** this very slow type of fermentation is very important in nature and leads to the formation of humus in farm manures and in the soil. It is caused by aerobic microorganisms (Cellvibrio, Cytophaga, Cellfalcicula, cellulomonas) or by anaerobic microorganisms (Clostridium cellulosolvens, for example).

The aerobic type are the most active. They also attack the majority of sugars. Anaerobic fermentation takes place principally at high temperatures (65 ºC).

These bacteria can be a source of trouble in the paper trade, by causing degradation of the woodpulp, due to localized lysis of the fibres: this is the well-known phenomenon of "slimes", due not only to the bacteria mentioned above, but also to waterborne microorganisms: Aerobacter, Pseudomonas, Achromobacter, Escherichia coli, Sphaerotilus, Desulfovibrio and certain of the Cyanophyceae.

● **Putrid fermentation.**

This comprises many types of fermentation, all of which lead to the decomposition of nitrogenous organic materials into simpler compounds. They are all accompanied by ammoniacal fermentation and are anaerobic. The products of such fermentation are:

— volatile acids of the $C_nH_{2n}O_2$ series (formic, acetic, butyric, etc., acids);

— ammoniacal compounds;

— gases: CH_4, CO_2, H_2, N_2;

— ptomaines, organic bases which are extremely toxic but which are destroyed by heat.

Sulphur fermentation (the conversion of sulphur compounds into H_2S) can also be placed in this category.

The ammonium compounds and H_2S will then undergo further biochemical transformations, before, once again, being assimilated by living creatures: in this way the nitrogen and sulphur cycles are closed.

4.6. The nitrogen cycle

With the exception of the fertilizer industry, built-up areas and industry discharge nitrogen in the reduced form of organic or ammoniacal compounds. Bacterial ammonization in waterways is very rapid and the ammoniacal nitrogen level correlates well with the organic pollution from the sewers.

Indeed, there is a risk that the state of equilibrium will be changed by the nitrogen which is thus introduced into the natural cycle. The nitrogen cycle is shown in diagrammatic form in fig. 575.

In an aerobic environment, such as a river containing dissolved oxygen, organic matter and ammonium salts are converted into nitrites and then into nitrates, with the consumption of oxygen. This is the process known as **nitrification** which covers two successive reactions. The first stage of nitrification is **the formation of nitrites** by bacteria: Nitrosomonas, ·Nitrosocystis, Nitrosospira, Nitrosoglea, etc... The second stage, **the formation of nitrates,** is the work of bacteria of the genera: Nitrobacter, Nitrocystis, Bactoderma, Microderma, etc...

All these bacteria are autotrophic (see page 16) and strictly aerobic. They

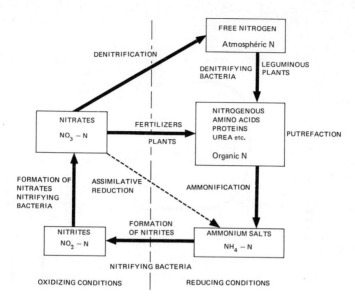

Fig. 575. — The nitrogen cycle.

use the energy produced by the oxidation of ammonia and of nitrites to reduce inorganic carbon, originating from carbon dioxide or carbonates.

Favourable conditions are frequently combined in surface waters: high level of dissolved oxygen, organic pollution diluted and thus removing the risk of the inhibition of autotrophic bacteria. The waste nitrogen products are rapidly converted to nitrates. 4.57 mg of oxygen is necessary per mg of nitrogen, if the oxidation is to be complete; the simplifed reaction is represented as:

$$NH_3 + 2O_2 \longrightarrow HNO_3 + H_2O$$

Thus the process of nitrification tends to reduce the oxygen content in the waterway in the same way as does assimilation of organic pollution.

It is true that the nitrates afford an oxygen reserve, which they can give back through denitrification when conditions again become reducing and anaerobic, but little hope can be placed on such conditions in a river. It can be seen that denitrification can assume two aspects (fig. 576):

— **dissimilative reduction** or the reduction of nitrates by respiration; a process by which heterotrophic bacteria reduce nitrates and nitrites to nitrogen gas; bacterial synthesis, therefore, requires the presence of carbonaceous organic matter and of traces of ammonia;

— **assimilative reduction** which, on the contrary, takes place in the complete absence of ammoniacal nitrogen ($NH_4 \cdot N$); nitrates and nitrites are thus reduced to ammonia, which is used to produce new cells.

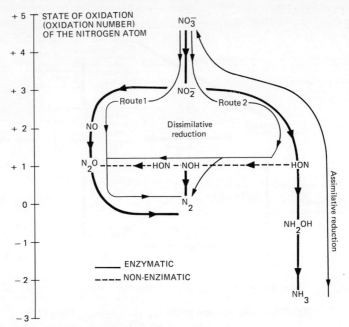

Fig. 576. — Routes for the reduction of nitrates.

Dissimilative reduction is the commonest reaction. This is what is understood when one speaks of *denitrification*, without qualification. It is particularly important in the treatment of waste water (see p. 107).

4.7. The sulphur cycle

It has been seen earlier (putrefaction fermentation) that anaerobic sulphur fermentation converts sulphur-containing organic compounds into H_2S.

Other aerobic bacteria can oxidize H_2S to sulphur (which they are sometimes capable of storing in the form of light-refracting granules, dispersed in their cytoplasm) and finally into sulphuric acid:

● **Photosynthetic** reactions in the **Rhodothiobacteriales** (or purple sulphur bacteria) such as Chromatium, Thiospirillum (see fig. 577) or Thiopedia, in the same way as in the case of the **Chlorothiobacteriales** (or green sulphur bacteria) such as Chlorobium or Chlorobacterium; elementary sulphur is formed first:

$$2\,H_2S + CO_2 \xrightarrow[\text{energy}]{\text{light}} (CH_2O)^1 + H_2O + 2\,S$$

1. In these reactions (CH_4O) represents synthesized organic matter.

The sulphur produced is, according to species, either stored in the bacterial cell or excreted. It may later be converted into sulphuric acid:

$$2\ S + 3\ CO_2 + 5\ H_2O \xrightarrow[\text{energy}]{\text{light}} 3\ (CH_2O) + 2\ H_2SO$$

The complete reaction may then be written:

$$H_2S + 2\ CO_2 + 2\ H_2O \xrightarrow[\text{energy}]{\text{light}} 2\ (CH_2O) + H_2SO_4$$

- Simple **oxidation-reduction** reactions in
— the **Leucothiobacteriales** (or colourless sulphur bacteria), such as Beggiatoa or Thiothrix:

$$2\ H_2S + O_2 \longrightarrow 2\ H_2O + 2\ S$$

— some **Protobacteriaceae.** such as Thiobacillus thiooxidans, which then oxidizes the sulphur to sulphuric acid:

$$2\ S + 3O_2 + 2\ H_2O \longrightarrow 2\ H_2SO_4$$

The final end-point, in an aerobic environment, may thus be the appearance of sulphates; on the other hand, in an anaerobic environment these may be reduced by other bacteria (Desulfovibrio or Sporovibrio desulfuricans, some Clostridium, etc...) which secrete sulfatoreductases, capable of catalysing the reaction:

$$H_2SO_4 + 4\ H_2 \longrightarrow H_2S + 4\ H_2O$$

Sulphite-reducing bacteria also exist (certain species of Clostridium and Welchia).

Some of these bacteria take part in the process of corrosion of cast iron or steel or concrete pipework (see Chapter 2, p. 38).

Fig. 577. — Thiospirillum × 1 000.

4.8. Bacterial oxidation of iron and manganese

The exothermic oxidation of iron can be catalysed by some bacteria, due to the oxidation-reduction enzymes which they excrete (flavoproteins, see page 974); trivalent iron, rendered insoluble in the form of the hydroxide, is then stored in mucilaginous secretions (sheaths, peduncles, capsules etc...) of these bacteria.

The organisms responsible for these phenomena are mainly the Siderobacteriales particularly:

— Chlamydobacteriales : Leptothrix (L. ochracea, L. crassa, L. discophora);

— Crenothricaceae   Crenothrix (Cr. polyspora), Clonothrix (Cl. ferruginea, Cl. fusca);
— Siderocapsaceae: Siderocapsa, Ferrobacillus, Sideromonas;
— Gallionellaceae: Gallionella (G. ferruginea, G. major).

This property is also shared by Protobacteriaceae (Thiobacillus ferrooxidans).

All these organisms can also cause the oxidation of manganese, if this latter element is distinctly more abundant than iron; besides these, other bacteria show a specific activity in this respect, for example:
— true bacteria: Pseudomonas (Ps. manganoxidans), Metallogenium (M. personatum, M. symbioticum);
— Siderobacteriales: Leptothrix (L. echinata, L. lopholes);
— Hypomicrobiales: Hypomicrobium H. vulgare).

The action of all these microorganisms can be very important in the processes of removal of iron and of manganese.   Their development appears to be encouraged by simultaneous nitrification.   Some of them are abundant in the acidic waters of mines.

4.9. Bacteria concerned in the obstruction and corrosion of pipework

A. Ferrous metal pipework.

Although the processes described in the preceding paragraph are beneficial in a deep-water treatment installation, they can, on the contrary, be very damaging in the interior of a cast iron or steel pipe.

Traces of iron in the water are sufficient to induce the development of the ferrobacteria mentioned above.   Three main genera are easily recognized under the microscope:
— **Leptothrix:** a filament (or trichome) containing a single line of cylindrical cells, surrounded by a sheath; this is at first thin and colourless, but becomes thicker and develops a brown colour, which becomes increasingly darker as it is impregnated with iron oxide (fig. 578).
— **Crenothrix:** the trichomes have openings at one end, through which cells escape, in several rows, to form new trichomes.   The development of the sheath is similar to that of Leptothrix.

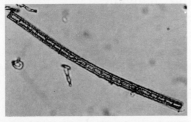

Fig. 578. — Leptothrix ochracea × 625.   *Fig. 579. — Gallionella ferruginea × 625.*

— **Gallionella:** isolated cells growing on spiral stalks (branched or not) which they secrete. The linkage is, in fact, fragile and the stalk only is frequently found (fig. 579).

The activity of these bacteria results in the formation of tubercles, which can block the pipe (the presence of these tubercles can be shown by partial solution in a concentrated acid and examination of the residue under the microscope, which reveals the bacterial filaments, sheathed with iron oxides). Areas coated with these tubercles become anaerobic, allowing the development of sulphate-reducing bacteria (a typical example being Desulfovibrio desulfuricans).

This bacterial growth promotes corrosion: their mechanism of action is described in Chapter 2.

B. *Concrete pipework.*

The main organisms responsible are:

— in anaerobic zones: Desulfovibrio (= Sporovibrio), Spirillum, Clostridium (production of H_2S);

— in aerobic zones: Thiobacillus thiooxidans (formation of H_2SO_4);

— the chemical reactions induced by these microorganisms are described on page 979.

5. VIROLOGY

● *General remarks:*

The viruses are minute pathogenic agents visible only under the electron microscope and capable of multiplying only inside a living cell. A virus consists of the association of a nucleic acid and a protein, components which have been successfully separated chemically and even crystallized. The reunion of these two components reconstitutes the virus with all its biological properties. When a living cell has been attacked by a virus, it is soon totally transformed into a granular mass of proliferating viruses ready to infect other cells. Viruses are responsible for colds in the head, influenza, measles, scarlet fever, smallpox, poliomyelitis, rabies and certain types of cancer.

● *Viral diseases transmissible by water:*

The following viruses are found in water:

a) **The Enteroviruses:**

— the virus of **poliomyelitis** which attacks the nerve centres;

— **echovirus (enteric cytopathic human orphan virus),** which is the cause of generally mild intestinal disease (diarrhoea in children); certain serological types can cause lymphocytic meningitis (reversible);

— **coxsackie A** and B can cause lymphocytic meningitis, myalgia and myocardia.

b) **The infectious hepatitis virus:** only type A is transmissible by water (type B is transmissible by blood transfusions). The disease is endemic in some hot regions.

c) **The Adenoviruses,** which attack the upper respiratory passages and the eyes, but which are also present in the intestine.

d) **Influenza** which can be passed on accidentally in swimming pools, as this virus, although rarely found in the intestine, is very abundant in nasal mucus.

e) **The Reo viruses,** still incompletely understood.

f) **The Papillomavirus,** which is responsible for the verrucas contracted in swimming pools.

In fact, the role played by water in the transmission of some viral diseases is very controversial. Moreover, if one refers to the reports of those rare cases in which it has been possible to attribute a water origin to viral epidemics, it will be found that the water has always been grossly contaminated from sewers: it seems that the risk of epidemic is present only in extreme conditions, but the present knowledge is still poor concerning minimal infective doses.

● *Precautions against viruses:*

The possible presence of viruses in the water must be taken into consideration when siting the drinking water intake points; these should be chosen in such a way as to avoid the presence of large quantities of sewer water, whether purified or not. .

Flocculation plus sedimentation is an effective method of eliminating 95 to 99 % of the viruses (especially in sludge blanket processes). Prior chlorination can improve these results, if well carried out.

Filtration is only successful if it is combined with a flocculation process. Filtration alone, even if carried out using very fine filter media (diatomaceous earth or porcelain), is ineffective.

Chlorine or ozone disinfection is of variable effectiveness, not all the viruses being equally resistant (regarding contact time and residual oxidant content, see page 403).

When the treatment plants are well designed and well operated, it is possible to eliminate all the viruses by a combination of the different procedures.

● *Virological examination of water:*

The viruses must, first of all, be concentrated; gauze, used for a long time, is now no longer used. The procedures at the present time are as follows: filtration on membranes (soluble or insoluble) or using cartridges, electrophoresis or electro-osmosis, coagulation-flocculation, centrifugation (with or without a non-water-miscible adsorbent phase), fluidized-bed or column microbead filtration (elution by pH control); this is followed by inoculation of cell cultures (usually monkey kidney cells) or of animals (young mice).

6. PATHOGENIC ORGANISMS OTHER THAN BACTERIA

6.1. Fungi

A microscopic fungus known as Histoplasma capsulatum, which causes histoplasmosis, sometimes infests mains piping.

The spores of this fungus are resistant to chlorine but can easily be eradicated by flocculation and filtration.

Another microscopic fungus occasionally develops on wet wood and can infect the skin of bathers at swimming pools.

6.2. Amoebae

These are unicellular animals of an average size of 50 μm.

As cysts, they can exist for over a month in water and are particularly resistant to the action of antispetics.

The species Entamoeba histolytica and E. tetragena are the causative agents of amoebic dysentery.

They can cause liver abscess and intestinal haemorrhages.

Furthermore, the species Naegleria gruberi may have been the cause of cases of encephalomeningitis, contracted from poorly chlorinated swimming pools.

6.3. Worms

Water may serve as the vehicle for many worm parasites of Man and animals. Disinfection with the dosage normally practised does not destroy these worms or their eggs. On the other hand, their size is usually sufficient to ensure their elimination by efficient filtration, thus suppressing contamination risks in practice. These worms generally pursue their development in several different hosts in succession:

— **Tenia solium** (tapeworm): spherical egg 35 μm in size forms a cyst in pigs;

— **Tenia saginata** (tapeworm): oval egg 25 × 35 μm forms a cyst in cattle;

— **Tenia echinococcus:** adult in dogs, forms hydratic cysts in sheep; eggs 25 × 35;

— **Bothriocephalus latus:** adult in man. The oval egg, 45 × 70 μm in size, develops in water and releases an embryo which infects a copepod. When the latter is swallowed by a fish, the embryo becomes a larva (8 to 30 mm long) and lodges in the muscles of the fish.

— **Distoma hepatica:** a flat worm, which is adult in man and sheep. The ovoid egg, 70 to 130 μm in size develops in water and produces a ciliated embryo which infects a water-snail, inside which it is transformed into a cercaria. This swims away and comes to rest on aquatic herbage before infecting shee pagain. The adult form of this worm lodges in the liver;

— **Distoma lanceolata:** similar to the above, the intermediate host being a planorbis snail;

— **Bilharzia:** a trematode worm the generic name of which is Schistosoma and which is the cause of a very serious disease, called bilharziosis or schistosomiasis

and widespread in the hotter parts of the world.   There are 2 forms of the disease and 3 species of this parasitic worm:
— vesical bilharziosis, due to S. haematobium (Egypt, tropical Africa, Madagascar);
— intestinal bilharziosis, caused by S. mansoni (Egypt, tropical Africa, Madagascar, South America) or by S. japonicum (Far East).

The adult lives in the blood vessels of man.   The egg is ovoid (50 × 150 μm), with a lateral spine of 25 μm, and is excreted in the urine.   It develops in water to produce a ciliated embryo (miracidium), which infects a snail living in still water.   There it turns into a cercaria (larva with a forked tail) which returns to the water, where it may enter a new human host, through the skin (or the mucous membranes of the mouth, if the water is swallowed).   The cercaria has a life span of 2 days.

This epidemic is fought by the destruction of the intermediary host, that is, the molluscs: using chemical methods, biological methods, changing the habitat of the molluscs.

The cercaria may be removed from drinking water by filtration through fine sand (0.35 mm maximum); but, above all, the guarantee of the destruction of these cercaria lies in effective prechlorination and final disinfection (chlorine or ozone), with the use of suitable dosages and times of contact.

— **Ascaris lumbricoides:** occurs frequently in the small intestine of man and pigs. The ovoid egg, 50 × 75 μm, develops in water or wet soil to form an embryo 0.3 mm long, which is transmitted to man direct.

— **Oxyuris vermicularis:** frequent in children.   Oval egg with one face slightly flattened, 20 × 50 μm.   Does not appear to live long in water.

— **Eustrongylus gigas:** infects the urinary passages.   The elliptical egg, 40 × 60 μm and paler at the two extremities, is expelled with the urine.   The embryo is 0.25 mm long and infects a fish as an intermediate host.

— **Ankylostoma duodenale** (hookworm): a small worm 6 to 20 mm long that lives in the intestine, buries into the mucosa and causes very persistent haemorrhages and diarrhoea (ankylostomiasis or miners' anaemia).   The egg, 30 × 60 μm, develops in water and needs a minimum temperature of 22 °C.   The embryo, 0.2 mm long, produces a larva capable of penetrating through the skin to infect a new victim.

— **Filaria medinensis** (Dracunculus, Guinea worm, Medina worm): a viviparous worm.   The embryo is 0.5 to 1 mm long and infects a copepod to form a larva. This copepod, if swallowed with water, enables the larva to develop in man.   The worm burrows through the intestinal wall and forms subcutaneous ulcers (dracunculosis).   The acult worm is 0.5 to 0.8 m long and 1 mm thick.

Transmission is also possible by direct passage of the larva through the skin.

— **Filaria sanguinis hominis:** this worm lives in the vessels in the lower part of the body, particularly in the bladder, and produces haematuria. It is transmitted by a mosquito as intermediate host.

— **Anguillula intestinalis:** a worm 2 to 3 mm long, which lives in the duodenum. The egg develops in water to form a worm (Anguillula Stercoralis) which lays eggs in the water, the larvae from which can reinfect man if swallowed in the water or by transmission through the skin.

6.4. Insects

The main aquatic insects that are a danger to health are the mosquitoes, whose larvae can live only in water. Mosquitoes act as carriers for certain diseases.

Malaria, or marsh fever, is transmitted by the anopheles mosquito (A. maculipennis, funestus or gambiae), the agent of the disease being Laveran's haematozoon. Yellow fever is transmitted by Aedes aegypti. The genus Culex can be the vector of certain diseases (viral encephalitis, filariasis).

7. A STUDY OF PLANKTON

GENERAL

To the above must be added those microscopic, non-pathogenic animals and plants whose natural habitat is water, and whose presence may either cause inconvenience (particularly from an aesthetic or organoleptic point of view) or be beneficial (cf. their function in the self-purifying of natural waters, in the biological membrane during slow filtration, the selective development of certain animal species in activated sludge, etc.).

Plankton constitute the mass of microscopic organisms which live suspended in water, and which can be divided into zooplankton (animal plankton) and phytoplankton (plant plankton). The majority of plankton species vary in size between a few millimetres and 20 μm (microplankton); those species less than 20 μm in size constitute what is called nanoplankton and are collected by centrifuging.

Simply by examining the plankton it is possible to ascertain certain chemical or physico-chemical properties of the untreated water; certain species are found only in water which is rich in dissolved organic matter, while others are characteristic of acid water.

7.1. Problems due to plankton

● **Algae (Phytoplankton):**

In surface waters, in temperate countries, algae, which are almost entirely absent in winter, undergo development, sometimes at an explosive rate, during periods of warm weather. In spring the diatoms generally develop first, and give way to the chlorophyceae in summer and to the cyanophyceae in the autumn.

Many species are able to live in fresh water. They are not generally of significance from the point of view of health, but they can cause trouble in filtration plants by clogging the filters very quickly, unless the latter are supplemented by a flocculation and sedimentation plant upstream.

Effective micro-straining enables the number of organisms to be reduced by 50 to 90 %, depending on the species.

In order to achieve maximum elimination of the microscopic algae by flocculation and sedimentation, it is often necessary, at the same time, to carry out pre-chlorination or a preliminary treatment with chlorine dioxide or ozone.

Among the algae most likely to cause clogging or choking must be mentioned: Melosira, Asterionella, Fragilaria (diatoms), Pediastrum (green algae). These algae gather together and form a very dense carpet on the surface of filters (especially slow filters), which obstructs the water from passing through.

The filters may also be clogged by the gases released by phytoplankton, in particular oxygen resulting from photosynthetic reactions; such occurrences, can also give rise to difficulties in settling.

Some algae are the cause of unpleasant taste and odours (earthy, musty, muddy, fishy, grassy, aromatic, etc...). These may be due to phenolic compounds which react with chlorine to give chlorophenols; they may also be due to metabolites, similar to those secreted by the Actinomycetes, especially geosmin. The principal organisms involved are the Cyanophyceae (Anabaena, Oscillatori, Aphanizomenon, Cylindrospermum, Rivularia), but Diatoms (such as Asterionella), Chlorophyceae (such as Cloadophora), Chrysophyceae (for example, Synura), etc., can be responsible.

Finally, some algae can synthesize toxic metabolites. The principal organisms responsible are, again, some Cyanophyceae: Microcystis, Anabaena, Lyngbya, Aphanizomenon, Nodularia, Nostoc, Gloeotrichia, etc... (mention should also be made of Prymnesium parvum, one of the Chrysophyceae, in brackish waters and of the "red tides" in sea water due to Peridinians). The toxic metabolites in fresh water do not present any danger to those drinking treated water, since they are not excreted; this means that they are eliminated with the algae during treatment (even so some cases of gastroenteritis have been attributed to them). However, they can cause the death of animals drinking the untreated water and can also be the cause of dermatitis and conjunctivitis in bathers.

On the other hand, there are some cyanophyceae which secrete substances of therapeutic value for the treatment of certain kinds of sores, ulcers, etc. These organisms can be cultivated in the laboratory and used in medicine.

● **Zooplankton:**

Animal plankton, too, may cause a certain number of problems:
— from the point of view of **health,** through the direct action of pathogenic organisms such as amoeba, worms, etc., which have been mentioned above,

or through the indirect action of certain crustaceans, insect larvae, or Nematoda, etc., the digestive tubes of which may contain bacteria or viruses, and which are, therefore, thus making them into disease vectors;

— from the **aesthetic** or **organoleptic** point of view. This concerns organisms which are able to develop in pipes: consumers' taps may deliver water containing either the animals themselves or the products of their metabolism, which can discolour or impart an unpleasant taste to drinking water;

— from the **technological** point of view: the fresh-water mussel (Dreissenia polymorpha), the larva of which is planktonic, for example, may succeed in completely blocking conduits and pipes; it can then only be removed by mechanical scraping to get rid of the adults, and by a shock chlorination to prevent further development of the larvae.

For all these reasons, plankton, even in their reproductive forms, must be completely eliminated from drinking water. (See Chapter 20, page 641— Elimination of plankton).

7.2. Quantitative estimate of plankton

For evaluation of the amount of plankton in a given sample of water, the following procedure may be used:

● **Samples:** a sample that is not to be examined for a few hours should have added to it 3 to 5 ml of a 40 % formaldehyde solution for each 100 ml of water, so as to prevent the multiplication of the living organisms (but it is preferable not to kill the organisms if a qualitative examination of zooplankton is to be made).

Having put the plankton in suspension by suitable means, carry out a microscopic examination, either of 1 ml of water if a counting cell is available, or of a drop placed between an ordinary slide and a cover glass.

If the water does not contain much plankton, the sample must be concentrated. This can be done by filtering it through very fine sand, but the most reliable method is to centrifuge it at 5 000 rev/min.

● *Expression of results:* count the number of individuals of each species and express it per ml of water, estimating if necessary the mean volume of each, in order to estimate the total biomass.

The planimetric standard unit is used in some countries; this is equal to 400 μm^2. Whipple's micrometer should be used: when placed in the eyepiece of the microscope, this marks out an area of 1mm^2 divided into squares with sides of 20 μm (with a 16 mm objective and a 10 × eyepiece).

● *Estimation of chlorophyll:* for the quantitative estimation of phytoplankton, it is necessary to concentrate them (centrifugation or filtration on filter paper) and to extract the chlorophyll with a solvent (methanol or acetone), in the cold or at boiling point. The concentration in the solution so made is estimated spectrophotometrically; a method of calculating the chlorophyll a, b and c values from the optical density at 663, 645 and 630 nm, will be found in "Standard Methods" (14th ed. 1976). The apparatus can also be standardized using a chlorophyll solution or with an artificial solution. But this overall measurement should not be made to the exclusion of the microscopic examination, as a qualitative examination should always be made.

7.3. Description of plankton

7.3.1. PHYTOPLANKTON

This consists mainly of algae, autotrophic chlorophyllic growths (which means that they are capable of reproducing by using CO_2, water and various mineral salts in conjunction with solar rays). When optimum light and temperature conditions and chemical composition of the water are found together, the proliferation of phytoplankton can be so great that the water takes on a deep colour, either green, blue-green or brown, according to the dominating species. In such water several thousand cells to the millilitre can be found. Under certain conditions planktonic algae float on the surface, forming an unbroken carpet called water bloom (e.g. cyanophyta). The water beneath is often depleted of oxygen.

Phytoplankton includes unicellular or colonial algae which are motile and flagellate. Most plankton species, however, are non-motile. Their density is similar to that ofthe water, and they are morphologically adapted to a drifting existence; this adaptation takes the form of spikes, fine threads (Micractinium, fig. 585), indentations (Pediastrum, fig. 584) and lobes, curves, etc. The secretion of mucilage, the presence of fatty cells, and the gas-vacuoles of certain Cyanophyceae (page 998) also contribute to this adaptation to a planktonic form of existence.

Certain species are purely planktonic (Asterionella, Pediatrum clathratum, Dinobryon) and never grow in filters, clarifiers or reservoirs; others are facultative-plankton (e.g. the filamentous diatoms of Melosira, fig. 598, Fragilaria, fig. 600 and Diatoma vulgare, fig. 604, which are capable of becoming attached by secreting a small amount of mucus, thus contributing to the constitution of the category known as Periphyton). Other species are attached in the first stage of their development and then free themselves to float in clusters more or less on the surface (filamentous green algae: Ulothrix, Spirogyra (fig. 592), Cladophora (fig. 591), Draparnaldia (fig. 589). Lastly, in certain cases, some species which are benthic (i.e. habitually attached to the shore or sea-bed) are moved by currents and found with the phytoplankton.

The classification of algae is based primarily on pigment (main phyla) and then on morphology, food substance, and flagella attachments. The pigments are carried by plastids (except in the case of the cyanophyta), where photosynthesis takes place. The green colour of chlorophyll which is always present, may be masked by other pigments (carotenes, phycobilins).

In each phylum dealt with, only the most common orders, characterized by one or two of the most frequently encountered species, will be mentioned here.

a) Chlorophyta.

This important phylum includes all green algae except the Euglena (Euglenophyta). The pure green plastids are of varying shapes. Starch is the food substance. The motile stages more often than not have two equal flagella.

— **Volvocales.** These planktonic algae are made up of small motile cells, with two or sometimes four flagella and one red eyespot (stigma).

— They are isolated:
e.g. genus *Chlamydomonas* (fig. 580);
— or colonial: in this case the motile cells of the Chlamydomonas type constitute a fairly thick jelly forming a coenobe, i.e. a colony within which individuals have no common physiological link.

The shape of the colony and the number of individuals composing it are characteristics of each species.

— Flat colony:
Gonium sociale (4 cells)
Gonium pectorale (16 cells) (fig. 581))
— Spherical colony:
Pandorina (fig. 582)
Eudorina (fig. 583).

Each cell gives birth to another colony of 8, 16 or 32 cells. Multiplication is very rapid.

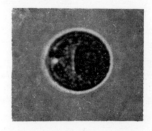

Fig. 580. — *Chlamydomonas*- × 750. *Numerous species, very common, cup-shaped chloroplast, two flagella, one stigma. Water rich in dissolved organic matter is particularly favourable to proliferation.*

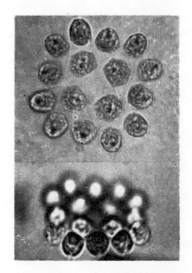

Fig. 581. — Gonium pectorale- × 700. Tabular colony formed of 16 cells (top and side view). Advances with flagella in front.

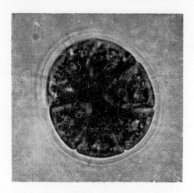

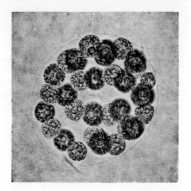

Fig. 582. — Pandorina morum × 750.
Spherical colony of 16 cells, grouped side by side; the two flagella of each cell pass through a thick jelly before being liberated.

Fig. 583. — Eudorina elegans- × 280.
The 32 cells are arranged fairly regularly in parallel circles on the edge of the spherical colony.

— **Tetrasporales.** They form green, gelatinous, irregular thalli which contain non-motile cells dispersed in the mucilage (palmelloid stage).

— **Chlorococcales.** Planktonic algae, very numerous in genera and species, non-motile in the vegetative state; the cells have neither flagella nor stigma. The organization is either unicellular or colonial, but never filamentous.
Pediastrum—appears as a small green disc, a flat coenobe, with a variable number of cells; these cells are either contiguous or have voids between them. The peripheral cells have extensions. Prolific in summer (fig. 584 and fig. 587).
Scenedesmus—small coenobe of two, four or eight cells. Numerous planktonic species (fig. 586).
Coelastrum—hollow spherical colony; the cells are joined by extensions. Frequent in alkaline water.
Micractinium—(fig. 585).

— **Ulothricales.** These are filamentous unbranched algae characterized by intercalary growth, forming a single thread of cells. The cells have a parietal chloroplast.
Ulothrix—filament fixed at the first stage of development, then free.
Hormidium—filament often separated into short elements (fig. 588).

— **Chaetophorales.** In the form of branched filaments with a main stem and branches. The terminal cells are elongated with a colourless tip (hairs).
Draparnaldia—found in fairly pure, quiet or slowly running water (fig. 589).

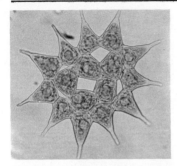

Fig. 584. — *Pediastrum clathratum*- × 200.
*Pure planktonic species. Proliferates in
summer. Clogs filters. Harmful to successful clarification. Very common in the
Seine river.*

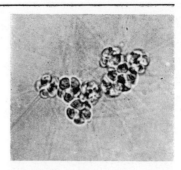

Fig. 585. — *Micractinium*- × 600.
*Spherical cells grouped in tetrahedra
clusters of four. Each cell has severa
very long fine threads.*

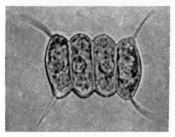

Fig. 586. — *Scenedesmus*- × 900.
*Very common, numerous planktonic species.
Four cells joined side by side in one plane;
the two ends are often spiked.*

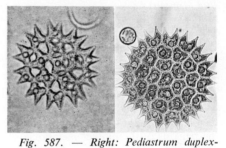

Fig. 587. — *Right: Pediastrum duplex*-
× 350.
*Peripheral cells with two cornua. Coenobe
cells not contiguous.*
Left: Pediastrum boryanum- × 250.
*Peripheral cells with two cornua but
coenobe cells contiguous.*

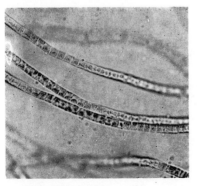

Fig. 588. — *Hormidium* × 300.
*Narrow filament (6 μm). Each cell has
a plastid half the circumference.*

Fig. 589. — *Draparnaldia* × 150. Branched
*filament with an axis of large cells and
hairy, fairly whorled branches.*

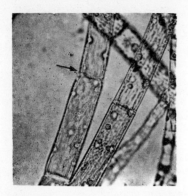

Fig. 590. — Oedogonium - × 300. Unbranched filaments. Each cell has a reticulated plastid. (Growth rings are visible on the upper part of the left cell.)

Fig. 591. — Cladophora - × 60. Branched filaments; the lateral divisions appear as buds made up of large coenocytes. The filaments are rough to the touch because of the chitinous nature of the superficial membrane.

— **Oedogoniales.** Algae very common in still water and consisting of single or branched filaments which are attached at the outset of their development and then free-floating. The upper part of the cell often has encased rings which are the remains of the mother membrane of the cell after each division.
Oedogonium—(fig. 590).

— **Cladophorales or Siphonocladales.** These are algae of single or branched filamentous structure, formed of multinucleate articulations with a parietal plastid in the network. Growth of the filaments is mainly terminal. They are often covered with epiphytic diatoms.
Cladophora—(fig. 591).

— **Conjugales.** This order of algae is characterized by its own method of reproduction—conjugation. There are never any motile spores.
They are divided into two classes.

— *Zygnemataceae*—unbranched filamentous algae. Numerous species.
Zygnema.
Mougeotia.
Spirogyra—(fig. 592 and 593).

— *Desmidiaceae*—unicellular or colonial. Numerous species exist in acid water.
Cosmarium—very common in fresh water (fig. 594 and 595).
Closterium—crescent-shaped cell with a special vacuole at each end in which grow crystals of calcium sulphate (fig. 596).

Fig. 592. — Spirogyra - Conjugation - × 150. Example of scalariform conjugation. The cells forming two parallel filaments are linked by copulatory tubes (one filament has the role of a ♂; the other, which plays the role of the ♀, contains the eggs or zygotes).

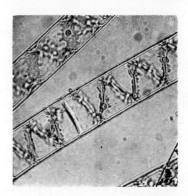

Fig. 593. — Spirogyra - × 300. Very common. Characteristic chloroplasts as bands running spirally along its length in the parietal layer of protoplasm. Found mainly in spring.

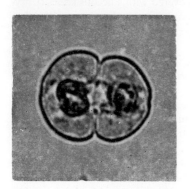

Fig. 594. — Cosmarium laeve - × 975.

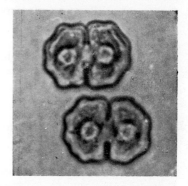

Fig. 595. — Cosmarium impressulum - × 975.

The cell is divided in two by a median constriction.

Fig. 596. — Closterium - × 250. Crescent-shaped cell with two symmetrical chloroplasts, the central area being occupied by the nucleus.

B) Euglenophyta.

This phylum is characterized by unicellular flagellate organisms, possessing pure green plastids but containing no starch; the food substance is paramylum, often in the form of rods.

Its various genera are mainly found in stagnant water which is rich in organic matter.

Euglena—the uniflagellate cell is spindle-shaped but may easily become distorted (metabole). Numerous species (fig. 597).

Phacus—flat organism, rigid membrane, no metabole, one flagellum.

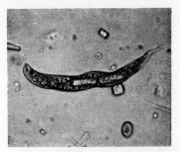

Fig. 597. — Euglena - × 250. In this photograph the flagellum is not visible, although its insertion and the eyespot can be clearly seen (right). Within the cell (left) can be seen two paramylum rods.

C) Chrysophyta.

This phylum comprises three large groups of algae, their food substance never being starch. The plastids are yellow-green, yellow or brown; they form leucosins or chyrsose and oils. These algae are either in the form of a siliceous skeleton (diatoms) or a membrane sometimes covered with siliceous scales (Chrysophyceae) or siliceous cysts (form of protection) (Chrysophyceae and Xanthophyceae).

Bacillariophyceae or diatoms.

These are isolated unicellular organisms, sometimes grouped in colonies, with brown plastids, characterized by their hard siliceous membrane, called the frustule or theca.

The frustule consists of two valves, one inter-linking into the other. Classification depends on the shape of this frustule, seen either from the valve or girdle aspect (and on the delicate structural ornamentation of the valves).

Diatoms play an important role in the plankton world. Intense proliferation generally takes place in spring and autumn.

Centrales order. Planktonic diatoms, axially symmetrical, with circular valves and numerous plastids.

Melosira—the cells are jointed to each other (valve side) by a cushion of mucilage forming long cylindrical filaments.

Numerous species; some are planktonic and tend to cause severe clogging (fig. 598).

Cyclotella—small isolated cells sometimes grouped in short filaments, easily dissociated.

They proliferate mainly in spring and by penetrating the sand can clog filters to a considerable depth (fig. 599).

Stephanodiscus—can be distinguished from Cyclotella by the presence of small marginal spikes (girdle view).

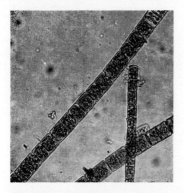

Fig. 598. — Melosira varians - × 150. Filaments consisting of large cells (girdle view) with numerous brown plastids. Considerable growth on the sand of slow filters (biological membrane) from spring to the end of autumn.

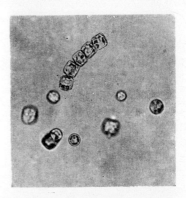

Fig. 599. — Cyclotella chaetoceras - × 300. Cells often joined in short filaments (girdle view) having long fine threads. Viewed from the valve side it is circular.

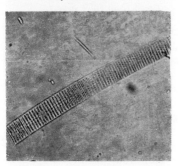

Fig. 600. — Fragilaria - × 150. Coupled rectangular cells forming long ribbons.

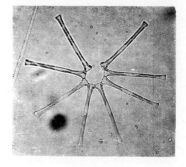

Fig. 601. — Asterionella formosa - × 300. Colony of eight tibia-shaped cells. Rapid proliferation in spring. Causes severe clogging and impedes clarification.

Pennales order. These diatoms have elongated valves of varying shapes which are symmetrical in one plane. This group includes species which move by gliding, as well as many benthic, fixed and epiphytic varieties. The classification of the Pennales is based on whether or not there is a raphe. The raphe is a seam running along the length of the valve and connecting the cell protosplasm with the outside medium.

— *Araphideae*—have no raphe.

Asterionella—a pure planktonic diatom; the cells are grouped in a star-shaped cluster by means of mucus excreted by one extremity of each cell (fig. 601).

Fragilaria—(fig. 602).

Fig. 602. — *Synedra acus* - × 250.

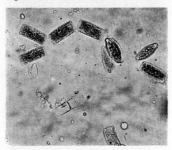

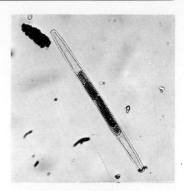

Fig. 603. — Synedra ulna - × 350.

Fig. 604. — Diatoma vulgare × 200. Zig-zag colony comprising four cells (girdle view); side view of a cell showing the two valves; cell seen from valve aspect.

Synedra—long, narrow cells, very often isolated, of varying size. Proliferation in spring and autumn. Certain species are the cause of serious clogging of slow filters. The very fine small species may escape clarification and pass through the primary filter (fig. 602 and 603).

Diatoma—species found in running water (fig. 604).

— *Monoraphideae*—one raphe on one valve.

Rhoicosphenia curvata—colonial species. Heteropolar cell attached by means of a peduncle of secreted mucilage (fig. 606).

— *Biraphideae*—one raphe on each valve.

Navicula—spindle-shaped cell with axial raphe. Numerous planktonic species, isolated, motile (fig. 605).

Nitzschia—the raphe is found in a protruding carina; numerous planktonic species (fig. 607).

Chrysophyceae.

A group of algae, mainly planktonic, generally found in cold water.

The cells have either one or two golden or yellow-brown plastids, one or two flagella and sometimes one stigma. The means of protection is a siliceous cyst.

Isolated cell:

Mallomonas—the membrane consists of overlapping siliceous scales, each bearing a long, thin siliceous spine; one flagellum (fig. 608).

Colonial structure:

Synura—spherical planktonic colony, prolific in spring (fig. 609).

Dinobryon—planktonic colony (fig. 610) in the form of a branched arbuscule.

Fig. 606. — Rhoicosphenia curvata - × 200.

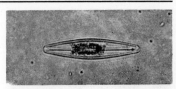

Fig. 605. — Navicula - × 450. Valve view.

Fig. 607. — Nitzschia sigmoidea × 200. Girdle view.

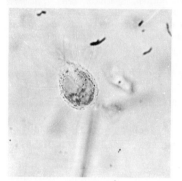

Fig. 608. — Mallomonas × 450. Unicellular algae, very mobile; the siliceous carapace is clearly visible.

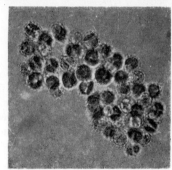

Fig. 609. — Synura - × 300. Biflagellate cells covered by a carapace of siliceous scales (visible in photograph). These species give an unpleasant flavour.

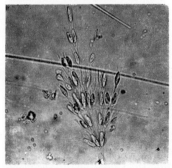

Fig. 610. — Dinobryon - × 200. Each cell occupies a small casing secreted by itself. Empty skeletons are often found due to the fragility of the species.

Fig. 611. — Vaucheria - × 50. Filamentous alga, siphoned (no septum) and branched.

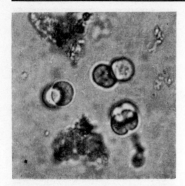

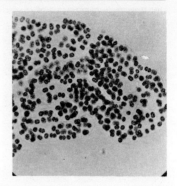

Fig. 612. — Chroococcus - × 650. Spherical cells, blue-green in colour; bipartite method of reproduction, often giving rise to associations of two or four cells.

Fig. 613. — Microcystis - × 100. An indeterminately shaped colony of small, closely packed, spherical cells contained in a gelatinous substance. The blackish appearance of the species is due to the presence of pseudo-gas vacuoles in each cell.

Xanthophyceae.

This phylum includes the yellow-green algae, which have the same structures as the Chlorophyceae but contain no starch.

Among the Heterotrichales, unbranched and filamentous, the genus *Tribonema* is often found in fresh water.

The Heterosiphonales include filamentous, branched, non-septated forms. The most common is *Vaucheria*, which possesses numerous small plastids (fig. 611).

D) Cyanophyta.

This phylum of "blue algae" contains organsims of a highly individual character. The cells have no nucelus indidualized by a membrane and no plastid; the pigments are evenly distributed in the protoplasm.

The normal colour of the cells is blue-green, although these algae may sometimes be yellowish or reddish; the presence of pseudovacuoles, or gas pockets, gives a blackish, bead-like appearance to some species. The food substance is glycogen. There is no flagellate stage.

The order of Chroococcales—includes unicellular forms, either isolated or colonial. *Chroococcus*—globular cells surrounded by a thin, colourless, gelatinous layer (fig. 612).

Microcystis—massed colony, comprising numerous closely packed cells in a gelatinous substance (fig. 613).

Coelosphaerium—the cells form spherical hollow colonies.

The order of Hormogonales comprises filamentous Cyanophyceae with intercalary growth, often surrounded by a casing of mucilage. The collection of cells inside this casing is called a trichome; the filament is made up of the casing and the trichome.

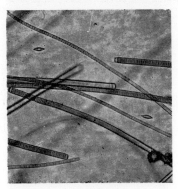

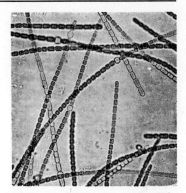

Fig. 614. — Oscillatoria and Lyngbia - × 100. The Oscillatoria are the small filaments which appear bent and often fuzzy. The Lyngbia are the filaments of larger diameter with a casing.

Fig. 615. — Anabaena - × 300. Filaments consisting of distinctly separated cells with round cells of a lighter colour - the heterocysts - at intervals.

This order includes numerous families. Many species are adapted to life in hot thermal springs.

— *Oscillatoriaceae family.*

The cells, all of which are identical, form cylindrical filaments, without branching and sometimes with a casing. No spores.

Oscillatoria—numerous species, no outer casing, usually an isolated filament subject to oscillation, which bends the filament (fig. 614).

Phormidium—numerous species, one gelatinous casing for each trichome, often barely visible. The filaments may group themselves together and become attached in the form of thalli.

Lyngbia—numerous species. Isolated filaments, with one trichome to each casing. The casing is firm and often yellow or brown in colour (fig. 614).

— *Nostocaceae family.*

These filamentous unbranched algae, often with a casing, possess cells which are not all identical. Here and there may be seen empty cells with a thickened membrane—the heterocysts. Larger pigmented cells—the spores—also exist.

Aphanizomenon—the trichomes are grouped in bundles. Planktonic. Often forms water bloom.

Anabaena—includes numerous species, some of which also form water bloom (fig. 615).

E) Peridinians or dinophyceae.

These are unicellular planktonic algae with numerous brown plastids; the body is often covered by a membrane consisting of cellulose plates.

The Peridinians are usually motile; two furrows cross the body, one transverse, with a flagellum which encircles the whole body, and the other longitudinal, containing another thicker flagellum which has the same point of insertion as the first.

The most important genera are:
Ceratium
Peridinium (fig. 616).

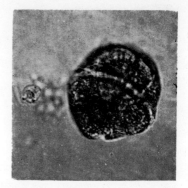

*Fig. 616. — Peridinium - × 600. Only
the transverse furrow encircling the body
is visible in this photograph.*

7.3.2. ZOOPLANKTON

Fresh water zooplankton is limited in genera and species. It is rarely represented by a large number of individuals in rivers and streams, although the contrary is true if lakes and ponds. It mainly embraces crustaceans, rotifers and
protozoa.

Planktonic animals are transparent. They swim fairly actively and possess
appendages which enable them to support themselves in the water. They sink
to varying depths according to the degree of turbulence on the surface of the
water. Daily vertical migrations take place under the influence of phototro-
pism. Animal plankton vary greatly in size. Certain phyla include only microscopic individuals (protozoa, rotifers), whereas others include individuals
measuring several millimetres long (crustaceans). They feed on algae, bacteria,
organic detritus or even on each other. Their multiplication is influenced by
seasonal changes and by the proliferation of phytoplankton.

As well as the three main groups already mentioned, there are other groups
of animals of which the larvae are planktonic, the adults living attached to a
support. Some species may form themselves into groups large enough to block
pipes carrying unfiltered water (e.g. the mollusc Dreissena polymorpha) or to
hinder filtration (molluscs, sponges, bryozoas). The eggs, cysts or larvae may
pass through filters and develop at a later stage (nematode worms).

A) Protozoa.

These are microscopic, unicellular, heterotrophic organisms. The cell
has a complex organization. We shall consider only the two principal groups.

● **The Ciliata** (formerly called Infusoria).

This class covers a large number of species, very varied in form and motile.
The body is wholly or partly covered by regularly distributed cilia. They are
classified in accordance with the distribution of the cilia.

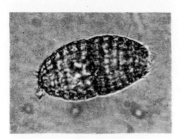

Fig. 617. — Coleps - × 600. The membrane, consisting of plaques of translucent organic matter, forms a characteristic chequered pattern.

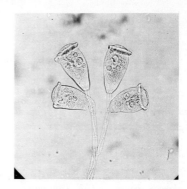

Fig. 618. — Vorticella - × 250. Cell attached to a support by a retractile filament which it has secreted.

Holotrichous ciliata—the cilia cover the body uniformly.
Examples: *Paramecium—Lionotus—Coleps* (fig. 617).
These three genera are abundant almost throughout the year and prefer to live among algae.

Heterotrichous ciliata—characterized by a fringe of cilia on the edge of the peristome, a depression at the bottom of which the mouth is situated.
Example: *Stentor:* very large, trumpet-shaped body, very wide at the peristome end.

Hypotrichous ciliata—flattened, with one flat ventral side having cirri (consisting of numerous agglutinated cilia), whose co-ordinated movements allow them to progress over the algae.
Example: *Stylonychia—Euplotes.*

Peritrichous ciliata—typified by *Vorticella* (fig. 618). The body is bell-shaped and non-ciliate; only the peristome has a spiral fringe of cilia. They are often attached by a peduncle, frequently very retractile, which quickly curls up into a spiral.

Acinetae or Tentaculifera—these organisms possess no cilia, but have hollow contractile tentacles. They feed on ciliata which they capture and suck with their tentacles.

● **The Rhizopods.**
These consists of a cell which becomes irregular in contour when it sends out extensions or pseudopods.

Amoeba—their shape is constantly changing by extension or contraction of the pseudopods, which enable them to move forward and take in food—algae or decomposing vegetable particles (fig. 619).

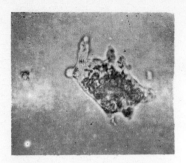

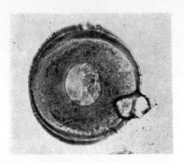

Fig. 619. — Amoeba - × *300. The pseu-dopods consist of a transparent proto-plasm; the body of the amoeba contains vegetable particles in process of digestion.*

Fig. 620. — Arcella - × *300. Theca seen from above. The central orifice can be seen by transparency.*

Thecamoeba are amoeba which secrete a chitinous shell or theca with only one orifice for the pseudopods.
Example: *Arcella* (fig. 620).

The heliozoa—have a spherical body which sends out fine radial protoplasmic extensions.
Example: *Actinophrys* (fig. 621).

B) Rotifers.

These are microscopic metazoa, very common among fresh-water plankton. In the apical region they possess a rotatory apparatus consisting of ciliary coronas. When the cilia beat, it is like a wheel turning. Their bodies can be divided into three sections: the head, crowned by the rotatory apparatus; the trunk, very often encased in a fairly thick but transparent cuticle, the lorica, with or without spikes; and the foot, fairly well developed and retractile, terminating either in two toes or in a type of sucker. It serves as an organ for temporary anchorage. The ciliary coronas are used for swimming and guiding food towards the mouth.
The majority of species of Rotifers belong to the order of **Ploima.**
Keratella cochlearis (fig. 622)—spiked lorica; no foot; species very common and abundant among plankton.
Asplanchna (fig. 623)—in the form of a transparent sac, large, no lorica, no foot.
Brachionus (fig. 624)—flattened, spiked lorica, long flexible retractile foot termi-nating in two minute toes. This is a large genus with numerous species.

The elongated rotifers with annulated cuticles belong to the order of **Bdelloidea.** They crawl on their support like leeches and also have short periods of free swimming.
Philodina—the body is inflated into a spindle; the rings of lesser diameter can telescope into the median rings (fig. 625).

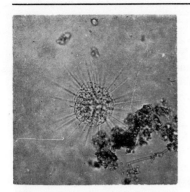

Fig. 621. — Actinophrys - × 250. Slender radial pseudopods.

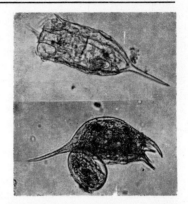

Fig. 622. — Keratella cochlearis - × 200. The upper photograph shows the rotifer seen from above. Below - the side view of the Keratella with an egg.

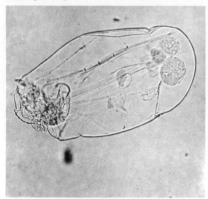

Fig. 623. — Asplanchna - × 80. Highly carnivorous rotifer. Very transparent organs.

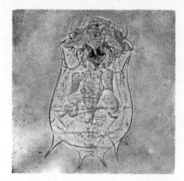

Fig. 624. — Brachionus - × 130. The retracted foot is not visible.

Fig. 625. — Philodina - × 150. The animal, stretched out from its point of attachment, spreads its two rotatory mechanisms.

C) Crustaceans.

The crustaceans are aquatic arthropods whose bodies are divided into three sections: head, thorax, and abdomen. The first two pairs of appendages carried by the head are the antennules and antennae.

Freshwater plankton includes three orders of lower crustaceans **(Entomo-straca).** These are small (about 0.1-3.5 mm).

Cladocera—the different segments of the body are not distinct and the body, except for the head, is enclosed in a transparent bivalve carapace. The beating of the heart, the branchiae, the movement of the intestines, and the five or six pairs of thoracic legs can all be seen. The eyes are united in one large, dark, uneven eye. The antennules are very small. The antennae are long and bira-mous and are used for swimming. Cladocera multiply very rapidly and serve as the staple diet of many fish. A large number of species exist.

Daphnia, or water flea, is very prevalent.

Bosmina—(fig. 626) is very common among plankton. The very large antennules are transformed into trunk-shaped cephalic extensions.

Copepoda—These have an elongated body composed of distinct segments. The antennule are long and bent and act as swimming organs. The antennae are much shorter. In the female the oviduct secretes a substance which binds the eggs into two ovoid masses (egg sacs) remaining attached to the first abdominal segment until hatching. The creature hatching from the egg is a larva known as the nauplius. The Copepoda prefer to remain near the surface at night; during the day they sink fairly deep in the water.

Cyclops—the cephalothorax is much wider than the abdomen; the latter ter-minates in two extensions bearing fine threads (fig. 627).

Ostracoda—the body consists of indistinct segments and is entirely enclosed in a bivalve carapace.

The two valves are provided with adductor muscles which enable the ani-mal to open or shut them.

Cypris—(fig. 628) are very fast swimmers.

D) Molluscs.

We shall cite only one bivalve mollusc; the freshwater mussel; *Dreissena polymorpha*, a species very widely distributed in running water, canals, streams, docks, and raw water pipes. The larvae are planktonic (fig. 630) and must be eliminated before they become attached and develop into adults, as they do very quickly in the dark. The diameter of water mains can be considerably reduced, by clusters of individuals which grow on top of one another.

E) Sponges.

Sponges form colonies, often shapeless and dirty yellow or greenish in colour. This mass, rough to the touch, is perforated by pores through which the water penetrates to the interior where the living cells or choanocytes are to be found. The skeleton is covered with minute spicules. The genus *Spongilla* is the most common.

F) Nematoda.

These are unsegmented filiform worms, enveloped in a transparent chitinous layer (fig. 629). The *Anguillulae* are to be found among this group.

Careful elimination is essential, otherwise the larvae may be found in filtered water pipes.

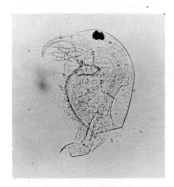

Fig. 626. — *Bosmina* - × 120. *The large eye and the antennules transformed into trunks are clearly visible.*

Fig. 627. — *Cyclops* - × 60. *The single median eye and well-developed antennules can be seen.*

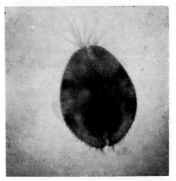

Fig. 628. — *Cypris* - × 50. *Side view; the hinge of the bivalve carapace can be distinguished. The ends of the cephalic appendages can be seen (top).*

Fig. 629. — *Nematode* × 50.

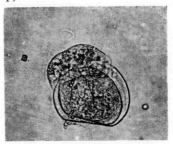

Fig. 630. — *Larva of Dreissena polymorpha* - × 200. *The two valves of the larval shell are clearly visible.*

8. THE MICROFLORA AND MICROFAUNA OF ACTIVATED SLUDGE

8.1. Purifying bacteria

Activated sludge consists, essentially, of bacteria and protozoons, with, from time to time, fungi, rotifers and nematodes.

The bacteria form the largest of the above groups and are the principal agents for the elimination of pollution, on the one hand, and for the formation of the floc, on the other.

A very large number of bacterial species can take part in the formation of activated sludge. The nature of the dominant genus is influenced by the nature of the organic compounds forming the pollution and by the characteristics of the environment: pH, temperature, dissolved oxygen, etc.

Thus a protein-rich effluent will promote the development of the genera Alcaligenes, Bacillus or Flavobacterium; glucide—or hydrocarbon—rich effluent will result in the predominance of the genus Pseudomonas. The presence of reducing sulphur will lead to the development of the genera Thiothrix, Microthrix, etc.

It was thought, for a long time, that activated sludge floc was composed of colonies of a particular bacterium, which was given the name Zooglea ramigera. It was shown later that many kinds of bacteria can be present in floc and that, while Zooglea ramigera certainly plays a role, it is only a partial one.

8.2. The bulking of sludge

Under certain conditions, i.e. glucide-rich pollution, a low pH and a deficiency of nitrogen and phosphorus, some fungi can take part in the formation of floc. Since they are the cause of a very filamentous floc, which does not settle and which constitutes one of the causes of the commonest "disease" of activated sludge, known as *bulking*, these fungi are undesirable.

Bulking affects the quality of activated sludge, especially its sedimentation capacity. The sludge volume index, or Mohlman's index, M.I. (see page 71), of such a sludge is high, greater than 150 cm³/g and may reach 500 cm³/g or higher. Such sludges may then accumulate in the clarifiers, resulting in the necessity to reduce the feed rate.

Bulking is a complex process, linked with the presence of filamentous microorganisms: bacteria such as Sphaerotilus natans, Thiothrix sp., Lactobacillus sp., Pelonemas or Peloploca sp.; fungi such as Leptomitus sp., Geotrichum candidum, etc.

The filamentous bacteria are microorganisms which develop in response to either unusual environmental conditions: temperature, pH, salinity, too low an oxygen level, an unusual substrate composition (excess glucides, lack of nutritive substances) or to sudden changes in the operating parameters of the installation: sudden surges of the polluting load, accidental loss of sludge, considerable variations in the recycling rate of sludge, etc. Another cause may be strains of bacteria or fungi contributed by some outside source such as effluent or from the sewers.

The fight against bulking is extremely difficult and prolonged and there is no one formula which is universally successful. The "flowchart" given in fig. 631 may be tried. If bulking continues, thought must be given to changing the treatment procedure

8.3. Predators

Besides the bacteria, the most important group is that of the Protozoa which do not act on the organic pollution directly, **but play the part of predators** on the bacterial flora.

Particularly sensitive to variations in the environmental conditions, the protozoons are valuable indicators to the plant operator, since the predominance of one species or another

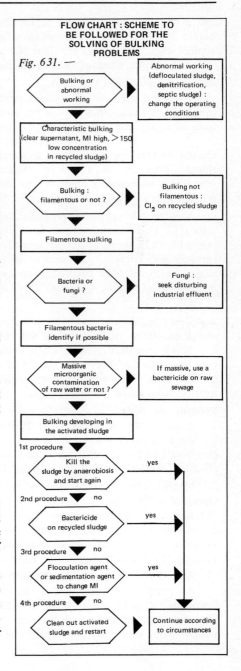

Fig. 631. —

FLOW CHART : SCHEME TO BE FOLLOWED FOR THE SOLVING OF BULKING PROBLEMS can give information on the adjustment level of the activated sludge to the nature of the substrate, on the oxygenation efficiency, the presence of toxic sub-stances, etc.   The rotifers are signs of a particularly stable biological system.

As examples, there follows a list of the principal predators in activated sludge.

Rhizopods or Amoeba (fig. 632).

Hyaline masses of protoplasm which are either unprotected or enclosed in a shell or theca—move about by extending pseudopods, which are evaginations of protoplasm.   Very resistant to anaerobiosis.

These forms make a brief appearance when treatment plants are started up. During normal operation, their appearance may indicate deterioration of the sludge.   They are also found in effluents from certain fermentation processes which have undergone a phase of anaerobiosis.

Zooflagellata (fig. 633).

Very mobile organisms propelled by one or more fine and very long flagella (f).   They can survive in an oxygen-deficient environment.   They are rarely found in domestic sewage, apart from that rich in fat and proteins (from canteens and kitchens).   Sometimes they are the only developed organisms in sludge which has been adapted to certain industrial waters containing phenols or products of organic synthesis.   They are the first developed species to appear in fresh sludge.

Aspidisca (fig. 634).

A euciliatum of the order of Spirotrichida and the family of Aspidiscidae.

This is a small ciliatum which looks like a crustacean and moves quite quickly over the surface of flocs with the aid of its frontal and anal cirri; it "browses" on thick flocs.   This ciliatum is very widely distributed in older activated sludges treating effluents of varying origins.

Epistylis (fig. 635).

A euciliatum of the order of Peritrichida and the family of Epistylidae.

An attached ciliatum that forms dense clusters, it is characteristically found in ageing and fairly well oxygenated activated sludges only.   It often takes the place of vorticella when the sewage contains a substantial proportion of various industrial effluents (phenols, etc.).

Opercularia (fig. 636).

A euciliatum of the order of Peritrichida and the family of Epistylidae.

An attached ciliatum in loose clusters.   It is characteristic of activated sludge that is reaching the phase of slower growth in certain industrial effluents.

Vorticella (fig. 637).

A euciliatum of the order of Petrichida and the family of Vorticellidae.

An attached ciliatum with a non-ramified peduncle.   Two species occur frequently: V. microstoma (small mouth) and V. macrostoma (large mouth). This very common predator is characteristic of fully-developed, well-oxygenated sludges in optimum condition in plants working under medium or heavy loading.

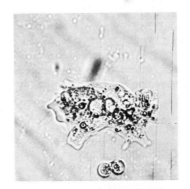

Fig. 632. — Amoeba × 400.

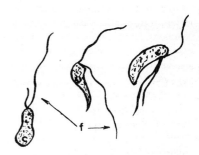

Fig. 633. — Zooflagellata × 1 000.

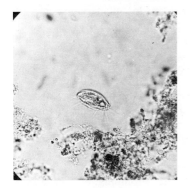

Fig. 634. — Aspidisca × 300.

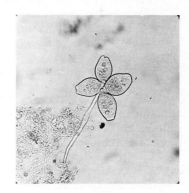

Fig. 635. — Epistylis × 150.

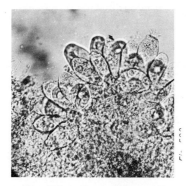

Fig. 636. — Opercularia × 150.

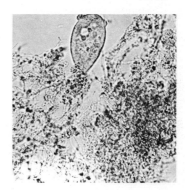

Fig. 637. — Vorticella × 150.

Didinium (fig. 638).

A euciliatum of the order of Holotrichida and the family of Didiniidae.

A free-swimming ciliatum, not very common in sewage, it is found most often in domestic waste containing a high proportion of fats and proteins (from kitchens, canteens, etc.). It is present as a transitional form, together with Coleps, Stentor, etc., and the attached ciliata, in environments where the variations in loading are frequent and substantial.

Acineta (fig. 639).

A ciliatum of the order of Tentaculiferida and the family of Acinetidae.

A resistant form which is found quite frequently in sludges used for the treatment of industrial sewage under heavy loading. Other genera in the same group: Podophrya.

Lionotus (fig. 641).

A euciliatum of the order of Holotrichida and the family of Amphileptidae.

A ciliatum that swins and crawls over the surface of flocs.

This genus is very common in sludges that are in full development, just before the attached ciliata appear, in plants operating under normal loading conditions. They require a high level of oxygenation (over 1 mg of dissolved oxygen per litre).

Related genera such as Acineria are also found occasionally.

Paramecium (fig. 640).

A euciliatum of the order of Holotrichida and the family of Paramecidae.

This is the typical ciliatum with a clearly defined buccal cavity surrounded by a peristome which is provided with a distinctive fringe of cilia; it has two more or less distinct nuclei and two fixed contractile vacuoles.

This free-swimming organism is quite a resistant form which predominates in poorly-oxygenated sludges in sewage or food industry effluents.

Other less common species, such as Colpidium, Frontonia, etc., have similar ecological characteristics.

Rotifers (fig. 641).

Multi-cellular organisms characterized by a head with a corona of cilia and which, being transparent, reveals a number of organs such as the cerebral ganglion and the masticatory organ or mastax. The trunk or abdomen comprises a large stomach supplied by the oesophagus, and also the genital organs. The abdomen terminates in a tail by the anus outlet. The typical feature of rotifers is their segmented foot, which terminates in "toes" and serves the animal for anchorage and movement.

Dorsal and lateral antennae can generally be seen.

These organisms "browse" on the surface of the flocs.

They are quite sensitive to poisons and loading variations but develop in ageing, well-flocculated and well-oxygenated sludges. They are thus characteristic of extended aeration.

Nematoda (fig. 629).

Multi-cellular worms (vermiforms) enveloped in a cuticle and provided with a well differentiated digestive system and genital organs. The buccal cavity incorporates an extensible probe which is used to seize suitable quarry.

These organisms are typical of old sludges or those in course of mineralization. They are frequently found in sludges undergoing aerobic digestion and also in the thick films on biological filters. Being more resistant than rotifers, they are often an indication of threshold conditions for aerobic life.

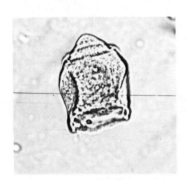

Fig. 638. — Didinium × 300.

Fig. 639. — Acineta × 500.

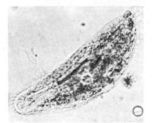

Fig. 640. — Paramecium × 250.

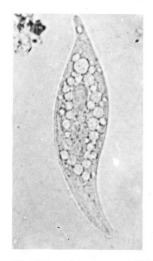

Fig. 641. — Lionotus × 1 000.

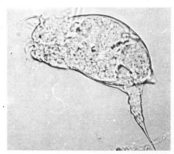

Fig. 642. — Rotifer × 200.

Part Five

Formulae

SCHEMA MATERIALIUM PRO LABORATORIO PORTATILI F.X

I	MINERÆ					
II	METALLA					
III	MINERALIA	Bismuth	Zinck	Marcasit Kobolt	Zaffra	Magnesia Magnes
IV	SALIA				Borax	Chrysocolla
V	DECOMPOSITA					
VI	TERRÆ	Crocus	Crocus	Vitrum / Vitrum	Minium Lithargirum	Cadmia Tutia / Ochra Schmalta
VII	DESTILLATA	Sp / Sp	Sp / Sp	Sp V	Sp	Sp
VIII	OLEA	Ol / Ol	Ol fœtid	Ol p deliq / Butyr	Liquor Silicum	Ol.Therebent
IX	LIMI	Fluxus Niger	Fluxus Albus	Arena Cineres	Creta Rubrica	Terra sigillata Bolus / Hæmatites Smiris	Talcum Granati	Asbestus
X	COMPOSITIONES		Cera Tinctoria Coloriza	Decoctio	Tirapelle	

1. SYSTEMS OF UNITS

The systems of units derived from the metric system have gradually given way to a single system, called the Système International d'Unités (SI) and the first part of this chapter is concerned with it. The cgs system (centimetre-gramme-second), whose units for the most part correspond to decimal multiples or sub-multiples of the International System, still remains important in certain fields.

Since a large number of countries continue to use British and American units, the second part of this chapter gives their equivalents against the SI units, with conversion tables.

2. INTERNATIONAL SYSTEM OF UNITS (S.I.)

This system, set up and steadily improved by the General Conferences of Weights and Measures and the International Committee of Weights and Measures, consists of basic units, derived units and supplementary units. It uses decimal multiples and submultiples, with the appropriate prefixes. In addition to the units listed below, it accepts units "used in conjunction with the International System" and "temporarily accepted" units. In principle, it advises against the use of cgs and other units.

The tables below are extracts from the International System relating to units which are used in various ways in water treatment. They were taken from various documents issued by the Association Française de Normalisation (AFNOR) Tour Europe, Cedex 7, 92080 Paris La Défense, France. Particular use is made of the diagrammatic table entitled "Le Système International d'Unités (SI)" to be found in document reference NF X 02-004.

2.1. General principles

● RULES FOR WRITING UNIT SYMBOLS

Unit symbols are only used after a number expressed in figures. Printed in upright characters, they are unchanged in the plural, and are written without a full stop, after the full numerical value, with a space between the numerical value and the symbol.

They are written in small letters; but a capital is used for the first letter when the name of the unit is derived from a surname.

Not more than one fraction bar must appear on the same line to represent a compound unit which is the quotient of two units, unless brackets are used to avoid any ambiguity.

● BASIC UNITS

Quantity	Unit	Symbol
Length	metre	m
Mass	kilogramme	kg
Time	second	s
Intensity of electric current	ampere	A
Thermodynamic temperature	kelvin	K
Amount of substance	mole	mol
Luminous intensity	candela	cd

Note : Celsius temperature t is connected to thermodynamic temperature T by the equation $t = T - 273.15$.

A difference of temperature may be expressed either in kelvin or in degrees Celsius. In this case, $1 \, ^{\circ}C = 1 \, K$.

● SUPPLEMENTARY UNITS
(these units can be used as basic units)

Quantity	Unit	Symbol
Plane angle	radian	rad
Solid angle	steradian	sr

● MULTIPLES AND SUBMULTIPLES

MULTIPLES			SUBMULTIPLES		
Factor	Prefix	Symbol	Factor	Prefix	Symbol
10^{18}	exa	E	10^{-1}	deci	d
10^{15}	peta	P	10^{-2}	centi	c
10^{12}	tera	T	10^{-3}	milli	m
10^{9}	giga	G	10^{-6}	micro	μ
10^{6}	mega	M	10^{-9}	nano	n
10^{3}	kilo	k	10^{-12}	pico	p
10^{2}	hecto	h	10^{-15}	femto	f
10	deca	da	10^{-18}	atto	a

● RULES FOR FORMING MULTIPLES AND SUBMULTIPLES

Multiples and submultiples of SI units are formed by adding one of the above prefixes to the name of the unit, e.g. centimetre.

However, for the word "kilogramme", since it already contains a prefix, the prefix is added to the word "gramme", e.g. milligramme.

The symbol of a prefix is considered to be combined with the symbol of the unit to which it is directly attached, thereby forming the symbol of a new unit, which can be raised to a power, e.g.

$$1 \text{ cm}^3 = (10^{-2} \text{ m})^6 = 10^{-3} \text{ m}^3.$$

Not more than one prefix must be juxtaposed: for example, nanometre (nm) but not millimicrometre (mμm).

A single prefix only is recommended for use with compound units: for example, millinewton metre and not decinewton centimetre.

Multiples are usually chosen so that the numerical value falls between 0.1 and 1 000.

2.2. Definitions of basic units and supplementary units

BASIC UNITS

metre: a metre is the length equal to 1 650 763.73 wavelengths in vacuo of the radiation corresponding to the transition between the levels $2p_{10}$ and $5d_5$ of the krypton-86 atom.

kilogramme: the kilogramme is the unit of mass equal to the mass of the International prototype kilogramme.

second: the second is the interval occupied by 9 192 631 770 cycles of the radiation corresponding to the transition between the two hyperfine levels of the ground state of the caesium-133 atom.

ampere: the ampere is the intensity of a constant current which, if maintained in two straight parallel conductors of infinite length, of negligible circular cross-section and placed at a distance of one metre apart in a vacuum, will produce between the conductors a force equal to 2×10^{-7} newton per metre length.

kelvin: the kelvin is the unit of thermodynamic temperature defined as the fraction 1/273.16 of thermodynamic temperature of the triple point of water.

mole: the mole is the amount of substance of a system which contains as many elementary units as there are atoms in 0.012 kg of carbon-12. When the mole is used the elementary units must be specified and may be atoms, molecules, ions, electrons, other entities or specified groups of such entities.

candela: the candela is the luminous intensity, in the perpendicular direction, of a surface of 1/60 square centimetre of black body radiation operating at the temperature of solidification of platinum under a pressure of 101 325 newtons per square metre.

SUPPLEMENTARY UNITS:

radian: the radian is the plane angle subtended at the centre of a circle by an arc equal in length to the radius of the circle.

steradian: the steradian is the solid angle which, with its vertex at the centre of a sphere, encloses an area on the surface of the sphere equal to the square of the radius.

2.3. Basic units. Derived units for frequently occurring quantities

Quantity	Unity		Expressed in other S.I. units	Expressed in basic units (B.U.) or supplementary units (S.U.)	Units used in conjunction with the SI or temporarily accepted (T)		
	Name	Symbol			Name	Symbol	Value in S.I. units
Space and time							
Length	Metre	m		B.U.	Ångström (T)	Å	$1\,\text{Å} = 10^{-10}$ m
					Nautical mile (T)		1 mile = 1852 m
Surface area	square metre	m²		m²	are (T)	a	1 are = 1 000 m²
					hectare (T)	ha	1 ha = 10^4 m²
Volume	cubic metre	m³		m³	litre	l	1 l = 1 dm³
Plane angle	radian	rad		S.U.	degree	°	$1° = (\pi/180)$ rad
					minute	′	$1′ = (1/60)°$
					second	″	$1″ = (1/60)′$
Solid angle	steradian	sr		S.U.			
Time	second	s		B.U.	minute	min	1 = min 60 s
					hour	h	1 h = 60 min
					day	d	1 d = 24 h
Angular velocity	radian per second	rad/s		s⁻¹ rad			
Speed	metre per second	m/s		m s⁻¹	knot (T)	kn	1 kn = 1852 m/h
Acceleration	metre per second squared	m/s²		m s⁻²			
Frequency	hertz	Hz	1/s	s⁻¹			
Mechanics							
Mass	kilogramme	kg		B.U.	tonne	t	1 t = 10^3 kg
Density	kilogramme per cubic metre	kg/m³		m⁻³ kg			
Masse rate of flow	kilogramme per second	kg/s		kg s⁻¹			
Volume rate of flow	metre cubed per second	m³/s		m³ s⁻¹			
Momentum	kilogramme metre per second	kg m/s		m kg s⁻¹			

Quantity	Unity		Expressed in other International units	Expressed in basic units (B.U.) or supplementary units	Units used in conjunction with the Si or temporarily accepted (T)		
	Name	Symbol			Name	Symbol	Value in S.I. units
Angular momentum	kilogramme metre squared per second	kg m²/s		m² kg s⁻¹			
Moment of inertia	kilogramme metre squared	kg m²		m² kg			
Force	newton	N	kg m/s²	m kg s⁻²			
Moment of force	newton metre	N m		m² kg s⁻²			
Pressure, stress	pascal	Pa	N/m²	m⁻¹kg s⁻²	bar (T) standard atmosphere (T)	bar atm	1 bar = 10⁵P 1 atm = 101235 Pa
Dynamic viscosity	pascal second	Pa s		m⁻¹ kg s⁻¹			
Kinematic viscosity	metre squared per second	m²/s		m² s⁻¹			
Surface tension	newton per metre	N/m					
Energy, work, amount of heat	joule	J	N m	m² kg s⁻²			
Power	watt	W	J/s	m² kg s⁻³			
Thermodynamics							
Thermodynamic temperature	kelvin	K		B.U.			
Linear expansion coefficient	kelvin to the power of minus one	K⁻¹					
Thermal conductivity	watt per metre kelvin	W/(m K)		m kg s⁻³ K⁻¹			
Specific heat capacity	Joule per kilogramme kelvin	J/(kg K)		m² s⁻² K⁻¹			
Entropy	joule per kelvin	J/K		m² kg s⁻² K⁻¹			
Internal energy, enthalpy	joule	J		m² kg s⁻²			
Optics							
Luminous intensity	candela	cd		B.U.			
Luminous flux	lumen	lm		cd sr			
Illumination	lux	lx	lm/m²	m⁻²cd sr			
Electricity and magnetism							
Intensity of electric current	ampere	A		B.U.			
Electric charge	coulomb	C	As	s A			
Potential, potential difference, voltage	volt	V	W/A	m² kg s⁻³ A⁻¹			
Electric field strength	volt per metre	V/m		m kg s⁻³ A⁻¹			
Capacitance	farad	F	C/V	m⁻² kg⁻¹ s⁴ A²			
Magnetic field strength	ampere per metre	A/m		m⁻¹ A			
Magnetic flux	weber	Wb	Vs	m² kg s⁻² A⁻¹			
Magnetic flux density	tesla	T	Wb/m²	kg s⁻² A⁻¹			

Quantity	Unity		Expressed in other International units	Expressed in basic units (B.U.) or supplementary units (S.U.)
	Name	Symbol		
Inductance, permeance	henry	H	Wb/a	m^2 kg s^{-2} A^{-2}
Resistance, impedance, reactance	ohm	Ω	V/A	m^2 kg s^{-3} A^{-2}
Conductance	siemens	S	A/V	m^{-2} kg^{-1} s^3 A^2
Resistivity	ohm metre	Ω m		m^3 kg s^{-3} A^{-2}
Conductivity	siemens per metre	S/m		m^{-3} kg^{-1} s^3 A^2
Chemistry, Physics, molecular physics				
Amount of substance	mole	mol		B.U.
Molar mass	kilogramme per mole	kg/mol		kg mol^{-1}
Molar volume	cubicmetre per mole	m^3/mol		m^3 mol^{-1}
Concentration	kilogramme per cubic metre	kg/m^3		m^{-3} kg
Molar concentration	mole per cubic metre	mol/m^3		m^{-3} mol
Molality	mole per kilogramme	mol/kg		kg^{-1} mol

2.4. Units not recommended or to be avoided

These units are listed in the last two columns below.

	Symbol of SI unit	cgs units	Other units (to be avoided)
Volume	m^3		stere (st); 1 st = 1m^3
Mass	kg		carat; 1 carat = 200 mg
Force	N	dyne; 1dyn = 10^5N	kg (f); 1 kg (f) = 9.81 N
Moment of force	N m		m kg(f); 1 m kg (f) = 9.81 N m
Pressure	Pa	barye; 1 barye = 10^{-1} Pa	kg(f)/cm^2 = metre of water column = 9.81. 10^4Pa
			mm Hg; 1 mm Hg = 74.10^{-4}Pa
Dynamic viscosity	Pa s	poise (P); 1 P = 10^{-1} Pa s	poiseuille (Pl); 1 Pl = 1 Pa s
Kinematic viscosity	m^2/s	stokes (St); 1 St = 10^{-4} m^2/s	
Energy, work, amount of heat	J	1 erg = 10^{-7} J	kg (f) m; 1 kg (f) m = 9.81 J
			calorie (cal); 1 cal = 4.187 J
			kilocalorie (kcal) = « millithermie » (mth); 1 kcal = 4187 J
Power	W	erg/s; 1 erg/s = 10^{-7} W	kg(f) m/s; 1 kg(f) m/s 9.81 W
Magnetic flux	Wb	maxwell; 1 Mx = 10^{-8} Wb	
Magnetic flux density	T	gauss; 1 Gs = 10^{-4}T	mho; 1 mho = 1 S
Conductance	S		

3. S.I. UNITS AND IMPERIAL OR U.S. UNITS CONVERSIONS

A. LENGTH:

Symbol

n or''	inch	=	0.0254 m
ft or'	foot = 12 in.................................	=	0.3048 m
yd	yard = 3 ft..................................	=	0.9144 m
mi	statute mile = 1 760 yd.......................	=	1.609 km
m	metre.......................................	≈	1.0936 yd
		≈	39.37 in
		≈	3.281 ft
		≈	$3'3\frac{3''}{8}$
km	kilometre	≈	0.6214 mi

Inches		Millimetres
1		25.40000
31/32	0.96875	24.60625
15/16	0.93750	23.81250
29/32	0.90625	23.01875
7/8	0.87500	22.22500
27/32	0.84375	21.43125
13/16	0.81250	20.63750
25/32	0.78125	19.84375
3/4	0.75000	19.05000
23/32	0.71875	18.25625
11/16	0.68750	17.46250
21/32	0.65625	16.66875
5/8	0.62500	15.87500
19/32	0.59375	15.08125
9/16	0.56250	14.28750
17/32	0.53125	13.49375
1/2	0.50000	12.70000
15/32	0.46875	11.90625
7/16	0.43750	11.11250
13/32	0.40625	10.31875
3/8	0.37500	9.52500
11/32	0.34375	8.73125
5/16	0.31250	7.93750
9/32	0.28125	7.14375
1/4	0.25000	6.35000
7/32	0.21875	5.55625
3/16	0.18750	4.76250
5/32	0.15625	3.96875
1/8	0.12500	3.17500
3/32	0.09375	2.38125
1/16	0.06250	1.5875
1/32	0.03125	0.79375

B. AREA

Symbol

Symbol			
in²	Square inch .... =	6.4516	cm²
ft²	Square foot (144 in²) .... ≈	9.2903	dm²
yd²	Square yard (9 ft) .... ≈	0.83613	m²
mile²	Square mile (640 acres) .... ≈	2.5900	km²
	Acre (4 roods) (4 840 yd²) .... ≈	0.40469	ha
cm²	Square centimetre .... ≈	0.1550	in²
m²	Square metre .... ≈	10.764	ft²
dam², a	Square decametre or are .... ≈	119.6	yd²
hm², ha	Square hectometre or hectare .... ≈	2.471	acres
km²	Square kilometre .... ≈	0.3861	mile²

The abbreviations sq. in, sq. yd are no longer used in British standards.

C. VOLUME AND CAPACITY

Symbol

Symbol			
in³	Cubic inch .... ≈	16.3871	cm³
ft³	Cubic foot .... ≈	28.317	dm³
yd³	Cubic yard .... ≈	0.7646	m³
U.K. pt	British pint (4 gills) .... ≈	0.5683	l
U.K. qt	British quart (2 U.K. pt) .... ≈	1.1365	l
U.K. gal	Imperial gallon (8 U.K. pt)................. ≈	4.5461	l
U.S. pt	U.S. liquid pint (4 gills) .... ≈	0.4732	l
U.S. qt	U.S. liquid quart (2 U.S. liq pt) .... ≈	0.9464	l
U.S. gal	U.S. gallon (4 U.S. liq pt) .... ≈	3.7854	l
bbl	U.S. barrel (petroleum) = 42 U.S. gal ...... ≈	158.987	l
bu	U.S. bushel (4 pecks) .... ≈	35.2391	l
	U.S. shipping ton = 40 ft³................. ≈	1.13267	m³
	Registered ton = 100 ft³ .... ≈	2.83168	m³
cm³ / ml	Cubic centimetre or Millilitre	≈ 0.0611	in³
dm³ / l	Cubic decimetre or Litre [1]	≈ 0.0353	ft³
		≈ 1.760	U.K. pt
		≈ 0.220	U.K. gal
		≈ 2.113	U.S. pt
		≈ 0.264	U.S. gal
m³ / st	Cubic metre or Stere	≈ 35.30	ft³
		≈ 1.3079	yd³
		≈ 220	U.K. gal
		≈ 264	U.S. gal
		≈ 6.293	U.S. bbl
		≈ 28.37	U.S. bu

1. 1 litre = 1 dm³ exactly and no longer the old definition in which 1 litre = the volume of 1 kg of water at 4 °C; this is equal to 1.000 028 dm³.

D. LINEAR SPEED
Symbol

in/s	Inch per second	=	91.44	m/h
ft/s	Foot per second	=	1.09728	km/h
yd/s	Yard per second	=	0.9144	m/s
mile/h	Mile per hour (statute)	≈	1.609	km/h
m/s	Metre per second	≈	3.280	ft per s.
m/h	Metre per hour	≈	3.280	ft per h.
km/h	Kilometre per hour	≈	0.622	mile per h.

E. FILTRATION RATE
Symbol

U.S. gal/ft² min	U.S. gpm per sq ft	≈	2.445 m/h
U.K. gal/ft² min	U.K. gpm per sq ft	≈	2.936 m/h
ft/min	cu ft/min sq ft (ft³/ft² min)	≈	18.29 m/h
m/h	Linear speed of 1 m per hour	≈	0.0547 ft/min
		≈	0.409 U.S. gpm per sq ft
	Cubic metre per hour and per square metre	≈	0.341 U.K. gpm per sq ft

F. MASS
Symbol

gr	Grain	64.799	mg
oz	Ounce	28.350	g
lb	Pound (livre = 16 oz = 700 gr)[1]	453.592	g
st	British stone (14 lb)	6.350	kg
qr	Quarter (British) (28 lb)	12.701	kg
cwt	Hundredweight (British) (112 lb)	50.802	kg
U.K. ton	Long Ton (British) (2 240 lb)...............	1.016	t
sh cwt	Hundredweight (U.S.A.) (100 lb)	45.359	kg
sh ton	Short ton (U.S.A.) (2 000 lb)	0.907	t
g	Gramme 1 g	15.432	gr
kg	Kilogramme	25.274	oz
		2.205	lb
t	Metric ton	19.684	cwt
		32.046	sh cwt
		1.1025	sh ton
		0.9842	U.K. ton

1. By definition, the pound is exactly 0.45359237 kg.

G. FORCE
Symbol

pdl	Poundal (foot-pound/s²)	0.0138	daN
lbf	Pound-force....................................	0.448	daN
tonf	Ton-force (British) (2 240 lbf)	996.402	daN
	Ton-force (U.S.A.) (2 000 lbf)	889.644	daN

H. PRESSURE, STRESS

Symbol

lbf/in² or psi	Pound-force per square inch	6 894.76	Pa
		0.0689476	bar or hpz
		or daN/cm²	
lbf/ft²	Pound-force per square foot	47.87	Pa
tonf/in²	Ton-force per square inch (British)........	154.44	bar
	Ton-force per square inch (U.S.A.)	137.90	bar
in H₂O	Inch of water	2.49	mbar
in Hg	Inch of mercury	33.86	mbar
Pa	Pascal (= 1 N/m²)	0.0209	lbf/ft²
bar	Bar (= 1 hpz)	14.504	lbf/in²
	Atmosphere	14.696	lbf/in²

● *Units of pressure*

The figures give the values of one unit read hereunder	Pieze	Bar	Atmo-sphere	mm of mercury	Metre of water	Pascal
Pascal SI	0.001	0.00001	9.87 × 10⁻⁶	0.0075	1.020 × 10⁻⁴	1
Pieze	1	0.01	0.00987	7.4975	0.10197	10³
Bar	100	1	0.98692	749.75	10.1972	10⁵
Standard atmo-sphere	101.325	1.01325	1	760	10.3323	101325
Kgf/cm²	98.066	0.98066	0.96784	735.514	10	9.81 × 10⁴
Metre of mercury	133.377	1.33377	1.316	1 000	13.596	1.33 × 10⁵
Metre of water at 4ºC	9.806	0.09807	0.09678	73.551	1	9.81 × 10³

I. VISCOSITY

Symbol

pdl s/ft²	● *Dynamic (or absolute) viscosity* Poundal second per square foot (= Pound per foot second) .... ≈ 1.4882 Pa/s
in²/s	● *Kinematic viscosity* Inch squared per second .... = 6.452.10⁻⁴ m²/s

J. DENSITY AND CONCENTRATION

Symbol

lb/in³	Pound per cubic inch................	27.6799	g/cm³
lb/ft³	Pound per cubic foot	16.0185	kg/m³
gr/U.K. gal	Grain per Imp gallon	14.25	mg/l
gr/U.S. gal	Grain per U.S. gallon	17.12	mg/l
lb/U.K. gal	Pound per Imp gallon	99.77	g/l
lb/U.S. gal	Pound per U.S. gallon...............	119.3	g/l
gr/in³	Grain per cubic inch	10.076	g/l
gr/ft³	Grain per cubic foot	2.296	mg/l
g/cm³	Gramme per cm³ = 1 kg/dm³ ....⟨	0.036127	lb/in³
kg/dm³	Kilogramme per dm³ = 1 g/cm³ ....⟨	62.427	lb/ft³
mg/l	Milligramme per litre ....⟨	0.0703	gr/U.K. gal
g/m³	Gramme per m³	0.0584	gr/U.S. gal
		0.4356	gr/ft³

K. ENERGY, WORK, HEAT

Symbol

ft pdl	Food poundal	0.042 14	J
ft lbf	Foot pound-force	1.356	J
hp h		2.685 × 10⁶	J
	Horsepower-hour ....⟨	0.746	kWh
		0.641	th
		1 055.06	J
BTU	British thermal unit ....⟨	0.293	Wh
		0.252	mth
	Therm (= 10⁵ BTU) ....⟨	105,500	kj
		25,200	kcal
J	Joule (= 10⁷ erg = 0.239 cal) ....⟨	23.72	ft pdl
		0.737	ft lbf
kJ	Kilojoule $\left(= \frac{1}{3.6}\,Wh\right)$ ....⟨	0.3725 × 10⁻	hp h
		0.948	BTU
Wh	Watt-hour (= 3 600 J = 0.860 mth) ...	3.41	BTU
kWh	Kilowatt-hour (= 3 600 kJ)	1.34	hp h
cal	Calorie (4.187 J)		
kcal	Kilocalorie	3.97	BTU
or	or = 1.163 Wh ....⟨		
mth	Millithermie	1.56 × 10⁻³	hp h
th	Thermie (= 10³ kcal)		

● *Corresponding values of the various units of energy*

The figures give the values of one unit read hereunder	Joule	kWh	kcal or mth	hp h	BTU
Joule	1	27.78×10^{-8}	239×10^{-6}	37.25×10^{-8}	948×10^{-6}
Kilowatt-hour	3.6×10^6	1	860	1.341	3 413
Kilocalorie	4 186	116×10^{-5}	1	156×10^{-5}	3.968
Horsepower-hour	2.68×10^6	0.746	641	1	2 545
British thermal unit	1 055	293×10^{-6}	0.252	393×10^{-6}	1

L. CALORIFIC POWER

Symbol

B.T.U./lb	British thermal unit per pound	2.326	J/g
		0.556	mth/kg
B.T.U./ft³	British thermal unit per cubic foot	37.259	kJ/m³
		8.901	mth/m³
kcal/m³	Kilocalorie per cubic metre		
or	or	0.1124	B.T.U./ft³
mth/m³	Millithermie per cubic metre		

M. POWER

Symbol

ft/ pdl/s	Foot poundal per second	0.042 14	W
ft lbf/s	Foot pound-force per second	1.355 82	W
hp or HP	Horsepower (= 550 ft lbf/s)	0.745 7	kW
B.T.U./h	British thermal unit per hour	0.2931	W
		0.252	mth/h
B.T.U./s	British thermal unit per second	1.055	kW
		0.252	mth/s
kW	Kilowatt	1.341	hp
		0.948	B.T.U./s
mth/h	Millithermie per hour	3.968	B.T.U./h
mth/s	Millithermie per second	3.968	B.T.U./s

N. RATES OF FLOW

Corresponding metric and American Units

Value of one unit given below	m³/h	m³/s	l/s	1 000 m³/d	ft³/sec	ft³/min	US gpm	US mgd
m³/h	1	278×10^{-6}	0.2778	0.024	9.81×10^{-3}	0.588	4.403	6.34×10^{-3}
m³/s	3 600	1	1 000	86.4	35.30	2 118	15 852	22.82
l/s	3.6	0.001	1	0.0864	35.3×10^{-3}	2.118	15.85	22.8×10^{-3}
1 000 m³/d	41.67	11.6×10^{-3}	11.575	1	0.4085	24.5	183.47	0.264
ft³/s (= cfs = cusec)	102	28.3×10^{-4}	28.317	2.448	1	60	449	0.647
ft³/min (= cfm)	1.70	472×10^{-6}	0.472	0.0408	0.0167	1	7.48	0.0108
US gal/min (US gpm)	0.2271	6.3×10^{-5}	0.0631	5.45×10^{-3}	2.223×10^{-3}	0.1336	1	1.44×10^{-3}
US mgd	157.7	43.8×10^{-3}	43.80	3.785	1.546	92.80	694	1

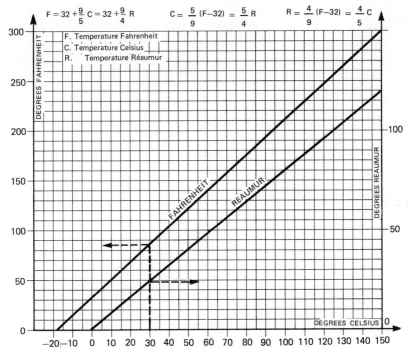

Fig. 643. — Thermometric scales for Fahrenheit, Celsius and Reaumur.

1. ALGEBRA AND ARITHMETIC

- **Proportions:**

  The equation $\dfrac{a}{b} = \dfrac{c}{d}$ may be expressed $\dfrac{a}{b} = \dfrac{a \pm c}{b \pm d}$

  or $\dfrac{a \pm b}{b} = \dfrac{c \pm d}{d}$

- **Progressions:**

  Arithmetic progression has the form:
  $$a;\ a + d;\ a + 2d;\ \ldots\ldots\ a + (n - 1)\,d$$
  where $a = $ first term; $d = $ common difference; $n = $ number of terms:
  $$a + (n - 1)\,d = l = n^{\text{th}}\text{ term}$$

  The sum of the first n terms $= \text{S} = \dfrac{(a + l)\,n}{2}.$

  Geometric progression has the form:
  $$a;\ ar;\ ar^2;\ \ldots\ldots\ ar^{n-1}$$
  where $a = $ first term; $r = $ common quotient; $n = $ number of terms; $l = ar^{n-1}$ $= n^{\text{th}}$ term.

  The sum of the first n terms $= \text{S}\,\dfrac{a\,(r^n - 1)}{r - 1}$

  The sum of an infinite number of terms when $r < 1 = \text{S} = \dfrac{a}{1 - r}.$

- **Arrangements:**

  Total number of possible arrangements of m things taken n at a time:
  $$\text{A}^n_m = m\,(m - 1)\,(m - 2)\,(m - 3)\,\ldots\ldots\,(m - n + 1)$$

- **Permutations:**

  Number of possible permutations with n things:
  $$\text{M}_n = 1 \times 2 \times 3 \times 4 \times 5 \times \ldots\ldots\ldots \times n = n!$$

- **Combinations:**

  Total number of possible combinations of n things taken p at a time
  $$\text{C}^p_n = \frac{n\,(n - 1)\,(n - 2)\,\ldots\ldots\,(n - p + 1)}{1.\,2.\,3.\,4.\,5.\,\ldots\ldots\,(p - 1)\,p} = \frac{\text{A}^p_n}{p!}$$

- **Compound interest:**

  P = principal; r = rate of interest
  C = capital obtained at compound interest after n years.

  $$C = P\,(1 + r)^n$$

- **Annuities:**

  Actual value A of a series of n equal annual repayments of a principal P:

  $$A = P\,\frac{(1 + r)^n - 1}{r}$$

- **Repayments:**

  P = principal.
  a = annual repayment of principal and interest:

  $$a = \frac{P\,r\,(1 + r)^n}{(1 + r)^n - 1}$$

- **Logarithms:**

  $$a^x = y$$

  x is the logarithm of y in a system which has a as its base.

  $$\log (a \times b \times c) = \log a + \log b + \log c$$

  $$\log \frac{a}{b} = \log a - \log b;\ \log a^n = n \log a$$

  $$\log \sqrt[n]{a} = \frac{1}{n} \log a$$

  Common logarithms have 10 as the base.
  Natural (Napierian) logarithms have the symbol e as the base such that:

  $$e = 1 + \frac{1}{1} \times \frac{1}{1 \times 2} + \frac{1}{1 \times 2 \times 3} + \frac{1}{1.2.3.4} + \ldots = 2.71828$$

  Common log $e = 0.43429 \ldots = M$

  Natural log a or $\log_e a = \dfrac{1}{M}$ common log $a = 2.30259 \log a$

  Common log a or $\log a = M \log_e a = 0.43429 \log_e a$

- **Quadratic equations:**

  $$ax^2 + bx + c = 0 \qquad x = \frac{-b \pm \sqrt{b^2 - 4ac}}{2a}$$

The two roots x' and x'' satisfy the equations:

$$x' + x'' = -\frac{b}{a}; \qquad x'x'' = \frac{c}{a}$$

- **Expansion of series:**

  Newton's Binomial Theorem:

  $$(a + b)^m = a^m + \frac{m}{1} a^{m-1} b + \frac{m(m-1)}{1.2} a^{m-2} b^2 + \dots$$

  $$+ \dots + \frac{m(m-1)\dots(m-n+1)}{1.2.3\dots n} a^{m-n} b^n + \dots + b^m$$

2. TRIGONOMETRICAL FORMULAE

The sine, cosine, and tangent of an angle are respectively equal to the cosine sine, and cotangent of the complementary angle.

$\sin a = \cos(90^\circ - a)$; $\cos a = \sin(90^\circ - a)$; $\tan a = \cot(90^\circ - a)$

Complementary arcs $a + a' = \dfrac{\pi}{2}$ or 90°

Supplementary arcs $a + a' = \pi$ or 180°

The sines of supplementary arcs are equal and of the same sign.   The cosines and tangents are equal and of opposite sign.

$\sin a = \sin(180^\circ - a)$ $\qquad \cos a = -\cos(180^\circ - a)$

$\tan a = -\tan(180^\circ - a)$

Fundamental trigonometrical formulae for any angle:

$$\sin^2 a + \cos^2 a = 1 \qquad\qquad \tan a = \frac{\sin a}{\cos a}$$

$$\cos a = \frac{1}{\pm\sqrt{1 + \tan^2 a}} \qquad \sin a = \frac{\tan a}{\pm\sqrt{1 + \tan^2 a}}$$

- **Addition, subtraction, multiplication, division of trigonometrical functions**

$$\sin(a + b) = \sin a \cos b + \cos a \sin b$$
$$\cos(a + b) = \cos a \cos b - \sin a \sin b$$
$$\sin(a - b) = \sin a \cos b - \cos a \sin b$$
$$\cos(a - b) = \cos a \cos b + \sin a \sin b$$
$$\sin 2a = 2 \sin a \cos a$$

$$\cos 2a = \cos^2 a - \sin^2 a \qquad \tan 2a = \frac{2 \tan a}{1 - \tan^2 a}$$

$$\sin \frac{a}{2} = \pm\sqrt{\frac{1 - \cos a}{2}}$$

$$\cos \frac{a}{2} = \pm\sqrt{\frac{1 + \cos a}{2}}$$

$$\sin a = \frac{2 \tan \frac{a}{2}}{1 + \tan^2 \frac{a}{2}} \qquad\qquad \cos a = \frac{1 - \tan^2 \frac{a}{2}}{1 + \tan^2 \frac{a}{2}}$$

● **Resolution of triangles**

Right angle triangles.

$B + C = 90^o; \ a^2 = b^2 + c^2$
$b = a \sin B = a \cos C$
$c = a \cos B = a \sin C$
$b = c \tan B = c \cot C$
$c = b \tan C = b \cot B$

All triangles.

$a^2 = b^2 + c^2 - 2 \, bc \cos A$

$$\frac{a}{\sin A} = \frac{b}{\sin B} = \frac{c}{\sin C}$$

● **Further formulae**

$$\sin a + \sin b = 2 \sin \frac{a + b}{2} \cos \frac{a - b}{2}$$

$$\sin a - \sin b = 2 \sin \frac{a - b}{2} \cos \frac{a + b}{2}$$

$$\cos a + \cos b = 2 \cos \frac{a + b}{2} \cos \frac{a - b}{2}$$

$$\cos a - \cos b = 2 \sin \frac{a + b}{2} \sin \frac{b - a}{2}$$

$$\tan a \pm \tan b = \frac{\sin (a \pm b)}{\cos a \cos b}$$

3. GEOMETRICAL FORMULAE

3.1. Plane surfaces

● **Regular polygon**

l, side; R, radius of circumscribed circle;
n, number of sides;
r, radius of inscribed circle;
A = area of the polygon.
Sum of the angles of a polygon = 2 (*n* − 2) right angles.

Polygon	R	r	l	A
Triangle	0.577 *l*	0.289 *l*	1.732 R or 3.463 *r*	0.433 *l*² or 1.299 R²
Square	0.707 *l*	0.500 *l*	1.414 R or 2.000 *r*	1.000 *l*² or 2.000 R²
Pentagon......	0.851 *l*	0.688 *l*	1.176 R or 1.453 *r*	1.721 *l*² or 2.378 R²
Hexagon	1.000 *l*	0.866 *l*	1.000 R or 1.155 *r*	2.598 *l*² or 2.598 R²
Heptagon	1.152 *l*	1.038 *l*	0.868 R or 0.963 *r*	3.634 *l*² or 2.736 R²
Octagon	1.307 *l*	1.208 *l*	0.765 R or 0.828 *r*	4.828 *l*² or 2.828 R²
Nonagon....∶.	1.462 *l*	1.374 *l*	0.684 R or 0.728 *r*	6.182 *l*² or 2.892 R²
Decagon......	1.618 *l*	1.540 *l*	0.618 R or 0.649 *r*	7.694 *l*² or 2.939 R²
Undecagon ...	1.775 *l*	1.703 *l*	0.563 R or 0.587 *r*	9.366 *l*² or 2.974 R²
Dodecagon ...	1.932 *l*	1.866 *l*	0.518 R or 0.536 *r*	11.19 *l*² or 3.000 R²

● Triangle

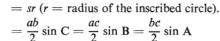

$$\text{Area} = \frac{b \cdot h}{2}$$

$$= \sqrt{s\,(s - a)\,(s - b)\,(s - c)}$$

$$\text{where } s = \frac{a + b + c}{2}$$

$$= \frac{a \cdot b \cdot c}{4\,R} \text{ where R = radius of the circumscribed circle.}$$

$$= sr \ (r = \text{radius of the inscribed circle}).$$

$$= \frac{ab}{2}\sin C = \frac{ac}{2}\sin B = \frac{bc}{2}\sin A$$

● Square, rectangle, parallelogram, trapezium, rhombus

A = Area

Square : A = a^2

Rectangle : A = ab

Parallelogram: A = bh

Trapezium: A = $\dfrac{b + b'}{2}\,h$

Rhombus: A = $\dfrac{d \cdot e}{2}$

● Irregular polygons

The polygon is divided into triangles.

Area = area *abe* + area *bce* + area *cde*.

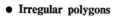

● **Circle, sector, segment, annulus, ellipse**

Circle   *Sector*   *Segment*

Circle: Area $= \pi R^2 = \dfrac{\pi D^2}{4}$ $\pi = 3.1416$ Circumference $= 2 \pi R$

Sector: Area $= \dfrac{arc\ AcB \times R}{2}$ $= \dfrac{\pi R^2}{360} a$ $a = arc$ Ac B expressed in degrees.

Segment: Area $= \dfrac{\pi R^2 \beta}{360} - \dfrac{c}{2} (R\text{-}s)$ β in degrees.

Chord : $C = 2\sqrt{S (2 R\text{-}S)} = 2 R \sin \dfrac{\beta}{2}$

Sagitta (perpendicular from mid points of chord to arc) $S = R \left(1 - \cos \dfrac{\beta}{2}\right)$

$$= R \pm \sqrt{R^2 - \dfrac{c^2}{4}}$$

Annulus   *Ellipse*

Area $= \pi (R^2\text{-}r^2)$
$= \pi (R\text{-}r) (R + r)$

Area $= \pi ab$

3.2. Surface areas and volumes of solids

● **Cube**
$A = 6 a^2$
$V = a^3$

A = Total surface area
V = volume

● **Rectangular parallelepiped**
$A = 2 (a.b + b.h + a.h)$
$V = B \times h = abh$

● **Right or oblique prism**
$V = B \times h$
B = area of base
h = perpendicular height

- **Frustum of a prism**

on a triangular base

$$V = abc \times \frac{h_1 + h_2 + h_3}{3}$$

abc = area of the base

on a parallelogram base

$$V = abcd \times \frac{h_1 + h_2 + h_3 + h_4}{4}$$

$abcd$ = area of the base

- **Body having parallel rectangular faces**

$$V = \frac{h}{6}\left[(2\,a + a')\,b + (2\,a' + a)\,b'\right]$$

h = perpendicular distance between parallel faces

- **Body having parallel polygonal faces**

$$V = \frac{h}{6}(B + B' + 4\,B'').$$

h = perpendicular distance between parallel faces.
B and B' = area of the two parallel faces.
B'' = area of the cross-section parallel to faces
    B and B' taken through mid-point of h.

- **Prism having oblique ends**

$V = abcde \times h$

    $abcde$ = area of one end
    h = perpendicular distance from the centre of gravity
       of the other end to the plane of the first.

- **Cylinder**

Right cylinder: $V = \dfrac{\pi\,d^2}{4}\,h$

Right cylinder with oblique section: $V = \dfrac{\pi\,d^2}{4} \times \dfrac{h_1 + h_2}{2}.$

Oblique cylinder with parallel ends: $V = A'h$
    A' = area of one end
    h = perpendicular distance between the ends

Cylinder with both ends oblique:
    $V = A \times gg_1$
    A = area of right cross-section.
    gg_1 = distance between centres of gravity of the ends.

- **Pyramid and frustum of pyramid with parallel section**

Pyramid:

$$V = \frac{1}{3}\,B \times h \qquad \begin{array}{l} B = \text{area of the base} \\ h = \text{perpendicular height to the apex.} \end{array}$$

Frustum of pyramid with parallel section:
B = area of the base.
b = area of parallel section.
h = perpendicular distance between faces.

$$V = \frac{1}{3} h (B + b + \sqrt{Bb})$$

● **Cone and frustum of cone**

Right of oblique circular cone: $V = \pi r^2 \dfrac{h}{3}$

Right or *oblique non-circular cone:* $V = B \times \dfrac{h}{3}$

    In both cases, h = perpendicular height to the apex.

Frustum of cone with parallel section: $V = \dfrac{1}{3} h (B + b + \sqrt{Bb})$

    B = area of the base.
    b = area of parallel section.
    h = perpendicular distance between parallel faces.

● **Sphere, and sector, segment and zone of a sphere**

1. *Sphere:* Surface area $= 4 \pi r^2$
    Volume $V = \dfrac{4}{3} \pi r^3 = \dfrac{\pi d^3}{6}$.

2. *Sector of a sphere:* Surface area $= \dfrac{\pi r}{2} (4 h + d)$
    Volume $= \dfrac{2}{3} \pi r^2 h = 2.0944 \, r^2 h$

3. *Segment of a sphere:* Lateral surface area $= 2 \pi rh = \dfrac{\pi}{4} (d^2 + 4 h^2)$
    Volume $V = \pi h^2 \left(r - \dfrac{1}{3} h\right) = \pi h \left(\dfrac{d^2}{8} + \dfrac{h^2}{6}\right)$

4. *Zone of a sphere:* Lateral surface area $= 2 \pi rh$
    Volume $V = \dfrac{1}{6} \pi h \left(3 R_1^2 + 3 R_2^2 + h^2\right)$

● **Circular ring**
Surface area A $= 4 \pi^2 R.r$
Volume   V $= 2 \pi^2 r^2 R$

● **Barrel**
Approximate volume
$V = 0.262 \, l \, (2 D^2 + d^2)$

● **Ellipsoid of revolution**

$V = \frac{4}{3} \pi a^2 b$ or $V = \frac{4}{3} \pi ab^2$ depending on whether the revolution takes place

around the major axis or the minor axis

$a = 1/2$ major axis. $b = 1/2$ minor axis.

● **Ellipsoid having three axes**

$V = \frac{4}{3} \pi abc$

● **Paraboloid of revolution**

$V = \frac{\pi d^2 h}{8}$

● **Area generated by a plane curve turning about an axis in the plane of the curve but not cutting it.**

Area $A = 2 \pi rl$

    l = actual length of the curve

    r = radius of the circle described by the centre of gravity.

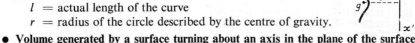

● **Volume generated by a surface turning about an axis in the plane of the surface but not cutting it**

Volume $V = 2 \pi r A$

    A = area

    r = radius of the circle described by the centre of gravity.

4. LENGTH OF ARCS, CHORDS, AND SAGITTA, AND AREAS OF SEGMENTS

When radius $R = 1$ and the angle at the centre is expressed in degrees.
For a radius $= R$, multiply by:
— R the values given below for arcs, chords and sagitta.
— R^2 the values given below for areas of segments.

● **General formulae**

Length of arc: $l = R \frac{a\pi}{180} = Ra \times 0.01745329$

Length of chord: $c = R \, 2 \sin\frac{a}{2} = 2\sqrt{2Rs - s^2}$

Sagitta: $s = R\left(1 - \cos\frac{a}{2}\right) = R \pm \sqrt{R^2 - \frac{c^2}{4}}$

Area of segment: $A = \frac{R^2}{2}\left(\frac{a\pi}{180} - \sin a\right)$

Angle in degrees a	Arc l	Chord c	Sagitta s	Area of Segment A	Angle in degrees a	Arc l	Chord c	Sagitta s	Area of Segment A
1	0.0175	0.0175	0.00004	0.00000	41	0.7156	0.7004	0.0633	0.02976
2	0.0349	0.0349	0.00015	0.00000	42	0.7330	0.7167	0.0664	0.03195
3	0.0524	0.0524	0.00034	0.00001	43	0.7505	0.7330	0.0695	0.03425
4	0.0698	0.0698	0.00061	0.00003	44	0.7679	0.7492	0.0728	0.03614
5	0.0873	0.0872	0.00095	0.00006	45	0.7854	0.7654	0.0761	0.03915
6	0.1047	0.1047	0.00137	0.00010	46	0.8029	0.7815	0.0795	0.04176
7	0.1222	0.1221	0.00187	0.00015	47	0.8203	0.7975	0.0829	0.04448
8	0.1396	0.1395	0.00244	0.00023	48	0.8378	0.8135	0.0865	0.04731
9	0.1571	0.1569	0.00308	0.00032	49	0.8552	0.8294	0.0900	0.05025
10	0.1745	0.1743	0.00381	0.00044	50	0.8727	0.8452	0.0937	0.05331
11	0.1920	0.1917	0.00460	0.00059	51	0.8901	0.8610	0.0974	0.05649
12	0.2094	0.2091	0.00548	0.00076	52	0.9076	0.8767	0.1012	0.05978
13	0.2269	0.2264	0.00643	0.00097	53	0.9250	0.8924	0.1051	0.06319
14	0.2443	0.2437	0.00745	0.00121	54	0.9425	0.9080	0.1090	0.06673
15	0.2618	0.2611	0.00856	0.00149	55	0.9599	0.9235	0.1130	0.07039
16	0.2793	0.2783	0.00973	0.00181	56	0.9774	0.9389	0.1171	0.07417
17	0.2967	0.2956	0.01098	0.00217	57	0.9948	0.9543	0.1212	0.07808
18	0.3142	0.3129	0.01231	0.00257	58	1.0123	0.9696	0.1254	0.08212
19	0.3316	0.3301	0.01371	0.00302	59	1.0297	0.9848	0.1296	0.08629
20	0.3491	0.3473	0.01519	0.00352	60	1.0472	1.0000	0.1340	0.09059
21	0.3665	0.3645	0.01675	0.00408	61	1.0647	1.0151	0.1384	0.09502
22	0.3840	0.3816	0.01837	0.00468	62	1.0821	1.0301	0.1428	0.09958
23	0.4014	0.3987	0.02008	0.00535	63	1.0996	1.0450	0.1474	0.10428
24	0.4189	0.4158	0.02185	0.00607	64	1.1170	1.0598	0.1520	0.10911
25	0.4363	0.4329	0.02370	0.00686	65	1.1345	1.0746	0.1560	0.11408
26	0.4538	0.4499	0.02563	0.00771	66	1.1519	1.0893	0.1613	0.11919
27	0.4712	0.4669	0.02763	0.00862	67	1.1694	1.1039	0.1661	0.12443
28	0.4887	0.4838	0.02969	0.00961	68	1.1868	1.1184	0.1710	0.12982
29	0.5061	0.5008	0.03185	0.01067	69	1.2043	1.1328	0.1759	0.13535
30	0.5236	0.5176	0.03407	0.01180	70	1.2217	1.1472	0.1808	0.14102
31	0.5411	0.5345	0.0363	0.01301	71	1.2392	1.1614	0.1859	0.14683
32	0.5585	0.5512	0.0387	0.01429	72	1.2566	1.1756	0.1910	0.15279
33	0.5760	0.5680	0.0411	0.01566	73	1.2741	1.1896	0.1961	0.15889
34	0.5934	0.5847	0.0437	0.01711	74	1.2915	1.2036	0.2014	0.16514
35	0.6109	0.6014	0.0462	0.01864	75	1.3090	1.2175	0.2066	0.17154
36	0.6283	0.6180	0.0489	0.02027	76	1.3256	1.2313	0.2120	0.17808
37	0.6458	0.6346	0.0516	0.02198	77	1.3439	1.2450	0.2174	0.18477
38	0.6632	0.6511	0.0544	0.02378	78	1.3614	1.2586	0.2229	0.19160
39	0.6807	0.6676	0.0573	0.02568	79	1.3788	1.2722	0.2284	0.19859
40	0.6981	0.6840	0.0603	0.02767	80	1.3963	1.2856	0.2340	0.20573

Angle in degrees a	Arc l	Chord c	Sagitta s	Area of Segment A	Angle in degrees a	Arc l	Chord c	Sagitta s	Area of Segment A
81	1.4137	1.2989	0.2396	0.21301	121	2.1118	1.7407	0.5076	0.62734
82	1.4312	1.3121	0.2453	0.22045	122	2.1293	1.7492	0.5152	0.64063
83	1.4486	1.3252	0.2510	0.22804	123	2.1468	1.7576	0.5228	0.65404
84	1.4661	1.3383	0.2569	0.23578	124	2.1642	1.7659	0.5305	0.66759
85	1.4835	1.3512	0.2627	0.24367	**125**	2.1817	1.7740	0.5388	0.68125
86	1.5010	1.3640	0.2686	0.25171	126	2.1991	1.7820	0.5460	0.65905
87	1.5184	1.3767	0.2746	0.25990	127	2.2166	1.7899	0.5538	0.70897
88	1.5359	1.3893	0.2807	0.26825	128	2.2340	1.7976	0.5616	0.72301
89	1.5533	1.4018	0.2867	0.27675	129	2.2515	1.8052	0.5695	0.73716
90	1.5708	1.4142	0.2929	0.28540	**130**	2.2689	1.8126	0.5774	0.75144
91	1.5882	1.4265	0.2991	0.29420	131	2.2864	1.8199	0.5853	0.76584
92	1.6057	1.4387	0.3053	0.30316	132	2.3038	1.8271	0.5933	0.78034
93	1.6232	1.4507	0.3116	0.31226	133	2.3213	1.8341	0.6013	0.79497
94	1.6406	1.4627	0.3180	0.39152	134	2.3387	1.8410	0.6093	0.80970
95	1.6580	1.4746	0.3244	0.33093	**135**	2.3562	1.8478	0.6173	0.82454
96	1.6755	1.4863	0.3309	0.34050	136	2.3736	1.8544	0.6254	0.83949
97	1.6930	1.4979	0.3374	0.35021	137	2.3911	1.8698	0.6335	0.85455
98	1.7104	1.5094	0.3439	0.36008	138	2.4086	1.8672	0.6416	0.86971
99	1.7279	1.5208	0.3506	0.37009	139	2.4260	1.8733	0.6498	0.88497
100	1.7453	1.5321	0.3572	0.38026	**140**	2.4435	1.8794	0.6580	0.90034
101	1.7628	1.5432	0.3639	0.39058	141	2.4609	1.8853	0.6662	0.91580
102	1.7802	1.5543	0.3707	0.40104	142	2.4784	1.8910	0.6744	0.93135
103	1.7977	1.5652	0.3775	0.41166	143	2.4958	1.8966	0.6827	0.94700
104	1.8151	1.5760	0.3843	0.42242	144	2.5133	1.9021	0.6910	0.96274
105	1.8236	1.5867	0.3912	0.43334	**145**	2.5307	1.9074	0.6993	0.97858
106	1.8500	1.5973	0.3982	0.44439	146	2.5482	1.9126	0.7076	0.99449
107	1.8675	1.6077	0.4052	0.45560	147	2.5656	1.9176	0.7160	1.01050
108	1.8850	1.6180	0.4122	0.46695	148	2.5831	1.9225	0.7244	1,02658
109	1.9024	1.6282	0.4193	0.47844	149	2.6005	1.9273	0.7328	1.04275
110	1.9199	1.6383	0.4264	0.49008	**150**	2.6180	1.9319	0.7412	1.05900
111	1.9373	1.6483	0.4336	0.50187	151	2.6354	1.9363	0.7496	1.07532
112	1.9548	1.6581	0.4408	0.51379	152	2.6529	1.9406	0.7581	1.09171
113	1.9722	1.6678	0.4481	0.52586	153	2.6704	1.9447	0.7666	1.10818
114	1.9897	1.6773	0.4554	0.53807	154	2.6878	1.9487	0.7750	1.12472
115	2.0071	1.6868	0.4627	0.55041	**155**	2.7053	1.9526	0.7836	1.14132
116	2.0246	1.6961	0.4701	0.56289	156	2.7227	1.9563	0.7921	1.15799
117	2.0420	1.7053	0.4775	0.57551	157	2.7402	1.9598	0.8006	1.17472
118	2.0595	1.7143	0.4850	0.68827	158	2.7576	1.9632	0.8092	1.19151
119	2.0769	1.7233	0.4925	0.60116	159	2.7751	1.9665	0.8178	1.20835
120	2.0944	1.7321	0.5000	0.61418	**160**	2.7925	1.9696	0.8264	1.22525

Angle in degrees a	Arc l	Chord c	Sagitta s	Area of Segment A	Angle in degrees a	Arc l	Chord c	Sagitta s	Area of Segment A
161	2.8100	1.9726	0.8350	1.24221	171	2.9845	1.9938	0.9215	1.41404
162	2.8274	1.9754	0.8436	1.25921	172	3.0020	1.9951	0.9302	1.43148
163	2.8449	1.9780	0.8522	1.27626	173	3.0194	1.9963	0.9390	1.44878
164	2.8623	1.9805	0.8608	1.29335	174	3.0369	1.9973	0.9477	1.46617
165	2.8798	1.9829	0.8695	1.31049	**175**	3.0543	1.9981	0.9564	1.48359
166	2.8972	1.9851	0.8781	1.32766	176	3.0718	1.9988	0.9651	1.50101
167	2.9147	1.9871	0.8868	1.34487	177	3.0892	1.9993	0.9738	1.51854
168	2.9322	1.9890	0.8955	1.36212	178	3.1067	1.9997	0.9825	1.53589
169	2.9496	1.9908	0.9042	1.37940	179	3.1241	1.9999	0.9913	1.55334
170	2.9671	1.9924	0.9128	1.39671	**180**	3.1416	2.0000	1.0000	1.57080

5. COMMON FACTORS IN CALCULATIONS

π	3.14159	$\sqrt{\dfrac{1}{2\pi}}$	0.39894	$\dfrac{1}{2g}$	0.05097
$\log \pi$	0.49715				
2π	6.28318	$\sqrt{\dfrac{3}{2\pi}}$	0.69099	$\dfrac{1}{g^2}$	0.01039
3π	9.42478				
$\dfrac{\pi}{2}$	1.57080	$\sqrt{\dfrac{2}{\pi}}$	0.79788	$\sqrt{\dfrac{1}{g}}$	0.31929
$\dfrac{\pi}{3}$	1.04720	π^2	9.86960	$e =$ base of Napierian logarithms	
		π^3	31.00628		
$\dfrac{\pi}{5}$	0.62832	$\sqrt{\pi}$	1.77245	$\ln a = \dfrac{\log a}{\log e}$	
$\dfrac{1}{\pi}$	0.31831	$\sqrt{2}$	1.41421	$e = 2.718282$	
		$\sqrt{3}$	1.73205	$\log e = 0.43429$	
$\dfrac{2}{\pi}$	0.63662	$\dfrac{1}{\sqrt{3}}$	0.57735	$\dfrac{1}{e} = 0.367879$	
$\dfrac{3}{\pi}$	0.95493				
$\sqrt{\pi}$	1.77245	$\dfrac{1}{\sqrt{2}}$	0.70710	$\dfrac{1}{\log e} = 2.30259$	
$\dfrac{1}{\sqrt{\pi}}$	0.56418	g	9.80896		
		$\dfrac{1}{g}$	0.10195		

6. SIMPLE INTEREST

l = Interest.
r = Rate of interest per annum.
C = Capital.
t = Time in years.
m = Time in months.
d = Time in days.

$$l = r\,Ct = r\,C\,\frac{m}{12} = r\,C\,\frac{d}{360}$$

Interest on 1 unit of currency for one day

Rate %	Year of 360 days	Year of 365 days	Rate %	Year of 360 days	Year of 365 days
0.5	0.000013888	0.000013699	3.5	0.000097221	0.000095890
1	0.000027777	0.000027397	4	0.000111111	0.000109589
1.5	0.000041665	0.000041096	4.5	0.000124999	0.000123288
2	0.000055555	0.000054795	5	0.000138888	0.000136986
2.5	0.000069443	0.000068493	6	0.000166667	0.000164384
3	0.000083333	0.000082192	7	0.000194444	0.000191871

7. COMPOUND INTEREST

Over a period of t years, the capital becomes:

$$C = P\,(1 + r)^t \text{ (where P is the principal)}$$

Table of values of $(1 + r)^t$ reached by one unit of currency placed at compound interest of r per cent after t years.

t yrs	Rate of Interest—r					
	4 %	5 %	6 %	7 %	8 %	9 %
1	1.04000	1.05000	1.06000	1.07000	1.08000	1.09000
2	1.08160	1.10250	1.12360	1.14490	1.16640	1.18810
3	1.12486	1.15762	1.19101	1.22504	1.25971	1.29503
4	1.16985	1.21550	1.26247	1.31079	1.36049	1.41158
5	1.21665	1.27628	1.33822	1.40255	1.46932	1.53862
6	1.26531	1.34009	1.41852	1.50073	1.58687	1.67710
7	1.31593	1.40710	1.50363	1.60578	1.71382	1.82804
8	1.36856	1.47745	1.59384	1.71818	1.85093	1.99256
9	1.43321	1.55132	1.68948	1.83846	1.99900	2.17189
10	1.48024	1.62889	1.79084	1.96715	2.15892	2.36736
11	1.53945	1.71033	1.89830	2.10485	2.33164	2.58042
12	1.60103	1.79585	2.01219	2.25219	2.51817	2.81266
13	1.66507	1.88564	2.13292	2.40984	2.71962	3.06580
14	1.73167	1.97993	2.26090	2.57853	2.93719	3.34172
15	1.80094	2.07892	2.39655	2.75903	3.17217	3.64248
16	1.87298	2.18287	2.54035	2.95216	3.42594	3.97030
17	1.94790	2.29201	2.69277	3.15881	3.70002	4.32763
18	2.02581	2.40661	2.85434	3.37993	3.99602	4.71712
19	2.10684	2.52695	3.02560	3.61653	4.31570	5.14166
20	2.19112	2.65329	3.20713	3.86968	4.66095	5.60441
21	2.27876	2.78596	3.39956	4.14056	5.03383	6.10881
22	2.36991	2.92526	3.60353	4.43040	5.43654	6.65860
23	2.46471	3.07152	3.81975	4.74053	5.87146	7.25787
24	2.56330	3.22510	4.04893	5.07236	6.34118	7.91108
25	2.66583	3.38635	4.29187	5.42743	6.84847	8.62308
30	3.2434	4.3219	5.7435	7.6123	10.0627	13.2677
35	3.9461	5.5160	7.6861	10.6766	14.8754	20.4140
40	4.8010	7.0400	10.2857	14.9745	21.7245	31.4094
45	5.8412	8.9850	13.7646	21.0025	31.9204	48.3273
50	7.1067	11.4674	18.4210	29.4571	46.9018	74.3575
60	10.5196	18.6792	32.9877	57.9466	101.2576	176.0313
70	15.5716	30.4264	59.0759	113.9894	218.6074	416.7303
80	23.050	49.561	105.796	224.235	471.958	986.552
90	34.119	80.730	189.464	441.105	1,018.915	2,325.528
100	50.505	131.501	339.302	867.720	2,199.783	5,529.043

8. REPAYMENTS

If a is the annual payment which will repay in t years the capital C at a rate of r compound interest.

$$C = \frac{a}{r}[(1 + r)^t - 1] \qquad a = C\frac{r}{(1 + r)^t - 1}$$

Rate of repayment: $\quad T = \dfrac{r}{(1 + r)^t - 1}$

Table of values of T

years	Rate of Interest—r				
	4 %	5 %	6 %	8 %	10 %
1	1.000	1.0000	1.0000	1.0000	1.0000
2	0.49019	0.48780	0.48543	0.480	0.476
3	0.32035	0.31720	0.31411	0.308	0.302
4	0.23549	0.23201	0.22859	0.221	0.215
5	0.18462	0.18097	0.17739	0.170	0.163
6	0.15076	0.14701	0.14336	0.136	0.129
7	0.12661	0.12282	0.11913	0.112	0.105
8	0.10852	0.10472	0.10103	0.094	0.087
9	0.09449	0.09069	0.08702	0.080	0.073
10	0.08329	0.07950	0.07586	0.069	0.062
11	0.07415	0.07038	0.06679	0.060	0.053
12	0.06655	0.06282	0.05927	0.052	0.048
13	0.06014	0.05645	0.05296	0.046	0.040
14	0.05467	0.05102	0.04758	0.041	0.035
15	0.04994	0.04634	0.04296	0.036	0.031
16	0.04582	0.04227	0.03895	0.032	0.027
17	0.04220	0.03869	0.03544	0.029	0.024
18	0.03899	0.03554	0.03235	0.026	0.021
19	0.03614	0.03274	0.02962	0.024	0.019
20	0.03358	0.03024	0.02718	0.021	0.017
21	0.01328	0.02799	0.02500	0.019	0.015
22	0.02920	0.02597	0.02304	0.018	0.014
23	0.02730	0.02413	0.02127	0.016	0.012
24	0.02558	0.02247	0.01967	0.015	0.011
25	0.02401	0.02095	0.01822	0.013	0.010
30	0.01783	0.01505	0.01264	0.008	0.006
35	0.01357	0.01107	0.00897	0.005	0.003
40	0.01052	0.00827	0.00646		
45	0.00826	0.00626	0.00470		
50	0.00655	0.00477	0.00344		
60	0.00420	0.00282	0.00187		
70	0.00274	0.00169	0.00103		
80	0.00181	0.00103	0.00057		
90	0.00120	0.00062	0.00031		
100	0.00080	0.00038	0.00017		

Annual payments of interest and repayments:

Every year the annual payment of interest and repayments for one unit of currency is $T + r$:

$$T + r = \frac{r(1 + r)^t}{(1 + r)^t - 1}$$

9. TRIGONOMETRICAL TABLES

Degrees	SINES						
	0′	10′	20′	30′	40′	50′	
0	0.00000	0.00291	0.00582	0.00873	0.01164	0.01454	89
1	0.01745	0.02036	0.02327	0.02618	0.02908	0.03199	88
2	0.03490	0.03781	0.04071	0.04362	0.04653	0.04943	87
3	0.05234	0.05524	0.05814	0.06105	0.06395	0.06685	86
4	0.06976	0.07266	0.07556	0.07846	0.08136	0.08426	85
5	0.08716	0.09005	0.09295	0.09585	0.09874	0.10164	84
6	0.10453	0.10742	0.11031	0.11320	0.11609	0.11898	83
7	0.12187	0.12476	0.12764	0.13053	0.13341	0.13629	82
8	0.13917	0.14205	0.14493	0.14781	0.15069	0.15356	81
9	0.15643	0.15931	0.16218	0.16505	0.16792	0.17078	80
10	0.17365	0.17651	0.17937	0.18224	0.18509	0.18795	79
11	0.19081	0.19366	0.19652	0.19937	0.20222	0.20507	78
12	0.20791	0.21076	0.21360	0.21644	0.21928	0.22212	77
13	0.22495	0.22778	0.23062	0.23345	0.23627	0.23910	76
14	0.24192	0.24474	0.24756	0.25038	0.25320	0.25601	75
15	0.25882	0.26163	0.26443	0.26724	0.27004	0.27284	74
16	0.27564	0.27843	0.28123	0.28402	0.28680	0.28959	73
17	0.29237	0.29515	0.29793	0.30071	0.30348	0.30625	72
18	0.30902	0.31178	0.31454	0.31730	0.32006	0.32282	71
19	0.32557	0.32832	0.33106	0.33381	0.33655	0.33929	70
20	0.34202	0.34475	0.34748	0.35021	0.35293	0.35565	69
21	0.35837	0.36108	0.36379	0.36650	0.36921	0.37191	68
22	0.37461	0.37730	0.37999	0.38268	0.38567	0.38805	67
23	0.39073	0.39341	0.39608	0.39875	0.40141	0.40408	66
24	0.40674	0.40939	0.41204	0.41469	0.41734	0.41998	65
25	0.42262	0.42525	0.42788	0.43051	0.43313	p.43575	64
26	0.43837	0.44098	0.44359	0.44620	0.44880	0.45140	63
27	0.45399	0.45658	0.45917	0.46175	0.46433	0.46690	62
28	0.46947	0.47204	0.47460	0.44716	0.47971	0.48226	61
29	0.48481	0.48735	0.48989	0.49242	0.49495	0.49748	60
30	0.50000	0.50252	0.50503	0.50754	0.51004	0.51254	59
31	0.51504	0.51753	0.52002	0.52250	0.52498	0.52745	58
32	0.52992	0.53238	0.53484	0.53730	0.53975	0.54220	57
33	0.54464	0.54708	0.54951	0.55194	0.55436	0.55678	56
34	0.55919	0.56160	0.56401	0.56641	0.56880	0.57119	55
35	0.57358	0.57596	0.57833	0.58070	0.58307	0.58543	54
36	0.58779	0.59014	0.59248	0.59482	0.59716	0.59949	53
37	0.60182	0.60414	0.60645	0.60876	0.61107	0.61337	52
38	0.61566	0.61795	0.62024	0.62251	0.62479	0.62706	51
39	0.62932	0.63158	0.63383	0.63608	0.63832	0.64056	50
40	0.64279	0.64501	0.64723	0.64945	0.65166	0.65386	49
41	0.65606	0.65825	0.66044	0.66262	0.66480	0.66697	48
42	0.66913	0.67129	0.67344	0.67559	0.67773	0.67987	47
43	0.68200	0.68412	0.68624	0.68835	0.69046	0.69256	46
44	0.69466	0.69675	0.69883	0.70091	0.70298	0.70505	45
45	0.70711						
	60′	50′	40′	30′	20′	10′	Degrees
	COSINES						

Degrees	COSINES						
	0′	10′	20′	30′	40′	50′	
0	1.00000	1.00000	0.99998	0.99996	0.99993	0.99989	89
1	0.99985	0.99979	0.99973	0.99966	0.99958	0.99949	88
2	0.99939	0.99929	0.99917	0.99905	0.99892	0.99878	87
3	0.99863	0.99847	0.99831	0.99813	0.99795	0.99776	86
4	0.99756	0.99736	0.99714	0.99692	0.99668	0.99644	85
5	0.99619	0.99594	0.99567	0.99540	0.99511	0.99482	84
6	0.99452	0.99421	0.99390	0.99357	0.99324	0.99290	83
7	0.99255	0.99219	0.99182	0.99144	0.99106	0.99067	82
8	0.99027	0.98986	0.98944	0.98902	0.98858	0.98814	81
9	0.98769	0.98723	0.98676	0.98629	0.98580	0.98531	80
10	0.98481	0.98430	0.98378	0.98325	0.98272	0.98218	79
11	0.98163	0.98107	0.98050	0.97992	0.97934	0.97875	78
12	0.97815	0.97754	0.97692	0.97630	0.97566	0.97502	77
13	0.97437	0.97371	0.97304	0.97237	0.97169	0.97100	76
14	0.97030	0.96959	0.96887	0.96815	0.96742	0.96667	75
15	0.96593	0.96517	0.96440	0.96363	0.96285	0.96206	74
16	0.96126	0.96046	0.95964	0.95882	0.95799	0.95715	73
17	0.95630	0.95545	0.95459	0.95372	0.95284	0.95195	72
18	0.95106	0.95015	0.94924	0.94832	0.94740	0.94646	71
19	0.94552	0.94457	0.94361	0.94264	0.94167	0.94068	70
20	0.93969	0.93869	0.93769	0.93667	0.93565	0.93462	69
21	0.93358	0.93253	0.93148	0.93042	0.92935	0.92827	68
22	0.92718	0.92609	0.92499	0.92388	0.92276	0.92164	67
23	0.92050	0.91936	0.91822	0.91706	0.91590	0.91472	66
24	0.91355	0.91236	0.91116	0.90996	0.90875	0.90753	65
25	0.90631	0.90507	0.90383	0.90259	0.90133	0.90007	64
26	0.89879	0.89752	0.89623	0.89493	0.89363	0.89232	63
27	0.89101	0.88968	0.88835	0.88701	0.88566	0.88431	62
28	0.88295	0.88158	0.88020	0.87882	0.87743	0.87603	61
29	0.87462	0.87321	0.87178	0.87036	0.86892	0.86748	60
30	0.86603	0.86457	0.86310	0.86163	0.86015	0.85866	59
31	0.85717	0.85567	0.85416	0.85264	0.85112	0.84959	58
32	0.84805	0.84650	0.84495	0.84339	0.84182	0.84025	57
33	0.83867	0.83708	0.83549	0.83389	0.83228	0.83066	56
34	0.82904	0.82741	0.82577	0.82413	0.82248	0.82082	55
35	0.81915	0.81748	0.81580	0.81412	0.81242	0.81072	54
36	0.80902	0.80730	0.80558	0.80386	0.80212	0.80038	53
37	0.79864	0.79688	0.79512	0.79335	0.79158	0.78980	52
38	0.78801	0.78622	0.78442	0.78261	0.78079	0.77897	51
39	0.77715	0.77531	0.77347	0.77162	0.76977	0.76791	50
40	0.76604	0.76417	0.76229	0.76041	0.75851	0.75661	49
41	0.75471	0.75280	0.75088	0.74896	0.74703	0.74509	48
42	0.74314	0.74120	0.73924	0.73728	0.73531	0.73333	47
43	0.73135	0.72937	0.72737	0.72537	0.72337	0.72136	46
44	0.71934	0.71732	0.71529	0.71325	0.71121	0.70916	45
45	0.70711						
	60′	50′	40′	30′	20′	10	Degrees
	SINES						

| Degrees | **TANGENTS** | | | | | | |
	0′	10′	20′	30′	40′	50′	
0	0.00000	0.00291	0.00582	0.00873	0.01164	0.01455	89
1	0.01746	0.02036	0.02328	0.02619	0.02910	0.03201	88
2	0.03492	0.03783	0.04075	0.04366	0.04658	0.04949	87
3	0.05241	0.05533	0.05824	0.06116	0.06408	0.06700	86
4	0.06993	0.07285	0.07578	0.07870	0.08163	0.08456	**85**
5	0.08749	0.09042	0.09335	0.09629	0.09923	0.10216	84
6	0.10510	0.10805	0.11099	0.11394	0.11688	0.11983	83
7	0.12278	0.12574	0.12869	0.13165	0.13461	0.13758	82
8	0.14054	0.14351	0.14648	0.14945	0.15243	0.15540	81
9	0.15838	0.16137	0.16435	0.16734	0.17033	0.17333	**80**
10	0.17633	0.17933	0.18233	0.18534	0.18835	0.19136	79
11	0.19438	0.19740	0.20042	0.20345	0.20648	0.20952	78
12	0.21256	0.21560	0.21864	0.22159	0.22475	0.22781	77
13	0.23087	0.23393	0.23700	0.24008	0.24316	0.24624	76
14	0.24933	0.25242	0.25552	0.25862	0.26172	0.26483	**75**
15	0.26795	0.27107	0.27419	0.27732	0.28046	0.28360	74
16	0.28675	0.28990	0.29305	0.29621	0.29938	0.30255	73
17	0.30573	0.30891	0.31210	0.31530	0.31850	0.32171	72
18	0.32492	0.32814	0.33136	0.33460	0.33783	0.34108	71
19	0.34433	0.34758	0.35085	0.35412	0.35740	0.36068	**70**
20	0.36397	0.36727	0.37057	0.37388	0.37720	0.38053	69
21	0.38386	0.38721	0.39055	0.39391	0.39727	0.40065	68
22	0.40403	0.40741	0.41081	0.41421	0.41763	0.42105	67
23	0.42447	0.42791	0.43136	0.43481	0.43828	0.44175	66
24	0.44523	0.44872	0.45222	0.45573	0.45924	0.46277	**65**
25	0.46631	0.46985	0.47341	0.47698	0.48055	0.48414	64
26	0.48773	0.49134	0.49495	0.49858	0.50222	0.50587	63
27	0.50953	0.51319	0.51688	0.52057	0.52427	0.52798	62
28	0.53171	0.53545	0.53920	0.54296	0.54673	0.55051	61
29	0.55431	0.55812	0.56194	0.56577	0.56962	0.57348	**60**
30	0.57735	0.58124	0.58513	0.58905	0.59297	0.59691	59
31	0.60086	0.60483	0.60881	0.61280	0.61681	0.62083	58
32	0.62487	0.62892	0.63299	0.63707	0.64117	0.64528	57
33	0.64941	0.65355	0.65771	0.66189	0.66608	0.67029	56
34	0.67451	0.67855	0.68301	0.68728	0.69157	0.69588	**55**
35	0.70021	0.70455	0.70891	0.71329	0.71769	0.72211	54
36	0.72654	0.73100	0.73547	0.73996	0.74447	0.74900	53
37	0.75355	0.75812	0.76272	0.76733	0.77196	0.77661	52
38	0.78129	0.78598	0.79070	0.79544	0.80020	0.80498	51
39	0.80978	0.81461	0.81946	0.82434	0.82923	0.83415	**50**
40	0.83910	0.84407	0.84906	0.85408	0.85912	0.86419	49
41	0.86929	0.87441	0.87955	0.88473	0.88992	0.89515	48
42	0.90040	0.90569	0.91099	0.91633	0.92170	0.92709	47
43	0.93252	0.93797	0.94345	0.94896	0.95451	0.96008	46
44	0.96569	0.97133	0.97700	0.98270	0.98843	0.99420	**45**
45	1.00000						
	60′	50′	40′	30′	20′	10′	Degrees
	COTANGENTS						

Degrees	COTANGENTS						Degrees
	0′	10′	20′	30′	40′	50′	
0	00	343.77371	171.88540	114.58865	85.93979	68.75009	89
1	57.28996	49.10888	42.96408	38.18846	34.36777	31.24158	88
2	28.63625	26.43160	24.54176	22.90377	21.47040	20.20555	87
3	19.08114	18.07498	17.16934	16.34986	15.60478	14.92442	86
4	14.30067	13.72674	13.19688	12.70621	12.25051	11.82617	85
5	11.43005	11.05943	10.71191	10.38540	10.07803	9.78817	84
6	9.51436	9.25530	9.00983	8.77689	8.55555	8.34496	83
7	8.14435	7.95302	7.77035	7.59575	7.42871	7.26873	82
8	7.11537	6.96823	6.82694	6.69116	6.56055	6.43484	81
9	6.31375	6.19703	6.08444	5.97576	5.87080	5.76937	80
10	5.67128	5.57638	5.48451	5.39552	5.30928	5.22566	79
11	5.14455	5.06584	4.98940	4.91516	4.84300	4.77286	78
12	4.70463	4.63825	4.57363	4.51071	4.44942	4.38969	77
13	4.33148	4.27471	4.21933	4.16530	4.11256	4.06107	76
14	4.01078	3.96165	3.91364	3.86671	3.82083	3.77595	75
15	3.73205	3.68909	3.64705	3.60588	3.56557	3.52609	74
16	3.48741	3.44951	3.41236	3.37594	3.34023	3.30521	73
17	3.27085	3.23714	3.20406	3.17159	3.13972	3.10842	72
18	3.07768	3.04749	3.01783	2.98868	2.96004	2.93189	71
19	2.90421	2.87700	2.85023	2.82391	2.79802	2.77254	70
20	2.74748	2.72281	2.69853	2.67462	2.65109	2.62791	69
21	2.60509	2.58261	2.56046	2.53865	2.51715	2.49597	68
22	2.47509	2.45451	2.43422	2.41421	2.39449	2.37504	67
23	2.35585	2.33693	2.31826	2.29984	2.28167	2.26374	66
24	2.24604	2.22857	2.21132	2.19430	2.17749	2.16090	65
25	2.14451	2.12832	2.11233	2.09654	2.08094	2.06553	64
26	2.05030	2.03526	2.02039	2.00569	1.99116	1.97680	63
27	1.96291	1.94858	1.93470	1.92098	1.90741	1.89400	62
28	1.88073	1.86760	1.85462	1.84177	1.82906	1.81649	61
29	1.80405	1.79174	1.77955	1.76749	1.75556	1.74375	60
30	1.73205	1.72047	1.70901	1.69766	1.68643	1.67530	59
31	1.66428	1.65337	1.64256	1.63185	1.62125	1.61074	58
32	1.60033	1.59002	1.57981	1.56969	1.55966	1.54972	57
33	1.53986	1.53010	1.52043	1.51084	1.50133	1.49190	56
34	1.48256	1.47330	1.46411	1.45501	1.44598	1.43703	55
35	1.42815	1.41934	1.41061	1.40195	1.39336	1.38484	54
36	1.37638	1.36800	1.35968	1.35142	1.34323	1.33511	53
37	1.32704	1.31904	1.31110	1.30323	1.29541	1.28764	52
38	1.27994	1.27230	1.26471	1.25717	1.24969	1.24227	51
39	1.23490	1.22758	1.22031	1.21310	1.20593	1.19882	50
40	1.19175	1.18474	1.17777	1.17085	1.16398	1.15715	49
41	1.15037	1.14363	1.13694	1.13029	1.12369	1.11713	48
42	1.11061	1.10414	1.09770	1.09131	1.08496	1.07864	47
43	1.07237	1.06613	1.05994	1.05378	1.04766	1.04158	46
44	1.03553	1.02952	1.02355	1.01761	1.01170	1.00583	45
45	1.00000						
	60′	50′	40′	30′	20′	10′	Degrees
	TANGENTS						

1. GUIDE-LINES FOR ESTIMATING WATER REQUIREMENTS

- **Domestic uses**

  (excluding watering of gardens and industrial uses) .... 125 to 250 l/d hd (33 to 66 gal/hd d)

  Watering of gardens (average of peak months) 300 to 600 l/are d (3 100 to 6 200 gal/acre d)

- **Water requirements of towns** (per inhabitant):

  Rural water supply .... 125 l/d hd (33 gal/hd d)

  Towns of less than 3 000 inhabitants 200 l/d hd (54 gal/hd d)

  Towns of 3 000 to 15 000 inhabitants ...... 115[1], 235[2], 315[3] l/d hd (30, 62, 34 gal/hd d)

  Towns of 15 000 to 60 000 inhabitants ..... 175[1], 285[2], 350[3] l/d hd (47, 76, 94 gal/hd d)

  Towns of more than 60 000 inhabitants .... 220[1], 345[2], 380[3] l/d hd (59, 92, 102 gal/hd d)

- **Fire-fighting** : 60 m³/h (16 000 gal/h) for 2 h. Rate to be doubled in towns. Minimum reserve 120 m³ (32 000 gal).

- **Agriculture:**

  Wheat .... 1 500 m³/tonne produced (396 000 gal)

  Rice .... 4 000 m³/tonne produced (1 056 000 gal)

  Rye and the like .... 1 000 m³/tonne produced (264 000 gal)

  Cotton .... 10 000 m³/tonne produced (2 640 000 gal)

- **Irrigation by continuous spraying** (in temperate climate) .... 1.5 m³/(h.ha) (160 gal/h acre)

- **Stock raising** (per head):

  Cattle .... 60 to 80 l/d (16 to 20 gal/d)

  Piggeries with hydraulic cleaning, French or Danish type .... 4 to 20 l/d (l to 5 gal/d)

  Piggeries with dry or mixed cleaning .... 2 to 6 l/d (0.5 to 1.6 gal/d)

  Sheep .... 5 l/d (1.3 gal/d)

● **Agricultural industries :**

Buttermaking	2 to 4 l/l of milk
Cheesemaking	6 to 10 l/l of milk
Milk powder	7 to 17 l/l of milk
Cider making (excluding bottling)..	4 m³/t of apples (1 060 gal/tonne
Bottle washing	2 to 6 l/bottle
Wine making	2 l/l of wine
Vinegar making	50 l/l of vinegar

Water requirements for the processing of 1 tonne (2 200 lbs)

	Raw material or product	m³ of water	gal
● **Agricultural industries :**			
Brewing (manufacture only)	malt	20 - 30	5 300 - 7 900
Malting	barley	1.5 - 3	400 - 800
Sugar refining	sugar beets	2 - 15	530 - 4 000
Yeast	yeast	150	40 000
Fruit canning	fruit	12 - 15	3 160 - 4 000
Vegetable canning	vegetables	6	1 600
Fish canning	fish	20	5 300
Meat canning	preserve	70	18 500
Starch manufacture	potatoes	15	4 000
Corn starch making	corn	15 - 20	4 000 - 5 300
Abattoirs:			
Cattle, pigs (excluding cooling circuits)	carcass	5 - 15	1 300 - 3 900
Poultry	carcass	10 - 20	2 650 - 5 300
Cooling circuits, maximum without economizer	carcass	30	79 000
● **Non-agricultural industries:**			
Tanning:	product	20 - 140	5 300 - 37 000
Paper making:			
Paper pulp	product	300	7 900
Packing—cardboard	product	40	10 600
Special papers	product	500	132 000
Textiles:			
Cotton (according to degree of preparation)	product	15 - 200	4 000 - 53 000
Wool (carding—bleaching)	product	165	43 600
Rayon	product	400 - 1 000	106 000 - 264 000
Chemical products	product	200 - 1 000	53 000 - 264 000
Oil refining	product	0.1 - 40	26 - 10 600
Steel	product	6 - 300	1 580 - 79 200
Rolled steel	product	400	106 000
Aluminium	product	1300	343 000

Thermal power generation, with   3 to 400 m³/MWh (800 to 106 000
water cooling ....   gal/MWh)

Note: The quantity of water in circulation may vary within wide limits for one and the same industry depending on the techniques used. Water consumption may also be considerably reduced by appropriate recycling.

2. HEAD LOSSES THROUGH FRICTION IN WATER PIPES

2.1. Empirical formulae:

Many authors, including Prony, Flamant, Darcy and Levy, have proposed empirical formulae for the calculation of these head losses, based on a number of practical tests with types of piping and joints which are now out of date. These formulae, which are of limited application, did not reflect the physical reality of the phenomena either, and the results obtained were sometimes very approximate. For these various reasons they are hardly used any more.

The empirical formula of *Williams and Hazen*, although old, is nevertheless still in use in the U.S.A. It takes the form (in metric units):

$$J = 6.815 \left(\frac{V}{C_{wh}} \right)^{1.852} D^{-1.167}$$

the coefficient C_{wh} varying with the diameter of the pipes and the state of their internal surface.

2.2. Colebrook formula based on the experiments by Nikuradze:

$$J = \frac{\lambda}{D} \cdot \frac{V^2}{2g}$$

$$\frac{1}{\sqrt{\lambda}} = -2 \log \left(\frac{k}{3.7\,D} + \frac{2.51}{Re\,\sqrt{\lambda}} \right)$$

J   = loss of head through friction, in m of water per m of pipe
λ = head loss coefficient.
D   = diameter of pipe, or hydraulic diameter (see p. 1080) for non-cylindrical conduits, in m.
V   = rate of flow, in m/s.
g   = acceleration due to gravity in m/s.s (= 9.81 in Paris)

k   = equivalent roughness coefficient of the wall, in m.

Re = Reynolds number = $\dfrac{VD}{v}$, where the value of the kinematic viscosity of water in m²/s at normal pressure [1] is:

t °C	0	5	10	15	20	30	40	50	60	70	80	90	100
$v \times 10^6$	1.792	1.52	1.31	1.14	1.006	0.80	0.66	0.56	0.48	0.41	0.36	0.33	0.30

A. Choice of roughness coefficient

This prior choice determines the accuracy of the calculation of the head losses through friction.   For pipes carrying water, it is linked both to the nature of the walls, their variation with time, and the chemical and physical characteristics of the water carried.

● **Smooth non-corroding piping with scaling unlikely**

These conditions are achieved with clear water passing through pipes made of plastic, asbestos-cement, centrifuged cement or any material which is non-corroding or has a perfectly smooth lining.   The roughness coefficient to be used in practice is k = 0.1 mm, because of the inevitable minimum ageing with time, although in theory k = 0.03 mm is assumed in the new condition.

For all the usual materials, the roughness coefficients k are as follows, under average service conditions, including joints:

Material	k (mm)	Material	k (mm)
Steel : new plastic lining non-porous } smooth lining }	0.1 0.03	Brass—copper—lead : new Aluminium : new	0.1 0.015-0.06
Cast iron : new bitumen lining cement lining	0.1-1 0.03-0.2 0.03-0.1	Concrete : new centrifuged new/smooth moulds new/rough moulds	0.03 0.2-0.5 1.0-2.0
Plastics	0.03-0.1	Asbestos-cement : new	0.03-0.1
		Glazed stoneware	0.1-1

● **Corrodable piping and probable scaling.**

When this type of piping is carrying water which is relatively aggressive, corrosive, scale-producing or turbid, it is assumed that the mean roughness coefficient will be about k = 2 mm.   For non-aggressive and non-scale-forming water it becomes k = 1 mm.   With untreated, fairly clear water and filtered

---

1. See for the viscosity of other liquids, see page 879.

water which is neither aggressive nor scale-producing and has undergone anti-algae treatment, we can assume k = 0.5 mm.

Under average water quality conditions it is also possible, as a first approximation, to adopt as the value J of the head loss in the tables on pages 1057 and foll., the arithmetic mean of those found in the "New pipes" and "Existing pipes" columns.

B. Calculation using the universal diagram

This diagram (page 1054), which is valid for industrial piping with walls of variable roughness, gives the values of the coefficient λ to be used in the Colebrook formula as a function of the Reynolds number Re corresponding to the true flow conditions and of the relative roughness $\frac{k}{D}$ of the walls.

The table (on page 1055) gives the values of the ratio $\frac{\lambda}{D}$ taken from the universal diagram for some usual values of the coefficient k.

It facilitates calculation by determining as a whole all the friction and local head losses Δh, expressed in m of water,

$$\Delta h = JL + K \frac{V^2}{2\,g} = \left(\frac{\lambda}{D} L + K\right) \frac{V^2}{2\,g}$$

where:

L = total length of section, in m, with flow rate V.

K = sum of the elementary head loss coefficients in the local transition zones of this section (see pages 1069 to 1075).

● **Universal pipe friction diagram.**

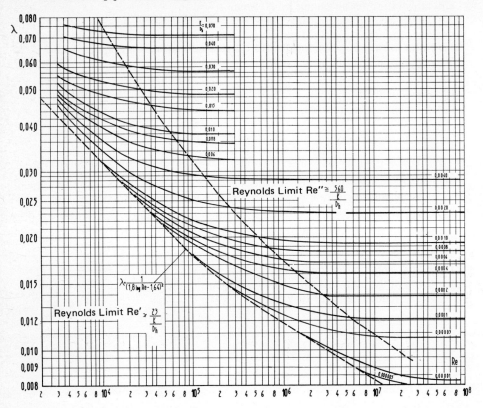

Diameter (m)	Coefficient $\frac{\lambda}{D}$ for roughness equal to:			
	k = 0.1 mm	k = 0.5 mm	k = 1.0 mm	k = 2.0 mm
0.030	1.02	1.54	2.00	2.71
0.040	0.700	1.04	1.34	1.80
0.050	0.528	0.78	0.985	1.30
0.070	0.35	0.500	0.615	0.80
0.080	0.290	0.413	0.512	0.660
0.100	0.222	0.310	0.380	0.490
0.125	0.168	0.232	0.284	0.360
0.150	0.133	0.182	0.223	0.280
0.175	0.110	0.150	0.180	0.229
0.200	0.0935	0.128	0.153	0.190
0.225	0.0813	0.110	0.129	0.162
0.250	0.0710	0.096	0.114	0.141
0.300	0.0573	0.076	0.090	0.110
0.350	0.0475	0.0625	0.0735	0.0900
0.400	0.0400	0.0530	0.0625	0.0758
0.450	0.0351	0.0460	0.0538	0.0650
0.500	0.0308	0.040	0.047	0.0566
0.600	0.0245	0.0322	0.0371	0.0477
0.700	0.0206	0.0266	0.0307	0.0368
0.800	0.0175	0.0225	0.0260	0.0310
0.900	0.0151	0.0194	0.0225	0.0267
1.000	0.0134	0.0170	0.0197	0.0234
1.250	0.0102	0.0130	0.0150	0.0177
1.500	0.00827	0.0104	0.0120	0.0140
1.750	0.00686	0.00857	0.0098	0.0116
2.000	0.00586	0.00735	0.0084	0.00980
2.500	0.00453	0.0056	0.0064	0.00745
Range of speeds with good approximation	1 to 3 m/s	1 to 3 m/s	$\geqslant$ 1 m/s	$\geqslant$ 0.5 m/s

2.3. Tables of values of the loss of head J.

In practice the roughness coefficients used most often are either 0.1 mm or 2 mm, or an intermediate value such that it is only necessary to take the arithmetic mean of the two values of J corresponding to each of these coefficients k (see page 1052). The diameter of the metal pipes is standardized. The following tables give for current diameters the value of the loss of head through friction J under the least favourable assumption of water at a temperature near 0 °C with the maximum viscosity.

● **Partially-filled circular piping.**

Let: q (l/s) be the flow rate in a pipe D in diameter, sloping at p (mm/m) and filled to X % of its diameter, and

Q (l/s) the flow rate in the same pipe D in diameter, running full with a head loss p (mm/m) here equal to the slope (see head loss tables pages 1057 to 1068). Knowing D and p (and therefore Q), the unknown flow q is given by the equation:

$$q = mQ$$

m being given by the following table as a function of X

X (%)	20	25	30	35	40	45	50	55	60	65	70	75
m	0.08	0.13	0.185	0.25	0.32	0.40	0.50	0.58	0.67	0.74	0.82	0.89

● **Piping of rectangular cross-section.**

The hydraulic diameter of the cylindrical pipe equivalent to a cross-section a × b is:

$$D_h = \frac{2\,ab}{a+b}$$

TABLE *giving the* loss of head J in metres

as a function of the diameter of the pipe and the mean velocity:

(1) For new pipes (k = 0.1 mm) and (2) for existing pipes
(k = 2 mm) (*the temperature of the water being at* 0 °C.)

Mean velocity in metres/sec.	PIPE DIAMETER: 0.04 m Pipe section: 0.0012566 m²			PIPE DIAMETER: 0.05 m Pipe section: 0.0019635 m²		
	Head per metre of pipe length		Flow in litres/sec.	Head per metre of pipe length		Flow in litres/sec.
	New pipes	Existing pipes		New pipes	Existing pipes	
0.01			0.0125			0.0196
0.05			0.0628			0.0982
0.10			0.1256			0.1963
0.15			0,1884			0.2945
0.20	0.002115		0.2513	0.001590		0.3927
0.25	0.003138		0.3140	0.002358		0.4909
0.30	0.004329		0.3769	0.003248		0.5890
0.35	0.005694		0.4396	0.004281	0.008237	0.6872
0.40	0.007242		0.5024	0.005451	0.010690	0.7854
0.45	0.008966	0.018576	0.5652	0.006708	0.013468	0.8836
0.50	0.010861	0.022868	0.6280	0.008115	0.016587	0.9817
0.55	0.012895	0.027640	0.6908	0.009668	0.020046	1.0790
0.60	0.015116	0.032856	0.7538	0.011340	0.023826	1.1781
0.65	0,017493	0.038512	0.9164	0.013118	0.027924	1.2763
0.70	0.020072	0.044652	0.8792	0.015013	0.032374	1.3744
0.75	0.022793	0.051212	0.9420	0.017030	0.037128	1.4726
0.80	0.025647	0.058227	1.0048	0.019213	0.042210	1.5708
0.85	0.028681	0.065742	1.0676	0.021509	0.047658	1.6690
0.90	0.031845	0.073703	1.1304	0.023948	0.053429	1.7671
0.95	0.035190	0.082110	1.1932	0.026496	0.059524	1.8653
1.00	0.038546	0.090981	1.2566	0.029155	0.065955	1.9635
1.05	0.042143	0.100299	1.3194	0.031916	0.072710	2.0617
1.10	0.046021	0.110081	1.3822	0.034782	0.079801	2.1598
1.15	0.050052	0.120327	1.4451	0.037750	0.087229	2.2580
1.20	0.054224	0.131091	1.5079	0.040884	0.094980	2.3562
1.25	0.058535	0.142157	1.5707	0.044152	0.103054	2.4555
1.30	0.063011	0.153670	1.6335	0.047549	0.111465	2.5525
1.35	0.067647	0.165809	1.6963	0.051090	0.120200	2.6507
1.40	0.072428	0.178322	1.7592	0.054745	0.129271	2.7489
1.45	0.077423	0.191281	1.9220	0.058509	0.138665	2.8471
1.50	0.082570	0.204704	1.8846	0.062386	0.148396	2.9452
1.55	0.087865	0.218591	1.9474	0.066373	0.158463	3.0434
1 60	0.093293	0.232907	2.0105	0.070459	0.168841	3.1416
1.65	0.098874	0.247704	2.0733	0.074658	0.179568	3.2397
1.70	0.104657	0.262931	2.1362	0.078953	0.190606	3.3379
1.75	0.110597	0.278639	2.1990	0.083420	0.210993	3.4361
1.80	0.116671	0.294775	2.2608	0.088020	0.213691	3.5343
1.85	0.122893	0.311375	2.3236	0.092732	0.225725	3.6324
1.90	0.129260	0.328440	2.3864	0.097557	0.238096	3.7306
1.95	0.135764	0.345951	2.4499	0.102487	0.250790	3.8288
2.00	0.142410	0.363926	2.5132	0.107526	0.263821	3.9270
2.05	0.149244	0.382347	2.5760	0.112669	0.277175	4.0251
2.10	0.156222	0.401232	2.6388	0.117920	0.290865	4.1233
2.15	0.163337	0.420564	2.7016	0.123271	0.304879	4.2215
2.20	0.170586	0.440342	2.7645	0.128772	0.319217	4.3197
2.25	0.178048	0.460601	2.8273	0.134336	0.333904	4.4179
2.30	0.185708	0.481290	2.8888	0.140046	0.348901	4.5160
2.35	0.193518	0.502442	2.9516	0.145863	0.364235	4.6142
2.40	0.201476	0.524058	3.0158	0.151786	0.379905	4.7124
2.45	0.209576	0.546121	3.0786	0.157870	0.395899	4.8106
2.50	0.217815	0.568630	3.1412	0.164058	0.412217	4.9087
3.00	0.307923	0.818833	3.7698	0.233035	0.593597	5.8905
3.50	0.414432	1.114518	4.3981	0.312190	0.807948	6.8723
4.00	0.536204	1.455703	5.0264	0.404498	1.055283	7.8540

TABLE *giving the* loss of head J in metres

as a function of the diameter of the pipe and the mean velocity:

(1) For new pipes (k = 0.1 mm) and (2) for existing pipes
(k = 2 mm) (*the temperature of the water being at 0 °C*)

Mean velocity in metres/sec.	PIPE DIAMETER: 0.06 m — Pipe section: 0.00282744 m²			PIPE DIAMETER: 0.08 m — Pipe section: 0.0050265 m²		
	Head per metre of pipe length		Flow in litres/sec.	Head per metre of pipe lenght		Flow in litres/sec.
	New pipes	Existing pipes		New pipes	Existing pipes	
0.01			0.0283			0.0503
0.05			0.1414			0.2513
0.10			0.2827	0.000256		0.5027
0.15	0.000753		0.4241	0.000520		0.7540
0.20	0.001249		0.5655	0.000863	0.001438	1.0053
0.25	0.001856		0.7069	0.001280	0.002213	1.2566
0.30	0.002557	0.004713	0.8482	0.001775	0.003154	1.5080
0.35	0.003364	0.006354	0.9896	0.002336	0.004263	1.7593
0.40	0.004277	0.008262	1.1310	0.002994	0.005539	2.0106
0.45	0.005289	0.010406	1.2723	0.003702	0.006969	2.2620
0.50	0.006412	0.012803	1.4137	0.004467	0.008568	2.5133
0.55	0.007633	0.015466	1.5551	0.005339	0.010347	2.7646
0.60	0.008961	0.018374	1.6965	0.006274	0.012290	3.0159
0.65	0.010388	0.021530	1.8378	0.007280	0.014396	3.2672
0.70	0.011907	0.024955	1.9792	0.008353	0.016680	3.5186
0.75	0.013523	0.028612	2.1206	0.009450	0.019120	3.7699
0.80	0.015223	0.032522	2.2619	0.010646	0.021733	4.0212
0.85	0.017034	0.036682	2.4033	0.011910	0.024515	4.2726
0.90	0.018959	0.041084	2.5447	0.013249	0.027458	4.5239
0.95	0.020968	0.045771	2.6861	0.014651	0.030590	4.7752
1.00	0.023064	0.050715	2.8274	0.016119	0.033895	5.0266
1.05	0.025257	0.055909	2.9688	0.017644	0.037367	5.2779
1.10	0.027556	0.061361	3.1102	0.019241	0.041011	5.5192
1.15	0.029941	0.067073	3.2516	0.020906	0.044828	5.7805
1.20	0.032418	0.073033	3.3929	0.022635	0.048811	6.0319
1.25	0.034975	0.079242	3.5343	0.024420	0.052961	6.2832
1.30	0.037615	0.085709	3.6757	0.026273	0.057283	6.5345
1.35	0.040392	0.092426	3.8170	0.028181	0.061772	6.7858
1.40	0.043257	0.099401	3.9584	0.030145	0.066434	7.0372
1.45	0.046204	0.106624	4.0998	0.032175	0.071261	7.2885
1.50	0.049255	0.114106	4.2412	0.034261	0.076262	7.5398
1.55	0.052392	0.121848	4.3825	0.036478	0.081436	7.7911
1.60	0.055606	0.129828	4.5239	0.038753	0.086769	8.0425
1.65	0.058908	0.138076	4.6653	0.041093	0.092283	8.2937
1.70	0.062308	0.146564	4.8066	0.043490	0.097955	8.5451
1.75	0.065796	0.155320	4.9480	0.045952	0.103807	8.7965
1.80	0.069359	0.164314	5.0894	0.048489	0.109818	9.0478
1.85	0.073003	0.173568	5.2368	0.051089	0.116003	9.2991
1.90	0.076759	0.183080	5.3721	0.053751	0.122360	9.5505
1.95	0.080625	0.192841	5.5135	0.056472	0.128884	9.8018
2.00	0.084576	0.202861	5.6549	0.059253	0.135580	10.0531
2.05	0.088607	0.213129	5.7963	0.062118	0.142443	10.3044
2.10	0.092722	0.223646	5.9376	0.065046	0.149479	10.5558
2.15	0.096914	0.234432	6.0790	0.068032	0.156680	10.8071
2.20	0.101266	0.245457	6.2204	0.071078	0.164049	11.0584
2.25	0.105710	0.256749	6.3617	0.074187	0.171597	11.3097
2.30	0.110234	0.268282	6.5031	0.077350	0.179304	11.5610
2.35	0.114844	0.280072	6.6445	0.080574	0.187184	11.8124
2.40	0.119540	0.292122	6.7859	0.083857	0.195238	12.0637
2.45	0.124318	0.304420	6.9272	0.087196	0.203457	12.3150
2.50	0.129176	0.316967	7.0686	0.090591	0.211842	12.5664
3.00	0.183110	0.456436	8.4823	0.128731	0.305056	15.0795
3.50	0.246110	0.621258	9.8960	0.172875	0.415213	17.5928
4.00	0.318732	0.811442	11.3098	0.224268	0.542321	20.1060

TABLE *giving the* loss of head J in metres

as a function of the diameter of the pipe and the mean velocity:

(1) For new pipes (k = 0.1 mm) and (2) for existing pipes
(k = 2 mm) (*the temperature of the water being at* 0 °C)

Mean velocity in metres/sec.	PIPE DIAMETER: 0.10 m Pipe section: 0.007854 m²			PIPE DIAMETER: 0.125 m Pipe section: 0.012272 m²		
	Head per metre of pipe length		Flow in litres/sec.	Head per metre of pipe length		Flow in litres/sec.
	New pipes	Existing pipes		New pipes	Existing pipes	
0.01			0.0785			0.1227
0.05			0.3927			0.6136
0.10	0.000191		0.7854	0.000144	0.000207	1.2272
0.15	0.000388	0.000604	1.1781	0.000291	0.000449	1.8408
0.20	0.000643	0.001054	1.5708	0.000486	0.000783	2.4544
0.25	0.000956	0.001622	1.9635	0.000726	0.001204	3.0680
0.30	0.001335	0.002312	2.3562	0.001009	0.001712	3.6816
0.35	0.001763	0.003120	2.7489	0.001330	0.002311	4.2952
0.40	0.002248	0.004060	3.1416	0.001701	0.003004	4.9088
0.45	0.002786	0.005111	3.5343	0.002104	0.003785	5.5224
0.50	0.003370	0.006281	3.9270	0.002548	0.004656	6.1360
0.55	0.004009	0.007584	4.3197	0.003037	0.005618	6.7496
0.60	0.004707	0.009006	4.7124	0.003560	0.006568	7.3632
0.65	0.005447	0.010543	5.1051	0.004120	0.007804	7.9768
0.70	0.006245	0.012215	5.4978	0.004726	0.009037	8.5904
0.75	0.007090	0.014000	5.8905	0.005369	0.010356	9.2040
0.80	0.007985	0.015911	6.1830	0.006059	0.011769	9.8176
0.85	0.008931	0.017951	6.6759	0.006765	0.013279	10.4312
0.90	0.009930	0.020108	7.0686	0.007531	0.014878	11.0448
0.95	0.010980	0.022402	7.4613	0.008332	0.016567	11.6584
1.00	0.012080	0.024822	7.8540	0.009166	0.018349	12.2720
1.05	0.013233	0.027365	8.2467	0.010047	0.020228	12.8866
1.10	0.014431	0.030033	8.6394	0.010962	0.022201	13.4992
1.15	0.015673	0.032829	9.0321	0.011913	0.024268	14.1128
1.20	0.016555	0.035786	9.4248	0.012091	0.026424	14.7264
1.25	0.018301	0.038785	9.8175	0.013921	0.028670	15.8400
1.30	0.019692	0.041950	10.2102	0.014988	0.031010	15.9536
1.35	0.021142	0.045237	10.6029	0.016089	0.033440	16.5672
1.40	0.022637	0.048651	10.9956	0.017231	0.035964	17.1808
1.45	0.024197	0.052197	11.3883	0.018406	0.038578	17.7944
1.50	0.025803	0.055849	11.7810	0.019615	0.041285	18.4080
1.55	0.027456	0.059638	12.1737	0.020857	0.044086	19,0216
1.60	0.029149	0.063544	12.5664	0.022140	0.046973	19.6352
1.65	0.030890	0.067581	12.9591	0.023458	0.049957	20.2488
1.70	0.032671	0.071735	13.3518	0.024805	0.053028	20.8624
1.75	0.034514	0.076021	13.7445	0.026200	0.056196	21.4760
1.80	0.036397	0.080423	14.1372	0.027625	0.059450	22.0896
1.85	0.038324	0.084952	14.5293	0.029097	0.062798	22.7032
1.90	0.040296	0.089608	14.9226	0.030588	0.066240	23.3168
1.95	0.042347	0.094385	15.3153	0.032126	0.069772	23.9304
2.00	0.044446	0.099290	15.7081	0.033714	0.073397	25.5440
2.05	0.046589	0.104315	16.1007	0.035335	0.077112	25.1576
2.10	0.048777	0.109468	16.4934	0.036990	0.080921	25.7712
2.15	0.051010	0.114742	16.8861	0.038678	0.084820	26.3848
2.20	0.053285	0.120138	17.2788	0.040437	0.088808	26.9984
2.25	0.055608	0.125665	17.6715	0.042236	0.092894	27.6120
2.30	0.057970	0.131310	18.0642	0.044068	0.097057	28.2256
2.35	0.060377	0.137081	18.4569	0.045960	0.101333	28.8392
2.40	0.062828	0.142978	18.8496	0.047890	0.105692	29.4538
2.45	0.065320	0.148998	19.2423	0.049858	0.110142	30.0664
2.50	0.065853	0.155139	19.6350	0.051862	0.114682	30.6800
3.00	0.096333	0.223402	23.5620	0.073580	0.165143	36.816
3.50	0.129559	0.304073	27.4890	0.098802	0.224777	42.952
4.00	0.167589	0.397152	31.4160	0.128004	0.293587	49.088

TABLE *giving the* **loss of head J in metres**

as a function of the diameter of the pipe and the mean velocity:

(1) For new pipes (k = 0.1 mm) and (2) for existing pipes
(k = 2 mm) (*the temperature of the water being at 0 °C*)

Mean velocity in metres/sec.	PIPE DIAMETER: 0.150 m Pipe section: 0.0176725 m²			PIPE DIAMETER: 0.200 m Pipe section: 0.031416 m²		
	Head per metre of pipe length		Flow in litres/sec.	Head per metre of pipe length		Flow in litres/sec.
	New pipes	Existing pipes		New pipes	Existing pipes	
0.01			0.1767			0.3142
0.05	0.000034		0.8836	0.000024	0.000030	1.5708
0.10	0.000114	0.000163	1.7671	0.000079	0.000110	3.1416
0.15	0.000232	0.000352	2.6507	0.000162	0.000238	4.7424
0.20	0.000387	0.000612	3.5343	0.000270	0.000413	6.2832
0.25	0.000578	0.000941	4.4179	0.000400	0.000636	7.8540
0.30	0.000801	0.001336	5.3014	0.000557	0.000903	9.4248
0.35	0.001059	0.001810	6.1850	0.000736	0.001217	10.9956
0.40	0.001351	0.002347	7.0686	0.000940	0.001581	12.5664
0.45	0.001674	0.002948	7.9522	0.001169	0.001989	14.1372
0.50	0.002031	0.003622	8.8357	0.001421	0.002445	15.7080
0.55	0.002421	0.004374	9.7193	0.001692	0.002945	17.2788
0.60	0.002842	0.005187	10.6029	0.001986	0.003491	18.8496
0.65	0.003293	0.006070	11.4865	0.002298	0.004080	20.4204
0.70	0.003777	0.007028	12.3700	0.002642	0.004734	21.9912
0.75	0.004289	0.008054	13.2536	0.002996	0.005433	23.5620
0.80	0.004834	0.009155	14.1372	0.003376	0.006181	25.1328
0.85	0.005411	0.010329	15.0208	0.003784	0.006979	26.7036
0.90	0.006017	0.011572	15.9043	0.004212	0.007824	28.2744
0.95	0.006652	0.012883	16.7879	0.004658	0.008717	29.8452
1.00	0.007316	0.014268	17.6715	0.005122	0.009659	31.4160
1.05	0.008009	0.015722	18.5550	0.005619	0.010648	32.9868
1.10	0.008732	0.017247	19.4386	0.006139	0.011686	34.5576
1.15	0.009487	0.018852	20.3222	0.006680	0.012774	36.1284
1.20	0.010271	0.020527	21.2058	0.007241	0.013909	37.6992
1.25	0.011086	0.022273	22.0893	0.007821	0.015092	39.2700
1.30	0.011933	0.024091	22.9729	0.008424	0.016324	40.8048
1.35	0.012813	0.025978	23.8565	0.009047	0.017603	42.4116
1.40	0.013726	0.027939	24.7401	0.009695	0.018931	43.9824
1.45	0.014667	0.029970	25.6237	0.010362	0.020307	45.5532
1.50	0.015642	0.032072	26.5072	0.011049	0.021737	47.1240
1.55	0.016646	0.034248	27.3908	0.011756	0.023206	48.6948
1.60	0.017684	0.036491	28.2744	0.012480	0.024726	50.2656
1.65	0.018752	0.038809	29.1580	0.013232	0.026297	51.8364
1.70	0.019846	0.041195	30.0415	0.014010	0.027913	53.4072
1.75	0.020970	0.043656	30.9251	0.014790	0.029581	54.9780
1.80	0.022129	0.046184	31.8087	0.015597	0.031294	56.5488
1.85	0.023317	0.048785	32.6922	0.016424	0.033056	58.1196
1.90	0.024533	0.051459	33.5758	0.017268	0.034868	59.6904
1.95	0.025777	0.054202	34.4594	0.018141	0.036727	61.2612
2.00	0.027062	0.057018	35.3430	0.019032	0.038635	62.8320
2.05	0.028374	0.059905	36.2265	0.019942	0.040591	64.4028
2.10	0.029716	0.062863	37.1101	0.020882	0.042596	65.9736
2.15	0.031085	0.065892	37.9937	0.021841	0.044548	67.5444
2.20	0.032497	0.068991	38.8772	0.022831	0.046748	69.1152
2.25	0.033941	0.072165	39.7608	0.023843	0.048899	70.6860
2.30	0.035411	0.075406	40.6444	0.024873	0.051095	72.2568
2.35	0.036911	0.078720	41.5279	0.025924	0.053340	73.8276
2.40	0.038441	0.082107	42.4115	0.026981	0.055635	75.3984
2.45	0.039998	0.085564	43.2951	0.028071	0.057978	76.9692
2.50	0.041583	0.089090	44.1787	0.029180	0.060367	78.5400
3.00	0.059023	0.128291	53.0145	0.041400	0.086929	94.2480
3.50	0.079296	0.174618	61.8503	0.055757	0.118320	109.956
4.00	0.102483	0.228073	70.6860	0.072051	0.154541	125.664

TABLE *giving the* loss of head J in metres

as a function of the diameter of the pipe and the mean velocity:

(1) For new pipes (k = 0.1 mm) and (2) for existing pipes
(k = 2 mm) (*the temperature of the water being at 0 °C*)

Mean velocity in metres/sec.	PIPE DIAMETER: 0.250 m Pipe section: 0.0490875 m²			PIPE DIAMETER: 0.300 m Pipe section: 0.070686 m²		
	Head per metre of pipe length		Flow in litres/sec.	Head per metre of pipe length		Flow in litres/sec.
	New pipes	Existing pipes		New pipes	Existing pipes	
0.01			0.4909			0.7069
0.05	0.000017	0.000022	2.4544	0.000014	0.000018	3.5343
0.10	0.000060	0.000081	4.9087	0.000048	0.000064	7.0686
0.15	0.000122	0.000175	7.3631	0.000097	0.000139	10.6029
0.20	0.000204	0.000305	9.8175	0.000163	0.000241	14.1372
0.25	0.000303	0.000469	12.2719	0.000244	0.000370	17.6715
0.30	0.000424	0.000668	14.7262	0.000339	0.000527	21.2058
0.35	0.000563	0.000902	17.1806	0.000450	0.000711	24.7401
0.40	0.000720	0.001173	19.6350	0.000574	0.000925	28.2744
0.45	0.000890	0.001477	22.0894	0.000712	0.001164	31.8087
0.50	0.001080	0.001815	24.5437	0.000864	0.001431	35.3430
0.55	0.001286	0.002188	26.9981	0.001031	0.001725	38.8773
0.60	0.001512	0.002594	29.4525	0.001215	0.002046	42.4116
0.65	0.001753	0.003034	31.9069	0.001411	0.002393	45.9459
0.70	0.002013	0.003511	34.3612	0.001622	0.002769	49.4802
0.75	0.002294	0.004024	36.8156	0.001845	0.003170	53.0145
0.80	0.002586	0.004573	39.2700	0.002079	0.003603	56.5488
0.85	0.002896	0.005159	41.7244	0.002326	0.004064	60.0831
0.90	0.003226	0.005781	44.1787	0.002588	0.004556	63.6174
0.95	0.003571	0.006440	46.6331	0.002866	0.005076	67.1517
1.00	0.003935	0.007136	49.0875	0.003157	0.005624	70.6860
1.05	0.004315	0.007867	51.5418	0.003461	0.006200	74.2203
1.10	0.004712	0.008634	53.9962	0.003778	0.006804	77.7546
1.15	0.005123	0.009437	56.4506	0.004110	0.007438	81.2889
1.20	0.005555	0.010276	58.9050	0.004453	0.008099	84.8232
1.25	0.006002	0.011150	61.3593	0.004808	0.008787	88.3575
1.30	0.006464	0.012060	63.8137	0.005177	0.009504	91.8918
1.35	0.006944	0.013005	66.2681	0.005561	0.010249	95.4261
1.40	0.007441	0.013986	68.7225	0.005957	0.011022	98.9604
1.45	0.007956	0.015002	71.1769	0.006365	0.011823	102.4947
1.50	0.008486	0.016055	73.6312	0.006785	0.012653	106.0290
1.55	0.009033	0.017144	76.0856	0.007217	0.013511	109.5633
1.60	0.009593	0.018267	78.5400	0.007659	0.014397	113.0976
1.65	0.010169	0.019428	80.9944	0.008123	0.015311	116.6319
1.70	0.010759	0.020622	83.4487	0.008602	0.016252	120.1662
1.75	0.011364	0.021854	85.9031	0.009090	0.017223	123.7005
1.80	0.011989	0.023120	88.3575	0.009595	0.018221	127.2348
1.85	0.012629	0.024422	90.8118	0.010106	0.019247	130.7691
1.90	0.013285	0.025760	93.2662	0.010635	0.020302	134.3034
1.95	0.013954	0.027133	95.7206	0.011170	0.021384	137.8377
2.00	0.014639	0.028543	98.1750	0.011723	0.022495	141.3720
2.05	0.015345	0.029988	100.6293	0.012288	0.023633	144.9063
2.10	0.016067	0.031469	103.0837	0.012865	0.024801	148.4406
2.15	0.016804	0.032985	105.5381	0.013461	0.025996	151.9749
2.20	0.017564	0.034537	107.9924	0.014070	0.027218	155.5092
2.25	0.018341	0.036126	110.4468	0.014691	0.028470	159.0435
2.30	0.019133	0.037748	112.9012	0.015324	0.029749	162.5778
2.35	0.019940	0.039407	115.3555	0.015969	0.031057	166.1121
2.40	0.020763	0.041103	117.8099	0.016627	0.032393	169.6464
2.45	0.021600	0.042833	120.2643	0.017296	0.033756	173.1807
2.50	0.022465	0.044598	122.7187	0.017988	0.035148	176.7150
3.00	0.031873	0.064222	147.2625	0.025490	0.050613	212.058
3.50	0.042907	0.087413	171.8063	0.034341	0.068890	247.401
4.00	0.055455	0.114173	196.3500	0.044527	0.089979	282.744

TABLE *giving the* **loss of head J in metres**

as a function of the diameter of the pipe and the mean velocity:

(1) For new pipes (k = 0.1 mm) and (2) for existing pipes
(k = 2 mm) (*the temperature of the water being at* 0 °C.)

Mean velocity in metres/sec.	PIPE DIAMETER: 0.350 m Pipe section: 0.0962115 m²			PIPE DIAMETER: 0.400 m Pipe section: 0.125664 m²		
	Head per metre of pipe length		Flow in litres/sec.	Head per metre of pipe length		Flow in litres/sec.
	New pipes	Existing pipes		New pipes	Existing pipes	
0.01			0.0621			1.2566
0.05	0.000011	0.000014	4.8106	0.000010	0.000012	6.2832
0.10	0.000039	0.000052	9.6211	0.000033	0.000044	12.5664
0.15	0.000081	0.000112	14.4317	0.000068	0.000094	18.8496
0.20	0.000135	0.000195	19.2423	0.000115	0.000164	25.1328
0.25	0.000203	0.000298	24.0529	0.000172	0.000253	31.4160
0.30	0.000282	0.000425	28.8634	0.000239	0.000360	37.6992
0.35	0.000374	0.000574	33.6740	0,000317	0.000485	43.9824
0.40	0.000477	0.000747	38.4846	0.000406	0.000631	50.2656
0.45	0.000594	0.000941	43.2952	0.000506	0.000795	56.5488
0.50	0.000721	0.001157	48.1057	0.000615	0.000978	62.8320
0.55	0.000860	0.001396	52.9163	0.000732	0.001180	69.1152
0.60	0.001009	0.001657	57.7269	0.000858	0.001400	75.3984
0.65	0.001172	0.001942	62.5375	0.000996	0.001640	81.6816
0.70	0.001348	0.002252	67.3480	0.001146	0.001899	87.9648
0.75	0.001533	0.002584	72.1586	0.001305	0.002177	94.2480
0.80	0.001730	0.002940	76.9692	0.001472	0.002473	100.5312
0.85	0.001936	0.003320	81.7798	0.001648	0.002790	106.8144
0.90	0.002153	0.003722	86.5903	0,001832	0.003128	113.0976
0.95	0.002383	0.004147	81.4009	0.002026	0.003485	119.3808
1.00	0.002626	0.004595	96.2115	0.002233	0.003861	125.6640
1.05	0.002878	0.005065	101.0221	0.002447	0,004257	131.9472
1.10	0.003142	0.005559	105.8326	0.002672	0.004672	138.2304
1.15	0.003417	0.006077	110.6432	0.002905	0.005106	144.5136
1.20	0.003701	0.006616	115.4538	0.003147	0.005560	150.7968
1.25	0.003998	0.007179	120.2644	0.003399	0.006033	157.0800
1.30	0.004304	0.007765	125.0749	0.003659	0.006525	163.3632
1.35	0.004623	0.008373	129.8855	0.003929	0.007037	169.6464
1.40	0.004952	0.009005	134.6961	0.004208	0.007567	175.9296
1.45	0.005291	0.009660	139.5067	0.004498	0.008117	182.2128
1.50	0.005642	0.010338	144.3172	0.004796	0.008687	188.4960
1.55	0.006004	0.011039	149.1278	0.005107	0.009276	194.7792
1.60	0.006375	0.011762	153.9384	0.005425	0.009884	201.0624
1.65	0.006760	0.012509	158.7490	0.005752	0.010512	207.3456
1.70	0.007155	0.013278	163.5595	0.006087	0.011158	213.6288
1.75	0.007560	0.014071	168.3701	0.006431	0.011825	219.9120
1.80	0.007979	0.014886	173.1807	0.006783	0.012509	226.1952
1.85	0.008403	0.015725	177.9913	0.007143	0.013214	232.4784
1.90	0.008852	0.016586	182.8018	0.007516	0.013938	238.7116
1.95	0.009286	0.017470	178.6124	0.007898	0.014781	245.0448
2.00	0.009745	0.018378	192.4230	0.008288	0.015444	251.3280
2.05	0.010214	0.019309	197.2336	0.008686	0.016226	257.6112
2.10	0.010693	0.020262	202.0441	0.009092	0.017027	263.8944
2.15	0.011188	0.021239	206.8547	0.009513	0.017848	269.1776
2.20	0.011693	0.022237	211.6653	0.009942	0.018687	276.4608
2.25	0.012209	0.023261	216.4759	0.010380	0.019547	282.7440
2.30	0.012734	0.024305	221.2864	0.010826	0.020425	289.0272
2.35	0.013270	0.025373	226.0970	0.011280	0.021322	295.3104
2.40	0.013816	0.026465	230.9076	0.011744	0.022240	301.5936
2.45	0.014371	0.027579	235.7182	0.012215	0.023176	307.8768
2.50	0.014945	0.028716	240.5287	0.012695	0.024131	314.1600
3.00	0.012167	0.041351	288.6345	0.017971	0.034749	376.992
3.50	0.028543	0.056283	336.7403	0.024273	0.047297	439.824
4.00	0.036908	0.073513	384.8460	0.031296	0.061276	502.656

TABLE *giving the* loss of head J in metres

as a function of the diameter of the pipe and the mean velocity:

(1) For new pipes (k = 0.1 mm) and (2) for existing pipes
(k = 2 mm) *(the temperature of the water being at 0 °C.)*

Mean velocity in metres/sec.	PIPE DIAMETER: 0.450 m Pipe section: 0.1590435 m²			PIPE DIAMETER: 0.500 m Pipe section: 0.19635 m²		
	Head per metre of pipe length		Flow in litres/sec.	Head per metre of pipe length		Flow in litres/sec.
	New pipes	Existing pipes		New pipes	Existing pipes	
0.01			1.5904			1.9635
0.05	0.000008	0.000010	7.9522	0.000007	0.000009	5.8175
0.10	0.000029	0.000037	15.9043	0.000025	0.000033	19.6350
0.15	0.000059	0.000081	23.8565	0.000052	0.000070	29.4525
0.20	0.000099	0.000141	31.8087	0.000088	0.000123	39.2700
0.25	0.000149	0.000217	39.7609	0.000131	0.000189	49.0875
0.30	0.000207	0.000309	47.7130	0.000182	0.000270	58.9056
0.35	0.000275	0.000418	55.6652	0.000242	0.000365	68.7225
0.40	0.000352	0.000543	63.6174	0.000310	0.000474	78.5400
0.45	0.000438	0.000684	71.5696	0.000386	0.000597	88.3575
0.50	0.000533	0.000841	79.5217	0.000469	0.000735	98.1750
0.55	0.000636	0.001016	87.4739	0.000560	0.000887	107.9925
0.60	0.000746	0.001205	95.4261	0.000658	0.001053	117.8100
0.65	0.000865	0.001412	103.3783	0.000763	0.001233	127.6275
0.70	0.000994	0.001634	111.3304	0.000875	0.001427	137.4450
0.75	0.001131	0.001872	119.2826	0.000995	0.001635	147.2625
0.80	0.001276	0.002127	127.2348	0.001123	0.001856	157.0800
0.85	0.001429	0.002399	135.1870	0.001258	0.002093	166.8975
0.90	0.001589	0.002688	143.1391	0.001400	0.002343	176.7150
0.95	0.001757	0.002991	151.0913	0.001548	0.002606	186.5325
1.00	0.001936	0.003313	159.0435	0.001704	0.002885	196.3500
1.05	0.002122	0.003652	166.9957	0.001869	0.003180	306.1675
1.10	0.002316	0.004008	174.9478	0.002040	0.003491	215.9850
1.15	0.002520	0.004382	182.9000	0.002219	0.003815	225.8025
1.20	0.002730	0.004771	190.8522	0.002405	0.004154	235.6200
1.25	0.002948	0.005177	198.8044	0.002596	0.004508	245.4375
1.30	0.003174	0.005599	206.7565	0.002794	0.004876	255.2550
1.35	0.003408	0.006038	214.7087	0.003000	0.005258	265.0725
1.40	0.003650	0.006494	222.6609	0.003213	0.005654	274.8900
1.45	0.003901	0.006965	230.6131	0.003436	0.006065	284.7075
1.50	0.004162	0.007454	238.5652	0.003665	0.006491	294.5250
1.55	0.004430	0.007960	246.5174	0.003902	0.006931	304.3425
1.60	0.004706	0.008481	254.4696	0.004144	0.007385	314.1600
1.65	0.004990	0.009020	262.4218	0.004393	0.007854	323.9775
1.70	0.005280	0.009574	270.3739	0.004649	0.008337	333.7950
1.75	0.005578	0.010147	278.3261	0.004911	0.008835	343.6125
1.80	0.005883	0.010734	286.2783	0.005179	0.009347	353.4300
1.85	0.006194	0.011338	294.2305	0.005456	0.009873	363.2475
1.90	0.006518	0.011960	302.1826	0.005741	0.010414	373.0650
1.95	0.006848	0.012598	310.1348	0.006031	0.010970	382.8825
2.00	0.007186	0.013252	318.0870	0.006328	0.011540	392.7000
2.05	0.007530	0.013923	326.0392	0.006632	0.012124	402.5175
2.10	0.007887	0.014611	333.9913	0.006946	0.012723	412.3350
2.15	0.008252	0.015315	341.9435	0.007266	0.013336	422.1525
2.20	0.008623	0.016035	349.8957	0.007593	0.013963	431.7900
2.25	0.009003	0.016773	357.8479	0.007927	0.014605	441.7875
2.30	0.009389	0.017526	365.8000	0.008267	0.015261	451.6050
2.35	0.009783	0.018296	373.7522	0.008613	0.015932	461.4225
2.40	0.010184	0.019083	381.5044	0.008966	0.016617	471.2400
2.45	0.010593	0.019887	389.6566	0.009315	0.017317	481.0575
2.50	0.011008	0.020706	397.6087	0.009697	0.018030	490.8750
3.00	0.015607	0.029817	477.1305	0.013762	0.025964	589.05
3.50	0.021035	0.040585	556.6523	0.018519	0.035340	687.225
4.00	0.027034	0.053009	636.174	0.023976	0.046158	785.4

TABLE *giving the* loss of head J in metres

as a function of the diameter of the pipe and the mean velocity:

(1) For new pipes (k = 0.1 mm) and (2) for existing pipes
(k = 2 mm) *(the temperature of the water being at 0 °C.)*

Mean velocity in metres/sec.	PIPE DIAMETER: 0.600 m Pipe section: 0.282744 m²			PIPE DIAMETER: 0.700 m Pipe section: 0.384646 m²		
	Head per metre of pipe length		Flow in litres/sec.	Head per metre of pipe length		Flow in litres/sec.
	New pipes	Existing pipes		New pieds	Existing pipes	
0.01			2.8274			3.8484
0.05	0.000006	0.000007	14.1372	0.000005	0.000006	19.2423
0.10	0.000020	0.000026	28.2744	0.000017	0.000022	38.4846
0.15	0.000041	0.000056	42.4116	0.000034	0.000047	57.7269
0.20	0.000068	0.000095	56.5488	0.000057	0.000080	76.9682
0.25	0.000105	0.000149	70.6860	0.000087	0.000123	96.2115
0.30	0.000146	0.000212	84.8232	0.000121	0.000175	115.4538
0.35	0.000193	0.000287	98.9604	0.000160	0.000236	134.6961
0.40	0.000247	0.000372	113.0976	0.000205	0.000308	153.9384
0.45	0.000307	0.000469	127.2348	0.000255	0.000387	173.1807
0.50	0.000372	0.000577	141.3720	0.000473	0.000473	192.4230
0.55	0.000443	0.000697	155.5092	0.000368	0.000571	211.6653
0.60	0.000521	0.000827	169.6464	0.000433	0.000679	230.9076
0.65	0.000605	0.000969	183.7836	0.000502	0.000795	250.1499
0.70	0.000695	0.001122	197.9208	0.000576	0.000921	269.2922
0.75	0.000790	0.001287	212.0580	0.000655	0.001057	288.6345
0.80	0.000890	0.001463	226.1952	0.000738	0.001202	307.8768
0.85	0.000996	0.001651	240.3324	0.000826	0.001358	327.1191
0.90	0.001107	0.001849	254.4696	0.000917	0.001521	346.3614
0.95	0.001221	0.002059	268.6068	0.001015	0.001681	365.6037
1.00	0.001341	0.002279	282.7440	0.001117	0.001880	384.8460
1.05	0.001472	0.002513	296.8812	0.001224	0.002068	404.0883
1.10	0.001609	0.002758	311.0184	0.001338	0.002272	423.3306
1.15	0.001750	0.003014	325.1556	0.001454	0.002482	442.5729
1.20	0.001897	0.003282	339.2928	0.001562	0.002701	461.8152
1.25	0.002049	0.003561	353.4300	0.001688	0.002934	481.0575
1.30	0.002208	0.003852	367.5672	0.001817	0.003175	500.3998
1.35	0.002372	0.004154	381.7044	0.001946	0.003420	519.5421
1.40	0.002541	0.004467	395.8416	0.002084	0.003680	538.7844
1.45	0.002715	0.004792	409.9788	0.002225	0.003950	558.0267
1.50	0.002896	0.005128	424.1160	0.002376	0.004223	577.2690
1.55	0.003082	0.005476	438.2532	0.002528	0.004512	596.5113
1.60	0.003273	0.005835	452.3904	0.002681	0.004884	615.7536
1.65	0.003469	0.006205	466.5276	0.002843	0.005115	634.9995
1.70	0.003673	0.006587	480.6648	0.003012	0.005437	654.2382
1.75	0.003879	0.006980	494.8020	0.003181	0.005750	673.4805
1.80	0.004090	0.007384	508.9392	0.003356	0.006079	692.7228
1.85	0.004309	0.007800	523.0764	0.003530	0.006424	711.9651
1.90	0.004533	0.008228	537.2136	0.003714	0.006775	731.2074
1.95	0.004761	0.008666	551.3508	0.003901	0.007146	750.4497
2.00	0.004995	0.009117	565.4882	0.004088	0.007508	769.6920
2.05	0.005234	0.009587	579.6250	0.004286	0.007895	788.9343
2.10	0.005477	0.010051	593.7624	0.004484	0.008277	808.1666
2.15	0.005729	0.010536	607.8996	0.004686	0.008673	827.4189
2.20	0.005986	0.011031	622.0368	0.004889	0.009089	846.6612
2.25	0.006249	0.011539	636.1740	0.005103	0.009502	865.9035
2.30	0.006516	0.012057	650.3112	0.005322	0.009944	885.1458
2.35	0.006788	0.012587	664.4484	0.005547	0.010451	904.3881
2.40	0.007066	0.013128	678.5856	0.005773	0.010806	923.6304
2.45	0.007353	0.013681	692.7228	0.006010	0.011268	942.8727
2.50	0.007645	0.014245	706.8600	0.006248	0.011733	962.1150
3.00	0.010841	0.020513	848.232	0.008867	0.016901	1,154.5380
3.50	0.014610	0.027920	989.604	0.011925	0.022997	1,346.9610
4.00	0.018893	0.036476	1,130.967	0.015436	0.030410	1,539.3840

TABLE *giving the* **loss of head J in metres**

as a function of the diameter of the pipe and the mean velocity:

(1) For new pipes (k = 0.1 mm) and (2) for existing pipes
(k = 2 mm) (*the temperature of the water being at 0 °C.*)

Mean velocity in metres/sec.	PIPE DIAMETER: 0.800 m Pipe section: 0.502656 m²			PIPE DIAMETER: 0.900 m Pipe section: 0.636174 m²		
	Head per metre of pipe length		Flow in litres/sec.	Head per metre of pipe length		Flow in litres/sec.
	New pipes	Existing pipes		New pipes	Existing pipes	
0.01			5.0205			6.3617
0.05	0.000004	0.000005	25.1328	0.000004	0.000005	31.8087
0.10	0.000014	0.000018	50.2656	0.000012	0.000015	63.6174
0.15	0.000029	0.000039	75.3984	0.000025	0.000034	95.4261
0.20	0.000049	0.000067	100.5312	0.000043	0.000058	127.2348
0.25	0.000074	0.000103	125.6640	0.000064	0.000087	159.0435
0.30	0.000103	0.000147	150.7968	0.000089	0.000124	190.8522
0.35	0.000137	0.000198	175.9296	0.000167	0.000167	222.6609
0.40	0.000174	0.000258	201.0624	0.000150	0.000218	254.4696
0.45	0.000216	0.000324	226.1952	0.000186	0.000274	286.2783
0.50	0.000262	0.000398	251.3280	0.000225	0.000336	318.0870
0.55	0.000312	0.000481	275.4608	0.000268	0.000406	349.8957
0.60	0.000367	0.000572	301.5936	0.000316	0.000483	381.7044
0.65	0.000425	0.000670	326.7264	0.000367	0.000565	413.5131
0.70	0.000489	0.000776	351.8592	0.000421	0.000654	445.3218
0.75	0.000557	0.000890	376.9920	0.000479	0.000749	477.1305
0.80	0.000628	0.001012	402.1248	0.000540	0.000852	508.9392
0.85	0.000703	0.001142	427.2576	0.000605	0.000961	540.7479
0.90	0.000781	0.001279	452.3904	0.000671	0.001077	572.5566
0.95	0.000864	0.001425	477.5232	0.000743	0.001199	604.3653
1.00	0.000952	0.001579	502.6560	0.000817	0.001327	636.1740
1.05	0.001044	0.001741	527.7888	0.000896	0.001461	667.9827
1.10	0.001139	0.001910	552.9216	0.000980	0.001606	699.7914
1.15	0.001239	0.002088	578.0544	0.001065	0.001752	731.6001
1.20	0.001341	0.002274	603.1872	0.001144	0.001910	763.4088
1.25	0.001448	0.002467	628.3200	0.001237	0.002073	795.2175
1.30	0.001559	0.002668	653.4528	0.001332	0.002241	827.0262
1.35	0.001673	0.002877	678.5856	0.001428	0.002420	858.8349
1.40	0.001791	0.003095	703.7184	0.001529	0.002604	890.6436
1.45	0.001914	0.003319	728.8512	0.001632	0.002787	922.4523
1.50	0.002041	0.003552	753.9840	0.001741	0.002983	954.2610
1.55	0.002174	0.003793	779.1168	0.001857	0.003186	986.0697
1.60	0.002309	0.004042	804.2496	0.001968	0.003398	1,017.8784
1.65	0.002449	0.004298	829.3824	0.002086	0.003610	1,049.6871
1.70	0.002593	0.004563	854.5152	0.002208	0.003837	1,081.4958
1.75	0.002740	0.004835	879.6480	0.002337	0.004061	1,113.3045
1.80	0.002890	0.005115	904.7808	0.002461	0.004299	1,145.1132
1.85	0.003044	0.005403	926.9136	0.002594	0.004538	1,176.9219
1.90	0.003202	0.005699	955.0464	0.002726	0.004792	1,208.7306
1.95	0.003363	0.006003	980.1792	0.002862	0.005044	1,240.5393
2.00	0.003530	0.006315	1,005.3120	0.003001	0.005307	1,272.3480
2.05	0.003700	0.006635	1,030.4448	0.003144	0.005578	1,304.1567
2.10	0.003875	0.006963	1,055.5776	0.003296	0.005850	1,335.9654
2.15	0.004052	0.007298	1,080.7104	0.003445	0.006136	1,367.7741
2.20	0.004234	0.007641	1,105.8432	0.003598	0.006424	1,399.5828
2.25	0.004419	0.007993	1,130.9760	0.003757	0.006712	1,421.3915
2.30	0.004611	0.008352	1,156.1088	0.003915	0.007025	1,463.2002
2.35	0.004806	0.008719	1,181.2416	0.004074	0.007319	1,495.0089
2.40	0.005006	0.009094	1,206.3744	0.004240	0.007641	1,526.8176
2.45	0.005209	0.009477	1,231.5072	0.004419	0.007960	1,558.6263
2.50	0.005416	0.009867	1,256.6400	0.004590	0.008288	1,590.4435
3.00	0.007695	0.014209	1,507.968.	0.006518	0.011923	1,908.5220
3.50	0.010357	0.019340	1,759.296	0.008782	0.016246	2,226.6090
4.00	0.013405	0.025261	2,010.624	0.011367	0.021204	2,544.6960

TABLE *giving the* loss of head J in metres

as a function of the diameter of the pipe and the mean velocity:

(1) For new pipes (k = 0.1 mm) and (2) for existing pipes
(k = 2 mm) (*the temperature of the water being at* 0 °C.)

Mean velocity in metres/sec.	PIPE DIAMETER: 1.000 m Pipe section: 0.785398 m²			PIPE DIAMETER: 1.250 m Pipe section: 1.22719 m²		
	Head per metre of pipe length		Flow in litres/sec.	Head per metre of pipe length		Flow in litres/sec.
	New pipes	Existing pipes		New pipes	Existing pipes	
0.01			7.8539			12.2715
0.05	0.000003	0.000004	39.2694	0.000003	0.000003	61.3575
0.10	0.000010	0.000013	78.5389	0.000008	0.000010	122.7150
0.15	0.000022	0.000029	117.8083	0.000017	0.000022	184.0725
0.20	0.000037	0.000051	157.0778	0.000028	0.000038	245.4300
0.25	0.000056	0.000078	196.3472	0.000043	0.000059	306.7875
0.30	0.000078	0.000111	235.6167	0.000060	0.000084	368.1450
0.35	0.000103	0.000150	274.8861	0.000079	0.000113	429.5025
0.40	0.000132	0.000195	314.1556	0.000101	0.000147	490.8600
0.45	0.000164	0.000246	353.4250	0.000125	0.000185	552.2175
0.50	0.000200	0.000308	392.6945	0.000152	0.000227	613.5755
0.55	0.000239	0.000365	431.9639	0.000182	0.000274	674.9325
0.60	0.000280	0.000433	471.2334	0.000213	0.000326	736.2900
0.65	0.000325	0.000507	510.5028	0.000248	0.000382	797.6475
0.70	0.000372	0.000587	549.7723	0.000285	0.000443	859.0050
0.75	0.000423	0.000673	589.0417	0.000324	0.000509	920.3625
0.80	0.000478	0.000765	628.3112	0.000366	0.000579	981.7200
0.85	0.000536	0.000863	667.5806	0.000409	0.000653	1,043.0775
0.90	0.000596	0.000966	706.8501	0.000456	0.000732	1,104.4350
0.95	0.000660	0.001076	746.1195	0.000505	0.000815	1,165.7925
1.00	0.000726	0.001193	785.3980	0.000556	0.000903	1,227.1500
1.05	0.000795	0.001315	824.6584	0.000609	0.000995	1,288.5075
1.10	0.000868	0.001443	863.9279	0.000665	0.001092	1,349.8650
1.15	0.000944	0.001577	903.1973	0.000723	0.001193	1,411.2215
1.20	0.001024	0.001718	942.4668	0.000783	0.001299	1,472.5800
1.25	0.001106	0.001864	971.7362	0.000846	0.001409	1,533.9375
1.30	0.001191	0.002016	1,021.0057	0.000911	0.001524	1,595.2950
1.35	0.001280	0.002174	1,050.2751	0.000979	0.001644	1,656.6525
1.40	0.001372	0.002338	1,099.5446	0.001049	0.001767	1,718.3010
1.45	0.001486	0.002508	1,138.8140	0.001121	0.001895	1,779.7267
1.50	0.001563	0.002684	1,178.0835	0.001196	0.002028	1,840.0825
1.55	0.001663	0.002866	1,217.3529	0.001274	0.002166	1,902.0005
1.60	0.001767	0.003053	1,256.6224	0.001353	0.002307	1,963.4405
1.65	0.001873	0.003247	1,295.8918	0.001434	0.002454	2,024.7975
1.70	0.001983	0.003447	1,335.1613	0.001518	0.002604	2,086.1550
1.75	0.002096	0.003653	1,374.4307	0.001603	0.002760	2,147.5125
1.80	0.002213	0.003864	1,413.7002	0.001691	0.002920	2,208.8700
1.85	0.002332	0.004082	1,452.9696	0.001782	0.003084	2,270.2275
1.90	0.002455	0.004306	1,492.2381	0.001875	0.003253	2,331.5850
1.95	0.002580	0.004535	1,531.5075	0.001971	0.003427	2,392.9425
2.00	0.002708	0.004771	1,570.7780	0.002068	0.003605	2,454.3000
2.05	0,002838	0.005012	1,610.0474	0.002168	0.003787	2,515.6575
2.10	0.002972	0.005260	1,649.3169	0.002269	0.003964	2,577.0150
2.15	0.003108	0.005513	1,688.5863	0.002375	0.004166	2,638.3725
2.20	0.003246	0.005773	1,727.8558	0.002483	0.004361	2,699.7300
2.25	0.003388	0.006038	1,767.1252	0.002593	0.004562	2,761.0875
2.30	0.003532	0.006309	1,806.3947	0.002705	0.004767	2,822.4450
2.35	0.003679	0.006587	1,845.6641	0.002819	0.004976	2,883.8025
2.40	0.003831	0.006870	1,884.9336	0.002936	0.005191	2,945.1600
2.45	0.003985	0.007159	1,924.2030	0.003055	0.005409	3,006.5175
2.50	0.004141	0.007454	1,963.4725	0.003178	0.005632	3,067.8750
3.00	0.005985	0.010734	2,356.194	0.004510	0.008110	3,681.57
3.50	0.007930	0.014610	2,748.893	0.006084	0.011039	4,295.165
4.00	0.010259	0.019083	3,141.592	0.007875	0.014418	4,908.76

TABLE *giving the* loss of head J in metres

as a function of the diameter of the pipe and the mean velocity:

(1) For new pipes (k = 0.1 mm) and (2) for existing pipes (k = 2 mm) (*the temperature of the water being at* 0 °C)

Mean velocity in metres/sec.	PIPE DIAMETER: 1.500 m Pipe section: 1.76715 m²			PIPE DIAMETER: 1.750 m Pipe section: 2.405281 m²		
	Head per metre of pipe length		Flow in litres/sec.	Head per metre of pipe length		Flow in litres/sec.
	New pipes	Existing pipes		New pipes	Existing pipes	
0.01			17.671			24.053
0.05	0.000002	0.000002	88.355	0.000002	0.000002	120.264
0.10	0.000006	0.000008	176.710	0.000005	0.000007	240.528
0.15	0.000013	0.000018	265.065	0.000011	0.000014	360.792
0.20	0.000023	0.000030	353.420	0.000019	0,000025	481,056
0.25	0.000034	0.000047	441.775	0.000028	0.000038	601.320
0.30	0.000048	0.000067	530.130	0.000040	0.000055	721.584
0.35	0.000063	0.000090	618.485	0.000053	0.000074	841.848
0.40	0.000081	0.000117	706.840	0.000068	0.000096	962.112
0.45	0.000101	0.000148	795.195	0.000084	0,000121	1,082.376
0.50	0.000122	0.000182	883.550	0.000102	0,000149	1,202.641
0.55	0.000146	0.000219	971.905	0.000122	0.000181	1,322.905
0.60	0,000172	0.000260	1,060.260	0.000144	0.000215	1,443.169
0.65	0.000200	0.000305	1,148.615	0.000167	0.000251	1,563.433
0.70	0.000230	0.000353	1,236.970	0.000191	0.000291	1,683.697
0.75	0.000261	0.000405	1,325.325	0.000217	0.000334	1,803.961
0.80	0.000295	0.000461	1,413.680	0.000246	0.000380	1,924.225
0.85	0,000330	0.000521	1,502.035	0.000276	0.000429	2,044.489
0.90	0,000368	0.000584	1,590.390	0.000306	0.000481	2,164.753
0.95	0.000406	0.000651	1,678.745	0.000339	0.000536	2,285.017
1.00	0.000447	0.000721	1,767.100	0.000373	0.000594	2,405.281
1.05	0.000490	0.000795	1,855.455	0.000409	0.000654	2,525.545
1.10	0.000535	0.000872	1,943.810	0.000447	0.000718	2,645.809
1.15	0.000582	0.000953	2,032.165	0.000486	0.000785	2,766.073
1.20	0.000631	0.001038	2,120.520	0.000527	0.000854	2,886.337
1.25	0.000682	0.001126	2,208.875	0.000570	0.000927	3,006.601
1.30	0.000735	0.001218	2,297.230	0.000614	0.001003	3,126.865
1.35	0.000789	0.001314	2,385.585	0.000659	0.001081	3,247.129
1.40	0.000845	0.001412	2,473.940	0.000706	0.001163	3,367.393
1.45	0.000903	0.001515	2,562.295	0.000754	0.001247	3,487.657
1.50	0.000963	0.001621	2,650.650	0.000805	0.001335	3,607.922
1.55	0.001025	0.001731	2,739.005	0.000857	0.001425	3,728.186
1.60	0.001089	0.001844	2,827.360	0.000911	0.001519	3,848.500
1.65	0.001155	0.001961	2,915.715	0.000966	0.001615	3,968.714
1.70	0.001223	0.002082	3,004.070	0.001023	0.001715	4,088.978
1.75	0.001292	0.002206	3,092.425	0.001080	0.001817	4,209.242
1.80	0.001363	0.002334	3,180.780	0.001140	0.001922	4,329.506
1.85	0.001436	0.002466	3,269.135	0.001201	0.002030	4,449.770
1.90	0.001512	0.002601	3,357.490	0.001264	0.002142	4,570.034
1.95	0.001589	0.002739	3,445.845	0.001329	0.002256	4,690.298
2.00	0.001669	0.002882	3,534.200	0.001396	0,002373	4,810.562
2.05	0.001749	0.003027	3,622.555	0.001463	0,002493	4,930.826
2.10	0.001833	0.003177	3,710.910	0.001531	0.002617	5,051.090
2.15	0.001916	0.003330	3,799.265	0.001602	0.002742	5,171.354
2.20	0.002003	0,003487	3,887.620	0.001675	0.002872	5,291.618
2.25	0.002092	0.003647	3,975.975	0.001749	0.003004	5,411.882
2.30	0.002182	0.003811	4,064.330	0.001824	0.003138	5,532.146
2.35	0.002274	0.003978	4,152.685	0.001901	0.003277	5,652.410
2.40	0.002378	0.004149	4,241.040	0.001980	0.003417	5,772.674
2.45	0.002462	0.004324	4,329.395	0.002058	0.003561	5,892.938
2.50	0.002557	0.004502	4,417.750	0.002139	0.003708	6,013.203
3.00	0.003633	0,006483	5,301.45	0.003035	0.005340	7,215.843
3.50	0.004891	0.008825	6,185.025	0.004092	0.007268	8,418.484
4.00	0.006350	0.011526	7,068.400	0.005303	0.009493	9,621.124

TABLE *giving the* loss of head J in metres

as a function of the diameter of the pipe and the mean velocity:

(1) For new pipes (k = 0.1 mm) and (2) for existing pipes (k = 2 mm) *(the temperature of the water being at 0 °C.)*

Mean velocity in metres/sec.	PIPE DIAMETER: 2.000 m Pipe section: 3.141592 m²			PIPE DIAMETER: 2.500 m Pipe section: 4.908738 m²		
	Head per metre of pipe length		Flow in litres/sec.	Head per metre of pipe length		Flow in litres/sec.
	New pipes	Existing pipes		New pipes	Existing pipes	
0.01			31.416			49.087
0.05	0.000001	0.000002	157.080	0.000001	0.000001	245.437
0.10	0.000005	0.000006	314.159	0.000003	0.000004	490.874
0.15	0.000009	0.000012	471.239	0.000007	0.000009	736.311
0.20	0.000016	0.000021	628.318	0.000012	0.000016	981.748
0.25	0.000024	0.000032	785.398	0.000018	0.000025	1,227.185
0.30	0.000034	0.000046	942.478	0.000026	0.000035	1,472.621
0.35	0.000045	0.000062	1,099.557	0.000035	0.000048	1,718.058
0.40	0.000058	0.000081	1,256.637	0.000044	0.000062	1,963.495
0.45	0.000072	0.000102	1,413.716	0.000055	0.000078	1,208.932
0.50	0.000087	0.000126	1,570.796	0.000067	0.000096	2,454.369
0.55	0.000104	0.000152	1,727.876	0.000080	0.000116	2,699.806
0.60	0.000122	0.000181	1,884.955	0.000094	0.000138	2,945.243
0.65	0.000142	0.000212	2,042.035	0.000109	0.000161	3,190.680
0.70	0.000163	0.000246	2,199.114	0.000125	0.000187	3,436.117
0.75	0.000186	0.000282	2,356.194	0.000143	0.000214	3,681.554
0.80	0.000210	0.000321	2,513.274	0.000161	0.000244	3,926.990
0.85	0.000235	0.000363	2,670.353	0.000181	0.000275	4,172.427
0.90	0.000261	0.000406	2,827.433	0.000202	0.000308	4,417.864
0.95	0.000289	0.000452	2,984.512	0.000223	0.000343	4,663.301
1.00	0.000319	0.000501	3,141.592	0.000246	0.000380	4,908.738
1.05	0.000349	0.000552	3,298.672	0.000270	0.000419	5,154.175
1.10	0.000381	0.000605	3,455.571	0.000295	0.000459	5,399.612
1.15	0.000415	0.000662	3,612.831	0.000321	0.000502	5,645.049
1.20	0.000450	0.000720	3,769.910	0.000348	0.000546	5,890.486
1.25	0.000487	0.000782	3,926.990	0.000376	0.000593	6,135.923
1.30	0.000524	0.000845	4,084.070	0.000405	0.000641	6,381.359
1.35	0.000563	0.000912	4,241.149	0.000435	0.000691	6,626.796
1.40	0.000603	0.000981	4,398.229	0.000467	0.000743	6,872.233
1.45	0.000645	0.001052	4,555.308	0.000498	0.000797	7,117.670
1.50	0.000688	0.001126	4,712.388	0.000531	0.000853	7,363.107
1.55	0.000733	0.001202	4,869.468	0.000566	0.000911	7,608.544
1.60	0.000779	0.001281	5,026.547	0.000601	0.000971	7,853.981
1.65	0.000826	0.001362	5,183.627	0.000638	0.001032	8,099.418
1.70	0.000874	0.001446	5,340.706	0.000675	0.001096	8,344.855
1.75	0.000923	0.001532	5,497.786	0.000714	0.001161	8,590.292
1.80	0.000974	0.001621	5,654.866	0.000753	0.001229	8,835.728
1.85	0.001027	0.001712	5,811.945	0.000794	0.001298	9,081.165
1.90	0.001080	0.001806	5,959.025	0.000836	0.001369	9,326.602
1.95	0.001136	0.001902	6,126.104	0.000878	0.001442	9,572.039
2.00	0.001193	0.002001	6,283.184	0.000922	0.001517	9,817.476
2.05	0.001250	0.002102	6,440.264	0.000966	0.001594	10,062.913
2.10	0.001308	0.002206	6,597.343	0.001011	0.001672	10,308.350
2.15	0.001369	0.002313	6,754.423	0.001057	0.001753	10,553.787
2.20	0.001431	0.002421	6,911.502	0.001105	0.001835	10,799.224
2.25	0.001494	0.002533	7,068.582	0.001154	0.001920	11,044.661
2.30	0.001559	0.002647	7,225.662	0.001204	0.002006	11,290.097
2.35	0.001634	0.002763	7,382.741	0.001254	0.002094	11,535.534
2.40	0.001691	0.002882	7,539.821	0.001307	0.002184	11,780.971
2.45	0.001759	0.003003	7,696.900	0.001361	0.002276	12,026.408
2.50	0.001829	0.003127	7,853.980	0.001416	0,002370	12,271.845
3.00	0.002592	0.004503	9,424.776	0.002015	0.003413	14,726.214
3.50	0.003497	0.006128	10,995.572	0.002712	0.004645	17,180.583
4.00	0.004526	0.008004	12,566.368	0.003517	0.006068	19,634.952

3. LOCAL HEAD LOSSES IN PIPES, UNIONS, VALVES, ETC. USED TO CARRY WATER

A. Sudden reduction in diameter.

$$\Delta h = \frac{1}{2}\left(1 - \frac{D_2^2}{D_1^2}\right)\frac{V^2}{2g}$$

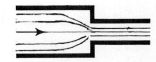

Δh = head loss in metres of water.
V^2 = mean velocity after reduction in diameter in metres per second.
g = acceleration due to gravity = 9.81 m/s².
D_1 = pipe diameter before reduction, in metres.
D_2 = pipe diameter after reduction, in metres.

● **Particular case: intake to a pipe leading from a large tank.**

a) $\Delta h = \dfrac{1}{2}\dfrac{V^2}{2g}$

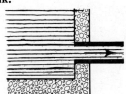

b) With protrusion inside tank (protrusion greater than 1/2 diameter):

$$\Delta h = \frac{V^2}{2g}$$

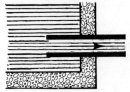

c) With rounded-edge connection:

$\left(\text{if } \dfrac{r}{D} > 0.18\right)\quad \Delta h = 0.05\,\dfrac{V^2}{2g}$

d) With round pipe connection at an angle:

$$\Delta h = K \frac{V^2}{2\,g}$$

where $K = 0.5 + 0.3 \cos \beta + 0.2 \cos^2 \beta$

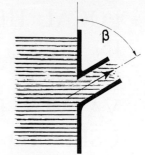

β	20°	30°	45°	60°	70°	80°	90°
K	0.96	0.91	0.81	0.70	0.63	0.56	0.50

e) With open-ended nozzle:

$$\Delta h = 1.5 \frac{V^2}{2\,g}$$

for $2\,D < 1 < 5\,D$

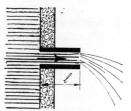

B. Sudden enlargement in diameter.

$$\Delta h = \frac{(V_1 - V_2)^2}{2\,g} = \frac{V_1^2}{2\,g}\left(1 - \frac{D_1^2}{D_2^2}\right)^2$$

$V_1 =$ mean velocity before enlargement, in metres per second.
$V_2 =$ mean velocity after enlargment, in metres per second.
$D_1 =$ pipe diameter before enlargement, in metres.
$D_2 =$ pipe diameter after enlargement, in metres.

● **Particular case: outlet from a pipe entering a large tank.**

$$\Delta h = \frac{V^2}{2\,g}$$

In actual practice use the following formula:

$$\Delta h = \alpha \frac{V^2}{2\,g} \quad \text{with } 1.06 < \alpha < 1.1$$

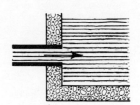

C. Converging cone.

$$\Delta h = \Delta h_1 + \Delta h_2$$

a) Friction loss (Δh_1):

Calculate the loss of head $\Delta h'_1$ in a cylindrical pipe of the same length, the cross-section of which is equal to the large cross-section;

$$\Delta h_1 = x \, \Delta h'_1$$

with $x =$
$$\frac{n(n^4 - 1)}{4(n - 1)}$$

where
$$n = \frac{D}{d}$$

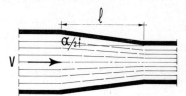

$D =$ inlet diameter
$d =$ outlet diameter

b) Loss by detachment (Δh_2):

$$\Delta h_2 = K \frac{V^2}{2g}$$

$V =$ calculated velocity in the large section, in metres per second.

Values of K:

$n = \dfrac{D}{d}$	1.15	1.25	1.50	1.75	2	2.5
Angle at apex						
6°	0.006	0.018	0.085	0.23	0.5	1.5
8°	0.009	0.028	0.138	0.373	0.791	2.42
10°	0.012	0.04	0.20	0.53	1,05	3.4
15°	0.022	0.07	0.344	0.934	1.98	6.07
20°	0.045	0.12	0.60	1.73	3.5	11
30°	0.280	0.25	1.25	3.4	7	

D. Divergent cone.

Lorenz' formula: $\Delta h = \left(\dfrac{4}{3} \tan \dfrac{\alpha}{2}\right) \dfrac{V_1^2}{2g}$

with:
$\alpha =$ angle at the apex of divergent cone.
$V_1 =$ velocity in the pipe before divergent cone.

E. Bends.

a) Rounded bends

$$\Delta h = K \frac{V^2}{2g}$$

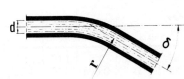

Values of K:
$r =$ radius of curvature of elbow, in metres.
$d =$ pipe diameter, in metres.

$\dfrac{r}{d}$	1	1.5	2	3	4
$\delta = 22°5$	0.11	0.10	0.09	0.08	0.08
$\delta = 45°$	0.19	0.17	0.16	0.15	0.15
$\delta = 60°$	0.25	0.22	0.21	0.20	0.19
$\delta = 90°$	0.33	0.29	0.27	0.26	0.26
$\delta = 135°$	0.41	0.36	0.35	0.35	0.35
$\delta = 180°$	0.48	0.43	0.42	0.42	0.42

Elbow opening into full tank (K total)

$\delta = 90°$	1.68	1.64	1.62	1.61	1.61

for a " 3d elbow":

$$2\,r = 3\,d, \text{ ie. } \frac{r}{d} = 1.5$$

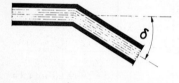

b) Sharp bends

$$\Delta h = K\,\frac{V^2}{2\,g}$$

δ	22°5	30°	45°	60°	75°	90°
K	0.17	0.20	0.40	0.70	1.00	1.50

F. T branches.

It is assumed that:
— the branch pipe has the same diameter as the main pipe;
— the edges of the joint are sharp.

a) Out-flowing:

$$\Delta h = K\,\frac{V^2}{2\,g}$$

Q = total flow in cubic metres per second.
Q_a = flow in branch pipe in cubic metres per second.
V = velocity of total flow in metres per second.
Kb = coefficient relating to the branch-pipe.
Kr = coefficient relating to the straight section.

$\dfrac{Qa}{Q}$	0	0.1	0.2	0.3	0.4	0.5	0.6	0.7	0.8	0.9	1
Kb	(1.0)	1.0	1.01	1.03	1.05	1.09	1.15	1.22	1.32	1.38	1.45
Kr	0	0.004	0.02	0.04	0.06	0.10	0.15	0.20	0.26	0.32	(0.40)

b) In-flowing:

$$\Delta h = K \frac{V^2}{2 g}$$

Q = total flow in cubic metres per second.
Qa = flow in branch-pipe in cubic metres per second.

$\frac{Qa}{Q}$	0	0.1	0.2	0.3	0.4	0.5	0.6	0.7	0.8	0.9	1
Kb	(− 0.60)	− 0.37	− 0.18	− 0.07	+ 0.26	0.46	0.62	0.78	0.94	1.08	1.20
Kr	0	0.16	0.27	0.38		0.46	0.53	0.57	0.59	0.60	0.59	0.55

c) Symmetrical T, separation of streams (welded steel T):

$$Kr_1 = 1 + 0.3 \left(\frac{Qa_1}{Q}\right)^2$$

$$Kr_2 = 1 + 0.3 \left(\frac{Qa_2}{Q}\right)^2$$

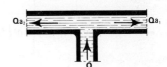

d) Symmetrical T, meeting of streams:

$$Kr_1 = 2 + 3 \left[\left(\frac{Qa_1}{Q}\right)^2 - \frac{Qa_2}{Q}\right]$$

$$Kr_2 = 2 + 3 \left[\left(\frac{Qa_2}{Q}\right)^2 - \frac{Qa_1}{Q}\right]$$

G. Valves.

$$\Delta h = K \frac{V^2}{2 g}$$

a) Rotating or butterfly valves

The head loss coefficient according to the degree to which the valve is opened depends on the hydrodynamic profile of the butterfly: the following table gives, as an indication, some normal values, but it is wise to refer to the manufacturers' tables for more accuracy.

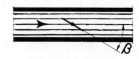

β	0° − 5°	10°	20°	30°	40°	45°	50°	60°	70°
K	0.25 to 0.35	0.52	1.54	3.91	10.8	18.7	32.6	118	751

b) Gate-valves:

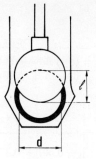

Gate-lowering value $\dfrac{l}{d}$	0	$\dfrac{1}{8}$	$\dfrac{2}{8}$	$\dfrac{3}{8}$	$\dfrac{4}{8}$	$\dfrac{5}{8}$	$\dfrac{6}{8}$	$\dfrac{7}{8}$
K	0.12	0.15	0.26	0.81	2.06	5.52	17	98

c) Ball valves:

β	10°	20°	30°	40°	45°	50°	55°
K	0.31	1.84	6.15	20.7	41	95.3	275

d) Clack-valves:

β	15°	20°	25°	30°	35°	40°	45°	50°	60°	70°
K	90	62	42	30	20	14	9.5	6.6	3.2	1.7

H. Open valves and couplings.

$$\Delta h = K \frac{V^2}{2g}$$

	Usual K	Variations of K
Parallel seat valve	0.12	0.08 to 0.2
Oblique seat valve		0.15 to 0.19
Angle valve		2.1 to 3.1
Needle valve		7.2 to 10.3
Straight-way screw-down valve	6	4 to 10
Right-angle screw-down valve		2 to 5
Float valve	6	
Ball valve		0.15 to 1.5
Non-return clack valve	2 to 2.5	1.3 to 2.9
Foot valve (excluding strainer)	0.8	
Sleeve coupling		0.02 to 0.07

Coefficients C_V of a valve

For certain valves, and especially control valves, the tendency now is not to give the head-loss in this form, but the flow coefficient C_V for the different openings. By definition, C_V is the flow of water of density 1 expressed in gallons per minute, which flows through the contracted cross-section for a head-loss of 1 p.s.i., which corresponds approximated to the flow of water in litres per minute creating a head-loss of 5 mbars, or 0.05 m of water. For water we therefore have:

$$C_V = \frac{Q}{\sqrt{\Delta h}}$$

With: Q = flow in gpm
Δh = head-loss in p.s.i.
or in metric units:

$$C_V = 13.3 \frac{Q}{\sqrt{\Delta h}}$$

With: Q = flow in l/s
Δh = head-loss in m of water.

4. DESIGN AND CALCULATION OF NEGATIVE PRESSURE SYSTEMS

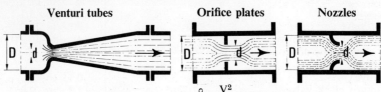

Venturi tubes   **Orifice plates**   **Nozzles**

A. Approximate calculation. $h = K \dfrac{\rho}{1\,000} \dfrac{V^2}{2\,g} (m^2 - 1)$, i.e.

With:
$$\frac{D^2}{d^2} = m = \sqrt{\frac{1\,000\ 2\ gh}{K\ \rho\ V^2} + 1}$$

h $=$ negative pressure created by the device, in metres of water at 4 °C (specific gravity 1 000 kg/m³).

K $=$ experimental coefficient (approx. 1).

ρ $=$ specific gravity of the fluid in the true flow conditions, in kg/m³.

V $=$ velocity of the fluid entering the device, in metres per second.

g $=$ acceleration due to gravity, 9.81 m/s².

D $=$ pipe diameter, in metres.

d $=$ diameter of the liquid jet at the narrowest point in metres.

m $=$ ratio of the pipe cross-section to the cross-section of the liquid jet at the narrowest point.

- **Calculation of the orifice size.** The actual diameter d_0 of the orifice plate is equal to $\dfrac{d}{0.8}$.

- **Head loss of an orifice plate** (for Re $> 10^5$):

$$h = \frac{K\ \rho}{1\,000} \frac{V^2}{2\,g} \quad \text{with } K = \left(1 + 0.707 \sqrt{1 - \frac{d_0^2}{D^2}} - \frac{d_0^2}{D^2}\right)^2 \left(\frac{D^2}{d_0^2}\right)^2$$

for an orifice plate with sharpened edges, the orifice diameter of which id d_0, expressed with the same unit as with the inside diameter of the pipe D.

B. Precise calculation of a negative-pressure measurement system:

Refer to the French standards NF X-10.101, NF X-10.102 and NF X 10.110.

Conditions of installation. The orifice plates, nozzles and Venturi nozzles must be located in a straight length of pipe ensuring that the upstream portion is at least equal to 10 D and the downstream portion is longer than 5 D, these minimum values being further increased for small contractions. For conventional Venturis, the minimum straight upstream length is only 1.5 to 6 D according to the degree of contraction (standard X-10.102, pp. 9 and 10).

The length of a Venturi tube is determined by the standardized shape coefficients (above standard) and by the choice of the contraction D — d.

5. FLOW RATES FOR ORIFICES AND NOZZLES

Flow rate $Q = kS \sqrt{2gh}$ (m³ per second).

Mean velocity: $V = k \sqrt{2gh}$ (metres per second).

With:

S = area of orifice measured at outer section (in square metres).

g = acceleration due to gravity, 9.81 m/s².

h = head on the orifice measured from the upstream level of the liquid to the centre of gravity of the orifice (in metres).

The relationship $k = K^{-2}$ must be allowed for between the coefficient k used here and the coefficient K used in sub-chapters 2 and 3 above pages 1053 and 1069 to 1075.

- Simplified formula for $k = 0.62$: $\underset{(m^3/h)}{Q} \simeq \underset{(cm^2)}{S} \underset{(m)}{\sqrt{h}}$

Orifice		k	Orifice		k
Orifice having exactly the same shape as the jet of liquid		1	Converging conical nozzle (12° angle)		0·94
Small orifice in thin wall		0·62	Diverging conical nozzle		1
Submerged orifice		0·62	Round re-entrant nozzle		0·5
Rectangular orifice in thin wall		0·62	Round nozzle projecting outwards $2\emptyset < 1 < 5\emptyset$		0·82

- **Pitot tube** — This method of flow measurement, although not standardized, is often used whenever it is difficult to design or fit a negative-pressure measuring element. For measurements in a pipe, the curved end of the Pitot tube sampling the pressure is generally placed along the axis of the pipe. The differential pressure h_c obtained represents the difference between the static pressure and the total pressure, and therefore the true dynamic pressure at the pressure sampling point. If V_o is the flow velocity along the axist at the sampling point, in m/s, and V_m is the mean flow velocity, in m/s, in the cross-section of diameter D, in m, for the rate of flow Q, in m³/s, of a fluid of specific gravity ρ, in kg/m³, the value of the differential pressure in mm of water is:

$$h_c = \rho \frac{V_c^2}{2\,g} = \rho \left[\frac{0.2874\,Q}{D^2 \dfrac{V_m}{V_c}} \right]^2$$

where the specific gravity ρ, in the absolute temperature T and absolute pressure P conditions of flow, is deduced from ρo under normal conditions by the formula:

$$\rho = \rho_0 \times \frac{P_0}{P} \times \frac{T_0}{T}$$

When the flow is symmetrically distributed in the cross-section because there are sufficient straight lengths, the following diagram gives the values of $\dfrac{V_m}{V_c}$ as a function of the Reynolds number Re:

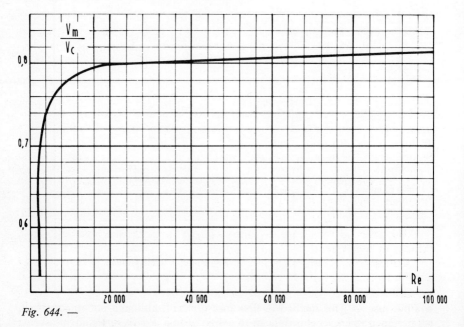

Fig. 644. —

For low values of Re, a mean velocity must be found by moving the Pitot tube in the flow cross-section.

6. FLOW OF WATER IN CHANNELS

A. *Empirical formulae for calculation of the head losses through friction.*

Only the following formulae are still in current use for the calculation of **head losses through friction**. Manning-Strickler's formulae becoming more wide-spread because of its simplicity and its general application to all forms of uniform flow in channels or rivers.

● **Bazin's formula:**

$$V = \frac{87 \sqrt{RI}}{1 + \dfrac{\gamma}{\sqrt{R}}}$$

● **Manning-Strickler's formula:**

$$V = K_s \, R^{23/} \, I^{1/2}$$

where:

V = mean velocity of flow in the cross-section, in m/s.

R = hydraulic or mean radius, in m, equal to the ratio of the liquid cross-section in the channel (m^2) to the wetted perimeter (m).

I = channel gradient, in metres per metre.

γ and K_s = wall roughness constants.

Nature of walls	γ	K_s
Very smooth walls (smooth cement rendering, planed wood)	0.06	100
Walls with ordinary cement rendering	—	90
Smooth walls (bricks, freestone, plain concrete)	0.16	70-80
Rough walls (rough stone)	0.46	60-70
Mixed walls (trimmed or stone pitched slopes)	0.85	50-60
Channels with earth walls (ordinary slopes)	1.30	40
Channels with earth walls with pebble bottoms and grassy sides	1.75	25-35

The nature and consistency of the walls may limit the maximum permissible velocity in their vicinity.

Critical conditions are reached in a channel of rectangular cross-section of a width l for a water head H_c such that $Q^2 = gl^2 H_c^3$ at a flow rate Q (i.e. a critical velocity $V_c = \sqrt{g \, H_c}$). At higher velocities the flow is torrential: it obeys complex laws and must form the subject of special studies (mathematical models, scale models, etc.). Below this, the flow is fluvial with $H < H_c$ and $V < V_c$. In water treatment works the flow is most often of the fluvial type; the two above conditions have therefore to be checked.

In uniform fluvial flow, the wetted cross-section and the velocity are constant in the successive profiles, the head losses through friction being exactly compensated by the gradient. The application of Bazin's or Mannin-Strickler's formula relating the velocity, the hydraulic radius and the gradient, makes it possible to calculate one of these values knowing the two others, i.e. three of the following four parameters: flow rate, wetted cross-section, wetted perimeter and gradient.

From the normal equilibrium level defined in this way, the local rises in level, or "jumps", resulting either from velocity increases or from restitution of energy in local transition sections must be calculated as indicated in para. below. In water treatment plant, where straight lengths are generally short, the variations in level in the various local sections are of great relative importance.

B. Use of the universal friction diagram.

This diagram (p. 1054) giving λ, the friction head-loss coefficient also applies to channels with walls of heterogeneous roughness. For concrete channels, the roughness coefficient k is on the average from 0.5 mm (smooth rendering) to 2 mm (plain concrete in average conditions). The method of calculation is the same as in pipes (p. 1051) using the hydraulic diameter.

$$\mathrm{Dh} = \frac{4\,\mathrm{S}}{\mathrm{p}_m}$$

S being the channel cross-section occupied by water and p_m the wetted perimeter, expressed in m² and m.

C. Calculation of local transition head losses.

The calculation is carried out as for pipes (p. 1069), from downstream and for the velocity of the uniform fluvial flow obtained. The local jumps upstream represent the head losses in the local transition sections.

D. Head loss through a bar screen.

$$\Delta\,\mathrm{h} = \mathrm{K}_1\,\mathrm{K}_2\,\mathrm{K}_3\,\frac{\mathrm{V}^2}{2\,\mathrm{g}}$$

V = approach velocity in the channel, in m/s.
— **Values of K_1** (clogging)

- clean screen $\mathrm{K}_1 = 1$
- clogged screen $\mathrm{K}_1 = \left(\dfrac{100}{\mathrm{m}}\right)^2$

where m is the percentage of the passage cross-section remaining at the maximum tolerated clogging. The latter, of the order of 60 to 90%, is related to the type of screen (manual or mechanical cleaning), the dimensions of the materials to be stopped and their nature. To prevent the carry-over of these materials between the screen bars, it is advisable to limit the true passage velocity through the clean screen at a value between 0.80 and 1.40 m/s (2.6 and 4.6 ft/s).

— **Values of K$_2$** (shape of the horizontal cross-section of the bars):

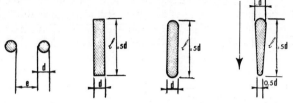

K$_2$ = 0.74   K$_2$ = 1   K$_2$ = 0.76   K$_2$ = 0.37

— **Values of K$_3$** (passage cross-section between bars):

$\frac{l}{4}\left(\frac{2}{e}+\frac{1}{h}\right)$	$\frac{e}{e+d}$									
	0.1	0.2	0.3	0.4	0.5	0.6	0.7	0.8	0.9	1
0	245	51.5	18.2	8.25	4.0	2.0	0.97	0.42	0.13	0
0.2	230	48	17.4	7.70	3.75	1.87	0.91	0.40	0.13	0.01
0.4	221	46	16.6	7.40	3.60	1.80	0.88	0.39	0.13	0.01
0.6	199	42	15	6.60	3.20	1.60	0.80	0.36	0.13	0.01
0.8	164	34	12.2	5.50	2.70	1.34	0.66	0.31	0.12	0.02
1	149	31	11.1	5.00	2.40	1.20	0.61	0.29	0.11	0.02
1.4	137	28.4	10.3	4.60	2.25	1.15	0.58	0.28	0.11	0.03
2	134	27.4	9.90	4.40	2.20	1.13	0.58	0.28	0.12	0.04
3	132	27.5	10.0	4.50	2.24	1.17	0.61	0.31	0.15	0.06

e = spacing between bars
d = width of bars
l = thickness of bars
h = submerged height of bars, vertical or oblique.
   These different values are to be expressed in the same units.

E. *Passage velocity of some materials.*
— Depth of water 1 m, straight channels:

	Diameter in mm	Mean velocity m/s
Silt	0.005— 0.05	0.15—0.20
Fine sand	0.05 — 0.25	0.20—0.30
Medium sand	0.25 — 1.00	0.30—0.55
Non-compacted clay	—	0.30—0.40
Coarse sand	1.00 — 2.5	0.55—0.65
Fine gravel	2.5 — 5	0.65—0.80
Medium gravel	5 —10	0.80—1.00
Coarse gravel	10 —15	1.00—1.20

— Correction for other depths of water:

H (m)	0.3	0.5	0.75	1.0	1.5	2.5
k	0.8	0.9	0.95	1.0	1.1	1.2

7. WEIRS

The flow rate of weirs is given by the general formula:

$$Q = \mu\, lh \sqrt{2\,gh}$$

where:
Q = flow rate, in m³/s (or l/s)
μ = flow coefficient of the weir
l = width of the weir, in m
h = height of sheet of water, in m (or cm)
h = height of sheet of water, in m (or cm)
g = acceleration due to gravity, in m/s² (= 9.81 in Paris).

In addition, p in m (or cm) is the height of the sill above the upstream bed and L is the width in m of the channel upstream of the weir.

A. *Rectangular thin-wall weir with low approach velocity*

$$\mu \simeq 0.40$$

in the case of a reservoir outlet for example.
● Particular case of the circular overflow weir

$$\mu \simeq 0.34$$

for an overflow of diameter 0.20 m $< \varnothing <$ 0.70 m with sufficient funnelling to avoid any downstream reaction.

B. *Rectangular thin-wall weir across a channel*

● **Weir with no lateral contraction** (l = L), with free fall.

A weir is so defined when the thickness e of the sill is less than half the head h, when the flow is such that it leaves a space ω filled with air at atmospheric pressure

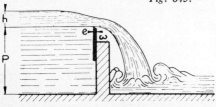

Fig. 645. — between the sheet of water and the downstream sill wall, and when the width of the overflowing sheet of water is exactly the same as that of the channel.

The flow coefficient μ is given by one of the following formulae:

— BAZIN'S formula (1898) in general use in France:

$$\mu_1 = 0.405 + \frac{0.003}{h}\left[1 + 0.55\,\frac{h^2}{(h + p)^2}\right]$$

— REHBOCK'S formula (1912):

$$\mu_2 = \frac{2}{3}\left[0.605 + \frac{1}{1\,050\,h - 3} + 0.08\,\frac{h}{p}\right]$$

— formula proposed by the S.I.A. (Société des Ingénieurs et Architectes Suisses):

$$\mu_3 = 0.410\left[1 + \frac{1}{1\,000\,h + 1.6}\right]\left[1 + 0.5\,\frac{h^2}{(h + p)^2}\right]$$

These formulae can be used for sheet depths h between 0.10 and 0.60 m for the Bazin formula, and between 0.025 m and 0.80 m for the Rehbock and S.I.A. formulae, the latter two giving practically identical results, but slightly lower than those obtained by the Bazin formula.

Other conditions of application

- for Bazin   : p between 0.20 and 2 m
- for Rehbock: p at least equal to h — 0.10 m
- for S.I.A.   : p greater than h.

Finally, h should be measured at a distance from the sill equal to at least five times the maximum height of the sheet. If the aeration under the waterfall is insufficient (depressed waterfall), the rate of flow is increased and its characteristic poorly defined, which is not admissible for a measuring weir.

Flow in l/s per m of sill length according to Bazin [1]

Sheet depth h m	Height of sill p in metres								
	0.20	0.30	0.40	0.50	0.60	0.80	1.00	1.50	2.00
0.10	64.7	63.0	62.3	61.9	61.6	61.3	61.2	61.1	61.0
0.12	85.3	82.7	81.5	80.8	80.4	79.9	79.7	79.7	79.3
0.14	108.2	104.4	102.6	101.5	100.9	100.1	99.8	99.3	99.2
0.16	133.2	128.1	125.5	124.0	123.0	122.0	121.4	120.7	120.5
0.18	160.2	153.7	150.2	148.1	146.8	145.3	144.5	143.5	143.2
0.20	189.3	181.0	176.6	173.9	172.1	170.0	168.9	167.7	167.1
0.22	220.2	210.2	204.6	201.2	198.9	196.2	194.8	193.1	192.4
0.24	253.0	241.0	234.2	230.0	227.2	223.8	221.9	219.7	218.8
0.26	287.6	273.6	265.5	260.3	256.9	252.7	250.3	247.5	246.4
0.28	323.9	307.8	298.2	292.1	288.0	282.9	280.0	276.5	275.1
0.30	361.8	343.6	332.5	325.4	320.5	314.4	310.9	306.6	304.9
0.32		380.9	368.3	360.1	354.3	347.2	343.0	337.9	335.7
0.34		419.8	405.6	396.1	389.5	381.2	376.2	370.2	367.2
0.36		460.1	444.2	433.5	426.0	416.4	410.7	403.6	400.5
0.38		502.0	484.3	472.3	463.8	452.8	446.3	438.0	434.4
0.40		545.2	525.8	512.4	502.9	490.5	483.0	473.5	469.3
0.45		659.4	635.3	618.3	606.0	589.6	579.6	566.5	560.6
0.50			752.9	732.1	716.7	696.0	682.9	665.7	657.8
0.55			878.2	853.4	834.8	809.2	792.9	770.9	760.5
0.60			1 011.1	982.1	960.0	929.2	909.3	881.9	868.7

1. Rectangular thin-wall weirs with no lateral contraction.

Flow in l/s per m of sill length according to Rehbock [1]

Sheet depth h (m)	Heigh of sill p in metres									
	0.10	0.20	0.30	0.40	0.50	0.60	0.80	1.00	2.00	3.00
0.02	5.7	5.6	5.6	5.6	5.5	5.5	5.5	5.5	5.5	5.5
0.04	15.7	15.3	15.1	15.1	15.0	15.0	15.0	15.0	14.9	14.9
0.06	29.1	28.0	27.7	27.5	27.4	27.3	27.2	27.2	27.1	27.0
0.08	45.5	43.4	42.7	42.3	42.1	42.0	41.8	41.7	41.5	41.4
0.10	64.9	61.1	59.9	59.3	58.9	58.7	58.3	58.2	57.8	57.7
0.12	87.0	81.2	79.2	78.2	77.6	77.2	76.7	76.8	75.8	75.7
0.14	112.0	103.3	100.4	99.0	98.1	97.5	96.8	96.4	95.5	95.2
0.16	139.7	127.6	123.5	121.5	120.3	119.5	118.5	117.9	116.7	116.3
0.18	170.1	153.9	148.5	145.8	144.1	143.1	141.7	140.9	139.3	138.7
0.20	203.3	182.2	175.1	171.6	169.5	168.1	166.3	165.3	163.2	162.5
0.22		212.5	203.6	199.1	196.4	194.6	192.4	191.0	188.4	187.5
0.24		244.8	233.7	228.1	224.8	222.5	219.8	218.1	214.8	213.7
0.26		279.0	265.4	258.6	254.6	251.9	248.5	246.4	242.4	241.0
0.28		315.2	298.9	290.7	285.8	282.5	278.4	276.0	271.1	269.5
0.30		353.3	333.9	324.2	318.4	314.5	309.7	306.7	300.9	299.0
0.32			370.6	359.2	352.4	347.8	342.1	338.7	331.8	329.5
0.34			408.9	395.6	387.7	382.4	375.7	371.7	363.8	361.1
0.36			448.8	433.5	424.3	418.2	410.5	405.9	396.8	393.7
0.38			490.3	472.8	462.3	455.3	446.5	441.2	430.7	427.2
0.40			533.4	513.5	501.5	493.6	483.6	477.6	465.7	461.7
0.45				621.4	605.3	594.7	581.3	573.3	557.2	551.9
0.50				738.0	717.1	703.2	685.8	675.4	654.5	647.5
0.55					836.8	819.1	797.0	783.8	757.3	748.4
0.60					964.2	942.2	914.8	898.3	865.4	854.4
0.65						1 072.6	1 039.0	1 018.9	978.7	965.3
0.70						1 210.0	1 169.7	1 145.5	1 097.0	1 080.9
0.75							1 306.6	1 277.8	1 220.3	1 201.1
0.80							1 449.8	1 416.0	1 348.4	1 325.9

1. Rectangular thin-wall weirs with no lateral contraction.

Flow in l/s per m of sill length according to S.I.A. [1]

Sheet depth h (m)	Height of sill p in metres									
	0.10	0.20	0.30	0.40	0.50	0.60	0.80	1.00	2.00	3.00
0.02	5.4	5.4	5.4	5.4	5.4	5.4	5.4	5.4	5.4	5.4
0.04	15.5	15.1	15.0	14.9	14.9	14.9	14.9	14.9	14.9	14.9
0.06	29.0	27.8	27.5	27.4	27.3	27.2	27.2	27.2	27.1	27.1
0.08	45.7	43.3	42.5	42.2	42.0	41.9	41.8	41.7	41.6	41.6
0.10		61.2	59.8	59.2	58.8	58.6	58.3	58.2	58.1	58.0
0.12		81.5	79.2	78.1	77.5	77.2	76.8	76.5	76.2	76.2
0.14		103.9	100.6	99.0	98.1	97.5	96.9	96.5	96.0	95,9
0.16		128.5	124.0	121.7	120.4	119.5	118.6	118.1	117.3	117.1
0.18		155.1	149.2	146.2	144.3	143.2	141.8	141.1	139.9	139.7
0.20			176.3	172.3	169.9	168.3	166.5	165.5	163.9	163.5
0.22			205.1	200.1	197.0	195.0	192.6	191.3	189.2	188.7
0.24			235.6	229.5	225.7	223.1	220.1	218.4	215.6	215.0
0.26			267.7	260.4	255.8	252.7	248.9	246.3	243.3	242.4
0.28			301.5	292.9	287.4	283.7	279.1	276.5	272.0	271.0
0.30				326.9	320.4	316.0	310.5	307.4	301.9	300.6
0.32				362.3	354.9	349.7	343.2	339.4	332.9	331.3
0.34				399.2	390.7	384.7	377.1	372.1	364.9	362.9
0.36				437.5	427.8	421.0	412.3	407.1	397.9	395.6
0.38				477.1	466.3	458.6	448.6	442.7	431.9	429.2
0.40					506.0	497.4	486.1	479.3	466.9	463.7
0.45					611.0	599.9	585.0	575.9	558.7	554.1
0.50						709.8	690.9	679.1	656.2	649.9
0.55						826.9	803.6	788.8	759.3	751.0
0.60							923.0	904.8	867.9	857.1
0.65							1 048.9	1 027.1	981.8	968.2
0.70							1 181.0	1 155.4	1 100.9	1 084.1
0.75							1 319.3	1 289.5	1 225.0	1 204.7
0.80								1 429.5	1 354.1	1 329.9

1. Rectangular thin-wall weirs with no lateral contraction.

● **Weir with lateral contraction**

The S.I.A. has proposed for μ the following formula:

$$\mu = \left[0.385 + 0.025 \left(\frac{l}{L}\right)^2 + \frac{2.410 - 2 \left(\frac{l}{L}\right)^2}{1\,000\,h + 1.6} \right] \cdot \left[1 + 0.5 \left(\frac{l}{L}\right)^4 \left(\frac{h}{h + p}\right)^2 \right]$$

when $p \geqslant 0.30$ m; $\quad 0.025 \frac{L}{l} \leqslant h \leqslant 0.80$ m; $\quad h \leqslant p$; $\quad l > 0.3\,L$

Simplified Francis formula: $\quad Q = 1.83\,(1 - 0.2\,h)\,h^{3/2}$

for which the extra width either side of the sill must be at least equal to 3 h, the sheet depth being measured at least 2 m upstream.

C. *Triangular weir with thin wall*

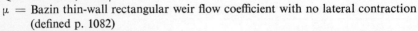

$$Q = \frac{4}{5}\, \mu\, h^2 \sqrt{2\ gh}\, \tan \frac{\theta}{2}$$

where

Q = flow rate, in m³/s

μ = Bazin thin-wall rectangular weir flow coefficient with no lateral contraction (defined p. 1082)

h = head in metres

θ = weir notch apex angle

The flow rate of a triangular weir can be deduced from the flow rate of a rectangular weir with no lateral contraction, with the same sheet and sill height, by multiplying this flow rate by $\frac{4}{5}\, h \tan \frac{\theta}{2}$.

For $\theta = 90°$, Thompson's formula is sometimes used:

$$Q = 1.42\ h^{5/2}$$

This formula is very approximate since it does not take into account the effect of the sill height.

8. HEAD LOSSES FOR ANY FLUID

The general formula for head losses, in pipes of any cross-section and in channels, takes the form:

$$\Delta h = \Delta h_0 + \Delta h_1 + \ldots$$

In turbulent flow:

$$\underbrace{\Delta h_0 = J_0 L_0}_{\text{(friction)}} \quad + \quad \underbrace{10^{-5}\ K_0\ \rho\ \frac{v_0^2}{2}}_{\substack{\text{(transition)}\\ \text{(points)}}}$$

with:

$$J_0 = 10^{-5}\ \rho\ \frac{\lambda}{D_h}\ \frac{v_0^2}{2}$$

In laminar flow, the formula giving Δh_0 is the same with $J_0 = 10^{-5}\ \dfrac{64}{Re\ D_h}\ \rho\ \dfrac{v^2}{2}$, K_0 then having particular values calculated with the aid of specialist reference books.

Notation:

Δh = total head-loss in bars.

Δh_1, Δh_2, etc. = elementary head losses in sections in which the respective velocities v_0, v_1, etc. are constant.

J_0 = friction head-loss coefficient in bars per metre of pipe (or channel) length, at velocity v_0.

L_0 = length of pipe (or channel) in m at velocity v_0.

K_0 = sum of the head-loss coefficients in transition sections at velocity v_0.

ρ = specific gravity of the fluid in the real flow temperature and pressure conditions in kg/m^3.

v_0, v_1, etc. = velocities of the fluid in the real flow conditions, in m/s.

λ = coefficient given by the universal diagram (see page 1054) as a function of the Reynolds number $Re = \dfrac{V_0 D_h}{\nu}$ (ν = kinematic viscosity of the fluid in m^2/s in the flow conditions (see pages 1052 and 1088) and as a function of the relative roughness $\dfrac{k}{D_h}$ (k = roughness coefficient of the wall in m, given on page 1052).

D_h = hydraulic diameter of the pipe (or channel) in metres $D_h = \dfrac{4\,S}{p_m}$

where S is the cross-section of pipe (or channel) occupied by the fluid in m^2, and p_m the 'perimeter wetted' by the fluid in this cross-section in m. D_h is the quadruple of the hydraulic radius or usual mean radius. In a circular pipe of diameter D, $D_h = D$. The calculation is carried out for the head losses by friction as indicated on page 1052 (pipes) and page 1079 (channels), and for the local head losses as indicated on pages 1069 and 1080.

It is often customary to give the values of the head loss in metres of water (of specific gravity 1 000 kg/m^3 at 4 °C). The above formulae then become:

— in turbulent flow:

$$\Delta h_0 = \frac{\rho}{1\,000}\,\frac{\lambda}{D_h}\,\frac{v_0^2}{2\,g}\,L_0 + \frac{\rho}{1\,000}\,K_0\,\frac{v_0^2}{2\,g}$$
$$= \frac{\rho}{1\,000}\left(\frac{\lambda}{D_h}\,L_0 + K_0\right)\frac{v_0^2}{2\,g}$$

— in laminar flow:

$$\Delta h_0 = \frac{\rho}{1\,000}\left(\frac{64}{Re\,D_h}\,L_0 + K_0\right)\frac{v_0^2}{2\,g},$$

where K_0 has particular values to be calculated with the aid of specialist reference books.

The case with the usual gases

• The kinematic viscosity ν as a function of the temperature, at the normal pressure of 760 mm of mercury, has a value in m^2/s:

t°C	0	20	40	60	80	100
Air	$13.20.10^{-6}$	$15.00.10^{-6}$	$16.98.10^{-6}$	$18.85.10^{-6}$	$20.89.10^{-6}$	$23.00.10^{-6}$
Steam	11.12. —	12.90. —	14.84. —	16.90. —	18.6. —	21.50. —
Cl_2	3.80. —	4.36. —	5.02. —	5.66. —	6.36. —	7.15. —
CH_4	14.20. —	16.50. —	18.44. —	20.07. —	22.90. —	25.40. —
CO_2	7.00. —	8.02. —	9.05. —	10.30. —	12.10. —	12.80. —
NH_3	12.00. —	14.00. —	16.00. —	18.10. —	20.35. —	22.70. —
O_2	13.40. —	15.36. —	17.13. —	19.05. —	21.16. —	23.40. —
SO_2	4.00. —	4.60. —				7.60. —

• This viscosity has to be corrected depending on the pressure in accordance with the following formula (not valid for steam):

$$\nu' = \nu \, \frac{P}{P'}$$

with:

ν' = corrected kinematic viscosity, in m^2/s

P' = real absolute pressure

P = normal absolute pressure expressed in the same units as P'.

The specific gravity ρ' of the fluid in kg/m^3, in the flow temperature t' (°C) and absolute pressure P' conditions, is deduced from the specific gravity ρ in the normal conditions in accordance with the formula:

$$\rho' = \rho \times \frac{P}{P'} \times \frac{273}{273 + t'}$$

9. MISCELLANEOUS INFORMATION

● TIME NEEDED TO EMPTY A TANK OF CONSTANT HORIZONTAL CROSS-SECTION IN THE BOTTOM OF WHICH AN ORIFICE IS MADE

Emptying time in seconds is:

$$t = \frac{2 S (\sqrt{h_1} - \sqrt{h_2})}{Ks \sqrt{2 g}}$$

S = tank surface area in cm²
s = area of the orifice in cm²
K = coefficient of contraction of the orifice (see page 1077)
g = acceleration due to gravity 981 cm/s²
h_1 = initial head of water above the orifice in cm
h_2 = final head of water above the orifice in cm.
($h_2 = 0$ for complete emptying)

● PUMPS

The power needed, in kW, is $P = \dfrac{Q (H + h)}{r \times 366}$

Q = required output, in m³/h
H = total static lifting head (metres)
h = loss of head in the pipes (metres)
r = efficiency of the pump (0.6 to 0.9).

In principle, h should be $< \dfrac{H}{10}$.

When the speed of rotation N becomes $N' = kN$, the characteristics of centrifugal pumps follow the formulae:

$$Q' = kQ; \ H' = k^2H; \ P' = k^3P$$

The efficiency is practically independent of the speeds of rotation.

● HYDRAULIC MACHINES

The power provided, in kW, is $P = \dfrac{QHr}{366}$.

Q = rate of flow in m³/h.
H = height of water fall in metres.
r = efficiency of the turbine.

	Values of r
Hydraulic wheel	0.70 to 0.75
Screw turbine and Francis turbine	0.70 to 0.88
Kaplan and Pelton turbines	0.70 to 0.92

● ECONOMIC DIAMETER OF PIPES FOR PUMPED SUPPLIES (see page 1091)

For this diameter, one obtains the minimum value for the sum of the pumping energy costs (directly related to the head losses), and the pipe amortization costs (cost of the pipe itself plus pipe-laying, the latter being related to the material cost).

In 1948, Vibert published a formula for cast-iron pipes, taking account of the latest economic conditions; he assumed for the pipe supply cost the formula $P_1 = k_1 D^{3/2}$, for the laying cost the formula $P_2 = C + k_2 D^{3/2}$, and for the power cost $P_3 = k_3 \dfrac{D^5}{Q^2}$. With an interest and amortization rate of 5 % over 50 years, Vibert's formula giving the minimum value for $P_1 + P_2 + P_3$ was:

$$D = 15.47 \left(\frac{ne}{f}\right)^{0.154} . \, Q^{0.46}$$

where:
D = economic diameter in metres.
n = number of daily pumping hours divided by 24.
e = mean cost per kWh of electricity used for pumping.
f = cost per kg of grey cast-iron pipe.
Q = flow delivered in m³/s.

The application of the latter formula gives rates of flow which are admissible in industrials systems or for lengths which are not excessive. For extensive systems or large diameters it is preferable to carry out a fresh detailed economic study for each individual case. Some authorities assume that the economic diameter is obtained as a first approximation for a flow velocity between 1 and 1.20 m/s.

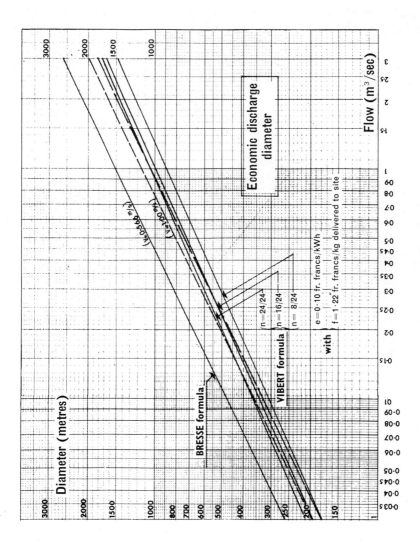

Fig. 646. —

ELECTRICITY

1. NOTATIONS, UNITS, SYMBOLS

I	Current..	Ampere	(A)
u	Voltage of a direct or single-phase alternating current (phase voltage)	Volt	(V)
U	Voltage composed of a three-phase alternating current (between two phases)	Volt	(V)
P	Active power	Watt	(W)
Pa	Apparent power	Volt-Ampere	(VA)
R	Resistance	Ohm	(Ω)
X	Reactance	Ohm	(Ω)
Z	Impedance	Ohm	(Ω)
cos φ	Power factor		
ρ	Efficiency		

2. USUAL DEFINITIONS AND FORMULAE

● **Current**

Direct current $I = \dfrac{P}{u}$

Single-phase alternating current : $I = \dfrac{P}{u \cos \varphi}$

Three-phase alternating current: $I = \dfrac{P}{U \sqrt{3} \cos \varphi}$

● **Resistance**

$$R = r \frac{l}{s} 10^{-2}$$

r = resistivity in microhm-cm
l = length in m
s = cross-section in mm²

● **Power factor or cos** φ

The strength and the voltage of an alternating current are rarely in phase. The cosine of the angle of the two vectors: current-voltage is called the power factor.

The self-induction causes the current to lag behind the voltage; the capacity causes it to lead. In both cases the active power is reduced.

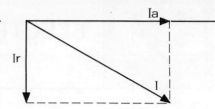

Fig. 647. — Case of a circuit comprising self-induction.

The current I is that which is read off the ammeter, and the diagram shows that it has two components I_a and I_r.

For industrial current of sinusoidal form (general case):

$I_a = I \cos \varphi$ is the active or watt current, in phase with the voltage u, contributing to the actual work.

$I_r = I \sin \varphi$ is the reactive or wattless current, in quadrature with the voltage, maintaining the magnetic field and not contributing to the actual work.

● **Apparent—active—reactive power**

I being the current read off the ammeter
U the mains voltage between phases, in three-phase,
u the mains voltage between phase and neutral, in single-phase,
the powers are given in the following table:

	single-phase	three-phase
Apparent power	uI	$U I \sqrt{3}$
Active power	u I cos φ	$U I \sqrt{3} \cos \varphi$
Reactive power	u I sin φ	$U I \sqrt{3} \sin \varphi$

The apparent power is expressed in volt-amperes   (VA)
The active power is expressed in watts   (W)
The reactive power is expressed in reactive volt-amperes   (VAr)

● **Voltage drops in the cables (copper conductor)**

The following formulae do not take account of the line self-induction.

— Single-phase or direct $\Delta u = \dfrac{I}{28.7} \times \dfrac{L}{s}$

— Three-phase $\Delta U = \dfrac{I}{28.7} \times \dfrac{L}{s} \times 0.866$

ΔU and Δu = voltage drop in volts
I = current in amperes
L = length of the line in metres
s = unit cross-section of the conductors in mm^2

- **Impedance of a circuit**

    This is the resultant of the ohmic resistance and reactance of the circuit.

$$Z = \sqrt{R^2 + X^2}$$

R = resistance

X = reactance (inductive, capacitive)

- **Quantity of heat emitted by a circuit** of resistance R through which passes a current I for one second:

$$Q \text{ (watt)} = RI^2$$

- **Temperature rise time constant**

    This is the time needed by a body emitting heat (e.g. a motor) to attain, without heat exchange with the ambient medium, the temperature which it attains in stabilized conditions when this exchange does exist.

    This constant, expressed in seconds, is given approximately by the formula:

$$t = 417 \times \frac{G \times T}{P}$$

G = mass of the body in kilogrammes,

T = stabilized maximum temperature in degrees Celsius,

P = power dissipated as heating expressed in watts.

- **Star connection—delta connection**

    Resistances and motor windings can be star connected or delta connected. The following diagrams show the details of these connections.

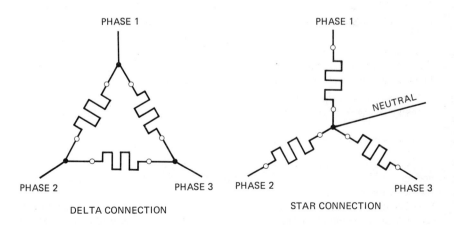

DELTA CONNECTION   STAR CONNECTION

Fig. 648. —

3. INDUSTRIAL APPLICATIONS

3.1. General installations

● **Determination of the average cos φ of an installation**

Let Q_a be the consumption of active energy noted on the active meter for a given period of time, and Q_r the consumption of reactive energy noted on the reactive meter for the same period, the average cos φ of the installation during this period is given by the formula:

$$\cos \varphi = \frac{Q_a}{\sqrt{Q_a^2 + Q_r^2}}$$

● **Improvement of the cos φ**

The line losses through the Joule effect are proportional to I^2, whereas the active energy consumed is only proportional to I cos φ, and therefore the electricity authorities penalize the consumers whose installations have a power factor lower than a certain value, of the order of 0.85.

When the reduction in cos φ is due to an inductive reaction of the installation, it is improved by a battery of capacitors.

The power of the batteries of capacitors, expressed in kilovars, which is needed to bring the cos φ to the desired value is determined in each case by calculation.

● **Lighting and heating**

Current consumed by incandescent lamps:

$$I = \frac{P}{u}$$

Current consumed by fluorescent lamps:

$$I = \frac{P}{u \cos \varphi}$$

Current consumed by electric fires:

— single-phase $I = \dfrac{P}{u}$

— three-phase $I = \dfrac{P}{U \sqrt{3}}$

Calculation of the power needed to heat a building takes many factors into account: volume of the room, orientation, thermal insulation, minimum outside temperature, etc.

It can be considered however that for a minimum outside temperature of — 10 °C and to obtain + 18 °C in the building, it is necessary to assume 200 to 300 watts per m².

3.2. Motors

The motors in current use are asynchronous motors; the information given below relates only to this type of motor.

● **Rating—Power consumed**

The **rating** is that indicated in the catalogue or on the motor plate. It corresponds to the **mechanical power** developed on the motor shaft and is expressed in kW.

The electric power **consumed** is given by the relation:

$$P \text{ consumed} = \frac{\text{mechanical power delivered}}{\text{efficiency}}$$

It is also expressed in kW.

It is the latter which is to be used in establishing the power balance of an installation.

Example: A motor rated at 15 kW driving a pump consuming 13 kW.

For this load the efficiency of the motor is 0.80.

Electric power consumed:

$$\frac{13}{0.80} = 16.2 \text{ kW}$$

A motor must be used at the frequency specified by the manufacturer. A motor designed for a 50 Hz supply will have a lower torque if it is supplied at 60 Hz.

● **Efficiency**

For motors of current manufacture, the greater the power of the motor, the higher the efficiency,

Example: 50 kW motor $\rho = 0.85$
          1 kW motor $\rho = 0.7$

For a given motor, the efficiency quoted by the manufacturer corresponds to operation at full load and decreases slightly with the load.

Example: 50 kW motor $\rho = 0.85$ at 4/4 load,
          $\rho = 0.82$ at 3/4 load
          $\rho = 0.80$ at 1/2 load

● **Determination of the power of a motor**

To determine the rating of a motor, it is recommended that the following allowances should be taken in relation to the mechanical power absorbed by the driven machine (except particular cases such as grinding mills, comminutors, etc.):

— 10 to 15 % in the case of direct coupling;

— 20 % in the case of belt transmission.

● **Supply voltage**

— As the power varies approximately as the square of the voltage, it is essential to design the motor for the exact mains voltage.

For example, a motor developing 15 kW at the shaft at a voltage of 380 V will only develop about 12.5 kW at a voltage of 350 V.

— The majority of manufacturers provide terminal boxes with six terminals allowing star and delta connection by moving straps:

● the first will be used for example at 380 V three-phase,

● the second at 220 V three-phase.

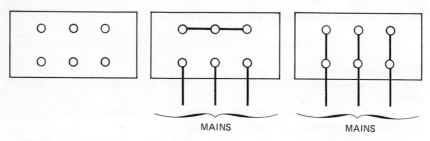

MAINS   MAINS

Fig. 649. — Box with six terminals. *Fig. 650. — Star connection 220/380. Motor supplied at 380 V.* *Fig. 651. — Delta connection 220/380. Motor supplied at 220 V.*

If the motor is to have star-delta starting, the six terminals must be brought out and designed for the following voltages:

Mains voltage	Motor voltage
220 V	220/380
380 V	380 delta
440 V	440 delta

For this method of starting, no strap needs be fitted in the terminal box.

● **Speed of asynchronous motors**

No-load: single-phase or three-phase asynchronous motors have a no-load speed which is practically equal to synchronism and is given by the formula:

$$N = \frac{60 \times F}{n}$$

N = number of revolutions per minute (rev/min),
F = frequency in cycles per second (Hz),
n = number of pairs of poles.
Example:
motor 2 poles 50 Hz 3 000 rev/min   2 poles 60 Hz 3 600 rev/min
     4 poles 50 Hz 1 500 rev/min   4 poles 60 Hz 1 800 tev/min
     6 poles 50 Hz 1 000 rev/min   6 poles 60 Hz 1 200 rev/min.

On load: : The speed is slightly lower than the no-load speed. The difference is due to the slip which is expressed by the relation:

$$g = \frac{\text{synchronous speed—speed on load}}{\text{synchronous speed}}$$

g is between 2 and 8 % of the synchronous speed.

- **Choice of motor and starting system**

These two points are dependent on the machine being driven and the requirements of the mains supply.

As regards the driven machine, whatever the starting method, the accelerating torque (difference between the motor torque and the resisting torque) must be sufficient to allow the set to come up to speed. There are two main factors:
— the moment of gyration Gd^2, expressed in Newtons per square metre;
— the torque needed for starting the machine.

Some machines start practically at no load, fans for example, but the mass and diameter of the rotating parts (elements characterizing the Gd^2) are such that a relatively high energy, converted almost solely into kinetic energy, is needed to bring them up to speed.

For other machines (sewage comminution pumps, compressors), as soon as it is switched on the motor has to allow them to do mechanical work, while accelerating; it is therefore the torque needed on starting which has to be taken into consideration.

As regards the mains supply, French legislation in general forbids direct starting of motors with short-circuited rotors when their rating exceeds 2.9 kW.

Apart from the legal conditions, it is sometimes necessary to provide for a starting system which allows the current called for from the mains to be reduced (example: supply from a generating set).

The following table gives the characteristics of the various starting methods:

Starting method	Torque on starting	Current called for
Direct	Cd	Id
Star delta	$0.3\ Cd$	$0.3\ Id$
Stator resistances	kCd	$\sqrt{k}Id$
Auto-transformer	kCd	kId

The value of k is to be fixed in accordance with the above considerations.

The ratios of the current taken and the torques on start-up, to the rated values, are given in the following tables (approximate values):

Type of motor	$\dfrac{Id}{In}$	$\dfrac{Cd}{Cn}$
Short-circuited rotor (direct starting) ...	6	1.6
Short-circuited rotor (star-delta starting)	2	0.6
Double-cage rotor (direct starting)	4	1.7
Double-cage rotor (star-delta starting) ..	1.33	0.55
Double-cage rotor (stator starting)	1.8	0.5 to 0.7
Ring-wound rotor	1.33	0.8 to 2
Rotor with centrifugal or similar starter	1.5 to 2	1.2 to 1.5

Id = current on starting
In = rated current at full load
Cd = torque on starting
Cn = rated torque
Cn is expressed in newton metres. If N is the speed in revolutions per minute, and P the rated power in kilowatts.

$$Cn = \frac{9564\ P}{N} \text{ newton metres}$$

the former unit was the metre kilogramme force, 1 newton = 9.81 kilogramme force).

● **Current consumed**

direct current $\qquad I = \dfrac{P_n\ 10^3}{u\ \rho}$

single-phase current $I = \dfrac{P_n\ 10^3}{u\ \rho \cos \varphi}$

three-phase current $I = \dfrac{P_n\ 10^3}{U \sqrt{3}\ \rho \cos \varphi}$

P_n is the rated power of the motor expressed in kilowatts.

● **Approximate values of the current consumed**
(motor of 1 to 10 kW)

	1 500 rev/min	3 000 rev/min
Single-phase 220 V	5.5 A per kW	5 A per kW
Three-phase 220 V	4.3 A per kw	3.8 A per kW
Three-phase 380 V	2.5 A per kW	2.2 A per kW

For a given power, the cos φ and the efficiency decrease as the number of poles increases, and therefore the lower the rated speed, the higher the current consumed, for a given power.

Thus a motor of 750 rev/min will consume about 20 % more than a motor of the same rating at 3000 rev/min and 10 % more than a motor of the same rating at 1 500 rev/min.

● **Supply cables**

The permissible voltage drop at the terminals of a motor at full load is 5%, and the cross-section of the supply cables must be calculated accordingly, taking into account in particular the current consumed at full load and the length of the cable.

The following table gives as an indication the characteristics of the supply cable for a three-phase voltage of 380 V, a maximum length of 25 m and the diameter of the gland to be used on the terminal box.

Motor rating	Number and cross-section of conductors (copper)	Gland diameter mm
up to 8 kW	4 × 2.5 mm²	13
8 to 14 kW	4 × 4 mm²	16
14 to 18 kW	4 × 6 mm²	16
18 to 25 kW	4 × 10 mm²	21

The fourth conductor is for earthing the motor, which is generally done inside the terminal box. The use of a sealing bolt to earth the motor is prohibited.

Do not forget that in the case of a star-delta starting, two cables are to be provided, and the cross-section of the conductors can be the same. One of the two cables will contain the fourth conductor for earthing the motor.

4. NUMERICAL VALUES AND ORDERS OF MAGNITUDE

● Principal metals and alloys which are conductors at 0 °C:

Nature of the conductor	Resistance in microhms cm (resistivity)	Temperature coefficient
Copper, electrolytic	1.593	0.00388
Copper, annealed	1.538	0.0045
Aluminium	2.9	0.0039
Silver	1.505	0.0039
Pure iron	9.065	0.00625
Iron wire	13.9	0.00426
Steel wire	15.8	0.0039
Silicon bronze (telephone)	3.84	0.00023
Ferro-nickel	78.3	0.00093
Nickel silver	30	0.00036
Constantan	50	0
Mercury	95	0.00099
Zinc	6	0.0037

Resistivity at t °C: $r_t = r_o (1 + at)$.
r_o resistivity at 0 °C.
a temperature coefficient.
t temperature in degrees Celsius.

● Voltage of a lead accumulator:
    2 volts per element

● Minimum insulation of a low-voltage motor
    per operating volt: 1 000 Ω.

33

HEAT

By the application of the basic principles of heat science, the water processer s able to solve various problems that confront him. A good knowledge of heat generation and thermodynamics is needed, for example, for a proper analysis of problems in drying, heat treatment, and incineration of waste-water sludge, degassing of water, or drying the air supply to an ozonizer. Likewise, in designing equipment for reheating sludge in anaerobic digestion, the laws of heat exchange have to be known. The following chapter gives some basic information for reference in the treatment of such problems.

1. GAS PHYSICS AND THERMODYNAMICS

1.1. Ideal gases

A. Ideal gas law:

$$pV_m = RT$$

p   = gas pressure in pascal (N/m²)
V_m = molar volume in m³
T   = absolute temperature in K
R   = molar constant of ideal gases = 8.314 J/(mol.K)

B. Gay-Lussac law:

$$\frac{m(1 + \alpha t)}{p} = \frac{m_1(1 + \alpha t_1)}{p_1}$$

m, m_1 = specific masses of a gas at pressures p and p_1 and temperatures t and t_1 (°C)

$$\alpha = \frac{1}{273} = 0.003\ 67$$

C. Avogadro-Ampère law:

$$M = 29 \, d$$

relates molar mass M of a gas in g/mol to its density d relative to air at standard temperature and pressure.

D. Specific heats:

   The specific heat is by definition the ratio of the heat capacity to the mass.

The heat capacity is the ratio dQ of a small quantity of heat $\dfrac{dQ}{dT}$ supplied to a system to the temperature increase dT that it causes.   For a gas, the specific heat cp at constant pressure is distinguished from the specific heat c_v at constant volume.

   Specific heats c_p at constant pressure, in kJ/(kg. °C) for some gases at 0 °C and 760 mm Hg:

air...................................... 1.
oxygen 0.92
nitrogen 1.06
ammonia............................... 2.09
carbon dioxide.......................... 0.88
chlorine................................ 0.48
sulphur dioxide......................... 0.63

The values of the ratio $\gamma = \dfrac{c_p}{c_v}$ are very nearly : 1.67 for monatomic gases
                                                                     1.41 for diatomic gases
                                                                     1.33 for polyatomic gases

1.2. Water vapour

A. Saturated vapour: is vapour in the presence of the liquid phase that produces it, and is said to be dry if it contains no water droplets.

● **Enthalpy of a vapour** (total heat): is the total amount of heat needed to convert 1 kg of water at 0 °C into saturated vapour at t °C.   It is the sum of the heat needed to raise the water from 0 to t °C (enthalpy of the water) and the heat of evaporation at t °C, corresponding to the energy needed to convert 1 kg of water at t °C into 1 kg of vapour.

   If vapour enthalpy tables are not available, one can, as a first approximation, use Regnault's formula for the enthalpy as a function of the temperature (°C):

$$H = 2\,538 + 1.276 \, t \quad kJ/kg,$$
or $\qquad\qquad\quad H = 606.5 + 0.305 \, t \quad kcal/kg.$

B. Wet vapour: contains droplets of water, and is described by the quantity x, the mass of vapour (in kg) present in 1 kg of the mixture.

C. Superheated vapour : is at a temperature exceeding that of saturated vapour at the pressure concerned.   As a first approximation, it may be regarded as behaving like an ideal gas.

The enthalpy of superheated vapour may be calculated as
$$H = 2\ 538 + 1.276\ t + c_p\ (t - t_1)\quad kJ/kg$$
or
$$H = 606.5 + 0.305\ t + c_p\ (t - t_1)\quad kcal/kg,$$
where $t - t_1$ is the temperature difference between the saturated vapour and the superheated vapour at constant pressure. As a first approximation, one can take $c_p = 2.1$ kJ/kg, or 0.5 kcal/kg.

In particular, this formula allows an approximate calculation of the enthalpy of water evaporated in an incinerating furnace from which the gases leave at a temperature t; in this case, $t_1 = 373$ K (100 °C).

D. Water vapour diagrams.

● **Density of water vapour as a function of temperature and pressure**

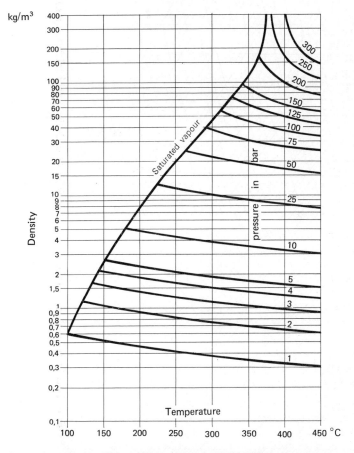

Fig. 652. — According to Chenais (French Standard NF 10101).

● **Mollier diagram**
for water vapour
by E. Schmidt (Munich)

| kcal, atm |

1 kcal = 4,1868 kJ
1 atm = 0,980665 bar

To convert enthalpy differences into flow velocities v, apply the
following equation:

$$v = \sqrt{2\Delta h}$$

or, if Δh is expressed in kcal/kg and v in m/s:

$$v = 91,5 \sqrt{\Delta h}$$

Fig. 653. —

● Boiling-point of water in vacuum

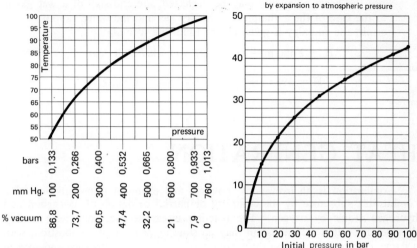

PERCENTAGE OF WATER VAPORIZED
by expansion to atmospheric pressure

Initial pressure in bar

1.3. Wet gases

A. Definitions.
● **Dry temperature:** temperature of an unsaturated wet gas, measured by a dry thermometer.
● **Wet temperature:** temperature of a wet gas saturated by contact with a layer of liquid.
● **Dew-point:** temperature at which the vapour present in a gas begins to condense on cooling.
● **Saturated vapour pressure:** partial pressure of vapour in the gas at the dew-point.
● **Relative humidity:** ratio of partial pressure of water vapour in the gas to the saturated vapour pressure corresponding to the dry temperature of the gas; usually expressed as a percentage.

B. Water vapour content of a gas.
If P is the total pressure of a gas with molar mass M, and p_v the partial pressure of vapour, the water content of the gas in kg per kg of dry gas is:

$$m = \frac{18}{M} \frac{p_v}{P - p_v},$$

or, for air,

$$m = 0.622 \frac{p_v}{P - p_v},$$

For example, with water-saturated air at 20 °C and standard atmospheric pressure P = 1.013 bar, we have $p_v = 0.023$ bar and m = 14.7 g/kg.

C. Enthalpy of wet gas.

Since the heats of mixing may be neglected, the enthalpy of a wet gas is equal to the sum of the enthalpies of the dry gas and the water vapour.

For **air,** the enthalpy is given by the following formula:

$$H = 2\ 490\ m + (1 + 1.97\ m)\ t \quad kJ/kg\ dry\ gas$$

or

$$H = 595\ m + (0.24 + 0.47\ m)\ t \quad kcal/kg.$$

2. HEAT GENERATION

Combustion is an irreversible physical and chemical transformation of a combustible substance by the complete or partial oxidation of its carbon and hydrogen atoms by oxygen. The reaction is exothermic, and yields carbon dioxide and water. Combustion may be slow or fast. Fast combustion, with a flame, cannot begin unless the amounts of combustible substance and oxygen are in a ratio lying between two **inflammability limits** and if the conditions for combustion to be initiated are satisfied.

In practice, combustion is initiated by raising the temperature at a point in the medium to a value equal to or exceeding the ignition temperature.

The **heat of combustion** is the quantity of heat released in the combustion of a body subjected to complete oxidation. For a compound, it is equal to the heat of formation plus the heats released by the combustion of its various elements.

2.1. Calorific value

The heat of combustion is expressed per mole of the oxidized substance. When expressed per unit mass or volume of the fuel, it becomes the **calorific value.**

For liquid or solid fuels, the calorific value is generally given per unit mass in standard atmospheric conditions (1 atmosphere pressure, 25 °C temperature). Then the fuel, oxidizing agent, and combustion products are reduced to the same conditions of pressure and temperature.

For gaseous fuels, the calorific value is generally given per unit standard volume of the gas (1 atmosphere pressure, 0 °C temperature).

The calorific value of a fuel is the same whether it burns in air or in oxygen, but the combustion temperature is much higher for the latter.

The calorific value per unit mass or volume of the fuel is greatly influenced by the presence of inert elements in it. Thus we define:

— **the pure calorific value** as the quantity of heat released in the combustion of unit mass or volume of fuel containing no moisture or inorganic matter or incombustible gas;

— **the dry calorific value** for unit mass or volume of fuel without moisture;

— **the crude calorific value** when all incombustible substances in unit mass or volume are taken into account.

If the fuel contains hydrogen or its compounds, these appear as water in the combustion products. The net and gross calorific value are defined according to the gas or liquid state of this water.

— **The net calorific value (N.C.V.)** does not include the heat of evaporation of the water present in the combustion products, which is thus regarded as remaining in the form of vapour. It is normally expressed in S.I. units in kJ/kg or kJ/m³, and also still commonly in kcal/kg or kcal/m³.

The gross calorific value (G.C.V) includes the heat of evaporation of the water formed during combustion, but not that of any water present in certain gaseous fuels. Its definition thus assumes that all the water from the moisture in the fuel and from the combustion is in the condensed state in the combustion products. The water present in the oxidizing agent (e.g. damp air) is assumed to remain as vapour.

The difference between the gross and net calorific values corresponds to the latent heat of evaporation of water.

If m (kg) is the mass of water vapour in the combustion products, then in S.I. units

$$G.C.V - N.C.V. = 2\,440\ m,$$

2 440 kJ being the latent heat of evaporation of 1 kg of water at 25 °C.

If m_1 (kg) is the mass of water in the fuel and m_2 that of hydrogen, then for a solid or liquid fuel

$$m = m_1 + 9\ m_2,$$

9 kg being the mass of water formed by the combustion of 1 kg of hydrogen in the reaction

$$H_2 + \frac{1}{2} O_2 = H_2O.$$

The G.C.V. is defined on the assumption of complete condensation of water, which hardly ever occurs, because the combustion gases are usually saturated with water vapour at ordinary temperature, and the final temperature of the gases may be below the dew-point. Unlike the N.C.V., the G.C.V. can be measured directly with a bomb calorimeter.

The N.C.V., the only one used in the design of installations, is found from the G.C.V. after determining the content of hydrogen and water in the fuel. The following are some average values of the N.C.V.

Fuel	N.C.V.			
	kJ/kg	kcal/kg	kJ/m³	kcal/m³
Bituminous coal (crude)	31 000	7 400		
Metallurgical coke (crude)	29 700	7 100		
Wastewater sludge (per kg volatile matter) (dry)	21 000	5 000		
Domestic fuel oil	43 000	10 000		
Household waste	5 000 - 8 000	1 200 - 1 900		
Commercial propane			45 800	11 000
Commercial butane			44 600	10 700
Coke oven gas			26 000 - 33 000	6 200 - 7 900
Blast furnace gas			22 000 - 37 000	5 300 - 8 800
Lacq gas			49 300	11 800
Digestion gas			22 000	5 300

The difference between the G.C.V. and the N.C.V. for dry organic matter is 10-15 %, and for domestic fuel oil 5-9 %.

Economic studies usually refer to the concept of **tons of coal equivalent.** This is based on a ton of coal with N.C.V. 6 500 kcal/kg (about 27 200 kJ/kg).

2.2. Combustion

There are two main types of combustion: **conventional,** in which the oxidizing agent is atmospheric air, and **oxygenated combustion,** where pure oxygen is supplied.

In conventional combustion, the following kinds are distinguished:

— **theoretical combustion,** where the amount of air used is equal to the combustion capacity of the fuel (defined below) and the combustion is incomplete;

— **oxidizing and semi-oxidizing combustion,** where the amount of air used exceeds the combustion capacity of the fuel, and the combustion is respectively complete and incomplete;

— **reducing and semi-reducing combustion,** where the amount of air used is less than the combustion capacity of the fuel, and it is respectively completely and partly absorbed;

— **mixed combustion,** giving smoke that contains unused oxygen and unburnt matter, as is sometimes found in practice because of technical difficulties.

Neutral combustion is complete and uses exactly the amount of oxidizing agent necessary and sufficient for it to be so. This is a theoretical concept and is difficult to achieve in practice. It serves, however, in the definition of various characteristic parameters of a combustion process.

The **combustion capacity** of a fuel is the amount of air just necessary for neutral combustion of unit amount of the fuel. For solid and liquid fuels, it may be taken, as a first approximation, to be about 1 m³ (STP) of air per kg when the N.C.V. is 4 000 kJ/kg. For gaseous fuels, the value is the same but per m³ (STP), not per kg. For sewage sludge it is about 6.5 m³ (STP) per kg of organic matter.

The **smoke-generating capacity** of a fuel is the quantity of smoke produced by neutral combustion of the fuel. The usual practical concept is the **wet-smoke-generating capacity,** the water vapour being assumed not condensed. As a first approximation, this is given by Véron's formulae:

— solid fuel : 1 m³ (STP) per kg per 3 500 kJ/kg of N.C.V.
— liquid fuel : 1 m³ (STP) per kg per 3 800 kJ/kg of N.C.V.
— gas fuel : 1 m³ (STP) per m³ (STP) per 3 500 kJ/m³ (STP) of N.C.V.
      (valid only for gases with N.C.V. exceeding 8 000 kJ/m³ (STP).

3. HEAT EXCHANGE

3.1. Definitions

Heat can be propagated by three different processes: conduction, convection, and radiation.

Conduction is the transfer of heat between two bodies in contact at different temperatures, or between two parts of the same body at different temperatures

The heat flux Φ carried by conduction over a length x across an area S perpendicular to the heat flux is given by Fourier's formula

$$\Phi = \lambda S \frac{(\theta_1 - \theta_2)}{x},$$

where $\theta_1 - \theta_2$ is the temperature drop over the distance x, and λ is the thermal conductivity of the material, expressed in kcal/(m.h. deg C.), or better in W/(m.K.)

For most solids, λ is an almost linear function of the temperature: $\lambda = \lambda_0 (1 + \alpha \theta)$; α is usually positive for insulators and negative for metals, except aluminium and brass. However, the thermal conductivity varies relatively little with the temperature. Between 0 and 100 °C, the following values may be taken as a first approximation :

	W/(m.K)	kcal/(m.h.deg C)
— Mild steel (1 % carbon)	15	39
— Stainless steel (Z2 CN 18-10)	45	13
— Pure copper	384	330
— Aluminium	200	175
— Brass (30 % zinc)	99	85
— Glass wool	0.038	0.033
— Expanded cork	0.040	0.035
— Expanded polystyrene	0.035	0.030
— Calm water at ambient temperature	0.58	0.5
— Still air at ambient temperature	0.027	0.023

Fourier's equation may also be written

$$\Phi = \frac{\Delta\theta}{R} \quad \text{where} \quad R = \frac{x}{\lambda S}$$

acts as a calorific resistance.

For conduction through several materials in series with a total temperature drop $\Delta\theta$, we can write

$$\Phi = \frac{\Delta\theta}{\Sigma R}$$

Convection is the transfer of heat within a fluid from a solid by movements due to density differences (natural convection) or caused by mechanical means (forced convection).

In practice, heat transfer between a solid at temperature θ and a fluid at temperature θ_1 involves both convection and conduction, which especially complicates the phenomenon. An overall transmission coefficient k is then defined by

$$\Phi = kS\,(\theta - \theta_1)$$

In given units, the value of k depends on various physical properties of the fluid, its flow rate, and the geometry of the solid. The values of k may therefore vary considerably, as shown by the following:

	$W/(m^2.K)$	$kcal/(m^2.h.deg\,C)$
— boiling water	1 700-50 000	1 500-45 000
— film of condensed water vapour	5 800-17 000	5 000-15 000
— heating or cooling of water	300-17 000	250-15 000
— heating or cooling of air................	1.2-45	1-40

Radiation is the transmission of heat in the form of radiant energy, occurring without material support and therefore possible even in a vacuum.

The Stefan-Boltzmann law gives as the heat flux carried by emitted radiation

$$\Phi = k\Sigma ST^4$$

where T is the absolute temperature of the radiating body, Σ an emission factor that is zero for a perfect reflector and unity for a black body, and k a dimensional constant.

In practice, heat exchange takes place by at least two of these three processes simultaneously.

In the field of water treatment, there may be radiative and convective heat transfer in sludge incineration furnaces, and in particular conductive and convective transfer in heat exchangers.

3.2. Heat exchangers

A. Definitions

A heat exchanger allows the transfer of heat between two fluids at different temperatures. It may have various forms, e.g. boilers, evaporators, condensers, and exchangers proper, where the two fluids are in the same state.

Exchangers may be divided into three classes according to the relative direction of motion of the fluids: concurrent with parallel flows in the same direction, countercurrent with parallel flows in opposite directions, and cross-flow with non-parallel flows, often at right angles.

In sludge treatment, countercurrent tube exchangers are the most common.

B. *Calculation of exchange area*

The quantity of heat passing through a wall is $Q = kSd_m$

— where S is the exchange area (m^2);
— d_m is the mean temperature difference across the wall, based on the logarithmic mean of the fluid entry and exit temperatures;
— and k is the overall transfer coefficient in W/(m^2. K) or kcal/(m^2.h.deg C), which depends on the nature and conditions of the flows and on the characteristics of the wall;
— Q is in watts or in kcal/h, according to the units adopted.

● **Determination of the overall transfer coefficient**

For heat transfer by conduction and forced convection (as with heat exchangers in sludge treatment), a mathematical formulation based on dimensional analysis gives a theoretical transfer coefficient for a cylindrical surface. Actually, because there is some variation in the nature of the sludge and the interstitial liquid, the transfer coefficient is mostly found by experiment.

Two examples:

— in sludge digestion, the exchange coefficient may have values up to 1 300 W/(m^2. K) or 1 100 kcal/(m^2.h.deg C) for flow rates between 1 and 2 m/s;
— in heat treatment of sludge with sludge-sludge exchangers, the exchange coefficient may have values up to 350 (W/m^2. K) or 300 kcal/(m^2.h.deg C) for flow rates between 0.5 and 1 m/s.

● **Determination of the logarithmic mean temperature**

Let a countercurrent heat exchanger have two fluids circulating as shown:

fluid 2 (hot)	
T_0 ⟶	T_1
fluid 1 (cold)	
t_1 ⟵	t_0

The logarithmic mean is

$$d_m = \frac{d_1 - d_2}{\log (d_1/d_2)} \qquad \text{with } d_1 = T_0 - t_1,\ d_2 = T_1 - t_2.$$

Hausband's table (below) gives $\dfrac{d_m}{d_1}$ as a function of $\dfrac{d_2}{d_1}$

when $d_1 > d_2$

Hausband's table

$\dfrac{d_2}{d_1}$	$\dfrac{d_m}{d_1}$	$\dfrac{d_2}{d_1}$	$\dfrac{d_m}{d_1}$	$\dfrac{d_2}{d_1}$	$\dfrac{d_m}{d_1}$
0.01	0.215	0.16	0.458	0.70	0.843
0.02	0.251	0.18	0.478	0.75	0.872
0.03	0.277	0.20	0.500	0.80	0.897
0.04	0.298	0.22	0.518	0.85	0.921
0.05	0.317	0.24	0.535	0.90	0.953
0.06	0.335	0.26	0.557	0.95	0.982
0.07	0.352	0.30	0.583	1.00	1.000
0.08	0.368	0.35	0.624		
0.09	0.378	0.40	0.658		
0.10	0.391	0.45	0.693		
0.11	0.405	0.50	0.724		
0.12	0.418	0.55	0.756		
0.13	0.430	0.60	0.786		
0.14	0.440	0.65	0.815		

N.B.: It can be shown mathematically that, if $0.5 < \dfrac{d_2}{d_1} < 2$, the logarithmic mean differs from the arithmetic mean by less than 5 %. This justifies the use of the arithmetic mean in the majority of heat exchangers employed for sludge treatment.

C. *Heat losses in an exchanger*

Every exchanger loses heat to the exterior, to an extent governed mainly by:
— the movement and temperature of the air around the equipment;
— the temperatures of the fluids circulating in it;
— the nature and colour of the exchanger materials.

These heat losses can be restricted by suitable lagging.

4. NUMERICAL DATA

A. Linear expansion coefficients of some solids.

Substance	Density kg/dm³	Linear expansion coefficient 20-100 ⁰C [m/(m . deg C)] × 10⁶ or 10⁶ K⁻¹
Metals		
Aluminium	2.70	23.8
Brass (35 % zinc)	8.45	20.3
Bronze	8.9	15.5
Cast iron (grey)	7.2	11-12
Cast iron (spheroidal graphitized)	7.4	17.5-19.5
Copper	8.9	16.8
Iron	7.87	11.4
Lead	11.4	28.6
Steel (ordinary carbon)	7.85	12.4
Steel (heat-resistant)	7.9	15.5
Steel (austenitic stainless)	7.9	16.5
Steel (ferritic stainless)	7.7	10.5
Tin	7.28	27
Titanium	4.5	8.35
Zinc	7.14	30
Plastics		
Acetal (resin) (Delrin)	1.4	130
Glass fabric reinforced epoxide (40 % resin)	1.7-1.8	10
Glass fabric reinforced polyester (40 % resin)	1.8	30
Polyamide 6 (Nylon)	1.12-1.15	70-140
Polyamide 11 (Rilsan)	1.04	110-150
Polycarbonate (Makrolon)	1.20	60-70
Polyvinyl chloride	1.35-1.45	50-180
Chlorinated polyvinyl chloride	1.50-1.55	60-80
Low-pressure polyethylene	0.95	110-140
Polymethyl metacrylate (Plexiglas)	1.17-1.20	50-90
Polypropylene	0.9	70-150
PTFE (Teflon)	2.1-2.3	80-120

B. Melting-points and enthalpies of fusion of some metals.

Metal	t_f (°C)	Enthalpy of fusion	
		J/mol	kcal/mol
Aluminium	660	10 660	2 550
Chromium	1 550	16 430	3 930
Copper	1 083	13 000	3 110
Iron	1 530	14 890	3 560
Lead	327	5 200	1 244
Nickel	1 455	17 570	4 200
Silver	960	11 290	2 700
Tin	231.8	7 190	1 720
Zinc	419	6 670	1 595

Part Six

Legislation

L'ARC DE TRIOMPHE

LEGISLATION AND REGULATIONS

Legislation and regulations with regard to water treatment, like the techniques whose development tends to be shaped by them, are constantly evolving in all countries and international organisations. The resultant documentation is very bulky and must continually be brought up to date.

It is therefore a rather hopeless task to try to summarize all this legislation in the limited number of pages of a work, the new editions of which are published at intervals of several years.

The designers and users of water treatment plant will nevertheless find in this chapter, in a condensed form, a large amount of useful facts and figures which will continue to be valid. These facts and figures can therefore be used as guidelines in the assessment of results anticipated or obtained in each particular case.

The present chapter deals with the following:

1. Bibliography and useful addresses (cf. below)
2. Legislation covering drinking water and swimming pool water (page 1124)
3. Legislation covering domestic sewage and industrial waste water (page 1140)
4. Miscellaneous legislation (page 1147).

1. GENERAL INFORMATION

1.1. Works to consult

(See also Bibliography on page 1151)

● *Code permanent Environnement et Nuisances*
Éditions Législatives et Administratives (up dated quarterly), 22, rue de Cronstadt - 75015 Paris.
● *Guide de l'Eau*
Éditions Johannet et Fils, Paris, 1976.
● *Le Moniteur des Travaux Publics et du Bâtiment*
L'eau et la propriété privée. Les loisirs et les sports. Les collectivités locales. L'organisation administrative. Textes officiels et Commentaires. 2ᵉ édition, Paris, September 1974.
● *Législation étrangère dans la lutte contre la pollution des eaux.*
D.I.P.P.N., Édition A.F.E.E., Paris, September 1974.

● *European Standards for Drinking Water.*
W.H.O. Publications, Geneva, Switzerland 1971.

● *International Standards for Drinking Water.*
W.H.O. Publications, Geneva, Switzerland, 1971-1972.
On Sale in:
France: Librairie Arnette, 6, rue Casimir-Delavigne - 75006 Paris.
Belgium: Office International de Librairie, 30, avenue Marnix - Brussels.
Spain: Libreria Diaz de Santos, Lagasca 95 - Madrid 6
U.S.A.: The American Public Health Association, Inc., 1740 Broadway - New
York 10019.
Italy: Edizioni Minerva Medica, Via Lamarmora 3 - Milano.
Netherlands: N.V. Martinus Nijhoff's Boekhandel en Uitgevers Maatschappij,
Lange Voorhout 9 - The Hague.
West Germany: Govi-Verlag GmbH, Beethovenplatz 1-3 - Frankfurt am Main 6.
U.K.: H.M. Stationery Office, 49 High Holborn - London WC1.
Sweden: Aktiebolaget C.E. Fritzed, Kungl. Hovbokhandel, Fredsgatan 2 - Stock-
holm 16.
Switzerland: Medizinischer Verlag, Hans Huber, Länggass Strasse 6 - 3000 Bern 9.

● *La pollution due à l'industrie des pâtes et papiers. Situation actuelle et tendances.*
O.C.D.E., Paris, 1973.

● *Problème de la pollution trans-frontière.*
O.C.D.E., Paris, 1974.

● The law and practice relating to pollution control in :
Volume 1 : In France: by C.A. COLLIARD for Environmental Resources Ltd.
Volume 2 : In EEC countries: by J. M. LOUGHLIN.
Volume 3 : In West Germany: by H. STEIGER and O. KIMMINICH.
Volume 4 : In Italy: by P. DELL'ANNO.
Volume 5 : In Great Britain: by J. McLOUGHLIN.
Volume 6 : In Belgium and Luxemburg: by J. M. DIDIER and others.
Volume 7 : In Holland: by J. J. DE GRAEFF and J.M. POLAK.
Graham & Trotman Limited, Great Britain, 1976.

● *Environmental Regulation Handbook.*
Environment Information Center, Washington, 1976.
A Guide to Environmental Legislation in the Fifty States and the Distirict of
Columbia.
Volume 1 : Government Aid.   Air pollution.   Land use, Mobil sources.
Volume 2 : Noise, Pesticides.   Radiactive materials.   Solid wastes.
Volume 3 : Toxic substances.   Water pollution.   States laws and regulations.

● *Brochures du Journal Officiel.*
26, rue Desaix - 75732 Paris Cedex 15.

1.2. Useful addresses

1.2.1. FRANCE

● Ministère de l'Environnement (Ministry of Environment), 14, boulevard du Général-Leclerc - 92151 Neuilly-sur-Seine - Tél : (1) 758-12-12.

● Ministère de la Santé (Ministry of Health), 1, place Fontenoy - 75007 Paris - Tél. : (1) 567-54-00.

● Conseil Supérieur d'Hygiène Publique de France, (Higher Council of Public Health of France), 1, place Fontenoy - 75007 Paris - Tél. : (1) 567-55-44.

● Centre de Formation et de Documentation sur l'Environnement Industriel (C.F.D.E.), (Training and Documentation Centre on the Industrial Environment), 11 *bis*, rue Léon-Jouhaux - 75010 Paris - Tél. : (1) 607-66-23.

● Association Française pour l'Étude des Eaux (A.F.E.E.), (French Association for the Study of Water), 23, rue de Madrid - 75008 Paris - Tél. : (1) 522-14-67 - (1) 522-99-61 : (1) 522-89-82.

● Association Française pour la Protection des Eaux, (French Association for the Protection of Water), 195, rue Saint-Jacques - 75005 Paris - Tél. : 326-70-53.

● Basin Financial Agencies:

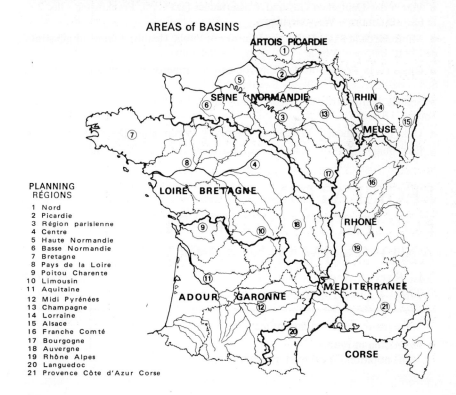

AREAS of BASINS

PLANNING RÉGIONS

1 Nord
2 Picardie
3 Région parisienne
4 Centre
5 Haute Normandie
6 Basse Normandie
7 Bretagne
8 Pays de la Loire
9 Poitou Charente
10 Limousin
11 Aquitaine
12 Midi Pyrénées
13 Champagne
14 Lorraine
15 Alsace
16 Franche Comté
17 Bourgogne
18 Auvergne
19 Rhône Alpes
20 Languedoc
21 Provence Côte d'Azur Corse

— Agence de Bassin « Artois-Picardie » 764, boulevard Lahure B.P. 818 - 59508 Douai - Tél. : (20) 87-01-94.

— Agence du Bassin « Seine-Normandie » 10-12, rue du Capitaine-Ménard - 75732 Paris Cedex 15 - Tél. : (1) 578-12-00.

— Agence de Bassin « Rhin-Meuse », Le Longeau, B.P. 36, Rozerieulles - 57160 Moulin-les-Metz - Tél. : (87) 60-48-88.

— Agence de Bassin « Loire-Bretagne », avenue Buffon - 45018 Orléans Cedex - Tél. : (38) 63-08-16.

— Agence de Bassin « Adour-Garonne », 84, rue Férétra - 31078 Toulouse Cedex - Tél. : (61) 53-21-51.

— Agence de Bassin « Rhône-Méditerranée-Corse », 31, rue Jules-Guesde - 69310 Pierre-Bénite - Tél. : (78) 50-16-40.

1.2.2. OTHER COUNTRIES

● Abwassertechnische Vereinigung e.V., Markt 1, (Stadthaus) - 5205 St-Augustin 1, West Germany.

● Institut für das Recht der Wasserwirtschaft der Universität Bonn, Lennestrasse 35 - D 5300 Bonn, West Germany,

● Verein des Deutschen Gas-und Wasserfaches (DVGW), Frankfurter Allee 27 - D 6236 Eschborn - West Germany.

● Ministère de la Santé Publique (Ministry of Public Health), Centre administratif de l'État, Bâtiment Vésale - 1010 Brussels - Belgium.

● Office de la Protection de l'Environnement (Environment Protection Board), Schwartztorstrasse 53 - Bern - Switzerland.

● Statens Naturwardsverk, Kommunbyran Vattenvardssektionen, Fach 17120, Solna 1 - Sweden.

● U.S. Environmental Protection Agency - Washington D.C. 20460, U.S.A.

● Environmental Protection Service, Environment Canada - Ottawa, Ontario K1A OH3 - Canada.

2. LEGISLATION FOR DRINKING WATER AND SWIMMING POOLS
2.1. Recommendations of the World Health Organization (1972) - (W.H.O.) Palace of Nations, Geneva

The main recommendations may be summarized as follows:

● **Expression of results**
The notation for expressing results is mg/l.

The expression 'parts per million' (p.p.m.) should be progressively abandoned. Wherever possible, chemical components should be expressed in ions. Turbidity should be expressed in units of turbidity, and colour in units of colour based on the platinum-cobalt scale. Volumes should be expressed in millilitres (ml) and the temperature in degrees Celsius (^{o}C). In bacteriological examinations, the total number of microorganisms developing on solid media should be expressed in significant figures as colonies per millilitre of water, the medium, time and temperature of incubation being stated. Estimates of the number of coliform organisms—*Escherichia coli* and other microorganisms indicative of pollution— should be given in terms of the 'Most Probable Number' per 100 ml (MPN/100 ml).

In reporting chemical analyses, the sensitivity, accuracy and precision of the method should be indicated. This includes the proper use of significant figures and the expression of reliability limits.

● **Examination and sampling frequency**

Treated water, as it enters the distribution system from each treatment point, should undergo the following examinations:
— bacteriological analysis once a day (at least once a week);
— a check several times a day on each stage of the chemical treatment, with recording of the results;
— *in situ* inspection at least twice a year by engineering and sanitation experts acting on behalf of the responsible authorities.

For untreated water entering the distribution system, the following maximum intervals between routine examinations are proposed:

Population served		Maximum interval between successive samplings
Up to	20 000	1 month
20 001 to	50 000	2 weeks
50 001 to	100 000	4 days
more than	100 000	1 day

For samples taken at several points on the distribution system, whether the water has previously been treated or not, the following collection procedure is proposed:

Population served	Maximum interval between successive samplings	Minimum number of samples to be taken from entire distribution system
Up to.......... 20 000	1 month	1 sample per 5 000 of population per month
20 001 .... 50 000	2 weeks	
50 001 to........ 100 000	4 days	
More than....... 100 000	1 day	1 sample per 10 000 of population per month

● **Bacteriological quality**

The following standards have been laid down for water distributed in pipe systems:

a) Water treated, for example, by chlorination: effective treatment followed by chlorination or another form of disinfection, will normally provide water free of coliforms, however polluted the initial raw water. In practice, this means that no random sample of 100 ml of water will reveal the presence of coliforms. Any sample taken at the inlet to the network and not conforming to this standard should call for an immediate inspection of the purification process and the sampling method. However, when examining chlorinated water, the samples assumed to be positive must always be subjected to an appropriate confirmation test.

b) Untreated water: in this case the water entering the distribution network will be considered unsatisfactory if Escherichia coli are shown to be present in a 100 ml sample. In the absence of E. coli, a maximum of 3 coliforms can be tolerated in a sample of 100 ml taken periodically from an undisinfected network, provided that the network is regularly and frequently examined and that the intakes and reservoirs are considered to be satisfactory. If sampling repeatedly shows that coliforms are present, steps must be taken to discover, and if possible to eliminate, the cause of the pollution. Where the number of coliforms exceeds 3 per 100 ml the water must be considered unusable without prior disinfection.

The following recommendations are made for samples taken from distribution networks:

1. In any one year, at least 95 % of the 100 ml samples must be free of coliforms.
2. No single 100 ml sample must contain E. coli.
3. No single sample must contain more than 10 coliforms per 100 ml.
4. Coliforms must not be detected in 2 successive 100 ml samples.

If examination of a sample reveals the presence of coliforms, at least one more sample must be taken. If 1 to 10 coliforms (or more in some samples) per 100 m are regularly found, there is reason to believe that undesirable substances are entering the water, and urgent measures are necessary to discover and eliminate the cause of the pollution.

● **Standards of chemical and physical quality—toxic substances**

The table below gives the maximum content of certain dangerous substances in drinking water:

Substances	Provisional maximum Concentrations permissible
Lead (Pb)	0.10 mg/l
Selenium (Se)	0.01 mg/l
Arsenic (As)	0.05 mg/l
Cyanide (CN)	0.05 mg/l
Cadmium (Cd)	0.01 mg/l
Total mercury (Hg)	0.001 mg/l

If the content of any one of these substances exceeds the limit indicated, the water cannot be distributed to the public for domestic use.

In addition, to the substances listed above, the presence of others (barium, beryllium, cobalt, tin, molybdenum, nitrilotriacetates, thiocyanates, uranium, and vanadium) should be monitored in drinking water, but at present there is not enough information to allow provisional maximum concentrations to be allocated. In the case of fluorides, it is recommended that the content be limited as follows:

Mean annual maximum daytime temperatures (°C)	Recommended maximum and minimum concentrations for fluorides (in F) (mg/l)	
	Min. concentration	Max. concentration
10.0 - 12.0	0.9	1.7
12.1 - 14.6	0.8	1.5
14.7 - 17.6	0.8	1.3
17.7 - 21.4	0.7	1.2
21.5 - 26.2	0.7	1.0
26.3 - 32.6	0.6	0.8

A nitrate content of 45 mg/l (as NO_3) is the safe limit for the health of some children under 12 months of age.

● **Chemical substances tolerated in water**

In view of the wide variations in the composition of water in different parts of the world, rigid standards of chemical quality cannot be established. The imits designated "desirable" below apply to water that would be generally acceptable to the consumer. They are given as an indication.

Substance or property	Maximum desirable concentration	Maximum permissible concentration
Total solids	500 mg/l	1 500 mg/l
Colour	5 units [1]	50 units [1]
Turbidity	5 units [2]	25 units [2]
Taste	acceptable	
Odour	acceptable	
Iron (Fe)	0.1 mg/l	1.0 mg/l
Manganese (Mn)	0.05 mg/l	0.5 mg/l
Copper (Cu)	0.05 mg/l	1.5 mg/l
Zinc (Zn)	5 mg/l	15 mg/l
Calcium (Ca)	75 mg/l	200 mg/l
Magnesium (Mg)	30 mg/l if the water contains at least 250 mg/l of sulphates	150 mg/l
Sulphates (SO_4)	200 mg/l	400 mg/l
Chlorides (Cl)	200 mg/l	600 mg/l
pH	7 to 8.5	6.5 to 9.2
Phenolic compounds (as phenol)	0.001 mg/l	0.002 mg/l
Anionic detergents	0.2 mg/l	1.0 mg/l
Mineral oils	0.01 mg/l	0.30 mg/l
Total hardness	2 meq/l (100 mg/l $CaCO_3$)	10 meq/l (500 mg/l $CaCO_3$)

1. Platinum-cobalt colour scale.
2. Turbidity units.

● **Maximum permissible concentrations of radioactive elements**

The radioactivity levels indicated below are considered to be the upper limits applicable to drinking water intended for normal use:
— total alpha radioactivity: 3 pCi/l;
— total beta radioactivity: 30 pCi/l.

These levels apply to the mean of all radioactivity analyses carried out over a period of 3 months.

A beta radioactivity level of $\leqslant$ 3 pCi/l is acceptable even if it is entirely imputed to Radium 226.

A beta radioactivity level of $\leqslant$ 30 pCi/l is acceptable even if it is entirely imputed to Strontium 90.

For radioactivity levels exceeding those indicated above, additional analysis is necessary.

2.2. French regulations governing the quality of water for human consumption

The Conseil Supérieur d'Hygiène Publique de France (Higher Council of Public Health of France) lays down the following standards, in accordance with the Public Health Code:

● **Quality control of water for human consumption**

Water intended for drinking purposes must have the characteristics defined by the Ministry of Public Health on advice from the Higher Council of Public Health of France.

Before distribution to the public, a water must satisfy the following conditions:
1. It must be free of all parasitic and pathogenic organisms;
2. If untreated, it must be free of *Escherichia coli* (in 100 ml of water), *Streptococcus faecalis* (in 50 ml of water) and sulphite-reducing *Clostridium* (in 20 ml of water).

If treated, it must not contain *E. coli* (in 100 ml of water) or *Streptococcus faecalis* (in 50 ml of water). The presence of small numbers of sulphite-reducing *Clostridium* in treated water may be tolerated and is not sufficient on its own to render the water unfit for drinking;
3. The degree of coloration must not exceed 20 units (platinum-cobalt color-imetric scale), and turbidity must not exceed 15 drops of a 1/1 000 alcoholic solution of gum mastic, under normal operating conditions. Under exceptional circumstances, however, 30 drops of mastic (in 50 ml of optically-clear water) may be tolerated for a limited period;
4. Its clogging capacity must not exceed 0.1 through the presence of elements in suspension; it must be free of algae and other live cells.
5. The degree of chemical pollution and concentrations of toxic and other undesirable substances must not exceed the figures shown in the following table:

	Maximum concentration (in milligrammes per litre)
Lead (as Pb)	0.1
Selenium (as Se)	0.05
Fluorides (as F)	1.0
Arsenic (as As)	0.05
Hexavalent chromium Cyanides	Less than the minimum quantity detectable by analysis
Copper (as Cu)	1.0
Iron (as Fe)	Total 0.3 comprising 0.2 Fe and 0.1 Mn
Manganese (as Mn)	
Zinc (as Zn)	5.0
Phenol compounds (as phenol)	Nil

6. The total mineral content must not exceed 2 grammes per litre. Furthermore, the water must be free from any unpleasant odour or taste.

7. The radioactivity of the water must not be greater than that defined by the regulations in force. The monitoring of radioactive elements will be carried out by laboratories approved by the Ministry of Social Affairs on the advice of the Central Department for Protection against Ionizing Radiation, which is responsible for the coordination of measurements and the centralization of results.

It is also preferable that the concentration of certain elements should not exceed the following figures:

$$
\begin{aligned}
&\text{Magnesium (as Mg)} && 125 \text{ mg/l} \\
&\text{Chlorides (as Cl)} && 250 \text{ mg/l} \\
&\text{Sulphates (as } SO_4) && 250 \text{ mg/l}
\end{aligned}
$$

As regards nitrates, account must be taken of the fact that a concentration higher than 10 mg/l (as N) or 44 mg/l (as NO_3) is liable to cause disorders, particularly in nursing infants.

Furthermore, it is desirable for the total hardness to be less than 30 (French) degrees, the optimum appearing to be about 12 to 15 (French) degrees.

When it is necessary for water to be treated chemically, steps must be taken to ensure that the introduction of the products necessary for the treatment does not react unfavourably upon its initial composition, except of course where such a modification is desirable in itself. In particular, water treated with aluminium sulphate must not contain a quantity of aluminium ions greater than those contained in the raw water.

In so far as concerns the treatment of water with chlorine or its compounds, the amount to free chlorine must not exceed 0.10 mg per litre in normal working conditions at the delivery point into the distribution network.

Bottled drinking water must comply with the following requirements :

1° It must be free of all parasites and pathogenic germs;

2º It must be free of *Escherichia coli* in 100 ml of water, *Streptococcus faecalis*, in 50 ml of water and sulphite-reducing *clostridium* in 20 ml of water;

3º It must be free from any unpleasant odour or taste;

4º It must be free of algae and other live cells;

5º Its degree of coloration must not exceed 5 units (platinum-cobalt colorimetric) scale), and its turbidity must not exceed 5 drops of mastic (in 50 ml of optically-clear water);

6º The degree of chemical pollution and concentrations of toxic or other undesirable substances must not exceed those specified for public water supplies, with the additional condition that the iron content (as Fe) must not exceed 0.1 mg/l and the manganese content (as Mn) must not exceed 0.05 mg/l.

Water to be used for the preparation or preservation of food, of consumable merchandise and ice for human consumption must also satisfy certain conditions. Ice creams and ices must not contain:

— more than 300 000 mesophilic aerobic germs per millilitre of compact product.

— more than 100 coliform bacteria per millilitre of compact product.

Furthermore, they must not contain any *Escherichia coli* per ml, nor any pathogenic germ, in particular *staphylococcus*, in 0.1 ml, nor any *salmonella* in 25 ml.

Testing must be carried out in accordance with the methods laid down in bye-laws.

The quality of drinking waters is controlled by means of periodic analyses carried out by laboratories specially approved by the Ministry of Public Health and Population.

● **Periodic analyses to inspect water for human consumption**

The number of periodic analyses is determined by the Prefet, with a minimum of three per year. There are three types :

Type I   : Complete analysis;

Type II   : Brief (or observational) analysis;

Type III : Short observational analysis.

An interministerial circular gives general instructions on **bacteriological analyses of water and ice for human consumption.** It indicates the procedure for taking samples for analysis, and for their transportation and preservation. It sets out in detail the microbe detection tests and counts to be carried out, as well as the required presentation of results. In addition, it shows how to interpret the results of the bacteriological analyses depending on the nature of the water analysed: naturally pure water, not treated by an antiseptic, water given some kind of corrective treatment, not using a disinfecting agent, water given corrective treatment by disinfection by physical means (ultraviolet rays) or by chemical means (chlorine, chloramines, hypochlorites, ozone).

2.3. Conditions for using various substances in drinking water treatment

A. In France:

● **Treatment of water with polyphosphates** (circular of 14 April 1962).

The Higher Council for Public Health of France advocates introducing into drinking water small doses of polyphosphates in order to prevent excessive scaling of pipes and distribution systems: the level of P_2O_5 in drinking water supplies must not exceed 5 mg/l. Treatment plans must be authorized by the Health Authorities. Before being treated with polyphosphates, the water must have all the chemical and bacteriological characteristics of drinking water and a temporary hardness at least equal to 10 French degrees.

● **Use of cation-resins in drinking water treatment and the production of foodstuffs** (circular of 3 May 1963).

Catio-carboxylic resins, and catio-sulphonic resins contained in appliances used on drinking water treatment, and authorized as such by the Higher Council of Public Health of France, are exempt from the regulations resulting from the inclusion in table C of poisonous substances.

They must be supplied in sealed packaging, covered by a label which conforms to the regulations governing dangerous substances, and mentioning that it is authorized by the Ministry of Public Health and Population.

This label is not required for resins supplied in domestic appliances, but such appliances must bear the date of authorization.

Cation-resins can only be authorized for treating drinking water if the soluble extract obtained after six months of recycling distilled water through these resins operating in a hydrogen cycle, is lower than 100 mg/l, and when they are used exclusively to exchange certain undesirable ions for those ions contained in the resins.

In accordance with the decree of 19 October 1976 projects for the treatment of drinking water by cation resins must be submitted to the Higher Council of Public Health of France when the treated water is distributed in a public drinking water system.

● **Treatment of drinking water with silicates** (circular of 5 June 1964).

The silicates used must not contain any substance which may affect the health of consumers.

The concentration used, expressed as SiO_2 must be lower than 10 mg/l.

Any project for setting up treatment plant involving cation-resins, silicates, possibly combined with polyphosphates or another authorized product, to prevent the precipitation of the iron and manganese in water to be supplied to a public drinking water system, must be accepted by the regional health authority which will automatically submit it for examination by the Higher Council of Public Health of France.

B. In Belgium: Law of 20 June 1964: decree of 18 May 1965 issued by the Ministry of Public Health and the Family.

Products	Conditions
Chlorine, Chloramines Chlorine dioxide Sodium or calcium hypochlorite	Free residual chlorine 0.25 ppm
Sodium thiosulphate SO_2	Conditions imposed by residual chlorine so that the pH of the acidic water is made approximately 7
$CuSO_4$	< 1 mg/l Cu
$Al_2(SO_4)_3$	< 1 mg/l Al
$KMnO_4$	< 0.3 mg/l Mn
$FeCl_3 - FeCl_2$	< 1 mg/l Fe
Sodium silicate	10 mgl/ SiO_2

C. In Germany: West German Official Gazette of 27.6.1960

Products	Conditions
Chlorine, Hypochlorites ClO_2, NH_4OH and ammonium salts	$\leqslant 0.3$ mg of active Cl_2 and 0.6 mg of ammonium ion. The content of chlorine in drinking water may be as high as 0.6 mg/l
Ozone	Authorized
K, Na, Ca salts of mono and poly-phosphoric acids Silicic acid and silicates	$\leqslant 5$ mg/l (as P_2O_5) $\leqslant 40$ mg/l (as SiO_2)
Ag, AgCl, Silver chloride—silver sulphate complexes	$\leqslant 0.1$ mg/l Ag
Thiosulphates	$\leqslant 0.5$ mg/l (as S_2O_3)
Clay, activated carbon	$\leqslant 0.5$ mg/l

D. In Great Britain: lists of authorized polyelectrolytes are published regularly in the journal *Water Treatment & Examination*, for example: (1970) 19, No. 4 p. 329; (1971) 20, No. 2 p. 94.

E. In the U.S.A.: lists of authorized polyelectrolytes are published in the papers of the E.P.A. and in the journal *J.A.W.W.A.*, for example: *J.A.W.W.A.* (1975) No. 8, August p. 468-470.

2.4. Table comparing standards for drinking wate

COUNTRIES	INTER-NATIONAL STANDARDS	EEC			UNITED STATES	CANADA
References	applied in Great Britain, Ireland, Denmark, Finland, Austria, Nigeria, South Africa W.H.O. 1972	Extract from Official Journal of EEC No. C 214/6 to 11, 18/9/75 Directives			EPA Federal Register 24/12/75 No. 51 and 248, Vol. 40, and 9/7/76, No. 133, Vol. 41 Environmental Regulation Handbook 1976	Standards and objectives for drinking water i Canada October 1969
Parameters	Max. or min.-max concentrations	Recom-mended level	max.	min.	Max. or min.-max. concentrations	Max. or min.-max concentrations
1. Organoleptic factors:						
Colour Pt　　　　mg/l		5	20		3-15	15
Turbidity silica　　mg/l		5	10		1-5	5
Temperature　　　°C		12	25			
2. Physico-chemical factors:						
pH		6.5-8.5	9.5	6		6.5-8.3
Conductivity　　μ S/cm		400	1 250			
Hardness　　　　Fr.°		35		10		18
Calcium　　　　mg/l		100		10	80-100	200
Magnesium　　　mg/l		30	50	5	80-100	150
Sodium　　　　mg/l		< 20	100		(1)	
Potassium　　　mg/l		⩽ 10	12			
Aluminium　　　mg/l			0.05			
Sulphates SO⁻₄　mg/l	200-400	5	250		(1)　　250	500
Chlorides Cl⁻　　mg/l		5	200		250	250
Free chlorine　　mg/l					0.2-0.3	
Nitrates NO₃　　mg/l	45		50		45	
NO₃ + NO₂ as N　mg/l					10	10
Nitrites NO₂　　mg/l			0.1	0.5		
Ammonia NH⁺₄　mg/l		0.05	0.5			0.5 (as N)
Kjeldahl Nitrogen　mg/l		0.00	0.5			
Silica　　　　mg/l		5 mg/l in excess of the initial content				
Chloroform Extract		0.1			0.2-0.7	0.2
3. Biological factors:						
Dissolved oxygen　mg/l		5				
Oxidizability KMnO₄　mg/l		1	5			
Dry extract at 110 °C　mg/l					0.2	0.5
4. Undesirable or toxic substances:						
Silver　　　　mg/l			0.01		0.05	0.05
Arsenic　　　mg/l	0.05		0.05		0.05	0.01-0.5
Barium　　　mg/l			0.1		1	1
Boron　　　　mg/l						5
Cadmium　　　mg/l	0.01		0.005		0.01	0.01
Cyanide　　　mg/l	0.05		0.05		(1)　0.01-0.2	0.01-0.2
Total chromium　mg/l			0.05			
Chromium VI　mg/l					0.05	0.05
Copper　　　mg/l			0.05		0.2-1	1
Fluorine　　　mg/l	1.5	0.1	0.7-1.5		(2)　1.4-2.4	
Iron　　　　mg/l			0.3		0.05-0.3	0.3
Mercury　　　mg/l	0.001		0.001		0.002	0.05
Manganese　　mg/l		0.02	0.05		0.01-0.05	
Nickel　　　mgl/		0.005	0.05			

Foot-notes: See page 1136.

uality in the E.E.C. and a few other countries

WEST GERMANY	SWITZER-LAND	BELGIUM	SPAIN	ITALY	SWEDEN	FRANCE
Recommendations DIN standard 2 000 1973 KfW Mitteilungen February 1975 Official Gazette 27/6/60	(Manuel suisse des denrées alimentaires) Swiss manual of foodstuffs) RS 817-20 and amendments of 9/4/75 RO 75,662	Decree of 6 May, 1977 Ministry of Public Health and Family	Official Gazette of Spain No. 253 23/8/67	L'Ultima Acqua Chimica Analitica depurazione e legislazione delle acque 1.2. A. Canuti 1974 AFEE 2482 1/2	Recommendations	Draft standards of 4 April 1973. Adour-Garonne report No. 7, September 1974
Max. or min.-max. concentrations	Max. or min.-max. concentrations	Max. concentrations	Max. or min.-max. concentrations	Examples of analysis	Max. or min.-max. concentrations	Max. or min.-max. concentrations
	1		5-15 5-10		20-40	(1) 5 12
8.5-9.5		6,5	7-9.2	7-8.3	7-9.5	7-8.5 2 000
		150	100-200 60-100	160 160	20-40 100	
0.2		0.1	tolerated		0.15	
240		250 1 000	200-400 250-350	100 35	100-200 100-300	250 200-600
0.3-0.6	10	0.25		0.2		
50-90	20	76	30	10		50-100
0.6 (as N)	0 0.02 (as N)			0 0 (as N)	0.05 (as N)	0.05
				2.5		
		1.5 (at 105° C)	12 0.75			1-1.5 0.1-0.2
0.1 0.04		0.05	0.2	0		0.5
0.006 0.005		0.01 0.01	0.01	0 0		0.005 0.05
0.05		0.05 1	0.05 1.5	0		0.05 0.05
1.5		1.5	1.5		0.05	1.5
0.2	0.1	0.3	0.2-0.3	0.5	1.5	0.1
0.1		0.1	(Fe + Mn) 1.5	0 0.2	0.2-0.4 0.1	0.001 0.05

oot-notes: See page 1136.

COUNTRIES	INTER-NATIONAL STANDARDS	EEC			UNITED STATES	CANADA
References	applied un Great Britain, Ireland, Denmark, Finland, Austria, Nigeria, South Africa W.H.O. 1972	Extract from Official Journal of EEC No. C 214/6 to 11, 18/9/75 Directives			EPA Federal Register 24/12/75 No. 51 and 248, Vol. 40, and 9/7/76, No. 133, Vol. 41 Environmental Regulation Handbook 1976	Standards and objectives for drinking water in Canada October 1969
Parameters	Max. of min.-max concentrations	Recommended level	max.	min.	Max. or min.-max. concentrations	Max. or min.-max. concentrations
Phosphorus mg/l		0.3	2			0.002 (PO$_4$)
Lead mg/l	0.1		0.05		0.05	0.05
Hydrogen sulphides mg/l			0			0.05
Thiosulphates mg/l						
Antimony mg/l			0.01			
Selenium mg/l	0.01		0.01		0.01	0.01
Zinc mg/l	1.5-5		0.1-2		5	2
Oils-greases mg/l			0.01			
Uranyl as UO$_2$ mg/l					5	
Hydrocarbons mg/l			0.0002			
Phenols mg/l			0.0005		0.001	0.002
Detergents (ABS) mg/l			0.1		0.2-0.5	0.5
Pesticides and { Total:			0.0004			
related matter { per substance:			0.0001			
Aldrin mg/l					0-0,001	0-0,017
Chlordane mg/l					(3) 0.003	0-0.003
DDT mg/l					0-0.05	0-0.042
Dieldrin mg/l					0-0.001	0.017
Endrin mg/l					0.0002	0-0.01
Heptachlor mg/l					(3) 0.0001	0-0.18
Lindane mg/l					0.004	0.056
Methoxychlor mg/l					0.1	0.035
Carbamates mg/l						0.1
Toxaphene mg/l					0.005	0.005
Chlorophenyoxy 2.4.D and						
245 TP Sibrex mg/l					0.1	0.1
Other organic chlorides mg/l					0.01	
Organic phosphates mg/l					0.01	
5. Microbiological factors:						
Total coliforms/100 ml	0	5	0			
Faecal coliforms/100 ml	0	0	0		1-4	
Aerobic bacteria/1 ml					1-100	
Pathogenic germs	0	0	0			
6. Radioactivivy:						
Radium 226-228 pCi/l					5	
Strontium 90 pCi/l					2	
Alpha pCi/l					15	
Beta } pCl/l					4	< 10
Photons }					4	

(1) Decision in December 1976
(2) Depending on ambient temperature
(3) Temporary withdrawal of standards

WEST GERMANY	SWITZERLAND	BELGIUM	SPAIN	ITALY	SWEDEN	FRANCE
Recommandations DIN standard 2 000 1973 fW Mitteilungen February 1975 Official Gazette 27/6/60	(Manuel suisse des denrées alimentaires) (Swiss manual of foodstuffs) RS 817-20 and amendments of 9/4/75 RO 75,662	Decree of 6 May 1966 Ministry of Public Heath and Family	Official Gazette of Spain No. 253 23/8/67	L'Ultima Acqua Chimica Analitica depurazione e legislazione delle acque 1.2. A. Canuti 1974 AFEE 2482 1/2	Recommendations	Draft standards of 4 April 1973 Adour-Garonne report No. 7. September 1974
Max. or min.-max. concentrations	Max or min.-max. concentrations	Max. concentrations	Max. or min.-max. concentrations	Example of analysis	Max. or min.-max. concentrations	Max. or min.-max. concentrations
0.04		0.05	0.1	0 (PO_4) 0 100		(3) 0.3 (PO_4) 0.05
0.5						
0.008		5	0.05 1,5	0	1	0.01
0.5		0.001	0.001		0.001 0.5	0.01-0.03 0.001 (2) 0-0.05
						0.001 0.001
0	0 20-300	0 0 0	0-2 65-100 0			0
		100				3 30 1 000
						(1) Drops of mastic (2) Cationic: 0, anionic and non-ionic: 0.05 (3) Polyphosphates: 5 mg/l

2.5. Swimming pools in France and the E.E.C

2.5.1. STANDARDS OF QUALITY REQUIRED FOR SWIMMING POOL WATER IN FRANCE

(Decree of 13 June 1969.)

A. Physico-chemical standards

● Main tests: Essentially, these involve, for safety reasons, an objective assessment of transparency, and for health reasons, measurement of neutrality and a search for toxic or undesirable substances.

— The transparency of the water should be measured by means of a plate placed at the bottom of the pool on a swimming line at the deepest part.

— The pH of the water, with the exception of untreated waters, must be maintained as far as possible between 7.5 and 8.2.

— The concentration of toxic or undesirable substances must always be lower than acceptable dilution limits for fish.

● Supplementary tests: these involve the oxidizability of the water and the development of nitrogenized substances.

B. Bacteriological standards

Bacteriological standards are summarized in the table below:

Description	Bacteria Count per millilitre at 37 °C	Coliforms per 100 ml at 37 °C	Escherichia coli per 100 ml at 44 °C	Faecal streptococci per 10 ml	Remarks
Open circuit pools and bathing places (without disinfection system):					
Very low pollution zone	optional	0 to 50	0 tp 20	—	Very good quality water
Low pollution zone	optional	50 to 500	20 to 200	—	Good quality water
Moderate pollution zone	optional	500 to 5 000	200 to 2 000	—	Moderate quality water
Supplied by surface waters (rivers and stagnant water):					
High pollution zone	optional	5 000 and over	— —	— —	Water unusable without disinfection treatment
Open circuit swimming pool (without disinfection system), supplied by public network or underground water	500 [1]	0 to 50 [1]	0 to 20 [1]	—	— —
Open circuit swimming pool (equipped with disinfection system), supplied by various sources	100 [1]	0 to 20 [1]	0 [1]	—	—
Closed circuit swimming pools	100 [1]	0 to 20 [1]	0 [1]	0 [1]	

1. In the pool and at the outet.

● Main tests:

— Essential and obligatory tests: detection and counting of coliforms with identification of Escherichia coli, total enumeration of the bacteria at 37 °C, and detection of faecal streptococci.

● Secondary and optional tests: the isolation of buccopharyngeal germs in the surface film. Periodical detection of these germs is, however, recommended.

● *Supplementary and optional tests*:

They ensure that the pools and bathing areas are kept clean, and provide the health authorities with important information so that, if necessary, the latter can recommend appropriate measures.

They concern the detection of mycobacteria and leptospirae; tests should be carried out only in laboratories or specialized institutes.

2.5.2. BATHING WATERS: EEC DIRECTIVES
EEC Official Gazette, 5.2.76, No. L 31/5:76/160/EEC

Parameters	Guide	Requirement	Minimum frequency of sampling
Microbiological: Total coliforms/100 ml	500	10 000	fortnightly
Faecal coliforms/100 ml	100	2 000	fortnightly
Faecal streptococci/100 ml	100	—	to be checked if quality of water deteriorates
Salmonellae/1 l	—	0	as above
Enterovirus PFU/10 l	—		as above
Physico-chemical: pH	—	6-9	as above
Colour	—	no abnormal change in colour —	fortnightly to be checked if quality of water deteriorates
Mineral oils mg/l	— ⩽ 0.3	no visible film on surface of water, no odour —	fortnightly to be checked if quality of water deteriorates
Surface-active substances reacting with methylene blue mg/l	— ⩽ 0.3	no persistent froth —	fortnightly to be checked if quality of water deteriorates
Phenols mg/l	— 0.005	no odour 0.05	fortnightly to be checked if quality of water deteriorates

Other parameters must be checked if the quality of the water deteriorates (pesticides, heavy metals), etc. or if eutrophication occurs (ammonia, nitrates, phosphates, etc.).

3. WASTE WATERS

3.1. Domestic sewage and industrial effluents in France

3.1.1. WASTE DISCHARGE REGULATIONS

A. General case

The basic document is the circular of 10 June 1976, entitled "Sanitation in large centres of population: protection of environment receiving discharged effluent". It insists on the need to link sanitation projects with measures aimed at protecting the receiving environment. It lays down characteristics for domestic effluent after treatment and indicates several possibilities: (1) general case, (2) partial purification, (3) supplementary treatments for "**proximity zones**"— zones in which there is less than 8 km between the place where the treated water is discharged and the place where it is used —, (4) exceptional cases.

The circular also gives guidelines concerning the choice of sanitation system to install, purification processes to use according to the individual case, and shows how to set out projects for submission to the health authorities.

It specifies how to fix discharge levels depending on how the receiving medium is used. The levels are laid down by the decree of 13 May 1975, and are only enforced if the effluent to be treated shows the characteristics of urban waste, in which industrial waste water may be present but not preponderant; the effluent in that case should have an organic and nitrogenized loading such as:

— COD/BOD < 2.5;

— COD < 750 mg/l;

— Kjeldahl nitrogen < 100 mg/l.

Where one or more of the above conditions are not met, the discharges causing the abnormal pollution should be identified.

● **Characteristics of treated effluent:** all treated effluent, regardless of its level of treatment, must have a temperature of < 30 °C (< 25 °C for a flow of more than 10 litres per second), a pH between 5.5 and 8.5 (discharge into the sea $5.5 \leqslant$ pH $\leqslant 9$). It must not destroy fish 50 m downstream of the discharge point. Its colour must not cause any visible coloration of the receiving medium.

The type of treatment to be applied depends on the pollution to be eliminated and the use of the medium receiving the treated water. The treatment must adhere to the discharge levels defined by the decree of 13 May 1975, and are shown in the following table.

TYPE of TREATMENT — Discharge levels fixed by the decree of 13 May 1975, and amended by the decree of 6 January 1977

Criteria for discharged effluent	I — Construction in successive stages	II — Partial physico-chemical treatment	III — Partial treatment including a biological treatment stage	IV — Standard treatment	V — Reinforced treatment, including nitrification	VI — Reinforced treatment, Exceptional, including tertiary treatment
Total suspended solids Average concentration over 2 hr.		⩽ 20 % in weight after 24 hr from a raw unsettled sample	⩽ 100 mg/l	⩽ 30 mg/l	⩽ 30 mg/l	⩽ 20 mg/l
COD Average over 24 hr Average over 2 hr	⩽ 10 % of the daily flow of settlable matter		⩽ 120 mg/l measured in filtered effluent	⩽ 90 mg/l ⩽ 120 mg/l	⩽ 90 mg/l ⩽ 120 mg/l	⩽ 50 mg/l ⩽ 80 mg/l
BOD Average over 24 hr Average over 2 hr		⩽ 50 % in weight after 24 hr from a raw unsettled sample	⩽ 40 mg/l measured in effluent filtered through membranes identical to those used for determining S.S.	⩽ 30 mg/l ⩽ 40 mg/l	⩽ 20 mg/l ⩽ 30 mg/l	⩽ 15 mg/l ⩽ 20 mg/l
Kjeldahl organic N Average over 24 hr Average over 2 hr				⩽ 40 mg/l ⩽ 50 mg/l	⩽ 10 mg/l	⩽ 7 mg/l
Putrescibility test			After 5 days' incubation at 20 °C the effluent does not give off any putrid or ammoniacal odour			

● **Remarks:**

— Under standard treatment (Level IV), if, for an average sample over 24 hours, $2 < \dfrac{COD}{BOD} < 2.5$ or $450 < COD < 750$, a reduction of only 75 % of the daily flow of COD is required, provided that this can be allowed in the receiving medium.

— Under exceptional treatment (Level VI), values of varying severity can be given to certain parameters where the receiving medium is suitable.

In addition, when the effluent is discharged upstream of an intake of water for human consumption, it is necessary to ensure that the concentration of chloroform-extractable substances (C.E.S.) in the purification plant effluent is such that after mixing with the receiving water, the C.E.S. concentration at the point of reuse never exceeds 0.2 mg/l. The same comments apply with regard to radioactivity and toxic substances.

B. Particular cases

● **Discharging into lakes, canals and ponds:** the rules adopted for proximity zones must be followed.

● **Discharge into deep-lying water tables:** only permissible in exceptional circumstances. When percolating wells are used, the water to be discharged underground must first undergo purification treatment.

● **Discharge into the sea:** it is forbidden to discharge domestic sewage or industrial effluent into the sea without giving it at least a preliminary treatment. Treatment plant must be designed to accept the varying loads and flows that it will be required to handle.

The following is considered as a negligible hazard: discharge into the sea of a pollution load less than the pollution caused by 500 inhabitants and containing less than 100 g/d of hydrocarbons and less than 10 g/d of cyclic, hydroxylated compounds. Such a discharge is only authorized at a minimum distance of 1 000 m from a water intake, a bathing place, a shellfish bed, or an oyster-farm. The pH must be between 5.5 and 9; the temperature must not exceed 30 °C, or 25 °C for a flow of over 10 litres per second.

Sludge produced in a purification plant should only be discharged into the sea under extremely exceptional circumstances, and then only after treatment making it harmless bacteriologically and biologically.

3.1.2. SLUDGE OF AGRONOMIC VALUE PRODUCED BY WATER TREATMENT PLANT

The **experimental** AFNOR standard U 44-041 [1] applies to sludges produced by wastewater treatment plants, valuable from an agronomic point of view because they are rich in certain organic and/or inorganic substances, and useful in agriculture.

---

1. See note page 905.

Sludges are classified according to their richness in organic solids and nitrogen, as shown in the table below:

organic solids/ total nitrogen ratio	Percentage of organic solids / dry solids			
	lower than 10 %	between 10 and 25 %	between 25 and 60 %	over 60 %
between 15 and 40	inorganic nitrogenous sludge	nitrogenous sludge with low organic content	organic nitrogenous sludge	nitrogenous sludge with high organic content
over 40	inorganic sludge with low nitrogen content	sludge with low nitrogen and organic content	organic sludge with low nitrogen content	sludge with high organic and low nitrogen content

At least 70 % of the effluent, in dry weather, should be domestic in origin. This restriction does not apply to sludges produced by the food industries.

● **Concentrations of elements:** the concentrations of certain elements in sludge, before they can be marketed as standardized sludge, must not exceed the following amounts per kilogramme of dry solids:

— zinc	3 g	— lead	0.3 g	— cobalt	20 mg
— copper	1.5 g	— chromium	200 mg	— cadmium	15 mg
— manganese	0.5 g	— nickel	100 mg	— mercury	8 mg

● **Additives:** the following substances may be added for the treatment of effluents and sludges:

Inorganic additives iron salts (chlorides, sulphates), aluminium salts (alum, chlorides, sulphates), silica activated by acidifying sodium silicate, activated carbon, lime, sodium hydroxide, sulphuric acid, hydrochloric acid, phosphoric acid, chlorates, thiosulphates, sodium bisulphites and sodium aluminate.

Organic additives: authorized for this purpose by law No. 72-1139 of 22 December 1972.

3.2.   Table comparing effluent discharge

Physical properties and chemical constituents (mg/l)	WEST GERMANY				BELGIUM			
	Discharge into water course			To a treatment plant	Discharge into river			Discharge into sewer
	A	B	C		1	2*	3	
Temperature in °C	20	20-28	28	35	30	30	—	45
pH	6-9	5-10	5-10	6.5-9.5	6.5-8.5	—		6.5-8.5
Suspended solids (mg/l)	20				100		1 000	20-100
BOD$_5$ (average over 2 h)	25				15	30	50	
COD (average over 2 h)	80					500		
KMnO$_4$ oxidizability	18	18-40	40					
N (Kjeldahl)								
Fluorides (as F$^-$)						10		
Chlorides (as Cl$^-$)	150	150-350	350					
Sulphides (as S$_2^-$)								
Sulphates (as SO$_3^{2-}$)				400		2 000		
Cyanides (as CN$^-$)						0.5		
Arsenic (As)								
Barium (Ba)								
Cadmium (Cd)								
Chromium (Cr^{3+})								
(Cr^{6+})								
Iron (Fe)	0.5	0.5-1.5	1.5			2		
Manganese (Mn)	0.25	0.25-0.5	0.5			1		
Mercury (Hg)								
Nickel (Ni)				5				
Lead (Pb)			3			1		
Copper (Cu)								
Zinc (Zn)			5		100	5		
Cd + Cr + Cu + Ni + Zn + Fe								
Oil and grease	0	trace	trace	20-100				500
Hydrocarbons					5-15			
Phenols	0.005	0.005-0.1	0.1	100				
Organic solvents								
Active chlorine (mg Cl/l)								

* Decree of 3/8/76: Discharge of waste water from the Iron and Steel Industry

Remarks

and

References

Galvanotechnik (1971), 62, No. 12.
L'ultima acqua, A. Çanuti, 1974, AFEE 2482/2

Decree of 23/1/74
Law of 26/5/71
Decree of 23/1/75: discharge of waste water in the dairy industry
La Technique de l'Eau (1974), N° 329, April

standards in several countries

FRANCE		GREAT BRITAIN						SWITZERLAND	
Discharges depending on treatment levels	Metal finishing industries	Examples of river discharges				Sewer discharge		Water-course discharge	Discharge into sewer or before purification
		1	2	3	4				
30 5.5-8.5 20-40 80-150		26 6-9 30 20	32 5-9 30	25 5-9	7-8.5 500	30-43 6-10 100-400		30 6.5-8.5 20 20 (+) 10 (+)	60 6.5-9 * * *
7-80 0	15	10	10 1		10			* * 10 * 0.1	* 10 10 * 1
	0.1-1 0 3	0.1	1 200	1200		1 200 2-10		* 0.1 0.1 5 0.1	300 0.5 0.1 * 0.1
0	0.05-0.1	0.5 4 0.01			0.3	2 10-20		2 0,1 2 0.01 2	2 0.5 20 0.01 2
	0.1 15	0.1 4	4	50	50	500		0.5 0.5 2 20	0.5 1 2 *
5-20 0.5-1 0		0.5-1 0.5 0 0.5	1 0 1	 0 1	0.01 0			10 0.05 * 0.05	20 1-5 * 0.5-3
		Regional regulations (examples)						* Determined by each canton + over 24 hours	
Official Gazette Brochures		Water Act 1973 Protection Handbook of Pollution Control P. Sutton, 1975 Publ. A. Osborne						Statute on the discharge of waste waters, 8/12/75	

3.3. Industrial Wastes (in France)

DISCHARGE CONDITIONS OF INDUSTRIAL EFFLUENTS

Discharge into natural medium		Discharge away from water intakes for towns, beaches, shellfish beds or Salmonidae reserves			Discharge near water intakes for towns, beaches, shellfish beds or Salmonidae reserves
		Industrial pollution load			
		low	Significant but not preponderant	preponderant	
Dilution d (ratio between watercourse flow and effluent flow) depending on the duration of industrial effluent discharge	discharge over 24 hr	d > 300	150 < d < 300	d < 150	
	discharge over 10 hr	d < 720	360 < d < 720	d < 360	
pH—general		5.5 < pH < 8.5	5.5 < pH < 8.5	5.5 < pH < 8.5	5.5 < pH < 8.5
pH in the case of lime neutralization		5.5 < pH < 9.5	5.5 < pH < 9.5	5.5 < pH < 9.5	5.5 < pH < 9.5
Cyclic, hydroxylated compounds and their halide derivatives		forbidden	forbidden	forbidden	forbidden
Substances of a type likely to cause the appearance of smells, tastes or abnormal coloration in natural water when used for human consumption		forbidden	forbidden	forbidden	forbidden
Total suspended solids		100 mg/l	50 mg/l	30 mg/l	30 mg/l
BOD_5		200 mg/l	100 mg/l	40 mg/l	40 mg/l
Total nitrogen (Kjeldahl method)		60 mgl/ of N 80 mg/l of NH_4	30 mg/l of N 40 mg/l of NH_4	10 mg/l of N 15 mg/l of NH_4	10 mg/l of N 15 mg/l of NH_4
Substances likely to destroy fish downstream of discharge		forbidden	forbidden	forbidden	forbidden
Maximum temperature		30 ºC	30 ºC	30 ºC	30 ºC

4. MISCELLANEOUS FRENCH LEGISLATION

4.1. Air pollution

The targets to aim for in combatting air pollution and odours were defined by a law on 2 August 1961. The maximum pollution limits are given in a circular (24 November 1970) concerning the construction of **stacks for furnaces,** giving off **sulphur dioxide,** with a maximum permissible level in the atmosphere of 0.25 mg/m³.

A circular (13 August 1971) on the construction of **stacks giving off fine dust,** states that the maximum level permissible in the atmosphere is 0.15 mg/m³.

In both cases the minimum outlet velocity for the gases is 2-8 m/s. The maximum levels permitted are calculated at ground level, with the mean annual concentration at ground level at the area in question being deducted. If the latter cannot be measured, it may be assessed as:

— 0.01 mg/m³ for SO_2 in a low pollution area;
— 0.05 mg/m³ for fine dust in a low pollution area;
— 0.11 mg/m³ for SO_2 in a moderately industrial zone or medium density area;
— 0.09 mg/m³ for fine dust in a moderately industrial zone or medium density area;
— 0.16 mg/m³ for SO_2 in a highly urbanized or highly industrialized zone;
— 0.11 mg/m³ for fine dust in a highly urbanized or highly industrialized zone.

A decree (20 June 1975) summarizes the previous regulations and lays down, inter alia, limits for the emission of particles by means of the *blackening index* which must be lower than 4 (French standard X 43-002) and the *ponderal index:* a generator, operating normally, must not emit
— more than 1 g of dust per thermie of solid fuel consumed;
— more than 0.25 g of dust per thermie of liquid or gaseous fuel consumed.

Incineration of urban refuse

A circular (12 June 1972) lays down the maximum levels for normal emission:
— 0.15 to 1 g/m³ of dust depending on the capacity of the furnace:
  under 1 t/h : 1 g/m³
  between 1 t/h and 4 t/h : 0.6 g/m³
  between 4 t/h and 7 t/h : 0.25 g/m³
  over 7 t/h : 1 g/m³
— 0.1% CO in volume
— $7 \% CO_2$ in volume
Outlet velocity of the flue-gases: 8 m/s.

4.2. Noise abatement

A circular (26 November 1971) sets out guidelines for noise abatement in workplaces.

4.2.1. ACCEPTABLE LIMITS FOR INDUSTRIAL NOISE are given below in accordance with the recommendations of the Technical Commission for studying noise of the Ministry of Public Health (17 March 1961).

A curve representing noise level as a function of frequency shows the limits which should not be exceeded for exposure to a complex noise during eight hours; hese limits are considered "not harmful for a normal, healthy ear".

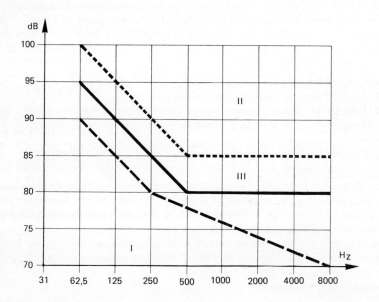

3 zones are defined:
— Zone 1: The levels included in this zone may be considered as not dangerous or possibly, not troublesome.
— Zone II: The levels in this zone are considered as dangerous.
— Zone III: In this zone there is a probability of danger.

The curve laid down by the Commission is approximately situated in the middle of zone III, and was based on physiological considerations. This may be described as the "warning level" put forward by the commission.

4.2.2. DANGEROUS NOISE LIMITS for workplaces were established as follows by the Technical Commission for studying noise:

A. For permanent exposure (40 hours per week) at a work station, to a level of steady noise, without impact or impulses, with the noise being measured at a weighted average level A with the instruments used for the standard AFNOR S 31-009, set to "slow response";

● 85 dB (A) must be taken as the warning level when monitoring noise;

● 90 dB (A) should be considered as the danger level; above this, there is a considerable risk of occupational deafness, which increases with the noise level and the number of years of work.

In this case, staff should be given a regular audiometric check-up, and all steps should be taken to reduce noise or the length of exposure, or to improve the protection of workers.

B. For intermittent exposure, to noise levels without impulses, of varying duration, for a standard period (week) at a work station, an equivalent noise level is calculated by the process given in the standard AFNOR S 31-013.

The noise level is measured in dB (A) at the approximate position of the worker's ear with instruments as used for the standard AFNOR S 31-009, set for "slow response". If the noise varies rapidly, it must be recorded for further analysis.

The various noise levels observed in the course of a week are classified in Li levels, at 5 dB intervals, Li being the mid-point of each category (for example: 92 dB (A) is in the 90 category; 93 dB (A) is in the 95 category); the total weekly duration ti is recorded for each Li category. If this duration is less than ten minutes, the minimum value of ten minutes is used. These Li and ti values are used to calculate an equivalent level of exposure to noise in accordance with the method given in the experimental standard AFNOR S 31-013.

● The equivalent noise level to 85 dB (A) must be considered as the warning level when monitoring noise;

● The equivalent noise level of 90 dB (A) must be considered as the danger level. It gives rise to the same consequences and necessitates the same precautions as for permanent exposure at 90 dB (A) level.

C. Impulsive noises:

The previous recommendations do not apply to impulsive noises or isolated, transitory phenomena. However, if these phenomena are present on top of an equivalent level calculated as above, the environmental noise is even more dangerous. In the case of hammering, for example, a first approximation would be to add 5 units to the corresponding dB (A) level.

D. Protection of individual workers:

The equivalent noise level is calculated as follows:

For a worker equipped with a protective helmet, the equivalent level should be calculated as it is heard from inside the helmet.   For this,

— the noise must be measured in octaves or thirds of an octave in accordance with the recommendations of the standard AFNOR S 31-009;

— the deadening effect given by the helmet corresponding to these frequency bands must be subtracted from these values;

— the average weighing corrections A are to be applied to the levels Ln thus obtained, in accordance with the above standards, Ln being the level found in the frequency band n;

— the level in dB (A) must be calculated in accordance with the recommendations of the standard AFNOR S 31-009.

If exposure is intermittent, the equivalent level is calculated as before from the L levels, divided into categories.

4.2.3. THE EQUIPMENT USED ON WORK SITES must not be so noisy as to cause excessive irritation,

Regulations exist for the following equipment:

A. Air compressors: when used less than 50 metres away from residential buildings or workplaces, they must not give rise, at a distance of 1 metre, to a noise above 85 decibels.

B. Internal combustion engines: noise measured 7 metres away must not exceed 80 decibels (A) or 90 decibels (A) in the case of engines of 200 H.P. and over.

C. Generating sets:
(Decrees of 26 November 1975 and 10 December 1976.)

D. Pneumatic road breakers and picks:
(Decree of 4 November 1975.)

For each category of machine and equipment, there are rules for agreement procedure.   See the decree of October 1975 on the limitation of noise levels.

A circular of June 1976 concerns noise from equipment installed in factories referred to as dangerous, unhealthy and uncomfortable by a December 1917 law.   Any discomfort is assessed in accordance with the provisions of the French standard NF S 31.010.

BIBLIOGRAPHIC DATA

WORKS ON WATER AND WATER TREATMENT

The lists for each subject are not exhaustive but are representative of the present state of technical knowledge.

GENERAL WORKS

Précis d'hydrologie, A. MORETTE (Masson, 1964).

Précis d'écologie, R. DAJOZ (Dunod, 1975).

Précis général des nuisances (6 volumes), J. A. TERNISIEN (G. Le Prat, 1971-72-73-74).

Livre de l'eau (5 volumes) (Centre Belge d'Étude et de Documentation des Eaux, Liège, 1964-65-66).

Guide de l'eau, R. COLAS (Pierre Johannet, 1977).

Water & Water Pollution Handbook, Vol. 1 Environmental systems (Leonardo L. Ciaccio, Marcel Dekker, N.Y., 1971-72-73).

Water Supply and Sewerage, E. W. STEEL (McGraw Hill, 1953, 3rd ed.).

Water Supply and Waste Water Disposal, G. M. FAIR, J. C. GEYER, J. C. MORRIS (John Wiley, Chapman & Hall, 1954).

Water : Examination, assessment, conditioning, chemistry, bacteriology, biology, K. HÖLL (W. de Gruyter, 1972).

Wasser. Untersuchung, Beurteilung, Aufbereitung von Wasser, K. HÖLL (W. de Gruyter & Co., Berlin, 1960).

Techniques et contrôle du traitement des eaux, Ch. R. COX (W.H.O., Geneva, 1967).

Théorie des eaux naturelles, L. LEGRAND and POIRIER (Eyrolles, Paris, 1972).

Chimie des eaux naturelles : agressivité, corrosivité, traitements, applications numériques, L. LEGRAND and POIRIER (Eyrolles, Paris, 1976).

Dureté de l'eau (2 volumes), E. LECLERC (Cebedoc, 1959).
Vol. I : Unités; méthodes de mesure.
Vol. II : Effets; dureté des eaux dans le monde; traitements d'adoucissement.

La corrosion des conduites d'eau et de gaz. Causes et remèdes, M. NEVEUX (Eyrolles, 1968).

Métal et eau, le bréviaire de la corrosion, H. E. HÖMIG (translated from the German; published by Dia-Prosim — Vulkan Verlag, Essen, 1966).

Active carbon manufacture, properties and applications, M. SMISEK, S. CERNY (Elsevier Publishing Co., 1970).

DRINKING WATER

L'alimentation en eau des agglomérations, P. KOCH (Dunod, 2nd ed., 1969).

Le traitement des eaux de distribution, C. GOMELLA and H. GUERREE (Eyrolles, 1973).

Les paramètres de la qualité des eaux, S.P.E.P.E. (La Documentation française, 1973).

Water quality and treatment; a Handbook of Public Water Supplies A.W.W.A. (McGraw Hill, 3rd ed. 1971).

SWIMMING POOL WATER

Le Moniteur des T.P. et du Bâtiment, numéro spécial hors série « Équipements sportifs et sociaux éducatifs » (January 1977).

OZONE, CHLORINE, DISINFECTION

Handbook of chlorination, G. C. WHITE (Van Nostrand Reinhold Co, 1972).

Ozone in water and waste water treatment, F. L. EVANS (Ann Arbor Science, 1972).

Disinfection water and waste water, J. D. JOHNSON (Ann Arbor Science, 1975).

Ozone chemistry and technology, a review of the literature, J. S. MURPHY, J. R. ORR (The Franklin Institute Press, 1975).

INDUSTRIAL PROCESS WATER AND BOILER WATER

Water treatment for industrial and other uses, NORDELL (Reinhold N.Y., 2nd ed., 1961).

Industrial water purification, L. F. MARTIN (Noyes Data Corp., 1974).

The chemical treatment of cooling water, I. McCOY (Chemical Publishing Co., 1974).

Richtlinien für die Aufbereitung von Kesselspeisewasser und Kühlwasser. Vereinigung der Grosskesselbesitzer (Vulkan Verlag, Essen, 1958).

Handbuch Wasser, VKW (Vulkan Verlag, Essen, 1974).

ION EXCHANGE

Les séparations par les résines échangeuses d'ions, TRÉMILLON (Gauthier-Villars, 1965).

L'échange d'ions et les échangeurs, G. V. AUSTERWEIL (Gauthier-Villars, 1955).

Demineralization by ion exchange, APPLEBAUM (Academic Press, New York, 1968).

Ion exchange separation in analytical chemistry, O. SAMUELSON (J. Wiley & Sons, N.Y., 1952).

Ion exchangers, F. HELFERICH (McGraw Hill, New York, 1959).

Ion exchange resins, R. KUNIN and R. J. MYERS (J. Wiley & Sons, N.Y., 2nd ed., 1958).

Ion exchange technology, F. C. NACHOD, J. SCHUBERT (Acad. Press Inc., 1956).

Ion exchangers. Properties and applications, K. DORFNER (Ann Arbor Science, 1972).

Ion exchange resins, C. PLACEK (Noyes Data Corp., 1970).

DESALINATION

La production d'eau potable par dessalement, A. CLERFAYT (Cébedoc, SPRL, 1967).

Les problèmes du dessalement de l'eau de mer et des eaux saumâtres, J. R. VAILLANT (Eyrolles, 1970).

Desalination and its role in water supply (United Kingdom Atomic Energy Authority, 1970).

Reverse osmosis, S. SOURIRAJAN (Logos Press Ltd., 1971).

Fundamentals of water desalination, E. D. HOWE (M. Deeker, 1974).

Industrial processing with membranes, R. E. LACEY and S. LOEB (Wiley Interscience, 1972).

Membrane technology and industrial separation techniques, P. R. KELLER (Noyes Data Corp., 1976).

MUNICIPAL WASTE WATER

Manuel de l'assainissement urbain, IMHOFF, trans. Koch (Dunod, 1970).

Pratique de l'assainissement des agglomérations urbaines et rurales, H. GUERRÉE (Eyrolles, 1970).

Mémento d'assainissement. Mise en service, entretien et exploitation des ouvrages d'assainissement, H. MONCHY (Eyrolles, 1975).

Ouvrages d'assainissement. Calcul et exécution, A. VALENTIN (Eyrolles, 1976).

L'analyse écologique des boues activées, B. VEDRY (Segetec, 1975).

Précis d'épuration biologique par boues activées, P. BROUZES (Technique et Documentation, 1973).

Biological treatment of sewage and industrial wastes, McCABE, ECKENFELDER (Reinhold).

    I : Aerobic oxydation (1956).
    II : Anaerobic digestion and solid-liquid separation (1958).

Sewerage and sewage treatment, H. E. BABBITT (J. Wiley, New York, Chapman & Hall, London, 7th ed., 1953).

Biological treatment of sewage, ECKENFELDER (Pergamon Press, 1961).

Aeration in waste water treatment plants, W.P.C.F. Manual of practice no. 5 (1971).

Operation of waste water treatment plants, W.P.C.F. Manual of practice no. 11 (1970).

Aeration of activated sludge in sewage treatment, D. L. GIBSON (Pergamon Press, 1974).

Sewage treatment, R. L. BOLTON and L. KLEIN (Butterworths Scientific Publications, 1971).

Community waste water collection and disposal, D. A. OKUN and G. PONGHIS (W.H.O., 1975).

Sewage treatment in hot climates, D. MARA (John Wiley and Sons, 1976).

Wastewater engineering, METCALF & EDDY INC. (McGraw Hill, 1972).

Lehr- und Handbuch der Abwassertechnik, O. PALLASCH, W. TRIEBEL (Wilhelm Ernst & Sohn, Berlin, Munich, 1967-1969).

    I : Grundlagen der Abwassertechnik.
    II : Abwasserbehandlung.
    III : Schlammbehandlung.

Die Abwassertechnik, K. R. DIETRICH (Dr. A. Hüttig Verlag, 1973).

Leitfaden für den Betrieb von Kläranlagen, W. TRIEBEL (Abwassertechnische Vereinigung, 1971).

PHYSICAL/CHEMICAL TREATMENT PROCESSES

Les procédés physico-chimiques d'épuration des eaux usées urbaines, A.F.E.E. (1975).

Les charbons actifs et les eaux usées, AGENCE SEINE NORMANDIE and A.F.E.E. (1973).

Purification with activated carbon, J. W. HASSLER (Chemical Publishing, 1974).

Advanced waste treatment, R. L. CULP, G. C. CULP (Van Nostrand Reinhold, 1971).

SLUDGE TREATMENT

Stabilisation non biologique des boues fraîches d'origine urbaine, AGENCE SEINE NORMANDIE and A.F.E.E. (1976).

La filtration industrielle des liquides (4 volumes) SOCIÉTÉ BELGE DE FILTRATION (Derouaux, 1975).

Le séchage et ses applications industrielles, A. DASCALESCU (Dunod, 1969).

Thermique appliquée aux fours industriels (volumes I and II), W. HEILIGENSTAEDT (Dunod, 1971).

Treatment and disposal of waste water sludge, A. VESILIND (Ann Arbor Science, 1974).

Sewage sludge treatment disposal, R. W. JAMES (Noyes Data Corp., 1976).

Anaerobic sludge digestion, W.P.C.F. Manual of practice no. 16 (1968).

Sludge dewatering, W.P.C.F. Manual of practice no. 20 (1969).

Utilization of municipal waste water sludge, W.P.C.F. Manual of practice no. 2 (1971).

Fluidization engineering, D. KUNII, O. LEVENSPIEL (J. Wiley & Sons, 1969).

INDUSTRIAL EFFLUENTS

L'eau dans l'industrie. Pollution, traitement, recherche de la qualité, W. W. ECKEN-FELDER (Entreprise Moderne d'Édition, 1972).

Analyse des eaux résiduaires industrielles, J. BORMANS (Eyrolles-Cebedoc, 1974).

Industrial water pollution control, W. W. ECKENFELDER (McGraw Hill, 1966).

Industrial wastes, their disposal and treatment, W. RUDOLFS (Reinhold N.Y., 1953).

Industrial waste treatment, E. B. BESSELIEVRE (McGraw Hill N.Y., 1969).

Industrial wastes and salvage (2 volumes), Ch. H. LIPSETT (Atlas Publishing Company, 1971).

Liquid waste of industry : Theories, practices and treatment, N. L. NEMEROW (Addison Wesley Publishing Company, 1971).

Process design techniques for industrial waste treatment, C. E. ADAMS, W. W. ECKEN-FELDER (Enviropress, 1974).

Industrial waste water management handbook, H. S. AZAD (McGraw Hill, 1976).

Industrie Abwässer, F. MEINCK, H. STOFF, H. KOHLSCHÜTTER (Gustav Fischer Verlag, Stuttgart, 1960). French translation : Les eaux résiduaires industrielles, 4th ed., André Gasser (Masson et Cie, 1970).

Die gewerblichen und industriellen Abwässer, SIERP (Springer Verlag, 1967).

Die Abwässer in der Metallindustrie (Metallverarbeitende und Galvanotechnische Betriebe), R. WEINER (Eugen G. Lenze Verlag, 1965).

Technik der industriellen Abwässerbehandlung, F. RÜB (Krausskopf Verlag GmbH Mainz, 1974).

PULP AND PAPER

Handbook of pulp and paper technology, K. W. BRITT (Van Nostrand Reinhold, 1970)

AGRICULTURAL AND FOOD INDUSTRIES

Pollution control in meat, poultry and sea food processing, H. R. JONES (Noyes Data Corporation, 1974).

Pollution control in the dairy industry, H. R. JONES (Joyes Data Corporation, 1974).

Waste disposal control in the fruit and vegetable industry, H. R. JONES (Noyes Data Corporation, 1973),

TEXTILE INDUSTRIES

Pollution control in the textile industry, H. R. JONES (Noyes Data Corporation, 1971).

TREATMENT OF METAL FINISHING EFFLUENTS

Épuration des eaux résiduaires dans la transformation et la galvanisation des métaux, R. WEINER (Eyrolles, 1975).

Pollution control in metal finishing, H. R. JONES (Noyes Data Corporation, 1973).

REFINERIES AND CHEMICAL INDUSTRIES

Manual on disposal of refinery wastes. Volume on liquid wastes. American Petroleum Institute (A.P.I., 1969, 1st ed.).

Aqueous wastes from petroleum and petrochemical plants, BEYCHOCK (Wiley, London, 1967).

Environmental control in the organic and petrochemical industries, H. R. JONES (Noyes Data Corporation, 1971).

Pollution control in the organic chemical industry, M. SITTIG (Noyes Data Corporation, 1974).

LEGISLATION

(See also the chapter on Legislation, p. 1121.)

Législation des nuisances. Aide-mémoire, P. GOUSSET (Dunod, 1973).

Tous les problèmes juridiques des pollutions et nuisances industrielles, A. GRANIER, SARGOS (J. Delmas, 1973).

Les résidus industriels (volumes I and II), M. MAES (Technique et Documentation, 1975-77).

METHODS OF ANALYSIS

L'analyse de l'eau : Eaux naturelles. Eaux résiduaires. Eaux de mer (2 volumes), J. RODIER (Dunod, 5th ed., 1975).

Les mesures physico-chimiques dans l'industrie : pH, potentiel d'oxydo-réduction, conductivité, ions spécifiques, P. BENOIT, E. DERANSART (Technique et Documentation, 1976).

Guide de l'aide biologiste, G. SIRJEAN.

Vol. I : Analyse physico-chimique des eaux de consommation (1951).
Vol. II : Analyse bactériologique des eaux de consommation (1952).

L'analyse bactériologique des eaux de consommation, R. BUTTIAUX (Flammarion, 1951).

Standard methods for the examination of water and waste water, AMERICAN PUBLIC HEALTH ASSOCIATION (APHA, AWWA, WPCF, N.Y., 14th ed., 1975).

Chemical analysis of industrial water, James W. McCOY (Chemical Publishing Co., N.Y., 1969).

The analysis of organic pollutants in water and waste water, W. LEITHE (Ann Arbor Science Pub., 1973).

The chemical analysis of water, A. L. WILSON (Society for Analytical Chemistry, 1974).

Methods of seawater analysis, K. GRASSHOFF (Verlag Chemie 1976).

BIOLOGY

Les algues d'eau douce, BOURELLY (Boubée), 3 vol.

Vol. I : Les algues vertes (1966).
Vol. II : Les algues jaunes et brunes (1968).
Vol. III : Les algues bleues et rouges (1970).

Limnologie. L'étude des eaux continentales, DUSSART (Gauthier-Villars, 1966).

La pollution des eaux continentales : incidences sur les biocénoses aquatiques, P. PESSON et al. (Gauthier-Villars, 1976).

Bioénergétique, A. L. LEHNINGER (Ediscience, 1969).

Fresh-water biology, WARD and WHIPPLE (John Wiley Science & Sons, N.Y., 2nd ed., 1959).

Handbuch der Frischwasser- und Abwasserbiologie, H. LIEBMANN (R. Oldenbourg, Munich).
Vol. I : 1951.
Vol. II : 1960.
Das Phytoplankton des Süsswassers, G. HUBER-PESTALOZZI (Schweizerbart, Stuttgart, 6 volumes, 1972).

BACTERIOLOGY

Traité de systématique bactérienne (2 volumes), A. R. PRÉVOT (Dunod, 1961).
Corrosion bactérienne, J. CHANTEREAU (Technique et Documentation, 1977).
Clé d'identification des bactéries hétérotrophes, N. GONTCHAROFF (Dunod, 1971).
Techniques d'enzymologie bactérienne, J. BRISON (Masson, 1971).
Mémento technique de microbiologie, J. P. and M. LARPENT (Technique et Documentation, 1975).
An introduction to the microbiology of water and sewage for engineering students, P. L. GAINEY and T. H. LORD (Burgess Publishing Co., 2nd ed., 1950).
Methods in aquatic microbiology, A. C. RODINA, R. F. COLWELL and M. S. ZAMBRUSKI, University Park Press, Baltimore (Butterworth, London, 1972).
Microbiology, DAVIS, DULBECCO, EISEN, GINSBERG, WOOD (Harper & Row, 1973).
BERGEY'S manual of determinative bacteriology, BERGEY, BREED, MURRAY and SMITH (Williams & Wilkins Company, Baltimore, 1974).
Microbiologie générale, H. LECLERC (Doin, Paris, 1975).
Microbiologie appliquée, H. LECLERC, R. BUTTIAUX, J. GUILLAUME and P. WATTRE (Doin, Paris, 1977).

HYDRAULICS, WATER COLLECTION AND DISTRIBUTION

Hydraulique générale, L. ESCANDE (E. Privat, 1948, 3 vol.).
Manuel d'hydraulique générale, A. LENCASTRE. Translated from the Portuguese by the author and J. VALEMBOIS (Eyrolles, 1966).
Mémento d'hydraulique pratique, J. VALEMBOIS (Eyrolles, 1958).
Hydraulique urbaine, A. DUPONT (Eyrolles, 1969).
Mémento des pertes de charge, I. E. IDEL'CIK. Translated from the Russian by Mme Meury (Eyrolles, 1969).
Distribution d'eau dans les agglomérations, A. CAUVIN, G. DIDIER (Eyrolles, 1960).
Les réseaux d'égout, données d'établissement et de calcul, P. KOCH (Dunod, 1954).
Les stations de pompage d'eau, A.G.H.T.M. (Collection I.P.E. — Industrie, Protection, Environnement, 1977).
Water supply engineering, H. E. BABBITT, J. J. DOLAND (McGraw Hill, 1955, 5th ed.).
Wasserversorgung, C. DAHLAUS and H. DAMRATH (B.G. Teubner, 1974).

PERIODICALS

ENGLISH-LANGUAGE JOURNALS

Water Research.
Pergamon Press, Headington Hill Hall, Oxford OX3 0BW, England.

Progress in Water Technology.
Pergamon Press, Headington Hill Hall, Oxford OX3 0BW, England.

Water Pollution Control.
Ledson House, 53, London Road, Maidstone, Kent, England.

Effluent & Water Treatment Journal.
Thunberbird Enterprises Ltd., 102 College Road, Harrow, Middlesex, HA1 1BQ, England.

Water & Waste Treatment.
Dale Reynolds, Craven House, 121 Kingsway, London WC2, England.

Journal of the Institution of Water Engineers and Scientists.
(Formerly Proceedings of the Society of Water Treatment and Examination), 6-8 Sackville Street, London W1X 1DD, England.

Water Services.
(Formerly Water & Water Engineering).
Fuel and Metallurgical Journals Ltd., Queensway House, 2 Queensway, Redhill, Surrey, RH4 1QS, England.

Environmental Pollution Management.
Polcon Publishing Ltd., 268 High Street, Uxbridge, Middlesex, UB8 1VA, England.

Journal of American Water Works Association.
J.A.W.W.A., 2, Park Avenue, New York, N.Y. 10016, U.S.A.

Water and Sewage Works (Formerly Industrial Water & Wastes).
4345 Wabash, Chicago III 60605, U.S.A.

Environmental Science & Technology.
American Chemical Society, 1115, 16th Street N.W., Washington, D.C. 20036, U.S.A.

Journal of Water Pollution Control Federation.
J.W.P.C.F., R. Canham, 3900 Wisconsin Av., Washington, D.C. 20016, U.S.A.

Analytical Chemistry.
Am. Chem. Society, 1155 16th Street, N.W., Washington, D.C. 20036, U.S.A.

Water & Wastes Engineering.
The Reuben H. Donneley Corp., 466 Lexington Av., New York, 20017, U.S.A.

Industrial Water Engineering.
Wakeman-Walworth Inc., Box 1144 Darien, Connecticut 06820, U.S.A.

Water S.A.
P.O. Box 824, Pretoria 0001, South African Republic.

FRENCH-LANGUAGE JOURNALS

Techniques et Sciences Municipales, l'Eau.
A.G.H.T.M., 9, rue de Phalsbourg, 75017 Paris.

Nuisances et Environnement.
Compagnie Française d'Éditions, 40, rue du Colisée, 75008 Paris.

Information Eaux.
A.F.E.E., 23, rue de Madrid, 75008 Paris.

L'Eau Pure.
A.N.P.E., 195, rue Saint-Jacques, 75005 Paris.

La Houille Blanche.
5, rue des Marronniers, 38008 Grenoble.

Revue Générale de Thermique.
2, rue des Tanneries, 75013 Paris.

L'Eau et l'Industrie.
P. Johannet et Fils, 7, avenue F.-D. Roosevelt, 75008 Paris.

Filtration et Techniques Séparatives.
C.F.E., 40, rue du Colisée, 75008 Paris.

Revue Technique Internationale de l'Eau.
Éd. Géographiques Professionnelles, 9, rue Coëtlogon, 75006 Paris.

Informations — Chimie.
5, rue Jules-Lefevre, 75009 Paris.

Journal Français d'Hydrologie.
A.P.F.H., 4, avenue de l'Observatoire, 75006 Paris.

L'Actualité Chimique.
Société Chimique de France, 250, rue Saint-Jacques, 75005 Paris.

Analusis.
Masson, 120, boulevard Saint-Germain, 75280 Paris Cedex 06.

La Technique de l'Eau et de l'Assainissement.
9, rue du Monastère, Bruxelles, Belgium.

La Tribune du Cebedeau.
CEBEDOC, 3, boulevard Frère-Orban, 4000 Liège, Belgium.

GERMAN-LANGUAGE JOURNALS

Korrespondenz Abwasser.
Markt A, 5205 Sankt-Augustin 1, Federal Republic of Germany.

GWF, Wasser, Abwasser. (Formerly Gas und Wasserfach (Section Wasser).
R. Oldenbourg Verlag, 8 München, Rosenheimer Strasse 145, (F.R.G.)

Wasser, Luft und Betrieb.
65, Mainz, Lessingstrasse 12-14, Federal Republic of Germany.

Kommunalwirtschaft.
Deutscher Kommunal Verlag GmbH, Düsseldorf, Federal Republic of Germany.

Zeitschrift für Wasser- und Abwasser Forschung.
Verlagsgesellschaft GmbH & Co., 8000 München 40, Federal Republic of Germany.

Forum Städte Hygiene.
2106 Bendestorf, Freudenthalweg 430, Federal Republic of Germany.

Vom Wasser.
Verlag Chemie GmbH, Diesbach, 694 Weinheim, Bergstrasse, (F.R.G.)

Wasserwirtschaft Wassertechnik.
VEB Verlag, 108 Berlin, Französischestrasse 13/14, German Democratic Republic.

Gaz, Eau, Eau usée.
Soc. suisse Ind. Gaz et Eaux, Zürich, Switzerland (German, French).

ITALIAN-LANGUAGE JOURNAL

Inquinamento Acqua, Aria, Suolo.
Etas Dompass, Via Mantegna 6, 20154 Milan, Italy.

ALPHABETICAL INDEX

N.B. *Entries relating to undesirable substances to be eliminated from water (manganese, micropollutants, cyanides, etc.) are printed in italics and indented.*

A

D

G

H

I

J

K

L

M

P

S

U

V

W

Z

The fifth English edition
of the
Water Treatment Handbook
is a translation
of the
"Mémento Technique de l'Eau"
prepared and edited in 1978 by
the following members
of the
Degrémont Company :

CHRISTIAN BARRAQUÉ
JEAN BÉBIN
JACQUES BERNARD
FRANÇOIS BERNÉ
JEAN BOUCHARD
DANIEL BOURGUIGNAT
JEAN DUROT
RENÉE FLAMION
GUY FROMONT
ANDRÉ HAUBRY
RADU HOLCA
JACQUES LEMAIRE
ROBERT LOUBOUTIN
JEAN MARMAGNE
JEAN MIGNOT
PIERRE MOUCHET
MAURICE PARÉ
BERNARD RAULT
YVES RICHARD
JEAN-MARIE ROVEL
VINCENT SAVALL
JACQUES TARDIVEL
ANDRÉ TICHIT
PIERRE TREILLE

whose work has been coordinated by
JACQUES BÉCHAUX

Translation revised and edited by
DONALD F. LONG
of Language Consultants (France) Ltd.

Printed in France par FIRMIN-DIDOT S.A.
Dépôt légal : 2e trimestre 1979
N° d'impression : 4491